STAT2
Modeling with Regression and ANOVA SECOND EDITION

Ann R. Cannon
Cornell College

George W. Cobb
Mount Holyoke College

Bradley A. Hartlaub
Kenyon College

Julie M. Legler
St. Olaf College

Robin H. Lock
St. Lawrence University

Thomas L. Moore
Grinnell College

Allan J. Rossman
California Polytechnic State University

Jeffrey A. Witmer
Oberlin College

New York

Vice President, STEM: Daryl Fox
Senior Program Manager: Karen Carson
Marketing Manager: Nancy Bradshaw
Marketing Assistant: Savannah DiMarco
Development Editor: Leslie Lahr
Executive Media Editor: Catriona Kaplan
Associate Editor: Andy Newton
Editorial Assistant: Justin Jones
Director of Content Management Enhancement: Tracey Kuehn
Senior Managing Editor: Lisa Kinne
Director of Design, Content Management: Diana Blume
Design Services Manager: Natasha Wolfe
Cover Designer: John Callahan
Text Designer: Lumina Datamatics, Ltd.
Director of Digital Production: Keri deManigold
Senior Media Project Manager: Alison Lorber
Media Project Manager: Hanna Squire
Senior Workflow Project Supervisor: Susan Wein
Senior Content Project Manager: Edgar Doolan
Senior Project Manager: Vanavan Jayaraman, Lumina Datamatics, Ltd.
Senior Photo Editor: Robin Fadool
Composition: Lumina Datamatics, Ltd.
Printing and Binding: LSC Communications
Cover and Back Cover Photo Credit: ©Gary M. Hart, Gary Hart Photography, www.EloquentImages.com

Library of Congress Control Number: 2018944738

Student Edition Hardcover:
ISBN-13: 978-1-319-05407-6
ISBN-10: 1-319-05407-2
Student Edition Loose-leaf:
ISBN-13: 978-1-319-05697-1
ISBN-10: 1-319-05697-0

Printed in the United States of America

1 2 3 4 5 6 23 22 21 20 19

W. H. Freeman and Company
One New York Plaza
Suite 4500
New York, NY 10004-1562
www.macmillanlearning.com

Brief Contents

Contents

To the Teacher

"Please, sir, I want some more."
— "Oliver Twist" in the novel *Oliver Twist* by Charles Dickens

This book introduces students to statistical modeling beyond what they learn in an introductory course. We assume that students have successfully completed a Stat 101 college course or an AP Statistics course. Building on basic concepts and methods learned in that course, we empower students to analyze richer datasets that include more variables and address a broader range of research questions.

Guiding Principles

Principles that have guided the development of this book include:

- **Modeling as a unifying theme.** Students will analyze many types of data structures with a wide variety of purposes throughout this course. These purposes include making predictions, understanding relationships, and assessing differences. The data structures include various numbers of variables and different kinds of variables in both explanatory and response roles. The unifying theme that connects all of these data structures and analysis purposes is *statistical modeling*. The idea of constructing statistical models is introduced at the very beginning, in a setting that students encountered in their Stat 101 course. This modeling focus continues throughout the course as students encounter new and increasingly more complicated scenarios.

  Basic principles of statistical modeling that apply in all settings, such as the importance of checking model conditions by analyzing residuals graphically and numerically, are emphasized throughout. Although it's not feasible in this course to prepare students for all possible contingencies that they might encounter when fitting models, we want students to recognize when a model has substantial faults. Throughout the book, we offer two general approaches for analyzing data when model conditions are not satisfied: data transformations and computer-intensive methods such as bootstrapping and randomization tests.

  Students will go beyond their Stat 101 experience by learning to develop and apply models with both quantitative and categorical response variables, with both quantitative and categorical explanatory variables, and with multiple explanatory variables.

- **Modeling as an interactive process.** Students will discover that the practice of statistical modeling involves applying an interactive process. We employ a four-step process in all statistical modeling: *Choose* a form for the model, *fit* the model to the data, *assess* how well the model describes the data, and *use* the model to address the question of interest.

As students gain more and more facility with the interplay between data and models, they will find that this modeling process is not as linear as it might appear. They will learn how to apply their developing judgment about statistical modeling. This development of judgment and the growing realization that statistical modeling is as much an art as a science are more ways in which this second course is likely to differ from students' Stat 101 experiences.

- **Modeling of real, rich datasets.** Students will encounter real and rich datasets throughout this course. Analyzing and drawing conclusions from real data are crucial for preparing students to use statistical modeling in their professional lives. Using real data to address genuine research questions also helps motivate students to study statistics. The richness stems not only from interesting contexts in a variety of disciplines, but also from the multivariable nature of most datasets.

  This multivariable dimension is an important aspect of how this course builds on what students learned in Stat 101 and prepares them to analyze data that they will see in our modern world that is so permeated with data.

Prerequisites

We assume that students using this book have successfully completed an introductory statistics course (Stat 101), including statistical inference for comparing two proportions and for comparing two means. No further prerequisites are needed to learn the material in this book. Some material on data transformations and logistic regression assumes that students are able to understand and work with exponential and logarithmic functions.

Overlap with Stat 101

We recognize that Stat 101 courses differ with regard to coverage of topics, so we expect that students come to this course with different backgrounds and levels of experience. We also realize that having studied material in Stat 101 does not ensure that students have mastered or can readily use those ideas in a second course. To help all students make a smooth transition to this course, we recommend introducing the idea of statistical modeling while presenting some material that students are likely to have studied in their first course.

Chapter 0 reminds students of basic statistical terminology and also uses the familiar two-sample t-test as a way to illustrate the approach of specifying, estimating, and testing a statistical model. Some topics in the early regression chapters (e.g., inference for the slope of a regression line) may be familiar to students from their first course. For a class of students with strong backgrounds, you may choose to move more quickly through the first chapters, treating that material mostly as review to help students get "up to speed."

Organization of Units/Chapters

After completing this course, students should be able to work with statistical models where the response variable is either quantitative or categorical and where explanatory/predictor variables are quantitative or categorical (or with both kinds of predictors). Chapters are grouped into units that consider models based on the type of response and type of predictors.

The three main units (A, B, and C) each follow a similar structure. The unit begins with a chapter dealing with the "simple" case with a single predictor/factor. This helps students become familiar with the basic ideas for that type of model (linear regression, analysis of variance, or logistic regression) in a relatively straightforward setting where graphical visualizations are most feasible. Later chapters of each unit extend these ideas to models with multiple predictors/factors.

Finally, each unit concludes with a special chapter of "additional topics" that extend ideas discussed earlier. The topics in these final chapters are relatively independent. We don't anticipate that you will choose to cover all of these additional topics, but rather will pick and choose the ones most appropriate for your course goals, and incorporate them as needed as you move through the other material in that unit.

Chapter 0: Introduction: We remind students about basic statistical terminology and present our four-step process for constructing statistical models in the context of a two-sample t-test.

Unit A (Chapters 1–4): Linear regression models. These four chapters develop and examine statistical models for a quantitative response variable, first with one quantitative predictor and then with multiple predictors of both quantitative and categorical types.

Unit B (Chapters 5–8): Analysis of variance models. These four chapters also consider models for a quantitative response variable, but specifically with categorical explanatory variables/factors. We start with a single factor (one-way ANOVA) and then move to models that consider multiple factors and interactions. We also include elements of experimental design that are important for choosing an appropriate ANOVA model.

Unit C (Chapters 9–11): Logistic regression models. These three chapters introduce models for a binary response variable with either quantitative or categorical predictors.

Unit D (Chapter 12): Time series. This "unit" is a single chapter that develops models (functions of time, seasonal, and ARIMA) that arise from time series data.

Features of *STAT2: Modeling with Regression and ANOVA*, Second Edition

Flexibility Within and Between Units

The units and chapters are arranged to promote flexibility regarding order and depth in which topics are covered. Within a unit, some instructors may choose to "splice" in an additional topic when related ideas are first introduced. For example, Section 5.7 in the first ANOVA chapter introduces techniques for conducting pairwise comparisons with one-way ANOVA using Fisher's LSD method. Instructors who prefer a more thorough discussion of pairwise comparison issues at this point, including alternate techniques such as the Bonferroni adjustment or Tukey's HSD method, can proceed to present those ideas from Topic 8.2. Other instructors might want to move immediately to two-way ANOVA in Chapter 6 and then study pairwise procedures later.

Instructors can also adjust the order of topics between the units. For example, some might prefer to consider logistic regression models (Unit C) before studying ANOVA models (Unit B). Others might choose to study all three types of models in the "simple setting" (Chapters 1–2, 5, 9), and then return to consider each type of model with multiple predictors. One could also move to the ANOVA material in Unit B directly after starting with a review of the two-sample t-test for means in Chapter 0 and simple linear regression in Chapters 1 and 2.

Technology

Modern statistical software is essential for doing statistical modeling. We assume that students will use statistical software for fitting and assessing the statistical models presented in this book. We include output and graphs from both Minitab and R throughout the book, but we do not include specific software commands or instructions.

Our goal is to allow students to focus on understanding statistical concepts, developing facility with statistical modeling, and interpreting statistical output while reading the text. Toward these ends, we want to avoid the distractions that often arise when discussing or implementing specific software instructions. This choice allows instructors to use other statistical software packages; e.g., SAS, SPSS (an IBM company)*, DataDesk, JMP (developed by SAS), etc..

Exercises

Developing skills of statistical modeling requires considerable practice working with real data. Homework exercises are an important component of this book and appear at the end of each chapter. These exercises are grouped into four categories:

- Conceptual exercises. These questions are brief and require minimal (if any) calculations. They give students practice with applying basic terminology and assess students' understanding of concepts introduced in the chapter.
- Guided exercises. These exercises ask students to perform various stages of a modeling analysis process by providing specific prompts for the individual steps.
- Open-ended exercises. These exercises ask for more complete analyses and reporting of conclusions, without much or any step-by-step direction.
- Supplemental exercises. Topics for these exercises go somewhat beyond the scope of the material covered in the chapter.

Exercises in the Additional Topics chapters (4, 8, and 11) are grouped to align with the independent topics in those chapters.

What's New in the Second Edition

New Content and Organization

New statistical topics. Two topics that were requested most consistently from first edition users were repeated measures designs and time series. We have added new material (Topic 8.6) to give a brief introduction to repeated measures designs, and for instructors who want more depth in this topic we have included three more sections (Topics 8.9–8.11) in the online material. We have also added Chapter 12, giving a brief introduction to working with time series data. In addition to these new sections and chapters, we have made numerous changes to include new ideas (e.g., effect sizes) and give more guidance to students (e.g., how to choose a transformation).

*SPSS was acquired by IBM in October 2009

New organization. We reorganized the material in Unit B to better integrate ideas of experimental design with the topics of ANOVA. Chapter 6 now focuses on block designs and the additive ANOVA model, with interaction coming in Chapter 7, and additional ANOVA topics in Chapter 8.

New exercises and examples. The second edition has 243 worked examples and 646 exercises for students, increases of 76% and 63% over the first edition. We have also updated and revised almost 100 examples and exercises that are carried over from the first edition.

New datasets. We have 64 new datasets dealing with real data, many from research studies. We have also updated datasets from the first edition to bring the total dataset count to 190. Datasets are available in various formats for different software packages, and a data index, which follows the general index, lists all datasets and where they are used in the text.

New Pedagogical Features

Chapter opening section lists give an at-a-glance look at the content therein.

Learning objectives outline goals and expectations that help instructors create their syllabi and students understand where they're headed.

New, full-color design incorporates all-new figures, charts, and graphs. In addition, important definitions and explanations are highlighted for emphasis. Our goal in creating this design was to make the reading and learning experience more approachable by instructors and students alike.

Key terms are highlighted in the margins to help students build a solid statistics vocabulary.

Caution icons and text signal common misconceptions and important ideas to help the student avoid pitfalls and grasp key concepts.

Cars17

Data icons highlight the dataset in use for each example and exercise.

Media and Supplements

Introducing SAPLINGPLUS for *STAT2*, Second Edition!

The new second edition of *STAT2: Modeling with Regression and ANOVA* is now in SaplingPlus, an extraordinary new online resource providing the richest, most completely integrated text/media learning experience yet. SaplingPlus combines Macmillan's powerful multimedia resources with an integrated e-book.

Assets Integrated into SaplingPlus Include the Following:

- Interactive e-book provides powerful study tools for students, multimedia content, and easy customization for instructors. Students can search, highlight, and bookmark specific information, making it easier to study and access key content.
- LearningCurve provides students and instructors with powerful adaptive quizzing, a gamelike format, and instant feedback. The quizzing system features questions tailored specifically to the text and adapts to students' responses, providing material at different difficulty levels and topics based on student performance.

- Data files are available in JMP, ASCII, Excel, TI, Minitab, SPSS, R, and CSV formats.
- Student's Solutions Manual (written by the *STAT2* author team) provides solutions to the odd-numbered exercises in the text.
- R Companion Manual (written by the *STAT2* author team) provides students with a short introduction to R and the commands necessary to accomplish the analyses presented in this text. The manual is organized so that each chapter corresponds to the equivalent chapter in the book. Many of the examples and figures from the book are replicated in the companion.
- Minitab Companion Manual (written by the *STAT2* author team) provides students with a short introduction to Minitab. There is one chapter in the manual for each unit of the textbook.

STAT2, Second Edition Book Companion Website:

www.macmillanlearning.com/stat22e

Instructor access to the Companion website requires user registration as an instructor and features all the open-access student web materials, plus:

- Instructor's Guide (written by the *STAT2* author team) provides instructors with guidance for teaching from this text. Discussion includes pros and cons of teaching topics in different orders, pitfalls to watch out for, and which exercises can be assigned after particular sections have been covered. The guide concludes with several essays written by instructors, describing how they have used this textbook in an appropriate *STAT2* course. These essays also include a collection of projects the instructors have used with their classes.

- Instructor's Solutions Manual (written by the *STAT2* author team) provides solutions to every exercise in the text.
- Test Bank (written by Marian Frazier, The College of Wooster) provides instructors with questions tailored to each chapter and unit of the text.
- Lecture PowerPoint Slides (created by the *STAT2* author team) provide instructors with ready-made PowerPoint slides for each section of the book.
- Lecture PowerPoint Slides (R-version; created by the *STAT2* author team) provide instructors with ready-made PowerPoint slides for each section of the book, including the R code necessary to accomplish the analysis in that section.
- R Markdown files (created by the *STAT2* author team) provide instructors with all of the R code necessary for each of the examples, as well as comments to help the user understand the commands. There are R Markdown files for each section of the text.
- Image Slides contain all textbook figures and tables.

Acknowledgments

We are grateful for the assistance of a great number of people in writing the original edition and this revision of the text.

First, we thank all the reviewers and classroom testers listed at the end of this section. This group of people gave us valuable advice, without which we would have not progressed far from early drafts of our book.

We thank the students in our *STAT2* classes who took handouts of rough drafts of chapters and gave back the insight, suggestions, and kind of encouragement that only students can truly provide.

We thank our publishing team at Macmillan Learning: Andy Dunaway, Karen Carson, Nancy Bradshaw, Elizabeth Simmons, Leslie Lahr, Andy Newton, Catriona Kaplan, Doug Newman, Justin Jones, Edgar Doolan, Susan Wein, Robin Fadool, Diana Blume, and Natasha Wolfe for this second edition. It has been a pleasure working with such a competent organization. Vanavan Jayaraman, with Lumina Datamatics, Ltd., was a great help throughout the production process. We thank Maki Wiering for offering her keen editorial skills in the copyediting phase. Many thanks are due to our accuracy checker John Samons for carefully reading every word and checking every solution.

Learning R takes time, and we continue to learn new tricks as we refine the R Markdown files. Our colleagues, listed below, provided valuable advice and guidance along the way. They provided help in different ways; we acknowledge their assistance and thank them for their help.

Nick Horton
Amherst College

Danny Kaplan
Macalester College

Randall Pruim
Calvin College

Marie Snipes
Kenyon College

Amy Wagaman
Amherst College

We extend our gratitude to Marian Frazier, of The College of Wooster, for revising the text-specific test bank. In addition, our thanks go to Nicole Dalzell of Wake Forest University for accuracy reviewing the test bank.

We thank the students, faculty colleagues, and other researchers who have generously provided their data for use in this project. Rich, interesting data are the lifeblood of statistics and critical to helping students learn and appreciate how to effectively model real-world situations.

We thank Emily Moore of Grinnell College, for giving us our push into the uses of LaTeX typesetting.

We thank our families for their patience and support. The list would be very long if eight authors listed all family members who deserve our thanks. But we owe them a lot and will continue to let them know this.

Finally, we thank all our wonderful colleagues in the Statistics in the Liberal Arts Workshop (SLAW). For more than 30 years, this group has met and supported one another through a variety of projects and life experiences. The idea for the *STAT2* project, and the author team, came from this group and the proceeds from the book fund its continued work. Other current members of the group who have shared their ideas, criticism, and encouragement include

Katherine Halvorsen of Smith College (who was very helpful in reviewing the new time series material), Nick Horton of Amherst College, Shonda Kuiper of Grinnell College, Nathan Tintle of Dordt College, and Kelly McConville of Reed College.

In the first edition we also thanked four retired SLAW participants who were active with the group when the idea for a *STAT2* textbook went from a wish to a plan. These are the late Pete Hayslett of Colby College, Gudmund Iversen of Swarthmore College, Don Bentley of Pomona College, and David Moore of Purdue University. Pete taught us about balance in one's life, and so a large author team allowed us to make the project more fun and more social. Gudmund taught us early about the place of statistics within the liberal arts, and we sincerely hope that our modeling approach will allow students to see our discipline as a general problem-solving tool worthy of the liberal arts. Don taught us about sticking to our guns and remaining proud of our roots in many disciplines, and we hope that our commitment to a wide variety of applications, well represented by many datasets, will do justice to his teaching. All of us in SLAW have been honored by David Moore's enthusiastic participation in our group until his retirement; his leadership in the world of statistics education and writing great textbooks will continue to inspire us for many years to come. His work and his teaching give us a standard to aspire to.

Since the first edition was published we have had four additional retirements from the SLAW group, including three members of the author team. Rosemary Roberts of Bowdoin College was a cofounder of the group and played a vital role in its early development. Julie Legler brought her boundless enthusiasm and was particularly important for developing the material on logistic regression in the first edition. Although retired from teaching and SLAW, George Cobb has remained very active in writing this second edition and was the guiding force behind the reorganization of the ANOVA chapters to emphasize the critical role of design in those models. His creativity is legendary in statistics education circles, so if you notice an apt analogy, particularly deft phrasing, or find yourself chuckling at a clever side comment as you read, George is probably the source. Finally, Tom Moore was the other co-founder of SLAW and served as our principal moderator, organizer, fund raiser, historian, and motivator for most of its existence. His unassuming style is a model for effective leadership that was instrumental in bringing this project to fruition. We wish him well as he now has time to pursue other interests in poetry, bird watching, and grandchildren.

Ann Cannon
Cornell College

George Cobb
Mount Holyoke College

Brad Hartlaub
Kenyon College

Julie Legler
St. Olaf College

Robin Lock
St. Lawrence University

Tom Moore
Grinnell College

Allan Rossman
Cal Poly San Luis Obispo

Jeff Witmer
Oberlin College

Reviewers of the Second Edition

Ali Arab	*Georgetown University*
Jeff Bay	*Maryville College*
KB Boomer	*Bucknell University*
Margaret Bryan	*University of Missouri–Columbia*
Andre Buchheister	*Humboldt State University*
Lisa Carnell	*High Point University*
Catherine Case	*University of Georgia*
Julie M. Clark	*Hollins University*
Linda E. Clark	*Central Connecticut State University*
Linda Brant Collins	*University of Chicago*
Amy Donley	*University of Central Florida*
Marian L. Frazier	*College of Wooster*
Ned Gandevani, PhD	*New England College of Business and Finance (NECB)*
Andrey Glubokov	*Ave Maria University*
Lisa Bloomer Green	*Middle Tennessee State University*
Abeer Hasan	*Humboldt State University*
Kevin Hastings	*Knox College*
Leslie Hendrix	*University of South Carolina*
Laura Hildreth	*Montana State University*
Staci Hepler	*Wake Forest University*
Elizabeth Johnson	*George Mason University*
Dominic Klyve	*Central Washington University*
Michael Kowalski	*University of Alberta*

Chris Lacke	*Rowan University*
Megan Lutz	*University of Georgia*
Wilmina M. Marget	*John Carroll University*
Monnie McGee	*Southern Methodist University*
Herle McGowan	*North Carolina State University*
Andrea Metts	*Elon University*
Alison Motsinger-Reif	*North Carolina State University*
Shyamala Nagaraj	*University of Michigan*
Robert Pearson	*Grand Valley State University*
Linda M. Quinn	*Cleveland State University*
Steven Rein	*California Polytechnic State University*
Aimee Schwab-McCoy	*Creighton University*
James C. Scott	*Colby College*
Richard Single	*University of Vermont*
Judith Smrha, Ph.D.	*Baker University*
Lori Steiner	*Newman University*
Michael Sutherland	*University of Massachusetts*
Jessica Utts	*University of California, Irvine*
Amy Wagaman	*Amherst College*
Daniel Wang	*Central Michigan University*
Jingjing Wu	*University of Calgary*
Zhibiao Zhao	*The Pennsylvania State University*
Faye X. Zhu	*Rowan University*
Laura Ziegler	*Iowa State University*

Second Edition Class Testers

Betsy Greenberg	*University of Texas, Austin*
Georgia Huang	*University of Minnesota*
Andre Lubecke	*Lander University*
Elizabeth Moliski	*University of Texas, Austin*

Reviewers of the First Edition

Carmen O. Acuna	*Bucknell University*
David C. Airey	*Vanderbilt School of Medicine*
Jim Albert	*Bowling Green State University*
Robert H. Carver	*Stonehill College*
William F. Christensen	*Brigham Young University*
Julie M. Clark	*Hollins University*
Phyllis Curtiss	*Grand Valley State University*
Lise DeShea	*University of Oklahoma Health Sciences Center*
Christine Franklin	*University of Georgia*
Susan K. Herring	*Sonoma State University*
Martin Jones	*College of Charleston*
David W. Letcher	*The College of New Jersey*
Ananda Manage	*Sam Houston State University*
John D. McKenzie, Jr.	*Babson College*
Judith W. Mills	*Southern Connecticut State University*
Alan Olinsky	*Bryant University*
Richard Rockwell	*Pacific Union College*
Laura Schultz	*Rowan University*
Peter Shenkin	*John Jay College of Criminal Justice*
Daren Starnes	*The Lawrenceville School*
Debra K. Stiver	*University of Nevada, Reno*
Linda Strauss	*Pennsylvania State University*
Dr. Rocky Von Eye	*Dakota Wesleyan University*
Jay K. Wood	*Memorial University*
Jingjing Wu	*University of Calgary*

First Edition Class Testers

Sarah Abramowitz	*Drew University*
Ming An	*Vassar College*
Christopher Barat	*Stevenson College*
Nancy Boynton	*SUNY, Fredonia*
Jessica Chapman	*St. Lawrence University*
Michael Costello	*Bethesda–Chevy Chase High School*
Michelle Everson	*University of Minnesota*
Katherine Halvorsen	*Smith College*
Joy Jordan	*Lawrence University*
Jack Morse	*University of Georgia*
Eric Nordmoe	*Kalamazoo College*
Ivan Ramler St.	*Lawrence University*
David Ruth	*U.S. Naval Academy*
Michael Schuckers	*St. Lawrence University*
Jen-Ting Wang	*SUNY, Oneonta*

To the Student

In your introductory statistics course, you saw many facets of statistics but you probably did little if any work with the formal concept of a statistical model. To us, modeling is a very important part of statistics. In this book, we develop statistical models, building on ideas you encountered in your introductory course. We start by reviewing some topics from Stat 101 but adding the lens of modeling as a way to view ideas. Then we expand our view as we develop more complicated models.

You will find a thread running through the book:

- Choose a type of model.
- Fit the model to data.
- Assess the fit and make any needed changes.
- Use the fitted model to understand the data and the population from which they came.

We hope that the Choose, Fit, Assess, Use quartet helps you develop a systematic approach to analyzing data.

Modern statistical modeling involves quite a bit of computing. Fortunately, good software exists that enables flexible model fitting and easy comparisons of competing models. We hope that by the end of your STAT2 course, you will be comfortable using software to fit models that allow for deep understanding of complex problems.

To David S. Moore, with enduring affection, admiration, and thanks:

Thank you, David, for all that your leadership has done for our profession, and thank you also for all that your friendship, support, and guidance have done for each of us personally.

CHAPTER 0

Marcus Miranda/Shutterstock

What Is a Statistical Model?

In this chapter you will learn to:

- Discuss statistical models using the correct terminology.
- Employ a four-step process for statistical modeling.

This book is unified by a single theme, statistical models for data analysis. Such models can help to answer all kinds of questions, for example:

- Used cars. Can miles on the odometer predict the price? If so, how does each 1000 miles affect the price? What about the car's age as a predictor? If you know the mileage, does age matter? What about make and model? Do age and mileage give different price predictions for a BMW and a Maxima?
- Walking babies. Can special exercises help young babies walk earlier?
- Hawks' tails. Can we tell the species of a hawk from the length of its tail?
- Medical school admission. If your GPA is 3.5 and your friend's GPA is 3.6, how does that difference affect each of your chances of getting into medical school? What if one of you is male and the other female? What if you take both sex and GPA into account?
- Migraines. Can magnetic pulses to your head reduce your pain from a migraine?

- Ice cream. When people help themselves, do they take more if they have a bigger bowl? What about a bigger spoon?
- Golfers. Golf, like baseball, is a game for patient data nerds. Both games are slow, and both offer lots of things to measure. For example, in golf we measure driving distance, driving accuracy, putting performance, iron play, etc. In the end, what matters most is average strokes per hole. Which predictor or set of predictors works "best" for modeling overall score?
- Burgers. A national chain cooks their hamburgers 24 at a time in a 6×4 array. Health regulations require that each burger reach a fixed minimum temperature before you can sell it. The grill heats unevenly, and you don't want to overcook the burgers that heat faster. What should you measure?

Models can serve a variety of goals. Here are just seven, based on the preceding examples.

a. **Prediction:** Predict the price of a used car or the chance of getting into medical school.

b. **Classification:** Tell the species of a hawk based on its tail length.

c. **Evaluating a treatment:** Can special exercises help a baby walk sooner? Can magnetic pulses relieve migraine pain?

d. **Testing a theory:** Theory says you take more ice cream if you have a bigger bowl or a bigger spoon. What does the evidence say?

e. **Summarizing a pattern:** How are strokes per hole in golf related to distance and accuracy for drives, distance and accuracy for iron play, and distance and accuracy for putts?

f. **Improving a process:** What variables have an effect on how quickly a burger heats up?

g. **Making a decision:** The same burger chain needs to decide: Should we make all our burgers at a central location and ship them frozen, or should we make them fresh on site at each franchise?

Before we go into some details about models, we offer an important caution: *Every model simplifies reality.* No one would mistake a model airplane for the real thing. No one should make that same mistake with a statistical model. Just because the model comes dressed up with equations is no reason you have to trust it. Insist that the model prove its value. As one of the most influential statisticians of the last century (G. E. P. Box) said, "All models are wrong; some are useful." Remember Box. Always ask two questions,

"How far wrong is this model?"
"How useful is it?"

0.1 Model Basics

This section will show you some common terms related to statistical models. It may help if you keep the used cars in mind as an example.

What Is a Model?

response

In this book a statistical model is an equation that relates an outcome Y, called the **response**, to one or more other variables X, or X_1, X_2, ...

predictors
model

These "X-variables" are called **predictors**, or **explanatory variables**, or **independent variables**.* The **model** is a mathematical equation

$$DATA = MODEL + ERROR$$
$$Y = f(X) + \epsilon$$

For the used cars, our goal is to predict the price (response) from the mileage (predictor) or perhaps from all three of mileage, age, and make (explanatory variables).

We assume you have already seen examples of models in your beginning statistics course.

- For comparing two groups, the explanatory variable is the group number, $X = 1$ or $X = 2$. The model tells the group mean, either μ_1 for group 1 ($X = 1$) or μ_2 for group 2 ($X = 2$).
- For fitting a line to the points (x, y) of a scatterplot, the model is the equation $Y = \beta_0 + \beta_1 X$, for a line with intercept β_0 and slope β_1.

What is the error term? The ϵ in the model is called "error," not in the sense of mistake, but instead meaning "deviation from the model value." Remember Box: "All models are wrong." The error term tells "How far wrong?" As you may know from introductory statistics, many statistical models assume that the "errors" follow a probability distribution, typically a normal distribution. (More in Chapter 1.)

"Explain" does not mean what you might think. The words of statistics can sometimes mislead. Just as "error" does not mean mistake, the terms *explain* and *account for* do not have their everyday meaning. They are technical terms that refer to "percent reduction in error variation." (If we tell you that for the 50 U.S. states, the variability in median income "explains" 70% of the variability in poverty rate, we haven't really explained anything.)

Underlying models for inference. By "inference" we mean "using the data we *can* see to *infer* facts about numbers or relationships we *can't* see." In statistics†, there are two main kinds of inference:

sample
population
cause and effect

- Inference from a **sample** (the data we see) to a **population** (the larger group we hope our sample represents); and
- Inference about **cause and effect**. This kind of inference is easiest with a *randomized experiment*.

For the used cars, we hope that the sample of cars in our data set is typical of all used cars with similar ages, mileage, and make.

randomized experiment
observational units
experimental units
subjects
observational study

The crucial distinction is whether a research study is a randomized experiment or an observational study. In a **randomized experiment**, the researcher manipulates the explanatory variable by assigning the explanatory group or value to the **observational units**. (These observational units may be called **experimental units** or **subjects** in an experiment.) In an **observational study**, the researchers do not assign the explanatory variable but rather passively observe and record its information. This distinction is important because the type of study determines the scope of conclusion that can be drawn. Randomized experiments allow for drawing *cause-and-effect* conclusions. Observational studies, on the other hand, almost always only allow for

*In economics the response Y is called **endogenous**, and the independent variables X are called **exogenous**.

†Modeling ideas and inference concepts that pertain to statistics also pertain to data science, which sits at the intersection of statistics, computer science, and application.

concluding that variables are *associated*. Ideally, an observational study will anticipate alternative explanations for an association and include the additional relevant variables in the model. These additional explanatory variables are then called **covariates**.

covariates

Regardless of the type of study, the theoretical requirements for inference are necessarily rigorous: samples must be chosen at random from the population; conditions must be assigned at random to subjects or other experimental units. Fortunately, inference is just one use of statistical models. Many uses do not demand so much of your data. Also, particularly with observational studies like the used cars, you may be able to offer nonstatistical reasons to regard the sample *as if* it had been chosen at random.

When we use models for inference, it is important to have different words for two kinds of numbers,

statistic

(1) a number we compute from the data is a **statistic** and

parameter

(2) the hidden number we really want to know is a **parameter**.

If our goal is to use a sample for inference about a population, for example, the average from the sample is our statistic—we can compute it from the data; the average for the whole population is a parameter—we can't compute it, but its value is what we really want to know.

Statistical Data

case
variable
quantitative
categorical
Cars17

Cases and variables. Table 0.1 shows the first few rows of the data for used cars found in the dataset **ThreeCars2017**. Each row is a **case**, here a single car. Each column is a **variable**: *Price*, *Mileage*, *Age*, and *CarType*. For the car data our goal is to predict price (response) using mileage, age, and model (the explanatory variables). *Price*, *Mileage*, and *Age* are **quantitative**. You can use arithmetic to compare values: "This car has twice the mileage and costs $2500 less than that one." The make of the car is **categorical**. You can classify a car as a Mazda3 or Maxima, but it makes no sense to subtract a Mazda3 from a Maxima. (As an aside, ZIP codes and sports jersey numbers are numerical, but as variables they are categorical, not quantitative. It makes no sense to subtract your ZIP code from mine.)

CarType	Age	Price	Mileage
Mazda3	3	15.9	17.8
Mazda3	2	16.4	19.0
Mazda3	1	18.9	20.9
Mazda3	2	16.9	24.0
Mazda3	2	20.5	24.0
Mazda3	1	19.0	24.2
Mazda3	2	17.5	30.1
Mazda3	3	18.0	32.0
Mazda3	3	13.6	34.8
⋮	⋮	⋮	⋮

TABLE 0.1 The first 9 cases in the **ThreeCars2017** dataset

EXAMPLE 0.1 **Response or explanatory, quantitative or categorical?**

binary

a. Medical school admissions. The response is admission (Yes/No), which is categorical with just two categories, or **binary**. One explanatory variable is GPA, which is quantitative. The other explanatory variable is gender (male/female), which is binary.

b. Ice cream. The response is amount of ice cream taken, which is quantitative. There are two explanatory variables. Bowl size (large or small) is binary. Spoon size (large or small) is also binary.

c. Golf scores. The response is average number of strokes per hole, which is quantitative. There are many possible predictors, depending on how you choose to measure accuracy. Driving distance is clearly quantitative. Putting accuracy might be quantitative (distance from the cup at the end of the putt), but could be binary (yes/no, i.e., make/miss).

d. Migraines. For this example, the explanatory variable is clear: electronic pulses, yes or no. Equally clearly, this variable is binary. The response is not so clear. We want the response to measure relief from pain, but it is far from clear what measure to use.

The pairs of terms (response/predictor, quantitative/categorical) offer a way to classify statistical models into three families. Our three main groups of chapters, Units A, B, and C, are devoted to these families.

Families of Models

This book will show you three main groups of models, plus a bonus:

- Fitting equations to data (**Regression**, Chapters 1–4).
  - The response is quantitative.
  - The predictors are often quantitative, but can also be categorical.
- Comparing groups (**Analysis of Variance**, Chapters 5–8).
  - The response is quantitative.
  - The explanatory variables are almost always categorical. The exception is Analysis of Covariance, which includes both categorical and quantitative predictors.
- Estimating the chances for Yes/No outcomes (**Logistic Regression**, Chapters 9–11).
  - The response must be Yes/No.
  - The predictor(s) can be of either type.

time series

The final bonus Chapter 12, on **time series**, will describe two kinds of models for a response variable that changes over time.

We now turn to an extended example. Our goal is to show you how we rely on a set of four steps to choose, to fit, to assess, and to use a statistical model. We rely on these same four steps throughout the book.

0.2 A Four-Step Process

We will employ a four-step process for statistical modeling throughout this book. These steps are:

- **Choose** a form for the model.

- **Fit** that model to the data.
- **Assess** how well the model fits the data.
- **Use** the model to address the question that motivated the data collection in the first place.

The specific details for how to carry out these steps will differ depending on the type of analysis being performed and, to some extent, on the context of the data being analyzed. But these four steps are carried out in some form in all statistical modeling endeavors. In more detail, the four steps are:

- **Choose** a form for the model. This involves identifying the response and explanatory variable(s) and their types. Typically, we plot the data and explore the patterns we see. A major goal is to describe those patterns and select a model for the relationship between the response and explanatory variable(s).*
- **Fit** that model to the data. This usually entails estimating model parameters based on the sample data. We will almost always use statistical software to do the necessary number-crunching to fit models to data.
- **Assess** how well the model fits the data and meets our needs. How we assess a model depends on how we plan to use it.

    1. For many studies, the main goal of a model is to describe or summarize patterns in the data, and our assessment should focus on how well the model captures those patterns. We can use a variety of **diagnostic plots** to assess

        diagnostic plots

        a. the patterns of **centers** (does our chosen equation track that pattern, or are there features of the data that our model has missed?);

        centers

        b. the pattern of the **spreads** (is the spread in response values roughly uniform across the values of the explanatory variable, so that a single standard deviation is all we need to describe the spread?); and

        spreads

        c. the **shape** of the residuals (i.e., the shape of deviations from the overall pattern).

        shape

        If possible, we look for a scale for which the pattern of centers is simple, the spreads are roughly uniform, residuals follow a normal curve, at least approximately, and our model equation captures all but the normal-shaped residuals.†

    2. For some studies, the goal may be more ambitious: formal (probability-based) inference from the sample data to the larger population or process from which the sample was collected. To justify the use of probability-based methods, there are two additional requirements above and beyond those listed in (1) above.

        a. Randomness. Is the use of a probability model justified?
        b. Independence. Is it reasonable to think of each of the data points as being independent of the other data points?

    We will examine several tools and techniques to use, but the process of assessing model adequacy can be considered as much art as it is science.

- **Use** the model to address the question that motivated collecting the data in the first place. This might be to make predictions, or explain relationships, or assess differences, bearing in mind possible limitations on the scope of

*This might involve transforming one or more of the variables. First choose a suitable scale, then choose an equation. We say more about this in Section 1.4.

†Many inference procedures are based on having errors that are normally distributed, or at least close to normally distributed.

inferences that can be made. For example, if the data were collected as a random sample from a population, then inference can be extended to that population. If treatments were assigned at random to subjects, then a cause-and-effect relationship can be inferred. But if the data arose in other ways, then we have little statistical basis for drawing such conclusions.

To illustrate the process, we consider an example in the familiar setting of a two-sample t-procedure.

EXAMPLE 0.2

WtLoss4

Financial incentives for weight loss Losing weight is an important goal for many individuals. One group of researchers[1] investigated whether financial incentives would help people lose weight more successfully. Some participants in the study were randomly assigned to a treatment group that offered financial incentives for achieving weight-loss goals, while others were assigned to a control group that did not use financial incentives. All participants were monitored over a four-month period, and the net weight change (*Before* − *After*, in pounds) was recorded for each individual. Note that a positive value corresponds to weight loss and a negative value indicates weight gain. The data are given in Table 0.2 and stored in **WeightLossIncentive4**.

Control	12.5	12.0	1.0	−5.0	3.0	−5.0	7.5	−2.5	20.0	−1.0
	2.0	4.5	−2.0	−17.0	19.0	−2.0	12.0	10.5	5.0	
Incentive	25.5	24.0	8.0	15.5	21.0	4.5	30.0	7.5	10.0	18.0
	5.0	−0.5	27.0	6.0	25.5	21.0	18.5			

TABLE 0.2 Weight loss after four months (pounds)

The response variable in this study (weight change) is quantitative and the explanatory variable of interest (control versus incentive) is categorical and binary. The subjects were assigned to the groups at random, so this is a randomized experiment. We may investigate whether there is a statistically significant difference in the distribution of weight changes due to the use of a financial incentive.

CHOOSE

When choosing a form for the model, we consider the question of interest and types of variables involved, then look at graphical displays, and compute summary statistics for the data. Because the weight-loss incentive study has a binary explanatory variable and quantitative response, we examine dotplots and boxplots of the weight losses for each of the two groups (Figure 0.1) and calculate the sample mean and standard deviation for each group.*

Variable	Group	N	Mean	StDev
WeightLoss	Control	19	3.92	9.11
	Incentive	17	15.68	9.41

The dotplots and boxplots show a pair of reasonably symmetric distributions with roughly the same variability in weight loss for the two groups. The mean weight loss for the incentive group is larger than the mean for the control group. One model for these data would be for the weight losses to

*We present vertical dotplots (and boxplots). Horizontal dotplots are common and work perfectly well (and some software packages do not have a simple option for creating vertical dotplots). We prefer vertical dotplots so as to match the vertical boxplots, which foreshadow graphs we will use in Chapter 1.

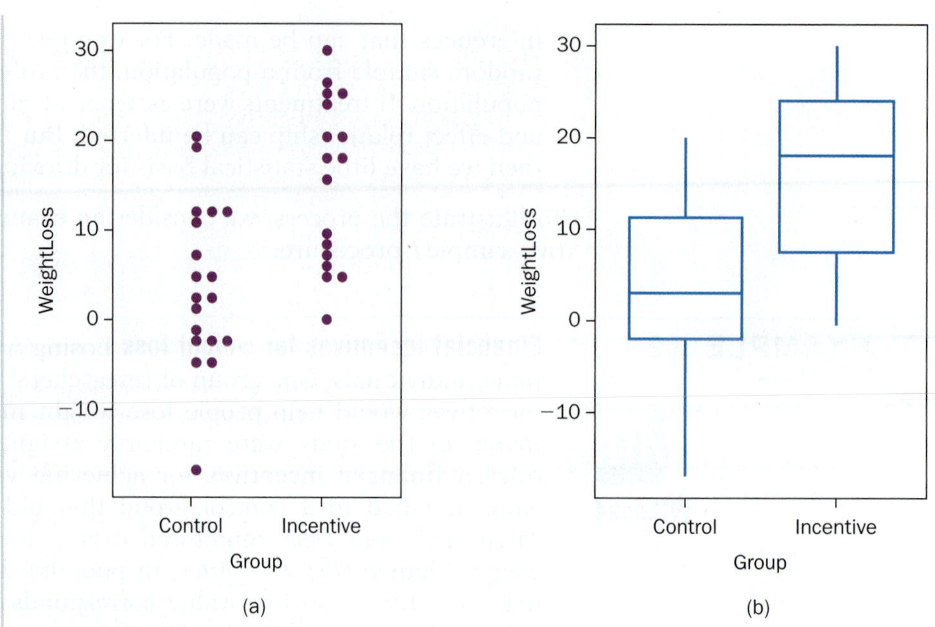

FIGURE 0.1 Weight loss for *Control* versus *Incentive* groups

come from a pair of normal distributions, with different means (and perhaps different standard deviations) for the two groups. Let the parameter μ_1 denote the mean weight loss after four months without a financial incentive, and let μ_2 be the mean with the incentive. If σ_1 and σ_2 are the respective standard deviations and we let the variable Y denote the weight losses, we can summarize the model as $Y \sim N(\mu_i, \sigma_i)$, where the subscript indicates the group and the symbol $\sim$ signifies that the variable has a particular probability distribution, such as normal (denoted in this case with $N(\mu_i, \sigma_i)$).* To see this in the $DATA = MODEL + ERROR$ format, this model could also be written as

$$Y = \mu_i + \epsilon$$

where μ_i is the population mean for the ith group and the random error term $\epsilon \sim N(0, \sigma_i)$. Because we only have two groups, this model says that

$$Y = (\mu_1 + \epsilon) \sim N(\mu_1, \sigma_1) \quad \text{for individuals in the control group}$$
$$Y = (\mu_2 + \epsilon) \sim N(\mu_2, \sigma_2) \quad \text{for individuals in the incentive group}$$

FIT

To fit this model, we need to estimate four parameters (the population means and standard deviations for each of the two groups) using the data from the experiment. The observed means and standard deviations from the two samples provide obvious estimates. We let $\bar{y}_1 = 3.92$ estimate the population mean weight loss for a population getting no incentive (control) and $\bar{y}_2 = 15.68$ estimate the mean for a population getting the incentive. Similarly, $s_1 = 9.11$ and $s_2 = 9.41$ estimate the respective (population) standard deviations. The fitted model (a prediction for the typical weight loss in either group) can then be expressed as

$$\hat{y} = \bar{y}_i$$

that is, that $\hat{y} = 3.92$ pounds for individuals without the incentive and $\hat{y} = 15.68$ pounds for those with the incentive.†

*For this example, an assumption that the variances are equal, $\sigma_1^2 = \sigma_2^2$, might be reasonable, but that would lead to the less familiar pooled variance version of the t-test.

†We use the carat symbol ˆ above a variable name to indicate predicted value, and refer to this as "y-hat."

Note that the error term does not appear in the fitted model since, when predicting a particular weight loss, we don't know whether the random error will be positive or negative. That does not mean that we expect there to be no error, just that the best guess for the *average* weight loss under either condition is the sample group mean, $\bar{y}_i$.

ASSESS

Our goal is to test a particular scientific hypothesis—that the financial incentive leads to greater weight loss—and to estimate the size of the effect of the incentive. The model we have presented expects that departures from the mean in each group (the random errors) should follow a normal distribution with mean zero. To assess this, we examine the sample *residuals* (deviations between the actual data weight losses and those predicted by the model):

$$residual = \text{observed} - \text{predicted} = y - \hat{y}$$

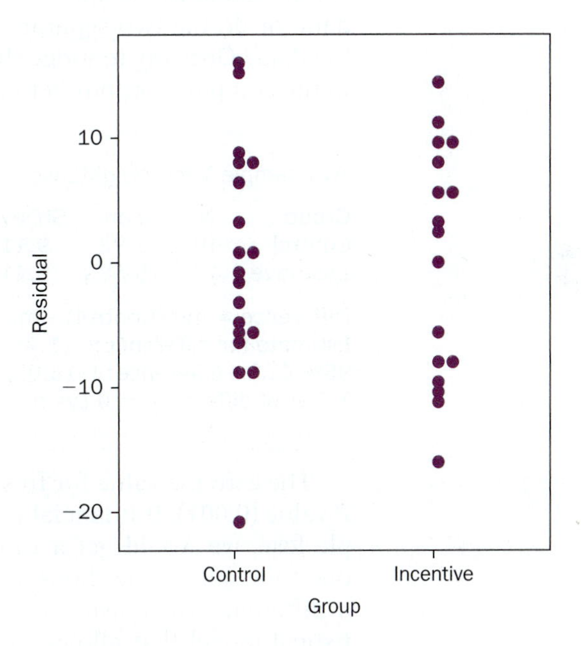

FIGURE 0.2 Residuals from group weight-loss means

The residuals are calculated by subtracting $\hat{y} = 3.92$ from each individual's weight loss in the control group and $\hat{y} = 15.68$ for each individual in the incentive group. Dotplots of the residuals for each group are shown in Figure 0.2. Note that the distributions of the residuals are the same as the original weight-loss data, except that both are shifted to have a mean of zero. We don't see any significant departures from normality in the dotplots, but it's difficult to judge normality from dotplots with so few observations. Normal quantile plots (as shown in Figure 0.3) are a more informative technique for assessing normality. Departures from a linear trend in such plots indicate a lack of normality. Figure 0.3 shows no substantial departure. Normal quantile plots will be examined in more detail in the next chapter.

As a second component of assessment, we consider whether a simpler model might fit the data essentially as well as our model with different means for each group. This is analogous to testing the standard hypotheses for a two-sample t-test:

$H_0 : \mu_1 = \mu_2$

$H_a : \mu_1 \neq \mu_2$

The null hypothesis (H_0) corresponds to the simpler model $Y = \mu + \epsilon$, which uses the same mean for both the control and incentive groups. The

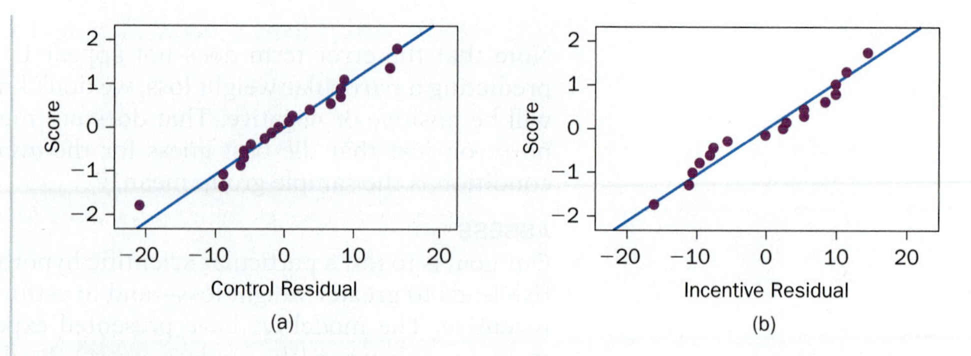

FIGURE 0.3 Normal quantile plots for residuals of weight loss

alternative (H_a) reflects the model that allows each group to have a different mean. Would the simpler (common mean) model suffice for the weight-loss data, or do the two separate group means provide a better explanation for the data? One way to judge this is with the usual two-sample *t*-test as shown in the computer output below.

```
Two-sample T for WeightLoss
Group       N    Mean   StDev  SE Mean
Control    19    3.92    9.11      2.1
Incentive  17   15.68    9.41      2.3

Difference = mu (Control) - mu (Incentive)
Estimate for difference: -11.76
95% CI for difference: (-18.05, -5.46)
T-Test of difference = 0 (vs not =): T-Value = -3.80 P-Value = 0.001 DF = 33
```

The extreme value for this test statistic ($t = -3.80$) produces a very small *P*-value (0.001). If financial incentive has no effect on how much weight people lose, we would get a value of *t* as large as 3.80 (in magnitude) about one time out of one thousand. The evidence that financial incentive makes a difference (increasing mean weight loss) is strong.* We prefer the statistical model that allows for different group means, despite its additional complexity, over the model that uses the same mean for both groups.†

Earlier (when first discussing the ASSESS step) we said that the formal (probability-based) inference requires the added conditions of randomness and independence. For the weight-loss data, the participants were randomly assigned to groups and each participant contributed exactly one data point, so the randomness and independence conditions are met.

USE

Because this was a controlled experiment with random assignment of the control and incentive conditions to the subjects, and because the data produced a very small *P*-value, we can infer that the financial incentives did produce a difference in the average weight loss over the four-month period; that is, the random allocation of conditions to subjects allows us to draw a cause-and-effect relationship.

This test based on a *P*-value addresses just one of three kinds of questions we often ask to summarize a fitted model.

*See Section 2.1 for a more complete discussion of interpreting *P*-values.

†A note about terminology: When the *P*-value is small, we say that there is "statistically significant" evidence that the population means are different. We might say "the *sample* means are (statistically) significantly different." That is *not* the same as saying that any difference between the two *population* means has practical importance.

Tests, Intervals, and Effect Sizes

- **Tests:** Is there an effect?
  Does the incentive have an effect? Does our model need separate means for the incentive and control groups?
- **Intervals:** How big is the effect?
  How precisely can we estimate the size of the effect? What is our margin of error?
- **Effect sizes:** Is the effect big enough to matter?

Tests: As you have seen, a t-test and P-value tell us yes, there is strong evidence of an effect due to the incentive. We *do* need two separate means in our model. *But—the* P-*value alone tells only the strength of the evidence that the means differ. The* P-*value tells nothing about the size of the difference between the two groups.*

Intervals: The difference in group means (incentive minus control) is 11.76 pounds. As you may remember (but we will show you later, in case you don't remember), there is a way to use the data to compute a margin of error, which here turns out to be 6.3. We use the margin of error to construct a 95% confidence interval, 11.76 ± 6.3. We can be 95% confident that the incentive is worth between 5.5 and 18.1 pounds of additional weight loss over four months on average. The interval tells you about precision, but not about importance.

Effect sizes: At this point we know from the P-value that the difference of 11.76 pounds is too big to be due just to the random assignment of subjects to groups. We know from the margin of error that with 95% confidence, the incentive is worth between 5.5 and 18.1 pounds. What we don't yet know is how much a difference of 11.76 pounds actually matters.

To address this question, we need a yardstick, a standard of comparison for judging the importance of 11.76. The usual choice is to use the typical size of person-to-person variation. We hope "typical size" and "variation" bring to mind "standard deviation." The only question left is "Which standard deviation?" There are several choices, and several definitions of effect size, but for now we just note the disagreement and leave it to others to argue about whose weeds matter and whose weeds don't.* To keep it simple, we use the pooled standard deviation of 9.25 as our yardstick.

Our effect size is the ratio of the observed difference (11.76) to the standard deviation (9.25) for person-to-person differences. This ratio, $11.76/9.25 = 1.27$, tells us that the effect of the incentive is 127% as big as the typical size of the difference between you and an average person.

$$\text{effect size} = \frac{|\bar{y}_1 - \bar{y}_2|}{s_p} = \frac{|15.68 - 3.92|}{9.25} = 1.27$$

Such a difference, 127%, is quite large. (General guidelines are that an effect size of 0.2 or 20% is small, an effect size of 0.5 or 50% is medium, and an effect size of 1.0 or 100% is large.)

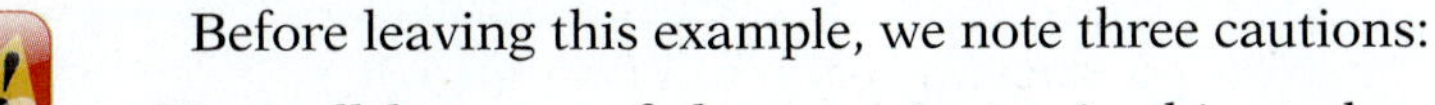

Before leaving this example, we note three cautions:

- First, all but two of the participants in this study were adult men, so we should avoid drawing conclusions about the effect of financial incentives on weight loss in women.

*There are many ways to calculate effect size $= \frac{|\bar{y}_1 - \bar{y}_2|}{s}$. Using the pooled standard deviation $s_p = \sqrt{\frac{(n_1-1)s_1^2 + (n_2-1)s_2^2}{n_1+n_2-2}}$ gives a number that is called Hedge's g (which is very closely related to a measure called Cohen's d). Some people prefer to use $s_{Control}$ in place of s_p; this is called Glass' Δ (delta). Others recommend using the larger of the two sample $s_i's$.

- Second, if the participants in the study were not selected by taking a random sample, we would have difficulty justifying a statistical basis for generalizing the findings to other adults. Any such generalization must be justified on other grounds (such as a belief that most adults respond to financial incentives in similar ways).
- Third, the experimenters followed up with subjects to see if weight losses were maintained seven months after the start of the study (and three months after any incentives expired).

The results from the follow-up study appear in Exercise 0.23.

CHAPTER SUMMARY

In this chapter, we reviewed basic terminology, introduced the four-step approach to modeling that will be used throughout the text, and revisited a common two-sample inference problem.

After completing this chapter, you should be able to distinguish between a **sample** and a **population**, describe the difference between a **parameter** and a **statistic**, and identify variables as **categorical** or **quantitative**. Prediction is a major component to modeling, so identifying **explanatory** (or predictor) **variables** that can be used to develop a model for the **response variable** is an important skill. Another important idea is the distinction between **observational studies** (where researchers simply observe what is happening) and **experiments** (where researchers impose "treatments").

We introduced the fundamental idea that a **statistical model** partitions data into two components, one for the model and one for error. Even though the models will get more complex as we move through the more advanced settings, this statistical modeling idea will be a recurring theme throughout the text. The error term and conditions associated with this term are important features in distinguishing statistical models from mathematical models. You saw how to compute **residuals** by comparing the observed data to predictions from a model as a way to begin quantifying the errors.

The **four-step process** of choosing, fitting, assessing, and using a model is vital. Each step in the process requires careful thought, and the computations will often be the easiest part of the entire process. Identifying the response and explanatory variable(s) and their types (categorical or quantitative) helps us **choose** the appropriate model(s). Statistical software will almost always be used to **fit** models and obtain estimates. Comparing models and **assessing** the adequacy of these models will require a considerable amount of practice, and this is a skill that you will develop over time. Try to remember that **using** the model to make predictions, explain relationships, or assess differences is only one part of the four-step process.

EXERCISES

Conceptual Exercises

0.1 Categorical or quantitative? Suppose that a statistics professor records the following for each student enrolled in her class:

- Gender
- Major
- Score on first exam
- Number of quizzes taken (a measure of class attendance)
- Time spent sleeping the previous night
- Handedness (left- or right-handed)
- Political inclination (liberal, moderate, or conservative)
- Time spent on the final exam
- Score on the final exam

For the following questions, identify the response variable and the explanatory variable(s). Also classify each variable as quantitative or categorical. For categorical variables, also indicate whether the variable is binary.

a. Do the various majors differ with regard to average sleeping time?

b. Is a student's score on the first exam useful for predicting his or her score on the final exam?

c. Do male and female students differ with regard to the average time they spend on the final exam?

d. Can we tell much about a student's handedness by knowing his or her major, gender, and time spent on the final exam?

0.2 More categorical or quantitative? Refer to the data described in Exercise 0.1 that a statistics professor records for her students. For the following questions, identify the response variable and the explanatory variable(s). Also, classify each variable as quantitative or categorical. For categorical variables, also indicate whether the variable is binary.

a. Do the proportions of left-handers differ between males and females on campus?

b. Are sleeping time, exam 1 score, and number of quizzes taken useful for predicting time spent on the final exam?

c. Does knowing a student's gender help predict his or her major?

d. Does knowing a student's political inclination and time spent sleeping help predict his or her gender?

0.3 Sports projects. For each of the following sports-related projects, identify observational units and the response and explanatory variables when appropriate. Also, classify the variables as quantitative or categorical.

a. Interested in predicting how long it takes to play a Major League Baseball game, an individual recorded the following information for all 14 games played on August 11, 2017: time to complete the game, total number of runs scored, margin of victory, total number of pitchers used, ballpark attendance at the game, and which league (National or American) the teams were in.

b. Over the course of several years, a golfer kept track of the length of all of his putts and whether or not he made the putt. He was interested in predicting whether or not he would make a putt based on how long it was.

c. Some students recorded lots of information about all of the football games played by Drew Brees during the 2016 season. They recorded his passing yards, number of pass attempts, number of completions, and number of touchdown passes.

0.4 More sports projects. For each of the following sports-related projects, identify observational units and the response and explanatory variables when appropriate. Also, classify the variables as quantitative or categorical.

a. A volleyball coach wants to see if a player using a jump serve is more likely to lead to winning a point than using a standard overhand serve.

b. To investigate whether the "home-field advantage" differs across major team sports, researchers kept track of how often the home team won a game for all games played in the 2016 and 2017 seasons in Major League Baseball, National Football League, National Basketball Association, and National Hockey League.

c. A student compared men and women professional golfers on how far they drive a golf ball (on average) and the percentage of their drives that hit the fairway.

0.5 Scooping ice cream. In a study reported in the *American Journal of Preventive Medicine*[2] 85 nutrition experts were asked to scoop themselves as much ice cream as they wanted. Some of them were randomly given a large bowl (34 ounces) as they entered the line, and the others were given a smaller bowl (17 ounces). Similarly, some were randomly given a large spoon (3 ounces) and the others were given a small spoon (2 ounces). Researchers then recorded how much ice cream each subject scooped for him- or herself. Their conjecture was that those given a larger bowl would tend to scoop more ice cream, as would those given a larger spoon.

a. Identify the observational units in this study.

b. Is this an observational study or a controlled experiment? Explain how you know.

c. Identify the response variable in this study, and classify it as quantitative or categorical.

d. Identify the explanatory variable(s) in this study, and classify it(them) as quantitative or categorical.

0.6 Diet plans. An article in the *Journal of the American Medical Association*[3] reported on a study in which 160 subjects were randomly assigned to one of four popular diet plans: Atkins, Ornish, Weight Watchers, and Zone. Among the variables measured were:

- Which diet the subject was assigned to
- Whether or not the subject completed the 12-month study
- The subject's weight loss after 2 months, 6 months, and 12 months (in kilograms, with a negative value indicating weight gain)

a. Classify each of these variables as quantitative or categorical.

b. The primary goal of the study was to investigate whether weight loss tends to differ significantly among the four diets. Identify the explanatory and response variables for investigating this question.

c. Is this an observational study or a controlled experiment? Explain how you know.

0.7 Wine model. In his book *SuperCrunchers: Why Thinking by Numbers Is the New Way to Be Smart*, Ian Ayres wrote about Orley Ashenfelter, who gained fame and generated considerable controversy by using statistical models to predict the quality of wine.

Ashenfelter developed a model based on decades of data from France's Bordeaux region, which Ayres reports as

$$\begin{aligned} WineQuality = 12.145 &+ 0.00117 WinterRain \\ &+ 0.0614 AverageTemp \\ &- 0.00386 HarvestRain + \epsilon \end{aligned}$$

where *WineQuality* is a function of the price, rainfall is measured in millimeters, and temperature is measured in degrees Celsius.

a. Identify the response variable in this model. Is it quantitative or categorical?

b. Identify the explanatory variables in this model. Are they quantitative or categorical?

c. According to this model, is higher wine quality associated with more or with less winter rainfall?

d. According to this model, is higher wine quality associated with more or with less harvest rainfall?

e. According to this model, is higher wine quality associated with more or with less average growing season temperature?

f. Are the data that Ashenfelter analyzed observational or experimental? Explain.

0.8 Predicting NFL wins. Consider the following model for predicting the number of games that a National Football League (NFL) team wins in a season:

$$Wins = 3.6 + 0.5PF - 0.3PA + \epsilon$$

where *PF* stands for average points a team scores per game over an entire season and *PA* stands for points allowed per game. Each team plays 16 games in a season.

a. Identify the response variable in this model. Is it quantitative or categorical?

b. Identify the explanatory variables in this model. Are they quantitative or categorical?

c. According to this model, how many more wins is a team expected to achieve in a season if they increase their scoring by an average of 3 points per game?

d. According to this model, how many more wins is a team expected to achieve in a season if they decrease their points allowed by an average of 3 points per game?

e. Based on your answers to parts (c) and (d), does it seem that a team should focus more on improving its offense or improving its defense?

f. Are the data analyzed for this study observational or experimental?

0.9 Measuring students. The registrar at a small liberal arts college computes descriptive summaries for all members of the entering class on a regular basis. For example, the mean and standard deviation of the high school grade point averages for all entering students in a particular year were 3.16 and 0.5247, respectively. The Mathematics Department is interested in helping all students who want to take mathematics to identify the appropriate course, so they offer a placement exam. A randomly selected subset of students taking this exam during the past decade had an average score of 71.05 with a standard deviation of 8.96.

a. What is the population of interest to the registrar at this college?

b. Are the descriptive summaries computed by the registrar (3.16 and 0.5247) statistics or parameters? Explain.

c. What is the population of interest to the Mathematics Department?

d. Are the numerical summaries (71.05 and 8.96) statistics or parameters? Explain.

0.10 Measuring pumpkins. A pumpkin farmer knows that to sell his pumpkins to the local grocery store, each pumpkin he sells must weigh at least 2 pounds. He keeps track of the proportion of pumpkins he is able to sell to the store and at the end of the season finds that 91% of his pumpkins met the 2-pound cutoff. A customer at the store is more interested in the number of pumpkin seeds inside the pumpkins at the store. She purchases 20 pumpkins over the season and carefully counts how many seeds are in each pumpkin. The mean number of seeds she records is 123.2.

a. What is the population of interest to the farmer?

b. Is the descriptive summary computed by the farmer (91%) a statistic or a parameter? Explain.

c. What is the population of interest to the customer?

d. Is the numerical summary (123.2) a statistic or a parameter? Explain.

Guided Exercises

0.11 Scooping ice cream. Refer to Exercise 0.5 on self-serving ice cream. The following table reports the average amounts of ice cream scooped (in ounces) for the various treatments:

	17-ounce bowl	34-ounce bowl
2-ounce spoon	4.38	5.07
3-ounce spoon	5.81	6.58

a. Does it appear that the size of the bowl had an effect on amount scooped? Explain.

b. Does it appear that the size of the spoon had an effect on amount scooped? Explain.

c. Which appears to have more of an effect: size of bowl or size of spoon? Explain.

d. Does it appear that the effect of the bowl size is similar for both spoon sizes, or does it appear that the effect of the bowl size differs substantially for the two spoon sizes? Explain.

0.12 Diet plans. The study discussed in Exercise 0.6 included another variable: the degree to which the subject adhered to the assigned diet, taken as the average of 12 monthly ratings, each on a 1–10 scale (with 1 indicating complete nonadherence and 10 indicating full adherence). As discussed in Exercise 0.6 the primary goal of the study was to investigate whether weight loss tends to differ significantly among the four diets. A secondary goal was to investigate whether weight loss is affected by the adherence level.

a. Identify the explanatory and response variables for investigating the secondary question.

b. If the researchers' analysis of the data leads them to conclude that there is a significant difference in weight loss among the four diets, can they legitimately conclude that the difference is because of the diet? Explain why or why not.

c. If the researchers' analysis of the data analysis leads them to conclude that there is a significant association between weight loss and adherence level, can they legitimately conclude that a cause-and-effect association exists between them? Explain why or why not.

0.13 Predicting NFL wins: Packers/Giants. Consider again, the model for predicting the number of games that a National Football League (NFL) team wins in a season:

$$Wins = 3.6 + 0.5PF - 0.3PA + \epsilon$$

where *PF* stands for average points a team scores per game over an entire season and *PA* stands for points allowed per game. Each team plays 16 games in a season.

a. The Green Bay Packers had a good regular season record in 2016, winning 10 games and losing 6. They averaged 27.0 points scored per game, while allowing an average of 24.25 points per game against them. How many wins does this model predict the Green Bay Packers had in 2016?

b. Find the residual for the Green Bay Packers in 2016 using this model.

c. The largest positive residual value from this model for the 2016 season belongs to the New York Giants, with a residual value of 3.04 games. The Giants actually won 11 games. Determine this model's predicted number of wins for the Giants.

0.14 Predicting NFL wins: Patriots/Chargers. Refer to the model in Exercise 0.13 for predicting the number of games won in a 16-game NFL season based on the average number of points scored per game (*PF*) and average number of points allowed per game (*PA*).

a. Use the model to predict the number of wins for the 2016 New England Patriots, who scored 441 points and allowed 250 points in their 16 games.

b. The Patriots actually won 14 games in 2016. Determine their residual from this model, and interpret what this means.

c. The largest negative residual value from this model for the 2016 season belongs to the San Diego Chargers, with a residual value of −3.48 games. Interpret what this residual means.

0.15 Roller coasters. The Roller Coaster Database (rcdb.com) contains lots of information about roller coasters all over the world. The following statistical model for predicting the top speed (in miles per hour) of a coaster was based on more than 100 roller coasters in the United States and data displayed on the database:

$$TopSpeed = 54 + 7.6TypeCode + \epsilon$$

where *TypeCode* = 1 for steel roller coasters and *TypeCode* = 0 for wooden roller coasters.

a. What top speed does this model predict for a wooden roller coaster?

b. What top speed does this model predict for a steel roller coaster?

c. Determine the difference in predicted speeds in miles per hour for the two types of coasters. Also identify where this number appears in the model equation, and explain why that makes sense.

0.16 Roller coasters: multiple predictors. Refer to the information about roller coasters in Exercise 0.15. Some other predictor variables available at the database include: age, total length, maximum height, and maximum vertical drop. Suppose that we include all of these predictor variables in a statistical model for predicting the top speed of the coaster.

a. For each of these predictor variables, indicate whether you expect its coefficient to be positive or negative. Explain your reasoning for each variable.

b. Which of these predictor variables do you expect to be the best single variable for predicting a roller coaster's top speed? Explain why you think that.

The following statistical model was produced from these data:

$$Speed = 33.4 + 0.10Height + 0.11Drop + 0.0007Length - 0.023Age - 2.0TypeCode + \epsilon$$

c. Comment on whether the signs of the coefficients are as you expect.

d. What top speed would this model predict for a steel roller coaster that is 10 years old, with a maximum height of 150 feet, maximum vertical drop of 100 feet, and length of 4000 feet?

Handwriting and gender. The file **Handwriting** contains survey data from a sample of 203 statistics students at Clarke University. Each student was given 25 handwriting specimens and they were to guess, as best they could, the gender (male or female) of the person who penned the specimen. (Each specimen was a handwritten address, as might appear on the envelope of a letter. There were no semantic clues in the specimen that could tip off the author's gender.) The survey was repeated two times—the second time was the same 25 specimens given in a randomly new order. Exercises 0.17 to 0.22 use the data from this study. **Handwrt**

0.17 CHOOSE/FIT: Survey1. One variable given in the **Handwriting** dataset is *Survey1*, which records the percent correct that each individual had on the first survey.

a. CHOOSE: Construct a histogram of the variable *Survey1*, and discuss what it tells you about the percent correct on the first survey.

b. CHOOSE: Do you think that the subjects are better in guessing the author's gender than mere 50-50 coin flipping would be? Write out the model that describes this situation.

c. FIT: Compute the summary statistics for the sample and give the fitted model.

0.18 CHOOSE/FIT: Survey2. One variable given in the **Handwriting** dataset is *Survey2*, which records the percent correct that each individual had on the second survey.

a. CHOOSE: Construct a histogram of the variable *Survey2*, and discuss what it tells you about the percent correct on the second survey.

b. CHOOSE: Do you think that the subjects' guesses are better in guessing the author's gender than mere 50-50 coin flipping would be? Write out the model that describes this situation.

c. FIT: Compute the summary statistics for the sample and give the fitted model.

0.19 ASSESS/USE: Survey1. Return to the variable *Survey1*.

a. ASSESS: Construct an appropriate plot of the residuals, and comment on whether the graph supports the idea that the residuals are normally distributed.

b. ASSESS: Do you think that the subjects are better in guessing the author's gender than mere 50-50 coin flipping would be? Assess whether our model is better than a coin-flipping model by performing a one-sample *t*-test.

c. USE: Give the conclusion in context to the test performed in part (b).

0.20 ASSESS/USE: Survey2. Return to the variable *Survey2*.

a. ASSESS: Construct an appropriate plot of the residuals and comment on whether the graph supports the idea that the residuals are normally distributed.

b. ASSESS: Do you think that the subjects are better in guessing the author's gender than mere 50-50 coin flipping would be? Assess whether our model is better than a coin-flipping model by performing a one-sample *t*-test.

c. USE: Give the conclusion in context to the test performed in part (b).

0.21 Two-sample Four-step: Survey1. Suppose we want to compare the ability of men and women students in guessing the gender identities of authors of handwriting samples. Consider again the measurements in *Survey1*.

a. CHOOSE: Construct an appropriate plot to compare the guessing abilities of men and women students, discuss what the plot tells you, and write out the model that allows for differences in guessing ability between the two groups.

b. FIT: Compute the appropriate summary statistics and give the fitted model.

c. ASSESS: Construct an appropriate plot of the residuals, and comment on whether the graph supports the idea that the residuals are normally distributed. Also, perform the appropriate two-sample *t*-test to compare the chosen model to the model of no difference between the two groups.

d. USE: Give your conclusion detailing which model you would use and why.

0.22 Two-sample Four-step: Survey2. Suppose we want to compare the ability of men and women students in guessing the gender identities of authors of handwriting samples. Consider again the measurements in *Survey2*.

a. CHOOSE: Construct an appropriate plot to visualize the differences between guessing abilities of men and women students, discuss what the plot tells you, and write out the model that allows for differences in guessing ability between the two groups.

b. FIT: Compute the appropriate summary statistics and give the fitted model.

c. ASSESS: Construct an appropriate plot of the residuals and comment on whether the graph supports the idea that the residuals are normally distributed. Also, perform the appropriate two-sample *t*-test to compare the chosen model to the model of no difference between the two groups.

d. USE: Give your conclusion detailing which model you would use and why.

Open-ended Exercises

0.23 Incentive for weight loss. The study (Volpp et al., 2008) on financial incentives for weight loss in Example 0.2 on page 7 used a follow-up weight check after seven months to see whether weight losses persisted after the original four months of treatment. The results are given in Table 0.3 and in the variable *Month7Loss* of the **WeightLossIncentive7** data file. Note that a few

Table 0.3 Weight loss after seven months (pounds)

Control	−2.0	7.0	19.5	−0.5	−1.5	−10.0	0.5	5.0	8.5	$\bar{y}_1 = 4.64$
	18.0	16.0	−9.0	4.5	23.5	5.5	6.5	−9.5	1.5	$s_1 = 9.84$
Incentive	11.5	20.0	−22.0	2.0	7.5	16.5	19.0	18.0	−1.0	$\bar{y}_2 = 7.80$
	5.5	24.5	9.5	10.0	−8.5	4.5				$s_2 = 12.06$

participants dropped out and were not reweighed at the seven-month point. As with the earlier example, the data are the change in weight (in pounds) from the beginning of the study and positive values correspond to weight losses. Using Example 0.2 as an outline, follow the four-step process to see whether the data provide evidence that the beneficial effects of the financial incentives still apply to the weight losses at the seven-month point. **WtLoss7**

0.24 Statistics student survey: resting pulse rate. An instructor at a small liberal arts college distributed a data collection card, similar to what is shown below, on the first day of class. The data for two different sections of the course are shown in the file **Day1Survey**. Note that the names have not been entered into the dataset. **Day1**

Data Collection Card

Directions: Please answer each question and return to me.

1. Your name (as you prefer): ____________
2. What is your current class standing? ____________
3. Sex: Male ________ Female ________
4. How many miles (approximately) did you travel to get to campus? ____________
5. Height (estimated) in inches: ____________
6. Handedness (Left, Right, Ambidextrous): ____________
7. How much money, in coins (not bills), do you have with you? $ ____________
8. Estimate the length of the white string (in inches): ___
9. Estimate the length of the black string (in inches): ___
10. How much do you expect to read this semester (in pages/week)? ____________
11. How many hours do you watch TV in a typical week? ____________
12. What is your resting pulse? ____________
13. How many text messages have you sent and received in the last 24 hours? ____________

a. Apply the four-step process to the survey data to address the question: "Is there evidence that the mean resting pulse rate for women is different from the mean resting pulse rate for men?"

b. Pick another question that interests you from the survey and compare the responses of men and women.

0.25 Statistics student survey: reading amount. Refer to the survey of statistics students described in Exercise 0.24 with data in **Day1Survey**. Use the survey data to address the question: "Do women expect to do more reading than men?" **Day1**

0.26 Marathon training: short versus long run. Training records for a marathon runner are provided in the file **Marathon**. The *Date*, *Miles* run, *Time* (in minutes:seconds:hundredths), and running *Pace* (in minutes:seconds:hundredths per mile) are given for a five-year period. The time and pace have been converted to decimal minutes in *TimeMin* and *PaceMin*, respectively. Use the four-step process to investigate if a runner has a tendency to go faster on short runs (5 or less miles) than long runs. The variable *Short* in the dataset is coded with 1 for short runs and 0 for longer runs. Assume that the data for this runner can be viewed as a sample for runners of a similar age and ability level. **Marath**

0.27 Marathon training: younger versus older runner. Refer to the data described in Exercise 0.26 that contains five years' worth of daily training information for a runner. One might expect that the running patterns change as the runner gets older. The file **Marathon** also contains a variable called *After*2004, which has the value 0 for any runs during the years 2002–2004 and 1 for runs during 2005 and 2006.* Use the four-step process to see if there is evidence of a difference between these two time periods in the following aspects of the training runs: **Marath**

a. The average running pace (*PaceMin*)

b. The average distance run per day (*Miles*)

0.28 Handwriting. Return to the data file **Handwriting** and suppose we decide upon an operational definition of ability to identify author gender that requires the subject make a correct identification on both surveys. Using the *Both* variable, compare the data to the coin-flipping model using the four-step procedure. **Handwrt**

0.29 Handwriting: males versus females. Consider the **Handwriting** dataset one last time. Suppose we want to compare the ability of men and women students in guessing the gender identities of authors of handwriting samples using their responses to both surveys. Using the *Both* variable, compare the data for men and women students using the four-step procedure. **Handwrt**

0.30 Student survey and marathon pace: effect sizes. Compute and interpret the effect size in each of the following situations.

a. Compare mean resting pulse rate for men and women students, as in Exercise 0.24, for the data in **Day1Survey**. **Day1**

b. Compare mean running pace for long and short training runs, as in Exercise 0.27, for the data in **Marathon**. **Marath**

0.31 Marathon training and handwriting: effect sizes. Compute and interpret the effect size in each of the following situations.

a. Compare mean miles per day for training before and after 2004, as in Exercise 0.27, for the data in **Marathon**. **Marath**

b. Compare mean guessing ability for men and women, as in Exercise 0.29, for the data in **Handwriting**. **Handwrt**

Supplementary Exercises

0.32 Pythagorean theorem of baseball. Renowned baseball statistician Bill James devised a model for predicting a team's winning percentage. Dubbed the "Pythagorean Theorem of Baseball," this model predicts a team's winning percentage as

*These data were collected by one of the authors, when he was training for marathons. Sadly, the data are old because he is old.

$$\text{Winning percentage} = \frac{(\text{runs scored})^2}{(\text{runs scored})^2 + (\text{runs against})^2} \times 100 + \epsilon$$

a. Use this model to predict the winning percentage for the Chicago Cubs, who scored 808 runs and allowed 556 runs in the 2016 season.

b. The Chicago Cubs actually won 103 games and lost 59 in the 2016 season. Determine the winning percentage and also determine the residual from the Pythagorean model (by taking the observed winning percentage minus the predicted winning percentage).

c. Interpret what this residual value means for the 2016 Cubs. (*Hints*: Did the team do better or worse than expected, given their runs scored and runs allowed? By how much?)

d. Repeat parts (a–c) for the 2016 San Diego Padres, who scored 686 runs and allowed 770 runs, while winning 68 games and losing 94 games.

e. Which team (Cubs or Padres) fell short of their Pythagorean expectations by more? Table 0.4 provides data,[4] predictions, and residuals for all 30 Major League Baseball teams in 2016.

f. Which team exceeded their Pythagorean expectations the most? Describe how this team's winning percentage compares to what is predicted by their runs scored and runs allowed.

g. Which team fell furthest below their Pythagorean expectations? Describe how this team's winning percentage compares to what is predicted by their runs scored and runs allowed.

Table 0.4 Winning percentage and Pythagorean predictions for baseball teams in 2016

TEAM	W	L	WinPct	RunScored	RunsAgainst	Predicted	Residual
Arizona Diamondbacks	69	93	42.59	752	890	41.65	0.94
Atlanta Braves	68	93	42.24	649	779	40.97	1.26
Baltimore Orioles	89	73	54.94	744	715	51.99	2.95
Boston Red Sox	93	69	57.41	878	694	61.55	−4.14
Chicago Cubs	103	58	63.98	808	556		
Chicago White Sox	78	84	48.15	686	715	47.93	0.22
Cincinnati Reds	68	94	41.98	716	854	41.28	0.70
Cleveland Indians	94	67	58.39	777	676	56.92	1.47
Colorado Rockies	75	87	46.30	845	860	49.12	−2.82
Detriot Tigers	86	75	53.42	750	721	51.97	1.45
Houston Astros	84	78	51.85	724	701	51.61	0.24
Kansas City Royals	81	81	50.00	675	712	47.33	2.67
Los Angeles Angels	74	88	45.68	717	727	49.31	−3.63
Los Angeles Dodgers	91	71	56.17	725	638	56.36	−0.18
Miami Marlins	79	82	49.07	655	682	47.98	1.09
Milwaukee Brewers	73	89	45.06	671	733	45.59	−0.53
Minnesota Twins	59	103	36.42	722	889	39.74	−3.32
New York Mets	87	75	53.70	671	617	54.19	−0.48
New York Yankees	84	78	51.85	680	702	48.41	3.44
Oakland Athletics	69	93	42.59	653	761	42.41	0.19
Philadelphia Phillies	71	91	43.83	610	796	37.00	6.83
Pittsburgh Pirates	78	83	48.45	729	758	48.05	0.40
San Diego Padres	68	94	41.98	686	770		
Seattle Mariners	86	76	53.09	768	707	54.13	−1.04
San Francisco Giants	87	75	53.70	715	631	56.22	−2.51
St. Louis Cardinals	86	76	53.09	779	712	54.48	−1.40
Tampa Bay Rays	68	94	41.98	672	713	47.04	−5.07
Texas Rangers	95	67	58.64	765	757	50.53	8.12
Toronto Blue Jays	89	73	54.94	759	666	56.50	−1.56
Washington Nationals	95	67	58.64	763	612	60.85	−2.21

© Gary M. Hart, Gary Hart Photography, www.EloquentImages.com

UNIT A

Unit A: Linear Regression

Response: Quantitative
Predictor(s): Quantitative

CHAPTER 1: SIMPLE LINEAR REGRESSION

Identify and fit a linear model for a quantitative response based on a quantitative predictor. Check the conditions for a simple linear model and use transformations when they are not met. Detect outliers and influential points.

CHAPTER 2: INFERENCE FOR SIMPLE LINEAR REGRESSION

Test hypotheses and construct confidence intervals for the slope of a simple linear model. Partition variability to create an ANOVA table and determine the proportion of variability explained by the model. Construct intervals for predictions made with the simple linear model.

CHAPTER 3: MULTIPLE REGRESSION

Extend the ideas of the previous two chapters to consider regression models with two or more predictors. Use a multiple regression model to compare two regression lines. Create and assess models using functions of predictors, interactions, and polynomials. Recognize issues of multicolinearity with correlated predictors. Test a subset of predictors with a nested F-test.

CHAPTER 4: ADDITIONAL TOPICS IN REGRESSION

Construct and interpret an added variable plot. Consider techniques for choosing predictors to include in a model. Identify unusual and influential points. Incorporate categorical predictors using indicator variables. Use computer simulation techniques (bootstrap and randomization) to do inference for regression parameters.

CHAPTER 1

Simple Linear Regression

In this chapter you will learn to:

- Identify and fit a linear model for a quantitative response based on a quantitative predictor.
- Check the conditions for a simple linear model and use transformations when they are not met.
- Detect outliers and influential points.

How is the price of a used car related to the number of miles it's been driven? Is the number of doctors in a city related to the number of hospitals? How can we predict the price of a textbook from the number of pages?

In this chapter, we consider a single quantitative predictor X for a quantitative response variable Y. A common model to summarize the relationship between two quantitative variables is the *simple linear regression model*. We assume that you have encountered simple linear regression as part of an introductory statistics course. Therefore, we review the structure of this model, the estimation and interpretation of its parameters, the assessment of its fit, and its use in predicting values for the response. Our goal is to introduce and illustrate many of the ideas and techniques of statistical model building that will be used throughout this book in a somewhat familiar setting. In addition to recognizing when a linear model may be appropriate, we also consider methods for dealing with relationships between two quantitative variables that are not linear.

1.1 The Simple Linear Regression Model

EXAMPLE 1.1

Accord

Prices for Honda Accords Suppose you want to buy a Honda Accord, but you can't afford a new one. What about used? How much should you expect to pay? Clearly the price will depend on many factors: age, condition, special features, etc. For this example we focus on the relationship between X = *Mileage* (1000s of miles) and Y = *Price* ($1000s). Note that both variables are quantitative.

Table 1.1 shows data for 30 used Accords downloaded from an Internet sales site.[1] This sample of 30 Accords was randomly chosen from recent listings of used Accords in a particular zip code. The data are also stored in the file named **AccordPrice**.

Car	Price ($1000s)	Mileage (1000s)	Car	Price ($1000s)	Mileage (1000s)
1	12.0	74.9	16	17.5	20.3
2	17.9	53.0	17	13.5	68.4
3	15.7	79.1	18	7.0	86.9
4	12.5	50.1	19	11.6	64.5
5	9.5	62.0	20	7.9	150.5
6	21.5	4.8	21	11.7	65.2
7	3.5	89.4	22	15.6	18.6
8	22.8	20.8	23	5.0	139.4
9	26.8	4.8	24	21.0	13.9
10	13.6	48.3	25	15.6	56.1
11	19.4	46.5	26	17.0	15.7
12	19.5	3.0	27	16.0	38.5
13	9.0	64.1	28	17.6	19.8
14	17.4	8.3	29	6.9	119.3
15	17.8	27.1	30	5.5	122.5

TABLE 1.1 *Price* and *Mileage* for used Honda Accords

We use plots to show how price and mileage are related. Although it's not standard when both X and Y are quantitative, we start with side-by-side box plots.* For the boxplots, we sort the 30 Accords into three groups of 10 based on mileage:

- Low: Fewer than 25,000 miles
- High: More than 65,000 miles
- Middle: In between

Figure 1.1 shows parallel boxplots for the low-mileage, medium-mileage, and high-mileage Accords, with each boxplot positioned above the midpoint of its mileage interval. We can see a decreasing linear relationship that seems to be fairly strong, and we also see similar variability across the three groups.

*Again, we remind you that normally we would start with a scatterplot of *Mileage* and *Price*. Using boxplots builds on comparing two means from Chapter 0 and anticipates the plots we use in Unit B.

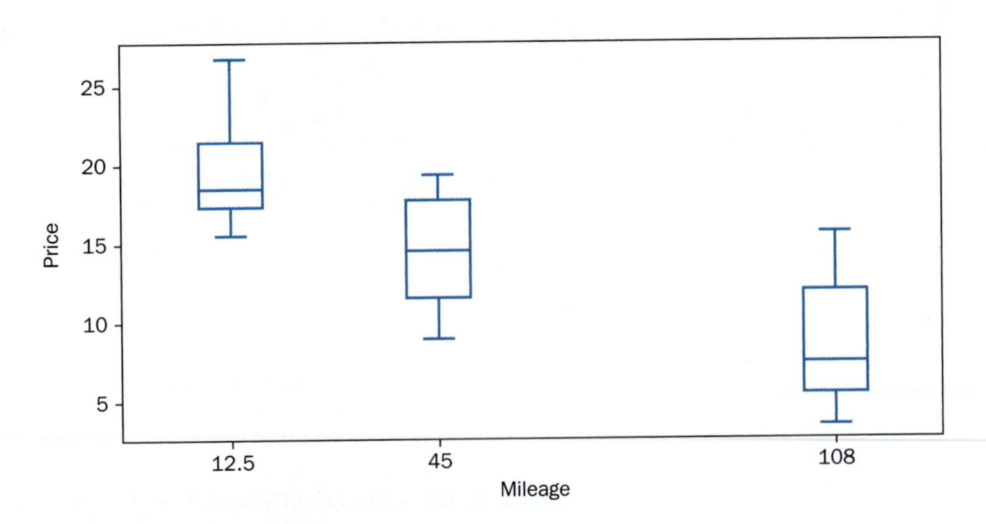

FIGURE 1.1 Parallel boxplots for Accords that have been sorted into three groups.

Choosing a Simple Linear Model

model
error

Recall that data can be represented by a **model** plus an **error** term:

$$Data = Model + Error$$

When the data involve a quantitative response variable Y and we have a single quantitative predictor X, the model becomes

$$Y = f(X) + \epsilon$$
$$= \mu_Y + \epsilon$$

where $f(X)$ is a function that gives the mean value of Y, μ_Y, at any value of X and ϵ represents the error (deviation) from that mean.*

simple linear regression

Scatterplots are the major tool for helping us choose a model when both the response and predictor are quantitative variables. If the scatterplot shows a consistent linear trend, then we use in our model a mean that follows a straight-line relationship with the predictor. This gives a **simple linear regression** model where the function, $f(X)$, is a linear function of X. If we let β_0 and β_1 represent the intercept and slope, respectively, of that line, we have

$$\mu_Y = f(X) = \beta_0 + \beta_1 X$$

and

$$Y = \beta_0 + \beta_1 X + \epsilon$$

EXAMPLE 1.2

Honda Accord prices—choosing a model

Accord

CHOOSE

We are interested in predicting the price of a used Accord based on its mileage, so the explanatory variable is *Mileage*, the response is *Price*. The scatterplot of price versus mileage for the sample of used Honda Accords is shown in Figure 1.2. The plot indicates a negative association between these two variables. It is generally understood that cars with lots of miles cost less, on average, than cars with only limited miles, and the scatterplot supports this understanding. Since the rate of decrease in the scatterplot is relatively constant as the mileage increases, a straight-line model might provide a good summary of the relationship between the average prices and mileages of used

*More formal notation for the mean value of Y at a given value of X is $\mu_{Y|X}$. To minimize distractions in most formulas, we will use just μ_Y when the role of the predictor is clear.

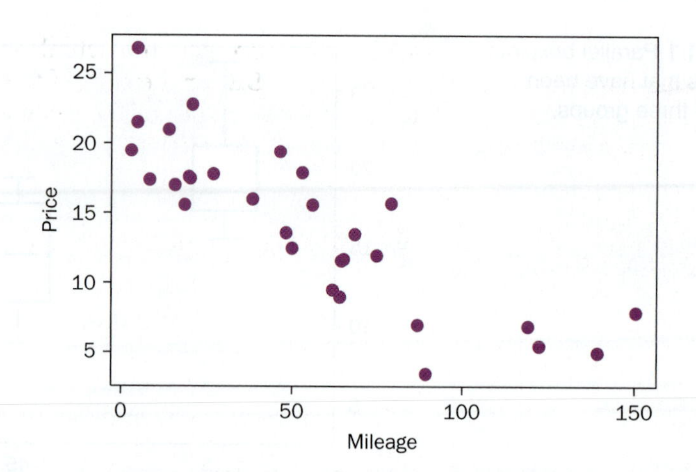

FIGURE 1.2 Scatterplot of Honda Accord *Price* versus *Mileage*

Honda Accords for sale on this Internet site. In symbols, we express the mean price as a linear function of mileage:

$$\mu_{Price} = \beta_0 + \beta_1\ Mileage$$

Thus, the model for actual used Honda Accord prices would be

$$Price = \beta_0 + \beta_1\ Mileage + \epsilon$$

This model indicates that Honda Accord prices should be scattered around a straight line with deviations from the line determined by the random error component, ϵ.

We now turn to the question of how to choose the slope and intercept for the line that best summarizes this relationship.

Fitting a Simple Linear Model

least squares regression

We want good estimates of the parameters β_0 and β_1. We use a technique called **least squares regression** to fit the model to the data. This method chooses estimates to minimize the sum of the squared prediction errors and leads to the best* results for predicting the response values in the sample. In practice, we rely on computer technology to compute the least squares estimates for the parameters. The fitted model is represented by

$$\hat{Y} = \hat{\beta}_0 + \hat{\beta}_1 X$$

In general, we use Greek letters (β_0, β_1, etc.) to denote parameters, and hats ($\hat{\beta}_0, \hat{\beta}_1$, etc.) are added to denote estimated (fitted) values of these parameters.

A key tool for fitting a model is to compare the values it predicts for the individual data cases† to the actual values of the response variable in the dataset.

residual

The discrepancy in predicting each response is measured by the **residual**:

$$\begin{aligned} residual &= observed\,y - predicted\,y \\ &= y - \hat{y} \end{aligned}$$

The magnitude of the residual measures the vertical distance from a point to the line. Summing the squares of the residuals provides a measure of how

*Here *best* is a technical term. There can be different criteria than "least squares" for assessing the fit of a line. In practice, "best" depends on your goal.

†We generally use a lowercase y when referring to the value of a variable for an individual case and an uppercase Y for the variable itself.

sum of squared errors (SSE)

least squares line

well the line predicts the actual responses for a sample. We often denote this quantity as **SSE** for the sum of the squared errors.* Statistical software calculates the fitted values of the slope and intercept so as to minimize this sum of squared residuals; hence, we call the fitted line the **least squares line.**

EXAMPLE 1.3

Accord

Honda Accord prices—fitting a model

FIT

For the ith car in the dataset, with mileage x_i, the model is

$$y_i = \beta_0 + \beta_1 x_i + \epsilon_i$$

The parameters, β_0 and β_1, in the model represent the true, population-wide intercept and slope for all Honda Accords for sale. The corresponding statistics, $\hat{\beta}_0$ and $\hat{\beta}_1$, are estimates derived from this particular sample of 30 Accords. These estimates are determined from statistical software, for example, in the output shown below.

```
Coefficients:
              Estimate Std.   Error    t value   Pr(>|t|)
(Intercept)    20.8096       0.9529    21.84     < 2e-16
Mileage        -0.1198       0.0141    -8.50     3.06e-09

Residual standard error: 3.085 on 28 degrees of freedom
Multiple R-squared: 0.7207, Adjusted R-squared: 0.7107
F-statistic: 72.25 on 1 and 28 DF, p-value: 3.06e-09

Analysis of Variance Table
            Df   Sum Sq   Mean Sq   F value     Pr(>F)
Mileage      1   687.66   687.66    72.253   3.055e-09 ***
Residuals   28   266.49     9.52
```

The least squares line is

$$\widehat{Price} = 20.81 - 0.1198\ Mileage$$

Thus, for every additional 1000 miles on a used Accord, the predicted price goes down by about $120. Also, if a (used!) Accord had zero miles on it, we would predict the price to be $20,810. In many cases, the intercept lies far from the data used to fit the model and has no practical interpretation.

Note that the first car in Table 1.1 had a mileage level of 74.9 (74,900 miles) and a price of 12.0 ($12,000), whereas the fitted line predicts a price of

$$\widehat{Price} = 20.81 - 0.1198(74.9) = 11.8$$

The residual here is

$$Price - \widehat{Price} = 12.0 - 11.8 = 0.2$$

If we do a similar calculation for each of the 30 cars, square each of the resulting residuals, and sum the squares, we get the SSE of 266.49 (as shown in the last line of the computer output). If you were to choose any other straight line to make predictions for these Accord prices based on the mileages, you could never obtain an SSE less than 266.49.

*Throughout this book we use the terms *error* and *residual* interchangeably to refer to the difference between a predicted and actual value.

Figure 1.3 shows the fitted least squares line added to the scatterplot. Points below the line correspond to cars in the sample with negative residuals, and points above the line have prices that are higher than what is predicted by the linear model.

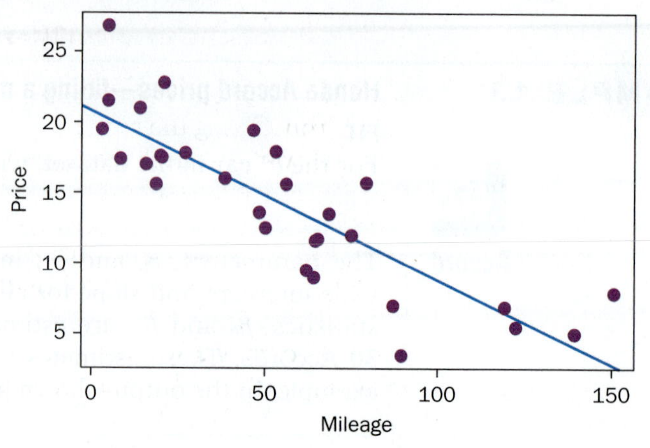

FIGURE 1.3 Linear regression to predict Accord *Price* based on *Mileage*

Centering a Predictor

In Example 1.3 we interpret the fitted value of the intercept, 20.8096, by saying that if a used Accord had zero miles on it, then we predict the price to be about \$20,810. In this dataset, zero is not a plausible value of the mileage predictor (because we are talking about "used" cars), so it is a bit silly to think about mileage = 0. However, we might have begun our analysis of the data by first **centering** the predictor variable, which simply involves subtracting a constant from each value of the predictor (X), where the constant is a number that is at or near the middle of the distribution of predictor values.

centering

EXAMPLE 1.4

Accord

Accord price with centered mileages Let the variable *MileageC* be *Mileage* − 50 for the Accord data. That is, shift the predictor *Mileage* by 50 units. The scatterplot with the centered data is given in Figure 1.4(b), while the original plot is Figure 1.4(a).

Centering the predictor variable at 50 does not change the pattern of the points nor the slope of the fitted regression line; it only changes how we interpret the intercept term. With the predictor being *MileageC* the fitted regression equation is

$$\widehat{Price} = 14.819 - 0.1198\ MileageC$$

which means that when *MileageC* is zero (so when a used Accord has 50,000 miles on it), we predict the price of a used Accord to be \$14,819.

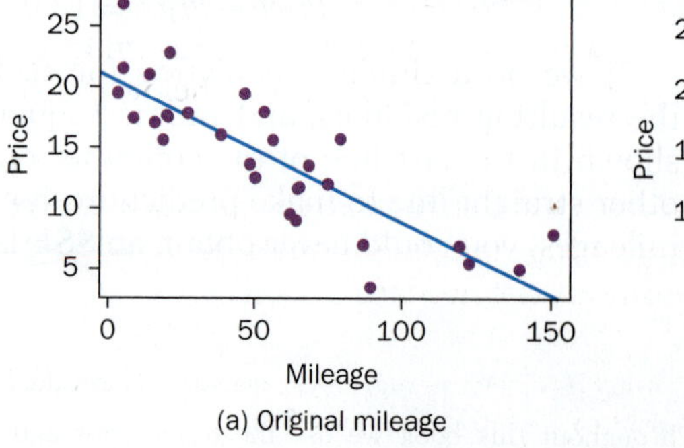

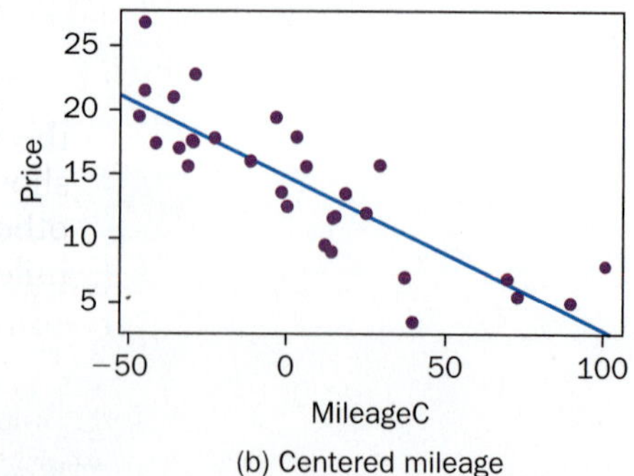

FIGURE 1.4 Plots comparing regressions to predict Accord prices

Note that we chose to center Mileage at 50 because 50 is a round number that is close to the mean of Mileage, but we could have centered using any number near the middle of the *Mileage* distribution.

1.2 Conditions for a Simple Linear Model

We know that our model won't fit the data perfectly. The discrepancies that result from fitting the model represent what the model did not capture in each case. We want to check whether our model is reasonable and captures the main features of the dataset. Are we justified in using our model? Do the conditions of the model appear to be reasonable? How much can we trust predictions that come from the model? Do we need to adjust or expand the model to better explain features of the data or could it be simplified without much loss of predictive power?

In specifying any model, certain conditions must be satisfied for the model to make sense. A key part of assessing any model is to check whether the conditions are reasonable for the data at hand. We hope that the residuals are small and contain no pattern that could be exploited to better explain the response variable. If our assessment shows a problem, then the model should be refined. Typically, we will rely heavily on graphs of residuals to assess the appropriateness of the model. In this section, we discuss the conditions that are commonly placed on a simple linear model. The conditions we describe here for the simple linear regression model are typical of those that will be used throughout this book. In the following section, we explore ways to use graphs to help us assess whether the conditions hold for a particular set of data.

When thinking about the conditions, keep in mind your goals for constructing a model. If your main goal is *description*, finding an equation to describe a pattern, all that matters is how well you capture the patterns in the sample you see. However, your goal may involve *inference*, for example to estimate features of an unseen population or to draw conclusions about a possible cause-and-effect relationship. For formal inference the conditions are more stringent.

One condition for a simple linear model is:

Linearity—The overall relationship between the variables has a linear pattern. If the relationship is not linear, you may need a change of scale to handle curvature (see Section 1.4) or additional predictors in the model (see Chapter 3).

The other model conditions deal with the distribution of the errors.

Zero Mean—The error distribution is centered at zero.* This means that the points are scattered at random above and below the line (determined by the linear part of the model, $\beta_0 + \beta_1 X$).

Uniform Spread—The variance of the response does not change as the predictor changes. If there is a pattern of changing variance, choosing a different scale might help (see Section 1.4).

Independence—The errors are assumed to be independent from one another. Thus, one point falling above or below the line has no influence on the location of another point.

When we are interested in using the model to make formal inferences (conducting hypothesis tests or providing confidence intervals), additional conditions are needed.

*By using least squares regression, we force the residual mean to be zero. Other techniques would not necessarily satisfy this condition.

Normality—For inference based on theory, the formulas for tests and intervals assume that the unseen errors (ϵ) in the model follow a normal probability distribution. (Even for description a normal shape is simplest to describe.)

Randomness—The data are obtained using a random process. This is a property of how you got your data. A random sample? A randomized experiment? These questions are crucial for determining the scope of inference, but they don't matter much if your only goal is description.

We can summarize these conditions for a simple linear model using the following notation.

SIMPLE LINEAR REGRESSION MODEL

For a quantitative response variable Y and a single quantitative explanatory variable X, the **simple linear regression model** is

$$Y = \beta_0 + \beta_1 X + \epsilon$$

where ϵ follows a normal distribution, that is, $\epsilon \sim N(0, \sigma_\epsilon)$, and the errors are independent from one another.

Note that the model comes in two parts:

1. The *form of the equation*, here $\beta_0 + \beta_1 X$
2. The *conditions for the error term*: errors are independent with normal shape and constant standard deviation

The first part of the model is important for *all* uses of the regression model. If a line does not capture the pattern, you should go back and start over. Once you have determined that the linear model describes the relationship well, then you will often want to make inferences based on this model. Before doing that you must check the conditions (noted in the second part of the model) carefully. The distinction between the two parts of the model—the equation and the error term—matters, but is not always clear.

Estimating the Standard Deviation of the Error Term

The simple linear regression model has three unknown parameters: β_1, the slope; β_0, the intercept; and σ_ϵ, the standard deviation of the errors around the line. We have already seen that software will find the least squares estimates of the slope and intercept. Now we consider how to estimate σ_ϵ, the standard deviation of the distribution of errors. Since the residuals estimate how much Y varies about the regression line, the sum of the squared residuals (SSE) is used to compute the estimate, $\hat{\sigma}_\epsilon$. The value of $\hat{\sigma}_\epsilon$ is referred to as the **standard error of regression** and is interpreted as the size of a "typical" error. The residuals, $y - \hat{y}$, estimate the errors (ϵ) for each data case and are used to compute the estimate of the standard error of regression.

standard error of regression

STANDARD ERROR OF REGRESSION

For a simple linear regression model, the estimated standard deviation of the error term based on the least squares fit to a sample of n observations is

$$\hat{\sigma}_\epsilon = \sqrt{\frac{\sum (y - \hat{y})^2}{n - 2}} = \sqrt{\frac{SSE}{n - 2}}$$

The predicted values and resulting residuals are based on a sample slope and intercept that are calculated from the data. Therefore, we have $n - 2$ **degrees of freedom** for estimating the standard error of regression, and the estimator follows the familiar form of a sum of squared deviations divided by degrees of freedom.* Looking ahead to Chapter 3 where we have more than one predictor, we lose an additional degree of freedom in the denominator for each new beta parameter that is estimated in the prediction equation.

degrees of freedom

EXAMPLE 1.5

Accord

Accord prices (continued) At the end of Example 1.3, we saw that the sum of squared residuals for the Accord data is 266.49. Thus, the standard error of regression is

$$\hat{\sigma}_\epsilon = \sqrt{\frac{266.49}{30 - 2}} = 3.085$$

Using mileage to predict the price of a used Accord, the typical error will be around \$3085. This gives us some feel for how far individual cases might spread above or below the regression line. Note that this value is labeled as *Residual standard error* in the output on page 25.

1.3 Assessing Conditions

A variety of plots are used to assess the conditions of the simple linear model. Scatterplots, histograms, and dotplots will be helpful to begin the assessment process. However, plots of residuals versus fitted values and normal plots of residuals will provide more detailed information, and these visual displays will be used throughout the text.

Residuals Versus Fits Plots

A scatterplot with the fitted regression line provides one visual method of checking linearity. Points will be randomly scattered above and below the line when the linear model is appropriate. Clear patterns, for example clusters of points above and below the line in a systematic fashion, indicate that the simple linear model is not appropriate.

A more informative way of looking at how the points vary about the regression line is a scatterplot of the residuals versus the fitted values for the prediction equation. This plot reorients the axes so that the regression line is represented as a horizontal line through zero. Positive residuals represent points that are above the regression line. The residuals versus fits plot allows us to focus on the estimated errors and look for any clear patterns without the added complexity of a sloped line.

The residuals versus fits plot is especially useful for assessing the linearity and constant variance conditions of a simple linear model. The ideal pattern will be random variation above and below zero in a band of relatively constant width. Figure 1.5 shows a typical residuals versus fits plot when these two conditions are satisfied.

Figure 1.6 shows examples of residuals versus fits plots that exhibit some typical patterns indicating a problem with linearity, constant variance, or both conditions.

*If a scatterplot has only 2 points, then it's easy to fit a straight line with residuals of zero, but we have no way of estimating the variability in the distribution of the error term. This corresponds to having zero degrees of freedom.

FIGURE 1.5 Residuals versus fitted values plot when linearity and constant variance conditions hold

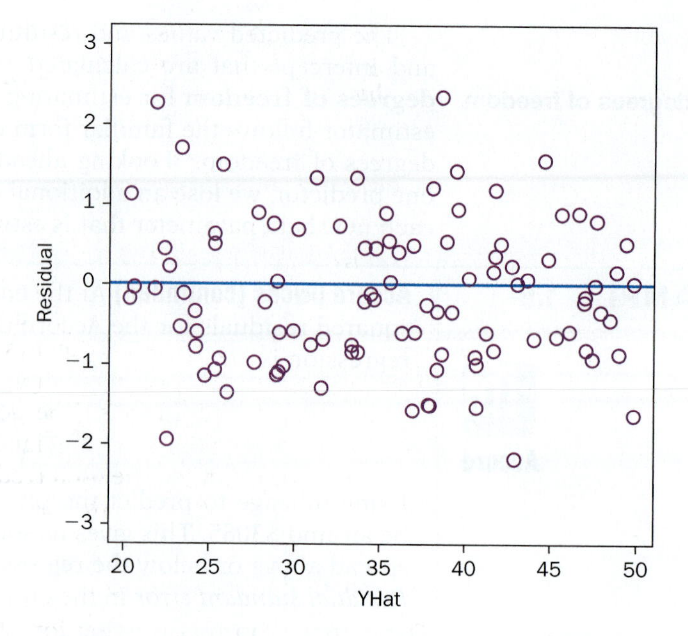

Figure 1.6(a) illustrates a curved pattern demonstrating a lack of linearity in the relationship. The residuals are mostly positive at either extreme of the graph and negative in the middle, indicating more of a curved relationship. Despite this pattern, the vertical width of the band of residuals is relatively constant across the graph, showing that the constant variance condition is probably reasonable for this model.

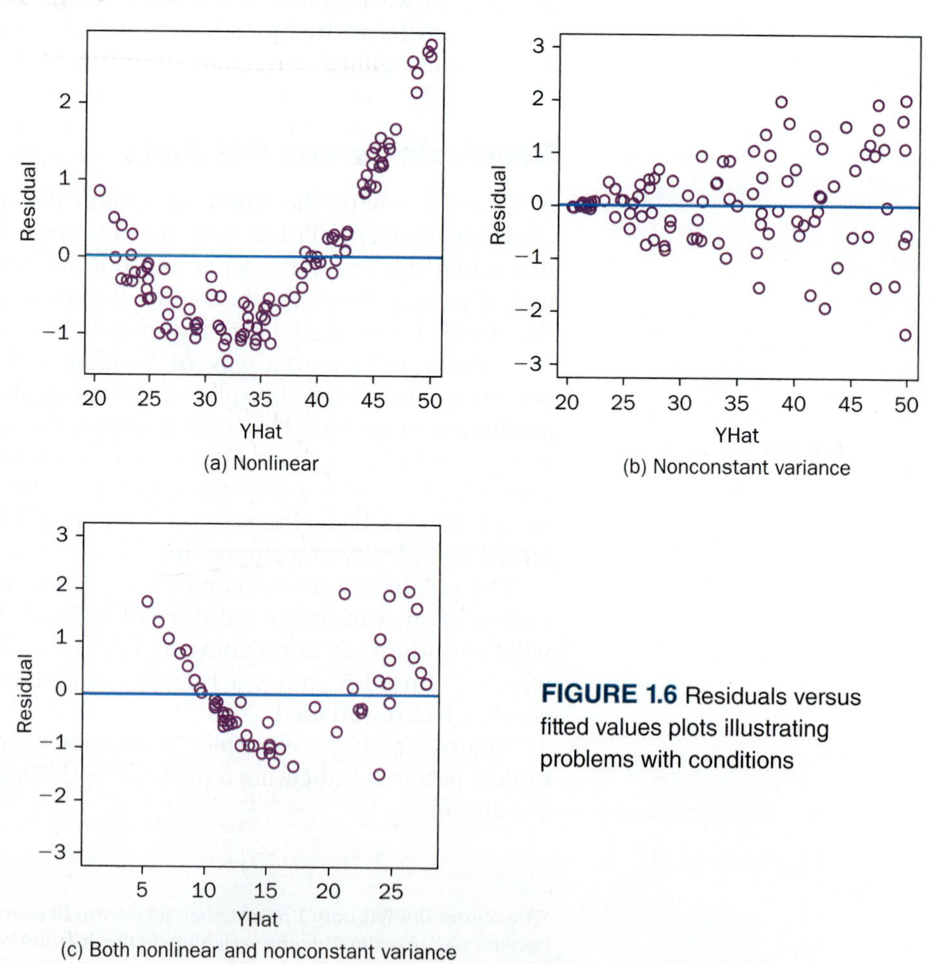

FIGURE 1.6 Residuals versus fitted values plots illustrating problems with conditions

Figure 1.6(b) shows a common violation of the equal variance condition. In many cases, as the predicted response gets larger, its variability also increases, producing a fan shape as in this plot. Note that a linearity condition might still be valid in this case since the residuals are still equally dispersed above and below the zero line as we move across the graph.

Figure 1.6(c) indicates problems with both the linearity and constant variance conditions. We see a lack of linearity due to the curved pattern in the plot and, again, variance in the residuals that increases as the fitted values increase.

In practice, the assessment of a residuals versus fits plot may not lead to as obvious a conclusion as in these examples. Remember that no model is perfect, and we should not expect to always obtain the ideal plot. A certain amount of variation is natural, even for sample data that are generated from a model that meets all of the conditions. The goal is to recognize when departures from the model conditions are sufficiently evident in the data to suggest that an alternative model might be preferred, or that we should use some caution when drawing conclusions from the model.

Normal Plots

Data from a normal distribution should exhibit a "bell-shaped" curve when plotted as a histogram or dotplot. However, we often need a fairly large sample to see this shape accurately, and even then it may be difficult to assess whether the symmetry and curvature of the tails are consistent with a true normal curve. As an alternative, a **normal plot** shows a different view of the data where an ideal pattern for a normal sample is a straight line. Although a number of variations exist, there are generally two common methods for constructing a normal plot.

normal plot

normal quantile plot

The first, called a **normal quantile plot**, is a scatterplot of the ordered observed data versus values (the theoretical quantiles) that we would expect to see from a "perfect" normal sample of the same size.* If the ordered data are increasing at the rate we would expect to see for a normal sample, the resulting scatterplot is a straight line. If the distribution of the data is skewed in one direction or has tails that are overly long due to some extreme outliers at both ends of the distribution, the normal quantile plot will bend away from a straight line.

Figure 1.7 shows several examples of normal quantile plots for residuals. The first (Figure 1.7(a)) was generated from residuals where the data were generated from a linear model with normal errors. The other two are from models with nonnormal errors.

normal probability plot

The second common method of producing a normal plot is to use a **normal probability plot**, such as those shown in Figure 1.8. Here, the residuals are plotted on the horizontal axis while the vertical axis is transformed to reflect the rate that normal probabilities grow. As with a normal quantile plot, the values increase as we move from left to right across the graph, but the revised scale produces a straight line when the values increase at the rate we would expect for a sample from a normal distribution. Thus, the interpretation is the same. A linear pattern (as in Figure 1.8(a)) indicates good agreement with normality, and curvature, or bending away from a straight line (as in Figure 1.8(b)), shows departures from normality.

Since both normal plot forms have similar interpretations, we will use them interchangeably. The choice we make for a specific problem often depends on the options that are most readily available in the statistical software we are using.

*The more common arrangement is to put the residuals on the vertical axis and theoretical normal quantiles on the horizontal axis (as in Figure 1.7), but some computer software reverses these locations.

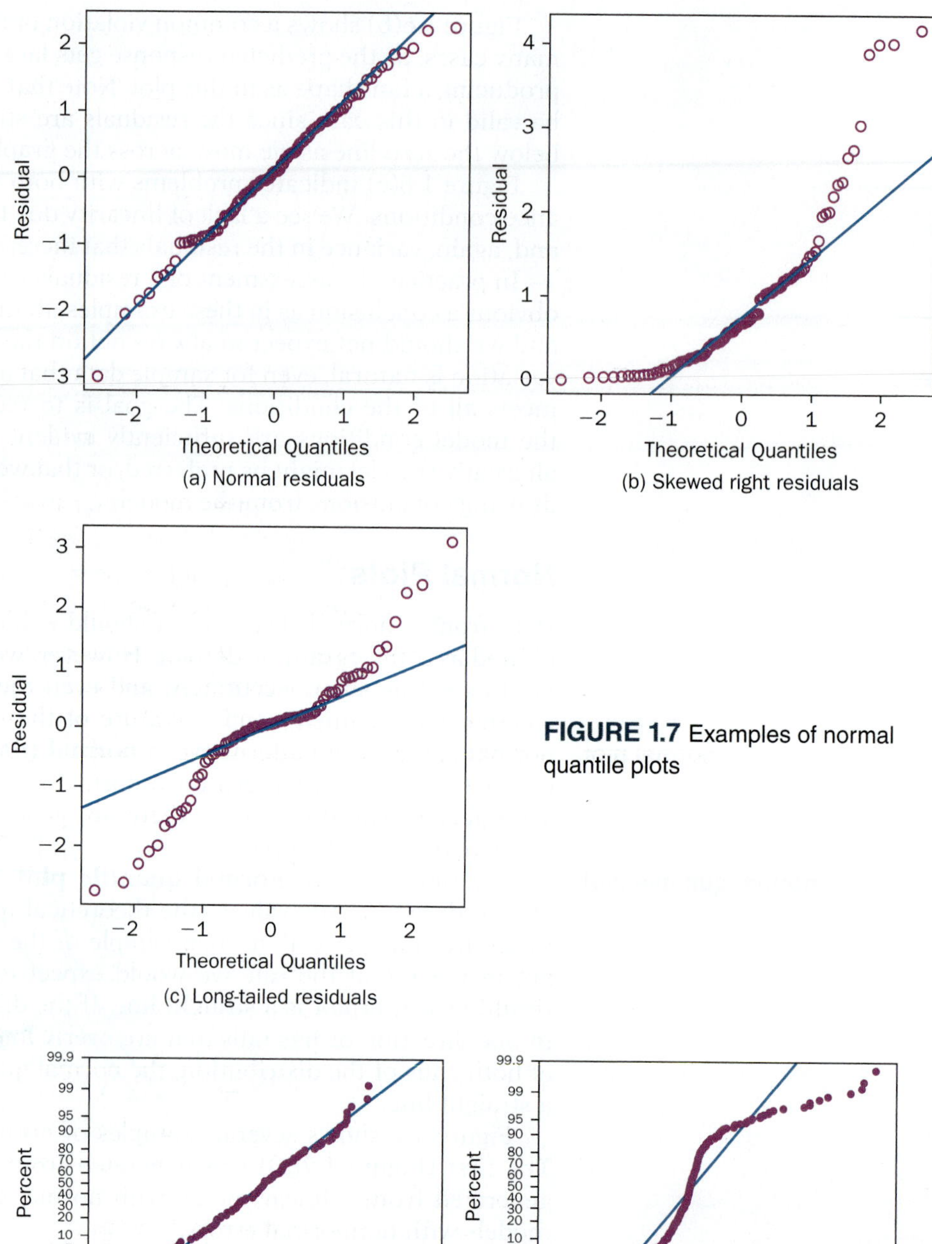

FIGURE 1.7 Examples of normal quantile plots

FIGURE 1.8 Examples of normal probability plots

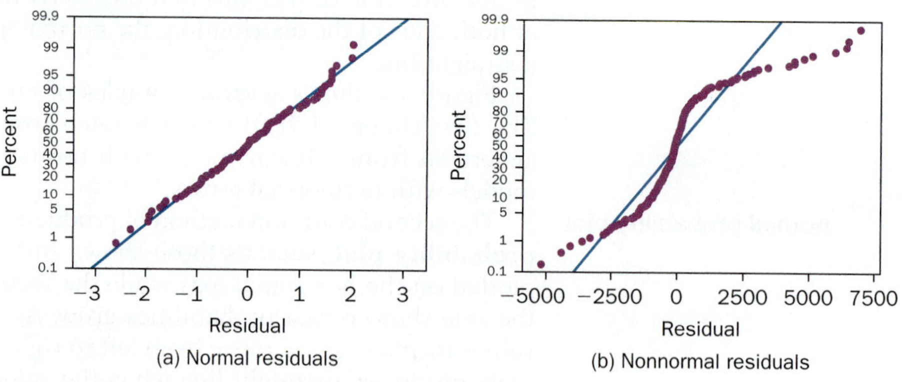

EXAMPLE 1.6

Accord

Accord prices—checking conditions

ASSESS

We illustrate these ideas by checking the conditions for the model to predict Honda Accord prices based on mileage.

Linearity: The scatterplot with regression line in Figure 1.3 shows that the linearity condition is reasonable as points show a consistent decline in prices with mileage and no obvious curvature. A plot of the residuals versus fitted values is shown in Figure 1.9. There may be some mild concern with a decreasing trend for the smaller fitted values, but there are only a few points in that region so it's hard to be too sure of a pattern. For the most part the horizontal band of points is generally scattered randomly above and below

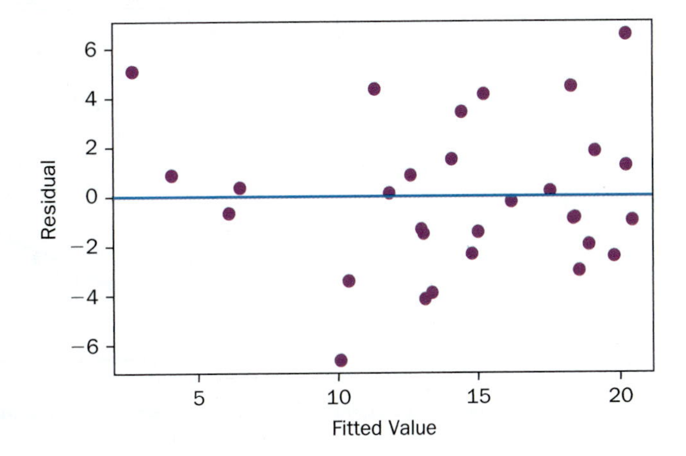

FIGURE 1.9 Plot of Accord residuals versus fitted values

the zero line, illustrating that a linear model is reasonably appropriate for describing the relationship between price and mileage.

Zero mean: We used least squares regression, which forces the sample mean of the residuals to be zero when estimating the intercept β_0. Also note that the residuals are scattered on either side of zero in the residual plot of Figure 1.9 and a histogram of the residuals, Figure 1.10, is centered at zero.

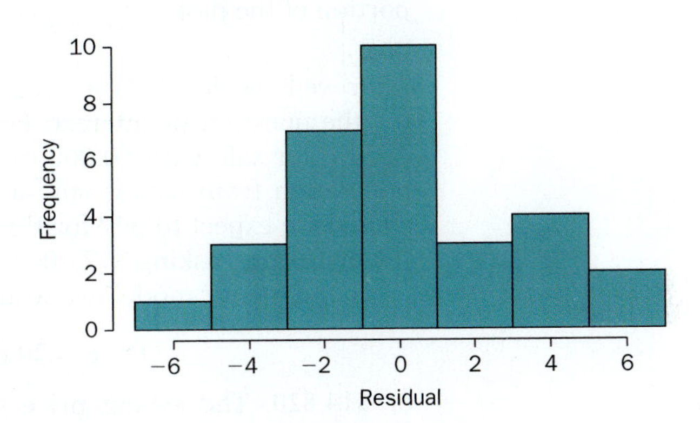

FIGURE 1.10 Histogram of Accord residuals

Constant variance: The fitted line plot in Figure 1.3 shows the data spread in roughly equal width bands on either side of the least squares line. Looking left to right in the plot of residuals versus fitted values in Figure 1.9 reinforces this finding as we see a fairly constant spread of the residuals above and below zero (where zero corresponds to prices that fall on the least squares regression line). This supports the constant variance condition.

Randomness and Independence: We cannot tell from examining the data whether these conditions are satisfied. However, the context of the situation and the way the data were collected make these reasonable assumptions. There is no reason to think that one seller changing the asking price for a used car would necessarily influence the asking price of another seller. We were also told that these data were randomly selected from the Accords for sale on the Cars.com website. So at the least we can treat it as a random sample from the population of all Accords on that site at the particular time the sample was collected. We might want to be cautious about extending the findings to cars from a different site, an actual used car lot, or a later point in time.

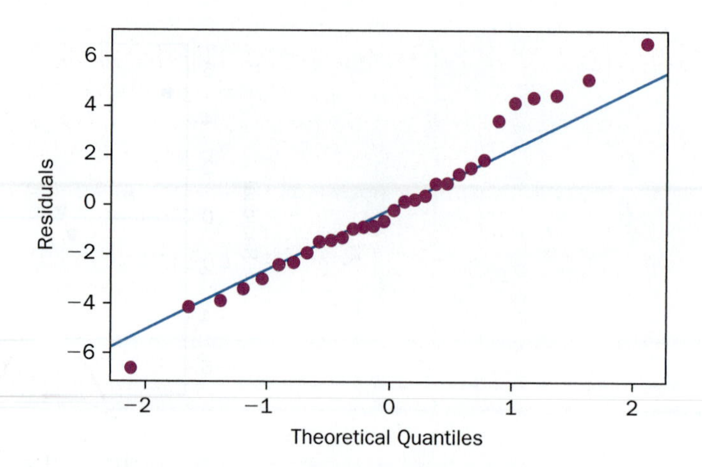

FIGURE 1.11 Normal quantile plot of residuals for Accord data

Normality: In assessing normality, we can refer to the histogram of residuals in Figure 1.10, where a reasonably bell-shaped pattern is displayed. However, a histogram based on this small sample may not be particularly informative and can change considerably depending on the bins used to determine the bars. A more reliable plot for assessing normality is the normal quantile plot of the residuals shown in Figure 1.11. This graph shows a fairly consistent linear trend that supports the normality condition, although we might have a mild concern about a slight right skew with the last few points in the upper portion of the plot.

USE

After we have decided on a reasonable model, we interpret the implications for the question of interest. For example, suppose we find a used Honda Accord for sale with 50,000 miles and we believe that it is from the same population from which our sample of 30 used Accords was drawn. What should we expect to pay for this car? Would it be an especially good deal if the owner was asking $12,000?

Based on our model, we would expect to pay

$$\widehat{Price} = 20.81 - 0.1198(50) = 14.82$$

or $14,820. The asking price of $12,000 is below the expected price of $14,820, but is this difference large relative to the variability in Accord prices? We might like to know if this is a really good deal or perhaps such a low price that we should be concerned about the condition of the car. This question will be addressed in Section 2.4, where we consider prediction intervals. For now, we can observe that the car's residual is a bit less than the size we called a "typical error" ($\hat{\sigma}_\epsilon = \$3.085$ thousand) below the expected price. Thus, it looks like a pretty good deal, but not an extremely low price.

1.4 Transformations/Reexpressions

Before settling on a model, it is critical to choose useful scales for the response and predictor variables. If one or more of the conditions for a simple linear regression model are not satisfied, we can consider transformations on one or both of the variables to address the problems. There is much more to come in this chapter about these choices. What comes next is just a preview.

EXAMPLE 1.7

CtyHlth

Doctors and hospitals in counties We expect the number of doctors in a county to be related to the number of hospitals, reflecting both the size of the county and the general level of medical care. Finding the number of hospitals in a given city is relatively easy, but counting the number of doctors is a more challenging task. Fortunately, the American Medical Associaton collects such data. The data in Table 1.2 show values for these two variables (and the *County* names) from the first few cases in the data file **CountyHealth**, which has a sample of 53 counties[2] that have at least two hospitals.

County	MDs	Hospitals
Bay, FL	351	3
Beaufort, NC	95	2
Beaver, PA	260	2
Bernalillo, NM	2797	11
Bibb, GA	769	5
Clinton, PA	42	2
Comanche, TX	13	2
Concordia, LA	20	2
Contra Costa, CA	2981	7
Coos, NH	83	3
⋮	⋮	⋮

TABLE 1.2 Number of MDs and hospitals for sample of n = 53 counties

CHOOSE

As usual, we start the process of finding a model to predict the number of doctors (*MDs*) from the number of hospitals (*Hospitals*) by examining a scatterplot of the two variables as seen in Figure 1.12. As expected, this shows an increasing trend with counties having more hospitals also tending to have more doctors, suggesting that a linear model might be appropriate.

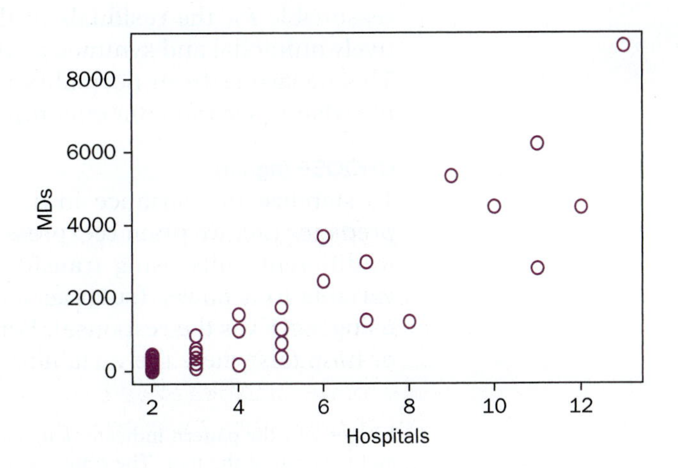

FIGURE 1.12 Scatterplot for number of doctors versus number of hospitals

FIT

Fitting a least squares line produces estimates for the slope and intercept as shown in the output below, giving the prediction equation $\widehat{MDs} = -1120.6 + 557.3\,Hospitals$. Figure 1.13(a) shows the scatterplot with this regression line as a summary of the relationship.

```
Coefficients:

(Intercept)   Hospitals
-1120.6       557.3
```

ASSESS

The line does a fairly good job of following the increasing trend in the relationship between number of doctors and number of hospitals. However, a closer look at the residual plots shows some considerable departures from our standard regression conditions. For example, the plot of residuals versus fitted values in Figure 1.13(b) shows a fan shape, with the variability in the residuals tending to increase as the fitted values get larger. This often occurs with count data like the number of MDs and number of hospitals, where variability increases as the counts grow larger. We can also observe this effect in a scatterplot with the regression line, Figure 1.13(a).

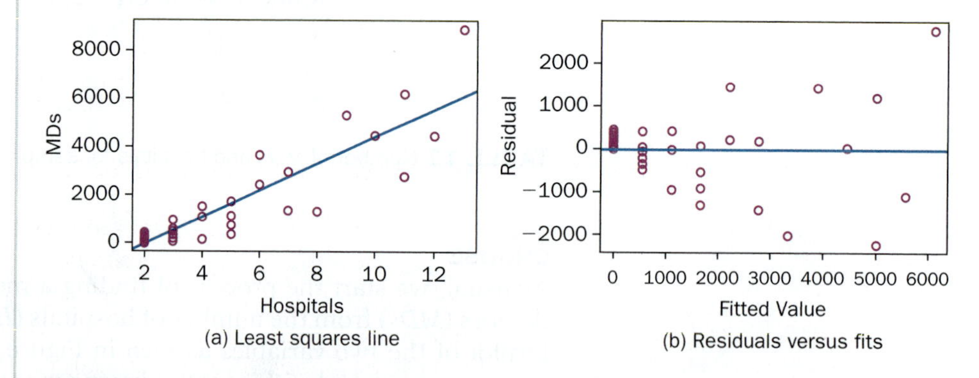

FIGURE 1.13 Regression for number of doctors based on number of hospitals

We also see from a histogram of the residuals, Figure 1.14(a), and normal quantile plot, Figure 1.14(b), that an assumption of normality would not be reasonable for the residuals in this model. Although the histogram is relatively unimodal and symmetric, the peak is quite narrow with very long tails. This departure from normality is seen more clearly in the normal quantile plot that has significant curvature away from a straight line at both ends.*

CHOOSE (again)

To stabilize the variance in a response (Y) across different values of the predictor (X), we often reexpress the response (Y) or predictor (X) (or both) in different units using transformations. Typical options include raising a variable to a power (such as $\sqrt{Y}$, Y^2, or $1/Y$) or taking a logarithm (e.g., using $\log(Y)$ as the response). For count data, such as the number of doctors or hospitals where the variability increases along with the magnitudes of the

*To see why the pattern indicates *long* tails rather than short tails, look at points at the far left and far right of the plot. The observed residuals are larger in magnitude than the theoretical line expects. A useful slogan is: "Long tails away, short tails toward." Here *away* means "away from the horizontal."

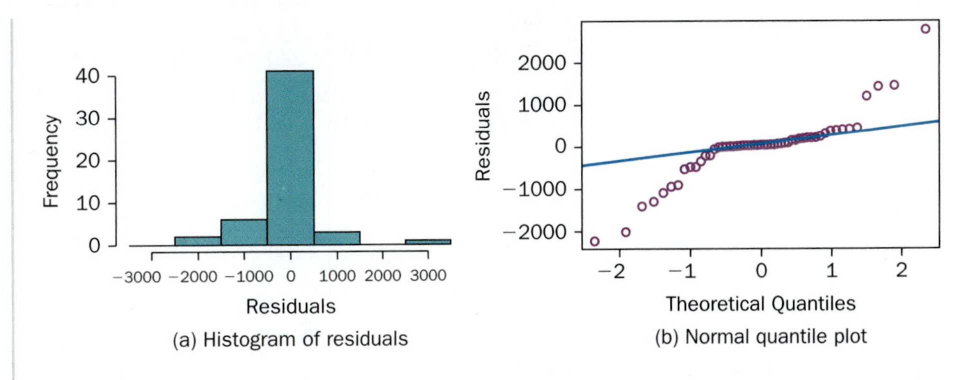

(a) Histogram of residuals (b) Normal quantile plot

FIGURE 1.14 Plots to check normality for residuals when predicting *MDs* with *Hospitals*

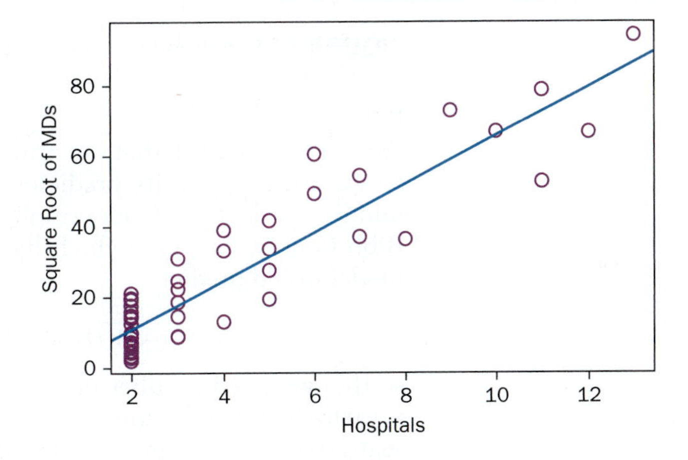

FIGURE 1.15 Least squares line for *Sqrt(MDs)* versus *Hospitals*

variables, a square root transformation is often helpful. Figure 1.15 shows the least squares line fit to the transformed data to predict the square root of the number of doctors based on the number of hospitals. The prediction equation is now $\widehat{\sqrt{MDs}} = -2.75 + 6.88\ Hospitals$.

When the equal variance condition holds, we should see roughly parallel bands of data spread along the line. Although there might still be slightly less variability for the smallest numbers of hospitals, the situation is much better than for the data on the original scale. The residuals versus fitted values plot for the transformed data in Figure 1.16(a) and normal quantile plot of the residuals in Figure 1.16(b) also show considerable improvement at meeting the constant variance and normality conditions of our simple linear model.

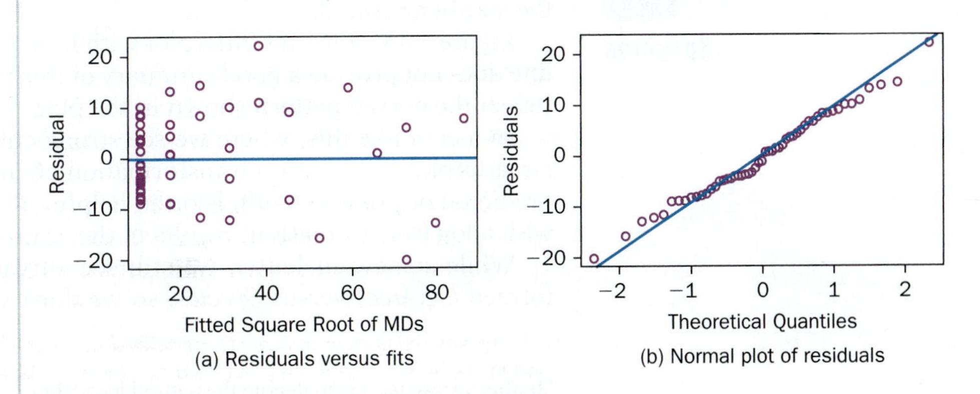

(a) Residuals versus fits (b) Normal plot of residuals

FIGURE 1.16 Residual plots when predicting *Sqrt(MDs)* with *Hospitals*

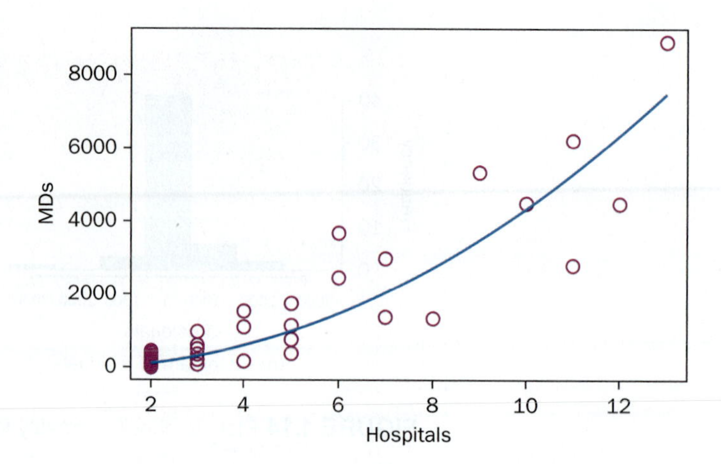

FIGURE 1.17 Predicted *MDs* from the linear model for *Sqrt(MDs)*

USE

We must remember that our transformed linear model is predicting $\sqrt{MDs}$, so we must square its predicted values to obtain estimates for the actual number of doctors. For example, if we consider the case from the data of Bibb County, Georgia, which has 5 community hospitals, the transformed model would predict

$$\widehat{\sqrt{MDs}} = -2.75 + 6.88(5) = 31.65$$

so the predicted number of doctors is $31.65^2 = 1001.7$, while Bibb County actually had 769 doctors at the time of this sample. Figure 1.17 shows the scatterplot with the predicted number of doctors after expressing the linear model for the square roots of the number of doctors back to the original scale so that

$$\widehat{NumMDs} = (-2.75 + 6.88\ NumHospitals)^2$$

Note that we could use this model to make predictions for other counties, but in doing so, we should feel comfortable only to the extent that we believe the sample to be representative of the larger population of counties with at least two hospitals.

EXAMPLE 1.8

SpecArea

Species by area The data in Table 1.3 (and the file **SpeciesArea**) show the number of mammal species and the area for 13 islands[3] in Southeast Asia. Biologists have speculated that the number of species is related to the size of an island and would like to be able to predict the number of species given the size of an island.

Figure 1.18 shows a scatterplot with least squares line added. Clearly, the line does not provide a good summary of this relationship, because it doesn't reflect the curved pattern shown in the plot.

In a case like this, where we see strong curvature and extreme values in a scatterplot, a logarithm transformation of either the response variable, the predictor, or possibly both, is often helpful. Reexpressing the *Area* predictor with a log transformation* results in the scatterplot of Figure 1.19(a).

While somewhat better, we still see curvature in the plot of the transformed *log*(*Area*) versus *Species*, so we don't yet have a linear relationship.

*In this text we use log to denote the natural logarithm.

Island	Area (km^2)	Mammal Species
Borneo	743244	129
Sumatra	473607	126
Java	125628	78
Bangka	11964	38
Bunguran	1594	24
Banggi	450	18
Jemaja	194	15
Karimata Besar	130	19
Tioman	114	23
Siantan	113	16
Sirhassan	46	16
Redang	25	8
Penebangan	13	13
Perhentian Besar	8	6

TABLE 1.3 *Species* and *Area* for Southeast Asian islands

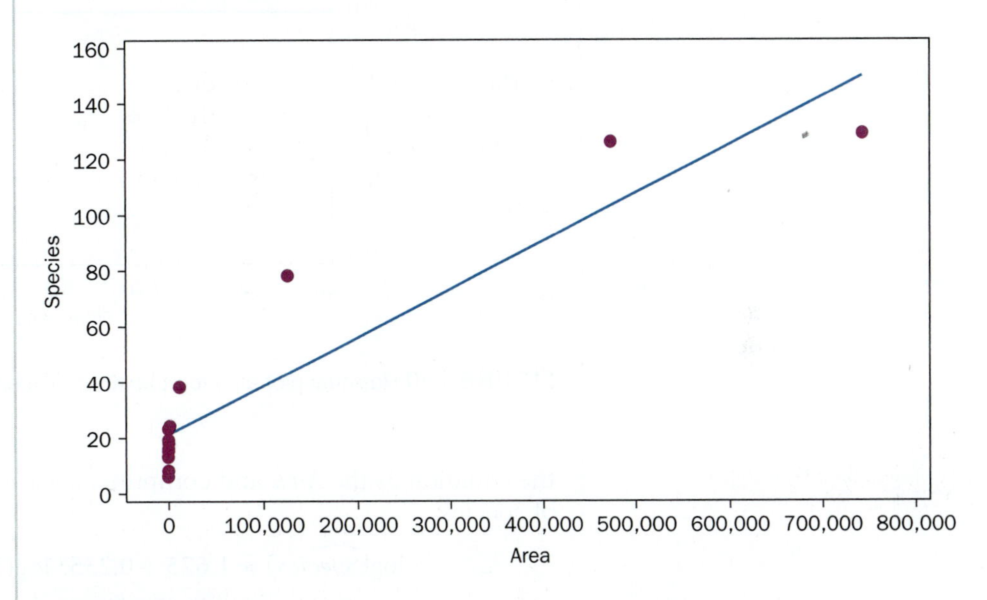

FIGURE 1.18 Number of *Mammal Species* versus *Area* for S.E. Asian islands

However, if we also take a log transformation of the *Species* (as well as the *Area*), we obtain the plot illustrated in Figure 1.19(b), which does show a relatively linear pattern. Figure 1.20 shows a residual plot from this regression. We might still have a little curvature, but this plot is much better behaved than the original scale.

We can predict log(*Species*) based on log(*Area*) for an island with the fitted model

$$\log(\widehat{Species}) = 1.625 + 0.2355\log(Area)$$

Suppose we wanted to use the model to find the predicted value for Java, which has an area of 125,628 square kilometers. We substitute 125,628 into

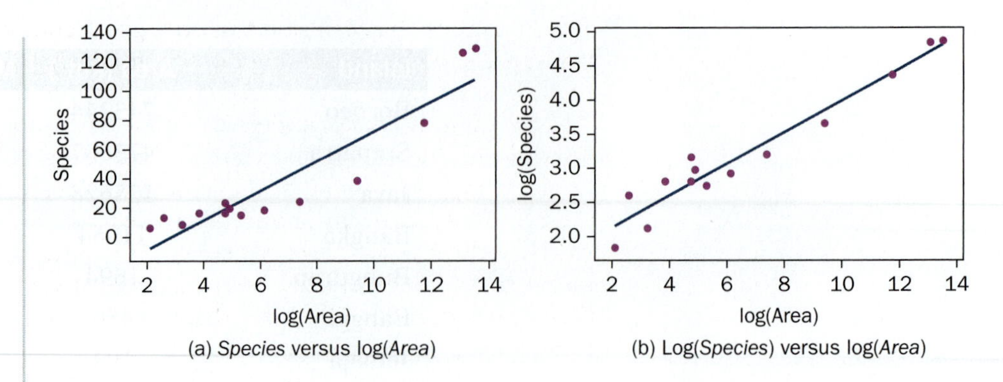

FIGURE 1.19 Log transformations of *Species* versus *Area* for S.E. Asian islands

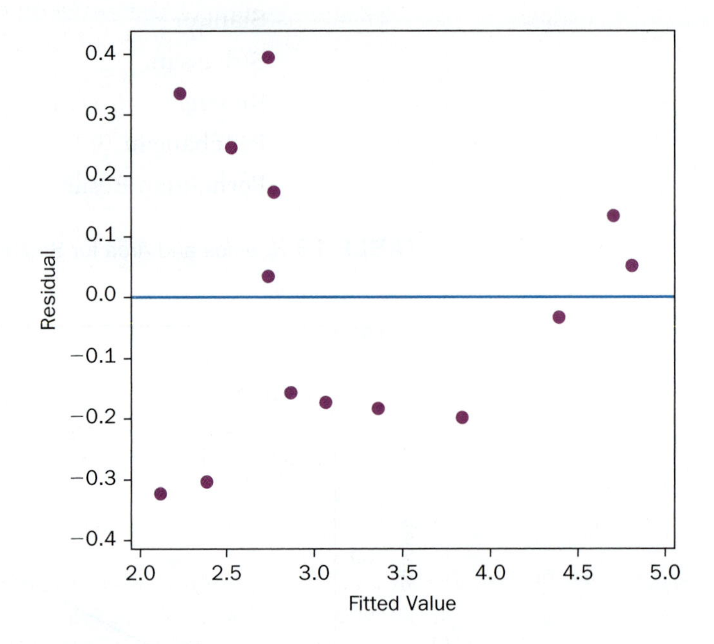

FIGURE 1.20 Residual plot after log transform of response and predictor

the equation as the *Area* and compute an estimate for the log of the number of species:

$$\widehat{\log(Species)} = 1.625 + 0.2355 \log(125{,}628) = 4.390$$

Our estimate for the number of species is then

$$e^{4.390} = 80.6 \text{ species}$$

The actual number of mammal species found on Java for this study was 78.

Choosing a Transformation

There is no guarantee that transformations will eliminate or even reduce the problems with departures from the conditions for a simple linear regression model. Often a logarithm transformation is very helpful, but not always. Finding an appropriate transformation can be as much an art as a science, yet there are guidelines that can point us to a good choice. Before discussing some of these, we start by addressing a frequently asked question about transforming: "Isn't it cheating?"

The idea of reexpression. We all learn about scales of measurement during our childhood. We learn to tell time using seconds, minutes, hours, days, weeks, and years. Some of us get a mark on the wall to record our height at each birthday; one way or another, we learn to measure distance in inches, feet, yards, and miles, or perhaps centimeters, meters, and kilometers. We learn to measure weight in ounces, pounds, and tons, or grams and kilograms. For all three basic quantities of time, distance, and mass, we grew up convinced that there is one "natural" way to measure.

But then we learn about decibels to measure "How loud?" for sounds. We learn about the Richter scale to measure "How strong?" for earthquakes. In chemistry, we learn to measure acidity using pH. All three of these—decibels, Richter scale, and pH—are based on the log transform. As a rule, we use them without needing to think about logarithms. For example, it makes sense if we are told that a toilet flushing registers at 75 decibels, riding in a subway car is at 95 decibels, and a rock band is 110. In the same way, we can use the Richter scale to compare earthquakes: The twin earthquakes in April 2017 near Mankato, Kansas, had magnitudes of 3.3 (April 7) and 2.9 (April 8). The 1906 San Francisco earthquake had a magnitude of 7.8, and the 1994 Northridge earthquake had a magnitude of 6.7. We take the logarithmic scale on its own terms, without trying to back-transform to find the amplitude of the seismic waves.

Bottom line: In some contexts, we take the log scale as "natural" in the same way we take minutes, inches, and pounds as "natural." It's not a matter of one scale being right and all others being wrong. In particular, measuring in the log scale doesn't change the data; logs just change the way we express the data. One of your authors is 71 inches tall. On the log scale, that same author registers as having log(height) = 4.26. Has the author suddenly been miniaturized? Of course not. His height is the same, but that height has been reexpressed in a new scale.*

Why logs? Perhaps you are thinking, "OK. Sometimes we want a log scale, sometimes not. But why logs?" One key is to recall that logarithms turn multiplication and division operations into addition and subtraction. In many data analysis situations, we are interested in differences: comparing one group with another, measuring change over time, or seeing how a change in one variable is related to the change in another variable. Differences are easy to see and compute. (Hold up ten fingers, put down three fingers, and you can see what's left. You can also see the difference between 10″ and 3″ on a ruler.) However, another important way to make comparisons is with ratios that express proportional relationships or measures such as percent difference. These may be harder to visualize directly. (You can't do ratios so well on your fingers because you can't hold up 5/16 of a finger.) However, a logarithm converts a ratio (which is harder to visualize) to a difference (which is easier to deal with).

The next example is artificially simple, in order to make the simple point that proportional relationships are common, but may be more easily handled when reexpressed with logarithms.

*That's why statistician John Tukey invented the term *reexpression* as an alternative to the mathematician's term *transformation*. The word *transform* suggests change, as in shrinking an author. The word *reexpress* better captures the true meaning.

EXAMPLE 1.9

Areas of circles For the sake of this example, imagine that you are a pioneer in the emerging field of circular science. Little is known, and your goal is to understand the relationship between the area of a circle (response) and its radius (predictor). Your lab produces the following data, which you plan to analyze using regression. You have a collection of circles, with radii ranging from 1″ to 50″. You know the area of each circle, but circular science does not know that area can be found by multiplying *radius*^{2} by π; or, as we learned in school, $A = \pi r^2$.

The data are shown in Figure 1.21(a). Notice the curved pattern: as radius increases, so does area, but not in a linear way.

With benefit of hindsight, we know that logs convert multiplication into addition: If $A = \pi r^2$ then $log(A) = log(\pi) + 2log(r)$. If we reexpress *Area* and *Radius* in a log scale, the curved pattern is gone, as can be seen in Figure 1.21(b).

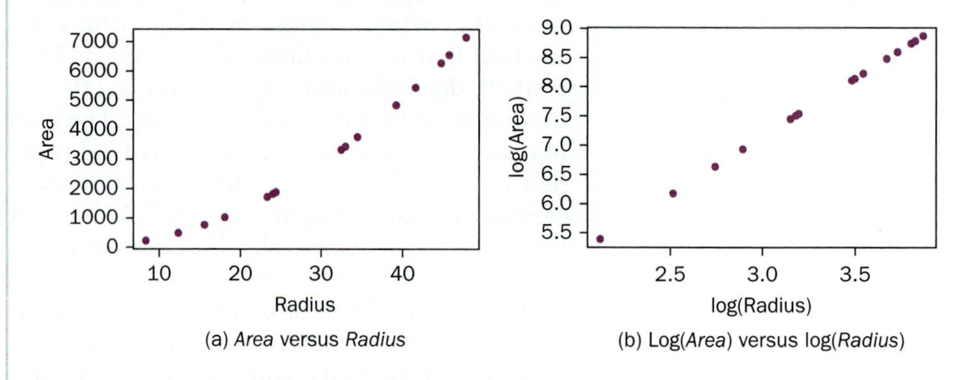

FIGURE 1.21 Circle *Area* versus *Radius* before and after log transformations

The last example was artificially simple, but you can use it as a stand-in for all those relationships in science that depend on proportional reasoning, that is, on multiplication.

- When moving at constant rate r, the distance traveled D is equal to rate times the travel time (t): $D = rt$.
- The volume V of a gas is inversely proportional to the pressure P: $V = cP^{-1}$.
- The gravitational force (F) acting on two bodies is inversely proportional to their distance (d) apart: $F = kd^{-2}$.

All of these relationships, and a host of others, are examples of "power law relationships" for which a response is proportional to some power of an explanatory variable. All power law relationships become linear in the log/log scale:

Power Law	log/log scale
$D = rt$	$log(D) = log(r) + (1)log(t)$
$V = cP^{-1}$	$log(V) = log(c) + (-1)log(P)$
$F = kd^{-2}$	$log(F) = log(k) + (-2)log(d)$

Power law relationships are also important in economics. The Cobb-Douglas formula* expresses the output of a production process as a power of the available resources.† For example,

$$Output \text{ is proportional to } Labor^{\alpha} \cdot Capital^{\beta}$$

*No relation to the Cobb who is one of this text's authors.

†This model goes beyond the single predictor models of this chapter, but we deal with multiple predictors in Chapter 3.

Taking logs gives

$$log(Output) = log(constant) + \alpha log(Labor) + \beta log(Capital)$$

The coefficients α and β are called "elasticities" and they have a useful interpretation: they tell the percent increase in output for a 1 percent increase in the resource. This interpretation and the metaphor of production can be a useful way to think about proportional relationships in general.

For the *Species* by *Area* example (page 38) we can think of island area as the resource, and number of mammal species as the output, and ask, "How much area does it take to produce a given number of species?" In Figure 1.19(b) the fitted relationship in the log/log scale is

$$log(Species) = 1.625 + 0.2355 log(Area)$$

which translates to a formula in the original *Species* and *Area* units of

$$Species = 5.0784 Area^{0.2355}$$

To summarize, here are some signs to consider a log transform:

SIGNS THAT A LOG TRANSFORMATION MIGHT BE USEFUL

- The distribution of one (or both) variables is skewed to the high end.
- The values of one (or both) variables range over two or more orders of magnitude.
- The scatterplot shows curvature.
- The plot of residuals versus fitted values shows a megaphone pattern, with more variability in residuals as the fitted values increase.
- Proportional relationships and the "production metaphor" are plausible.

Is the logarithm a magic bullet, or at least a magic button on your calculator? Of course not. For many situations, if you need to transform to get a linear plot you will be able to find a suitable choice by considering transformations from the "power family"—those of the form "*x*-to-the-power." These transformations take each value of x and change it to x^{power} for some fixed value of *power*. Squares ($power = 2$) transform x to x^2. Square roots ($power = 1/2$) transform x to $\sqrt{x}$. Reciprocals ($power = -1$) transform x to $1/x$. For technical reasons, the log transform corresponds to $power = 0$.

Any tips to choose the power? While you can use a trial-and-error approach with various reexpressions (as we did in Example 1.18), you can sometimes get a hint from the shape of a plot of the relationship. If your scatterplot is concave down, you should "transform down"; that is, try powers of the predictor less than 1. If your plot is concave up, you should "transform up"; that is, try predictor powers greater than 1. If transforming the response variable, you should reverse this advice (use a power greater than 1 for concave down and less than 1 for concave up). Note that Figure 1.13 on page 36 comparing number of doctors and hospitals for a sample of counties shows a slightly concave-up shape. This suggests either transforming the predictor to $Hospitals^2$ or the response to $\sqrt{MDs}$. We chose $\sqrt{MDs}$ in that example and it did a good job of "straightening" the relationship as well as stabilizing the variability.

1.5 Outliers and Influential Points

Sometimes, a data point just doesn't fit within a linear trend that is evident in the other points. Recall that one commonly used rule when we have just a single variable is to label a point as an *outlier* if it is above the third quartile, or below the first quartile, by more than 1.5 IQR. When we have (x, y) data, the situation is more interesting, and more complicated, as a point could be an outlier with respect to X, with respect to Y, or both. We will find it helpful to differentiate between two kinds of extreme values in a regression setting.

We might find an unusual point that doesn't fit with the other points in a scatterplot vertically—this is considered an *outlier*—or a point may differ from the others horizontally and vertically so that it is an *influential* point. In this section, we examine some quick methods for detecting outliers and influential points using graphs and summary statistics. More detailed methods for identifying unusual points in regression appear in Section 4.4.

Outliers

In the setting of simple linear regression, we call a point an *outlier* if the magnitude of the residual is unusually large. That is, we focus on extreme values in the vertical direction, relative to the linear regression line. (We'll think about unusual points in the horizontal direction later.) How large must a residual be for a point to be called an outlier? That depends on the variability of all the residuals, as we see in the next example.

EXAMPLE 1.10

Jump16

Olympic long jump During the 1968 Olympics, Bob Beamon shocked the track and field world by jumping 8.9 meters (29 feet 2.5 inches), breaking the world record for the long jump by 0.65 meter (more than 2 feet). Figure 1.22 shows the winning men's Olympic long jump distance[4] (labeled as *Gold*) versus *Year*, together with the least squares regression line, for the $n = 28$ Olympics held during the period 1900–2016. The data are stored in **LongJumpOlympics2016**.

The 1968 point clearly stands above the others and is far removed from the regression line. Because this point does not fit the general pattern in the scatterplot, it is an outlier. The unusual nature of this point is perhaps even more evident in Figure 1.23, a residual plot for the fitted least squares model.

The fitted regression model is

$$\widehat{Gold} = -16.47 + 0.01251 Year$$

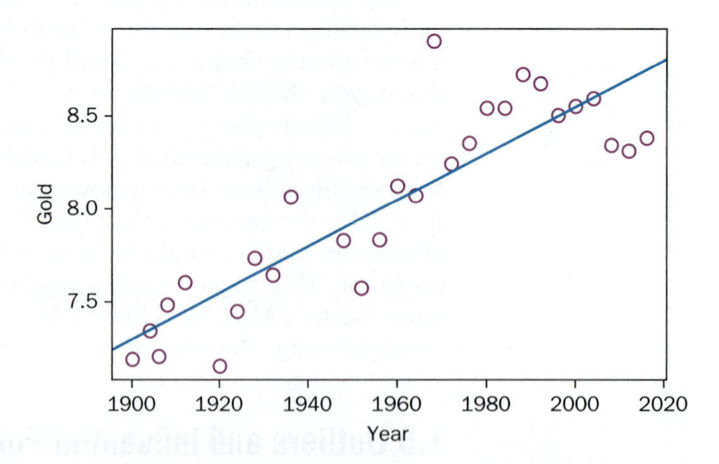

FIGURE 1.22 Gold-medal-winning distances (m) for the men's Olympic long jump, 1900–2016

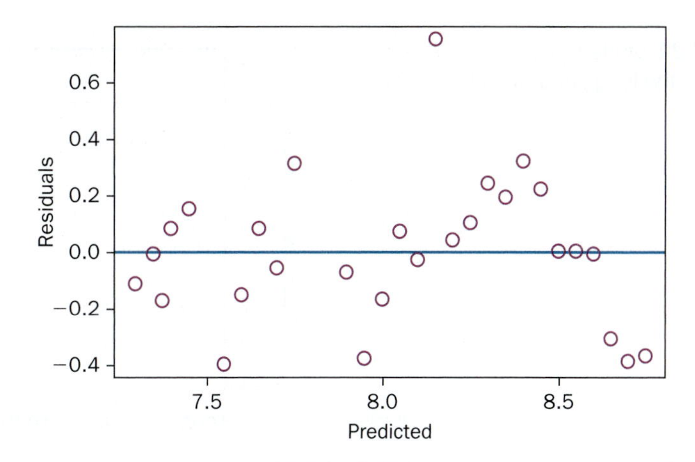

FIGURE 1.23 Residual plot for long jump model

Thus, the predicted 1968 winning long jump is $\widehat{Gold} = -16.47 + 0.01251(1968) = 8.15$ meters. The 1968 residual is $8.90 - 8.15 = 0.75$ meter.

Even when we know the context of the problem, it can be difficult to judge whether a residual of 0.75 m is unusually large. One method to help decide when a residual is extreme is to put the residuals on a standard scale. For example, since the estimated standard deviation of the regression error, $\hat{\sigma}_\epsilon$, reflects the size of a typical error, we could standardize* each residual using

$$\frac{y - \hat{y}}{\hat{\sigma}_\epsilon}$$

In practice, most statistical packages make some modifications to this formula when computing a **standardized** residual to account for how unusual the predicted value is for a particular case. Since an extreme outlier might have a significant effect on the estimation of σ_ϵ, another common adjustment is to estimate the standard deviation of the regression error using a model that is fit after omitting the point in question. Such residuals are often called **studentized**[†] (or **deleted-*t***) residuals.

standardized

studentized

If the conditions of a simple linear model hold, approximately 95% of the residuals should be within 2 standard deviations of the residual mean of zero, so we would expect most standardized or studentized residuals to be less than 2 in absolute value. We may be slightly suspicious about points where the magnitude of the standardized or studentized residual is greater than 2 and even more wary about points beyond ±3. For example, the standardized residual for Bob Beamon's 1968 jump is 2.96, indicating this point is an outlier. Figure 1.24 shows the studentized residuals for the long jump data plotted against the predicted values. The studentized residual for the 1968 jump is 3.57, while none of the other studentized residuals are beyond ±2, clearly pointing out the exceptional nature of that performance.

*Recall that *standardize* means "convert to standard units (*z*-score)." Question: "How many standard deviations from the mean is that?" Answer: "Subtract the mean and divide by the standard deviation."

[†]You may recall from an introductory statistics course that the *t*-distribution is sometimes called Student's *t*.

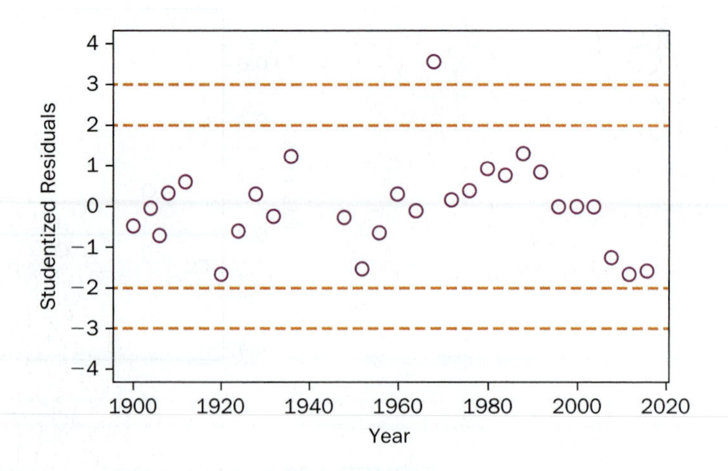

FIGURE 1.24 Studentized residuals for the long jump model

Influential Points

What should we do if a point has an unusually large or small x-value? If we see an unusual point in the horizontal direction, then we need to think about how that point influences the overall fit of the regression model. When we fit a regression model and make a prediction, we combine information across several observations, or cases, to arrive at the prediction, or fitted value, for a particular case. For example, we use the mileages and prices of many cars to arrive at a predicted price for a particular car. In doing this, we give equal weight to all of the cases in the dataset; that is, every case contributes equally to the creation of the fitted regression model and to the subsequent predictions that are based on that model.

Usually, this is a sensible and useful thing to do. Sometimes, however, this approach can be problematic, especially when the data contain one or more extreme cases that might have a significant impact on the coefficient estimates in the model.

EXAMPLE 1.11

PalmBch

Butterfly ballot The race for the presidency of the United States in the fall of 2000 was very close, with the electoral votes from Florida determining the outcome. Nationally, George W. Bush received 47.9% of the popular vote, Al Gore received 48.4%, and the rest of the popular vote was split among several other candidates. In the disputed final tally in Florida, Bush won by just 537 votes over Gore (48.847% to 48.838%) out of almost 6 million votes cast. About 2.3% of the votes cast in Florida were awarded to other candidates. One of those other candidates was Pat Buchanan, who did much better in Palm Beach County than he did anywhere else. Palm Beach County used a unique "butterfly ballot" that had candidate names on either side of the page with "chads" to be punched in the middle. This nonstandard ballot seemed to confuse some voters, who punched votes for Buchanan that may have been intended for a different candidate. Figure 1.25 shows the number of votes that Buchanan received plotted against the number of votes that Bush received for each county, together with the fitted regression line ($\widehat{Buchanan} = 45.3 + 0.0049\ Bush$). The data are stored in **PalmBeach**.

The data point near the top of the scatterplot is Palm Beach County, where Buchanan picked up over 3000 votes. Figure 1.26 is a plot of the residuals versus fitted values for this model; clearly, Palm Beach County stands out from the rest of the data. Using this model, statistical software computes the standardized residual for Palm Beach to be 7.65 and the studentized residual to be 24.08! No question that this point should be considered an outlier. Note also that the data point at the far right on the plots (Dade

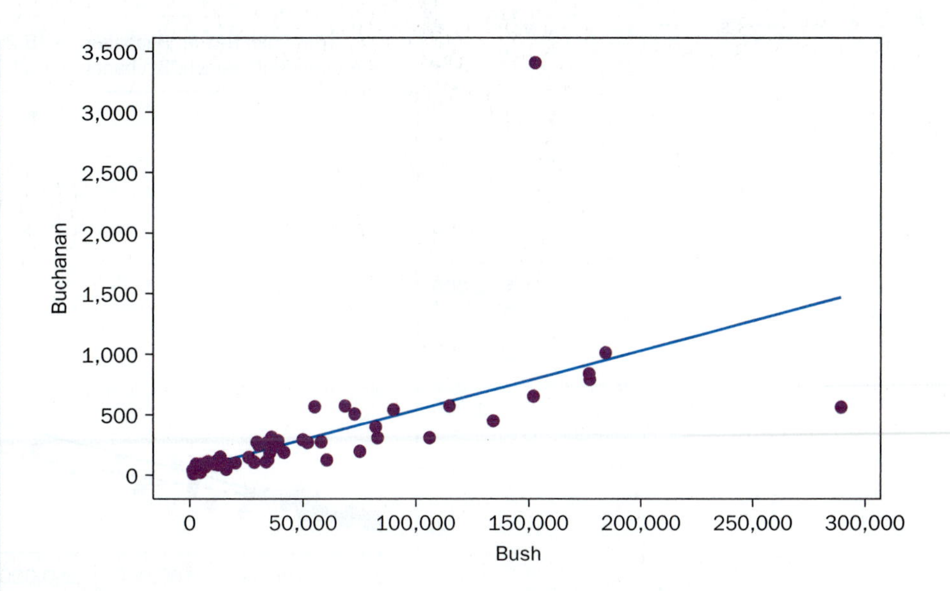

FIGURE 1.25 2000 presidential election totals in Florida counties

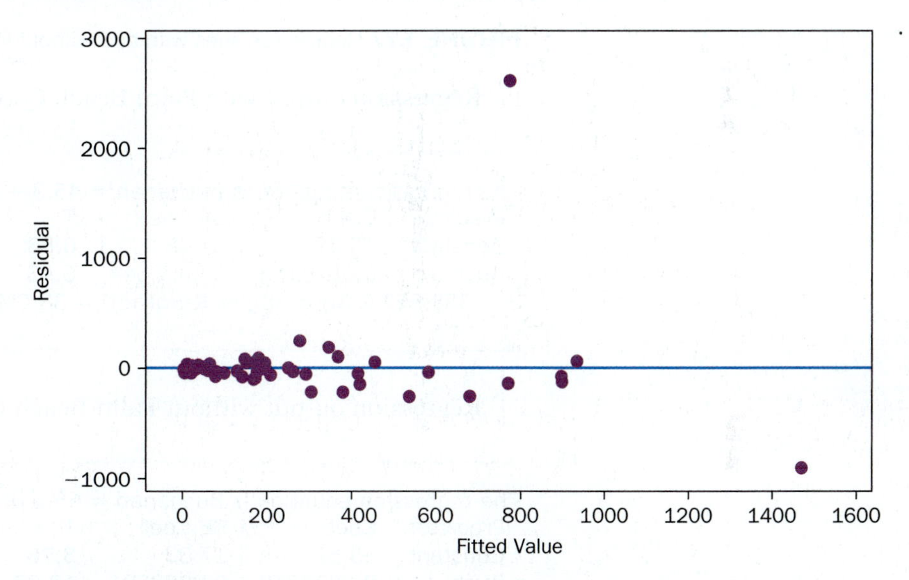

FIGURE 1.26 Residual plot for the butterfly ballot data

County) has a large negative residual of −907.5, which gives a standardized residual of −3.06; certainly something to consider as an outlier, although not as dramatic as Palm Beach.

Other than recognizing that the model does a poor job of predicting Palm Beach County (and to a lesser extent Dade County), should we worry about the effect that such extreme values have on the rest of the predictions given by the model? Would removing Palm Beach County from the dataset produce much change in the regression equation? Portions of computer output for fitting the simple linear model with and without the Palm Beach County data point are shown as follows. Figure 1.27 shows both regression lines, with the steeper slope (0.0049) occurring when Palm Beach County is included and the shallower slope (0.0035) when that point is omitted. Notice that the effect of the extreme value for Palm Beach is to "pull" the regression line in its direction.

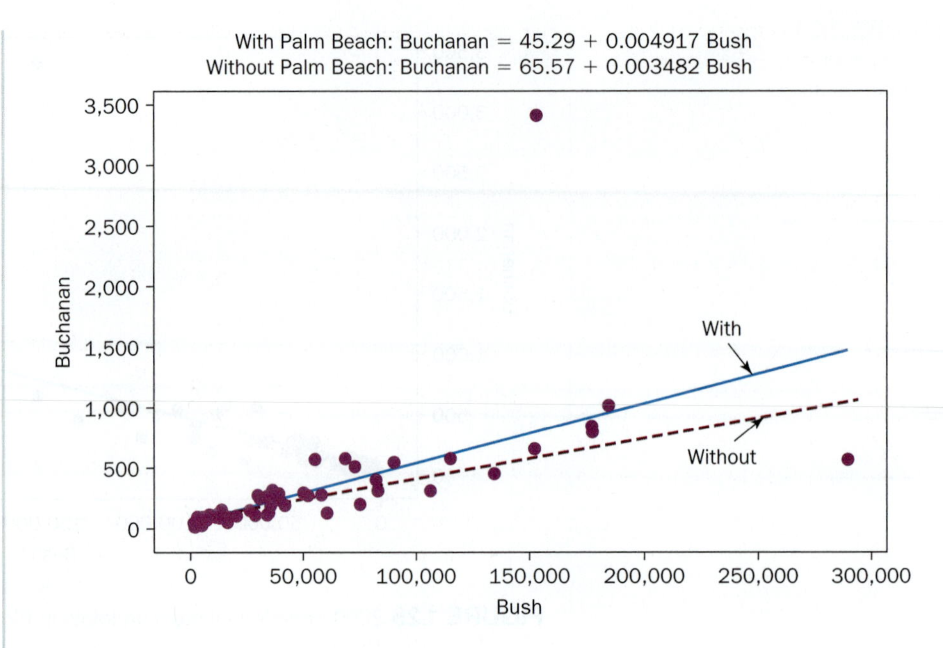

FIGURE 1.27 Regression lines with and without Palm Beach

Regression output with Palm Beach County:

```
The regression equation is Buchanan = 45.3 + 0.00492 Bush
 Predictor   Coef         SE Coef     T     P
 Constant   45.29         54.48       0.83  0.409
 Bush        0.0049168    0.0007644   6.43  0.000
S = 353.922 R-Sq = 38.9% R-Sq(adj) = 38.0%
```

Regression output without Palm Beach County:

```
The regression equation is Buchanan = 65.6 + 0.00348 Bush
 Predictor   Coef         SE Coef     T      P
 Constant   65.57         17.33       3.78   0.000
 Bush        0.0034819    0.0002501   13.92  0.000
S = 112.453 R-Sq = 75.2% R-Sq(adj) = 74.8%
```

influence The amount of **influence** that a single point has on a regression fit depends on how well it aligns with the pattern of the rest of the points and on its value for the predictor variable. Figure 1.28 shows the regression lines we would have obtained if the extreme value (3407 Buchanan votes) had occurred in Dade County (with 289,456 Bush votes), Palm Beach County (152,846 Bush votes), Clay County (41,745 Bush votes), or not occurred at all. Note that the more extreme values for the predictor (large Bush counts in Dade or Palm Beach) produced a bigger effect on the slope of the regression line than when the outlier was placed in a more "average" Bush county such as Clay.

Generally, points farther from the mean value of the predictor ($\overline{x}$) have greater potential to influence the slope of a fitted regression line. This concept

leverage is known as the **leverage** of a point. Points with high leverage have a greater capacity to pull the regression line in their direction than do low leverage points near the predictor mean. Although in the case of a single predictor, we could

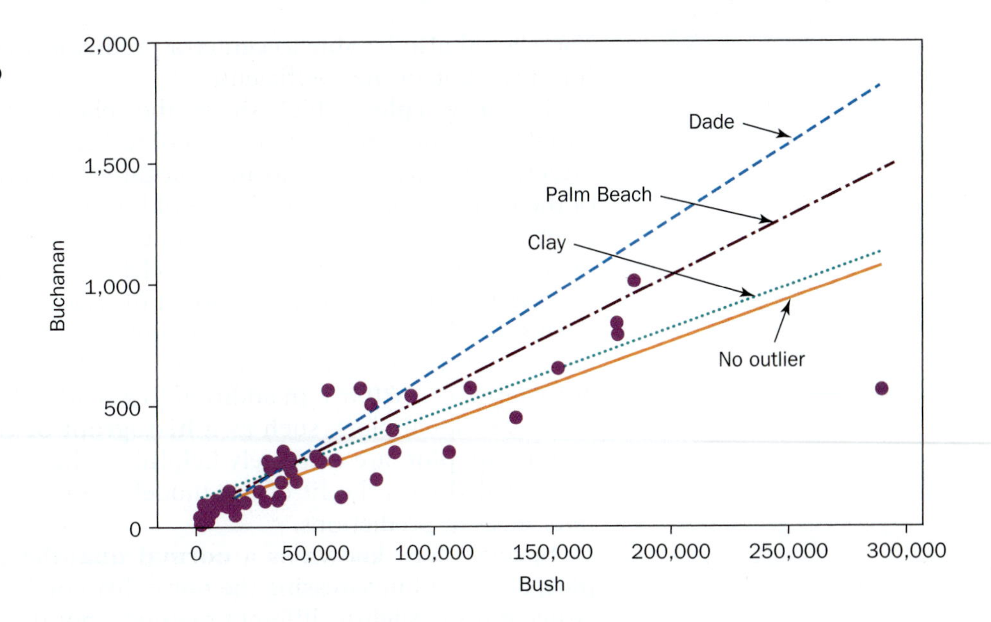

FIGURE 1.28 Regression lines with an outlier of 3407 "moved" to different counties

measure leverage as just the distance from the mean, we introduce a somewhat more complicated statistic in Section 4.4 that can be applied to more complicated regression settings. Notice that an unusual predictor value *might* indicate that a case greatly affects the regression line; however, it could also fall on the line determined by the other points and have little effect on the fit.

Extrapolation

extrapolation

Another potential problem with extreme values is trying to make a prediction for an unusually large or small value of the explanatory variable. This is known as **extrapolation**. Results may be very unreliable when trying to predict for a value that is far from the data that is used to build the model. If we tried to use the model of Example 1.3 to predict the price of a Honda Accord with 200 thousand miles, we would see $\widehat{Price} = 20.81 - 0.1198(200) = -3.15$. Clearly we shouldn't expect to get money and the car!

CHAPTER SUMMARY

In this chapter, we considered a **simple linear regression model** for predicting a single quantitative response variable Y from a single quantitative predictor X:

$$Y = \beta_0 + \beta_1 X + \epsilon$$

You should be able to use statistical software to estimate and interpret the slope and intercept for this model to produce a least squares regression line:

$$\hat{Y} = \hat{\beta}_0 + \hat{\beta}_1 X$$

The coefficient estimates, $\hat{\beta}_0$ and $\hat{\beta}_1$, are obtained using a method of least squares that selects them to provide the smallest possible sum of squared residuals (SSE). A **residual** is the observed response (y) minus the predicted response ($\hat{y}$), or the vertical distance from a point to the line. You should be able to interpret, in context, the intercept and slope, as well as residuals.

You should also be able to center a predictor and describe the impact on the interpretation of the coefficients.

The **scatterplot**, which shows the relationship between two quantitative variables, is an important visual tool to help choose a model. We look for the direction of association (positive, negative, or a random scatter), the strength of the relationship, and the degree of linearity. Assessing the fit of the model is a very important part of the modeling process. The conditions to check when using a simple linear regression model include **linearity**, **zero mean** (of the residuals), **constant variance** (about the regression line), **independence**, **random selection** (or random assignment), and **normality** (of the residuals). We can summarize several of these conditions by specifying the distribution of the error term, $\epsilon \sim N(0, \sigma_\epsilon)$. In addition to a scatterplot with the least squares line, various residual plots, such as a **histogram of the residuals** or a **residuals versus fits plot**, are extremely helpful in checking these conditions. Once we are satisfied with the fit of the model, we use the estimated model to make inferences or predictions.

Special plots, known as a **normal quantile plot** or **normal probability plot**, are useful in assessing the normality condition. These two plots are constructed using slightly different methods, but the interpretation is the same. A linear trend indicates that normality is reasonable, and departures from linearity indicate trouble with this condition. Be careful not to confuse linearity in normal plots with the condition of a linear relationship between the predictor and response variable.

You should be able to estimate the **standard deviation of the error** term, σ_ϵ, the third parameter in the simple linear regression model. The estimate is based on the sum of squared errors (SSE) and the associated degrees of freedom $(n - 2)$:

$$\hat{\sigma}_\epsilon = \sqrt{\frac{SSE}{n-2}}$$

and is the typical error, often referred to as the **standard error of regression**.

If the conditions are not satisfied, then **transformations** on the predictor, response, or both variables should be considered. Typical transformations include the square root, reciprocal, logarithm, and raising the variable(s) to another power. Identifying a useful transformation for a particular dataset (if one exists at all) is as much art as science. Trial and error is often a good approach.

You should be able to identify obvious **outliers** and **influential points**. Outliers are points that are unusually far away from the overall pattern shown by the other data. Influential points exert considerable impact on the estimated regression line. We will look at more detailed methods for identifying outliers and influential points in Section 4.4. For now, you should only worry about recognizing very extreme cases and be aware that they can affect the fitted model and analysis. One common guideline is to tag all observations with **standardized** or **studentized residuals** smaller than −2 or larger than 2 as possible outliers. To see if a point is influential, fit the model with and without that point to see if the coefficients change very much. In general, points far from the average value of the predictor variable have greater potential to influence the regression line.

EXERCISES

Conceptual Exercises

1.1 Equation of a line. Consider the fitted regression equation $\hat{Y} = 100 + 15X$. Which of the following is *false*?

a. The sample slope is 15.

b. The predicted value of Y when $X = 0$ is 100.

c. The predicted value of Y when $X = 2$ is 110.

d. Larger values of X are associated with larger values of Y.

1.2 Residual plots to check conditions. For which of the following conditions for inference in regression does a residual plot *not* aid in assessing whether the condition is satisfied?

a. Linearity

b. Constant variance

c. Independence

d. Zero mean

1.3 Sparrows slope. Priscilla Erickson from Kenyon College collected data on a stratified random sample of 116 Savannah sparrows at Kent Island. The weight (in grams) and wing length (in mm) were obtained for birds from nests that were reduced, controlled, or enlarged. The data[5] are in the file **Sparrows**. Based on the following computer output (which you will also use for the odd exercises through Exercise 1.11), what is the slope of the least squares regression line for predicting sparrow weight from wing length? **Sparrow**

The regression equation is Weight = 1.37 + 0.467 WingLength

Predictor	Coef	SE Coef	T	P
Constant	1.3655	0.9573	1.43	0.156
WingLength	0.4674	0.03472	13.46	0.000

S = 1.39959 R-Sq = 61.4% R-Sq(adj) = 61.1%

Analysis of Variance

Source	DF	SS	MS	F	P
Regression	1	355.05	355.05	181.25	0.000
Residual Error	114	223.31	1.96		
Total	115	578.36			

1.4 Olympic long jump distances slope. We recorded the length of the gold-winning long jump distance in meters for all of the Olympic Games from 1900 to 2016. We are interested in using the year to predict the winning jump length. The data are given in the file **LongJumpOlympics2016**. Based on the computer output that follows (which you will also use for the even exercises through Exercise 1.12), what is the slope of the least squares regression line for predicting the winning Olympic long jump length from year? **Jump16**

Regression Equation
Gold = -16.47 + 0.01251 Year

Term	Coef	SE Coef	T-Value	P-Value
Constant	-16.47	2.67	-6.18	0.000
Year	0.01251	0.00136	9.19	0.000

S	R-sq	R-sq(adj)	R-sq(pred)
0.259522	76.46%	75.56%	72.98%

Analysis of Variance

Source	DF	Adj SS	Adj MS	F-Value	P-Value
Regression	1	5.689	5.68939	84.47	0.000
Error	26	1.751	0.06735		
Total	27	7.441			

1.5 Sparrows intercept. Refer to the output in Exercise 1.3. Based on the regression output, what is the intercept of the least squares regression line for predicting sparrow weight from wing length? **Sparrow**

1.6 Olympic long jump intercept. Refer to the output in Exercise 1.4. Based on the regression output, what is the intercept of the least squares regression line for predicting winning Olympic long jump length from year? **Jump16**

1.7 Sparrows slope interpretation. Refer to the value reported as the answer to Exercise 1.3. Interpret this coefficient in the context of this setting. **Sparrow**

1.8 Olympic long jump slope interpretation. Refer to the value reported as the answer to Exercise 1.4. Interpret this coefficient in the context of this setting. **Jump16**

1.9 Sparrows regression standard error. Refer again to the output in Exercise 1.3. Based on the regression output, what is the size of the typical error when predicting weight from wing length? **Sparrow**

1.10 Olympic long jump regression standard error. Refer again to the output in Exercise 1.4. Based on the regression output, what is the size of the typical error when predicting winning jump length from year? **Jump16**

1.11 Sparrow degrees of freedom. Refer one more time to the output in Exercise 1.3. What are the degrees of freedom associated with the regression standard error when predicting weight from wing length for these sparrows? **Sparrow**

1.12 Olympic long jump degrees of freedom. Refer one more time to the output in Exercise 1.4. What are the degrees of freedom associated with the regression standard error when predicting winning jump length from year? **Jump16**

1.13 Computing a residual. Consider the fitted regression equation $\hat{Y} = 25 + 7X$. If $x_1 = 10$ and $y_1 = 100$, what is the residual for the first data point?

1.14 Computing a residual. Consider the fitted regression equation $\hat{Y} = 78 - 0.5X$. If $x_1 = 30$ and $y_1 = 60$, what is the residual for the first data point?

Guided Exercises

1.15 Leaf size. Biologists know that the leaves on plants tend to get smaller as temperatures rise. The file **LeafWidth** has data on samples of leaves from the species *Dodonaea viscosa* subsp. *angustissima*, which have been collected in a certain region of South Australia for many years. The variable *Width* is the average width, in mm, of leaves, taken at their widest points, that were collected in a given year. **LeafW**

a. Fit the regression of *Width* on *Year*. What is the fitted regression model?

b. Interpret the coefficient of *Year* in the context of this setting.

c. What is the predicted width of these leaves in the year 1966?

1.16 Glow-worms. The paper "I'm Sexy and I Glow It: Female Ornamentation in a Nocturnal Capital Breeder" is about glow-worms. Female glow-worms attract males by glowing, and flying males search for females. The authors write, "We found brightness to correlate with female fecundity." The file **GlowWorms** has data on 26 female glow-worms captured in Finland. The variable *Lantern* is the size, in mm, of the part of the female abdomen that glows. The variable *Eggs* is number of eggs laid by the glow-worm. **Glow**

a. Fit the regression of *Eggs* on *Lantern*. What is the fitted regression model?

b. Interpret the coefficient of *Lantern* in the context of this setting.

c. Suppose a glow-worm has a lantern size of 14 mm. What is the predicted number of eggs she will lay?

1.17 Grip strength and attractiveness. The file **Faces** has data on grip strength (*MaxGripStrength*) and facial attractiveness (*Attractive*), as rated by female college students, for each of 38 men.[6] Grip strength, measured in kilograms, is the maximum of three readings for each hand from the man squeezing a handheld dynamometer. **Faces**

a. Fit the regression of *MaxGripStrength* on *Attractive*. What is the fitted regression model?

b. Interpret the coefficient of *Attractive* in the context of this setting.

c. Predict the *MaxGripStrength* of a man with *Attractive* equal to 3.

1.18 Male body measurements. The file **Faces** has data on grip strength (*MaxGripStrength*) and shoulder-to-hip ratio (*SHR*) for each of 38 college men. Grip strength, measured in kilograms, is the maximum of three readings for each hand from the man squeezing a handheld dynamometer. *SHR* is the ratio of shoulder circumference to hip circumference. **Faces**

a. Fit the regression of *MaxGripStrength* on *SHR*. What is the fitted regression model?

b. Interpret the coefficient of *SHR* in the context of this setting.

c. Predict the *MaxGripStrength* of a man with *SHR* equal to 1.5.

1.19 Breakfast cereal, CHOOSE/FIT. The number of calories and number of grams of sugar per serving were measured for 36 breakfast cereals. The data are in the file **Cereal**. We are interested in trying to predict the calories using the sugar content. **Cereal**

a. Make a scatterplot and comment on what you see.

b. Find the least squares regression line for predicting calories based on sugar content.

c. Interpret the value (not just the sign) of the slope of the fitted model in the context of this setting.

1.20 Houses in Grinnell, CHOOSE/FIT. The file **GrinnellHouses** contains data from 929 house sales in Grinnell, Iowa, between 2005 and into 2015. In this question we investigate the relationship of the list price of the home (what the seller asks for the home) to the final sale price. One would expect there to be a strong relationship. In most markets, including Grinnell during this period, list price almost always exceeds sale price. **Grinnell**

a. Make a scatterplot with ListPrice on the horizontal axis and SalePrice on the vertical axis. Comment on the pattern.

b. Find the least squares regression line for predicting sale price of a home based on list price of that home.

c. Interpret the value (not just the sign) of the slope of the fitted model in the context of this setting.

1.21 Breakfast cereal, ASSESS. Refer to the data on breakfast cereals that is described in Exercise 1.19. The number of calories and number of grams of sugar per serving were measured for 36 breakfast cereals. The data are in the file **Cereal**. We are interested in trying to predict the calories using the sugar content. **Cereal**

a. How many calories would the fitted model predict for a cereal that has 10 grams of sugar?

b. Cheerios has 110 calories but just 1 gram of sugar. Find the residual for this data point.

c. Does the linear regression model appear to be a good summary of the relationship between calories and sugar content of breakfast cereals?

1.22 Grinnell houses, ASSESS. Refer to the data on sales of houses in Grinnell, Iowa, that is described in Exercise 1.20. The list price and the sales price of 929 houses are in the file **GrinnellHouses**. We are interested in trying to predict the sales price using the list price. **Grinnell**

a. What sales price would the fitted model predict for a house that has a $99,500 listing price?

b. The house at 1317 Prince Street has a listing value of $99,500 and a sales price of $95,000. Find the residual for this data point.

c. Does the linear regression model appear to be a good summary of the relationship between list price and sales price for houses in Grinnell, Iowa?

1.23 Sparrow residuals. Exercise 1.3 introduces you to a model for the weight (in grams) of sparrows using the wing length as a predictor and the data in **Sparrows**. Construct and interpret the following plots for this model. In each case, discuss what the plot tells you about potential problems (if any) with the regression conditions. **Sparrow**

a. Scatterplot that includes the least squares line. Are there any obvious outliers or influential points in this plot?

b. Histogram of the residuals.

c. Normal probability plot of the residuals.

1.24 Olympic long jump residuals. Exercise 1.4 introduced you to a model for the length (in meters) of the winning long jump in Olympic competition using the year as a predictor and the data in **LongJumpOlympic2016**. Construct and interpret the following plots for this model. In each case, discuss what the plot tells you about potential problems (if any) with the regression conditions. **Jump16**

a. Scatterplot that includes the least squares line. Are there any obvious outliers or influential points in this plot?

b. Histogram of the residuals.

c. Normal probability plot of the residuals.

1.25 Rails to Trails distance. The file **RailsTrails** contains data from a sample of 104 homes that were sold in 2007 in the city of Northampton, Massachusetts. The goal was to see if a home's proximity to a biking trail would relate to its selling price. Perhaps, for example, proximity to the trail would add value to the home. And to see if any such effect has changed over time, the researchers took estimated prices for these homes at four different time points using the website Zillow.com. Here we focus on the estimated price of homes in 2007 using the price in thousands of 2014 dollars (variable name *Adj2007*). The variable *Distance* measures the distance in miles to the nearest entry point to the rail trail network. **Rails**

a. Create a scatterplot of *Adj2007* versus *Distance* and describe the relationship.

b. Fit a simple linear regression model and interpret the results, making a clear interpretation of the slope parameter.

c. Report the regression standard error and discuss its meaning.

d. Comment on the conditions of the model.

1.26 Rails to Trails size. Consider again the dataset **RailsTrails** concerning a sample of 104 homes that sold in 2007 in Northamption, Massachusetts. It is commonly the case that larger homes will sell for more. Note: units are thousands of dollars for home price (*Adj2007*) and thousands of square feet for the *SquareFeet* variable. **Rails**

a. Use a scatterplot of *Adj2007* (the price in 2007 adjusted to 2014 dollars) against *SquareFeet* (the total interior area of the home) to decide if that common knowledge holds here. Describe the association (if any) between price and size.

b. Find a simple linear regression model for part (a), and interpret the results, making a clear interpretation of the slope.

c. Report and interpret the regression standard error.

d. Comment on the conditions of the model.

1.27 Capacitor voltage. A capacitor was charged with a 9-volt battery and then a voltmeter recorded the voltage as the capacitor was discharged. Measurements were taken every 0.02 second. The data are in the file **Volts**. **Volts**

a. Make a scatterplot with *Voltage* on the vertical axis versus *Time* on the horizontal axis. Comment on the pattern.

b. Create a residuals versus fits plot for predicting *Voltage* from *Time*. What does this plot tell you about the idea of fitting a linear model to predict *Voltage* from *Time*? Explain.

c. Transform *Voltage* using a log transformation and then plot log(*Voltage*) versus *Time*. Comment on the pattern.

d. Regress log(*Voltage*) on *Time* and write down the prediction equation.

e. Make a plot of residuals versus fitted values from the regression from part (c). Comment on the pattern.

1.28 Arctic sea ice. Climatologists have been measuring the amount of sea ice in both the Arctic and Antarctic regions for a number of years. The datafile **SeaIce** gives information about the amount of sea ice in the arctic region as measured in September (the time when the amount of ice is at its least) since 1979. The basic research question is to see if we can use time to model the amount of sea ice.

In fact, there are two ways to measure the amount of sea ice: *Area* and *Extent*. *Area* measures the actual amount of space taken up by ice. *Extent* measures the area inside the outer boundaries created by the ice. If there are areas inside the outer boundaries that are not ice (think about a slice of Swiss cheese), then the *Extent* will be a larger number than the *Area*. In fact, this is almost always true. Both *Area* and *Extent* are measured in 1,000,000 square km.

We will focus on the *Extent* of the sea ice in this exercise and see how it has changed over time since 1979. Instead of using the actual year as our explanatory variable, to keep the size of the coefficients manageable, we will use a variable called *BaseYear*, which measures time since 1978 (the value for 1979 is 1). **SeaIce**

a. Produce a scatterplot that could be used to predict *Extent* from *BaseYear*. Comment on the pattern.

b. Create a residuals versus fits plot for predicting *Extent* from *BaseYear*. What does this plot tell you about the idea of fitting a linear model to predict *Extent* from *BaseYear*? Explain.

c. Transform *Extent* by squaring it. Plot $Extent^2$ versus *BaseYear*. Comment on the pattern.

d. Create a residuals versus fits plot for this new response variable using *BaseYear*. Discuss whether there is improvement in this plot over the one you created in part (b).

e. Redo parts (c) and (d) using the cube of *Extent*.

f. Would you be comfortable using a linear model for any of these three response variables? Explain.

1.29–1.32 Caterpillars. Student and faculty researchers at Kenyon College conducted numerous experiments with *Manduca Sexta* caterpillars to study biological growth.[7] A subset of the measurements from some of the experiments is in the file **Caterpillars**. **Caterp**

Exercises 1.29–1.32 deal with some relationships in these data. The variables in the dataset include:

Variable	Description
Instar	A number from 1 (smallest) to 5 (largest) indicating stage of the caterpillar's life
Active Feeding	An indicator (Y or N) of whether or not the animal is actively feeding
Fgp	An indicator (Y or N) of whether or not the animal is in a free growth period
Mgp	An indicator (Y or N) of whether or not the animal is in a maximum growth period
Mass	Body mass of the animal in grams
Intake	Food intake in grams/day
WetFrass	Amount of frass (solid waste) produced by the animal in grams/day
DryFrass	Amount of frass, after drying, produced by the animal in grams/day
Cassim	CO_2 assimilation (ingestion–excretion)
Nfrass	Nitrogen in frass
Nassim	Nitrogen assimilation (ingestion–excretion)

Log (base 10) transformations are also provided as *LogMass*, *LogIntake*, *LogWetFrass*, *LogDryFrass*, *LogCassim*, *LogNfrass*, and *LogNassim*.

1.29 Caterpillar waste versus mass. We might expect that the amount of waste a caterpillar produces per day (*WetFrass*) is related to its size (*Mass*). Use the data in **Caterpillars** to examine this relationship as outlined below.

a. Produce a scatterplot for predicting *WetFrass* based on *Mass*. Comment on any patterns.

b. Produce a similar plot using the log (base 10) transformed variables, *LogWetFrass* versus *LogMass*. Again, comment on any patterns.

c. Would you prefer the plot in part (a) or part (b) to predict the amount of wet frass produced for caterpillars? Fit a linear regression model for the plot you chose and write down the prediction equation.

d. Add a plotting symbol for the grouping variable *Instar* to the scatterplot that you chose in (c). Does the linear trend appear consistent for all five stages of a caterpillar's life? (*Note:* We are not ready to fit more complicated models yet, but we will return to this experiment in Chapter 3.)

e. Repeat part (d) using plotting symbols (or colors) for the groups defined by the free growth period variable *Fgp*. Does the linear trend appear to be better when the caterpillars are in a free growth period? (Again, we are not ready to fit more complicated models, but we are looking at the plot for linear trend in the two groups.)

1.30 Caterpillar nitrogen assimilation versus mass. The *Nassim* variable in the **Caterpillars** dataset measures nitrogen assimilation, which might be associated with the size of the caterpillars as measured with *Mass*. Use the data to examine this relationship as outlined below.

a. Produce a scatterplot for predicting nitrogen assimilation (*Nassim*) based on *Mass*. Comment on any patterns.

b. Produce a similar plot using the log (base 10) transformed variables, *LogNassim* versus *LogMass*. Again, comment on any patterns.

c. Would you prefer the plot in part (a) or part (b) to predict the nitrogen assimilation of caterpillars with a linear model? Fit a linear regression model for the plot you chose and write down the prediction equation.

d. Add a plotting symbol for the grouping variable *Instar* to the scatterplot that you chose in (c). Does the linear trend appear consistent for all five stages of a caterpillar's life? (*Note:* We are not ready to fit more complicated models yet, but we will return to this experiment in Chapter 3.)

e. Repeat part (d) using plotting symbols (or colors) for the groups defined by the free-growth period variable *Fgp*. Does the linear trend appear to be better when the caterpillars are in a free-growth period? (Again, we are not ready to fit more complicated models, but we are looking at the plot for linear trend in the two groups.)

1.31 Caterpillar body mass and food intake. We might expect that larger caterpillars would consume more food. Use the data in **Caterpillars** to look at using food intake to predict *Mass* as outlined below.

a. Plot body mass (*Mass*) as the response variable versus food intake (*Intake*) as the explanatory variable. Comment on the pattern.

b. Plot the log (base 10) transformed variables, *LogMass* versus *LogIntake*. Again, comment on any patterns.

c. Do you think the linear model should be used to model either of the relationships in part (a) or (b)? Explain.

1.32 Caterpillar body mass and food intake, grouping variable. This exercise is similar to Exercise 1.31 except that we will reverse the roles of predictor and response, using caterpillar size (*Mass*) to predict food intake (*Intake*) for the data in **Caterpillars**.

a. Plot *Intake* as the response variable versus *Mass* as the explanatory variable. Comment on the pattern.

b. Plot the log (base 10) transformed variables, *LogIntake* versus *LogMass*. Again, comment on any patterns.

c. Would you prefer the plot in part (a) or (b) to fit with a linear model? Fit a linear regression model for the plot you chose and write down the prediction equation.

d. Add plotting symbols (or colors) for the grouping variable *Instar* to the scatterplot that you chose in (c). Is a linear model more appropriate for this relationship during some of the stages of caterpillar development?

1.33 U.S. stamp prices. Historical prices[8] of U.S. stamps for mailing a letter weighing less than 1 ounce are provided in the file **USStamps**. **USStamp**

a. Plot *Price* (in cents) versus *Year* and comment on any patterns.

b. Regular increases in the postal rates started in 1958. Remove the first four observations from the dataset and fit a regression line for predicting *Price* from *Year*. What is the equation of the regression line?

c. Plot the regression line along with the prices from 1958 to 2012. Does the regression line appear to provide a good fit?

d. Analyze appropriate residual plots for the linear model relating stamp price and year. Are the conditions for the regression model met?

e. Identify any unusual residuals.

1.34 Enrollment in mathematics courses. Total enrollments in mathematics courses at a small liberal arts college[9] were obtained for each semester from Fall 2001 to Spring 2012. The academic year at this school consists of two semesters, with enrollment counts for *Fall* and *Spring* each year as shown in Table 1.4. The variable *AYear* indicates the year at the beginning of the academic year. The data are also provided in the file **MathEnrollment**. **MthEnr**

Table 1.4 Math enrollments

AYear	Fall	Spring
2001	259	246
2002	301	206
2003	343	288
2004	307	215
2005	286	230
2006	273	247
2007	248	308
2008	292	271
2009	250	285
2010	278	286
2011	303	254

a. Plot the mathematics enrollment for each semester against time. Is the trend over time roughly the same for both semesters? Explain.

b. A faculty member in the Mathematics Department believes that the fall enrollment provides a very good predictor of the spring enrollment. Do you agree?

c. After examining a scatterplot with the least squares regression line for predicting spring enrollment from fall enrollment, two faculty members begin a debate about the possible influence of a particular point. Identify the point that the faculty members are concerned about.

d. Fit the least squares line for predicting math enrollment in the spring from math enrollment in the fall, with and without the point you identified in part (c). Would you tag this point as influential? Explain.

1.35 Pines. The dataset **Pines** contains data from an experiment conducted by the Department of Biology at Kenyon College at a site near the campus in Gambier, Ohio.[10] In April 1990, student and faculty volunteers planted 1000 white pine (*Pinus strobes*) seedlings at the Brown Family Environmental Center. These seedlings were planted in two grids, distinguished by 10- and 15-foot spacings between the seedlings. Several variables, described as follows, were measured and recorded for each seedling over time. **Pines**

Variable	Description
Row	Row number in pine plantation
Col	Column number in pine plantation
*Hgt*90	Tree height at time of planting (cm)
*Hgt*96	Tree height in September 1996 (cm)
*Diam*96	Tree trunk diameter in September 1996 (cm)

(Continued)

Variable	Description
Grow96	Leader growth during 1996 (cm)
Hgt97	Tree height in September 1997 (cm)
Diam97	Tree trunk diameter in September 1997 (cm)
Spread97	Widest lateral spread in September 1997 (cm)
Needles97	Needle length in September 1997 (mm)
Deer95	Type of deer damage in September 1995: 0 = none, 1 = browsed
Deer97	Type of deer damage in September 1997: 0 = none, 1 = browsed
Cover95	Thorny cover in September 1995: 0 = none; 1 = some; 2 = moderate; 3 = lots
Fert	Indicator for fertilizer: 0 = no, 1 = yes
Spacing	Distance (in feet) between trees (10 or 15)

a. Construct a scatterplot to examine the relationship between the initial height in 1990 and the height in 1996. Comment on any relationship seen.

b. Fit a least squares line for predicting the height in 1996 from the initial height in 1990.

c. Are you satisfied with the fit of this simple linear model? Explain.

1.36 Pines: 1997 versus 1990. Refer to the **Pines** data described in Exercise 1.35. Examine the relationship between the initial seedling height and the height of the tree in 1997. **Pines**

a. Construct a scatterplot to examine the relationship between the initial height in 1990 and the height in 1997. Comment on any relationship seen.

b. Fit a least squares line for predicting the height in 1997 from the initial height in 1990.

c. Are you satisfied with the fit of this simple linear model? Explain.

1.37 Pines: 1997 versus 1996. Refer to the **Pines** data described in Exercise 1.35. Consider fitting a line for predicting height in 1997 from height in 1996. **Pines**

a. Before doing any calculations, do you think that the height in 1996 will be a better predictor than the initial seedling height in 1990? Explain.

b. Fit a least squares line for predicting height in 1997 from height in 1996.

c. Does this simple linear regression model provide a good fit? Explain.

1.38 Fluorescence experiment. Suzanne Rohrback used a novel approach in a series of experiments to examine calcium-binding proteins. The data from one experiment[11] are provided in **Fluorescence**. The variable *Calcium* is the log of the free calcium concentration and *ProteinProp* is the proportion of protein bound to calcium. **Fluor**

a. Find the regression line for predicting the proportion of protein bound to calcium from the transformed free calcium concentration.

b. What is the regression standard error?

c. Plot the regression line and all of the points on a scatterplot. Does the regression line appear to provide a good fit?

d. Analyze appropriate residual plots. Are the conditions for the regression model met?

1.39 Caterpillar CO_2 assimilation and food intake. Refer to the data in **Caterpillars** that is described on page 54 for Exercises 1.29–1.32. Consider a linear model to predict CO_2 assimilation (*Cassim*) using food intake (*Intake*) for the caterpillars. **Caterp**

a. Plot *Cassim* versus *Intake* and comment on the pattern.

b. Find the least squares regression line for predicting CO_2 assimilation from food intake.

c. Are the conditions for inference met? Comment on the appropriate residual plots.

1.40 Caterpillar nitrogen assimilation and wet frass. Repeat the analysis described in Exercise 1.39 for a model to predict nitrogen assimilation (*Nassim*) based on the amount of solid waste (*WetFrass*) in the **Caterpillars** data. **Caterp**

1.41 Goldenrod galls 2003. Biology students collected measurements on goldenrod galls at the Brown Family Environmental Center.[12] The file **Goldenrod** contains the gall diameter (in mm), stem diameter (in mm), wall thickness (in mm), and codes for the fate of the gall in 2003 and 2004. **Golden**

a. Are stem diameter and gall diameter positively associated in 2003?

b. Plot wall thickness against stem diameter and gall diameter on two separate scatterplots for the 2003 data. Based on the scatterplots, which variable has a stronger linear association with wall thickness? Explain.

c. Fit a least squares regression line for predicting wall thickness from the variable with the strongest linear relationship in part (b).

d. Find the fitted value and residual for the first observation using the fitted model in (c).

e. What is the value of a typical residual for predicting wall thickness based on your linear model in part (c)?

1.42 Goldenrod galls 2004. Refer to the data on goldenrod galls described in Exercise 1.41. Repeat the analysis in that exercise for the measurements made in 2004 instead of 2003. The value of *Wall04* is missing for the first observation, so use the second case for part (e). **Golden**

1.43 Oysters. The dataset **Oysters** contains information from an experiment to compare two completely mechanical systems for sorting oysters for commercial purposes. Oysters are classified as small, medium, or large based upon volume, where small is less than 10 cc, medium is greater than or equal to 10 but less than 13 cc, and large is 13 cc or greater. In 2001, engineers at an R&D lab Agri-Tech, Inc., in Woodstock, Virginia, designed a 3-D system that they hoped would improve on the existing 2-D system. The 3-D system used computer scanning to estimate an oyster volume, whereas the old 2-D system estimated a cross-sectional area.

The data contain 30 cases on 30 oysters and 5 variables from a calibration experiment to see how well both the 2-D and 3-D machines work and compare their performance to one another. **Oyster**

a. Use scatterplots to see which of *ThreeD* or *TwoD* looks like the better predictor for *Volume*. Is there a clear choice and how much better do you think it will be?

b. Compute the two simple linear models related to (a). Comment on their fit, including a comparison of the typical size error each model will make when predicting an oyster volume.

Open-ended Exercises

1.44 Textbook prices. Two undergraduate students at Cal Poly took a random sample[13] of 30 textbooks from the campus bookstore in the fall of 2006. They recorded the price and number of pages in each book in order to investigate the question of whether the number of pages can be used to predict price. Their data are stored in the file **TextPrices** and appear in Table 1.5. **TxtPrc**

Table 1.5 Pages and price for textbooks

Pages	Price	Pages	Price	Pages	Price
600	95.00	150	16.95	696	130.50
91	19.95	140	9.95	294	7.00
200	51.50	194	5.95	526	41.25
400	128.50	425	58.75	1060	169.75
521	96.00	51	6.50	502	71.25
315	48.50	930	70.75	590	82.25
800	146.75	57	4.25	336	12.95
800	92.00	900	115.25	816	127.00
600	19.50	746	158.00	356	41.50
488	85.50	104	6.50	248	31.00

a. Produce the relevant scatterplot to investigate the students' question. Comment on what the scatterplot reveals about the question.

b. Determine the equation of the regression line for predicting price from number of pages.

c. Produce and examine relevant residual plots, and comment on what they reveal about whether the conditions for inference are met with these data.

1.45 Baseball game times, CHOOSE/FIT/ASSESS. What factors can help predict how long a Major League Baseball game will last? The data in Table 1.6 were collected at www.baseball-reference.com for the 14 games played on August 11, 2017, and stored in the file named **BaseballTimes** 2017. The *Time* is recorded in minutes. *Runs* and *Pitchers* are totals for both teams combined. *Margin* is the difference between the winner's and loser's scores. **BballT**

Table 1.6 Major League Baseball game times

Game	League	Runs	Margin	Pitchers	Attendance	Time
CHC-ARI	NL	11	5	10	39131	203
KCR-CHW	AL	9	3	7	18137	169
MIN-DET	AL	13	5	10	29733	201
SDP-LAD	NL	7	1	6	52898	179
COL-MIA	NL	9	3	10	20096	204
CIN-MIL	NL	21	1	10	34517	235
BOS-NYY	AL	9	1	8	46509	220
BAL-OAK	AL	9	1	8	14330	168
NYM-PHI	NL	13	1	11	26925	205
LAA-SEA	AL	11	1	9	38206	184
ATL-STL	NL	13	3	10	41928	188
CLE-TBR	AL	5	5	6	16794	164
HOU-TEX	AL	10	2	7	33897	202
PIT-TOR	IL	6	2	6	35965	161

a. First, analyze the distribution of the response variable (*Time* in minutes) alone. Use a graphical display (dotplot, histogram, boxplot) as well as descriptive statistics. Describe the distribution.

b. Examine scatterplots to investigate which of the quantitative predictor variables appears to be the best single predictor of time. Comment on what the scatterplots reveal.

c. Choose the one predictor variable that you consider to be the best predictor of time. Determine the regression equation for predicting time based on that predictor. Also, interpret the slope coefficient of this equation.

d. Analyze appropriate residual plots and comment on what they reveal about whether the conditions for inference appear to be met here.

1.46 Baseball game times, outliers and influential points. Refer to the previous Exercise 1.45 on the playing time of baseball games. The game between the Cincinnati Reds and the Milwaukee Brewers had an unusually large number of runs, suggesting that it might be an influential point. **BballT**

a. Examine the scatterplot predicting time from the number of runs. Find the CIN-MIL point on this plot and discuss whether you think it has a strong influence on the linear relationship between these two variables or not.

b. Omit the CIN-MIL point and find the least squares regression line. Is there much change from the equation you found in the previous exercise? What does that mean about the influence of this particular point on this relationship?

c. Continuing to omit the CIN-MIL point, examine scatterplots between *Time* and all of the other variables. Is *Runs* still the best variable for prediction purposes?

1.47 Retirement SRA. A faculty member opened a supplemental retirement account in 1997 to investment money for retirement. Annual contributions were adjusted downward during sabbatical years in order to maintain a steady family income. The annual contributions are provided in the file **Retirement**. **Retire**

a. Fit a simple linear regression model for predicting the amount of the annual contribution (*SRA*) using *Year*. Identify the two sabbatical years that have unusually low SRA residuals and compute the residual for each of those cases. Are the residuals for the two sabbatical years outliers? Provide graphical and numerical evidence to support your conclusion.

b. Sabbaticals occur infrequently and are typically viewed by faculty to be different from other years. Remove the two sabbatical years from the dataset and refit a linear model for predicting the amount of the annual contribution (*SRA*) using *Year*. Does this model provide a better fit for the annual contributions? Make appropriate graphical and numerical comparisons for the two models.

1.48 Metabolic rate of caterpillars. Marisa Stearns collected and analyzed body size and metabolic rates for Manduca Sexta caterpillars.[14] The data are in the file **MetabolicRate** and the variables are: **MetRate**

Variable	Description
Computer	Number of the computer used to obtain metabolic rate
BodySize	Size of the animal in grams
CO2	Carbon dioxide concentration in parts per million
Instar	Number from 1 (smallest) to 5 (largest) indicating stage of the caterpillar's life
Mrate	Metabolic rate

The dataset also has variables *LogBodySize* and *LogMrate* containing the logs (base 10) of the size and metabolic rate variables. The researchers would like to build a linear model to predict metabolic rate (either *Mrate* directly or on a log scale with *LogMrate*) using a measure of body size for the caterpillars (either *BodySize* directly or on a log scale with *LogBodySize*).

a. Which variables should they use as the response and predictor for the model? Support your choice with appropriate plots.

b. What metabolic rate does your fitted model from (a) predict for a caterpillar that has a body size of 1 gram?

1.49 Metabolic rate of caterpillars: grouping variable. Refer to Exercise 1.48 that considers linear models for predicting metabolic rates (either *Mrate* or *LogMrate*) for caterpillars using a measure of body size (either *BodySize* or *LogBodySize*) for the data in **MetabolicRate**. Produce a scatterplot for the model you chose in Exercise 1.48, and add a plotting symbol for the grouping variable *Instar* to show the different stages of development. Does the linear trend appear to be consistent across all five stages of a caterpillar's life? (*Note:* We are not ready to fit more complicated models yet, but we will return to this experiment in Chapter 3.) **MetRate**

1.50 County health care, logarithms. In Example 1.7 on page 35 we consider reexpressing the response variable (*MDs* = number of doctors in a county) with a square root transformation when fitting a model to predict with the number of *Hospitals* in the county. We've also seen several examples where a log transformation is helpful for addressing problems with the regression conditions. Would that also be helpful for the relationship between *MDs* and *Hospitals* in the **CountyHealth** dataset? Experiment with reexpressing the response (*MDs*), the predictor (*Hospitals*), or both with logarithms to see which works best—and whether the result would be preferred over the *sqrt*(*MDs*) transformation in Example 1.7. **CtyHlth**

Supplemental Exercises

1.51 Zero mean. One of the neat consequences of the least squares line is that the sample means $(\bar{x}, \bar{y})$ always lie on the line so that $\bar{y} = \hat{\beta}_0 + \hat{\beta}_1\bar{x}$. From this, we can get an easy way to calculate the intercept if we know the two means and the slope:

$$\hat{\beta}_0 = \bar{y} - \hat{\beta}_1\bar{x}$$

We could use this formula to calculate an intercept for *any* slope, not just the one obtained by least squares estimation. See what happens if you try this. Pick any dataset with two quantitative variables, find the mean for both variables, and assign one to be the predictor and the other the response. Pick any slope you like and use the formula above to compute an intercept for your line. Find predicted values and then residuals for each of the data points using this line as a prediction equation. Compute the sample mean of your residuals. What do you notice?

Reflexus/Deposit Photos

CHAPTER 2

Inference for Simple Linear Regression

In this chapter you will learn to:

- Test hypotheses and construct confidence intervals for the slope of a simple linear model.
- Partition variability to create an ANOVA table and determine the proportion of variability explained by the model.
- Construct intervals for predictions made with the simple linear model.

Recall that in Example 1.1 (page 22) we considered a simple linear regression model to predict the price of used Honda Accords based on mileage. Now we turn to questions of assessing how well that model works. Are prices significantly related to mileage? How much does the price drop as mileage increases? How much of the variability in Accord prices can we explain by knowing their mileages? If we are interested in a used Accord with about 50,000 miles, how accurately can we predict its price?

These are all questions of statistical inference. Remember the important difference between description and inference. The goal of description is to summarize the patterns *we can actually see*. The goal of inference is to use those patterns to draw conclusions about patterns *we can't see*: either about the population our sample came from, or about a possible cause-and-effect relationship between the predictor and the response. These goals of inference are ambitious. They rely on mathematical theory to bridge the gap between what we can see and what we can't see. Be hopeful, but be careful.

Recall that the simple linear model is $Y = \beta_0 + \beta_1 X + \epsilon$ where $\epsilon \sim N(0, \sigma_\epsilon)$. Many of the important inference questions deal with the slope parameter β_1. These include

- **Tests**: Is the predictor related to the response? Is it believable that the slope could be equal to zero? Note that if $\beta_1 = 0$ in the model, there is no linear relationship between the predictor and the response.
- **Intervals**: How precisely can we estimate the value of β_1? What are plausible values for how much the average response changes as the predictor changes?
- **Effect size**: Suppose we come to believe that the predictor and the response are associated. Is the association strong enough that we care? Does the predictor have a practical effect on the response?

We introduce a formal hypothesis test and a confidence interval for the slope, both based on the t-distribution, in Section 2.1. In Section 2.2, we examine an alternate test, based on the F-distribution. This depends on dividing the variability in the response variable into two parts: one attributed to the linear relationship and the other accounting for the random error term. We use the correlation coefficient, r, in Section 2.3 as another way to measure (and test) the strength of linear association between two quantitative variables. In Section 2.4, we consider two forms of intervals that are important for quantifying the accuracy of predictions based on a regression model. Finally, we present a case study in Section 2.5 that illustrates many of these inference ideas by looking at a possible relationship between summer temperatures and sizes of butterflies.

2.1 Inference for Regression Slope

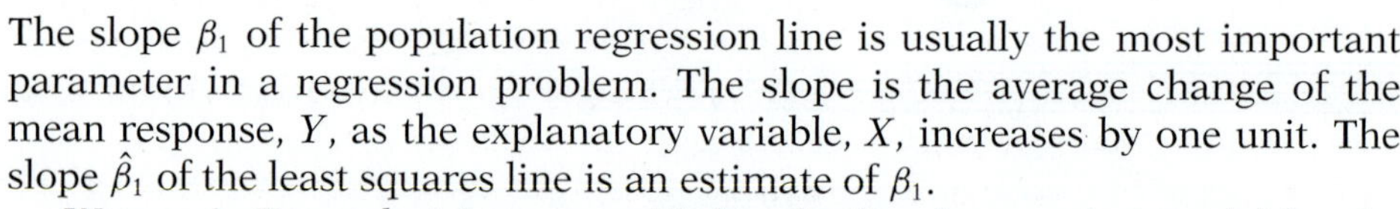

The slope β_1 of the population regression line is usually the most important parameter in a regression problem. The slope is the average change of the mean response, Y, as the explanatory variable, X, increases by one unit. The slope $\hat{\beta}_1$ of the least squares line is an estimate of β_1.

Accord

We saw in Example 1.3 on page 25 that the fitted regression model for the Accord cars dataset (**AccordPrice**) is

$$\widehat{Price} = 20.8 - 0.1198\ Mileage$$

This suggests that for every additional 1000 miles on a used car the price goes down by \$119.80, on average. A skeptic might claim that price and mileage have no linear relationship. Is the value of −0.1198 in our fitted regression model just a fluke? Is the corresponding parameter value in the population really zero? The skeptic might have in mind that the population scatterplot looks something like Figure 2.1 and that our data, shown as the dark points in the plot, arose as an odd sample. Even if there is no true relationship between Y and X, any particular sample of data will show some positive or negative fitted regression slope just by chance. After all, it would be extremely unlikely that a random sample would give a slope of exactly zero between X and Y.

Next we examine a formal mechanism for assessing when the population slope is likely to be different from zero. Then we discuss a method to put confidence bounds around the sample slope.

t-Test for Slope

We want to assess whether the slope for sample data provides significant evidence that the slope for the population differs from zero. To do that we need to estimate its variability. The standard error of the slope, $SE_{\hat{\beta}_1}$, measures how much we expect the sample slope (i.e., the fitted slope) to vary from one

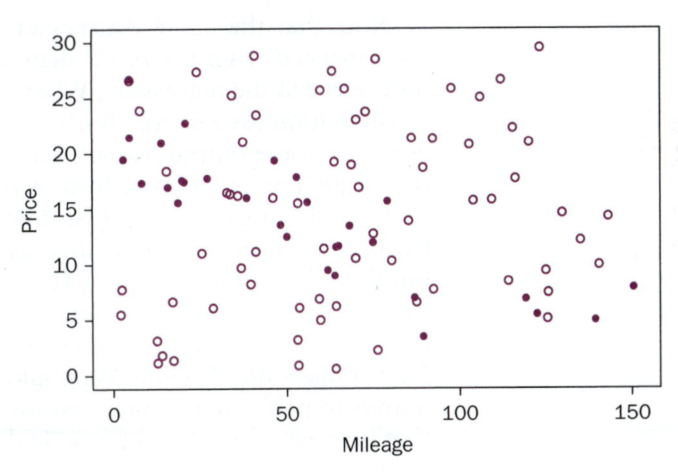

FIGURE 2.1 Hypothetical population with no relationship between *Price* and *Mileage*. Solid points represent our particular sample.

random sample to another.* The standard errors for coefficients (slope or intercept) of a regression model are generally provided by statistical software such as in the following output.

Predictor	Coef	SE Coef	T	P
Constant	20.8096	0.95286	21.839	0.000
Mileage	-0.11981	0.014095	-8.500	0.000

The ratio of the slope to its standard error is one test statistic for this situation:

$$t = \frac{\hat{\beta}_1}{SE_{\hat{\beta}_1}} = \frac{-0.11981}{0.014095} = -8.50$$

Notice that the test statistic is far from zero (the sample slope is more than 8 standard errors below a slope of zero). Given this, we can reject the hypothesis that the true slope, β_1, is zero.

Think about the null hypothesis of no linear relationship and the regression conditions. Under this hypothesis, the test statistic follows a ***t*-distribution** with $n-2$ degrees of freedom. Software will provide the *P*-value we need, based on this distribution. The *P*-value measures the probability of getting a statistic at least as extreme as the observed test statistic. Our statistical inference about the linear relationship depends on the magnitude of the *P*-value. That is, when the *P*-value is below our significance level α, we reject the null hypothesis and conclude that the slope differs from zero.

***t*-distribution**

t-TEST FOR THE SLOPE OF A SIMPLE LINEAR MODEL

To test whether the population slope is different from zero, the hypotheses are

$H_0 : \beta_1 = 0$
$H_a : \beta_1 \neq 0$

and the test statistic is

$$t = \frac{\hat{\beta}_1}{SE_{\hat{\beta}_1}}$$

If the conditions for the simple linear model, including normality, hold, we may compute a *P*-value for the test statistic using a *t*-distribution with $n-2$ degrees of freedom.

*Be careful to avoid confusing the **standard error of the slope**, $SE_{\hat{\beta}_1}$, with the **standard error of regression**, $\hat{\sigma}_\epsilon$.

Note that the $n - 2$ degrees of freedom for the t-distribution in this test are inherited from the $n - 2$ degrees of freedom in the estimate of the standard error of the regression. When the t-statistic is extreme in either tail of the t-distribution (as it is for the Accord data), the P-value will be small (reported as 0.000 in the output above). This provides strong evidence that the slope in the population is different from zero.

In some cases, we might be interested in testing for a relationship in a particular direction (e.g., a positive association between number of doctors and number of hospitals in a city). Here we would use a one-tailed alternative (such as $H_a : \beta_1 > 0$) in the t-test and compute the P-value using only the area in that tail. Software generally provides the two-sided P-value, so we need to divide by 2 for a one-sided P-value. We could also use the information in the computer output to perform a similar test for the intercept in the regression model, but that is rarely a question of practical importance.

The *P*-Value and What It Claims to Mean: A Way to Use (or Abuse) Your Model

To start, notice that we write *"your"* model, not *"the"* model. Whatever the model, it is *you* who chose to buy it, and if you now proceed to break it, the damage will be yours to own. All models should be handled with care.

That said, we turn to the P-value as a useful way to summarize the strength of evidence. The P in P-value stands for *probability*: "Suppose that there is no 'true' relationship between mileage and price of a used Accord. What is the *probability* that you would observe a value of t as extreme as the one you actually got?"

The meaning of the P-value is hard to unpack. Even though you have seen it before, it may take some additional time and effort to use it with skill and confidence, because its meaning is counterfactual—the meaning depends on assuming something that may not be true: "Suppose ...". The first challenge is to understand what the P-value claims to tell you. A second challenge is to understand the logic that connects the way the P-value gets computed and what it is that the P-value claims to tell you.

What the P-value claims to tell you. Abstractly, the Accord study seeks to answer the concrete question: "Does the mileage of a used Accord make a difference in price?" The abstract version of the question is: "How strong is the evidence from the data that there is in fact an association between mileage and price of used Accords?" More formally, suppose there is no "real" association between the two variables. That is, *pretend* for the sake of argument that any apparent association between the variables is just coincidence, an artifact of the random selection process. As you know, this pretend scenario has a formal name, the "null hypothesis." It is called "null" because it asks you to pretend that there is "no" association between the two variables.

A useful mental image, attributed to statistician Nick Maxwell, can help the P-value come to life. According to Maxwell, the P-value answers the question, "How embarrassed should the null hypothesis be in the face of the data?" A tiny P-value says that the evidence of association between the variables is very strong, and the null hypothesis should be very embarrassed. The larger the P-value, the more secure the null hypothesis can be. A large P-value doesn't mean that the null hypothesis is right, only that it has no reason to worry about being chased out of town by the data.

Although the P-value is computed using mathematical theory, you can understand its logic without the mathematics.*

*The approach we describe here is based on computer simulation instead of mathematical theory. This randomization-based approach can be useful when your dataset fails to satisfy the conditions for inference based on normal theory. For more, see Topic 4.6.

- Start with a list of the prices for each of the Accords in the dataset; call this list 1. Then, separately, make a list of the mileages of the Accords; call this list 2. Then scramble the list of prices, list 1, and attach them to the mileages, list 2, forming random pairings of price with mileage. Any association between the price and mileage variables will now be purely a matter of chance.
- Fit the regression model and compute the test statistic (in this case a t-statistic that compares the slope of the regression line to its standard error).
- Because you scrambled the data, the slope should be close to zero and you can expect the t-statistic to be close to 0. But, of course, random quantities will vary randomly. Some scrambled pairings, just by chance, will give t-statistics that are far from zero. The key question is: What percent of the time will these scrambled pairings give you a large t-statistic (large in absolute value, that is)? In particular, what is the chance of a t-statistic as large as the one from the actual data? That chance is the P-value.

As a concept and a construct for data analysis, the P-value is controversial. Thanks to computers, the number itself has become as easy to get as shoving coins into a vending machine. Just insert your data and model, push the button, and your P-value comes rolling down the chute. If you use statistical software as your vending machine, just type in the appropriate command and the magic number P will appear in the output, like a bottle clunking into the bin. Getting the bottle is easy. The hard question, the one that matters, is this: "What is it you are about to swallow?"

In 1977, in Waterloo, Ontario, there was a conference devoted to the theory of how to learn from data. One of the papers presented was by A. P. Dempster of Harvard University. His title was "Model searching and estimation in the logic of inference." In his presentation, Dempster argued that we use formal inference, that is, inference based on a probability model, to answer questions of two different kinds, which he called IS questions and IT questions. Of course Dempster knew he was oversimplifying, but more important, he knew that his oversimplification to IS/IT had roots that ran deep.

Dempster's first question, "IS there an effect?" is Yes/No. If your goal is to choose a model, you ask, "Is there evidence that a (nonzero) slope between X and Y belongs in the model?" If your goal is to test a null hypothesis, you ask, "Is there evidence that the association between price and mileage in the data is too big to be due to chance?" Whichever your question, you can use the P-value as a guide. The lower the P-value, the stronger the evidence that the association you see is not just the result of chance variation.

Testing a null hypothesis using a P-value is the traditional way to answer Dempster's IS question, "Is there an association?" At best, the P-value tells only whether there is detectable evidence of association between the two variables. This might be what you most want to know. *But: The* P*-value doesn't tell how strong the association is nor how well we can estimate it.* The P-value doesn't tell whether the association matters in real life. To address these two other important issues, Dempster's IT questions ("How big is it?"), we need the correlation coefficient, which measures the strength of the association, and a confidence interval for the slope, which tells the uncertainty we have about how large the slope really is. We will look at correlation soon, but first we consider confidence intervals.

CAUTION

Confidence Interval for Slope

A confidence interval for the slope may be more useful than a test because it tells us more than just whether or not the slope is zero. It shows how accurate the estimate $\hat{\beta}_1$ is likely to be.

CONFIDENCE INTERVAL FOR THE SLOPE OF A SIMPLE LINEAR MODEL

The confidence interval for β_1 has the form

$$\hat{\beta}_1 \pm t^* SE_{\hat{\beta}_1}$$

where t^* is the critical value for the t_{n-2} distribution corresponding to the desired confidence level.

EXAMPLE 2.1

Accord

Confidence interval for slope in the model for Accord prices In our simple linear regression example to predict Accord prices, the sample slope, $\hat{\beta}_1$, is −0.1198. The computer output (page 61) shows us that the standard error of this estimate is 0.0141. There are 30 cars so we have 28 degrees of freedom. For a 95% confidence level, the t^* value is 2.05. Thus, a 95% confidence interval for the true (population) slope is $-0.1198 \pm 2.05(0.0141)$ or -0.1198 ± 0.0289, which gives an interval of $(-0.149, -0.091)$. Thus we are 95% confident that as mileage increases by 1000 miles, the average price decreases by between \$91 and \$149 in the population of all used Accords.

Although it is not as widely used, we can use similar methods to make inferences about the intercept for a regression model. For example, the confidence interval for the intercept would be $\hat{\beta}_0 \pm t^* SE_{\hat{\beta}_0}$, where we can find the sample intercept and its standard error from the computer output.

2.2 Partitioning Variability—ANOVA

Another way to assess the effectiveness of a model is to measure how much of the variability in the response variable is explained by the predictions based on the fitted model. This general technique is known in statistics as **analysis of variance**, abbreviated as **ANOVA**. Although we will illustrate ANOVA in the context of the simple linear regression model, this approach could be applied to any situation in which a model is used to obtain predictions, as we demonstrate in later chapters.

ANOVA

The basic idea is to partition the total variability in the responses into two pieces. One piece summarizes the variability explained by the model. The other piece summarizes the variability due to error and captured in the residuals. In short, we have

TOTAL variation in Response Y	=	Variation explained by the MODEL	+	Unexplained variation in the RESIDUALS

In order to partition this variability, we start with deviations for individual cases. Note that we can write a deviation $y - \bar{y}$ as

$$y - \bar{y} = (\hat{y} - \bar{y}) + (y - \hat{y})$$

so that $y - \bar{y}$ is the sum of two deviations. The first deviation corresponds to the model and the second is the residual. We then sum the squares of these deviations to obtain the following relationship, known as the ANOVA sum of squares identity:*

$$\sum (y - \bar{y})^2 = \sum (\hat{y} - \bar{y})^2 + \sum (y - \hat{y})^2$$

We summarize the partition with the following notation:

$$SSTotal = SSModel + SSE$$

Some computer output showing the analysis of variance output for the Accord price model is shown below.*

Analysis of Variance

Source	DF	SS	MS	F	P
Regression	1	687.66	687.66	72.253	0.000
Residual Error	28	266.49	9.52		
Total	29	954.15			

Examining the "SS" column in the output shows the following values for this partition of variability for the Accord regression:

$$SSModel = \sum (\hat{y} - \bar{y})^2 = 687.66$$
$$SSE = \sum (y - \hat{y})^2 = 266.49$$
$$SSTotal = \sum (y - \bar{y})^2 = 954.15$$

and note that $954.15 = 687.66 + 266.49$.

ANOVA Test for Simple Linear Regression

How do we tell if the model explains a significant amount of variability or if the explained variability is due to chance alone? The relevant hypotheses would be the same as those for the *t*-test for the slope

$$H_0 : \beta_1 = 0$$
$$H_a : \beta_1 \neq 0$$

In order to compare the *explained* (model) and *error* (residual) variabilities, we need to adjust the sums of squares by appropriate degrees of freedom. We have already seen in the computation of the standard error of regression $\hat{\sigma}_\epsilon$ that the sum of squared residuals (*SSE*) has $n - 2$ degrees of freedom. When we partition variability as in the ANOVA identity, the degrees of freedom of the components add in the same way as the sums of squares. The total sum of squares (*SSTotal*) has $n - 1$ degrees of freedom when estimating the variance of the response variable (recall that we divide by $n - 1$ when computing a basic variance or standard deviation). Thus for a simple linear model with a single predictor, the sum of squares explained by the model (*SSModel*) has just 1 degree of freedom. We partition the degrees of freedom so that $n - 1 = 1 + (n - 2)$.

*There is some algebra involved in deriving the ANOVA sum of squares identity that reduces the right-hand side to just the two sums of squares terms, with no cross-product term.
*It is unusual to report a *P*-value of zero. Most statistical software packages include more decimal places or say that the *P*-value is < 0.001. We know the *P*-value is not exactly equal to zero!

mean squares

We divide each sum of squares by the appropriate degrees of freedom to form **mean squares**:

$$MSModel = \frac{SSModel}{1} \quad \text{and} \quad MSE = \frac{SSE}{n-2}$$

If the null hypothesis of no linear relationship is true, then both of these mean squares will estimate the variance of the error.* However, if the model is effective, the sum of squared errors gets small and the *MSModel* will be large relative to *MSE*. The test statistic is formed by dividing *MSModel* by *MSE*:

$$F = \frac{MSModel}{MSE}$$

F-distribution

Under H_0, we expect F to be approximately equal to 1. But under H_a, the F-statistic should be larger than 1. If the normality condition holds, then the test statistic F follows an ***F*-distribution** under the null hypothesis of no relationship. The F-distribution represents a ratio of two variance estimates so we use degrees of freedom for both the numerator and the denominator. In the case of simple linear regression, the numerator degree of freedom is 1 and the denominator degrees of freedom are $n-2$.

We summarize all of these calculations in an ANOVA table as follows. Statistical software provides the ANOVA table as a standard part of the regression output.

ANOVA FOR A SIMPLE LINEAR REGRESSION MODEL

To test the effectiveness of the simple linear model, the hypotheses are

$H_0 : \beta_1 = 0$
$H_a : \beta_1 \neq 0$

and the **ANOVA table** is

Source	Degrees of Freedom	Sum of Squares	Mean Square	*F*-statistic
Model	1	SSModel	MSModel	$F = \frac{MSModel}{MSE}$
Error	$n-2$	SSE	MSE	
Total	$n-1$	SSTotal		

If the conditions for the simple linear model, including normality, hold, the P-value is obtained from the upper tail of an F-distribution with 1 and $n-2$ degrees of freedom.

EXAMPLE 2.2

Accord

ANOVA for Accord price model Using the ANOVA output (page 65) for the Accord linear regression fit, we see the $MSModel = 687.66/1 = 687.66$ and $MSE = 266.49/(30-2) = 9.518$. Thus the test statistic is $F = 687.66/9.518 = 72.25$. Comparing this to an F-distribution with 1 numerator and 28 denominator degrees of freedom, we find a P-value very close to 0.000 (as seen in the computer output). Thus we conclude that price and mileage have some relationship and that this model based on mileage is effective for predicting the price of used Accords.

*We have already seen that *MSE* is an estimate for $\sigma_\epsilon{}^2$. It is not so obvious that *MSModel* is also an estimate of $\sigma_\epsilon{}^2$ when the true slope is zero, but deriving this fact is beyond the scope of this book.

After partitioning the variability in the response variable, we now return to modeling the association between the two quantitative variables.

2.3 Regression and Correlation

Recall that the correlation coefficient, r, is a number between -1 and $+1$ that measures the direction and strength of the linear association between two quantitative variables. Thus the correlation coefficient is also useful for assessing the significance of a simple linear model. In this section, we make a connection between correlation and the partitioning of variability from the previous section. We also show how the slope can be estimated from the sample correlation and how to test for a significant correlation directly.

Coefficient of Determination, r^2

coefficient of determination

Another way to assess the fit of the model using the ANOVA table is to compute the ratio of the *Model* variation to the *Total* variation. This statistic, known as the **coefficient of determination**, tells us how much variation in the response variable Y we explain by using the explanatory variable X in the regression model. Why do we call it r^2? An interesting feature of simple linear regression is that the square of the correlation coefficient happens to be exactly the coefficient of determination. We explore a t-test for r in more detail in the next section.

COEFFICIENT OF DETERMINATION

The coefficient of determination for a model is

$$r^2 = \frac{\text{Variability explained by the model}}{\text{Total variability in } y} = \frac{\sum(\hat{y}-\bar{y})^2}{\sum(y-\bar{y})^2} = \frac{SSModel}{SSTotal}$$

Using our ANOVA partition of the variability, this formula can also be written as

$$r^2 = \frac{SSTotal - SSE}{SSTotal} = 1 - \frac{SSE}{SSTotal}$$

Statistical software generally provides a value for r^2 as a standard part of regression output.

If the model fit perfectly, the residuals would all be zero and r^2 would equal 1. If the model does no better at predicting the response variable than simply using the mean $\bar{y}$ for every prediction, then *SSModel* and r^2 would both be equal to zero. The coefficient of determination is a useful tool for assessing the fit of the model. It can also be used for comparing competing models.

EXAMPLE 2.3

Accord

r^2 for the Accord price model Refer to the ANOVA output on page 65 for the model to predict price based on mileage for Honda Accords. We see that

$$r^2 = \frac{687.66}{954.15} = 0.721$$

We can confirm this value in the original computer output for fitting this regression model on page 25. Since r^2 is the fraction of the response variability that is explained by the model, we often convert the value to a percentage. In this context, we find that 72.1% of the variability in the prices of the Accords in this sample can be explained by the linear model based on their mileages.

Inference for Correlation

The least squares slope $\hat{\beta}_1$ is closely related to the correlation r between the explanatory and response variables X and Y in a sample. In fact, the sample slope can be obtained from the sample correlation together with the standard deviation of each variable as follows:

$$\hat{\beta}_1 = r\frac{s_Y}{s_X}$$

Testing the null hypothesis $H_0 : \beta_1 = 0$ is the same as testing that there is no correlation between X and Y in the population from which we sampled our data. Because correlation also makes sense when there is no explanatory-response distinction, it is handy to be able to test correlation without doing regression.

t-TEST FOR CORRELATION

If we let ρ denote the population correlation, the hypotheses are

$$H_0 : \rho = 0$$
$$H_a : \rho \neq 0$$

and the test statistic is

$$t = \frac{r\sqrt{n-2}}{\sqrt{1-r^2}}$$

If the conditions for the simple linear model, including normality, hold, we find the *P*-value using the *t*-distribution with $n - 2$ degrees of freedom.

EXAMPLE 2.4

Accord

Test correlation for the Accord data We only need the sample correlation and sample size to compute this test statistic. Note that this test statistic has the same *t*-value that we saw in Section 2.1. For the prices and mileages of the $n = 30$ Accords in Table 1.1 (page 22), the correlation is $r = -0.8489$. The test statistic is

$$t = \frac{-0.8489\sqrt{30-2}}{\sqrt{1-(-0.8489)^2}} = -8.50$$

This test statistic is far in the tail of a *t*-distribution with 28 degrees of freedom, so we have a *P*-value near 0.0000. Thus we conclude, once again, that there is significant evidence showing a correlation between the prices and mileages of used Accords.

Three Tests for a Linear Relationship?

We now have three distinct ways to test for a significant linear relationship between two quantitative variables.* The *t-test for slope*, the *ANOVA for regression*, and the *t-test for correlation* can all be used. Which test is best? If the conclusions differed, which test would be more reliable? Surprisingly, these three procedures are exactly equivalent in the case of simple linear regression. In fact, the *t*-test statistics for slope and correlation are equal, as we just

*Note that finding a significant linear relationship between two variables doesn't mean that the linear relationship is necessarily an adequate summary of the situation. For example, it might be that there is a curved relationship but that a straight line tells part of the story.

saw, and the F-statistic is the square of the t-statistic. For the Accord cars, the t-statistic is -8.50 and the ANOVA F-statistic is $(-8.50)^2 = 72.25$.

Why do we need three different procedures for one task? While the results are equivalent in the simple linear case, we will see that these tests take on different roles when we consider multiple predictors in Chapter 3.

Terminology Cautions

We pause here to note that some terms we use in a statistical sense may have somewhat different meanings than common usage.

- **Predicted**: The *predicted value* does *not* mean "the value we expect to get." Instead it means "the fitted value given by the model we have chosen to use." We might "expect" the actual value to be somewhere around the predicted value, but would probably be quite surprised if they matched exactly.
- **Explained**: *Explained* does *not* mean explained in the usual sense. Instead, it means "accounted for by the fitted model."
- **Significant**: *Significant* does not (necessarily) mean "important." Instead it means "too extreme to be just due to random variation from one response value to the next." More on this follows.

Practical Versus Statistical Significance

No matter which test we use to check for a linear relationship, the only question such a test can answer is, "Do we have evidence that the relationship we are seeing between X and Y is stronger than what would plausibly happen simply by chance?" A test does not address the more important question, *"Even if we suppose that* X *and* Y *are related, what difference does it make?"*

It might be that there is a link between X and Y, but the link is so small that we don't really care about it in practice. The correlation coefficient, r (and likewise, r^2), gives us a measure of how closely X and Y are related. Take a look at the t-test formula for testing a correlation, in the box on page 68. You can see that as the sample size n goes up, so does the magnitude of the test statistic. For a large enough sample size, even a moderate or small correlation coefficient will give a statistically significant t-test result. Again we ask "So what? Is r large enough that anyone should care?" The same reasoning applies to the slope of a fitted regression line. Especially for a very large sample, the tests for slope might show significant results, even when the actual value of the slope is small and the average responses change very little as the predictor changes.

EXAMPLE 2.5

Kershaw

Kershaw fastballs Clayton Kershaw won the Cy Young award in 2013 as the best baseball pitcher in the National League. The datafile **Kershaw** contains information on every pitch that he threw that season.[1] In this example we focus on only the $n = 2060$ fastballs (*PitchType*="*FF*") that Kershaw threw. We look at the speed the pitch was traveling when it reached the batter (*EndSpeed* in mph) and the number of batters he had already faced in the game (*BatterNumber*). Conventional baseball wisdom says that pitch speed should slow down as pitchers tire later in a game.

Figure 2.2 shows a scatterplot of the relationship between *EndSpeed* and *BatterNumber*. The fitted least squares line, $\widehat{EndSpeed} = 85.16 - 0.00924\ BatterNumber$, has been added to the scatterplot. The correlation is $r = -0.155$, which when squared is 0.024. This indicates that about 2.4% of the variability in pitch end speed can be explained by the batter number. While this seems like a fairly small amount, the t-test is highly significant ($t = -7.11$, P-value $\approx 10^{-12}$). Clearly Kershaw's average fastball end speed decreases as the game goes along, but is the difference enough to matter? The fitted model predicts the average end speed decreases by less than one hundredth of a mile per hour for each additional batter faced. This means predictions drop from 85.15 mph for the first batter at the beginning of a game to 84.61 mph for batter number 60 near the end. Whether a decrease of a little over half a mile per hour in a fastball's speed *actually* means a lot to a pitcher's effectiveness is a question for baseball experts to answer. Most would probably say it's not a big deal—certainly not as *practically* significant as the tiny *P*-value is *statistically* significant.

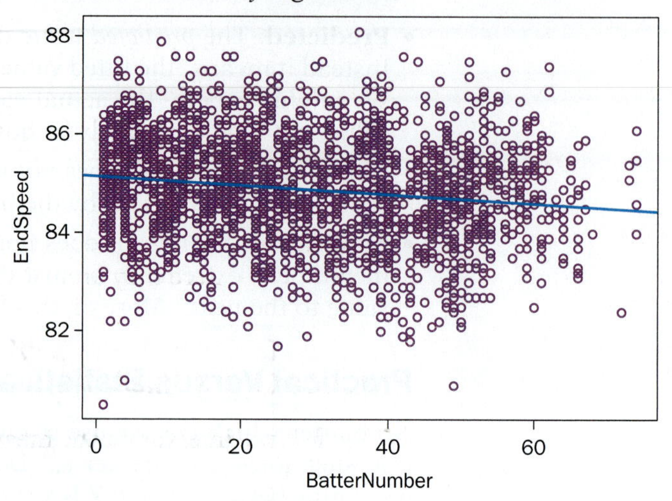

FIGURE 2.2 Pitch *EndSpeed* versus *BatterNumber* for Kershaw fastballs

As this example illustrated, large sample sizes make it easy to detect smaller differences and effects. Statistical significance and practical significance are very different concepts.

2.4 Intervals for Predictions

One of the most common reasons to fit a line to data is to predict the response for a particular value of the explanatory variable. In addition to a prediction, we often want a margin of error that describes how accurate the prediction is likely to be. This leads to forming an interval for the prediction, but there are two important types of intervals in practice.

To decide which interval to use, we must answer this question: Do we want to predict the *mean response* for a particular value x^* of the explanatory variable or do we want to predict the response y for an *individual case*? Both of these predictions may be interesting, but they are two different problems. The value of the prediction is the same in both cases, but the margin of error is different. The distinction between predicting a single outcome and predicting the mean of all outcomes when $X = x^*$ determines which margin of error is used.

For example, if we would like to know the average price for all Accords with 50,000 miles, then we would use an interval for the mean response. On the other hand, if we want an interval to contain the price of a particular Accord with 50,000 miles, then we need the interval for a single prediction. To emphasize the distinction, we use different terms for the two intervals.

confidence interval

To estimate the *mean* response, we use a **confidence interval for** μ_Y. It has the same interpretation and properties as other confidence intervals, but it estimates the mean response of Y when X has the value x^*. This mean response

$$\mu_Y = \beta_0 + \beta_1 x^*$$

is a parameter, a fixed number whose value we don't know.

prediction interval

To estimate an *individual* response y, we use a **prediction interval**. A prediction interval estimates a single random response y rather than a parameter such as μ_Y. The response y is not a fixed number. If we took more observations with $X = x^*$, we would get different responses. The meaning of a prediction interval is very much like the meaning of a confidence interval. A 95% prediction interval, like a 95% confidence interval, is right 95% of the time in repeated use. "Repeated use" now means that we take an observation on Y for each of the n values of X in the original data, and then take one more observation y with $X = x^*$. We form the prediction interval, then see if it covers the observed value of y for $X = x^*$. The resulting interval will contain y in 95% of all repetitions.

The main point is that it is harder to predict one response than to predict a mean response. Both intervals have the usual form

$$\hat{y} \pm t^* SE$$

but the prediction interval is wider (has a larger SE) than the confidence interval. You will rarely need to know the details because software automates the calculation, but here they are.

CONFIDENCE AND PREDICTION INTERVALS FOR A SIMPLE LINEAR REGRESSION RESPONSE

A confidence interval for the mean response μ_Y when X takes the value x^* is

$$\hat{y} \pm t^* SE_{\hat{\mu}}$$

where the standard error is

$$SE_{\hat{\mu}} = \hat{\sigma}_\epsilon \sqrt{\frac{1}{n} + \frac{(x^* - \bar{x})^2}{\sum (x - \bar{x})^2}}$$

A prediction interval for a single observation of Y when X takes the value x^* is

$$\hat{y} \pm t^* SE_{\hat{y}}$$

where the standard error is

$$SE_{\hat{y}} = \hat{\sigma}_\epsilon \sqrt{1 + \frac{1}{n} + \frac{(x^* - \bar{x})^2}{\sum (x - \bar{x})^2}}$$

The value of t^* in both intervals is the critical value for the t_{n-2} density curve to obtain the desired confidence level.

There are two standard errors: $SE_{\hat{\mu}}$ for estimating the mean response μ_Y and $SE_{\hat{y}}$ for predicting an individual response y. The only difference between the two standard errors is the extra 1 under the square root sign in the standard error for prediction. The extra 1, which makes the prediction interval wider, reflects the fact that an individual response will vary from the mean response μ_Y with a standard deviation of σ_ϵ. Both standard errors are multiples of the standard error of regression $\hat{\sigma}_\epsilon$. The degrees of freedom are again $n - 2$, the degrees of freedom of $\hat{\sigma}_\epsilon$.

EXAMPLE 2.6

Accord

Intervals for Accord prices Returning once more to the Accord prices, suppose that we are interested in a used Accord with 50,000 miles. The predicted price according to our fitted model is

$$\widehat{Price} = 20.81 - 0.1198(50) = 14.82$$

or an average price of about $14,820. Most software has options for computing both types of intervals using specific values of a predictor. This produces output such as what follows.

```
Predicted Values for New Observations
New
Obs     Fit   SE Fit          95% CI           95% PI
  1   14.82     0.57  (13.66, 15.98)  (8.39, 21.24)

Values of Predictors for New Observations
New
Obs  Mileage
  1     50.0
```

This confirms the predicted value of 14.82 (thousand) when $x^* = 50$. The "SE Fit" value of 0.57 is the result of the computation for $SE_{\hat{\mu}}$. The "95% CI" is the confidence interval for the mean response. Thus we can be 95% confident that the average price of all used Accords with 50,000 miles for sale at cars.com (at the time the sample was chosen) was somewhere between $13,660 and $15,980. The "95% PI" tells us that we should expect about 95% of those Accords with 50,000 miles to be priced between $8,390 and $21,240. So if we find an Accord with 50,000 miles on sale for $12,000, we know that the price is slightly better than average, but not unreasonably low.

Figure 2.3 shows both the confidence intervals for mean Accord prices μ_Y and prediction intervals for individual Accord prices (y) for each mileage. These have been added to the scatterplot with the regression line. Note how the intervals get wider as we move away from the mean mileage ($\bar{x} = 55,000$ miles). Many data values lie outside of the narrower confidence bounds for μ_Y; those bounds are only trying to capture the "true" line (mean Y at each value of X). Only one of the 30 Accords in this sample (the least expensive car at $3.5 thousand with 89,400 miles) has a price that falls outside of the 95% prediction bounds.

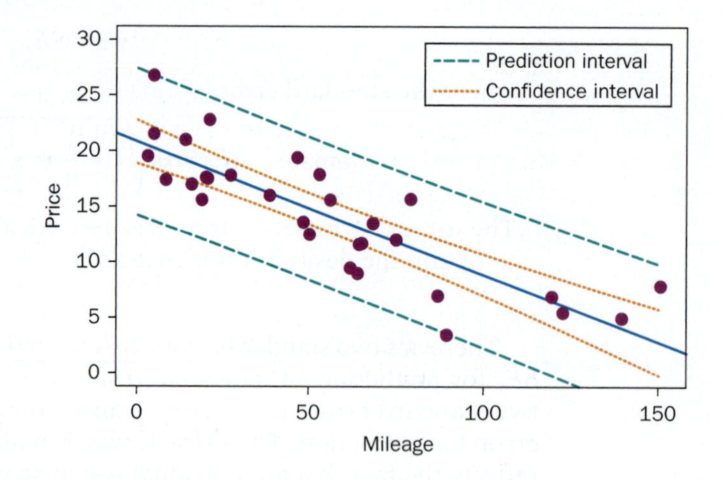

FIGURE 2.3 Confidence intervals and prediction intervals for Accord prices

2.5 Case Study: Butterfly Wings

Is butterfly growth related to temperature?

There is a species of butterfly called *Boloria chariclea* that is found in Greenland. Scientists know that body size is related to reproductive success and to "dispersal capacity," which is the ability of butterflies to move from one habitat to another. In short, butterflies don't want to be short (or narrow or small

Temp	Wing
0.9	19.1
1.1	18.8
1.4	19.5
1.6	19.0
1.6	18.9
1.6	18.6
2.3	18.9
2.4	18.9
2.4	18.8
2.8	19.1
2.7	18.9
2.6	18.7
2.6	18.6
2.9	18.6
2.7	18.4
4.0	18.2

TABLE 2.1 Previous summer temperature and average wing length for female butterflies in Greenland

in general). The climate is changing in Greenland, and this could affect the survival of the butterflies, as temperature could affect the development of a butterfly.

To study this, scientists measured average wing length for butterflies that were captured in each of many years. This variable (*Wing* in mm) is the response we will try to model. They also recorded the average temperature during the larval growing season of the previous summer. This variable (*Temp* in degrees Celsius) is the predictor. It is only slightly above freezing in Greenland during the summer, so the temperatures are quite small. The data[2] are in the file **ButterfliesBc**, and are shown in Table 2.1. The dataset has values for both male and female butterflies, but in this case study we focus only on the female butterflies.

ButterBc

CHOOSE

To study whether temperature and wing length are related, we start with a graph of the data. Figure 2.4 is a scatterplot of wing length versus temperature. There is a trend in the data: higher temperatures are associated with smaller wing lengths. This suggests trying a linear model to capture the trend.

$$Wing = \beta_0 + \beta_1\ Temp + \epsilon$$

FIT

Figure 2.5 shows the regression line added to the scatterplot. The equation of the fitted line is

$$\widehat{Wing} = 19.344 - 0.239\ Temp$$

ASSESS

There is a lot of variability in the data, with points well above and well below the line. But a test of the relationship shows that the linear relationship is stronger than would happen by chance if temperature and wing length were not related. Here is computer output from fitting the regression line.

FIGURE 2.4 Scatterplot of wing length versus temperature for female butterflies in Greenland

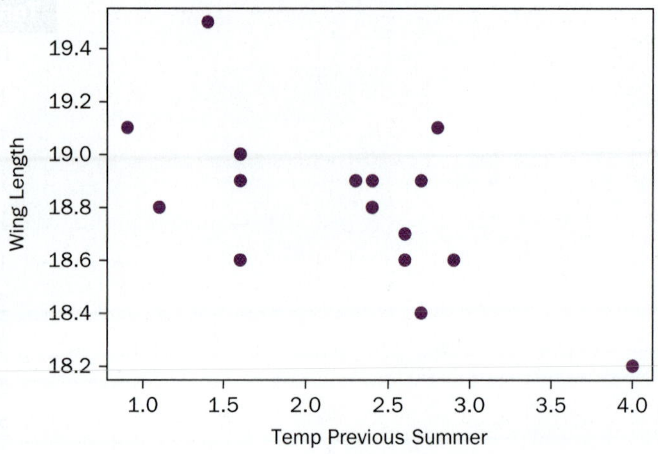

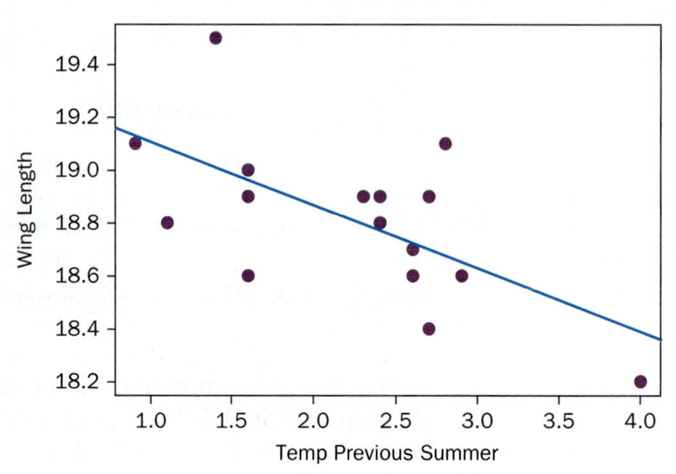

FIGURE 2.5 Scatterplot of wing length versus temperature with regression line

```
Coefficients:

            Estimate Std.   Error  t value  Pr(>|t|)
(Intercept) 19.3439        0.1872  103.33   <2e-16 ***
Temp        -0.2388        0.0795   -3.01   0.0094 **

Residual standard error: 0.246 on 14 degrees of freedom
Multiple R-squared: 0.392, Adjusted R-squared: 0.349
F-statistic: 9.03 on 1 and 14 DF, p-value: 0.00945
```

The *P*-value for the *Temp* coefficient is 0.0094, which is quite small, providing strong evidence that the pattern in the scatterplot is not due to chance. As expected, the ANOVA *F*-statistic has the same *P*-value, showing that *Temp* is an effective predictor of *Wing*. The R^2 value of 0.392 indicates that about 39% of wing length variability is explained by the previous summer's temperature.

Figure 2.6(a) is a residuals versus fits plot, which has no unusual pattern. This is good; the linear model tells us what is happening in the data. Figure 2.6(b) is a normal quantile plot of the residuals with a bit of curvature at the high end. There are a couple of points with residuals somewhat larger than we might expect from truly normal data. Should we be concerned? Probably not. We don't expect points to completely "hug the line" of a normal quantile plot. Moreover, the *t*-test is robust, so some mild departure from normality isn't a problem. Also, the *P*-value for the *t*-test is quite small, so the evidence of a link between temperature and wing length is strong, even if the calculated *P*-value of 0.0094 might be slightly inaccurate.

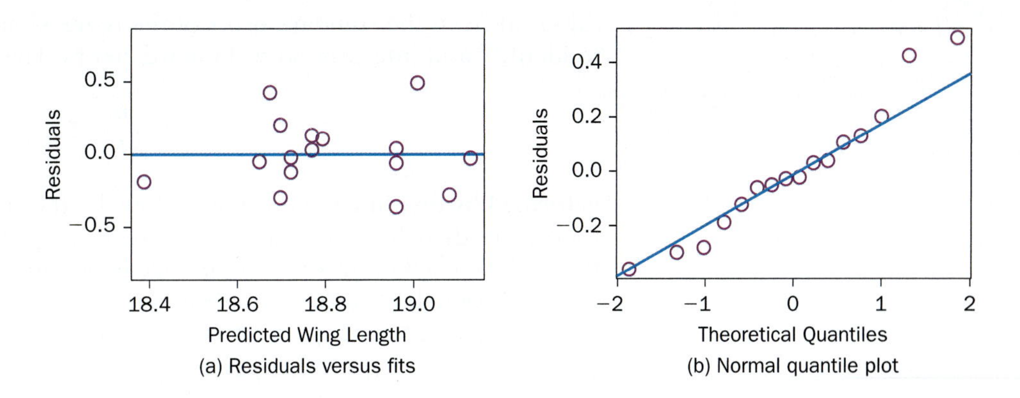

FIGURE 2.6 Residual plots for the butterfly model

We are convinced that the pattern in the scatterplot is not the result of mere chance variation. But how much does average wing length change as temperature varies? The standard error of the *Temp* coefficient is 0.0795, so a 95% confidence interval for the slope is

$$-0.239 \pm 2.145(0.0795) = -0.239 \pm 0.1705 = (-0.410 \text{ to } -0.069)$$

We are 95% confident that a one degree increase in summer temperature is associated with a decrease in average wing length of between 0.07 mm and 0.41 mm. Note that we do not conclude that increasing temperature *causes* a decrease in wing size. While this might be the case, these data were obtained from an observational study (the temperatures were merely recorded, not controlled by the researchers). There could be other factors (like cloud cover and rainfall) that affect both temperatures and butterfly growth.

Suppose we learn that the temperature in Greenland this summer is 3.0 degrees (Celsius). What size of female butterfly (of this species) do we expect to see next year? Our model gives a predicted average wing length of

$$\widehat{Wing} = 19.344 - 0.239(3.0) = 18.627$$

We expect average wing length to be 18.6 mm, but we don't expect our model to make perfect predictions. Given the variability in Figure 2.5, we shouldn't be surprised if wing length next year is quite a bit above 18.6 mm or quite a bit below 18.6 mm. How far off might our prediction of 18.6 mm be?

We can use technology to do the computational work. Here is some output:

```
Fit      SE Fit      95% CI              95% PI
18.6274  0.0870869  (18.4406, 18.8142)  (18.0671, 19.1878)
```

We see that a 95% confidence interval is (18.44, 18.81), based on a summer temperature of $3.0^{o}C$. Thus, there is a 95% chance that the average wing length of the female butterflies next year will be between 18.4 mm and 18.8 mm.

In this case study we have only used data for female butterflies, but there are also male butterfly data in **ButterfliesBc**. Males tend to have shorter wings than females, yet they show the same linear trend with temperature. We will look at males and females together, in one model, in Chapter 3. Stay tuned.

CHAPTER SUMMARY

In this chapter, we focused on statistical inference for a simple linear regression model. To test whether there is a nonzero linear relationship between a response variable Y and a predictor variable X, we use the estimated slope and the **standard error of the slope**, $SE_{\hat{\beta}_1}$. The standard error of the slope is different from the standard error of the regression line, so you should be able to identify and interpret both standard errors. The **slope test statistic**

$$t = \frac{\hat{\beta}_1}{SE_{\hat{\beta}_1}}$$

is formed by dividing the estimated slope by the standard error of the slope. It follows a t-distribution with $n - 2$ degrees of freedom.* The two-sided P-value that will be used to make our inference is provided by statistical software.

You should be able to compute and interpret **confidence intervals** for both the slope and intercept parameters. The intervals have the same general form:

$$\hat{\beta} \pm t^* SE_{\hat{\beta}}$$

where the coefficient estimate and standard error are provided by statistical software and t^* is the critical value from the t-distribution with $n - 2$ degrees of freedom.

Partitioning the total variability provides an alternative way to assess the effectiveness of our linear model. The **total variation** (*SSTotal*) is partitioned into one part that is **explained by the model** (*SSModel*) and another unexplained part **due to error** (*SSE*):

$$SSTotal = SSModel + SSE$$

This general idea of partitioning the variability is called **ANOVA** and will be used throughout the rest of the text. Each sum of squares is divided by its corresponding degrees of freedom (1 and $n - 2$, respectively) to form a mean sum of squares. An F-test statistic is formed by dividing the mean squares for the model by the mean squares for error:

$$F = \frac{MSModel}{MSE}$$

This statistic is compared to the upper tail of an F-distribution with 1 and $n - 2$ degrees of freedom. For the simple linear regression model, the inferences based on this F-test and the t-test for the slope parameter will always be equivalent.

The simple linear regression model is connected with the correlation coefficient that measures the strength of linear association through a statistic called the **coefficient of determination**. There are many ways to compute the coefficient of determination r^2, but you should be able to interpret this statistic in the context of any regression setting:

$$r^2 = \frac{SSModel}{SSTotal}$$

In general, r^2 tells us how much variation in the response variable is explained by using the explanatory variable in the regression model. This is often useful for comparing competing models.

The connection between regression and correlation leads to another test to see if our simple linear model provides an effective fit. The slope of the least squares regression line can be computed from the correlation coefficient and the standard deviations for each variable. That is,

$$\hat{\beta}_1 = r\frac{s_Y}{s_X}$$

*The normality condition is required for this test statistic to follow the t-distribution. The normality condition is also needed for all other inference procedures in Unit A.

The test statistic for the null hypothesis of no association against the alternative hypothesis of nonzero correlation is

$$t = \frac{r\sqrt{n-2}}{\sqrt{1-r^2}}$$

The P-value is obtained from a t-distribution with $n-2$ degrees of freedom.

Prediction is an important aspect of the modeling process in most applications, and you should be aware of the important differences between a **confidence interval for a mean response** and a **prediction interval for an individual response**. The standard error for predicting an individual response will always be larger than the standard error for predicting a mean response; therefore, prediction intervals will always be wider than confidence intervals. Statistical software will provide both intervals, which have the usual form, $\hat{y} \pm t^* SE$, so you should focus on picking the correct interval and providing the appropriate interpretation. Remember that the predicted value, $\hat{y}$, and either interval both depend on a particular value for the predictor variable (x^*). This should be a part of your interpretation of a confidence or prediction interval.

EXERCISES

Conceptual Exercises

2.1–2.7 ***True or False?*** Each of the statements in Exercises 2.1–2.7 is either true or false. For each statement, indicate whether it is true or false and, if it is false, give a reason.

2.1 If dataset A has a larger correlation between Y and X than dataset B, then the slope between Y and X for dataset A will be larger than the slope between Y and X for dataset B.

2.2 The degrees of freedom for the model is always 1 for a simple linear regression model.

2.3 The magnitude of the critical value (t^*) used to compute a confidence interval for the slope of a regression model decreases as the sample size increases.

2.4 The variability due to error (SSE) is always smaller than the variation explained by the model ($SSModel$).

2.5 If the size of the typical error increases, then the prediction interval for a new observation becomes narrower.

2.6 For the same value of the predictor, the 95% prediction interval for a new observation is always wider than the 95% confidence interval for the mean response.

2.7 If the correlation between X_1 and Y is greater (in magnitude) than the correlation between X_2 and Y, then the coefficient of determination for regressing Y on X_1 is greater than the coefficient of determination for regressing Y on X_2.

2.8 Using correlation. A regression equation was fit to a set of data for which the correlation, r, between X and Y was 0.6. Which of the following must be true?

a. The slope of the regression line is 0.6.

b. The regression model explains 60% of the variability in Y.

c. The regression model explains 36% of the variability in Y.

d. At least half of the residuals are smaller than 0.6 in absolute value.

2.9 Interpreting the size of r^2.

a. Does a high value of r^2, say, 0.90 or 0.95, indicate that a linear relationship is the best possible model for the data? Explain.

b. Does a low value of r^2, say, 0.20 or 0.30, indicate that some relationship other than linear would be the best model for the data? Explain.

2.10 Effects on a prediction interval. Describe the effect (if any) on the width (difference between the upper and lower endpoints) of a prediction interval if all else remains the same except:

a. the sample size is increased.

b. the variability in the values of the predictor variable increases.

c. the variability in the values of the response variable increases.

d. the value of interest for the predictor variable moves further from the mean value of the predictor variable.

Guided Exercises

2.11 Inference for slope. A regression model was fit to 40 data cases and the resulting sample slope, $\hat{\beta}_1$, was 15.5, with a standard error $SE_{\hat{\beta}_1}$ of 3.4. Assume that the conditions for a simple linear model, including normality, are reasonable for this situation.

a. Test the hypothesis that $\beta_1 = 0$.

b. Construct a 95% confidence interval for β_1.

2.12 Inference for slope, again. A regression model was fit to 82 data cases and the resulting sample slope, $\hat{\beta}_1$, was 5.3, with a standard error $SE_{\hat{\beta}_1}$ of 2.8. Assume that the conditions for a simple linear model, including normality, are reasonable for this situation.

a. Test the hypothesis that $\beta_1 = 0$.

b. Construct a 95% confidence interval for β_1.

2.13 Brain pH. Scientists measured the pH in brain tissue shortly after the deaths of 54 persons. One possible predictor of brain tissue pH is age at death of the person. The variables pH and Age are in the datafile **BrainpH**. Is Age a good predictor of pH? **BrainpH**

a. Produce a scatterplot of the data. Does there appear to be a linear relationship? Explain.

b. Test the hypothesis that there is a linear relationship between pH and age. Does this test confirm your suspicions from part (a)? Explain.

2.14 Racial animus. Professor Seth Stephens-Davidowitz studies the level of racial animus across different areas in America by measuring the percent of Google search queries that include racially charged language.[3] In his work, he finds the percentage of Google searches that included the "n" word (with or without an *s* at the end). He then applies a sophisticated algorithm, but for our purposes we will work with a measurement, *Animus*, that is derived from his numbers and is scaled to be between 0 (low racial animus) and 250 (high racial animus). The file **RacialAnimus** has data for each of the 196 media markets (we can think of a media market as a city and the surrounding area) on *Animus* and other variables. We wish to explore a linear model using *Animus* as the response variable.

One variable of interest is *ObamaKerry*, which is the percentage of the vote that Barack Obama won in 2008 minus the percentage of the vote that John Kerry won in 2004. **RacAnim**

a. Produce a scatterplot of the data. Does there appear to be a linear relationship? Explain.

b. Test the hypothesis that there is a linear relationship between *ObamaKerry* and *Animus*. Does this test confirm your suspicions from part (a)? Explain.

2.15 Breakfast cereal. The number of calories and number of grams of sugar per serving were measured for 36 breakfast cereals. The data are in the file **Cereal**. We are interested in trying to predict the calories using the sugar content. **Cereal**

a. Test the hypothesis that sugar content has a linear relationship with calories (in an appropriate scale). Report the hypotheses, test statistic, and *P*-value, along with your conclusion.

b. Find and interpret a 95% confidence interval for the slope of this regression model. Also explain what this slope coefficient means in the context of these data.

2.16 Textbook prices. Exercise 1.44 examined data on the price and number of pages for a random sample of 30 textbooks from the Cal Poly campus bookstore. The data are stored in the file **TextPrices** and appear in Table 1.5. **TxtPrc**

a. Perform a significance test to address the students' question of whether the number of pages is a useful predictor of a textbook's price. Report the hypotheses, test statistic, and *P*-value, along with your conclusion.

b. Find and interpret a 95% confidence interval for the slope of this regression model. Also explain what this slope coefficient means in the context of these data.

2.17 Dementia. Two types of dementia are Dementia with Lewy Bodies and Alzheimer's disease. Some people are afflicted with both of these. The file **LewyDLBad** has data from 20 such subjects. The variable *APC* gives annualized percentage change in brain gray matter. The variable *MMSE* measures change in functional performance on the Mini Mental State Examination. **LewyDL**

a. Is there a statistically significant linear relationship between *MMSE* (response) and *APC* (predictor)?

b. Assess the use of a regression model by making an appropriate residual plot.

c. Identify the estimated slope and its standard error.

d. Provide a 90% confidence interval for the slope of this model. Does your interval contain zero? How does this relate to your answer to part (a)?

2.18 Sparrows. In Exercises 1.3–1.11 on page 51, we consider a model for predicting the weight in grams from the wing length (in millimeters) for a sample of Savannah sparrows found at Kent Island, New Brunswick, Canada. The data are in the file **Sparrows**. **Sparrow**

a. Is the slope of the least squares regression line for predicting *Weight* from *WingLength* significantly different from zero? Show details to support your answer.

b. Construct and interpret a 95% confidence interval for the slope coefficient in this model.

c. Does your confidence interval in part (b) contain zero? How is this related to part (a)?

2.19 Real estate near Rails to Trails: distance. Consider the dataset **RailsTrails**, that looks at 104 homes sold in Northampton, Massachusetts, in 2007. Recall from Exercise 1.25 that a goal of the study was to ascertain if distance to a bike trail (*distance*) would positively impact the selling price of the home. Use *adj2007* as your response variable for this exercise. **Rails**

a. Fit a simple linear regression of *adj2007* (the 2007 selling price adjusted to 2014 dollars) on *distance*. Report the equation and the estimated mean increase in price associated with each foot closer a home is to a bike trail.

b. Give a 90% confidence interval for the estimate found in (a) along with a statement interpreting what the confidence interval means.

c. Comment on adherence to the simple linear regression model in this case and how this might affect our answer to (b).

2.20 Real estate nears Rails to Trails: size of home. Consider the dataset **RailsTrails** that looks at 104 homes sold in Northampton, Massachusetts, in 2007 (first discussed in Exercise 1.25). We can use the data to ascertain how much higher the selling price is for larger homes. **Rails**

a. Fit a simple linear regression of *adj2007* (the 2007 selling price adjusted to 2014 dollars) on *squarefeet*. Report the equation and the estimated increase in selling price associated with each additional square foot of home size.

b. Give a 90% confidence interval for the estimate found in (a) along with a statement interpreting what the confidence interval means.

c. Comment on the simple linear regression model conditions and how this affects our answer to (b).

2.21 Partitioning variability. The sum of squares for the regression model, *SSModel*, for the regression of Y on X was 110, the sum of squared errors, *SSE*, was 40, and the total sum of squares, *SSTotal*, was 150. Calculate and interpret the value of r^2.

2.22 Partitioning variability, again. The sum of squares for the regression model, *SSModel*, for the regression of Y on X was 38, the sum of squared errors, *SSE*, was 64, and the total sum of squares, *SSTotal*, was 102. Calculate and interpret the value of r^2.

2.23 U.S. stamp prices. Use the file **USstamps** to continue the analysis begun in Exercise 1.33 on page 55 to model regular postal rates from 1958 to 2012. Delete the first four observations from 1885, 1917, 1919, and 1932 before answering the questions below. **USStamp**

a. What percent of the variation in postal rates (*Price*) is explained by *Year*?

b. Is there a significant linear association between postal rates (*Price*) and *Year*? Justify your answer.

c. Find and interpret the ANOVA table that partitions the variability in *Price* for a linear model based on *Year*.

2.24 Metabolic rates. Use the file **MetabolicRate** to examine the linear relationship between the log (base 10) of metabolic rate and log (base 10) of body size for a sample of caterpillars. **MetRate**

a. Fit a least squares regression line for predicting *LogMrate* from *LogBodySize*. What is the equation of the regression line?

b. Is the slope parameter significantly different from zero? Justify your answer.

c. Find and interpret the ANOVA table for partitioning the variability in the transformed metabolic rates.

d. Calculate the ratio of the model sum of squares and the total sum of squares. Provide an interpretation for this statistic.

2.25 More sparrows. Refer to the **Sparrows** data and the model to predict weight from wing length that is described in Exercise 2.18. **Sparrow**

a. Is there a significant association between weight and wing length? Use the correlation coefficient between *Weight* and *WingLength* to conduct an appropriate hypothesis test.

b. What percent of the variation in weight is explained by the simple linear model with *WingLength* as a predictor?

c. Provide the ANOVA table that partitions the total variability in weight and interpret the F-test.

d. Compare the square root of the F-statistic from the ANOVA table with the t-statistic from testing the correlation.

2.26 Leaf width over time. Biologists know that the leaves on plants tend to get smaller as temperatures rise. The file **LeafWidth** has data on a sample of 252 leaves from the species *Dodonaea viscosa* subspecies *angustissima* have been collected in a certain region of South Australia for many years. The variable *Width* is the average width, in mm, of leaves, taken at their widest points, that were collected in a given year. **LeafW**

a. Is there a significant association between leaf width and year? Use the correlation coefficient between *Width* and *Year* to conduct an appropriate hypothesis test.

b. What percent of the variation in weight is explained by the simple linear model with *Year* as a predictor and *Width* as response?

c. Provide the ANOVA table that partitions the total variability in width and interpret the F-test.

d. Compare the square root of the F-statistic from the ANOVA table with the t-statistic from testing the correlation.

2.27 Brees passes in the NFL. The data in Table 2.2 (stored in **BreesPass**) are the passing yards and number of pass attempts for Drew Brees in each game[4] of the 2016 National Football League season. **Brees**

Table 2.2 Drew Brees passing for NFL games in 2016

Game	Opponent	Attempts	Yards	Game	Opponent	Attempts	Yards
1	OAK	42	423	9	DEN	29	303
2	NYG	44	263	10	CAR	44	285
3	ATL	54	376	11	LAR	36	310
4	SD	36	207	12	DET	44	326
5	CAR	49	465	13	TB	41	257
6	KC	48	367	14	ARI	48	389
7	SEA	35	265	15	TB	34	299
8	SF	39	323	16	ATL	50	350

a. Produce a scatterplot and determine the regression equation for predicting yardage from number of passes.

b. Is the slope coefficient equal to Brees's average number of yards per pass? Explain how you know.

c. How much of the variability in Brees's yardage per game is explained by knowing how many passes he threw?

2.28 Enrollment in mathematics courses. Exercise 1.34 on page 55 introduces data on total enrollments in mathematics courses at a small liberal arts college where the academic year consists of two semesters, one in the fall and another in the spring. Data for Fall 2001 to Spring 2012 are shown in Table 1.4 with the earlier exercise and stored in **MathEnrollment**. **MthEnr**

a. In Exercise 1.34, we found that the data for 2003 were unusual due to special circumstances. Remove the data from 2003 and fit a regression model for predicting spring enrollment (*Spring*) from fall enrollment (*Fall*). Prepare the appropriate residual plots and comment on the slight problems with the conditions for inference. In particular, make sure that you plot the residuals against order (or *AYear*) and comment on the trend.

b. Even though we will be able to improve on this simple linear model in the next chapter, let's take a more careful look at the model. What percent of the variability in spring enrollment is explained by using a simple linear model with fall enrollment as the predictor?

c. Provide the ANOVA table for partitioning the total variability in spring enrollment based on this model.

d. Test for evidence of a significant linear association between spring and fall enrollments.

e. Provide a 95% confidence interval for the slope of this model. Does your interval contain zero? Why is that relevant?

2.29 Goldenrod galls in 2003. In Exercise 1.41 on page 56, we introduce data collected by biology students with measurements on goldenrod galls at the Brown Family Environmental Center. The file **Goldenrod** contains the gall diameter (in mm), stem diameter (in mm), wall thickness (in mm), and codes for the fate of the gall in 2003 and 2004. **Golden**

a. Is there a significant linear relationship between wall thickness (response) and gall diameter (predictor) in 2003?

b. Identify the estimated slope and its standard error.

c. What is the size of the typical error for this simple linear regression model?

d. A particular biologist would like to explain over 50% of the variability in the wall diameter. Will this biologist be satisfied with this simple model?

e. Provide a 95% interval estimate for the mean wall thickness when the gall diameter is 20 mm.

f. Use the correlation coefficient to test if there is a significant association between wall thickness and gall diameter.

2.30 Goldenrod galls in 2004. Repeat the analysis in Exercise 2.29 using the measurements made in 2004, instead of 2003. **Golden**

2.31 Pines: 1990–1997. The dataset **Pines** introduced in Exercise 1.34 on page 55 contains measurement from an experiment growing white pine trees conducted by the Department of Biology at Kenyon College at a site near the campus in Gambier, Ohio. **Pines**

a. Test for a significant linear relationship between the initial height of the pine seedlings in 1990 and the height in 1997.

b. What percent of the variation in the 1997 heights is explained by the regression line?

c. Provide the ANOVA table that partitions the total variability in the 1997 heights based on the model using 1990 heights.

d. Verify that the coefficient of determination can be computed from the sums of squares in the ANOVA table.

e. Are you happy with the fit of this linear model? Explain why or why not.

2.32 Pines: 1996–1997. Repeat the analysis in Exercise 2.31 using the height in 1996 to predict the height in 1997. **Pines**

2.33 Pines regression intervals. Refer to the regression model in Exercise 2.32 using the height in 1996 to predict the height in 1997 for pine trees in the **Pines** dataset. **Pines**

a. Find and interpret a 95% confidence interval for the slope of this regression model.

b. Is the value of 1 included in your confidence interval for the slope? What does this tell you about whether or not the trees are growing?

c. Does it make sense to conduct inference for the intercept in this setting? Explain.

2.34 Moth eggs. Researchers[5] were interested in looking for an association between body size (*BodyMass* after taking the log of the measurement in grams) and the number of eggs produced by a moth. *BodyMass* and the number of eggs present for 39 moths are in the file **MothEggs**. **Moth**

a. Before looking at the data, would you expect the association between body mass and number of eggs to be positive or negative? Explain.

b. What is the value of the correlation coefficient for measuring the strength of linear association between *BodyMass* and *Eggs*?

c. Is the association between these two variables statistically significant? Justify your answer.

d. Fit a linear regression model for predicting *Eggs* from *BodyMass*. What is the equation of the least squares regression line?

e. The conditions for inference are not met, primarily because there is one very unusual observation. Identify this observation.

2.35 Moth eggs: subset. Use the data in the file **MothEggs** to continue the work of Exercise 2.34 to model the relationship between the number of eggs and body mass for this type of moth. **Moth**

a. Remove the moth that had no eggs from the dataset, and fit a linear regression model for predicting the number of eggs. What is the equation of the least squares regression line?

b. Prepare appropriate residual plots and comment on whether or not the conditions for inference are met.

c. Compare the estimated slope with and without the moth that had no eggs.

d. Compare the percent of variability in the number of eggs that is explained with and without the moth that had no eggs.

2.36–2.39 *When does a child first speak?* The data in Table 2.3 (stored in **ChildSpeaks**) are from a study about whether there is a relationship between the age at which a child first speaks (in months) and his or her score on a Gesell Aptitude Test taken later in childhood. Use these data to examine this relationship in Exercises 2.36–2.39 that follow. **ChldSpk**

Table 2.3 Gesell Aptitude Test and first speak age (in months)

Child #	Age	Gesell	Child #	Age	Gesell	Child #	Age	Gesell
1	15	95	8	11	100	15	11	102
2	26	71	9	8	104	16	10	100
3	10	83	10	20	94	17	12	105
4	9	91	11	7	113	18	42	57
5	15	102	12	9	96	19	17	121
6	20	87	13	10	83	20	11	86
7	18	93	14	11	84	21	10	100

2.36 Child first speaks: full data. Use the data in Table 2.3 and **ChildSpeaks** to consider a model to predict age at first speaking using the Gesell score.

a. Before you analyze the data, would you expect to see a positive relationship, negative relationship, or no relationship between these variables? Provide a rationale for your choice.

b. Produce a scatterplot of these data and comment on whether age of first speaking appears to be a useful predictor of the Gesell aptitude score.

c. Report the regression equation and the value of r^2. Also, determine whether the relationship between these variables is statistically significant.

d. Which child has the largest (in absolute value) residual? Explain what is unusual about that child.

2.37 Child first speaks: one point removed. Refer to Exercise 2.36 where we consider a model to predict age at first speech using the Gesell score for the data in **ChildSpeaks**. Remove the data case for the child who took 42 months to speak and produce a new scatterplot with a fitted line. Comment on how removing the one child has affected the line and the value of r^2.

2.38 Child first speaks: a second point removed. Refer to Exercise 2.36, where we consider a model to predict age at first speech using the Gesell score for the data in **ChildSpeaks**, and Exercise 2.37, where we remove the case where the child took 42 months to speak. Now remove the data for the child who took 26 months to speak, in addition to the child who took 42 months. Produce a new scatterplot with a fitted line when both points have been removed. Comment on how removing this child (in addition to the one identified in Exercise 2.37) has affected the line and the value of r^2.

2.39 Child first speaks: a third point removed. Refer to Exercise 2.36, where we consider a model to predict age at first speech using the Gesell score for the data in **ChildSpeaks**. In Exercises 2.37 and 2.38, we removed two cases where the children took 42 months and 26 months, respectively, to speak. Now also remove the data for a third child who is identified as an outlier in part (d) of Exercise 2.36. As in the previous exercises, produce a new scatterplot with fitted line and comment on how removing this third child (in addition to the first two) affects the analysis.

2.40 Predicting Grinnell house prices. Refer to the GrinnellHouses data introduced in Exercise 1.20 and use the simple regression of *SalePrice* on *ListPrice* to: **Grinnell**

a. Find a 90% confidence interval for the mean *SalePrice* of a home that is listed at $300,000.

b. Find a 90% prediction interval for the *SalePrice* of a home that is listed at $300,000.

c. Give interpretative sentences for your answers to (a) and (b) so that the distinction between what they mean is clear.

2.41 Leaf Width: regression intervals. The datafile **LeafWidth** was introduced in Exercise 2.26. In this exercise we will investigate what we would predict for a width if the linear trend continues for future years. **LeafW**

a. Find a 95% confidence interval for the mean *LeafWidth* in the year 2020.

b. Find a 95% prediction interval for the *LeafWidth* of one plant in the year 2020.

c. Give interpretative sentences for your answers to (a) and (b) so that the distinction between what they mean is clear.

2.42 Dementia: regression intervals. Use the data in the file **LewyDLBad** to continue the analysis of the relationship between *MMSE* and *APC* from Exercise 2.17. **LewyDL**

a. What would you predict the MMSE score to be for someone with an *APC* measurement of −1?

b. Provide a 95% confidence interval for mean *MMSE* when *APC* is −1.

c. Provide a 95% prediction interval for *MMSE* when *APC* is −1.

d. Tania wants to know why the interval from part (c) is wider than the interval from part (b). She doesn't want to see a bunch of mathematics; she wants to understand the reason. What would you say to her?

2.43 Sparrows: regression intervals. Refer to the **Sparrows** data and the model to predict weight from wing length that is described in Exercise 2.18. **Sparrow**

a. What would you predict the weight of a Savannah sparrow to be with a wing length of 20 mm?

b. Find a 95% confidence interval for the mean weight for Savannah sparrows with a wing length of 20 mm.

c. Find a 95% prediction interval for the weight of a Savannah sparrow with a wing length of 20 mm.

d. Without using statistical software to obtain a new prediction interval, explain why a 95% prediction interval for the weight of a sparrow with a wing length of 25 mm would be narrower than the prediction interval in part (b).

2.44 Textbook prices: regression intervals. Refer to Exercise 2.16 on prices and numbers of pages in textbooks. **TxtPrc**

a. Determine a 95% confidence interval for the mean price of a 450-page textbook in the population.

b. Determine a 95% prediction interval for the price of a particular 450-page textbook in the population.

c. How do the midpoints of these two intervals compare? Explain why this makes sense.

d. How do the widths of these two intervals compare? Explain why this makes sense.

e. What value for number of pages would produce the narrowest possible prediction interval for its price? Explain.

f. Determine a 95% prediction interval for the price of a particular 1500-page textbook in the population. Do you really have 95% confidence in this interval? Explain.

2.45 Racial animus: regression intervals. Refer to Exercise 2.14 on the difference in votes for Obama and Kerry, and the amount of racial animus in the areas. **RacAnim**

a. Determine a 95% confidence interval for the mean *ObamaKerry* for a region with a racial animus measurement of 150.

b. Determine a 95% confidence interval for *ObamaKerry* for a region with a racial animus measurement of 150.

c. How do the midpoints of these two intervals compare? Explain why this makes sense.

d. How do the widths of these two intervals compare? Explain why this makes sense.

e. What value for *Animus* would produce the narrowest possible prediction interval for *ObamaKerry*? Explain.

f. Determine a 95% prediction interval for *ObamaKerry* of a particular region with a racial animus measurement of 0. Do you really have 95% confidence in this interval? Explain.

2.46 Real estate near Rails to Trails: home size transformation. Consider the **RailsTrails** data, first discussed in Exercise 1.25, which give information about 104 homes sold in Northampton, Massachusetts, in 2007. In this exercise, use *adj2007* (the selling price in adjusted 2014 dollars) as the response variable and *squarefeet* (the home's floor space) as the explanatory variable. **Rails**

a. Use a simple linear regression model to predict the selling price of a home (in *adj2007* units, of course) if the home is 1500 square feet in size. (Note: the *squarefeet* units are in thousands.)

b. Give a 95% prediction interval and interpret the interval.

c. Comment on adherence to the model conditions and any effect on answer (b).

d. Redo the regression using log(*adj*2007) and log(*squarefeet*) as response and explanatory variables. Comment on the logged model, comparing its fit and its model conditions to the original.

e. Redo the prediction interval for the 1500 square foot home. Translate the prediction interval back to dollars—from log-dollars—and compare the two intervals.

2.47 Real estate near Rails to Trails: distance transformation. Consider, again, the **RailsTrails** data, which give information about 104 homes sold in Northampton, Massachusetts, in 2007. In this exercise, use *adj2007* (the 2007 selling price in adjusted 2014 thousands of dollars) as the response variable and *distance* (the home's distance to the nearest bike trail, in miles) as the explanatory variable. **Rails**

a. Use the simple linear regression model to predict the selling price for a home that is half a mile (*distance* = 0.5) from a bike trail.

b. Find a 90% prediction interval.

c. Assess model conditions and comment on the implications for parts (a) and (b).

d. Redo the prediction and the 90% interval using log(*adj*2007) and log(*distance*) for response and explanatory variables, respectively. Report the results giving an interpretive statement for the interval.

e. Redo the prediction interval for a home that is half a mile from the bike trail. Translate the prediction interval back to dollars—from log-dollars—and compare the two intervals.

2.48 Brees passes in the NFL: use of model. In Exercise 2.27, we fit a linear model to predict the number of yards Drew Brees passes for in a football game based on his number of passing attempts using the data in **BreesPass** and Table 2.2. **Brees**

a. Which game has the largest (in absolute value) residual? Explain what the value of the residual means for that game. Was this a particularly good game, or a particularly bad one, for Brees?

b. Determine a 90% prediction interval for Brees's passing yards in a game in which he throws 40 passes.

c. Would you feel comfortable in using this regression model to predict passing yardage for a different player? Explain.

2.49 Enrollment in mathematics courses: use of model. Use the file **MathEnrollment** to continue the analysis of the relationship between fall and spring enrollment from Exercise 2.28. Fit the regression model for predicting *Spring* from *Fall*, after removing the data from 2003. **MthEnr**

a. What would you predict the spring enrollment to be when the fall enrollment is 290?

b. Provide a 95% confidence interval for mean spring enrollment when the fall enrollment is 290.

c. Provide a 95% prediction interval for spring enrollment when the fall enrollment is 290.

d. A new administrator at the college wants to know what interval she should use to predict the enrollment next spring when the enrollment next fall is 290. Would you recommend that she use your interval from part (b) or the interval from part (c)? Explain.

2.50 Metabolic rates: comparing models. Use the file **MetabolicRate** to continue the work of Exercise 2.24 on inferences for the relationship between the log (base 10) of metabolic rate and log (base 10) of body size for a sample of caterpillars. **MetRate**

a. Use the linear model for *LogMrate* based on *LogBodySize* to predict the log metabolic rate and then convert to predicted metabolic rate when *LogBodySize* = 0. What body size corresponds to a caterpillar with *LogBodySize* = 0?

b. How much wider is a 95% prediction interval for *LogMrate* than a 95% confidence interval for mean *LogMrate* when *LogBodySize* = 0?

c. Repeat part (b) when *LogBodySize* = −2.

2.51 Handwriting Analysis. The file **Handwriting** contains survey data from a sample of 203 statistics students at Clarke University. Each student was given 25 handwriting specimens and they were to guess, as best they could, the gender (male or female) of the person who penned the specimen. (Each specimen was a handwritten address, as might appear on the envelope of a letter. There were no semantic clues in the specimen that could tip off the author's gender.) The survey was repeated two times—the second time was the same 25 specimens given in a randomly new order.

One might hope that subjects would be consistent in their guesses between the two surveys. To investigate the salience of this hope, we look at the relationship between the variables *Survey1* and *Survey2*, which give, respectively, the percentage of successes for the two surveys. **Handwrt**

a. Make a scatterplot with *Survey1* on the horizontal axis and *Survey2* on the vertical axis. Comment on the pattern. Note: We are arbitrarily designating the first survey as producing the explanatory variable and the second as the response.

b. Regress *Survey2* on *Survey1* and write down the prediction equation.

c. Include appropriate plots and comment on the adequacy of the model conditions for simple linear regression.

d. If gender could be predicted well by handwriting, we would expect that *Survey*1 = *Survey*2. How well does this model fit the situation described by these data? Enhance your discussion by referring to the results of significance tests on the regression coefficients.

2.52 Grinnell houses: use of model. The file **GrinnellHouses** contains data from 929 house sales in Grinnell, Iowa, between 2005 and into 2015. In Exercise 1.20 we investigated the relationship of the list price of the home (what the seller asks for the home) to the final sale price. We found that *ListPrice* was a good predictor for *SalePrice*, albeit with some concerns about the model conditions for simple linear regression. We will ignore those concerns for this exercise, although we will come back to them in later exercises.

Refit the simple linear regression model for predicting *SalePrice* from *ListPrice* and answer these questions. **Grinnell**

a. Give a 95% confidence interval for the slope, and write a sentence that interprets this interval in context.

b. Test the hypothesis that the true intercept for the regression line is 0.

c. Given the result in part (b), we might consider a simple model that *SalePrice* is predicted as a simple fraction of the *ListPrice*, where the estimate of that fraction is the line's slope. If this model works, we could also look for this fraction by computing a 95% confidence interval for the collection of ratios of *SalePrice* / *ListPrice*. Find this interval and compare to your answer in (a).

2.53 Grinnell houses: transformations. Refer to the **GrinnellHouses** data and define variables called *CbrtLP* and *CbrtSP*, which are, respectively, cube-roots of *ListPrice* and *SalePrice*. Fit a regression of *CbrtSP* on *CbrtLP* and answer these questions. **Grinnell**

a. Find the predicted Sale Price for a home listed at $300,000 using the cube-root model and put a 90% prediction interval around this.

b. Compare the answer from (a) to the prediction interval obtained in the untransformed model found in Exercise 2.40.

Open-ended Exercises

2.54 Baseball game times. In Exercise 1.45, you examined factors to predict how long a Major League Baseball game will last. The data in Table 1.6 were collected for the 14 games played on August 11, 2017, and stored in the file named **BaseballTimes2017**. **BballT**

a. Calculate the correlations of each predictor variable with the length of a game (*Time*). Identify which predictor variable is most strongly correlated with time.

b. Choose the one predictor variable that you consider to be the best predictor of time. Determine the regression equation for predicting time based on that predictor. Also, interpret the slope coefficient of this equation.

c. Perform the appropriate significance test of whether this predictor is really correlated with time in the population.

d. Analyze appropriate residual plots for this model, and comment on what they reveal about whether the conditions for inference appear to be met here.

2.55 Nitrogen in caterpillar waste. Exercises 1.29–1.32 starting on page 54 introduce some data on a sample of 267 caterpillars. Use the untransformed data in the file **Caterpillars** to answer the questions below about a simple linear regression model for predicting the amount of nitrogen in frass (*Nfrass*) of these caterpillars. Note that 13 of the 267 caterpillars in the sample are missing values for *Nfrass*, so those cases won't be included in the analysis. **Caterp**

a. Calculate the correlations of each possible predictor variable (*Mass*, *Intake*, *WetFrass*, *DryFrass*, *Cassim*, and *Nassim*) with *Nfrass*. Identify which predictor variable is the most strongly correlated with *Nfrass*.

b. Choose the one predictor variable that you consider to be the best predictor of *Nfrass*. Determine the regression equation for predicting *Nfrass* based on that predictor and add the regression line to the appropriate scatterplot.

c. Perform a significance test of whether this predictor is really correlated with *Nfrass*.

d. Are you satisfied with the fit of your model in part (b)? Explain by commenting on the coefficient of determination and residual plots.

2.56 CO_2 assimilation in caterpillars. Repeat the analysis in Exercise 2.55 to find and assess a single predictor regression model for CO_2 assimilation (*Cassim*) in these caterpillars. Use the untransformed variables *Mass*, *Intake*, *WetFrass*, *DryFrass*, *Nassim*, and *Nfrass* as potential predictors. **Caterp**

2.57 Horses for sale. Undergraduate students at Cal Poly collected data on the prices of 50 horses advertised for sale on the Internet.[6] Predictor variables include the age and height of the horse (in hands), as well as its sex. The data appear in Table 2.4 and are stored in the file **HorsePrices**.

Analyze these data in an effort to find a useful model that predicts price from one predictor variable. Be sure to consider transformations, and you may want to consider fitting a model to males and females separately. Write a report explaining the steps in your analysis and presenting your final model. **Horse**

Table 2.4 Horse prices

Horse ID	Price	Age	Height	Sex	Horse ID	Price	Age	Height	Sex
97	38000	3	16.75	m	132	20000	14	16.50	m
156	40000	5	17.00	m	69	25000	6	17.00	m
56	10000	1	*	m	141	30000	8	16.75	m
139	12000	8	16.00	f	63	50000	6	16.75	m
65	25000	4	16.25	m	164	1100	19	16.25	f
184	35000	8	16.25	f	178	15000	0.5	14.25	f
88	35000	5	16.50	m	4	45000	14	17.00	m
182	12000	17	16.75	f	211	2000	20	16.00	f
101	22000	4	17.25	m	89	20000	3	15.75	f
135	25000	6	15.25	f	57	45000	5	16.50	m
35	40000	7	16.75	m	200	20000	12	17.00	m
39	25000	7	15.75	f	38	50000	7	17.25	m
198	4500	14	16.00	f	2	50000	8	16.50	m
107	19900	6	15.50	m	248	39000	11	17.25	m
148	45000	3	15.75	f	27	20000	11	16.75	m
102	45000	6	16.75	m	19	12000	6	16.50	f
96	48000	6	16.50	m	129	15000	2	15.00	f
71	15500	12	15.75	f	13	27500	5	16.00	f
28	8500	7	16.25	f	206	12000	2	*	f
30	22000	7	16.50	f	236	6000	0.5	*	f
31	35000	5	16.25	m	179	15000	0.5	14.50	m
60	16000	7	16.25	m	232	60000	13	16.75	m
23	16000	3	16.25	m	152	50000	4	16.50	m
115	15000	7	16.25	f	36	30000	9	16.50	m
234	33000	4	16.50	m	249	40000	7	17.25	m

2.58 Infant mortality rate versus year. Table 2.5 shows infant mortality (deaths within one year of birth per 1000 births) in the United States from 1920–2010. The data[7] are stored in the file **InfantMortality2010**. InfMort

Table 2.5 U.S. Infant *Mortality*

Mortality	Year
85.8	1920
64.6	1930
47.0	1940
29.2	1950
26.0	1960
20.0	1970
12.6	1980
9.2	1990
6.9	2000
6.2	2010

a. Make a scatterplot of *Mortality* versus *Year* and comment on what you see.

b. Fit a simple linear regression model and examine residual plots. Do the conditions for a simple linear model appear to hold?

c. If you found significant problems with the conditions, find a transformation to improve the linear fit; otherwise, proceed to the next part.

d. Test the hypothesis that there is a linear relationship in the model you have chosen.

e. Use the final model that you have selected to make a statistical statement about infant mortality in the year 2010.

2.59 Caterpillars: nitrogen assimilation and mass. In Exercise 1.30 on page 54, we examine the relationship between nitrogen assimilation and body mass for a sample of caterpillars with data stored in **Caterpillars**. The log (base 10) transformations for both predictor and response (*LogNassim* and *LogMass*) worked well in this setting, but we noticed that the linear trend was much stronger during the free growth period ($Fgp = \text{Y}$) than at other times ($Fgp = \text{N}$). Compare and contrast the simple linear model for the entire dataset, with the simple linear model only during the free growth period. Caterp

Supplemental Exercise

2.60 Linear fit in a normal probability plot. A random number generator was used to create 100 observations for a particular variable. A normal probability plot from Minitab is shown in Figure 2.7.

Notice that this default output from Minitab includes a regression line and a confidence band. The band is narrow near the mean of 49.54 and gets wider as you move away from the center of the randomly generated observations.

a. Do you think Minitab used confidence intervals or prediction intervals to form the confidence band? Provide a rationale for your choice.

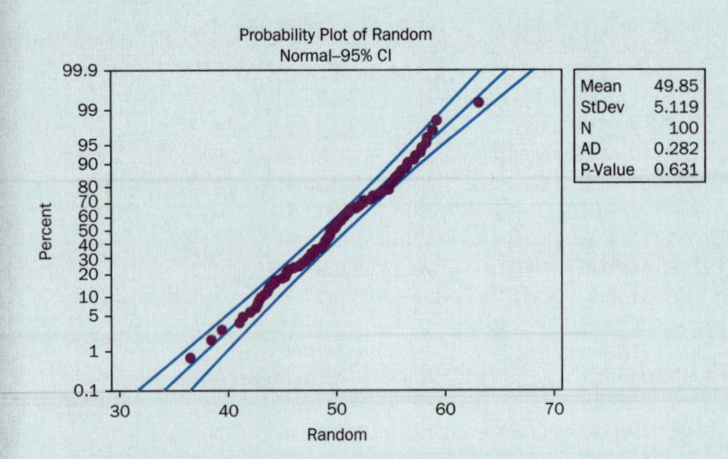

FIGURE 2.7 Normal probability plot for random numbers

b. The Anderson-Darling test statistic $AD = 0.282$ and corresponding P-value = 0.631 are provided in the output. The null hypothesis is that the data are normal and the alternative hypothesis is that the data are not normal. Provide the conclusion for the formal Anderson-Darling procedure.

c. Do the departures from linear trend in the probability plot provide evidence for the conclusion you made in part (b)? Explain.

2.61 Gate count: computing the least squares line from descriptive statistics. Many libraries have gates that automatically count persons as they leave the building, thus making it easy to find out how many persons used the library in a given year. (Of course, someone who enters, leaves, and enters again is counted twice.) Researchers conducted a survey of liberal arts college libraries and found the following descriptive statistics on enrollment and gate count:

	N	MEAN	MEDIAN	TRMEAN	STDEV	SEMEAN
Enroll	17	2009	2007	2024	657	159
Gate	17	247235	254116	247827	104807	25419

Correlation of Enroll and Gate = 0.701

The least squares regression line can be computed from five descriptive statistics, as follows:

$$\hat{\beta}_1 = r\frac{s_y}{s_x} \quad \text{and} \quad \hat{\beta}_0 = \bar{y} - \hat{\beta}_1\bar{x}$$

where $\bar{x}$, $\bar{y}$, s_x, and s_y are the sample means and standard deviations for the predictor and response, respectively, and r is their correlation.

a. Find the equation of the least squares line for predicting the gate count from enrollment.

b. What percentage of the variation in the gate counts is explained by enrollments?

c. Predict the number of persons who will use the library at a small liberal arts college with an enrollment of 1445.

d. One of the reporting colleges has an enrollment of 2200 and a gate count of 130,000. Find the value of the residual for this college.

2.62 Correlation using Bayesian analysis. Most statistical analyses, including those throughout this book, use what are broadly known as frequentist methods. However, there is another school of thought within the statistics world: Bayesian reasoning. In this exercise we provide a glimpse into Bayesian methods. The file **Cereal** contains data on 36 breakfast cereals. We want to consider the relationship between *Calories* and *Sugar* (the number of grams of sugar per serving). **Cereal**

a. Make a scatterplot of $Y = Calories$ versus $X = Sugar$. Find the correlation coefficient.

b. Conduct a test of the null hypothesis that there is no correlation between *Calories* and *Sugar* in the population of all breakfast cereals. What is the P-value? What is the 95% confidence interval?

c. The R package BayesianFirstAid* has a command bayes.cor.test that uses a technique called Markov chain Monte Carlo to estimate the population (true) correlation coefficient and to provide a measure of uncertainty about the estimate. Run the command BayesCorTestCereal < − bayes.cor.test(Sugar+Calories, data=Cereal). Then enter BayesCorTestCereal to see the results.

The Bayesian analysis gives the probability, conditional on the data, that the true correlation is positive and that it is negative.† It also provides a 95% credible interval, which is similar to a confidence interval, but which is interpreted differently. A Bayesian analysis concludes that there is a 95% probability that the true correlation is within the 95% credible interval.

Finally, run the command plot(BayesCorTestCereal) to see a plot of the results. At the top of the graph you see a histogram of the "posterior distribution" of the true correlation. This histogram represents our belief about the range of possible values for the true correlation, conditional on having seen the 36 data points. This histogram is centered near 0.48 and shows that there is only about a 0.2% chance, given the data, that the true correlation is negative and a 99.8% chance that it is positive. (We also see a scatterplot of the data and, on the margins, spikes corresponding to the data with a smattering of representative t-distributions that are fit to the variable.)

How does the P-value from part (b) compare to the Bayesian probability that the true correlation is negative?

d. How does the confidence interval from part (b) compare to the Bayesian 95% credible interval?

*This package is available via github. In R, type install.package("devtools") and then type devtools::install_github("rasmuxas/Bayesian_first_aid") to install the package. Then type library(BayesianFirstAid) to make the package active.

†The analysis conducted here uses a "non-informative prior" that says that before seeing the data the true correlation might have been any number between −1 and 1.

Dann Tardif/LWA/Blend Images/Alamy

CHAPTER 3

Multiple Regression

In this chapter you will learn to:

- Extend the ideas of the previous two chapters to consider regression models with two or more predictors.
- Use a multiple regression model to compare two regression lines.
- Create and assess models using functions of predictors, interactions, and polynomials.
- Recognize issues of multicollinearity with correlated predictors. Test a subset of predictors with a nested *F*-test.

When a scatterplot shows a linear relationship between a quantitative explanatory variable X and a quantitative response variable Y, we fit a regression line to the data in order to describe the relationship. We can also use the line to predict the value of Y for a given value of X. For example, Chapter 1 uses regression lines to describe relationships between:

- The price Y of a used Honda Accord and its mileage X
- The number of doctors Y in a county and the number of hospitals X
- The abundance of mammal species Y and the area of an island X

In all of these cases, other explanatory variables might improve our understanding of the response and help us better predict Y:

- The price Y of a used Accord may depend on its mileage X_1 and also its age X_2
- The number of doctors Y in a county may depend on the number of hospitals X_1, the number of beds in those hospitals X_2, and the number of Medicare recipients X_3
- The abundance of mammal species Y may depend on the area of an island X_1, the maximum elevation X_2, and the distance to the nearest island X_3

multiple linear regression

In Chapters 1 and 2, we studied simple linear regression with a single quantitative predictor. This chapter introduces the more general case of **multiple linear regression**, which allows several explanatory variables to combine in explaining a response variable.

EXAMPLE 3.1

NFL16

scatterplot matrix

NFL winning percentage Is offense or defense more important in winning football games? The data in Table 3.1 (stored in **NFLStandings2016**) contain the records for all NFL teams during the 2016 regular season,[1] along with the total number of points scored (*PointsFor*) and points allowed (*PointsAgainst*). We are interested in using the two scoring variables in a model to predict winning percentage (*WinPct*).

Figure 3.1 is a **scatterplot matrix** that shows separate scatterplots for each pair of variables. In the middle plot of the first row, we can see that *PointsFor* has a positive, linear relationship with *WinPct*, the relationship is fairly strong, and the variability in *WinPct* is consistent across different values of *PointsFor*. Looking at the last graph of the first row, the story with *PointsAgainst* is similar, except that the relationship with *WinPct* is negative. Finally, in the last graph of the second row, we see that *PointsFor* and *PointsAgainst* have no apparent linear relationship with each other.

Winning percentage Y could be related to points scored, X_1, and points allowed, X_2. The simple linear regressions for both of these relationships

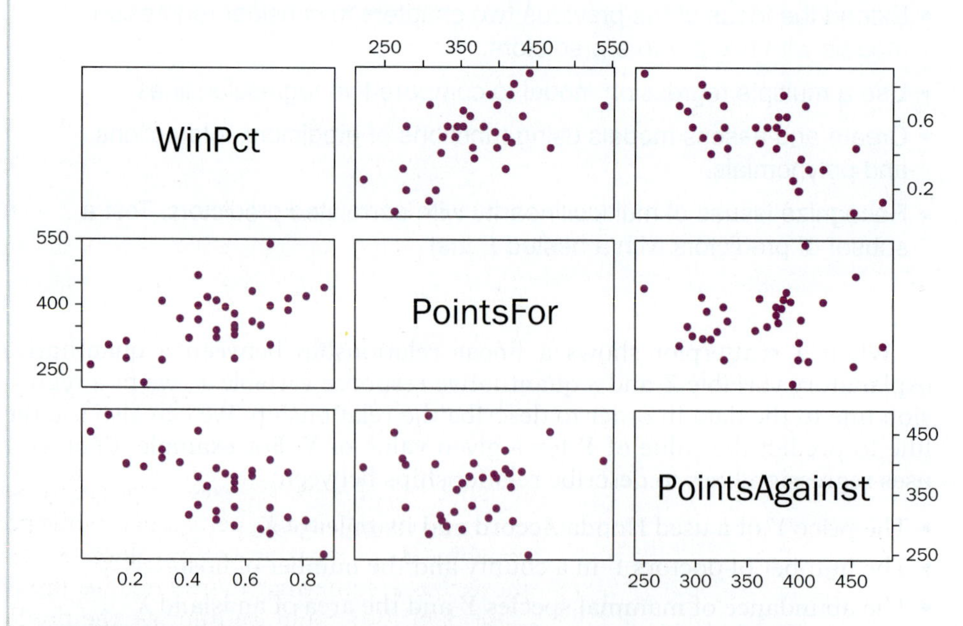

FIGURE 3.1 Scatterplot matrix for NFL winning percentages and scoring

Team	Wins	Losses	Ties	WinPct	PointsFor	PointsAgainst
New England Patriots	14	2	0	0.875	441	250
Dallas Cowboys	13	3	0	0.813	421	306
Kansas City Chiefs	12	4	0	0.750	389	311
Oakland Raiders	12	4	0	0.750	416	385
Pittsburgh Steelers	11	5	0	0.688	399	327
New York Giants	11	5	0	0.688	310	284
Atlanta Falcons	11	5	0	0.688	540	406
Seattle Seahawks	10	5	1	0.656	354	292
Miami Dolphins	10	6	0	0.625	363	380
Green Bay Packers	10	6	0	0.625	432	388
Houston Texans	9	7	0	0.563	279	328
Tennessee Titans	9	7	0	0.563	381	378
Denver Broncos	9	7	0	0.563	333	297
Detroit Lions	9	7	0	0.563	346	358
Tampa Bay Buccaneers	9	7	0	0.563	354	369
Washington Redskins	8	7	1	0.531	396	383
Baltimore Ravens	8	8	0	0.500	343	321
Indianapolis Colts	8	8	0	0.500	411	392
Minnesota Vikings	8	8	0	0.500	327	307
Arizona Cardinals	7	8	1	0.469	418	362
Buffalo Bills	7	9	0	0.438	399	378
Philadelphia Eagles	7	9	0	0.438	367	331
New Orleans Saints	7	9	0	0.438	469	454
Cincinnati Bengals	6	9	1	0.406	325	315
Carolina Panthers	6	10	0	0.375	369	402
New York Jets	5	11	0	0.313	275	409
San Diego Chargers	5	11	0	0.313	410	423
Los Angeles Rams	4	12	0	0.250	224	394
Jacksonville Jaguars	3	13	0	0.188	318	400
Chicago Bears	3	13	0	0.188	279	399
San Francisco 49ers	2	14	0	0.125	309	480
Cleveland Browns	1	15	0	0.063	264	452

TABLE 3.1 Records and points for NFL teams in 2016 season

are shown in Figure 3.2. Not surprisingly, scoring more points is positively associated with winning percentage, while points allowed has a negative relationship. By comparing the r^2 values (33.2% to 46.9%), we see that points allowed is a somewhat more effective predictor of winning percentage for these data. But could we improve the prediction of winning percentage by using both variables in the same model?

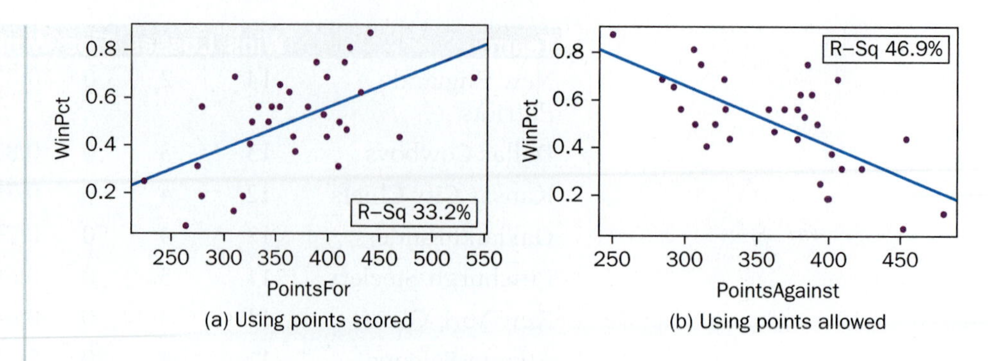

FIGURE 3.2 Linear regressions to predict NFL winning percentage

3.1 Multiple Linear Regression Model

Recall from Chapter 1 that the model for simple linear regression based on a predictor X is

$$Y = \beta_0 + \beta_1 X + \epsilon \quad \text{where } \epsilon \sim N(0, \sigma_\epsilon) \text{ and the errors are independent from one another.}$$

Choosing a Multiple Linear Regression Model

Moving to the more general case of multiple linear regression, we have k explanatory variables $X_1, X_2, \ldots, X_k$. The model now assumes that the mean response μ_Y for a particular set of values of the explanatory variables is a linear combination of those variables:

$$\mu_Y = \beta_0 + \beta_1 X_1 + \beta_2 X_2 + \cdots + \beta_k X_k$$

As with the simple linear regression case, an actual response value (Y) will deviate around this mean by some random error (ϵ). We continue to assume that these errors are **independent** of each other and have a **constant variance**. When we need to do formal inference for regression parameters, we also continue to assume that the distribution of ϵ follows a **normal distribution**. These conditions are summarized by assuming the errors in a multiple regression model are independent values from a $N(0, \sigma_\epsilon)$ distribution. The conditions imply that the responses Y at any combination of predictor values are independent, and the distribution of Y has mean given by μ_Y and constant variance σ_ϵ^2.

independent
constant variance
normal distribution

THE MULTIPLE LINEAR REGRESSION MODEL

We have n observations on k explanatory variables $X_1, X_2, \ldots, X_k$ and a response variable Y. Our goal is to study or predict the behavior of Y for the given set of the explanatory variables. The multiple linear regression model is

$$Y = \beta_0 + \beta_1 X_1 + \beta_2 X_2 + \cdots + \beta_k X_k + \epsilon$$

where $\epsilon \sim N(0, \sigma_\epsilon)$ and the errors are independent from one another.

This model has $k+2$ unknown parameters that we must estimate from data: the $k+1$ coefficients $\beta_0, \beta_1, \beta_2, \ldots, \beta_k$ and the standard deviation of the error σ_ϵ. Some of the X_i's in the model may be interaction terms, which are products of two explanatory variables. Others may be squares, higher powers, or other functions of quantitative explanatory variables. We can also include information from categorical predictors by coding the categories with $(0, 1)$ variables.

Thus the model can describe quite general relationships. We discuss each of these types of predictors later in this chapter. The main restriction is that the model is linear because each term $\beta_i X_i$ is a constant multiple of a predictor.

Fitting a Multiple Linear Regression Model

Once we have chosen a tentative set of predictors as the form for a multiple linear regression model, we need to estimate values for the coefficients based on data and then assess the fit. The estimation uses the same procedure of computing the sum of squared residuals, where the residuals are obtained as the differences between the actual Y values and the values obtained from a prediction equation of the form

$$\hat{Y} = \hat{\beta}_0 + \hat{\beta}_1 X_1 + \hat{\beta}_2 X_2 + \cdots + \hat{\beta}_k X_k$$

As in the case of simple linear regression, statistical software chooses estimates for the coefficients, $\beta_0, \beta_1, \beta_2, \ldots, \beta_k$, that minimize the sum of the squared residuals.

EXAMPLE 3.2

NFL16

NFL winning percentage: Fitting the model Figure 3.3 gives some computer output for fitting a multiple linear regression model to predict the winning percentages for NFL teams based on both the points scored and points allowed.

The fitted prediction equation in this example is

$$\widehat{WinPct} = 0.785 + 0.001699\ PointsFor - 0.002482\ PointsAgainst$$

If we consider the Green Bay Packers, who scored 432 points while allowing 388 points during the 2016 regular season, the predicted winning percentage is

$$\widehat{WinPct} = 0.785 + 0.001699(432) - 0.002482(388) = 0.557$$

Since the Packers' 10-6 record produced an actual winning percentage of 0.625, the residual in this case is $0.625 - 0.557 = 0.068$.

```
The regression equation is
WinPct = 0.785 + 0.001699 PointsFor - 0.002482 PointsAgainst

Predictor           Coef     SE Coef      T      P
Constant           0.785       0.154   5.11  0.000
PointsFor       0.001699    0.000263   6.47  0.000
PointsAgainst  -0.002482    0.000320  -7.74  0.000

S = 0.0965319    R-Sq =78.24%    R-Sq(adj) = 76.74%

Analysis of Variance

Source       DF      SS        MS       F      P
Regression    2  0.9715  0.485728   52.13  0.000
Error        29  0.2702  0.009318
Total        31  1.2417
```

FIGURE 3.3 Computer output for a multiple regression

Interpreting the coefficients is a bit trickier in the multiple regression setting because the predictors might be related to each other (for more details see Section 3.5). For example, in the model for NFL *WinPct* we see scoring 10 more points increases the predicted winning percentage by $0.001699(10) = 0.01669$

or about 1.7%. Of course, this would assume that the points allowed did not change, which might not be reasonable for an actual team. More formally, we say that we expect the winning percentage to increase by about 1.7% for each extra 10 points scored *when allowing for simultaneous changes in points against*.

In addition to the estimates of the regression coefficients, the other parameter of the multiple regression model that we need to estimate is the standard deviation of the error term, σ_ϵ. Recall that for the simple linear model we estimated the variance of the error by dividing the sum of squared residuals (SSE) by $n-2$ degrees of freedom. For each additional predictor we add to a multiple regression model, we have a new coefficient to estimate and thus lose 1 more degree of freedom. In general, if our model has k predictors (plus the constant term), we lose $k+1$ degrees of freedom when estimating the error variability, leaving $n-k-1$ degrees of freedom. This gives the estimate for the **standard error of the multiple regression model** with k predictors as

standard error of the multiple regression model

$$\hat{\sigma}_\epsilon = \sqrt{\frac{SSE}{n-k-1}}$$

From the multiple regression output for the NFL winning percentages in Figure 3.3, we see that the sum of squared residuals for the $n = 32$ NFL teams is $SSE = 0.2702$ and so the standard error of this two-predictor multiple regression model is

$$\hat{\sigma}_\epsilon = \sqrt{\frac{0.2702}{32-2-1}} = \sqrt{0.00932} = 0.0965$$

Note that the error degrees of freedom (29) and the estimated variance of the error term ($MSE = 0.009318$) are also given in the Analysis of Variance table of the computer output and the standard error is labeled as $S = 0.0965319$.

3.2 Assessing a Multiple Regression Model

Issues for assessing a multiple regression model are similar to what you have already seen for single predictor models (see Section 1.3 and Sections 2.1 and 2.2). Are the conditions for the model reasonably met? Is the predictor related to the response variable? Does the model do a good job of explaining variability in the response? The ways we address these questions when dealing with multiple regression will be familiar, with some additional distinctions to account for having more than one predictor.

For a single predictor, we use the correlation between the predictor and the response, and a scatterplot with regression line, as tools to judge the strength of a predictor. The scatterplot may also suggest possible transformations to help with linearity. These ideas still apply in the multiple regression setting, particularly for picking predictors. However, we tend to rely more on residuals versus fitted values plots or normal quantile plots of residuals for checking conditions for the multiple regression model as a whole. Fortunately, the methods for constructing these plots and their interpretation are the same when dealing with multiple predictors.

EXAMPLE 3.3

NFL16

NFL winning percentage: Checking conditions Figure 3.2 (page 90) shows scatterplots for NFL *WinPct* versus both predictors, *PointsFor* and *PointsAgainst*. Those show clear trends for both predictors and no strong evidence of curvature. Figure 3.4 shows (a) a plot of the residuals versus the fitted values for the multiple regression model in Example 3.2 and (b) a normal quantile plot of the residuals.

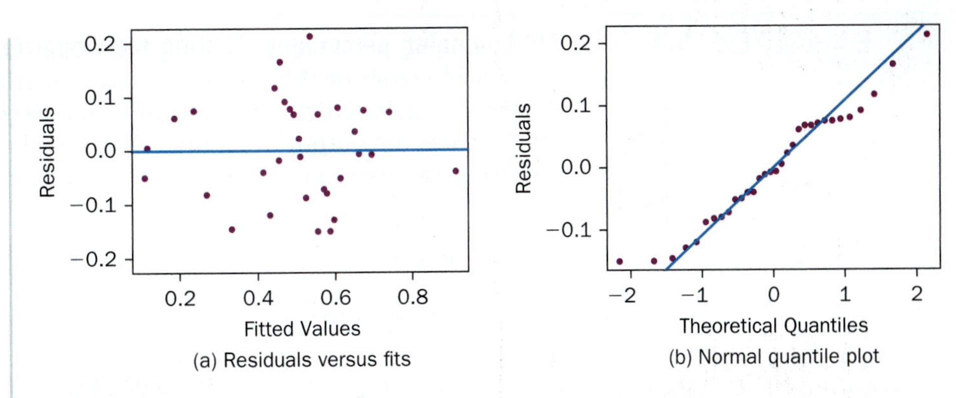

(a) Residuals versus fits

(b) Normal quantile plot

FIGURE 3.4 Residual plots for NFL winning percentage model

The residuals versus fitted values plot shows a fairly consistent band of residuals on either side of the zero line. We see no evidence of strong curvature or consistently changing variability of residuals. The normal quantile plot of residuals also looks fairly straight, with only a slight departure for the smallest residual. Thus we see no strong evidence to cause concern with the condition of normally distributed errors.

We have seen that for a simple linear model the t-test for slope and ANOVA F-test for regression are equivalent. If the slope is significantly different from zero (t-test), the model will have some effectiveness (ANOVA). These tests get more interesting with multiple predictors. t-tests will assess the importance of each predictor individually in the model, while the ANOVA F-test will check how the predictors do as a group in explaining variability in the response. We present these next in the order they were introduced in Chapter 2, individual t-test and then ANOVA. In practice, we often look at the ANOVA first ("Is the model effective?") before checking the t-tests ("Which predictors are helpful?").

t-Tests for Coefficients

With multiple predictors in the model, we need to ask the question of whether or not an individual predictor is helpful to include in the model. We can test this by seeing if the coefficient for the predictor is significantly different from zero.

INDIVIDUAL *t*-TESTS FOR COEFFICIENTS IN MULTIPLE REGRESSION

To test the coefficient for one of the predictors, X_i, in a multiple regression model, the hypotheses are

$$H_0 : \beta_i = 0$$
$$H_a : \beta_i \neq 0$$

and the test statistic is

$$t = \frac{\text{parameter estimate}}{\text{standard error of estimate}} = \frac{\hat{\beta}_i}{SE_{\hat{\beta}_i}}$$

If the conditions for the multiple linear model hold (including normality of errors), we compute the P-value for the test statistic using a t-distribution with $n - k - 1$ degrees of freedom.

EXAMPLE 3.4

NFL16

NFL winning percentage: Testing individual terms The parameter estimates, standard errors of the estimates, test statistics, and P-values for the t-tests for individual predictors appear in standard regression output. For example, we can use the output in Figure 3.3 to test the *PointsAgainst* predictor in the multiple regression to predict NFL winning percentages:

$$H_0 : \beta_2 = 0$$
$$H_a : \beta_2 \neq 0$$

From the line in the computer output for *PointsAgainst*, we see that the test statistic is

$$t = \frac{-0.002482}{0.000320} = -7.74$$

and the P-value (based on 29 degrees of freedom) is essentially zero. This provides very strong evidence that the coefficient of *PointsAgainst* is different from zero and thus the *PointsAgainst* variable has some predictive power in this model for helping explain the variability in the winning percentages of the NFL teams. In other words, the *PointsAgainst* variable has some additional predictive value over and above that contributed by the other independent variable. Note that the other predictor, *PointsFor*, also has a tiny individual P-value and can be considered an important contributor to this model.

Confidence Intervals for Coefficients

In addition to the estimate and the hypothesis test for regression coefficients, we may be interested in producing a confidence interval for one or more of the coefficients. As usual, we can find a margin of error as a multiple of the standard error of the estimator. Assuming the normality condition, we use a value from the t-distribution based on our desired level of confidence.

CONFIDENCE INTERVAL FOR A MULTIPLE REGRESSION COEFFICIENT

A confidence interval for the actual value of any multiple regression coefficient, β_i, has the form

$$\hat{\beta}_i \pm t^* \, SE_{\hat{\beta}_i}$$

where the value of t^* is the critical value from the t-distribution with degrees of freedom equal to the error df in the model ($n - k - 1$, where k is the number of predictors). The value of the standard error of the coefficient, $SE_{\hat{\beta}_1}$, is obtained from computer output.

EXAMPLE 3.5

NFL16

NFL winning percentage: CI for coefficients For the data on NFL winning percentages, the standard error of the coefficient of the *PointsFor* is 0.000263 and the error term has 29 degrees of freedom, so a 95% confidence interval for the size of the average improvement in winning percentage for every extra point scored during a season (assuming the same number of points allowed) is

$$0.00170 \pm 2.045(0.000263) = 0.00170 \pm 0.000538 = (0.00116 \text{ to } 0.00224)$$

The coefficient and confidence limits in this situation are very small since the variability in *WinPct* is small relative to the variability in points scored over an entire season. To get a more practical interpretation, we might consider what happens if a team improves its scoring by 50 points over the entire season. Assuming no change in points allowed,* we multiply the limits by 50 to find the expected improvement in winning percentage to be somewhere between 0.058 (5.8 percentage points) and 0.112 (11.2 percentage points).

ANOVA for Multiple Regression

In addition to assessing the individual predictors one-by-one, we are generally interested in testing the effectiveness of the model as a whole. To do so, we return to the idea of partitioning the variability in the data into a portion explained by the model and unexplained variability in the error term:

$$SSTotal = SSModel + SSE$$

where

$$SSModel = \sum (\hat{y} - \bar{y})^2$$
$$SSE = \sum (y - \hat{y})^2$$
$$SSTotal = \sum (y - \bar{y})^2$$

Note that although it takes a bit more effort to compute the predicted $\hat{y}$ values with multiple predictors, the formulas are exactly the same as we saw in the previous chapter for simple linear regression. To test the overall effectiveness of the model, we again construct an ANOVA table. The primary adjustment from the simple linear case is that we are now testing k predictors simultaneously, so we have k degrees of freedom for computing the mean square for the model in the numerator and $n - k - 1$ degrees of freedom left for the mean square error in the denominator.

ANOVA FOR A MULTIPLE REGRESSION MODEL

To test the effectiveness of the multiple linear regression model, the hypotheses are

$H_0 : \beta_1 = \beta_2 = \cdots = \beta_k = 0$
$H_a :$ at least one $\beta_i \neq 0$

and the ANOVA table is

Source	Degrees of Freedom	Sum of Squares	Mean Square	F-statistic
Model	k	$SSModel$	$MSModel = \frac{SSModel}{k}$	$F = \frac{MSModel}{MSE}$
Error	$n - k - 1$	SSE	$MSE = \frac{SSE}{n - k - 1}$	
Total	$n - 1$	$SSTotal$		

If the conditions for the multiple linear regression model, including normality, hold, we compute the P-value using the upper tail of an F-distribution with k and $n - k - 1$ degrees of freedom.

The null hypothesis that the coefficients for all the predictors are zero is consistent with an ineffective model in which none of the predictors has any

*More precisely, we should say "allowing for simultaneous linear change in points allowed."

linear relationship with the response variable.* If the model does explain a statistically significant portion of the variability in the response variable, the MSModel will be large compared to the MSE and the *P*-value based on the ANOVA table will be small. In that case, we conclude that one or more of the predictors is effective in the model, but the ANOVA analysis does not identify which predictors are significant. That is the role for the individual *t*-tests.

EXAMPLE 3.6

NFL16

NFL winning percentage: ANOVA for overall fit The ANOVA table from the computer output in Figure 3.3 for the multiple regression model to predict winning percentages in the NFL is reproduced here. The normal quantile plot of the residuals in Figure 3.4 indicates that the normality condition is fairly reasonable so we proceed with interpreting the *F*-test.

Analysis of Variance

Source	DF	SS	MS	F	P
Regression	2	0.9715	0.485728	52.13	0.000
Error	29	0.2702	0.009318		
Total	31	1.2417			

Using an *F*-distribution with 2 numerator and 29 denominator degrees of freedom produces a *P*-value of approximately 0.000 for the *F*-statistic of 52.13. This gives strong evidence to reject a null hypothesis that $H_0: \beta_1 = \beta_2 = 0$ and conclude that at least one of the predictors, *PointsFor* or *PointsAgainst*, is effective for explaining variability in NFL winning percentages. From the plots of Figure 3.2 and individual *t*-tests, we conclude that, in fact, both of these predictors are effective in this model.

Coefficient of Multiple Determination

coefficient of determination

In the previous chapter, we also encountered the use of the **coefficient of determination**, r^2, as a measure of the percentage of total variability in the response that is explained by the regression model. This concept applies equally well in the setting of multiple regression so that

$$R^2 = \frac{\text{Variability explained by the model}}{\text{Total variability in } y} = \frac{SSModel}{SSTotal} = 1 - \frac{SSE}{SSTotal}$$

Using the information in the ANOVA table for the model to predict NFL winning percentages, we see that

$$R^2 = \frac{0.9715}{1.2417} = 0.782$$

Thus we can conclude that 78.2% of the variability in winning percentage of NFL teams for the 2016 regular season can be explained by the regression model based on the points scored and points allowed.

Recall that in the case of simple linear regression with a single predictor, we used the notation r^2 for the coefficient of determination because that value happened to be the square of the correlation between the predictor X and response Y. That interpretation does not translate directly to the multiple regression setting, since we now have multiple predictors, $X_1, X_2, \ldots, X_k$, each of which has its own correlation with the response Y. So there is no longer a single correlation between predictor and response. However, we can consider the correlation between the predictions $\hat{y}$ and the actual y values for all of the data cases. The

*Note that the constant term, β_0, is not included in the hypotheses for the ANOVA test of the overall regression model. Only the coefficients of the predictors in the model are being tested.

square of this correlation is the coefficient of determination for the multiple regression model. For example, if we use statistical software to save the predicted values (fits) from the multiple regression model to predict NFL winning percentages and then compute the correlation between those fitted values and the actual winning percentages for all 32 teams, we find

```
Pearson correlation of WinPct and FITS = 0.8845
```

If we then compute $r^2 = (0.8845)^2 = 0.782$, we match the coefficient of determination for the multiple regression. In the case of a simple linear regression, the predicted values are a linear function of the single predictor, $\hat{Y} = \hat{\beta}_0 + \hat{\beta}_1 X$, so the correlation between X and Y must be the same as between $\hat{Y}$ and Y, up to possibly a change in $\pm$ sign. Thus in the simple case, the coefficient of determination could be found by squaring r computed either way. Since this doesn't work with multiple predictors, we generally use a capital R^2 to denote the coefficient of determination in a multiple regression setting.

As individual predictors in separate simple linear regression models, both *PointsFor* ($r^2 = 33.2\%$) and *PointsAgainst* ($r^2 = 46.9\%$) were less effective at explaining the variability in winning percentages than they are as a combination in the multiple regression model. This will always be the case. Adding a new predictor to a multiple regression model can never decrease the percentage of variability explained by that model (assuming the model is fit to the same data cases). At the very least, we could put a coefficient of zero in front of the new predictor and obtain the same level of effectiveness. In general, adding a new predictor will decrease the sum of squared errors and thus increase the variability explained by the model. But does that increase reflect important new information provided by the new predictor or just extra variability explained due to random chance? That is one of the roles for the t-tests of the individual predictors.

adjusted coefficient of determination

Another way to account for the fact that R^2 tends to increase as new predictors are added to a model is to use an **adjusted coefficient of determination** that reflects the number of predictors in the model as well as the amount of variability explained. One common way to do this adjustment is to divide the total sum of squares and sum of squared errors by their respective degrees of freedom and subtract the result from one. This subtracts the ratio of the estimated error variance, $MSE = \hat{\sigma}_\epsilon^2 = SSE/(n - k - 1)$, to the ordinary sample variance of the responses $S_Y^2 = \sum(y - \bar{y})^2/(n - 1)$, rather than just $SSE/SSTotal$.

ADJUSTED COEFFICIENT OF DETERMINATION

The **adjusted** R^2, which helps account for the number of predictors in the model, is computed as

$$R^2_{adj} = 1 - \frac{SSE/(n-k-1)}{SSTotal/(n-1)} = 1 - \frac{\hat{\sigma}_\epsilon^2}{S_Y^2}$$

Note that the denominator stays the same for all models fit to the same response variable and data cases. However, the numerator in the term subtracted can actually increase when a new predictor is added to a model if the decrease in the SSE is not sufficient to offset the decrease in the error degrees of freedom. Thus, the R^2_{adj} value might go down when a weak predictor is added to a model. Adjusted R^2 is especially useful, as we will see in later examples in this chapter, when comparing models with different numbers of predictors.

EXAMPLE 3.7

NFL16

NFL winning percentage: Adjusted R^2 For our two-predictor model of NFL winning percentages, we can use the information in the ANOVA table to compute

$$R^2_{adj} = 1 - \frac{0.2702/29}{1.2417/31} = 1 - \frac{0.00932}{0.04005} = 1 - 0.233 = 0.767$$

and confirm the value listed as $R\text{−}Sq(Adj) = 76.74\%$ for the computer output in Figure 3.3. While this number reveals relatively little new information on its own, it is particularly useful when comparing competing models based on different numbers of predictors.

Confidence and Prediction Intervals

Just as in Section 2.4, we can obtain interval estimates for the mean response or a future individual case given any combination of predictor values. We do not provide the details of these formulas since computing the standard errors is more complicated with multiple predictors. However, the overall method and interpretation are the same, and we rely on statistical software to manage the calculations.

EXAMPLE 3.8

NFL16

NFL winning percentage: Prediction interval Suppose an NFL team in the upcoming season scores 400 points and allows 350 points. Using the multiple regression model for *WinPct* based on *PointsFor* and *PointsAgainst*, some computer output for computing a prediction interval in this case is shown below:

```
Variable        Setting
PointsFor           400
PointsAgainst       350

   Fit      SE Fit         95% CI              95% PI
0.5965     0.01995   (0.5557, 0.6373)   (0.3949, 0.7981)
```

Based on the output, we would expect this team to have a winning percentage near 59.7%. Since we are working with an individual team, we interpret the "95% PI" provided in the output to say with reasonable confidence that the winning percentage for this team will be between 39.5% and 79.8% (or between 7 and 12 wins for a 16-game season).

We consider the multiple regression model to be one of the most powerful and general statistical tools for modeling the relationship between a set of predictors and a quantitative response variable. Having discussed in this section how to extend the basic techniques from simple linear models to fit and assess a multiple regression, we move in the next few sections to consider several examples that demonstrate the versatility of this procedure. Using multiple predictors in the same model also raises some interesting challenges for choosing a set of predictors and interpreting the model, especially when the predictors might be related to each other as well as to the response variable. We consider some of these issues later in this chapter.

3.3 Comparing Two Regression Lines

In Chapter 1, we considered a simple linear regression model to summarize a linear relationship between two quantitative variables. Suppose now that we want to investigate whether such a relationship changes between groups determined by some categorical variable. For example, is the relationship between price and mileage different for Honda Accords offered for sale at physical car lots compared to those for sale on the web? Does the relationship between the number of pages and price of textbooks depend on the field of study or perhaps on whether the book has a hard or soft cover? We can easily fit separate linear regression models by considering each categorical group as a different dataset. However, in some circumstances, we would like to formally test whether some aspect of the linear relationship (such as the slope, the intercept, or possibly both) is significantly different between the two groups or use a common parameter for both groups if it would work essentially as well as two different parameters. To help make these judgments, we examine multiple regression models that allow us to fit and compare linear relationships for different groups determined by a categorical variable.

EXAMPLE 3.9

ButterBc

Butterfly size: Females and males Section 2.5 (page 72) introduced a case study looking at a possible relationship between average wing length of *Boloria chariclea* butterflies in Greenland and previous summer temperatures. In the case study we only considered the female cases from the **ButterfliesBc** dataset. In this example we also account for the wing lengths of male butterflies of the same species. Some values for both sexes are shown in Table 3.2.

Temp	Wing	Sex
0.9	18.1	Male
1.1	18.2	Male
1.4	18.4	Male
1.6	18.1	Male
⋮	⋮	⋮
0.9	19.1	Female
1.1	18.8	Female
1.4	19.5	Female
⋮	⋮	⋮

TABLE 3.2 Previous summer temperature and average wing length for male and female butterflies

A quick look at the data shows that female butterflies are larger than male butterflies. Also, over time temperatures have tended to go up and butterfly size has tended to go down. But is the relationship between temperature and size the same for males as for females? The scatterplot in Figure 3.5 of wing length versus temperature helps show these relationships by using different symbols for males and females. For each sex there is a reasonably linear trend. Since the females appear to be quite a bit larger than males, we should not use the same linear model to describe the decreasing trend with temperature for each sex. If we consider each sex as its own dataset, we can fit separate linear regression models for each sex:

$$\text{For males:}\quad \widehat{Wing} = 18.396 - 0.231 \cdot Temp$$

$$\text{For females:}\quad \widehat{Wing} = 19.344 - 0.239 \cdot Temp$$

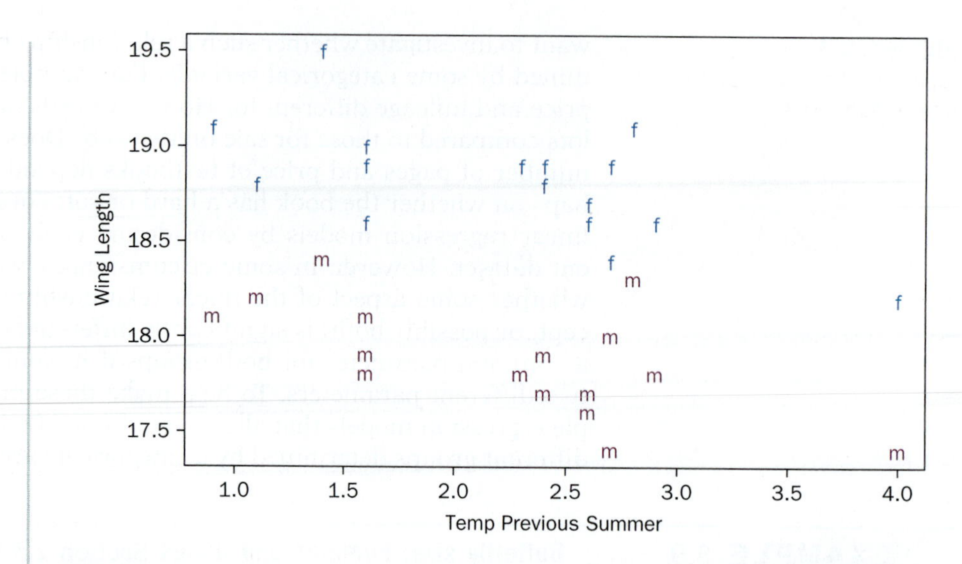

FIGURE 3.5 Scatterplot of wing length versus temperature for male and female butterflies

Note that the separate regression lines for the two sexes have similar slopes (roughly a 0.23 mm drop in wing length for every 1 degree increase in temperature), but have different intercepts (18.40 for males versus 19.34 for females). Since the two lines are roughly parallel, the difference in the intercepts gives an indication of how much larger females are than males, after taking into account the effect of temperature.

Indicator Variables

CHOOSE

indicator variable

Using multiple regression, we can examine the *Wing* versus *Temp* relationship for both sexes in a single model. The key idea is to use an **indicator variable** that distinguishes between the two groups, in this case males and females. We generally use the values 0 and 1 for an indicator variable so that 1 indicates that a data case does belong to a particular group ("yes") and 0 signifies that the case is not in that group ("no"). We can assign 1 to either of the groups, so an indicator for males would be defined as

$$IMale = \begin{cases} 0 & \text{if Sex = female} \\ 1 & \text{if Sex = male} \end{cases}$$

INDICATOR VARIABLE

An **indicator variable** uses two values, usually 0 and 1, to indicate whether a data case does (1) or does not (0) belong to a specific category.

Using the sex indicator, *IMale*, we consider the following multiple regression model:

$$Wing = \beta_0 + \beta_1 Temp + \beta_2 IMale + \epsilon$$

For data cases from females (where $IMale = 0$), this model becomes

$$\begin{aligned} Wing &= \beta_0 + \beta_1 Temp + \beta_2(0) + \epsilon \\ &= \beta_0 + \beta_1 Temp + \epsilon \end{aligned}$$

which looks like the ordinary simple linear regression, although the slope will be determined using data from both males and females.

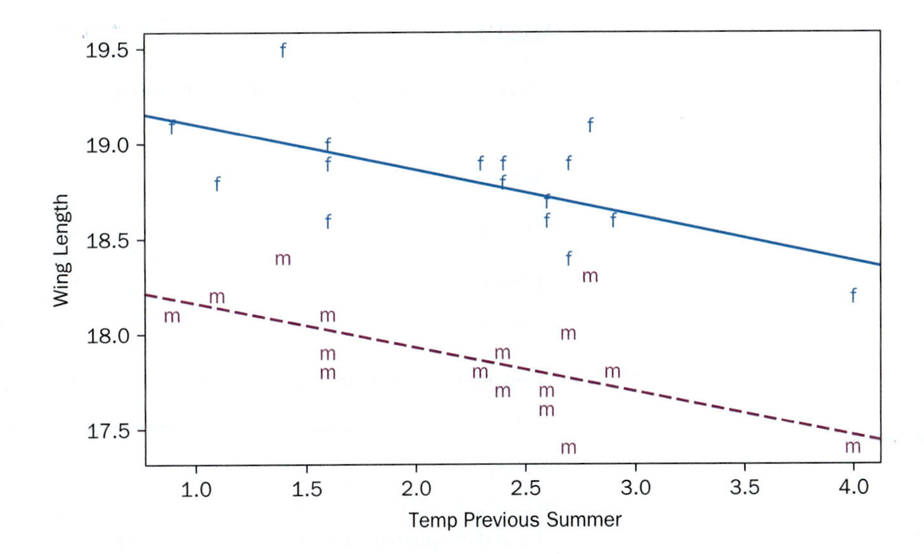

FIGURE 3.6 Separate regression lines for male and female butterflies

For data cases from males (where *IMale* = 1), this model becomes

$$\begin{aligned} Wing &= \beta_0 + \beta_1 Temp + \beta_2(1) + \epsilon \\ &= (\beta_0 + \beta_2) + \beta_1 Temp + \epsilon \end{aligned}$$

which also looks similar to a simple linear regression model, although the intercept is adjusted by the amount β_2. Thus the coefficient of the *IMale* indicator variable measures the difference in the intercepts between regression lines for males and females that have the same slope. This provides a convenient summary in a single model for the situation illustrated in Figure 3.6.

FIT

If we fit this multiple regression model to the **ButterfliesBc** data, we obtain the following output:

```
Coefficients:
 (Intercept)  Temperature      IMale
 19.3355          -0.2350   -0.9313
```

which gives the following prediction equation.

$$\widehat{Wing} = 19.34 - 0.235\ Temp - 0.931 IMale$$

Comparing Intercepts

By substituting the two values of the indicator variable into the prediction equation, we can obtain a least squares line for each sex:

$$\text{For males: } \widehat{Wing} = 19.3355 - 0.235\ Temp - 0.9313 = 18.4042 - 0.235\ Temp$$

$$\text{For females: } \widehat{Wing} = 19.3355 - 0.235\ Temp$$

Note that the intercepts (18.4042 for males and 19.3355 for females) are not quite the same as when we fit the two simple linear models separately (18.396 for males and 19.344 for females). That happens since this multiple regression model forces the two prediction lines to be parallel (common slope = −0.235) rather than allowing separate slopes (−0.231 for males and −0.239 for females). In the next example, we consider a slightly more complicated multiple regression model that allows either the slope or the intercept (or both) to vary between the two groups.

ASSESS

A summary of the multiple regression model fit includes the following output:

```
Coefficients:
              Estimate  Std. Error  t value  Pr(>|t|)
(Intercept)   19.33547     0.13372   144.60  < 2e-16 ***
Temp          -0.23504     0.05391    -4.36  0.00015 ***
IMale         -0.93125     0.08356   -11.14  5.35e-12 ***

Residual standard error: 0.2364 on 29 degrees of freedom
Multiple R-squared: 0.8316, Adjusted R-squared: 0.82
F-statistic: 71.6 on 2 and 29 DF, p-value: 6.059e-12
```

The *P*-values for the coefficients of both *Temp* and *IMale* as well as the overall ANOVA are all very small, indicating that the overall model is effective for explaining wing length and that each of the predictors is important in the model. The R^2 value shows that approximately 83% of the variability in wing length is explained by these two predictors.

If we fit a model with just the *Temp* predictor alone, only 11% of the variability is explained (and $R^2_{adj} = 8.1\%$), while *IMale* alone would explain approximately 72% ($R^2_{adj} = 71.2\%$). The model with both predictors does better than both of these with $R^2 = 83\%$ ($R^2_{adj} = 82.0\%$). When we fit the simple linear regression models to the data for each sex separately, we obtain values of $R^2 = 40\%$ ($R^2_{adj} = 35.7\%$) for males and $R^2 = 39\%$ ($R^2_{adj} = 34.9\%$) for females. Does this mean that our multiple regression model does a much better job at explaining wing length? Unfortunately, no, since the combined model introduces extra variability in the wing lengths by including both sexes, which the *IMale* predictor then helps successfully explain. The standard error of regression for the combined model ($\hat{\sigma}_\epsilon = 0.236$), which reflects how well wing lengths are predicted, is close to what is found in the separate regressions for each sex ($\hat{\sigma}_\epsilon = 0.235$ for males, $\hat{\sigma}_\epsilon = 0.246$ for females). So the multiple regression model predicts as well as (or slightly better than) the two separate regressions and uses one fewer parameter. We would prefer the multiple regression model in this situation.

The normal quantile plot in Figure 3.7(a) is relatively straight (with a bit of departure at the high end) and the histogram in Figure 3.7(b) has a single peak with a bit of a right skew. These indicate some mild, but not too strong, concern with the normality condition of the residuals in the multiple regression model to predict wing length using temperature and sex. The plot of residuals versus predicted values, Figure 3.7(c), shows no clear patterns or trends. Thus we don't see any serious concerns with the regression conditions for this model.

USE

In this example, we are especially interested in the coefficient (β_2) of the indicator variable since that reflects the magnitude of the difference in wing length between female and male butterflies. From the prediction equation, we can estimate the average female wing length is 0.93 mm longer than the average male wing length, and the very small *P*-value (5.35×10^{-12}) for this coefficient in the regression output gives strong evidence that the observed difference is not due to chance.

In addition to the estimate and the hypothesis test for the difference in intercepts, we may be interested in producing a confidence interval for the size of the female − male difference. The standard error of the coefficient of the indicator *IMale* is 0.0836 and the error term has 29 (32 − 2 − 1) degrees

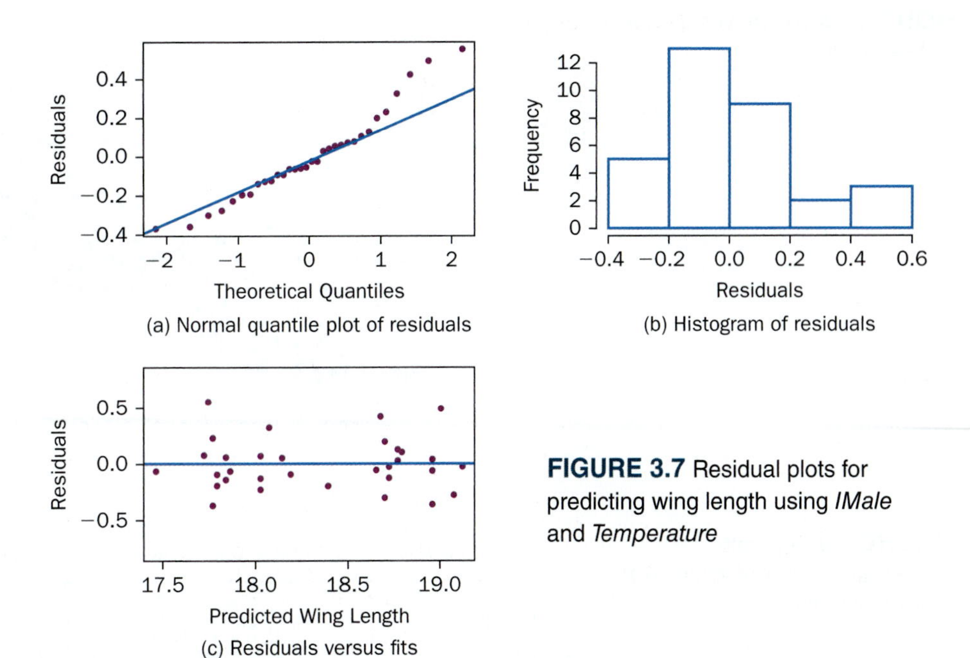

FIGURE 3.7 Residual plots for predicting wing length using *IMale* and *Temperature*

of freedom, so a 95% confidence interval for the size of the female − male difference (after adjusting for the temperature) is

$$-0.93125 \pm 2.045(0.0836) = -0.93125 \pm 0.171 = (-1.10, -0.76)$$

Thus we may conclude with reasonable confidence that the average wing length of females is between 0.76 mm and 1.10 mm greater than the average wing length of males.

EXAMPLE 3.10

Kids198

Growth rates of kids We all know that children tend to get bigger as they get older, but we might be interested in how growth rates compare. Do boys and girls gain weight at the same rates? The data displayed in Figure 3.8 show the ages (in months) and weights (in pounds) for a sample of 198 kids who were part of a larger study of body measurements of children.* The data are in the file **Kids198**. The plot shows a linear trend of increasing weights as the ages increase, and the variability in weights is fairly constant across ages. The dots for boys and girls seem to indicate that weights are similar at younger ages but that boys tend to weigh more at the older ages. This would suggest a larger growth rate for boys than for girls during these ages.

Comparing Slopes

Figure 3.9 shows separate plots for boys and girls with the regression line drawn for each case. The line is slightly steeper for the boys (slope = 0.91 pound per month) than it is for girls (slope = 0.63 pound per month). Figure 3.10 shows both regression lines on the same plot. Does the difference in slopes for these two samples indicate that the typical growth rate is really larger for boys than it is for girls, or could a difference this large reasonably occur due to random chance when selecting two samples of these sizes?

*These are cross-sectional data—a snapshot in time. If each child were followed for many years and measured repeatedly, then the data would be dependent and we would need other methods to conduct a proper analysis. See Chapter 12.

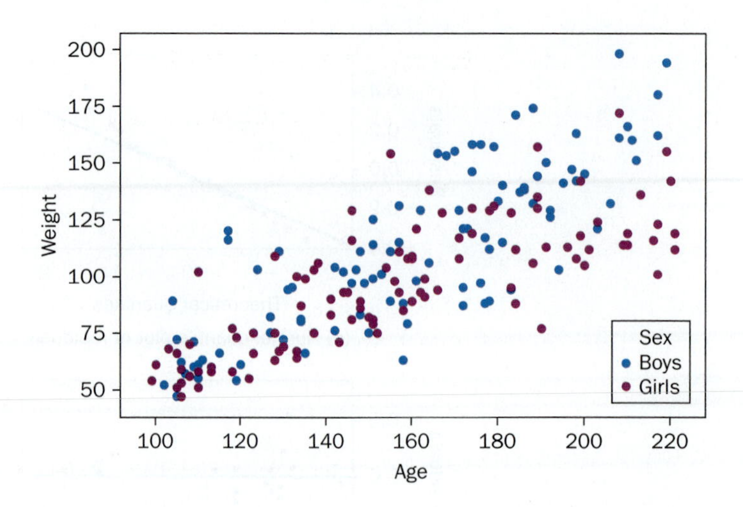

FIGURE 3.8 *Weight* versus *Age* by *Sex* for kids

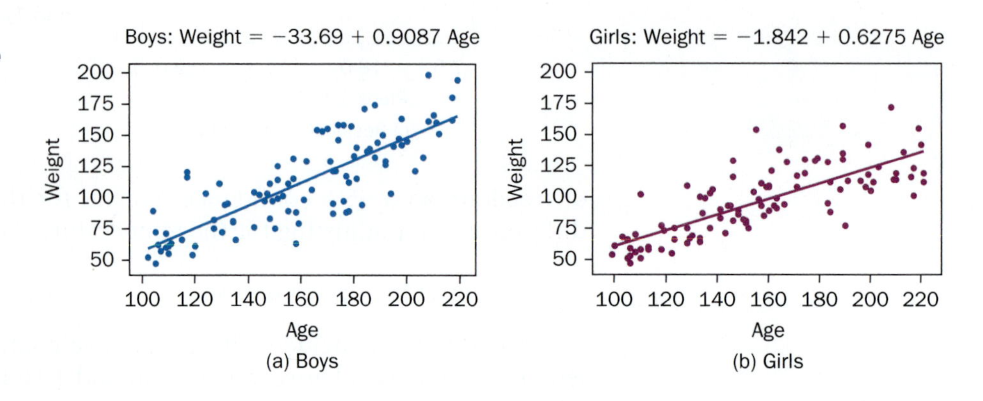

FIGURE 3.9 Separate regressions of *Weight* versus *Age* for boys and girls

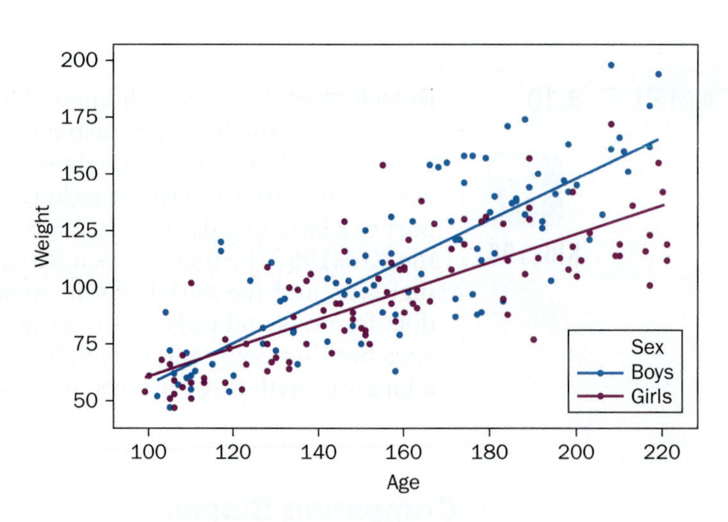

FIGURE 3.10 Compare regression lines by *Sex*

CHOOSE

To compare the two slopes with a multiple regression model, we add a new term* to the model considered in the previous example. Define the indicator variable *IGirl* to be 0 for the boys and 1 for the girls. You have seen that adding an indicator, such as *IGirl*, to a simple linear model of *Weight* versus *Age* allows

*As we build more complicated models with combinations of variables, we use *terms*, *predictors*, and *explanatory variables* interchangeably to refer to the components of a model.

for different intercepts for the two groups. The new predictor we add now is just the product of *IGirl* and the *Age* predictor.* This gives the model

$$Weight = \beta_0 + \beta_1 Age + \beta_2 IGirl + \beta_3 Age \cdot IGirl + \epsilon$$

For the boys in the study ($IGirl = 0$), the model becomes

$$Weight = \beta_0 + \beta_1\ Age + \epsilon$$

while the model for girls ($IGirl = 1$) is

$$Weight = \beta_0 + \beta_1\ Age + \beta_2(1) + \beta_3\ Age(1) + \epsilon$$
$$Weight = (\beta_0 + \beta_2) + (\beta_1 + \beta_3)\ Age + \epsilon$$

As in the previous model, the coefficients β_0 and β_1 give the slope and intercept for the regression line for the boys and the coefficient of the indicator variable, β_2, measures the difference in the intercepts between boys and girls. The new coefficient, β_3, shows how much the slopes change as we move from the regression line for boys to the line for girls.

FIT

Using statistical software to fit the multiple regression model produces the following output:

```
The regression equation is
Weight = - 33.7 + 0.909 Age + 31.9 IGirl - 0.281 Age*IGirl

Predictor       Coef   SE Coef      T      P
Constant      -33.69     10.01  -3.37  0.001
Age          0.90871   0.06106  14.88  0.000
IGirl          31.85     13.24   2.41  0.017
Age*IGirl   -0.28122   0.08164  -3.44  0.001
```

The first two terms of the prediction equation match the slope and intercept for the boys' regression line and we can obtain the least squares line for the girls by increasing the intercept by 31.9 and decreasing the slope by 0.281:

$$\text{For Boys: } \widehat{Weight} = -33.7 + 0.909\ Age$$
$$\text{For Girls: } \widehat{Weight} = -33.7 + 0.909\ Age + 31.9(1) - 0.281(1)\ Age$$
$$= -1.8 - 0.628\ Age$$

ASSESS

Figure 3.11 shows a normal probability plot of the residuals from the multiple regression model and a scatterplot of the residuals versus the predicted values. The linear pattern in the normal plot indicates that the residuals are reasonably normally distributed. The scatterplot shows no obvious patterns and a relatively consistent band of residuals on either side of zero. Neither plot raises any concerns about significant departures from conditions for the errors in a multiple regression model.

*This product is known as an interaction term and will be discussed further in Section 3.4.

It is reasonable to think that all perch have similar shapes, that for flat fish like perch, the weight might be proportional to its flat-side area. Moreover, as an admittedly simple approximation, we might regard the shape of a perch as roughly rectangular,[3] with area proportional to length $\times$ width. This logic suggests a model with the interaction *Length* $\cdot$ *Width* as a predictor.

Obs	Weight	Length	Width
104	5.9	8.8	1.4
105	32.0	14.7	2.0
106	40.0	16.0	2.4
107	51.5	17.2	2.6
108	70.0	18.5	2.9
109	100.0	19.2	3.3
110	78.0	19.4	3.1
⋮	⋮	⋮	⋮

TABLE 3.3 First few cases of fish measurements in the **Perch** datafile

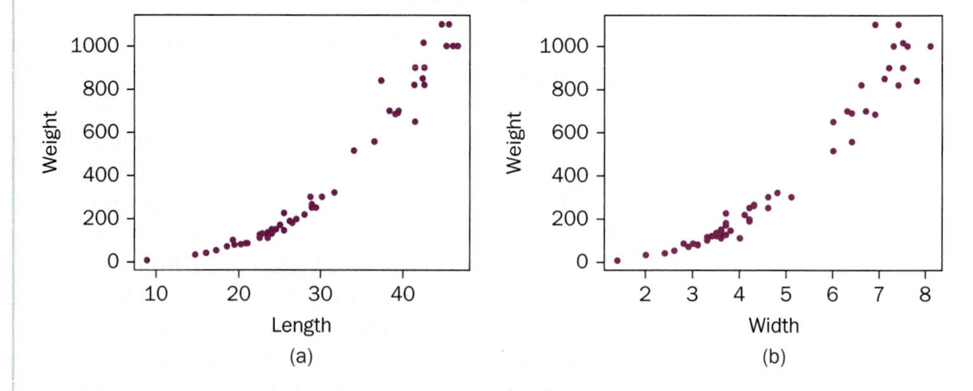

FIGURE 3.12 Individual predictors for perch weights

We create the new product variable (naming it *Length* $\cdot$ *Width*) and consider a simple model in which weight depends just on the interaction term:

$$Weight = \beta_0 + \beta_1 Length \cdot Width + \epsilon$$

Figure 3.13 shows that there is a strong relationship between *Weight* and this interaction product.

If we run a regression model with the interaction term, we get the following output:

```
              Estimate   Std. Error   t value   Pr(>|t|)
(Intercept)   -136.926       12.728     -10.8   4.8e-15 ***
Length*Width     3.319        0.068      48.8   < 2e-16 ***

Residual standard error: 52.3 on 54 degrees of freedom
Multiple R-squared: 0.978, Adjusted R-squared: 0.977
F-statistic: 2.38e+03 on 1 and 54 DF, P-value: <2e-16
```

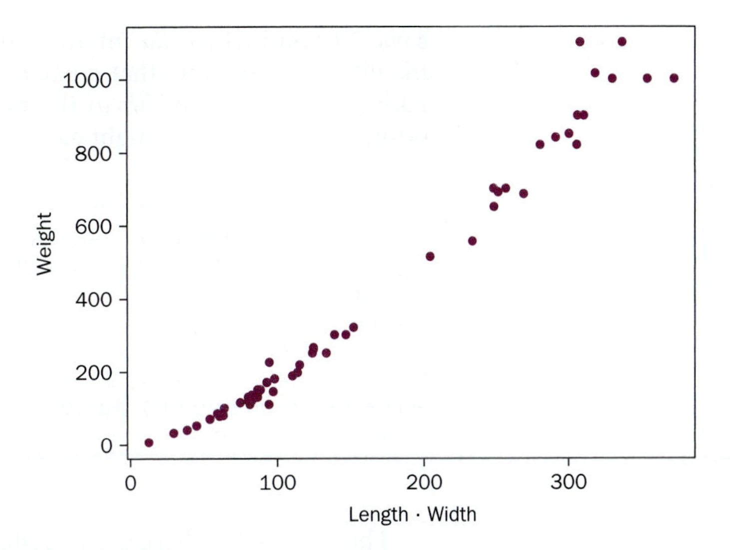

FIGURE 3.13 Perch weights plotted against the product of length and width

The fit here looks impressive; *Length · Width* is a strong predictor of the weight of a fish and the adjusted R^2 is 0.977. At the same time, the plot of *Weight* versus *Length · Width* shows the same two patterns as before: both curvature and a pattern of increasing spread as level increases. As before, these two patterns suggest that a change of scale may lead to a simpler, better-fitting model.

Moreover, the residual plot of Figure 3.14 strongly confirms these suggestions by magnifying both patterns. Indeed, if we look back at Figure 3.13 we can detect curvature in that scatterplot (although the pattern is much more subtle than in the residual plot of Figure 3.14).

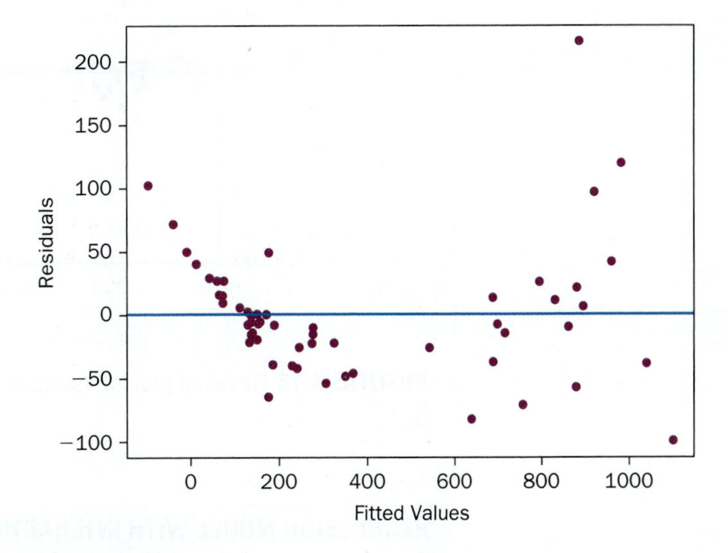

FIGURE 3.14 Residual plot for the one-term interaction model for perch weights

Having a highly significant predictor and a large R^2 does not mean that we have reached the end of the story. There are a couple of ways forward from this point. One thing we could do would be to follow the advice that says, "If you include an interaction term in a model, then you should also include the predictors that created the interaction term," which here means adding *Length* and *Width* to the model. Doing this produces the following output, from what we might call the "full interaction model":

	Estimate	Std. Error	t value	Pr(>|t|)
(Intercept)	113.935	58.784	1.94	0.058 .
Length	-3.483	3.152	-1.10	0.274
Width	-94.631	22.295	-4.24	0.000091 ***
Length*Width	5.241	0.413	12.69	< 2e-16 ***

Residual standard error: 44.2 on 52 degrees of freedom
Multiple R-squared: 0.985, Adjusted R-squared: 0.984
F-statistic: 1115 on 3 and 52 DF, *P*-value: <2e-16

The adjusted R^2 here is somewhat better than in the previous model and the *Width* term stands out as being statistically significant. Figure 3.15 is a residual plot, which is much improved over Figure 3.14. The problem with curvature has been mostly eliminated, but there is still an issue with variability increasing as the predicted perch weights increase. We return to this situation in Example 3.20 on page 131 where we consider an alternative model based on log transformations.

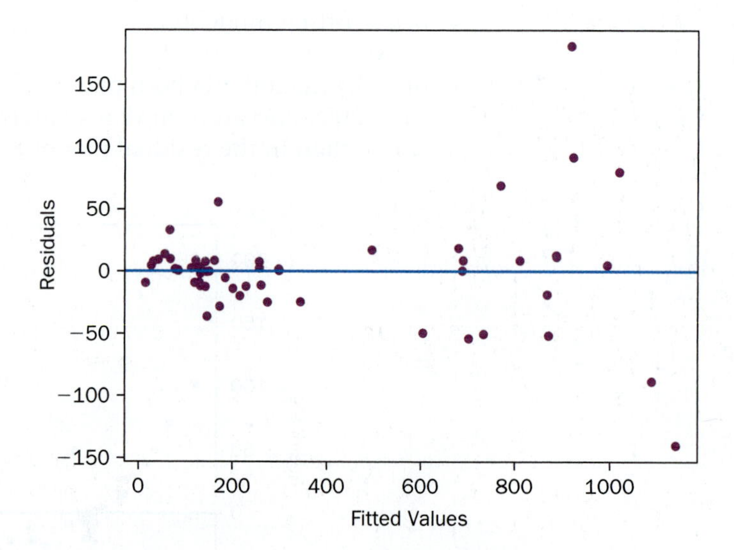

FIGURE 3.15 Residual plot for the full interaction model for perch weights

REGRESSION MODEL WITH INTERACTION

For two variables, X_1 and X_2, the full **interaction** model has the form

$$Y = \beta_0 + \beta_1 X_1 + \beta_2 X_2 + \beta_3 X_1 \cdot X_2 + \epsilon$$

The interaction product term, $X_1 \cdot X_2$, allows the slope with respect to one predictor to change for values of the second predictor.

EXAMPLE 3.12

IQ

Guessing IQ Can you tell how smart someone is just by looking at a photo of the person? This question was put to the test by researchers who showed photos of 40 college women to each of several raters. The file **IQGuessing** contains perceived intelligence (*GuessIQ*, computed from ratings of a panel of judges viewing only the photos), as well as actual intelligence (*TrueIQ*, as measured by an IQ test), and age for each woman.[4]

Person	Age	GuessIQ	TrueIQ
1	20	134	83
2	20	127	121
3	21	135	114
4	19	125	129
5	22	126	111
⋮	⋮	⋮	⋮

TABLE 3.4 First five cases of age and IQ data in **IQGuessing**

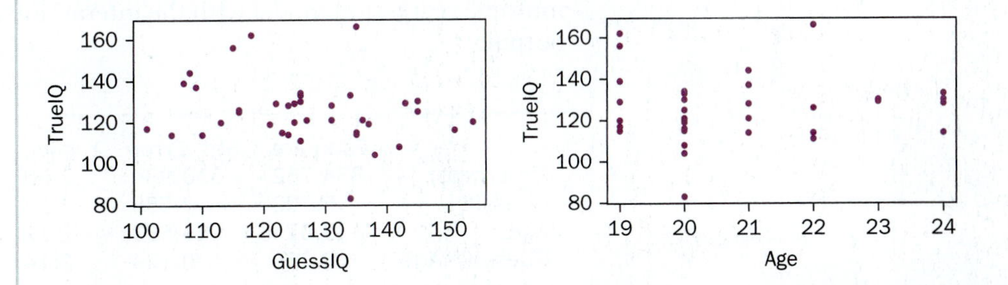

FIGURE 3.16 Scatterplots of individual predictors of *TrueIQ*

Table 3.4 shows the data for the first five women. Figure 3.16 shows that neither predictor, alone, has a strong relationship with *TrueIQ*. Each scatterplot shows a linear, but very weak, link between variables. But we want to explore how the predictors work in combination.

Figure 3.17 is a scatterplot of *TrueIQ* vs *GuessIQ*, but with the points color-coded according to whether age is low (younger) or high (older). The data points shown as dots are (younger) women in the bottom half of the age distribution and data points shown as triangles are (older) women in the top half. Regression lines are added for the two groups, using the same modeling approach as in Example 3.10. We can see that for the younger women there is a negative relationship between *TrueIQ* and *GuessIQ*, but for the older women the relationship is positive.

But *Age* is a quantitative variable and we don't have to limit ourselves to an interaction model with just two groups, younger and older. Instead, we can use age as a continuous variable and fit an interaction model that allows the relationship between *GuessIQ* and *TrueIQ* to change continuously as age changes. To do this, we create the interaction product between *GuessIQ* and *Age* and include it in the regression model.

The interaction model is

$$TrueIQ = \beta_0 + \beta_1 GuessIQ + \beta_2 Age + \beta_3 GuessIQ \cdot Age + \epsilon$$

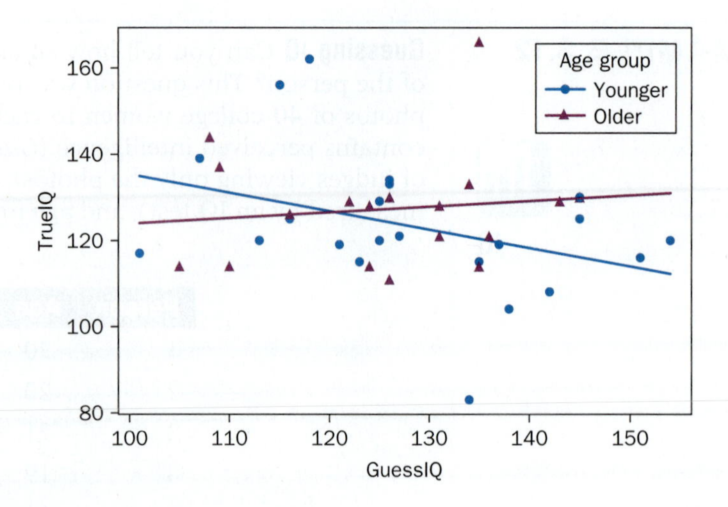

FIGURE 3.17 Regression lines for predicting *TrueIQ* from *GuessIQ* for younger and older women

FIT

We create the new product variable (calling it *GuessIQ* ∗ *Age*) and run the multiple regression model with the interaction term to produce the following output:

	Estimate	Std. Error	t value	Pr(>|t|)
(Intercept)	834.762	320.934	2.60	0.013 *
GuessIQ	-5.702	2.558	-2.23	0.032 *
Age	-33.023	15.553	-2.12	0.041 *
GuessIQ*Age	0.265	0.124	2.14	0.039 *

```
Residual standard error: 14.4 on 36 degrees of freedom
Multiple R-squared: 0.15, Adjusted R-squared: 0.0794
F-statistic: 2.12 on 3 and 36 DF, P-value: 0.115
```

Within this fitted model, the relationship between *TrueIQ* and *GuessIQ* if *Age* = 19 is

$$\widehat{TrueIQ} = 834.762 - 5.702\ GuessIQ - 33.023(19) + 0.265(19)\ GuessIQ = 207.325 - 0.667\ GuessIQ$$

while the relationship between *TrueIQ* and *GuessIQ* if *Age* = 23 is

$$\widehat{TrueIQ} = 834.762 - 5.702\ GuessIQ - 33.023(23) + 0.265(23)\ GuessIQ = 75.233 + 0.393\ GuessIQ$$

Because of the interaction term in this model, the slope between *TrueIQ* and *GuessIQ* changes as *Age* changes. If *Age* is small, then the *GuessIQ* effect is negative; but if *Age* is larger, then the *GuessIQ* effect is positive.

ASSESS

The *t*-tests for the individual predictors show that the interaction term, *GuessIQ* · *Age*, is important in the model, as are *GuessIQ* and *Age* separately. The R^2 value of 15% is rather small, meaning that the interaction model explains only 15% of the variability in *TrueIQ* (but see below for the

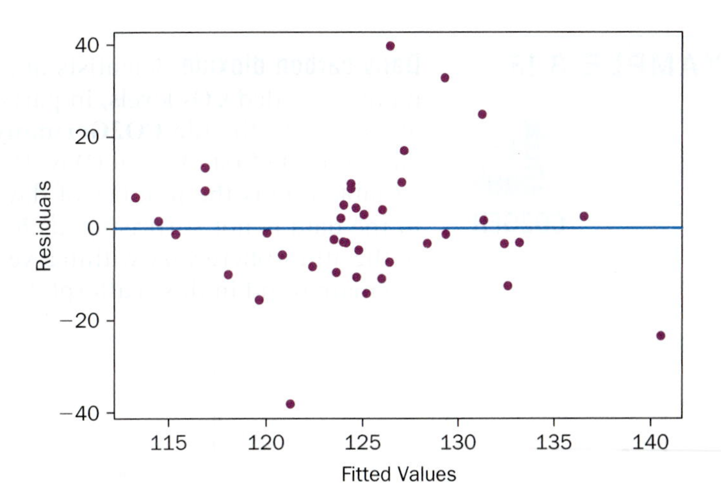

FIGURE 3.18 Residual plot for the regression of *TrueIQ* on *GuessIQ*, *Age*, and their interaction

heteroscedasticity

model without the interaction term). Figure 3.18 shows a residual plot with no extreme pattern; it also shows no evidence of **heteroscedasticity** (i.e., of nonconstant variance in the residuals).

As a closing note, consider the (additive) model that uses just *GuessIQ* and *Age* to predict *TrueIQ*. The output from fitting the model with no interaction is:

	Estimate	Std. Error	t value	Pr(>|t|)
(Intercept)	151.659	37.640	4.03	0.00027 ***
GuessIQ	-0.235	0.185	-1.27	0.21130
Age	0.153	1.485	0.10	0.91841

Residual standard error: 15.1 on 37 degrees of freedom
Multiple R-squared: 0.0419, Adjusted R-squared: -0.00989
F-statistic: 0.809 on 2 and 37 DF, *P*-value: 0.453

This model suggests that neither *GuessIQ* (*P*-value = 0.21) nor *Age* (*P*-value = 0.92) is very useful in predicting *TrueIQ* by themselves. The R^2 for this model is only 4.2%. It is only when we add the interaction term to the model that we see statistical significance for all three terms: *GuessIQ*, *Age*, and their interaction. Sometimes the important relationship to be found is one that involves an interaction. When dealing with two predictors, we suggest that you always start with the full model that includes the interaction term and then consider an additive model if the interaction term is not needed.

Polynomial Regression

In Section 1.4, we saw methods for dealing with data that showed a curved rather than linear relationship by transforming one or both of the variables. Now that we have a multiple regression model, another way to deal with curvature is to add powers of one or more predictor variables to the model.

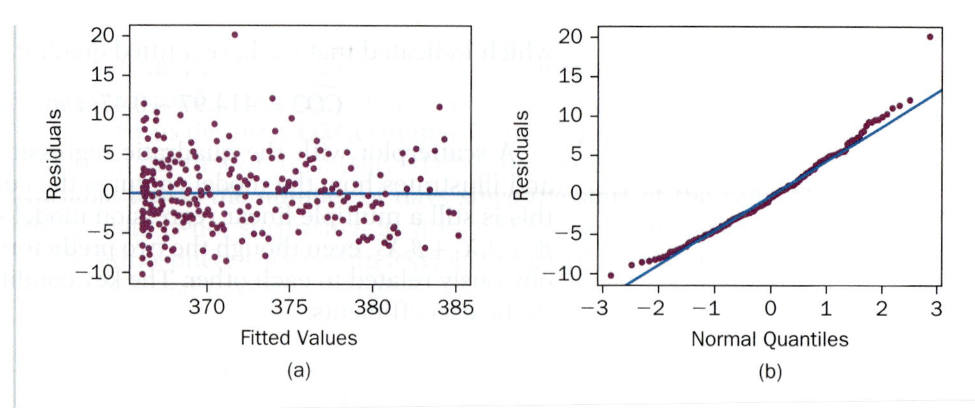

FIGURE 3.21 Diagnostic plots from the quadratic regression fit for CO_2 levels

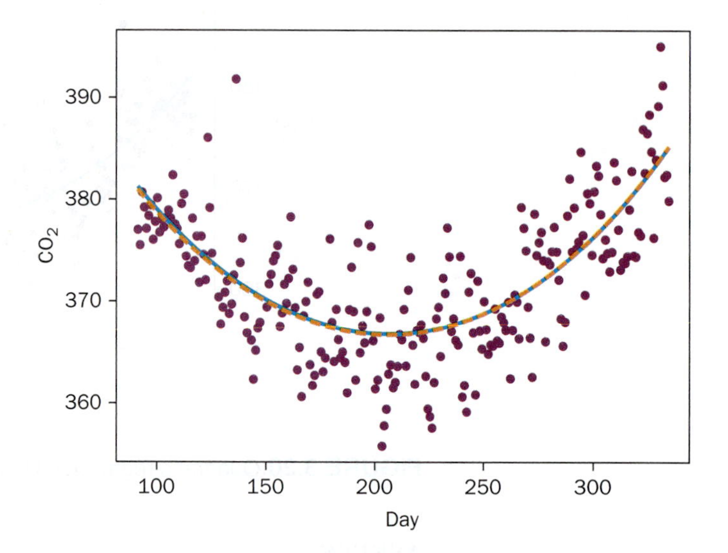

FIGURE 3.22 Quadratic regression fit for CO_2 levels with all data (solid line) and with outlier removed (dashed line)

Figure 3.21 shows that there is no unusual pattern in the residual plot. The residuals follow a normal distribution reasonably well, although there is an outlier on the high end. That is, there was one day with an unusually high *CO2* reading. Looking at Figure 3.20, we can spot that day as being around Day 140, so a day in May.*

There are a lot of points here ($n = 237$) and the t-tests are highly significant, so there is no worry that this day in May has much of an effect. However, we can delete this point and refit the quadratic regression model. This results in almost no change: Figure 3.22 is a repeat of Figure 3.20 but with a dashed line that shows what the quadratic model predicts if we delete the outlier.

USE

The graphs of the data show that CO_2 levels tend to decrease until around Day 205 and increase afterward.† So our summary is that a quadratic model fits well; CO_2 levels decrease smoothly and rather consistently in the spring, bottom out in July, and increase smoothly after July.‡

*To be precise, this was May 16.

†If we had data spanning a wider range of days, we would see a cyclical pattern that we would want to model using time series methods, as discussed in Chapter 12.

‡The minimum fitted value from the quadratic regression happens on Day 206, which is July 28.

We can easily generalize the idea of quadratic regression to include additional powers of a single quantitative predictor variable. However, note that additional polynomial terms may not improve the model much.

POLYNOMIAL REGRESSION MODEL

For a single quantitative predictor X, a **polynomial regression** model of degree k has the form

$$Y = \beta_0 + \beta_1 X + \beta_2 X^2 + \cdots + \beta_k X^k + \epsilon$$

For example, adding a $Day3 = Day^3$ predictor to the quadratic model to predict $CO2$ levels based on a third-degree polynomial yields the following summary statistics:

```
Coefficients:
               Estimate     Std. Error   t value    Pr(>|t|)
 (Intercept)  406.69882838  8.84848652    45.96    <2e-16 ***
 Day           -0.33964711  0.14097576    -2.41     0.017 *
 DaySq          0.00047026  0.00069887     0.67     0.502
 Day3           0.00000108  0.00000109     0.99     0.324

Residual standard error: 4.62 on 233 degrees of freedom
Multiple R-squared: 0.575, Adjusted R-squared: 0.57
F-statistic: 105 on 3 and 233 DF, P-value: <2e-16
```

The new cubic term in this model is clearly not significant (P-value $= 0.324$), the value of R^2 shows no improvement over the quadratic model (in fact, the adjusted R^2 doesn't change), and a plot of the fitted cubic equation with the scatterplot in Figure 3.23 is essentially indistinguishable from the quadratic fit shown in Figure 3.20.

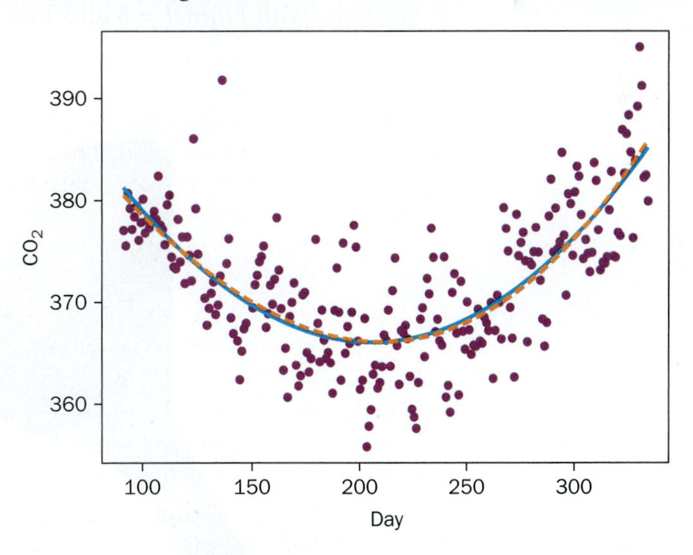

FIGURE 3.23 Quadratic regression for CO_2 levels (solid line) and cubic regression (dashed line)

Complete Second-Order Model

In some situations, we find it useful to combine the ideas from both of the previous examples. That is, we might choose to use quadratic terms to account for curvature and one or more interaction terms to handle effects that occur at particular combinations of the predictors. When doing so, we should take care that we don't overparameterize the model, making it more complicated than needed, with terms that aren't important for explaining the structure of the data. Examining t-tests for the individual terms in the model and checking how much additional variability is explained by those terms are two ways we can guard against including unnecessary complexity. In Section 3.6, we also consider a method for assessing the contribution of a set of predictors as a group.

COMPLETE SECOND-ORDER MODEL

For two predictors, X_1 and X_2, a **complete second-order model** includes linear and quadratic terms for both predictors along with the interaction term:

$$Y = \beta_0 + \beta_1 X_1 + \beta_2 X_2 + \beta_3 X_1^2 + \beta_4 X_2^2 + \beta_5 X_1 \cdot X_2 + \epsilon$$

This extends to more than two predictors by including all linear, quadratic, and pairwise interactions.

EXAMPLE 3.14

FunDrop

Funnel Swirling The diagram in Figure 3.24 shows the setup for an experiment where a steel ball is rolled through a tube at an angle into a funnel.[5] The ball swirls around as it travels down the funnel until it comes out the bottom end and hits the table. We are interested in maximizing the time it takes for the ball to make this trip. We can adjust the relative steepness by changing the angle of the funnel (as reflected in the funnel height) or the height of the drop tube.

In a class experiment students did 10 runs at each of four funnel heights (8, 11, 14, and 16 inches) combined with each of three tube heights (8, 11, and 14 inches). This produced a total of 120 drop times (measured in seconds with a stopwatch). They used a random order for the settings, although for efficiency they did two runs at each setting before adjusting the heights. The results are stored in the file **FunnelDrop**. The longest drop time was 23.6 seconds (with *Funnel* = 11 and *Tube* = 11) while the fastest trip was 8.2 seconds (with *Funnel* = 8 and *Tube* = 14).

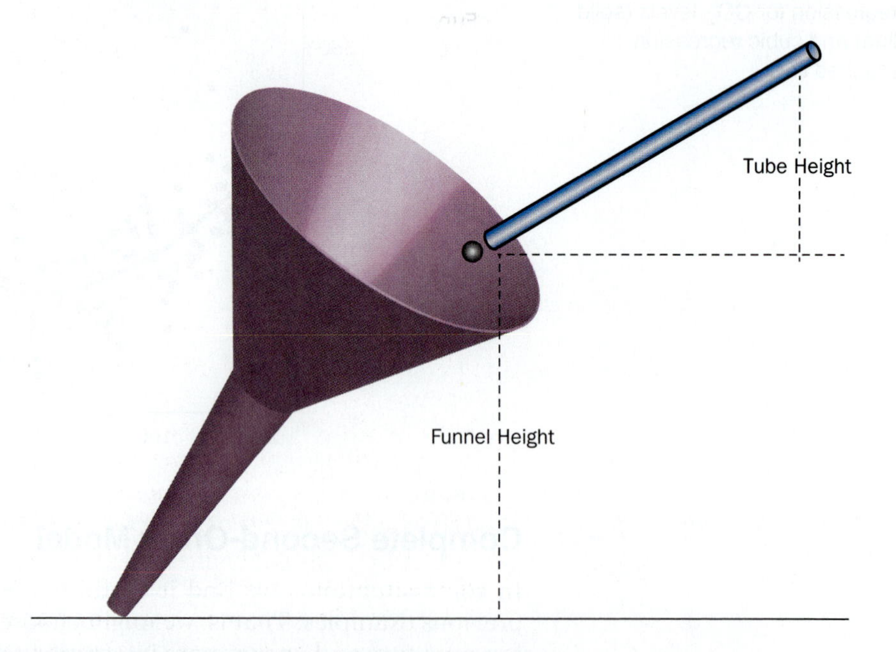

FIGURE 3.24 Setup for the funnel swirling experiment

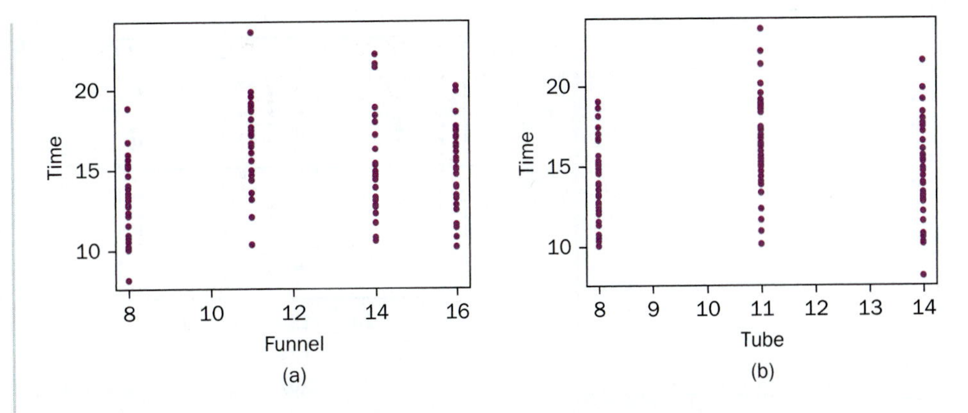

FIGURE 3.25 Scatterplots of drop times versus funnel and tube heights

CHOOSE

Figure 3.25 shows scatterplots of the drop *Time* versus both (a) *Funnel* height and (b) *Tube* height. In both cases, we see that the longer drop times tend to occur at the middle heights, while the times tend to be somewhat smaller at the more extreme (high or low) heights. This curvature suggests possibly adding quadratic terms for both the *Funnel* and *Tube* measurements. We might also expect that the angle of the tube relative to the funnel (as reflected by their respective heights) could have some impact on drop times. This would suggest trying an interaction term, as the product of the two heights. This gives the second-order model

$$Time = \beta_0 + \beta_1 Funnel + \beta_2 Tube + \beta_3 Funnel^2 + \beta_4 Tube^2 + \beta_5 Funnel \cdot Tube + \epsilon$$

FIT

Some output from fitting this model is given below.

```
Coefficients:
              Estimate  Std. Error  t value       Pr(>|t|)
(Intercept)  -26.80000     9.09938   -2.945   0.003913 **
Funnel         3.13797     0.90140    3.481   0.000708 ***
Tube           4.48722     1.28788    3.484   0.000701 ***
Funnelsq      -0.15691     0.03455   -4.541    1.40e-05 ***
Tubesq        -0.23528     0.05563   -4.229    4.75e-05 ***
Funnel*Tube    0.06719     0.03179    2.114   0.036724 *

Residual standard error: 2.585 on 114 degrees of freedom
Multiple R-squared: 0.2934, Adjusted R-squared: 0.2624
F-statistic: 9.465 on 5 and 114 DF, P-value: 1.433e-07
```

ASSESS

The residual plots in Figure 3.26 show no concerns with linearity, constant variance, or normality of the residuals. You may have an initial concern about the gaps between fitted values in the residuals versus fitted values plot, but remember that we have only four distinct values of the *Funnel* height and three for the *Tube*, so there are only 12 different values for the predicted drop times, each repeated 10 times. In the computer output, we see that the *P*-values are quite small for each of the terms (even the largest $p = 0.037$ for the interaction term is still significant at a 5% level). The ANOVA *P*-value (1.4×10^{-7}) is very small, indicating an effective fit. You might think the R^2 value of 29.3% is not very big, but you can check that a model without the second-order terms would only give $R^2 = 2.7\%$.

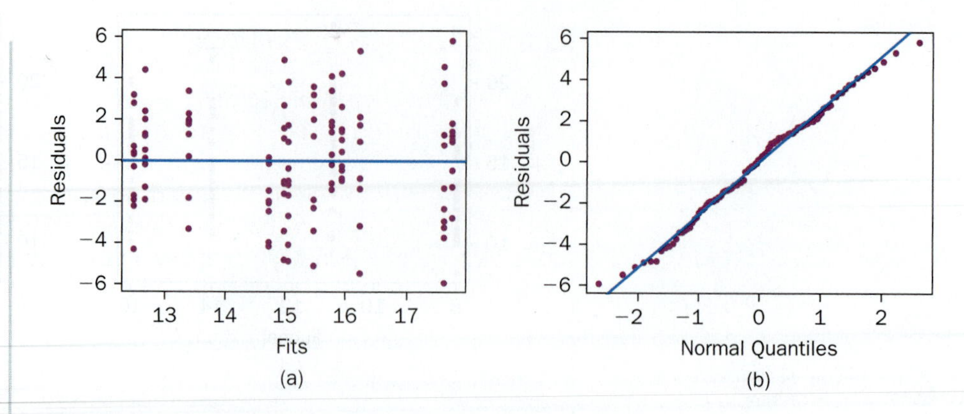

FIGURE 3.26 Residual plots for the second-order model for funnel drops

USE

Recall that one goal for doing this experiment was to try to find a combination of *Funnel* and *Tube* heights that would provide a long drop time. Figure 3.27 shows a plot of the predicted drop *Time*, based on the fitted second-order model, for various *Funnel* and *Tube* heights. Such models produce a surface for the response variable as a function of the two predictors (which is sometimes called a *response surface*). In this case the negative values for the coefficients of the squared terms yield a surface that is curving down from a single maximum point. A closer look at the plot, some guess/check using the prediction equation, or a bit of calculus will show that the maximum of this surface is a time of just over 18 seconds, which is the predicted drop time when the *Funnel* height is set at 12.4 inches and the *Tube* height is 11.3 inches.

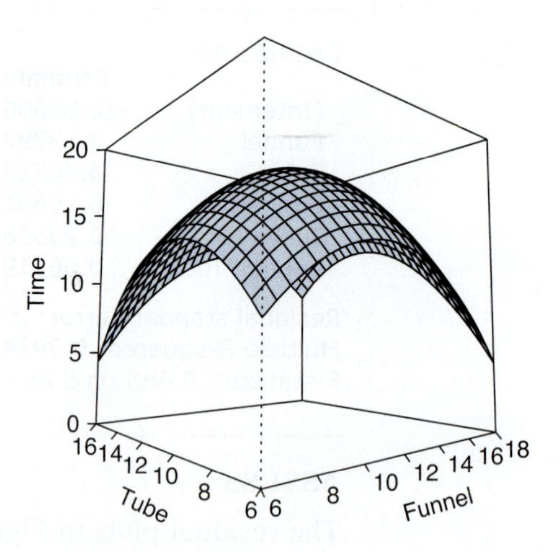

FIGURE 3.27 Predicted times from the fitted model for the funnel drop experiment

Although each of the terms, including the linear terms, in the second-order model for the funnel drops are statistically significant, this is not always the case. In general, when a higher-order term, such as a quadratic term or an interaction product, is important in the model, we usually keep the lower-order terms in the model, even if the coefficients for these terms are not significant.

3.5 Correlated Predictors

When fitting a multiple regression model, we often encounter predictor variables that are correlated with one another. We shouldn't be surprised if predictors that are related to some response variable Y are also related to each other. This is not necessarily a bad thing, but it can lead to difficulty in fitting and interpreting the model. The next example illustrates the sort of somewhat counterintuitive behavior that we might see when correlated predictors are added to a multiple regression model.

EXAMPLE 3.15

HseNY

House prices in NY The file **HousesNY** contains estimated prices (in \$1,000s) for a sample of 53 houses in the small, upstate NY town of Canton.[6] The file also contains data on the size of each house (in 1,000s of square feet), the number of bedrooms, bathrooms, and the size of the lot (in acres).

CHOOSE

Figure 3.28(a) shows that *Price* is positively associated with both house *Size* ($r = 0.512$) and *Beds* ($r = 0.419$). First, we fit each predictor separately to predict *Price* to get the fitted lines shown in Figure 3.28.

FIT

Using statistical software to regress *Price* on *Size* gives the following output:

```
            Estimate  Std. Error  t value      Pr(>|t|)
(Intercept)   54.078      14.832    3.646   0.000625 ***
Size          35.489       8.335    4.258   8.87e-05 ***
---

Residual standard error: 35.93 on 51 degrees of freedom
Multiple R-squared: 0.2622, Adjusted R-squared: 0.2478
F-statistic: 18.13 on 1 and 51 DF, P-value: 8.865e-05
```

Regressing *Price* on *Beds* gives similar output:

```
            Estimate  Std. Error  t value    Pr(>|t|)
(Intercept)   39.239      23.161    1.694   0.09632 .
Beds          21.905       6.644    3.297   0.00179 **
---

Residual standard error: 37.98 on 51 degrees of freedom
Multiple R-squared: 0.1757, Adjusted R-squared: 0.1595
F-statistic: 10.87 on 1 and 51 DF, P-value: 0.001785
```

Both of the predictors *Size* and *Beds* produce useful regression models for predicting *Price* on their own, with very small *P*-values when testing for significance of the slope. The model based on size of the house is slightly more effective (explaining 26.2% of the variability in prices) compared to the number of bedrooms ($R^2 = 17.6\%$). But now that we have the tools of multiple regression, we should consider putting both predictors in the same model to perhaps explain prices even more effectively. Here is the output for fitting a multiple regression model based on both *Size* and *Beds*.

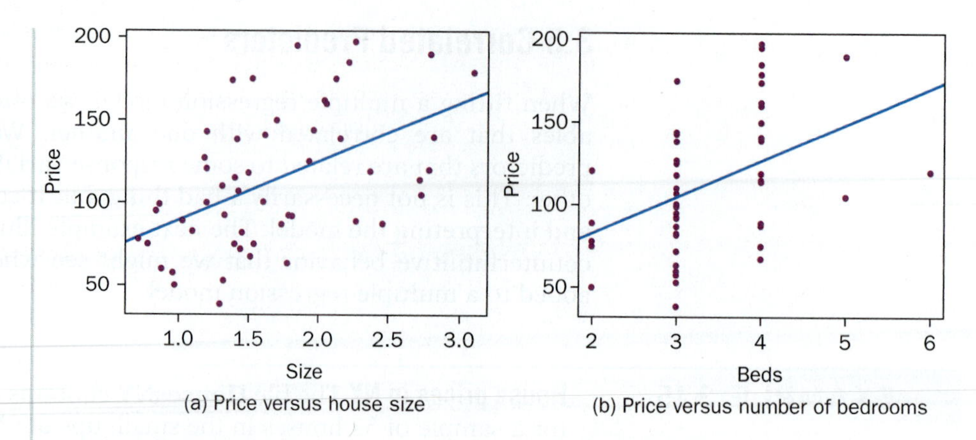

(a) Price versus house size

(b) Price versus number of bedrooms

FIGURE 3.28 Scatterplots of house selling price versus two predictors

```
            Estimate  Std. Error  t value  Pr(>|t|)
(Intercept)   46.498      22.277    2.087    0.042 *
Size          31.169      12.617    2.470    0.017 *
Beds           4.367       9.515    0.459    0.648
---
Residual standard error: 36.21 on 50 degrees of freedom
Multiple R-squared: 0.2653, Adjusted R-squared: 0.236
F-statistic: 9.03 on 2 and 50 DF, P-value: 0.0004489
```

The two-variable prediction equation is

$$\widehat{Price} = 46.5 + 31.17\ Size + 4.37\ Beds$$

ASSESS

Note a couple of interesting features of the combined multiple regression model. The most striking is that the *P*-value for the *Beds* predictor (0.648) is not at all significant. Does that mean that the number of bedrooms has suddenly become irrelevant for determining the price of a house? We also see that the the *P*-value for the *Size* predictor (0.017), while still significant at a 5% level, is not nearly as strong as when it was in a model by itself (*P*-value = 0.000089). When assessing the overall fit of the multiple regression model, we see that the *P*-value for the ANOVA (0.00045) is quite small—much smaller than the *P*-values of either of the two predictors in the model. Yet the amount of variability explained ($R^2 = 26.53\%$) is barely more than when using *Size* alone ($R^2 = 26.22\%$); and the adjusted R^2 goes down when *Beds* is added to a model that has *Size* in it. What might account for the somewhat apparent contradictions in these models? The key is the fact that the predictors, *Size* and *Beds*, are related to each other.

What might account for the possible contradictions in the previous example? Is the *Beds* variable related to *Price* or not? One key to understanding what is happening is the fact that the predictors, *Size* and *Beds*, in addition to being related to the response variable (*Price*), are also related to each other. It is not surprising that larger houses tend to have more bedrooms. The correlation between *Size* and *Beds* for this sample of 53 houses is $r = 0.746$. It is not an unusual situation for predictors that are each associated with a common response variable to also be associated with each other. The previous example illustrates that we need to take care not to make hasty conclusions based on the individual coefficients or their t-tests when predictors in a model are related to one another. One of the challenges of dealing with multiple predictors is accounting for such relationships between predictors.

MULTICOLLINEARITY

We say that a set of predictors exhibits **multicollinearity** when one or more of the predictors is strongly correlated with some combination of the other predictors in the set.

If one predictor has an *exact* linear relationship with one or more other predictors in a model, the least squares process to estimate the coefficients in the model does not have a unique solution. Most statistical software routines will either delete one of the perfectly correlated predictors when running the model or produce an error message. In cases where the correlation between predictors is high, but not 1, the coefficients can be estimated, but interpretation of individual terms can be problematic.

The individual t-test for a term in a multiple regression model assesses how much that predictor contributes to the model *after accounting for the other predictors in the model*. Thus while *Size* and *Beds* are both strong predictors of house prices on their own, they carry similar information (bigger houses tend to have more bedrooms and be more expensive). If *Size* is already in a model to predict *Price*, we don't really need to add *Beds*. We see that doing so only increases the R^2 by a tiny 0.31% in the two-predictor model. Thus the individual t-test for *Beds* in the multiple regression model is quite insignificant—it doesn't add much predictive power to what *Size* already does. On the other hand, the R^2 jumps considerably (from 17.6% to 26.5%) when *Size* is added as a second predictor over one with just *Beds*. Thus *Size* contains more additional information that is useful for predicting *Price*, that is not revealed by *Beds*. *Size* still has a significant *P*-value in the multiple regression model (but not as strong as the evidence when *Size* is the only predictor in the model).

Detecting Multicollinearity

How do we know when multicollinearity might be an issue with a set of predictors? One quick check is to examine scatterplots of pairs of variables to look for associations among the predictors. We can also look at pairwise correlations between the predictors to see which might be strongly related.

EXAMPLE 3.16

HseNY

House prices in NY: Correlations The data in **HousesNY** also shows the number of bathrooms (*Baths*) and size of the *Lot* (in acres) for each of the houses. A correlation matrix of the *Price* and four potential predictors is shown below and pairwise scatterplots for all variables are shown in Figure 3.29.

	Price	Beds	Baths	Size	Lot
Price	1.000	0.419	0.558	0.512	-0.011
Beds	0.419	1.000	0.356	0.746	-0.211
Baths	0.558	0.356	1.000	0.418	-0.039
Size	0.512	0.746	0.418	1.000	-0.214
Lot	-0.011	-0.211	-0.039	-0.214	1.000

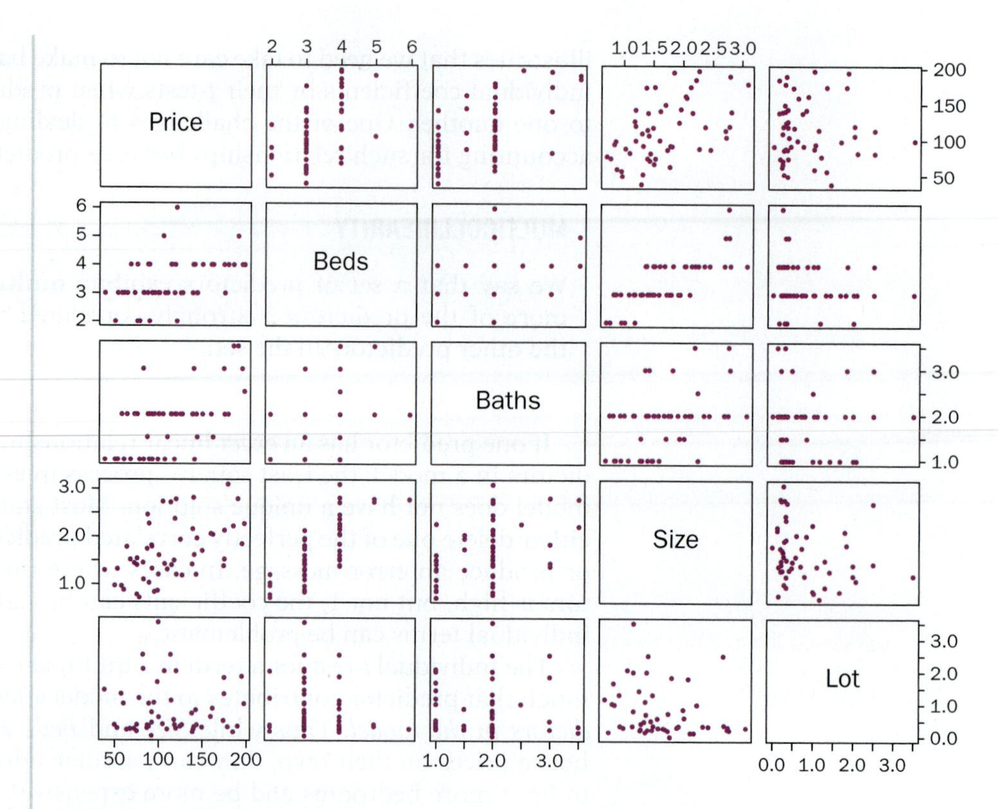

FIGURE 3.29 Matrix of scatterplots for variables in **HousesNY**

We see that *Baths* is actually the strongest individual predictor of *Price* ($r = 0.558$), but it is also related to *Beds* ($r = 0.356$) and *Size* ($r = 0.418$). These are not as strong as the relationship between *Beds* and *Size*, but are still significant. For this sample size, a single correlation beyond ± 0.27 will be significant at a 5% level. We also note that the size of the *Lot* is not strongly related to any of the other predictors or to the price of the house.

Here is some output for a multiple regression model using all four predictors of *Price*.

	Estimate	Std. Error	t value	Pr(>|t|)	
(Intercept)	14.590	23.266	0.627	0.5336	
Size	22.155	11.931	1.857	0.0695	.
Beds	2.771	8.730	0.317	0.7523	
Baths	26.238	7.844	3.345	0.0016	**
Lot	4.621	6.184	0.747	0.4585	

Residual standard error: 33.03 on 48 degrees of freedom
Multiple R-squared: 0.4134, Adjusted R-squared: 0.3645
F-statistic: 8.457 on 4 and 48 DF, *P*-value: 3.01e-05

We see here that only *Baths* has a tiny *P*-value in this model, and the *P*-value for *Size* is quite a bit larger than for the previous model. Nonetheless, this model still shows evidence that *Size* and *Baths* are useful predictors. However, the $R^2 = 41.3\%$ is a sizable increase over the two-predictor model based on *Size* and *Beds* ($R^2 = 25.6\%$) or a single predictor model based on *Baths* alone ($R^2 = 31.2\%$). We give some advice for dealing with these sorts of issues in picking a model at the end of this section.

When the predictors are related, we should also take care when interpreting the individual coefficients. For example, one might be tempted to interpret the model above to conclude that adding a new bathroom would increase the value of a house by around \$26,238 while an extra bedroom would be worth only about \$2,771. But in the model with *Beds* alone, we see an increase in expected price of \$21,905 for each additional bedroom. In the multiple regression model, with four predictors, the coefficient on *Beds* of 2.771 tells us that as the number of bedrooms increases by 1 we expect price to increase by \$2,771 *allowing for simultaneous changes in the other predictors*. A house with more bedrooms is likely to have more bathrooms, more total square footage, and to be on a larger lot. The interplay among all of these variables matters—and can make multiple regression coefficients difficult to assess on their own.

In some cases, a dependence between predictors can be more subtle. We may have some combinations of predictors that, taken together, are strongly related to another predictor. To investigate these situations, we can consider regression models for each individual variable in the model using all of the other variables as predictors. Any measure of the effectiveness of these models (e.g., R^2) could indicate which predictors might be strongly related to others in the model. One common form of this calculation available in many statistical packages is the **variance inflation factor**, which reflects the association between a predictor and all of the other predictors.

variance inflation factor

VARIANCE INFLATION FACTOR

For any predictor X_i in a model, the **variance inflation factor** (VIF) is computed as

$$VIF_i = \frac{1}{1 - R_i^2}$$

where R_i^2 is the coefficient of multiple determination for a model to predict X_i using the other predictors in the model. The VIF is the factor by which the variance of an estimated regression coefficient is inflated.

As a rough rule, we suspect multicollinearity with predictors for which the $VIF > 5$, which is equivalent to $R_i^2 > 80\%$.

EXAMPLE 3.17

CtyHlth

Doctors and hospitals in counties: VIF In Example 1.7, we used a square root transformation when modeling how the number of doctors (*MDs*) in a county is related to the number of *Hospitals*. The data in the datafile **CountyHealth** contains *MDs*, *Hospitals*, and a third variable *Beds*, which is the number of hospital beds in the county. Figure 3.30 shows a scatterplot matrix for the three variables. As expected, *sqrt(MDs)* is clearly related to both *Hospitals* and *Beds*, and those two predictors are related to each other.

Consider a model in which the square root of the number of doctors depends on *Hospitals*, *Beds*, and the interaction between *Hospitals* and *Beds*. Fitting this model produces the output shown below (after a software option for the variance inflation factor has been selected):

The regression equation is
sqrtMDs = -0.625 + 3.142 Hospitals + 0.0219 Beds - 0.00098 Hospitals*Beds

Predictor	Coef	SE Coef	T	P	VIF
Constant	-0.625	1.729	-0.36	0.72	
Hospitals	3.142	0.610	5.15	0.000	6.02
Beds	0.0219	0.0026	8.43	0.000	16.09
Hospitals*Beds	-0.00098	0.00021	-4.61	0.000	14.42

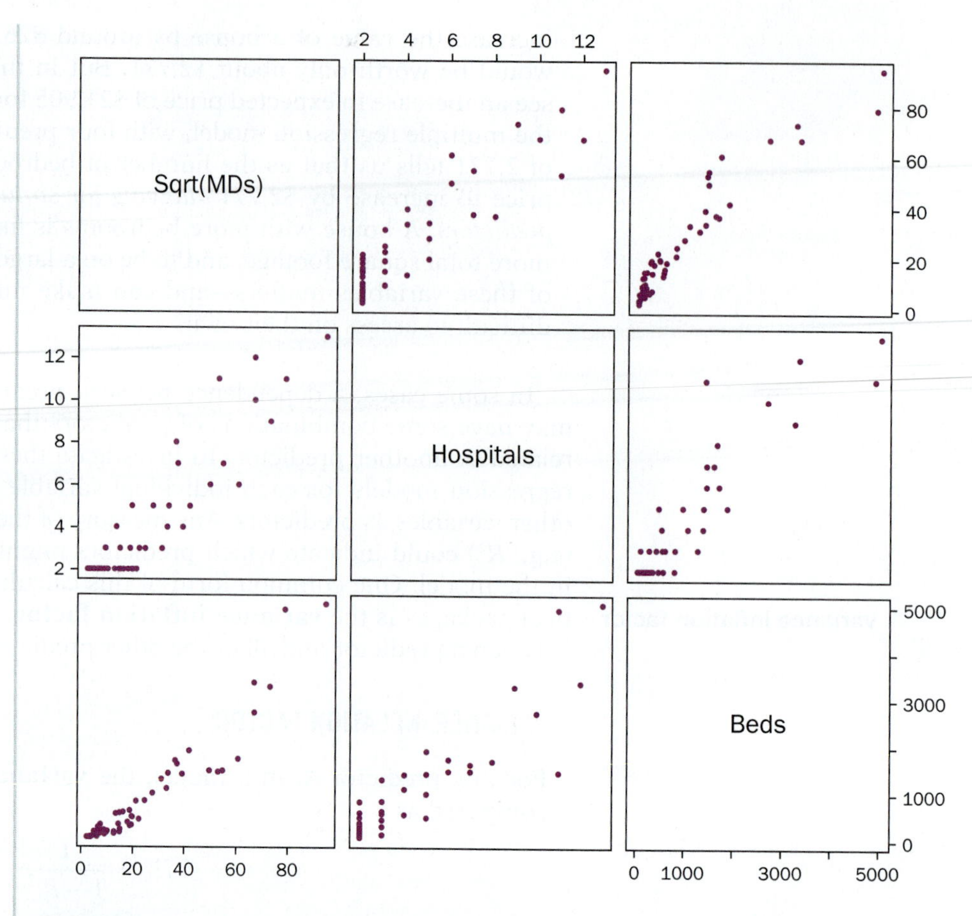

FIGURE 3.30 Scatterplot matrix of sqrt(MDs), hospitals, and hospital beds for 53 counties

As we should expect, the VIF values for all three predictor terms are quite large because those variables have obvious relationships.

What should we do if we detect multicollinearity in a set of predictors? First, realize that multicollinearity does not necessarily produce a poor model. As the previous example shows, certain types of models (such as ones with interactions or polynomials) will naturally lead to multicollinearity. In many cases, related predictors might all be important in the model. For example, when trying to predict house prices (Example 3.16), we see considerable improvement in the model when we add the *Size* variable to one containing *Baths* and *Beds*, even though it is strongly related to both predictors.

Here are some options for dealing with correlated predictors:

1. *Drop some predictors.* If one or more predictors are strongly correlated with other predictors, try the model with those predictors left out. If those predictors are really redundant and the reduced model is essentially as effective as the bigger model, you can probably leave them out of your final model. However, if you notice a big drop in R^2 or problems appearing in the residual plots, you should consider keeping one or more of those predictors in the model.
2. *Combine some predictors.* Suppose that you are working with data from a survey that has many questions on closely related topics that produce highly correlated variables. Rather than putting each of the predictors in a model individually, you could create a new variable with some formula

(e.g., a sum) based on the group of similar predictors. This would allow you to assess the impact of that group of questions without dealing with the predictor-by-predictor variations that could occur if each was in the model individually.

3. *Discount the individual coefficients and* t*-tests.* As we've seen in previous examples, multicollinearity can produce what initially looks like odd behavior when assessing individual predictors. A predictor that is clearly important in one model might produce a clearly insignificant t-test in another model, when the second model includes a strongly correlated predictor. We might even find a model that is highly effective overall, yet *none* of its individual t-tests are significant, or see a predictor that we expect to have a positive association with the response get a significant negative coefficient in a model due to other predictors with similar information. Since a model can still be quite effective with correlated predictors, we might choose to just ignore some of the individual t-tests. For example, we should probably keep both *Baths* and *Size* (and perhaps even *Beds*) in a model to predict house prices, but we should use caution when interpreting their individual coefficients.

3.6 Testing Subsets of Predictors

Individual t-tests for regression predictors ($H_0: \beta_i = 0$) allow us to check the importance of terms in the model *one at a time*. The overall ANOVA for regression ($H_0: \beta_1 = \beta_2 = \cdots = \beta_k = 0$) allows us to test the effectiveness of *all* of the predictors in the model as a group. Is there anything in between these two extremes? As we develop more complicated models with polynomial terms, interactions, and different kinds of predictors, we may encounter situations where we want to assess the contribution of some subset of predictors as a group. In this section we describe a general procedure for comparing the effectiveness of a relatively simple model to a more complicated alternative.

Nested *F*-Test

nested *F*-test

The procedure we use for testing a subset of predictors is called a **nested *F*-test**. We say that one model is nested inside another model if its predictors are all present in the larger model. For example, an interaction model $Y = \beta_0 + \beta_1 X_1 + \beta_2 X_2 + \beta_3 X_1 \cdot X_2 + \epsilon$ is nested within a complete second-order model using the same two predictors. Note that the smaller (nested) model should be entirely contained in the larger model, so models such as $Y = \beta_0 + \beta_1 X_1 + \beta_2 X_2 + \epsilon$ and $Y = \beta_0 + \beta_1 X_2 + \beta_2 X_3 + \epsilon$ have a predictor in common but neither is nested within the other.

The essence of a nested F-test is to compare the full (larger) model with a reduced (nested) model that eliminates the group of predictors that we are testing. If the full model does a (statistically significantly) better job of explaining the variability in the response, we may conclude that at least one of those predictors being tested is important to include in the model.

EXAMPLE 3.18

HseNY

House prices: Comparing models Let's look again at models to predict house prices introduced in Example 3.15 using the data in **HousesNY**. From the correlation matrix in Example 3.16 on page 123 it is apparent that the *Lot* variable is not very helpful for predicting house price, so let's drop it and consider a model based on the remaining three predictors.

$$\text{Full model:} \qquad Price = \beta_0 + \beta_1 Size + \beta_2 Beds + \beta_3 Baths + \epsilon$$

Here is some output for fitting this model:

```
Coefficients:
             Estimate  Std. Error  t value      Pr(>|t|)
(Intercept)  20.929       21.566    0.970      0.33659
Size         21.258       11.817    1.799      0.07818 .
Beds          2.230        8.661    0.257      0.79787
Baths        26.610        7.793    3.415     0.00129 **
---
Residual standard error: 32.88 on 49 degrees of freedom
Multiple R-squared: 0.4066, Adjusted R-squared: 0.3702
F-statistic: 11.19 on 3 and 49 DF, P-value: 1.042e-05

ANOVA Table
Model: Price ~ Size + Beds + Baths
        Df   Sum Sq   Mean Sq   F value          P(>F)
Model    3    36288     12096     11.19   1.042e-05 ***
Error   49    52967      1081
Total   52    89255
```

We see from the individual t-tests that both *Size* (P-value $= 0.078$) and *Beds* (P-value $= 0.798$) are not significant at a 5% level in this model. Suppose that we want to assess their impact in the model together with a single test. That is, we want to compare the effectiveness of the three-predictor model to a simpler model with just the *Baths* predictor. The relevant hypotheses are

$H_0: \beta_1 = \beta_2 = 0$

$H_a:$ at least one $\beta_i \neq 0$

and the simpler model specified by the null hypothesis is

$$\text{Reduced model:} \qquad Price = \beta_0 + \beta_3 Baths + \epsilon$$

Here is some output for fitting the model based on *Baths* alone.

```
Coefficients:
             Estimate  Std. Error  t value      Pr(>|t|)
(Intercept)   47.069      14.648    3.213   0.00228 **
Baths         35.815       7.453    4.806   1.4e-05 ***
---
Residual standard error: 34.71 on 51 degrees of freedom
Multiple R-squared: 0.3117, Adjusted R-squared: 0.2982
F-statistic: 23.1 on 1 and 51 DF, P-value: 1.399e-05

ANOVA Table
Model: Price ~ Baths
        Df   Sum Sq   Mean Sq   F value          P(>F)
Model    1    27821   27821.1    23.096   1.399e-05 ***
Error   51    61434    1204.6
Total   52    89255
```

One quick way to compare these models is to look at their R^2 values and see how much is "lost" when the *Size* and *Beds* terms are dropped from the model, with R^2 going from 40.7% in the full model down to 31.2% in the reduced model. This seems like a lot, but is it statistically significant? We now develop a formal test see if a substantial amount of variability explained is lost when we drop those two predictors.

Comparing the SSModel values for the two models, we see that

$$SSModel_{full} - SSModel_{reduced} = 36,288 - 27,821 = 8,467$$

Is that a "significant" amount of extra variability? As with the ANOVA procedure for the full model, we obtain a mean square by dividing this change in explained variability by the number of extra predictors needed to obtain it. To compute an F-test statistic, we then divide that mean square by the MSE for the full model:

$$F = \frac{8467/2}{52967/49} = \frac{4233.5}{1081} = 3.92$$

The degrees of freedom should be obvious from the way this statistic is computed; in this case, we should compare this value to the upper tail of an $F_{2,49}$ distribution. Doing so produces a P-value of 0.026, which is significant at a 5% level. This would imply that we might want to keep either *Size* or *Beds* together with *Baths* in the model for these house prices.

NESTED *F*-TEST

To test a subset of predictors in a multiple regression model,

$H_0 : \beta_i = 0$ for all predictors in the subset
$H_a : \beta_i \neq 0$ for at least one predictor in the subset

Let the **full** model denote one with all k predictors and the **reduced** model be the nested model obtained by dropping the predictors that are being tested. The test statistic is

$$F = \frac{(SSModel_{full} - SSModel_{reduced})/\#\text{ predictors tested}}{SSE_{full}/(n-k-1)}$$

The P-value is computed from an F-distribution with numerator degrees of freedom equal to the number of predictors being tested and denominator degrees of freedom equal to the error degrees of freedom for the full model.

An equivalent way to compute the amount of new variability explained by the predictors being tested is to use the fact that

$$SSModel_{full} - SSModel_{reduced} = SSE_{reduced} - SSE_{full}$$

Note that if we apply a nested F-test to all of the predictors at once, we get back to the original ANOVA test for an overall model. If we test just a single predictor using a nested F-test, we get a test that is equivalent to the individual t-test for that predictor. The nested F-test statistic will be the square of the test statistic from the t-test (see Exercise 3.47).

EXAMPLE 3.19

NFL16

NFL winning percentage: Nested *F*-test We return to the problem of predicting the winning percentage for NFL teams that was introduced in Example 3.1 on page 88. The **NFLStandings2016** dataset has additional information for each team that season, including the total yards the team gained (*YardsFor*), number of yards for their opponents (*YardsAgainst*), net points (*NetPts* = *PointsFor* − *PointsAgainst*), and number of touchdowns scored (*TDs*). To keep the coefficients from being too small, we convert the winning percentage to $WinPct100 = (100)WinPct$ and consider the five-predictor model.

$$WinPct100 = \beta_0 + \beta_1 PointsFor + \beta_2 PointsAgainst + \beta_3 YardsFor + \beta_4 YardsAgainst + \beta_5 TDs + \epsilon$$

Note that we don't include the *NetPts* variable as a predictor, since it is an exact function of *PointsFor* and *PointsAgainst*. Some regression output for fitting this model is given as follows.

```
The regression equation is
WinPct100 = 48.7 + 0.157 PointsFor - 0.3103 PointsAgainst - 0.00302 YardsFor
            + 0.01236 YardsAgainst+ 0.116 TDs

Coefficients
 Term              Coef     SE Coef   T-Value   P-Value     VIF
 Constant          48.7       31.6       1.54     0.135
 PointsFor         0.157      0.115      1.37     0.182    20.79
 PointsAgainst    -0.3103     0.0412    -7.54     0.000     1.80
 YardsFor         -0.00302    0.00629   -0.48     0.635     3.92
 YardsAgainst      0.01236    0.00547    2.26     0.032     1.89
 TDs               0.116      0.737      0.16     0.877    16.47

Analysis of Variance
 Source        DF        SS        MS   F-Value   P-Value
 Regression     5   10198.6   2039.73     23.91     0.000
 Error         26    2218.3     85.32
 Total         31   12416.9

       S     R-sq   R-sq(adj)
 9.23678   82.14%     78.70%
```

We see that the *P*-values for each of the offensive variables (*PointsFor*, *YardsFor*, *TDs*) are not statistically significant, while the two defensive variables (*PointsAgainst* and *YardsAgainst*) are significant at a 5% level. But notice the high VIF values for *PointsFor*, *TDs*, and (to a lesser extent) *YardsFor*. It is reasonable (at least to a football fan) that the three variables related to offensive scoring should be related. Would it really make sense to drop all three from the model? To examine this we should test

$H_0: \beta_1 = \beta_3 = \beta_5 = 0$

$H_a:$ at least one $\beta_i \neq 0$

Here is some output for comparing the sum of squared errors (denoted *RSS* in the output) between the reduced model (based on *PointsAgainst* and *YardsAgainst*) and the full model with *PointsFor*, *YardsFor*, and *TDs* added.

```
Model 1: WinPct100 ~ PointsAgainst + YardsAgainst
Model 2: WinPct100 ~ PointsFor + PointsAgainst + YardsFor + YardsAgainst + TDs
  Res.Df     RSS  Df  Sum of Sq       F     Pr(>F)
1     29  5085.9
2     26  2218.3   3     2867.7  11.204  6.67e-05 ***
```

Comparing the sum of squared errors to compute the *F*-statistic we have

$$F = \frac{(SSE_{reduced} - SSE_{full})/3}{MSE_{full}} = \frac{(5085.9 - 2218.3)/3}{85.32} = 11.2$$

which (as shown in the output) gives a *P*-value $= 0.000067$ from an $F_{3,26}$-distribution. This gives very strong evidence that at least one of *PointsFor*, *YardsFor*, and *TDs* is useful (with *PointsAgainst* and *YardsAgainst*) for predicting winning percentage of NFL teams. In fact, any one of the three will help improve the R^2 by a fair amount. Adding *PointsFor* increases it to 81.9%, *TDs* to 80.8%, and *YardsFor* to slightly less at 73.8%. The first two are very close to the $R^2 = 82.1\%$ of the full five-predictor model. But if you put any pair of these three in the same model, one of the pair will have an insignificant *P*-value (due to multicollinearity).

More Thoughts on Modeling

In this chapter we have introduced many modeling ideas, including t-tests for individual predictors, the nested F-test for considering several predictors together, R^2 and adjusted R^2, indicator variables, interaction terms, quadratic terms, and multicollinearity. Given that technology makes it easy to fit models with lots of features, it is tempting to create sophisticated models, but sometimes we can view data from a different perspective and see something new that simplifies our work.

EXAMPLE 3.20

Perch

Perch weights revisited In Example 3.11 on page 107 we analyzed the perch weight data (**Perch**) using a regression model with an interaction term. Recall that R^2 for the three-predictor model was 98.5%, which is rather high, but all of the scatterplots and residual plots in that example showed a clear increase in variability as *Weight* increases.

There are many indicators that a log transformation should be considered in this sort of situation:

a. a curved plot of Y against X

b. a pattern of increasing spread in Y as X increases

c. a variable that ranges over two orders of magnitude (e.g., the perch weights range from 5.9 to 1100 grams)

d. a predictor that is multiplicative (e.g., *Length* · *Width*)

Suppose we had taken logs of the three variables (*Weight*, *Length*, and *Width*).

Figure 3.31 shows the data in log scale. There are nice, linear trends here (although we might still notice a hint of increasing variability in each plot).

If we fit the model

$$log(Weight) = \beta_0 + \beta_1 log(Length) + \beta_2 log(Width) + \epsilon$$

we get the following output:

```
Coefficients:
              Estimate  Std. Error  t value   Pr(>|t|)
(Intercept)     -3.151       0.409    -7.70   3.3e-10 ***
log(Length)      2.199       0.198    11.11   1.9e-15 ***
log(Width)       0.864       0.174     4.96   7.7e-06 ***
---
Residual standard error: 0.0965 on 53 degrees of freedom
Multiple R-squared: 0.992, Adjusted R-squared: 0.992
F-statistic: 3496 on 2 and 53 DF, P-value: <2e-16
```

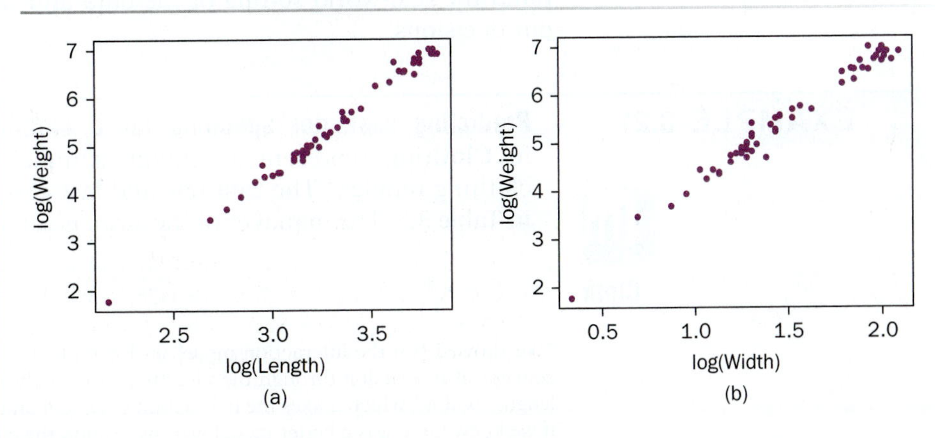

FIGURE 3.31 Individual predictors for perch weights

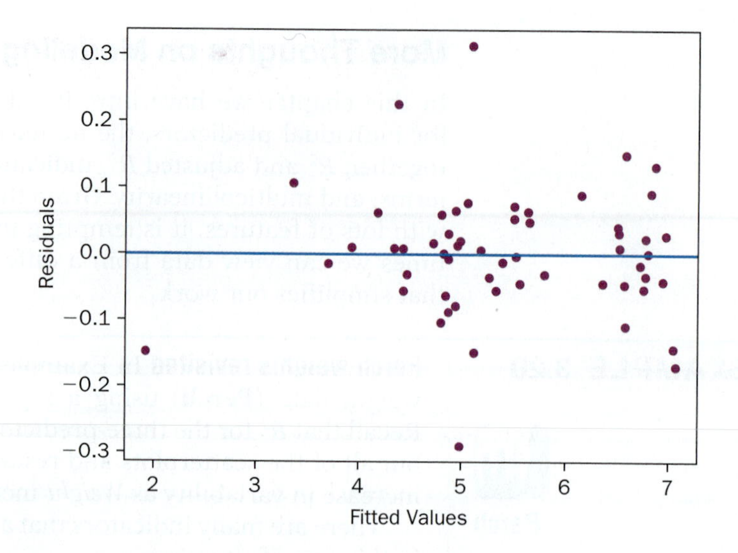

FIGURE 3.32 Residual plot of perch model that uses logs

This model fits very well; notice that R^2 is 99.2%, which is even better than the R^2 we obtained with the interaction model (without using logs) in Example 3.11. In the log scale, if we add an interaction term to the model, the P-value for that term is 0.19, so we don't need to use the interaction after converting to logs. The residual plot shown in Figure 3.32 for the two-predictor model in the log scale also looks quite nice—even the problem with increasing variability has been addressed.

*The lesson here is that there is more than one way to analyze a dataset and often our first analysis won't be the best.**

3.7 Case Study: Predicting in Retail Clothing

We will now look at a set of data with several explanatory variables to illustrate the process of arriving at a suitable multiple regression model. We first examine the data for outliers and other deviations that might unduly influence our conclusions. Next, we use descriptive statistics, especially correlations, to get an idea of which explanatory variables may be most helpful to choose for explaining the response. We fit several models using combinations of these variables, paying attention to the individual t-tests to see if any variables contribute little in a particular model, checking the regression conditions, and assessing the overall fit. Once we have settled on an appropriate model, we use it to answer the question of interest. Throughout this process, we keep in mind the real-world setting of the data and use common sense to help guide our decisions.

EXAMPLE 3.21

Cloth

Predicting customer spending for a clothing retailer The data provided in **Clothing** represent a random sample of 60 customers from a large clothing retailer.[7] The first few and last cases in the dataset are reproduced in Table 3.5. The manager of the store is interested in predicting how much a

*We showed you the interaction model, in Example 3.11, in the hope that you would grasp the concept of interaction through the idea that fish weight is proportional to size and that "area = length · width," which makes the interaction of *Length* and *Width* a natural thing to consider (even if we knew there was a better model waiting around the corner).

ID	Amount	Recency	Freq12	Dollar12	Freq24	Dollar24	Card
1	0	22	0	0	3	400	0
2	0	30	0	0	0	0	0
3	0	24	0	0	1	250	0
4	30	6	3	140	4	225	0
5	33	12	1	50	1	50	0
6	35	48	0	0	0	0	0
7	35	5	5	450	6	415	0
8	39	2	5	245	12	661	1
9	40	24	0	0	1	225	0
⋮	⋮	⋮	⋮	⋮	⋮	⋮	⋮
60	1,506,000	1	6	5000	11	8000	1

TABLE 3.5 First few cases of the **Clothing** data

customer will spend on his or her next purchase based on one or more of the available explanatory variables that are described below.

Variable	Description
Amount	The net dollar amount spent by customers in their latest purchase from this retailer
Recency	The number of months since the last purchase
Freq12	The number of purchases in the last 12 months
Dollar12	The dollar amount of purchases in the last 12 months
Freq24	The number of purchases in the last 24 months
Dollar24	The dollar amount of purchases in the last 24 months
Card	1 for customers who have a private-label credit card with the retailer, 0 if not

The response variable is the *Amount* of money spent by a customer. A careful examination of Table 3.5 reveals that the first three values for *Amount* are zero because some customers purchased items and then returned them. We are not interested in modeling returns, so these observations will be removed before proceeding. The last row of the data indicates that one customer spent $1,506,000 in the store. A quick consultation with the manager reveals that this observation is a data entry error, so this customer will also be removed from our analysis. We can now proceed with the cleaned data on 56 customers.

CHOOSE

We won't go through all of the expected relationships among the variables, but we would certainly expect the amount of a purchase to be positively associated with the amount of money spent over the last 12 months (*Dollar12*) and the last 24 months (*Dollar24*). Speculating about how the frequency of purchases over the last 12 and 24 months is related to the purchase amount is not as easy. Some customers might buy small amounts on a regular basis, while others might purchase large amounts of clothing at less frequent intervals because they don't like to shop. Other people like shopping and clothing, so they might purchase large amounts on a regular basis.

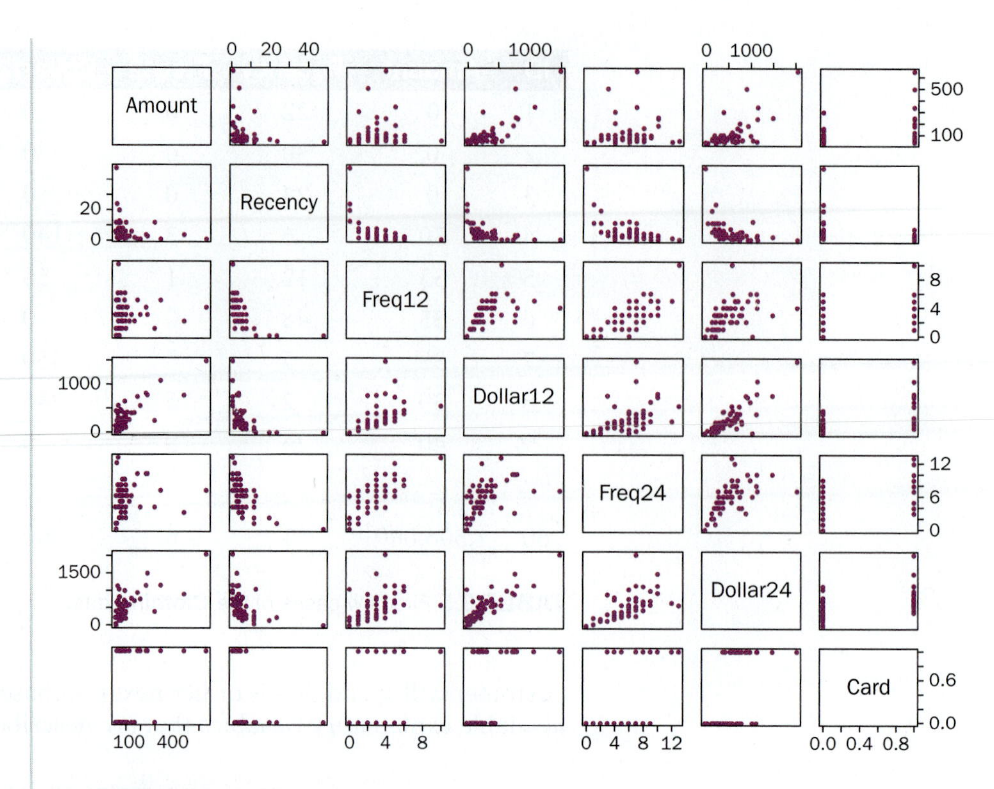

FIGURE 3.33 Matrix of scatterplots for variables in **Clothing**

Figure 3.33 is a scatterplot matrix that shows how *Amount* is related to the six predictor variables. We see that *Amount* is positively associated with all of the predictors except *Recency* and that it has a strong relationship with *Dollar12* and *Dollar24*. We also see multicollinearity, as *Freq12* is positively associated with *Dollar12*, *Freq24*, and *Dollar24*. The strongest association in the scatterplot matrix is between *Dollar12* and *Dollar24*. There are a few points that stand somewhat apart from the bulk of the data, but no points that are obvious outliers demanding special attention.

A matrix of correlation coefficients for the six quantitative variables is shown as follows. As expected, *Amount* is strongly correlated with past spending: $r = 0.804$ with *Dollar12* and $r = 0.677$ with *Dollar24*. However, the matrix also reveals that these explanatory variables are correlated with one another. Since the variables are dollar amounts in overlapping time periods, there is a strong positive association, $r = 0.827$, between *Dollar12* and *Dollar24*.

	Amount	Recency	Freq12	Dollar12	Freq24	Dollar24
Amount	1.000	-0.221	0.052	0.804	0.102	0.677
Recency	-0.221	1.000	-0.584	-0.454	-0.549	-0.432
Freq12	0.052	-0.584	1.000	0.556	0.710	0.421
Dollar12	0.804	-0.454	0.556	1.000	0.485	0.827
Freq24	0.102	-0.549	0.710	0.485	1.000	0.596
Dollar24	0.677	-0.432	0.421	0.827	0.596	1.000

Recency (the number of months since the last purchase) is negatively associated with the purchase *Amount* and with the four explanatory variables that indicate the number of purchases or the amount of those purchases. Perhaps recent customers (low *Recency*) tend to be regular customers

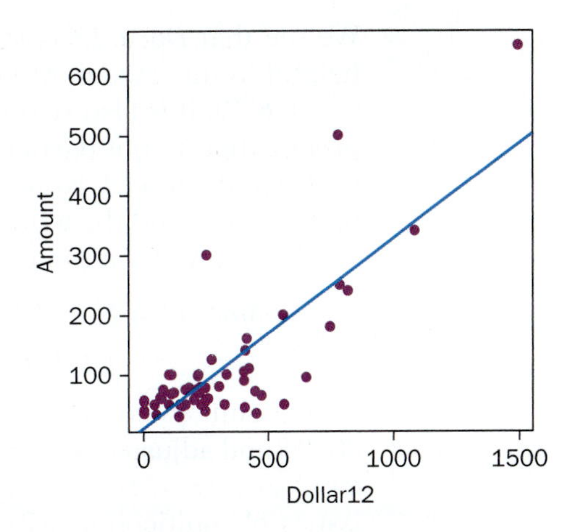

FIGURE 3.34 Regression of *Amount* on *Dollar12*

who visit frequently and spend more, whereas those who have not visited in some time (high *Recency*) include customers who often shop elsewhere.

To start, let's look at a simple linear regression model with the single explanatory variable most highly correlated with *Amount*. The correlations above show that this variable is *Dollar12*.

FIT and ASSESS

Some least squares regression output predicting the purchase *Amount* based on *Dollar12* is shown below and a scatterplot with the least squares line is illustrated in Figure 3.34:

	Estimate	Std. Error	t value	Pr(>|t|)	
(Intercept)	10.0756	13.3783	0.753	0.455	
Dollar12	0.3176	0.0320	9.925	8.93e-14	***

Residual standard error: 67.37 on 54 degrees of freedom
Multiple R-squared: 0.6459, Adjusted R-squared: 0.6393
F-statistic: 98.5 on 1 and 54 DF, *P*-value: 8.93e-14

The simple linear model based on *Dollar12* shows a clear increasing trend with a very small *P*-value for the slope. So this is clearly an effective predictor and accounts for 64.6% of the variability in purchase *Amounts* in this sample. That is a significant amount of variability, but could we do better by including more of the predictors in a multiple regression model?

Of the remaining explanatory variables, *Dollar24* and *Recency* have the strongest associations with the purchase amounts, so let's add them to the model:

	Estimate	Std. Error	t value	Pr(>|t|)	
(Intercept)	-23.05236	21.59290	-1.068	0.2906	
Dollar12	0.32724	0.05678	5.764	4.53e-07	***
Dollar24	0.02151	0.04202	0.512	0.6110	
Recency	2.86718	1.37573	2.084	0.0421	*

Residual standard error: 65.91 on 52 degrees of freedom
Multiple R-squared: 0.6736, Adjusted R-squared: 0.6548
F-statistic: 35.78 on 3 and 52 DF, *P*-value: 1.097e-12

We see that *Dollar12* is still a very strong predictor, but *Dollar24* is not so helpful in this model. Although *Dollar24* is strongly associated with *Amount* ($r = 0.677$), it is also strongly related to *Dollar12* ($r = 0.827$), so it is not surprising that it's not particularly helpful in explaining *Amount* when *Dollar12* is also in the model. Note that the R^2 value for this model has improved a bit, up to 67.4%, and the adjusted R^2 has also increased from 63.9% to 65.5%.

What if we try a model with all six of the available predictors?

$$Amount = \beta_0 + \beta_1 Dollar12 + \beta_2 Freq12 + \beta_3 Dollar24 + \beta_4 Freq24 + \beta_5 Recency + \beta_6 Card + \epsilon$$

We omit the output for this model here, but it shows that R^2 jumps to 88.2% and adjusted R^2 is 86.8%. However, only two of the six predictors (*Dollar12* and *Freq12*) have significant individual t-tests. Again, we can expect issues of multicollinearity to be present when all six predictors are in the model.

Since adding in all six predictors helped improve the R^2 and we are primarily interested in predicting *Amount*, perhaps we should consider adding quadratic terms for each of the quantitative predictors (note that $Card^2$ is the same as *Card* since the only values are 0 and 1). This 11-predictor model increases the R^2 to 89.3%, but the adjusted R^2 actually drops to 86.6%. Clearly, we have gone much too far in constructing an overly complicated model. Also, the output for this 11-predictor model shows once again that only the coefficients of *Dollar12* and *Freq12* are significant at a 5% level.

Since *Dollar12* and *Freq12* seem to be important in most of these models, it makes sense to try a model with just those two predictors:

$$Amount = \beta_0 + \beta_1 Dollar12 + \beta_2 Freq12 + \epsilon$$

Here is some output for fitting that two-predictor model:

```
Coefficients:
             Estimate  Std. Error  t value  Pr(>|t|)
(Intercept)  73.89763   10.46860    7.059   3.62e-09 ***
Dollar12      0.44315    0.02337   18.959    < 2e-16 ***
Freq12      -34.42587    3.56139   -9.666   2.72e-13 ***

Residual standard error: 40.91 on 53 degrees of freedom
Multiple R-squared: 0.8718, Adjusted R-squared: 0.867
F-statistic: 180.3 on 2 and 53 DF, P-value: < 2.2e-16
```

The R^2 drops by only 1% compared to the six-predictor model and the adjusted R^2 is essentially the same. The tests for each coefficient and the ANOVA for overall fit all have extremely small P-values, so evidence exists that these are both useful predictors and, together, they explain a substantial portion of the variability in the purchase amounts for this sample of customers. Although we should also check the conditions on residuals, if we were restricted to models using just these six individual predictors, the two-predictor model based on *Dollar12* and *Freq12* would appear to be a good choice to balance simplicity and explanatory power.

CHOOSE (again)

We have used the explanatory variables that were given to us by the clothing retailer manager and come up with a reasonable two-predictor model for explaining purchase amounts. However, we should think carefully about the data and our objective. Our response variable, *Amount*, measures the spending of a customer on an individual visit to the store. To predict this quantity,

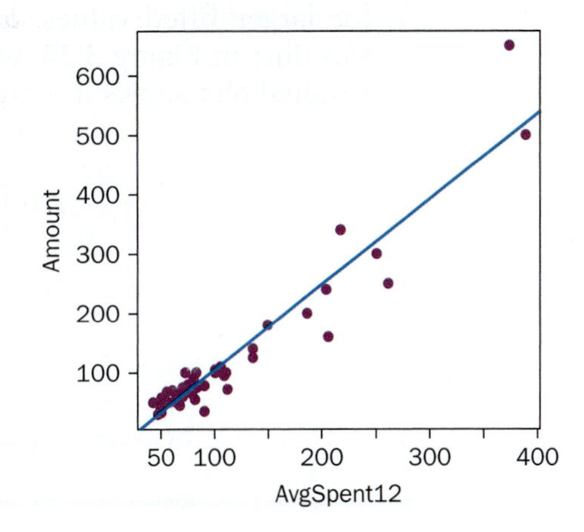

FIGURE 3.35 *Amount* versus *AvgSpent12* with regression line

the "typical" or average purchase for that customer over a recent time period might be helpful. We have the total spending and frequency of purchases over 12 months, so we can create a new variable, *AvgSpent12*, to measure the average amount spent on each visit over the past 12 months:

$$AvgSpent12 = \frac{Dollars12}{Freq12}$$

Unfortunately, four cases had no record of any sales in the past 12 months ($Freq12 = 0$), so we need to drop those cases from the analysis as we proceed. This leaves us with $n = 52$ cases.

FIT

We compute values for *AvgSpent12* for every customer in the reduced sample and try a simple linear regression model for *Amount* using this predictor. The results for this model are shown below, and a scatterplot with regression line appears in Figure 3.35:

```
Coefficients:
              Estimate  Std. Error  t value   Pr(>|t|)
 (Intercept)  -38.8254      8.3438   -4.653   2.43e-05 ***
 AvgSpent12     1.4368      0.0642   22.380    < 2e-16 ***

Residual standard error: 35.02 on 50 degrees of freedom
Multiple R-squared: 0.9092, Adjusted R-squared: 0.9074
F-statistic: 500.9 on 1 and 50 DF, P-value: < 2.2e-16
```

This looks like a fairly good fit for a single predictor and the R^2 value of 90.9% is even better than $R^2 = 90.2\%$ when the 11-term quadratic model using all of the original variables is applied to the reduced sample with $n = 52$ cases!

ASSESS

The linear trend looks clear in Figure 3.35, but we should still take a look at the residuals to see if the regression conditions are reasonable. Figure 3.36 shows a plot of the residuals versus fitted values for the simple linear regression of *Amount* on *AvgSpent12*. In this plot, we see a bit of curvature with positive residuals for most of the small fitted values, then mostly negative residuals for the middle fitted values with a couple of large positive residuals for larger fitted values. Looking back at the scatterplot with the regression line in Figure 3.35, we can also see this slight curvature, although the residual plot shows it more clearly:

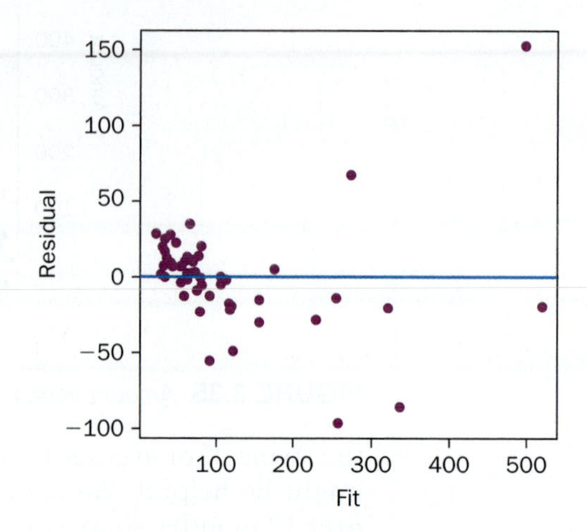

FIGURE 3.36 Residuals versus fits for the regression of *Amount* on *AvgSpent12*

This suggests adding a quadratic term $AvgSpent12Sq = (AvgSpent12)^2$ to the model. The output for fitting this model is shown below:

```
               Estimate   Std. Error  t value      Pr(>|t|)
(Intercept)   1.402e+01   1.457e+01     0.963   0.340464
AvgSpent12    5.709e-01   2.145e-01     2.661   0.010498 *
AvgSpent12Sq  2.289e-03   5.477e-04     4.180   0.000120 ***

Residual standard error: 30.37 on 49 degrees of freedom
Multiple R-squared: 0.9331, Adjusted R-squared: 0.9304
F-statistic: 341.7 on 2 and 49 DF, P-value: < 2.2e-16
```

The *t*-test for the new quadratic term shows it is valuable to include in the model. The values for R^2 (93.3%) and adjusted R^2 (93.0%) are both improvements over any model we have considered so far. We should take care when making direct comparisons to the earlier models since the models based on *AvgSpent12* were fit using a reduced sample that eliminated the cases for which $Freq12 = 0$. Perhaps those cases were unusual in other ways and the earlier models would also look better when fit to the smaller sample. However, a quick check of the model using *Dollar12* and *Freq12* with the reduced sample shows that the R^2 value is virtually unchanged at 87.5%.

The plot of residuals versus fitted values in Figure 3.37(a) shows that the issue with curvature has been addressed, as the residuals appear to be more randomly scattered above and below the zero line. We may still have some problem with the equal variance condition as the residuals tend to be larger when the fitted values are larger. We might expect it to be more difficult to predict the purchase amounts for "big spenders" than more typical shoppers. The normal plot in Figure 3.37(b) also shows some departures from the residuals following a normal distribution. Again, the issue appears to be with those few large residuals at either end of the distribution that

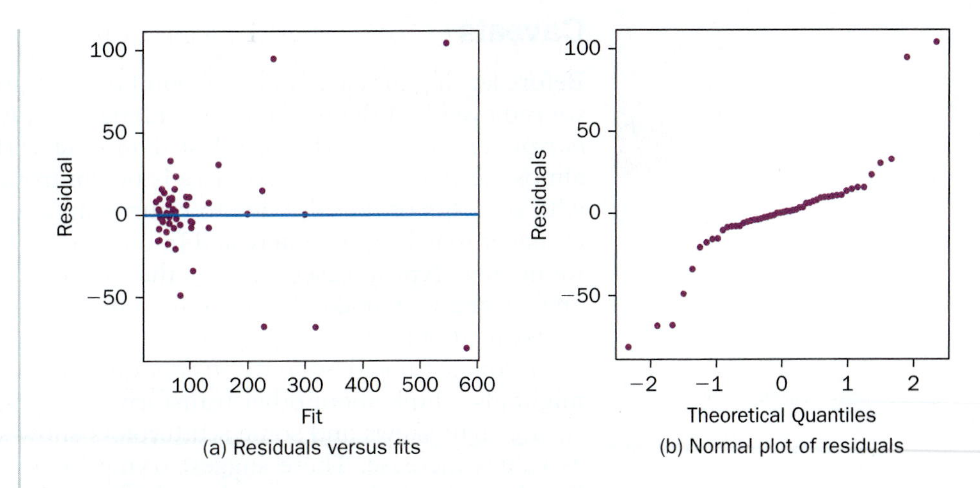

FIGURE 3.37 Residuals plots for quadratic model to predict *Amount* based on *AvgSpent12*

produce somewhat longer tails than would be expected based on the rest of the residuals.

Remember that no model is perfect! While we might have some suspicions about the applicability of the equal variance and normality conditions, the tests that are based on those conditions, especially the t-test for the coefficient of the quadratic term and overall ANOVA F-test, show very small P-values in this situation; moreover, we have a fairly large sample size ($n = 52$).

USE

We can be reasonably confident in recommending that the manager use the quadratic model based on average spending per visit over the past 12 months (*AvgSpent12*) to predict spending amounts for individual customers who have shopped there at least once in the past year. We might also recommend further sampling or study to deal with new or infrequent customers who haven't made any purchases in the past year ($Freq12 = 0$). Our final fitted model is given below and shown on a scatterplot of the data in Figure 3.38:

$$\widehat{Amount} = 14.02 + 0.5709 AvgSpent12 + 0.002289 (AvgSpent12)^2$$

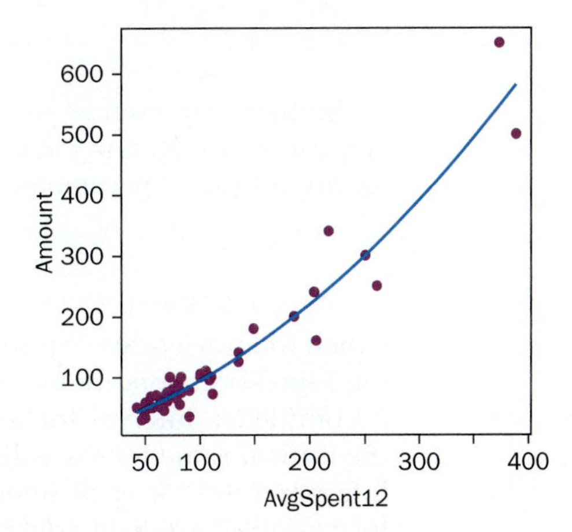

FIGURE 3.38 Quadratic regression fit of *Amount* on *AvgSpent12*

Caveats

Before leaving the case study, we point out two areas of possible concern. First, we removed 8 of the 60 data cases along the way. *Excluding cases may bias or narrow an analysis.* The very last data case with the absurdly large value is almost certainly an error, so there is no controversy in excluding it. The cases with no visits or no sales due to returns may be more problematic. The store manager may be quite interested in turning such cases into paying customers. Removing atypical cases restricts the range of customers to which the model applies and may make the fit to the remaining cases give an overly optimistic assessment of its effectiveness.

Second, although we did consider quadratic terms to address curvature, we might also think about other transformations. Several of the variables exhibit strong right skews and some scatterplots show signs of increasing variability as values increase. These suggest trying log or square root transformations. For this case study, we are primarily focused on the process of selecting predictors, so we have chosen to skip over some issues that would complicate the discussion. You are welcome to give them a try on your own and see if you can improve the model.

CHAPTER SUMMARY

In this chapter, we introduced the **multiple regression model**, where two or more predictor variables are used to explain the variability in a response variable:

$$Y = \beta_0 + \beta_1 X_1 + \beta_2 X_2 + \cdots + \beta_k X_k + \epsilon$$

where $\epsilon \sim N(0, \sigma_\epsilon)$ and the errors are independent from one another.

The procedures and conditions for making inferences are very similar to those used for simple linear models in Chapters 1 and 2. The primary difference is that we have more parameters ($k+2$) to estimate: Each of the k predictor variables has a coefficient parameter, plus we have the intercept parameter and the standard deviation of the errors. As in the case with a single predictor, the coefficients are estimated by minimizing the sum of the squared residuals. The computations are more tedious, but the main idea of using least squares to obtain the best estimates is the same. One of the major differences is that the degrees of freedom for error are now $n - k - 1$, instead of $n - 2$, so the **multiple standard error of regression** is

$$\hat{\sigma}_\epsilon = \sqrt{\frac{SSE}{n - k - 1}}$$

Statistical inferences for the coefficient parameters are based on the parameter estimates, $\hat{\beta}_i$, and standard errors of the estimates, $SE_{\hat{\beta}_i}$. The test statistic for an individual parameter is

$$t = \frac{\hat{\beta}_i}{SE_{\hat{\beta}_i}}$$

which follows a t-distribution with $n - k - 1$ degrees of freedom, provided that the regression conditions, including normality of the errors, are reasonable. A confidence interval for an individual parameter is $\hat{\beta}_i \pm t^* SE_{\hat{\beta}_i}$, where t^* is the critical value for the t-distribution with $n - k - 1$ degrees of freedom. The interpretation for each interval assumes that the values of the other predictor variables are held constant and the individual t-tests assess how much a predictor contributes to a model after accounting for the other predictors in the model. In short, the individual t-tests or intervals look at the effect of each predictor variable, one-by-one in the presence of others.

Partitioning the total variability into two parts, one for the multiple regression model (*SSModel*) and another for error (*SSE*), allows us to test the effectiveness of all of the predictor variables as a group. The **multiple regression ANOVA** table provides the formal inference, based on an F-distribution with k numerator degrees of freedom and $n - k - 1$ denominator degrees of freedom.

The **coefficient of multiple determination**, R^2, provides a statistic that is useful for measuring the effectiveness of a model: the amount of total variability in the response (*SSTotal*) that is explained by the model (*SSModel*). In the multiple regression setting, this coefficient can also be obtained by squaring the correlation between the fitted values (based on all k predictors) and the response variable. Making a slight adjustment to this coefficient to account for the fact that adding new predictors will never decrease the variability explained by the model, we obtain an **adjusted** R^2 that is extremely useful when comparing different multiple regression models based on different numbers of predictors.

After checking the conditions for the multiple regression model, as you have done in the previous chapters, you should be able to obtain and interpret a confidence interval for a mean response and a prediction interval for an individual response.

One of the extensions of the simple linear regression model is the multiple regression model for two regression lines. This extension can be expanded to compare three or more simple linear regression lines. In fact, the point of many of the examples in this chapter is that multiple regression models are extremely flexible and cover a wide range of possibilities. Using **indicator variables**, interaction terms, squared variables, and other combinations of variables produces an incredible set of possible models for us to consider. The regression model with **interaction**, **quadratic regression**, a **complete second-order model**, and **polynomial regression** were used in a variety of different examples. You should be able to choose, fit, assess, and use these multiple regression models.

When considering and comparing various multiple regression models, we must use caution, especially when the predictor variables are associated with one another. Correlated predictor variables often create **multicollinearity** problems. These multicollinearity problems are usually most noticeable when they produce counterintuitive parameter estimates or tests. An apparent contradiction between the overall F-test and the individual t-tests is also a warning sign for multicollinearity problems. The easiest way to check for association in your predictor variables is to look at pairwise scatterplots of the variables and to create a correlation matrix of pairwise correlation coefficients. Another way to detect multicollinearity is to compute the **variance inflation factor** (VIF), a statistic that measures the association between a predictor and all of the other predictors. As a rule of thumb, a VIF larger than 5 indicates multicollinearity. Unfortunately, there is no hard and fast rule for dealing with multicollinearity.[8] Dropping some predictors and combining predictors to form another variable may be helpful.

The **nested *F*-test** is a way to assess the importance of a subset of predictors. The general idea is to fit a full model (one with all of the predictor variables under consideration) and then fit a reduced model (one without a subset of predictors). The nested F-statistic provides a formal way to assess if the full model does a significantly better job explaining the variability in the response variable than the reduced model. In short, the nested F-test sits between the overall ANOVA test for all of the regression coefficients and the individual t-tests for each coefficient and you should be able to apply this method for comparing models.

Finally, the case study for the clothing retailer provides a valuable modeling lesson. Think about your data and the problem at hand before blindly applying complicated models with lots of predictor variables. Keeping the model as simple and intuitive as possible is appealing for many reasons.

EXERCISES

Conceptual Exercises

3.1 Predicting a statistics final exam grade. A statistics professor assigned various grades during the semester, including a midterm exam (out of 100 points) and a logistic regression project (out of 30 points). The prediction equation below was fit, using data from 24 students in the class, to predict the final exam score (out of 100 points) based on the midterm and project grades:

$$\widehat{Final} = 11.0 + 0.53\ Midterm + 1.20\ Project$$

a. What would this tell you about a student who got perfect scores on the midterm and project?

b. Michael got a grade of 87 on his midterm, 21 on the project, and an 80 on the final. Compute his residual and write a sentence to explain what that value means in Michael's case.

3.2 Breakfast cereals. A regression model was fit to a sample of breakfast cereals. The response variable Y is calories per serving. The predictor variables are X_1, grams of sugar per serving, and X_2, grams of fiber per serving. The fitted regression model is

$$\widehat{Calories} = 109.3 + 1.0\ Sugar - 3.7\ Fiber$$

a. How many calories would you predict for a breakfast cereal that had 1 gram of fiber and 11 grams of sugar per serving?

b. Frosted Flakes is a breakfast cereal that has 1 gram of fiber and 11 grams of sugar per serving. It also has 110 calories per serving. Compute the residual for Frosted Flakes and explain what this value means.

3.3 Predicting a statistics final exam grade: understanding estimates. Does the prediction equation for final exam scores in Exercise 3.1 suggest that the project score has a stronger relationship with the final exam than the midterm exam? Explain why or why not.

3.4 Breakfast cereals: understanding estimates. Does the prediction equation for number of calories per serving in Exercise 3.2 suggest that the amount of sugar has a weaker relationship with the number of calories than the amount of fiber? Explain why or why not.

3.5 Predicting a statistics final exam grade: interpreting estimates. In Exercise 3.1 the prediction equation below was fit, using data from 24 students in the class, to predict the final exam score (out of 100 points) based on the midterm and project grades:

$$\widehat{Final} = 11.0 + 0.53\ Midterm + 1.20\ Project$$

In the context of this setting, interpret 1.20, the coefficient of *Project*. That is, describe how the project grade is related to the final grade, in the presence of the midterm score.

3.6 Breakfast cereals: interpreting estimates. In Exercise 3.2 a regression model was fit to a sample of breakfast cereals. The response variable Y was calories per serving. The predictor variables were X_1, grams of sugar per serving, and X_2, grams of fiber per serving. The fitted regression model was

$$\widehat{Calories} = 109.3 + 1.0\ Sugar - 3.7\ Fiber$$

In the context of this setting, interpret −3.7, the coefficient of X_2. That is, describe how fiber is related to calories per serving, in the presence of the sugar variable.

3.7 Adjusting R^2. Decide if the following statements are true or false, and explain why:

a. For a multiple regression problem, the adjusted coefficient of determination, $R^2{}_{adj}$, will always be smaller than the regular, unadjusted R^2.

b. If we fit a multiple regression model and then add a new predictor to the model, the adjusted coefficient of determination, $R^2{}_{adj}$, will always increase.

3.8 More adjusting R^2. Decide if the following statements are true or false, and explain why:

a. If we fit a multiple regression model and then add a new predictor to the model, the (unadjusted) R^2 will never decrease.

b. Suppose we have two possible predictors, X_1 and X_2, and we use them to fit two separate regression models: Model1 uses only X_1 and yields R_1^2 while Model2 uses only X_2 and yields R_2^2. Then if we fit the multiple regression model that uses both X_1 and X_2, the adjusted coefficient of determination, R^2(adj), will always be greater than the larger of the R_1^2 and R_2^2.

3.9 Body measurements. Suppose that you are interested in predicting the percentage of body fat (*BodyFat*) on a man using the explanatory variables waist size (*Waist*) and *Height*.

a. Do you think that *BodyFat* and *Waist* are positively correlated? Explain.

b. For a fixed waist size (say, 38 inches), would you expect *BodyFat* to be positively or negatively correlated with a man's *Height*? Explain why.

c. Suppose that *Height* does not tell you much about *BodyFat* by itself, so that the correlation between the two variables is near zero. What sort of coefficient on *Height* (positive, negative, or near zero) would you expect to see in a multiple regression to predict *BodyFat* based on both *Height* and *Waist*? Explain your choice.

3.10 Car correlations. Suppose that you are interested in predicting the price of used cars using the explanatory variables current mileage (*Mileage*) and year of manufacture (*Year*).

a. Do you think that *Year* and *Mileage* are positively or negatively correlated? Explain.

b. For a fixed manufacture year, would you expect *Mileage* to be positively or negatively correlated with *Price*?

3.11 Modeling prices to buy a car. An information technology specialist used an interesting method for negotiating prices with used car sales representatives. He collected data from the entire state for the model of car that he was interested in purchasing. Then he approached the salesmen at dealerships based on the residuals of their prices in his regression model.

a. Should he pick dealerships that tend to have positive or negative residuals? Why?

b. Write down a two-predictor regression model that would use just the *Year* of the car and its *Mileage* to predict *Price*.

c. Why might we want to add an interaction term for *Year* · *Mileage* to the model? Would you expect the coefficient of the interaction to be positive or negative? Explain.

3.12 Free growth period. Caterpillars go through free growth periods during each stage of their life. However, these periods end as the animal prepares to molt and then moves into the next stage of life. A biologist is interested in checking to see whether two different regression lines are needed to model the relationship between metabolic rates and body size of caterpillars for free growth and no free growth periods.

a. Identify the multiple regression model for predicting metabolic rate (*Mrate*) from size (*BodySize*) and an indicator variable for free growth ($Ifgp = 1$ for free growth, 0 otherwise) that would allow for two different regression lines (slopes and/or intercepts) depending on free growth status.

b. Identify the multiple regression model for predicting *Mrate* from *BodySize* and *Ifgp*, when the rate of change in the mean *Mrate* with respect to size is the same during free growth and no free growth periods.

c. Identify the full and reduced models that would be used in a nested *F*-test to check if one or two regression lines are needed to model metabolic rates.

3.13 Models for well water. An environmental expert is interested in modeling the concentration of various chemicals in well water over time. Identify the regression models that would be used to:

a. Predict the amount of arsenic (*Arsenic*) in a well based on *Year*, the distance (*Miles*) from a mining site, and the interaction of these two variables.

b. Predict the amount of lead (*Lead*) in a well based on *Year* with two different lines depending on whether or not the well has been cleaned (*Iclean*).

c. Predict the amount of titanium (*Titanium*) in a well based on a possible quadratic relationship with the distance (*Miles*) from a mining site.

d. Predict the amount of sulfide (*Sulfide*) in a well based on *Year*, distance (*Miles*) from a mining site, depth (*Depth*) of the well, and any interactions of pairs of explanatory variables.

3.14 Degrees of freedom for free growth period. Suppose the biologist in Exercise 3.12 gives you data on 53 caterpillars. Identify the degrees of freedom for error in each of the models in Exercise 3.12 parts (a) and (b).

3.15 Degrees of freedom for well water models. Suppose that the environmental expert in Exercise 3.13 gives you data from 198 wells. Identify the degrees of freedom for error in each of the models from Exercise 3.13.

3.16 Predicting faculty salaries. A dean at a small liberal arts college is interested in fitting a multiple regression model to try to predict salaries for faculty members. If the residuals are unusually large for any individual faculty member, then adjustments in the person's annual salary are considered.

a. Identify the model for predicting *Salary* from age of the faculty member (*Age*), years of experience (*Seniority*), number of publications (*Pub*), and an indicator variable for gender (*IGender*). The dean wants this initial model to include only pairwise interaction terms.

b. Do you think that *Age* and *Seniority* will be correlated? Explain.

c. Do you think that *Seniority* and *Pub* will be correlated? Explain.

d. Do you think that the dean will be happy if the coefficient for *IGender* is significantly different from zero? Explain.

Guided Exercises

3.17 Active pulse rates. The following computer output comes from a study to model *Active* pulse rates (after climbing several flights of stairs) based on resting pulse rate (*Rest* in beats per minute), weight (*Wgt* in pounds), and amount of *Exercise* (in hours per week). The data were obtained from 232 students taking Stat2 courses in past semesters.

```
The regression equation is Active = 11.8 + 1.12 Rest +
0.0342 Wgt - 1.09 Exercise
  Predictor      Coef    SE Coef      T       P
  Constant      11.84      11.95   0.99   0.323
  Rest         1.1194     0.1192   9.39   0.000
  Wgt         0.03420    0.03173   1.08   0.282
  Exercise     -1.085      1.600  -0.68   0.498
S = 15.0452    R-Sq = 36.9%    R-Sq(adj) = 36.1%
```

a. Test the hypotheses that $\beta_2 = 0$ versus $\beta_2 \neq 0$ and interpret the result in the context of this problem. You may assume that the conditions for a linear model are satisfied for these data.

b. Construct and interpret a 90% confidence interval for the coefficient β_2 in this model.

c. What active pulse rate would this model predict for a 200-pound student who exercises 7 hours per week and has a resting pulse rate of 76 beats per minute?

3.18 Real estate near Rails-to-Trails. Consider the **RailsTrails** data, which gives information about 104 homes sold in Northampton, Massachusetts, in 2007. In this exercise, use *adj2007* (the 2007 selling price in adjusted 2014 thousands of dollars) as the response variable and *distance* (distance from a bike trail) and *squarefeet* (size of home) as two explanatory variables. The primary interest of the researchers was to ascertain the selling price impact of being close to a bike trail. By adding squarefeet to the model, we hope to address this impact, controlling for the size of the home. **Rails**

a. First fit the simple linear regression model of *adj2007* on *distance*.

b. Next, fit a regression model with both explanatory variables. Has the addition of *squarefeet* changed our estimate of the *distance* relationship to *adj2007*? Use both the estimated coefficient and R^2 in your answer.

c. Find 95% confidence intervals for the distance coefficient from each model and compare them.

d. Using the two-predictor model, give a prediction for the price of a particular home that goes on the market and that has 1500 square feet of space and is 0.5 mile from a bike trail.

3.19 Racial Animus. Consider the data described in Exercise 2.14. The file **RacialAnimus** has data for each of the 196 media markets on *Animus* and four predictor variables: *Black*, which is the percentage of African Americans living in the media market, *Hispanic*, which is the percentage of Hispanics living in the media market, *BachPlus*, which is the percentage of persons in the media market who have at least a bachelor's degree, and *Age65Plus*, which is the percentage of persons living in the media market who are at least 65 years old. **RacAnim**

a. Fit a multiple regression model with *Animus* depending on the four predictors. Report the fitted prediction equation.

b. Prepare appropriate residual plots and comment on the conditions for inference.

3.20 Enrollments in mathematics courses: FIT/ASSESS. In Exercise 2.28 on page 80, we consider a model to predict spring enrollment in mathematics courses based on the fall enrollment. The residuals for that model showed a pattern of growing over the years in the data. Thus it might be beneficial to add the academic year variable *AYear* to our model and fit a multiple regression. The data are provided in the file **MathEnrollment**. **MthEnr**

a. Fit a multiple regression model for predicting spring enrollment (*Spring*) from fall enrollment (*Fall*) and academic year (*AYear*), after removing the data from 2003 that had special circumstances. Report the fitted prediction equation.

b. Prepare appropriate residual plots and comment on the conditions for inference. Did the slight problems with the residual plots (e.g., increasing residuals over time) that we noticed for the simple linear model disappear?

3.21 Enrollments in mathematics courses: ASSESS. Refer to the model in Exercise 3.20 to predict *Spring* mathematics enrollments with a two-predictor model based on *Fall* enrollments and academic year (*AYear*) for the data in **MathEnrollment**. **MthEnr**

a. What percent of the variability in spring enrollment is explained by the multiple regression model based on fall enrollment and academic year?

b. What is the size of the typical error for this multiple regression model?

c. Provide the ANOVA table for partitioning the total variability in spring enrollment based on this model and interpret the associated *F*-test.

d. Are the regression coefficients for *both* explanatory variables significantly different from zero? Provide appropriate hypotheses, test statistics, and *P*-values in order to make your conclusion.

3.22 Breakfast cereal: ASSESS/USE. The regression model in Exercise 3.6 was fit to a sample of 36 breakfast cereals with calories per serving as the response variable. The two predictors were grams of sugar per serving and grams of fiber per serving. The partition of the sums of squares for this model is

$$\begin{aligned} SSTotal &= SSModel + SSE \\ 17190 &= 9350 + 7840 \end{aligned}$$

a. Calculate R^2 for this model and interpret the value in the context of this setting.

b. Calculate the regression standard error of this multiple regression model.

c. Calculate the *F*-ratio for testing the null hypothesis that neither sugar nor fiber is related to the calorie content of cereals.

d. Assuming the regression conditions hold, the *P*-value for the *F*-ratio in (c) is about 0.000002. Interpret what this tells you about the variables in this situation.

3.23 Brain pH. In Exercise 2.13 we introduced a study of brain tissue. The data are in the file **BrainpH**. Are *Age* and *Sex* a good combination of predictors of *pH*? **BrainpH.**

a. Make a scatterplot of $Y = pH$ versus $X = Age$ and with *Sex* as a grouping variable (i.e., use different colors or different plotting symbols for the two levels of *Sex*). Comment on the plot.

b. Fit the regression of *pH* on *Age* and test whether there is a linear association between the two variables.

c. Fit a model that produces parallel regression lines for the two levels of *Sex*. Write down the fitted prediction equation for each level of *Sex*.

3.24 Dementia. Two types of dementia are Dementia with Lewy Bodies and Alzheimer's disease. Some people are afflicted with both of these. The file **LewyBody2Groups** includes the variable *Type*, which has two levels: "DLB/AD" for the 20 subjects with both types of dementia and "DLB" for the 19 subjects with only Lewy Body dementia. The variable *APC* gives annualized percentage change in brain gray matter. The variable *MMSE* measures change in functional performance on the Mini Mental State Examination. **LewyB2**

a. Make a scatterplot of $Y = MMSE$ versus $X = APC$ and with *Type* as a grouping variable (i.e., use different colors or different plotting symbols for the two levels of Type). Comment on the plot.

b. Fit the regression of *MMSE* on *APC* and test whether there is a linear association between the two variables.

c. Fit a model that produces parallel regression lines for the two levels of *Type*. Write down the fitted prediction equation for each level of *Type*.

3.25 2016 Democratic nomination for president. In 2016 Hillary Clinton won the Democratic nomination for president over Bernie Sanders. A paper was circulated that claimed to show evidence of election fraud based, among other things, on Clinton doing better in states that don't have a paper trail for votes cast in a primary election than she did in states that have a paper trail. The file **ClintonSanders** has data from that paper for the 31 states that held primaries before June. The variable *Delegates* gives the percentage of delegates won by Clinton for each state. The variable *AfAmPercent* gives the percentage of residents in the state who are African American. *PaperTrail* indicates whether or not the voting system in the state includes a paper trail. **ClinSan**

a. Conduct a regression of *Delegates* on *PaperTrail*. What does this regression say about how Clinton did in states with and without a paper trail?

b. Conduct a regression of *Delegates* on *PaperTrail* and *AfAmPercent*. What does this regression say about how Clinton did in states with and without a paper trail? What is the effect of *AfAmPercent*?

c. Repeat parts (a) and (b) but in place of *Delegates* as the response variable, use *PopularVote*, which is the percentage of the popular vote that Clinton received.

3.26 Tadpoles. Biologists wondered whether tadpoles can adjust the relative length of their intestines if they are exposed to a fungus called *Batrachochytrium dendrobatidis (Bd)*. The file **Tadpoles** includes the variable *Treatment*, which has two levels: "Bd" for 14 exposed tadpoles and "Control" for 13 control tadpoles. The variable *GutLength* is the length (in mm) of the intestinal tract. The variable *Body* is the length (in mm) of the body of the tadpole. The variable *MouthpartDamage* is a measure of damage to the mouth (missing teeth, etc.) and is expected to have a positive association with *GutLength*, because more damage means reduced food intake, which means the tadpole needs to increase nutrient absorption. The biological question is whether *GutLength* will also have a positive association with *Treatment*; the predicted answer is "Yes." **Tad**

a. Make a scatterplot of $Y = GutLength$ versus $X = Body$ and with *Treatment* as a grouping variable (i.e., use different colors or different plotting symbols for the two levels of *Treatment*). Comment on the plot.

b. Fit the regression of *GutLength* on *Body* and test whether there is a linear association between the two variables.

c. Fit a model that produces parallel regression lines for the two levels of *Treatment*. Write down the fitted prediction equation for each level of *Treatment*.

d. Expand the model from part (c) by adding *MouthpartDamage* as a predictor. How does this fitted model relate to the question that the biologists had?

3.27 British trade unions: parallel lines. The British polling company Ipsos MORI conducted several opinion polls in the United Kingdom between 1975 and 1995 in which it asked people whether they agreed or disagreed with the statement, "Trade unions have too much power in Britain today." Table 3.6 shows the dates of the polls; the agree and disagree percentages; the *NetSupport* for unions that is defined *DisagreePct* – *AgreePct*; the number of months after August 1975 that the poll was conducted; and a variable, *Late*, that indicates whether an observation is from the early (0) or late (1) part of the period spanned by the data. The last variable is the unemployment rate in the United Kingdom for the month of each poll. The data are also stored in **BritishUnions**. **BritU**

November 2002–2004. A goal of the study was to investigate the fertility of fish that had been stocked in the lake. One measure of the viability of fish eggs is *percent dry mass(PctDM)*, which reflects the energy potential stored in the eggs by recording the percentage of the total egg material that is solid. Values of the *PctDM* for a sample of 35 lake trout (14 in September and 21 in November) are given in Table 3.7 along with the age (in years) of the fish. The data are stored in three columns in a file called **FishEggs**. **FshEgg**

Table 3.7 Percent dry mass of eggs and age for female lake trout

September

Age	7	7	7	7	9	9	11
PctDM	34.90	37.00	37.90	38.15	33.90	36.45	35.00
Age	11	12	12	12	16	17	18
PctDM	36.15	34.05	34.65	35.40	32.45	36.55	34.00

November

Age	7	8	8	9	9	9	9
PctDM	34.90	37.00	37.90	38.15	33.90	36.45	35.00
Age	10	10	11	11	12	12	13
PctDM	36.15	34.05	34.65	35.40	32.45	36.55	34.00
Age	13	13	14	15	16	17	18
PctDM	36.15	34.05	34.65	35.40	32.45	36.55	34.00

Ignore the month at first and fit a simple linear regression to predict the *PctDM* based on the *Age* of the fish.

a. Write down an equation for the least squares line and comment on what it appears to indicate about the relationship between *PctDM* and *Age*.

b. What percentage of the variability in *PctDM* does *Age* explain for these fish?

c. Is there evidence that the relationship in (a) is statistically significant? Explain how you know that it is or is not.

d. Produce a plot of the residuals versus the fits for the simple linear model. Does there appear to be any regular pattern?

e. Modify your plot in (d) to show the points for each *Month* (Sept/Nov) with different symbols or colors. What (if anything) do you observe about how the residuals might be related to the month? Now fit a multiple regression model, using an indicator (*Sept*) for the month and interaction product, to compare the regression lines for September and November.

f. Do you need both terms for a difference in intercepts and slopes? If not, delete any terms that aren't needed and run the new model.

g. What percentage of the variability in *PctDM* does the model in (f) explain for these fish?

h. Redo the plot in (e) showing the residuals versus fits for the model in (f) with different symbols for the months. Does this plot show an improvement over your plot in (e)? Explain why.

3.35 Elephants: firstborn. If you are an elephant, you are worse off if there is a drought during the first two years of your life. But does maternal experience matter? That is, is it helpful or harmful if you are firstborn? The file **ElephantsFB** has data on 138 male African elephants that lived through droughts in the first two years of life.[11] The variable *Height* records shoulder height in cm, *Age* is age in years, and *Firstborn* is 1 for firstborn and 0 for not-firstborn. **ElephFB**

a. Plot *Height* against *Age* and comment on the pattern.

b. What is the fitted quadratic regression model for using *Age* to predict *Height*?

c. Use the fitted model from part (b) to predict the height of a 15-year-old elephant.

3.36 Elephants, gender. How does growth of elephants differ between males and females? The file **ElephantsMF** has data on 288 African elephants that lived through droughts in the first two years of life.[12] The variable *Height* records shoulder height in cm, *Age* is age in years, and *Sex* is M for males and F for females. **ElephMF**

a. Plot *Height* against *Age* and comment on the pattern.

b. What is the fitted quadratic regression model for using *Age* to predict *Height*?

c. Use the fitted model from part (b) to predict the height of a 10-year-old elephant.

3.37 Diamond prices: carat only. The dataset found in **Diamonds** has information on several variables for 351 diamonds. The quantitative variables measured are size (*Carat*), price (*PricePerCt* and *TotalPrice*), and the *Depth* of the cut. The categorical variables measured are *Color* and *Clarity*. A young couple are shopping for a diamond and are interested in learning more about how these gems are priced. They have heard about the four C's: carat, color, cut, and clarity. Now they want to see if there is any relationship between these diamond characteristics and the price. **Diamond**

a. Produce a scatterplot with $Y = TotalPrice$ and $X = Carat$. What does this plot suggest to you about a model predicting price from carat size?

b. What is the fitted quadratic model using *Carat* to predict *TotalPrice*? What are the values for R^2 and adjusted R^2?

c. Use plots to check the conditions of the model in part (b). Are the conditions met?

d. Fit the cubic model using *Carat* to predict *TotalPrice*? What are the values for R^2 and adjusted R^2?

e. Use plots to check the conditions of the model in part (d). Are the conditions met?

3.38 Diamond prices: carat and depth. In Exercise 3.37, we looked at quadratic and cubic polynomial models for the price of diamonds (*TotalPrice*) based on the size (*Carat*). Another variable in the **Diamonds** datafile gives the *Depth* of the cut for each stone (as a percentage of the diameter). Run each of the models listed below, keeping track of the values for R^2, adjusted R^2, and which terms (according to the individual t-tests) are important in each model: **Diamond**

a. A quadratic model using *Depth*

b. A two-predictor model using *Carat* and *Depth*

c. A three-predictor model that adds interaction for *Carat* and *Depth*

d. A complete second-order model using *Carat* and *Depth*

Among these four models as well as the quadratic and cubic models you looked at in Exercise 3.37, which would you recommend using for *TotalPrice* of diamonds? Explain your choice.

3.39 Diamond prices: transformation. One of the consistent problems with models for the *TotalPrice* of diamonds in Exercise 3.38 was the lack of a constant variance in the residuals. As often happens, when we try to predict the price of the larger, more expensive diamonds, the variability of the residuals tends to increase. **Diamond**

a. Using the model you chose in Exercise 3.38, produce one or more graphs to examine the conditions for homoscedasticity (constant variance) and normality of its residuals. Do these standard regression conditions appear to be reasonable for your model?

b. Transform the response variable to be *logPrice* as the natural log of the *TotalPrice*. Is your "best" choice of models from Exercise 3.38 still a reasonable choice for predicting *logPrice*? If not, make adjustments to add or delete terms, keeping within the options offered within a complete second-order model.

c. Once you have settled on a model for *logPrice*, produce similar graphs to those you found in (a). Has the log transformation helped with either the constant variance or normality conditions on the residuals?

3.40 Diamond prices: USE. The young couple described in Exercise 3.37 has found a 0.5-carat diamond with a depth of 62% that they are interested in buying. Suppose that you decide to use the quadratic regression model for predicting the *TotalPrice* of the diamond using *Carat*. The data are stored in **Diamonds**. **Diamond**

a. What average total price does the quadratic model predict for a 0.5-carat diamond?

b. Find a 95% confidence interval for the mean total price of 0.5-carat diamonds. Write a sentence interpreting the interval in terms that will make sense to the young couple.

c. Find a 95% prediction interval for the total price when a diamond weighs 0.5 carat. Write a sentence interpreting the interval in terms that will make sense to the young couple.

d. Repeat the previous two intervals for the model found in part (b) of Exercise 3.39, where the response variable was *logPrice*. You should find the intervals for the log scale, but then exponentiate to give answers in terms of *TotalPrice*.

3.41 Fluorescence experiment: quadratic. Exercise 1.38 on page 56 describes a novel approach in a series of experiments to examine calcium-binding proteins. The data from one experiment are provided in **Fluorescence**. The variable *Calcium* is the log of the free calcium concentration and *ProteinProp* is the proportion of protein bound to calcium. **Fluor**

a. Fit a quadratic regression model for predicting *ProteinProp* from *Calcium* (if needed for the software you are using, create a new variable $CalciumSq = Calcium \cdot Calcium$). Write down the fitted regression equation.

b. Add the quadratic curve to a scatterplot of *ProteinProp* versus *Calcium*.

c. Are the conditions for inference reasonably satisfied for this model?

d. Is the parameter for the quadratic term significantly different from zero? Justify your answer.

e. Identify the coefficient of multiple determination and interpret this value.

3.42 Fluorescence experiment: cubic. In Exercise 3.41, we examine a quadratic model to predict the proportion of protein binding to calcium (*ProteinProp*) based on the log free calcium concentration (*Calcium*). Now we will check if a cubic model provides an improvement for describing the data in **Fluorescence**. **Fluor**

a. Fit a cubic regression model for predicting *ProteinProp* from *Calcium* (if needed for the software you are using, create *CalciumSq* and *CalciumCube* variables). Write down the fitted regression equation.

b. Add the cubic curve to a scatterplot of *ProteinProp* versus *Calcium*.

c. Are the conditions for inference reasonably satisfied for this model?

d. Is the parameter for the cubic term significantly different from zero? Justify your answer.

e. Identify the coefficient of multiple determination and interpret this value.

3.43 2008 U.S. presidential polls. The file **Pollster08** contains data from 102 polls that were taken during the 2008 U.S. presidential campaign. These data include all presidential polls reported on the Internet site pollster.com that were taken between August 29, when John McCain announced that Sarah Palin would be his running mate as the Republican nominee for vice president, and the end of September. The variable *MidDate* gives the middle date of the period when the poll was "in the field" (i.e., when the poll was being conducted). The variable *Days* measures the number of days after August 28 (the end of the Democratic convention) that the poll was conducted. The variable *Margin* shows Obama%–McCain% and is a measure of Barack Obama's lead. *Margin* is negative for those polls that showed McCain to be ahead.

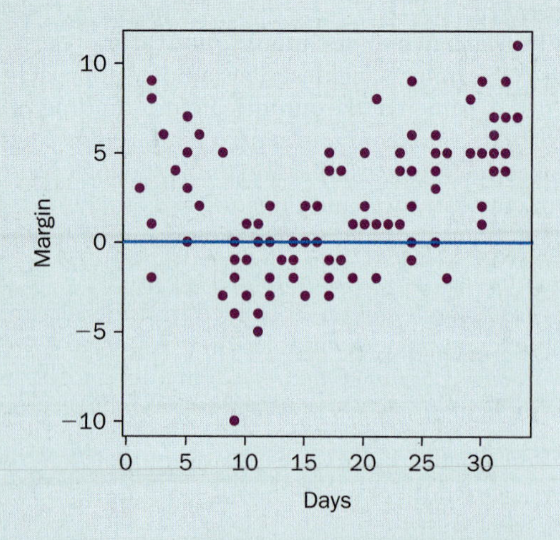

FIGURE 3.39 *Obama–McCain* margin in 2008 presidential polls

The scatterplot in Figure 3.39 of *Margin* versus *Days* shows that Obama's lead dropped during the first part of September but grew during the latter part of September. A quadratic model might explain the data. However, two theories have been advanced as to what caused this pattern, which you will investigate in this exercise.

The **Pollster08** datafile contains a variable *Charlie* that equals 0 if the poll was conducted before the telecast of the first ABC interview of Palin by Charlie Gibson (on September 11) and 1 if the poll was conducted after that telecast. The variable *Meltdown* equals 0 if the poll was conducted before the bankruptcy of Lehman Brothers triggered a meltdown on Wall Street (on September 15) and 1 if the poll was conducted after September 15. **Poll08**

a. Fit a quadratic regression of *Margin* on *Days*. What is the value of R^2 for this fitted model? What is the value of SSE?

b. Fit a regression model in which *Margin* is explained by *Days* with two lines: one line before the September 11 ABC interview (i.e., $Charlie = 0$) and one line after that date ($Charlie = 1$). What is the value of R^2 for this fitted model? What is the value of SSE?

c. Fit a regression model in which *Margin* is explained by *Days* with two lines: one line before the September 15 economic meltdown (i.e., $Meltdown = 0$) and one line after September 15 ($Meltdown = 1$). What is the value of R^2 for this fitted model? What is the value of SSE?

d. Compare your answers to parts (a–c). Which of the three models best explains the data?

3.44 Metropolitan doctors. In Example 1.7, we considered a simple linear model to predict the number of doctors (*MDs*) from the number of hospitals (*Hospitals*) in a metropolitan area. In that example, we found that a square root transformation on the response variable, *Sqrt*(*MDs*), produced a more linear relationship. In this exercise, use this transformed variable, in **CountyHealth**, as the response variable. **CtyHlth**

a. Either the number of hospitals (*Hospitals*) or number of beds in those hospitals (*Beds*) might be good predictors of the number of doctors in a city. Find the correlations between each pair of the three variables, *SqrtMDs*, *Hospitals*, *Beds*. Based on these correlations, which of the two predictors would be a more effective predictor of *SqrtMDs* in a simple linear model by itself?

b. How much of the variability in the *SqrtMDs* values is explained by *Hospitals* alone? How much by *Beds* alone?

c. How much of the variability in the *SqrtMDs* values is explained by using a two-predictor multiple regression model with both *Hospitals* and *Beds*?

d. Based on the two separate simple linear models (or the individual correlations), which of *Hospitals* and/or *Beds* have significant relationship(s) with *SqrtMDs*?

e. Which of these two predictors are important in the multiple regression model? Explain what you use to make this judgment.

3.45 British trade unions: interaction model. Consider the data on opinion polls on the power of British trade unions in Exercise 3.27. **BritU**

a. Create an interaction term between *Months* (since August 1975) and *Late*. Fit the regression model that produces two lines for explaining the *NetSupport*, one each for the early ($Late = 0$) and late ($Late = 1$) parts of the dataset. What is the fitted model?

b. Use a t-test to test the null hypothesis that the interaction term is not needed and parallel regression lines are adequate for describing these data.

c. Use a nested F-test to test the null hypothesis that neither of the terms involving *Late* is needed and a common regression line for both periods is adequate for describing the relationship between *NetSupport* and *Months*.

3.46 Diamond prices: nested *F*-test. Refer to the complete second-order model you found for diamond prices in Exercise 3.38(d). Use a nested F-test to determine whether all of the terms in the model that involve the information on *Depth* could be removed as a group from the model without significantly impairing its effectiveness. **Diamond**

3.47 Driving fatalities and speed limits. In Exercise 3.32, you considered a multiple regression model to compare the regression lines for highway fatalities versus year between years before and after states assumed control of setting speed limits on interstate highways. The data are in the file **Speed** with variables for *FatalityRate*, *Year*, and an indicator variable *StateControl*. **Speed**

a. Use a nested F-test to determine whether there is a significant difference in either the slope and/or the intercept of those two lines.

b. Use a nested F-test to test only for a difference in slopes ($H_0 : \beta_3 = 0$). Compare the results of this test to the t-test for that coefficient in the original regression.

3.48 Real estate near Rails to Trails: nested *F*-test. Consider the **RailsTrails** data, which give information about 104 homes sold in Northampton, Massachusetts, in

2007. In this exercise, use *adj2007* (the 2007 selling price in adjusted 2014 thousands of dollars) as the response variable and *distance* and *garagegroup* as two explanatory variables. The primary interest of the researchers was to ascertain the selling price impact of being close to a bike trail. By adding *garagegroup* to the model, we hope to address the distance-to-trail impact, controlling for whether the home has any garage spaces. The variable *garagegroup* is a binary variable, yes or no, on whether the home has any garage spaces at all. **Rails**

a. Use comparative boxplots and a t-test to decide if having a garage is related to the price of a home. Summarize your results.

b. Fit (again) the simple linear regression model of *adj2007* on *distance* and summarize the impact of distance to a trail on home price.

c. Now fit a model using both *garagegroup* and *distance* as predictors. Summarize the results.

d. Finally, fit a model with an interaction term between *distance* and *garagegroup*. Use the results of this model to estimate the two rates at which housing price decreases as distance increases. Is this difference in rates statistically significant? Explain.

e. Use a nested F-test to ascertain whether the terms involving garage space add significantly to the model of price on distance.

3.49 Real estate near Rails to Trails: transformation. Consider the **RailsTrails** data again, which give information about 104 homes sold in Northampton, Massachusetts, in 2007. In this exercise, use *adj2007* (the 2007 selling price in adjusted 2014 thousands of dollars) as the response variable and *distance* and *garagegroup* as two explanatory variables. The primary interest of the researchers was to ascertain the selling price impact of being close to a bike trail.

In this exercise we will begin with logged variables and we will look at the impact of using the number of full bathrooms on the model. So our response variable will be *logadj2007* and our predictors will be *logdistance*, *logsquarefeet*, and *no_full_baths*. **Rails**

a. Fit the linear model using just these three predictors. Report the model estimates and write a summary.

b. Assess model conditions for part (a) using basic residual plots.

c. Now fit a model that adds each of the three two-way interactions and a three-way interaction term. Summarize the model and compare it to (a).

d. Use a nested F-test to ascertain if the more complex model (b) adds significantly to the simple model (a) with only linear terms. Summarize your findings.

3.50 Elephants: firstborn quadratic. Exercise 3.35 introduced you to the file **ElephantsFB**, which has data on 138 male African elephants that lived through droughts in the first two years of life. The variable *Height* records shoulder height in cm, *Age* is age in years, and *Firstborn* is 1 for firstborn and 0 for not-firstborn. **ElephFB**

a. What is the fitted quadratic regression model for using *Age* to predict *Height*?

b. Now modify the model from part (a) by including an additive effect of *Firstborn*. Is the *Firstborn* effect statistically significant? Interpret the coefficient of *Firstborn* in the context of this setting.

c. Is the quadratic relationship between *Age* and *Height* different for firstborn elephants than for others? Augment the model from part (b) by adding two interaction terms (one for $Age \cdot Firstborn$ and one for $Age^2 \cdot Firstborn$). Use a nested-models F-test to choose between this model and the part (b) model.

3.51 Elephants: gender quadratic. Exercise 3.36 introduced you to the file **ElephantsMF** which has data on 288 African elephants that lived through droughts in the first two years of life. The variable *Height* records shoulder height in cm, *Age* is age in years, and *Sex* is M for males and F for females. **ElephMF**

a. What is the fitted quadratic regression model for using *Age* to predict *Height*?

b. Now modify the model from part (a) by including an additive effect of *Sex*. Is the *Sex* effect statistically significant? Interpret the coefficient of *Sex* in the context of this setting.

c. Is the quadratic relationship between *Age* and *Height* different for male than for female elephants? Augment the model from part (b) by adding two interaction terms (one for $Age \cdot Sex$ and one for $Age^2 \cdot Sex$). Use a nested-models F-test to choose between this model and the part (b) model.

3.52 Dementia: nested F-test. Consider the data on two types of dementia in Exercise 3.24. **LewyB2**

a. Fit an interaction model that produces two regression lines for predicting *MMSE* from *APC*, one for each of the two levels of *Type*. Write down the fitted prediction equation for each level of *Type*.

b. Use a t-test to test the null hypothesis that the interaction term is not needed and parallel regression lines are adequate.

c. Use a nested F-test to test the null hypothesis that neither of the terms involving *Type* is needed and a common regression line for both levels of *Type* is adequate for modeling how *MMSE* depends on *APC*.

3.53 First-year GPA. The data in **FirstYearGPA** contain information from a sample of 219 first-year students at a midwestern college that might be used to build a model to predict their first-year GPA. Suppose that you decide to use high school GPA (*HSGPA*), Verbal SAT score (*SATV*), number of humanities credits (*HU*), and an indicator for whether or not a student is white (*White*) as a four-predictor model for *GPA*. **Y1GPA**

a. A white student applying to this college has a high school GPA of 3.20 and got a 600 score on his Verbal SAT.

If he has 10 credits in humanities, what GPA would this model predict?

b. Produce an interval that you could tell an admissions officer at this college would be 95% sure to contain the GPA of this student after his first year.

c. How much would your prediction and interval change if you added the number of credits of social science (*SS*) to your model? Assume this student also had 10 social science credits.

3.54 Combining explanatory variables. Suppose that X_1 and X_2 are positively related with $X_1 = 2X_2 - 4$. Let $Y = 0.5X_1 + 5$ summarize a positive linear relationship between Y and X_1.

a. Substitute the first equation into the second to show a linear relationship between Y and X_2. Comment on the direction of the association between Y and X_2 in the new equation.

b. Now add the original two equations and rearrange terms to give an equation in the form $Y = aX_1 + bX_2 + c$. Are the coefficients of X_1 and X_2 both in the direction you would expect based on the signs in the separate equations? Combining explanatory variables that are related to each other can produce surprising results.

Open-Ended Exercises

3.55 Porsche versus Jaguar prices. Two students collected data on used cars from Internet sites. The first student collected information about used Porsche sports cars. A second student collected similar data for 30 used Jaguar sports cars. Both datasets show a decrease in price as the mileage increases. Does the nature of the price versus mileage relationship differ between the two types of sports cars? For example, does one tend to be consistently more expensive? Or does one model experience faster depreciation in price as the mileage increases? Use a multiple regression model to compare these relationships and include graphical support for your conclusions. The data for all 60 cars of both models are stored in the file **PorscheJaguar**. Note the variable *IPorsche* that is 1 for the 30 Porsches and 0 for the 30 Jaguars. **PorJag**

3.56 Major League Baseball winning percentage. The **MLBStandings2016** file records many variables for Major League Baseball (MLB) teams from the 2016 regular season. Suppose that we are interested in modeling the winning percentages (*WinPct*) for the teams. Find an example of a set of predictors to illustrate the idea that adding a predictor (or several predictors) to an existing model can cause the adjusted R^2 to *decrease*. **MLB16**

Supplemental Exercises

3.57 Caterpillar metabolic rates. In Exercise 1.49 on page 58, we learned that the transformed body sizes (*LogBodySize*) are a good predictor of transformed metabolic rates (*LogMrate*) for a sample of caterpillars with data in **MetabolicRate**. We also notice that this linear trend appeared to hold for all five stages of the caterpillar's life. Create five different indicator variables, one for each level of *Instar*, and fit a multiple regression model to estimate five different regression lines. Only four of your five indicator variables are needed to fit the multiple regression model. Can you explain why? You should also create interaction variables to allow for the possibility of different slopes for each regression line. Does the multiple regression model with five lines provide a better fit than the simple linear regression model? **MetRate**

3.58 Caterpillar nitrogen assimilation and mass. In Exercise 1.30 on page 54, we explored the relationship between nitrogen assimilation and body mass (both on log scales) for data on a sample of caterpillars in **Caterpillars**. In an exploratory plot, we notice that there may be a linear pattern during free growth periods that is not present when the animal is not in a free growth period. Create an indicator variable for free growth period (*Ifpg* = 1 for free growth, *Ifpg* = 0 for no free growth) and indicator variables for each level of *Instar*. Choose *Ifpg* and four of the five *Instar* indicators to fit a multiple regression model that has response variable *LogNassim* using predictors: *LogMass*, *Ifpg*, any four of the five *Instar* indicators, and the interaction terms between *LogMass* and each of the indicators in the model. Does this multiple regression model provide a substantially better fit than the simple linear model based on *LogMass* alone? Explain. **Cater**

Andy Dean Photography/Shutterstock

CHAPTER 4

Additional Topics in Regression

In this chapter you will learn to:

- Construct and interpret an added variable plot.
- Consider techniques for choosing predictors to include in a model.
- Use cross-validation techniques to assess a regression model.
- Identify unusual and influential points.
- Incorporate categorical predictors using indicator variables.
- Use computer simulation methods to perform randomization tests for regression parameters.
- Use computer simulation methods to compute bootstrap confidence intervals for regression parameters.

TOPICS

In Chapters 1, 2, and 3, we introduced the basic ideas of simple and multiple linear regression. In this chapter, we consider some more specialized ideas and techniques involved with fitting and assessing regression models that extend some of the core ideas of the previous chapters. We prefer to think of the sections in this chapter as separate topics that stand alone and can be read in any order. For this reason, the exercises at the end of the chapter have been organized by these topics and we do not include the usual chapter summary.

TOPIC 4.1 Added Variable Plots

added variable plot

An **added variable plot** is a way to investigate the effect of one predictor in a model after accounting for the other predictors.* The basic idea is to remove the predictor in question from the model and see how well the remaining predictors work to model the response. The residuals from that model represent the variability in the response that is not explained by those predictors. Next, we use the same set of predictors to build a multiple regression model for the predictor we had left out. The residuals from that model represent the information in that predictor that is not related to the other predictors. Finally, we construct the added variable plot by plotting one of the sets of residuals against the other. This shows what the "unique" information in that predictor can tell us about what was unexplained in the response. The next example illustrates this procedure.

EXAMPLE 4.1

HseNY

House prices in NY In the model to predict NY house prices based on the size of the house (*Size*) and number of bedrooms (*Beds*) in Example 3.5, we saw that the two predictors were correlated ($r = 0.746$). Thus to a certain extent, *Size* and *Beds* are redundant: When we know one of them, we know something about the other. But they are not completely redundant, as there is information in the *Size* variable beyond what the *Beds* variable tells. We use the following steps for the data in **HousesNY** to construct a plot to investigate what happens when we add the variable *Size* to a model that predicts *Price* by *Beds* alone:

1. Find the residuals when predicting *Price* without using *Size*.
   The correlation between *Price* and *Beds* is 0.558, so *Beds* tells us something about *Price*, but not everything. Figure 4.1 shows the scatterplot with a regression line and residual plot from the regression of *Price* on *Beds*. These residuals represent the variability in *Price* that is *not* explained by *Beds*. We let a new variable called *Resid*1 denote the residuals for the regression of *Price* on *Beds*.

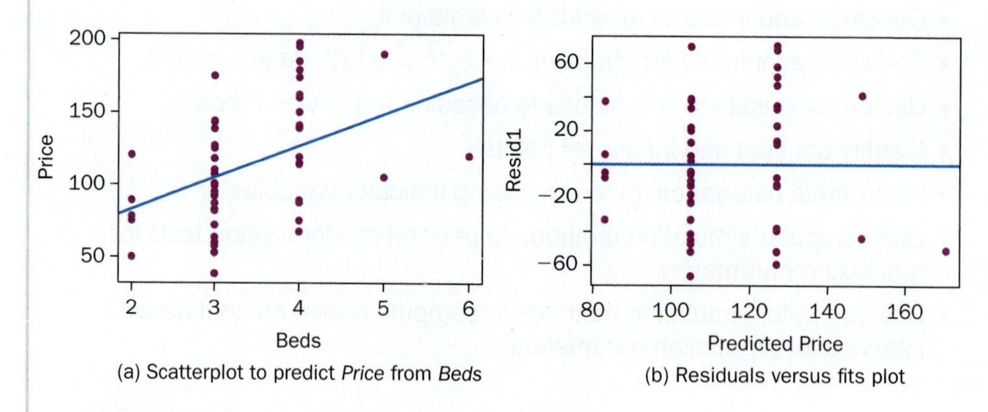

(a) Scatterplot to predict *Price* from *Beds*

(b) Residuals versus fits plot

FIGURE 4.1 Plots from regression of *Price* on *Beds*

2. Find the residuals when predicting *Size* using the other predictor *Beds*.
   If we regress *Size* on *Beds*, then the residuals capture the information in *Size* that is not explained by *Beds*. Figure 4.2 shows the scatterplot with a regression line and residual plot from the regression of *Size* on *Beds*. We store the residuals from the regression of *Size* on *Beds* in *Resid*2.

*Such a plot is sometimes called a **partial regression plot**.

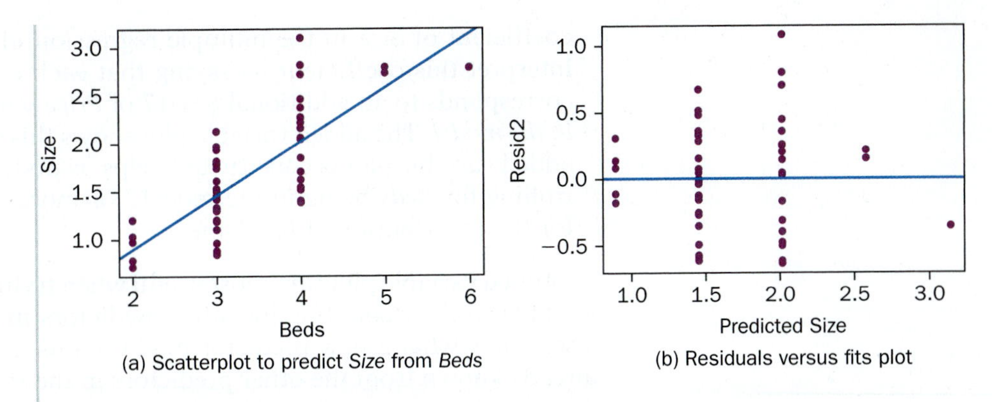

(a) Scatterplot to predict *Size* from *Beds*

(b) Residuals versus fits plot

FIGURE 4.2 Plots from regression of *Size* on *Beds*

3. Plot the residuals from the first model versus the residuals from the second model to create the added variable plot.

Figure 4.3 shows a scatterplot with a regression line to predict *Resid*1 (from the regression of *Price* on *Beds*) using *Resid*2 (from the regression of *Size* on *Beds*). The correlation between these two sets of residuals is $r = 0.33$, which indicates that the unique information in *Size* does explain some of the variation in *Price* that is not already explained by *Beds*. This helps explain why the *Size* variable is significant at a 5% level when combined with *Beds* in the multiple regression model. Some summary output for the multiple regression model is shown below:

```
Coefficients:
            Estimate  Std. Error  t value  Pr(>|t|)
(Intercept)   46.498      22.277    2.087     0.042
Beds           4.367       9.515    0.459     0.648
Size          31.169      12.617    2.470     0.017
```

The equation of the regression line in Figure 4.3 is

$$\widehat{Resid1} = 0 + 31.169\ Resid2$$

The intercept of the line must be zero since both sets of residuals have mean zero and thus the regression line must go through $(0, 0)$. Note that the slope of the line in the added variable plot is 31.169, exactly the same as the

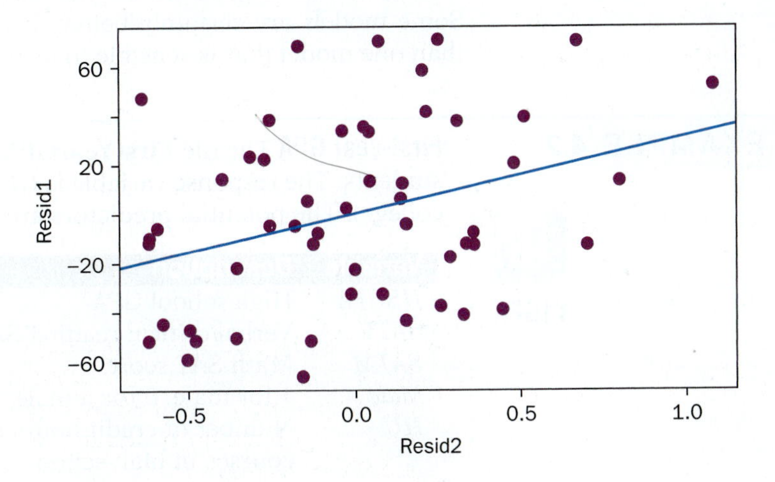

FIGURE 4.3 Added variable plot for adding *Size* to *Beds* when predicting *Price*

coefficient of *Size* in the multiple regression of *Price* on *Beds* and *Size*. We interpret this coefficient by saying that each additional square foot of *Size* corresponds to an additional \$31.17 of *Price* while *controlling for Beds being in the model*. The added variable plot shows this graphically. The way that the added variable plot is constructed helps us understand what the words "controlling for *Beds* being in the model," or more accurately, "after controlling for the contribution of *Beds*" mean.

Added variable plots are most useful when trying to identify the effect of one variable while accounting for other predictors in multiple regression models. They show what a new predictor might tell us about the response that is not already known from the other predictors in the model.

TOPIC 4.2 Techniques for Choosing Predictors

Sometimes, there are only one or two variables available for predicting a response variable and it is relatively easy to fit any possible model. In other cases, we have many predictors that might be included in a model. If we include quadratic terms, interactions, and other functions of the predictors, the number of terms that can be included in a model can grow very rapidly. It is tempting to include every candidate predictor in a regression model, as the more variables we include, the higher R^2 will be. But there is a lot to be said for creating a model that is simple, that we can understand and explain, and that we can expect to hold up in similar situations. That is, we don't want to include in a model a predictor that is related to the response only due to chance for a particular sample.

term

Since we can create new "predictors" using functions of other predictors in the data, we often use the word **term** to describe any predictor, a function of a predictor (like X^2), or a quantity derived from more than one predictor (like an interaction). When choosing a model, a key question is "Which terms should be included in a final model?" If we start with k candidate terms, then there are 2^k possible models (since each term can be included or excluded from a model). This can be quite a large number. For example, in the next example we have nine predictors so we could have $2^9 = 512$ different models, even before we consider any second-order terms.

Before going further, we note that often there is more than one model that does a good job of predicting a response variable. It is quite possible for different statisticians who are studying the same dataset to come up with somewhat different regression models. Thus, we are not searching for the one true, ideal model, but for a good model that helps us answer the question of interest. Some models are certainly better than others, but often there may be more than one model that is sensible to use.

EXAMPLE 4.2

Y1GPA

First-year GPA The file **FirstYearGPA** contains measurements on 219 college students. The response variable is *GPA* (grade point average after one year of college). The potential predictors are:

Variable	Description
HSGPA	High school GPA
SATV	Verbal/critical reading SAT score
SATM	Math SAT score
Male	1 for male, 0 for female
HU	Number of credit hours earned in humanities courses in high school

SS	Number of credit hours earned in social science courses in high school
FirstGen	1 if the student is the first in her or his family to attend college
White	1 for white students, 0 for others
CollegeBound	1 if attended a high school where ≥ 50% of students intend to go on to college

Figure 4.4 shows a scatterplot matrix of these data, with the top row of scatterplots showing how *GPA* is related to the quantitative predictors. Figure 4.5 shows boxplots of *GPA* by group for the four categorical predictors.

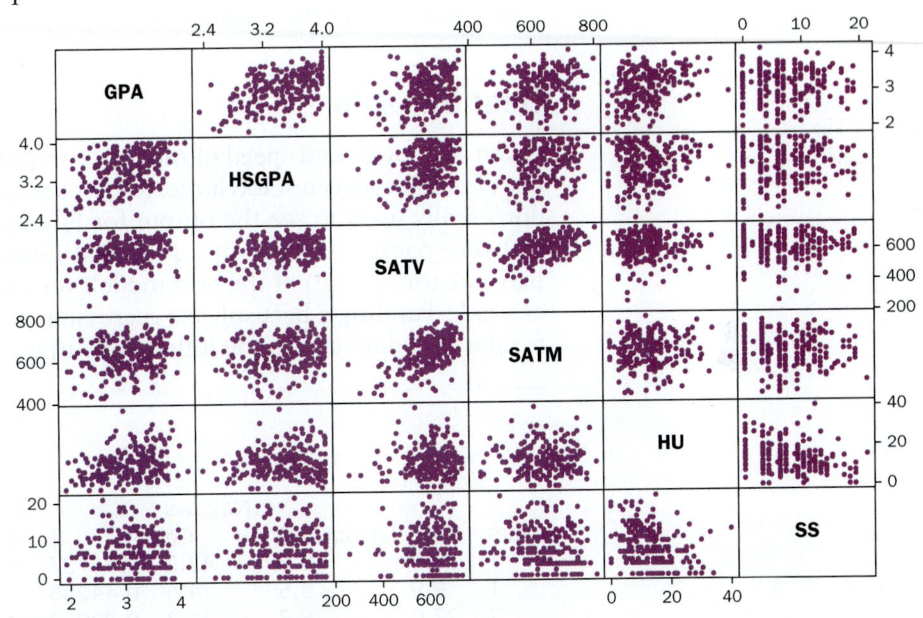

FIGURE 4.4 Scatterplot matrix for first-year GPA data

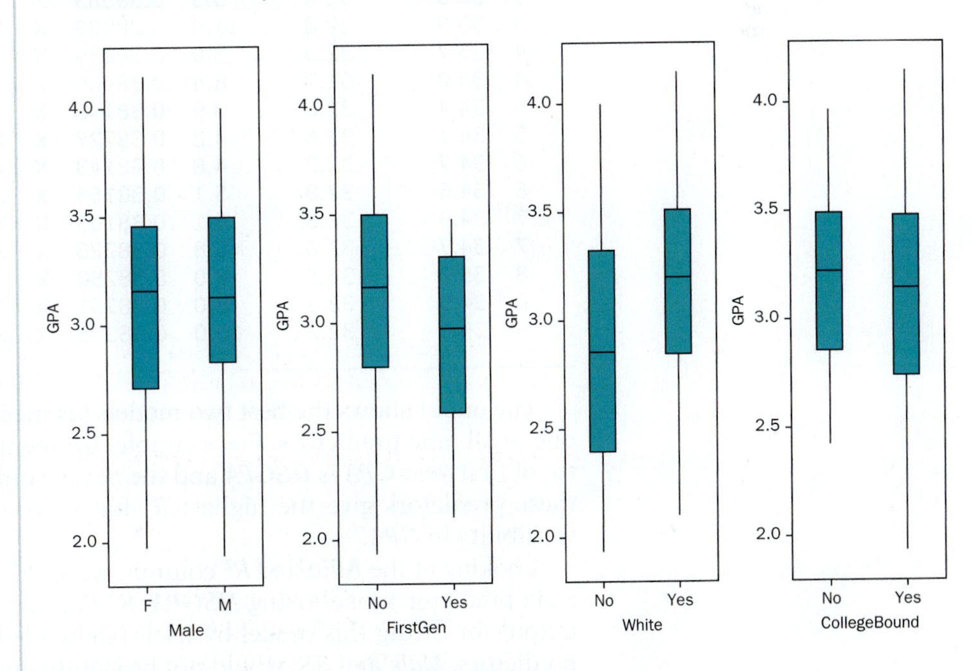

FIGURE 4.5 *GPA* versus categorical predictors

In order to keep the explanation fairly simple, we will not consider interactions and higher-order terms, but will just concentrate on the nine predictors in the dataset. As we consider which of them to include in a regression model, we need a measure of how well a model fits the data. Whenever a term is added to a regression model, R^2 goes up (or at least it doesn't go down), so if our goal were simply to maximize R^2, then we would include all nine predictors (and get $R^2 = 35.0\%$ for this example). However, some of the predictors may not add much value to the model. As we saw in Section 3.2, the adjusted R^2 statistic "penalizes" for the addition of a term, which is one way to balance the conflicting goals of predictive power and simplicity. Suppose that our goal is to maximize the adjusted R^2 using some subset of these nine predictors: How do we sort through the 512 possible models in this situation to find the optimal one?

Best Subsets

Given the power and speed of modern computers, it is feasible, when the number of predictors is not too large, to check all possible subsets of predictors. We don't really want to see the output for hundreds of models, so most statistical software packages (including R and Minitab) offer procedures that will display one (or several) of the best models for various numbers of predictors. For example, running a best subsets regression procedure using all nine predictors for the *GPA* data shows the following output:

Vars	R-Sq	R-Sq(adj)	Mallows Cp	S	HSGPA	SATV	SATM	Male	HU	SS	FGen	White	CBnd
1	20.0	19.6	42.2	0.41737	X								
1	9.9	9.5	74.5	0.44285					X				
2	27.0	26.3	21.7	0.39962	X				X				
2	26.8	26.1	22.2	0.40007	X							X	
3	32.3	31.4	6.5	0.38563	X				X			X	
3	30.8	29.8	11.4	0.38993	X	X			X				
4	33.7	32.5	3.9	0.38239	X	X			X			X	
4	33.0	31.7	6.4	0.38466	X				X		X	X	
5	34.4	32.8	3.9	0.38148	X	X			X	X		X	
5	34.1	32.6	4.8	0.38227	X	X		X	X			X	
6	34.7	32.9	4.8	0.38143	X	X		X	X	X		X	
6	34.6	32.8	5.1	0.38164	X	X			X	X	X	X	
7	34.9	32.8	6.1	0.38163	X	X		X	X	X	X	X	
7	34.7	32.6	6.8	0.38226	X	X		X	X	X		X	X
8	35.0	32.5	8.0	0.38250	X	X		X	X	X	X	X	X
8	34.9	32.5	8.0	0.38251	X	X	X	X	X	X	X	X	
9	35.0	32.2	10.0	0.38338	X	X	X	X	X	X	X	X	X

The ouput shows the best two models (as measured by R^2) of each size from one to all nine predictors. For example, we see that the best individual predictor of first-year *GPA* is *HSGPA* and the next strongest predictor is *HU*. Together, these predictors give the highest R^2 for a two-variable model for explaining variability in *GPA*.

Looking at the Adjusted R^2 column, we see that value is largest at 32.9% for a six-predictor model using *HSGPA*, *SATV*, *Male*, *HU*, *SS*, and *White*. But the output for fitting this model by itself (shown below) indicates that two of the predictors, *Male* and *SS*, would not be significant if tested at a 5% level. Would a simpler model be as effective?

The regression equation is
GPA = 0.547 + 0.483 HSGPA + 0.000694 SATV + 0.0541 Male + 0.0168 HU + 0.00757 SS + 0.205 White

Predictor	Coef	SE Coef	T	P
Constant	0.5467	0.2835	1.93	0.055
HSGPA	0.48295	0.07147	6.76	0.000
SATV	0.0006945	0.0003449	2.01	0.045
Male	0.05410	0.05269	1.03	0.306
HU	0.016796	0.003818	4.40	0.000
SS	0.007570	0.005442	1.39	0.166
White	0.20452	0.06860	2.98	0.003

S = 0.381431   R-Sq = 34.7%   R-Sq(adj) = 32.9%

Analysis of Variance

Source	DF	SS	MS	F	P
Regression	6	16.3898	2.7316	18.78	0.000
Residual Error	212	30.8438	0.1455		
Total	218	47.2336			

The method of best subsets is most useful when the number of predictors is relatively small. When we have a large number of predictors, then stepwise regression methods, discussed shortly, might be more useful.

Mallows's C_p

Most of the criteria we have used so far to evaluate a model (e.g., R^2, adjusted R^2, MSE, $\hat{\sigma}_\epsilon^2$, overall ANOVA, individual t-tests) depend only on the predictors in the model being evaluated. None of these measures takes into account what information might be available in the other potential predictors that aren't in the model. A new measure that does this was developed by statistician Colin Mallows.

MALLOWS'S C_P

When evaluating a regression model for a subset of m predictors from a larger set of k predictors using a sample of size n, the value of Mallows's C_p is computed by

$$C_p = \frac{SSE_m}{MSE_k} + 2(m+1) - n$$

where SSE_m is the sum of squared residuals from the model with just m predictors and MSE_k is the mean square error for the full model with all k predictors.

We prefer models where C_p is small.

As we add terms to a model, we drive down the first part of the C_p statistic, since SSE_m goes down while the scaling factor MSE_k remains constant. But we also increase the second part of the C_p statistic, $2(m+1)$, since $m+1$ measures the number of coefficients (including the constant term) that are being estimated in the model. This acts as a kind of penalty in calculating C_p. If the reduction in SSE_m is substantial compared to this penalty, the value of C_p will decrease, but if the new predictor explains little new variability, the value of C_p will increase. A model that has a small C_p value is thought to be a good compromise between making SSE_m small and keeping the model simple. Generally, any model for which C_p is less than $m+1$ or not much greater than $m+1$ is considered to be a model worth considering.

Note that if we compute C_p for the full model with all k predictors we get,

$$C_p = \frac{SSE_k}{MSE_k} + 2(k+1) - n = \frac{SSE_k}{SSE_k/(n-k-1)} + 2(k+1) - n$$
$$= (n-k-1) + (2k+2) - n = k+1$$

so the C_p when all predictors are included is always 1 more than the number of predictors (as you can verify from the last row in the best subsets output on page 158).

When we run the six-predictor model for all $n = 219$ cases using *HSGPA*, *SATV*, *Male*, *HU*, *SS*, and *White* to predict *GPA*, we see that $SSE_6 = 30.8438$ and for the full model with all nine predictors we have $MSE_9 = 0.14698$. This gives

$$C_p = \frac{30.8438}{0.14698} + 2(6+1) - 219 = 4.85$$

which is less than $6+1=7$ so this is a reasonable model to consider, but could we do better?

Fortunately, the values of Mallows's C_p are displayed in the best subsets regression output on page 158. The minimum $C_p = 3.9$ occurs at a four-predictor model that includes *HSGPA*, *SATV*, *HU*, and *White*. Interestingly, this model omits both the *Male* and *SS* predictors that had insignificant t-tests in the six-predictor model with the highest adjusted R^2. When running the smaller four-predictor model, all four terms have significant t-tests at a 5% level. A five-predictor model (putting *SS* back in) also has a C_p value of 3.9, but we would prefer the smaller model that attains essentially the same C_p.

AIC *and* BIC

Akaike's information criterion

Bayesian information criterion

Other selection criteria, similar to Mallows's C_p, are provided in some software packages and preferred by some practicing statisticians. Two common examples are **Akaike's information criterion (AIC)** and **Bayesian information criterion (BIC)**. Like C_p, smaller values of AIC and BIC indicate preferred models. These measures account for the number of predictors in the model as well as how well the response is explained. The details for computing these values are beyond the scope of this section,* so we rely on statistical software for finding and using them to help select models.

Backward Elimination

Another model selection method that is popular and rather easy to implement is called **backward elimination**:

1. Start by fitting the full model (the model that includes all terms under consideration).
2. Identify the term for which the individual t-test produces the largest P-value:
   a. If that P-value is large (say, greater than 5%), eliminate that term to produce a smaller model. Fit that model and return to the start of Step 2.
   b. If the P-value is small (less than 5%), stop since all of the predictors in the model are "significant."

Note: This process can be implemented with, for example, the goal of minimizing C_p at each step, rather than the criterion of eliminating all non-significant predictors. At each step eliminate the predictor that gives the largest

*AIC and BIC are based on modified likelihood functions that add a term depending on the number of predictors, p.

drop in C_p, until we reach a point that C_p does not get smaller if any predictor left in the model is removed.

If we run the full model to predict *GPA* based on all nine predictors, the output includes the following *t*-tests for individual terms:

Predictor	Coef	SE Coef	T	P
Constant	0.5269	0.3488	1.51	0.132
HSGPA	0.49329	0.07456	6.62	0.000
SATV	0.0005919	0.0003945	1.50	0.135
SATM	0.0000847	0.0004447	0.19	0.849
Male	0.04825	0.05703	0.85	0.398
HU	0.016187	0.003972	4.08	0.000
SS	0.007337	0.005564	1.32	0.189
FirstGen	-0.07434	0.08875	-0.84	0.403
White	0.19623	0.07002	2.80	0.006
CollegeBound	0.0215	0.1003	0.21	0.831

The largest *P*-value for a predictor is 0.849 for *SATM*, which is certainly not significant, so that predictor is dropped and we fit the model with the remaining eight predictors:

Predictor	Coef	SE Coef	T	P
Constant	0.5552	0.3149	1.76	0.079
HSGPA	0.49502	0.07384	6.70	0.000
SATV	0.0006245	0.0003548	1.76	0.080
Male	0.05221	0.05298	0.99	0.325
HU	0.016082	0.003925	4.10	0.000
SS	0.007177	0.005487	1.31	0.192
FirstGen	-0.07559	0.08830	-0.86	0.393
White	0.19742	0.06958	2.84	0.005
CollegeBound	0.0212	0.1001	0.21	0.833

Now *CollegeBound* is the "weakest link" (*P*-value = 0.833), so it is eliminated and we continue. To avoid showing excessive output, we summarize below the predictors that are identified with the largest *P*-value at each step:

9 terms:

Predictor	Coef	SE Coef	T	P
SATM	0.0000847	0.0004447	0.19	0.849

8 terms:

Predictor	Coef	SE Coef	T	P
CollegeBound	0.0212	0.1001	0.21	0.833

7 terms:

Predictor	Coef	SE Coef	T	P
FirstGen	-0.07725	0.08775	-0.88	0.380

6 terms:

Predictor	Coef	SE Coef	T	P
Male	0.05410	0.05269	1.03	0.306

5 terms:

Predictor	Coef	SE Coef	T	P
SS	0.007747	0.005440	1.42	0.156

4 terms:

Predictor	Coef	SE Coef	T	P
SATV	0.0007372	0.0003417	2.16	0.032

After we get a four-term model consisting of *HSGPA*, *SATV*, *HU*, and *White*, we see that the weakest predictor (*SATV*, *P*-value = 0.032) is still significant at a 5% level, so we would stop and keep this model. This produces the same model that minimized the C_p in the best subsets procedure. This is a nice coincidence when it happens, but is certainly not something that will occur for every set of predictors. Statistical software allows us to automate this process even further to give the output tracing the backward elimination process such as that shown below:

Backward elimination. Alpha-to-Remove: 0.05
Response is GPA on 9 predictors, with N = 219

Step	1	2	3	4	5	6
Constant	0.5269	0.5552	0.5825	0.5467	0.5685	0.6410
HSGPA	0.493	0.495	0.492	0.483	0.474	0.476
T-Value	6.62	6.70	6.81	6.76	6.68	6.70
P-value	0.000	0.000	0.000	0.000	0.000	0.000
SATV	0.00059	0.00062	0.00063	0.00069	0.00075	0.00074
T-Value	1.50	1.76	1.79	2.01	2.19	2.16
P-value	0.135	0.080	0.075	0.045	0.029	0.032
SATM	0.00008					
T-Value	0.19					
P-Value	0.849					
Male	0.048	0.052	0.053	0.054		
T-Value	0.85	0.99	1.00	1.03		
P-Value	0.398	0.325	0.316	0.306		
HU	0.0162	0.0161	0.0161	0.0168	0.0167	0.0151
T-Value	4.08	4.10	4.10	4.40	4.39	4.14
P-Value	0.000	0.000	0.000	0.000	0.000	0.000
SS	0.0073	0.0072	0.0071	0.0076	0.0077	
T-Value	1.32	1.31	1.30	1.39	1.42	
P-Value	0.189	0.192	0.194	0.166	0.156	
FirstGen	-0.074	-0.076	-0.077			
T-Value	-0.84	-0.86	-0.88			
P-Value	0.403	0.393	0.380			
White	0.196	0.197	0.196	0.205	0.206	0.212
T-Value	2.80	2.84	2.84	2.98	3.00	3.09
P-Value	0.006	0.005	0.005	0.003	0.003	0.002
CollegeBound	0.02	0.02				
T-Value	0.21	0.21				
P-Value	0.831	0.833				
S	0.383	0.383	0.382	0.381	0.381	0.382
R-Sq	34.96	34.95	34.94	34.70	34.37	33.75
R-Sq(adj)	32.16	32.47	32.78	32.85	32.83	32.51
Mallows Cp	10.0	8.0	6.1	4.8	3.9	3.9

Backward elimination has an advantage in leaving us with a model in which all of the predictors are significant and it requires fitting relatively few models (in this example, we needed to run only 6 of the 512 possible models). One disadvantage of this procedure is that the initial models tend to be the most complicated. If multicollinearity is an issue (as it often is with large sets of predictors), we know that the individual *t*-tests can be somewhat unreliable for correlated predictors, yet that is precisely what we use as a criterion in making decisions about which predictors to eliminate. In some situations, we might eliminate a single strong predictor at an early stage of the backward elimination process if it is strongly correlated with several other predictors. Then we have no way to "get it back in" the model at a later stage when it might provide a significant benefit over the predictors that remain.

Forward Selection and Stepwise Regression

The difficulties we see when starting with the most complicated models in backward elimination suggest that we might want to build a model from the other direction, starting with a simple model using just the best single predictor and then adding new terms. This method is known as **forward selection**.

forward selection

1. Start with a model with no predictors and find the best single predictor (the largest correlation with the response gives the biggest initial R^2).
2. Add the new predictor to the model, run the regression, and find its individual *P*-value:
   a. If that *P*-value is small (say, less than 5%), keep that predictor in the model and try each of the remaining (unused) predictors to see which would produce the most benefit (biggest increase in R^2) when added to the existing model.
   b. If the *P*-value is large (over 5%), stop and discard this predictor. At this point, no (unused) predictor should be significant when added to the model and we are done.

The forward selection method generally requires fitting many more models. In our current example, we would have nine predictors to consider at the first step, where *HSGPA* turns out to be the best with $R^2 = 20.0\%$. Next, we would need to consider two-predictor models combining *HSGPA* with each of the remaining eight predictors. Statistical software automates this process to give the output such as that shown below:

```
Forward selection. Alpha-to-Enter: 0.05
Response is GPA on 9 predictors, with N = 219
Step          1        2        3        4
Constant      1.1799   1.0874   0.9335   0.6410

HSGPA         0.555    0.517    0.507    0.476
T-Value       7.36     7.11     7.23     6.70
P-Value       0.000    0.000    0.000    0.000

HU                     0.0172   0.0153   0.0151
T-Value                4.55     4.18     4.14
P-Value                0.000    0.000    0.000

White                           0.266    0.212
T-Value                         4.12     3.09
P-Value                         0.000    0.002

SATV                                     0.00074
T-Value                                  2.16
P-Value                                  0.032

S              0.417    0.400    0.386    0.382
R-Sq          19.97    26.97    32.31    33.75
R-Sq(adj)     19.60    26.30    31.36    32.51
Mallows Cp    42.2     21.7      6.5      3.9
```

The forward selection procedure arrives at the same four-predictor model (*HSGPA*, *SATV*, *HU*, and *White*) as we obtained from backward elimination and minimizing C_p with best subsets. In fact, the progression of forward steps mirrors the best one-, two-, three-, and four-term models in the best subsets output. While this may happen in many cases, it is not guaranteed to occur.

In some situations (but not this example), we may find that a predictor that was added early in a forward selection process becomes redundant when more predictors are added at later stages. Thus, X_1 may be the strongest individual predictor, but X_2 and X_3 together contain much of the same information that X_1 carries about the response. We might choose X_1 to add at the first step but later, when X_2 and X_3 have been added, discover that X_1 is no longer needed and has an insignificant P-value. To account for this, we use **stepwise regression**, which combines features of both forward selection and backward elimination. A stepwise procedure starts with forward selection, but after any new predictor is added to the model, it uses backward elimination to delete any predictors that have become redundant in the model. In our current example, this wasn't necessary so the stepwise output would be the same as the forward selection output.

stepwise regression

In the first-year GPA example, we have not considered interaction terms or power terms. However, the ideas carry over. If we are interested in, say, interactions and quadratic terms, we simply create those terms and put them in the set of candidate terms before we carry out a model selection procedure such as best subsets regression or stepwise regression. If we decide that an interaction or quadratic term should be in the model, then we include the lower-order terms as well.

Caution about Automated Techniques

Model selection procedures have been implemented in computer programs such as Minitab and R, but they are not a substitute for thinking about the data and the modeling situation. At best, a model selection process should suggest to us one or more models to consider. Looking at the order in which predictors are entered in a best subsets or stepwise procedure can help us understand which predictors are relatively more important and which might be more redundant. Do not be fooled by the fact that several of these procedures gave us the same model in the first-year GPA example. This will not always be the case in practice! Even when it does occur, we should not take that as evidence that we have found *the* model for our data.

It is always the responsibility of the modeler to think about the possible models, to conduct diagnostic procedures (such as plotting residuals against fitted values), and to use only models that make sense.

Moreover, as we noted earlier, there may well be two or more models that have essentially the same quality. In practical situations, a sightly less optimal model that uses variables that are much easier (or cheaper) to measure might be preferred over a more complicated one that squeezes an extra 0.5% of R^2 for a particular sample.

No matter what procedure we use to arrive at a model, it is good practice to set aside part of the data, when searching for a model, and to use those set-aside data to validate the chosen model. This idea, of "cross-validation," is explored in the next section.

TOPIC 4.3 Cross-validation

In Chapter 0 we introduced several purposes for fitting a model, including to make predictions, understand relationships, and assess differences. When the first of these is a primary goal, we have an easy way to check how well the model works. Just use the model to make a prediction, then wait until the response variable can be measured to see how close the prediction comes to the actual value. But wouldn't it be nice to judge how well the model works *before* using it on new cases?

Of course, we already have measures like R^2, *SSE*, and *MSE* that measure how well a model fits, but those only assess the accuracy of predicting data cases *that were used to build and fit the model*. Should we always expect the same level of accuracy to hold when using the model to predict completely new data cases? No.* The coefficient estimates are chosen to minimize quantities such as *SSE* for the original sample data, so we shouldn't expect them to work just as well for new cases. But can we get some idea, in advance, for how the model might do when applied to new data? One technique for doing this is **cross-validation**.

cross-validation

Here is one fairly simple approach to cross-validation:

training sample

- Build and fit the model with one set of data, called the **training sample**.
- Use the fitted model to predict responses for a second set of data, called the **testing** or **holdout sample**.
- Compare the accuracy of the fit in the training sample to what is observed for the holdout sample.

testing sample

Generally, we expect the performance of the model to be worse for predicting new data, but we hope not too much worse. If it is much worse, we should suspect that the original model might just reflect individual features of the training sample, rather than patterns that hold in the more general population.

Where do the training and testing samples come from? While we might go out, after fitting a model, and collect a new sample for testing purposes, a more common approach is to split the original sample cases into two groups to create the training and holdout samples. Generally we split the full data by taking a random sample of cases for training, leaving the rest for testing. If the dataset is already a random sample in no particular order, we can simply choose a convenient block of cases for training. We use *only* the data in the training sample to build and fit a model. Then we calculate predictions for the holdout sample using its predictor values and the model fitted from the training sample. Finally, we compare the actual responses for the holdout sample to the predictions based on the training model. Here is a simple example.

EXAMPLE 4.3

HseNY

Houses in NY: Cross-validation In Example 3.15 on page 121 we use the data in **HousesNY** to look at regression models to predict prices for houses based on a sample of 53 homes in Canton, New York. This is a fairly small dataset, so we'll stick to a simple linear model, using just *Size* of the house (in thousands of square feet) to predict its *Price* (in thousands of dollars). Since the data are a random sample in no particular order, we'll use the first 35 cases as the training sample and reserve the final 18 cases for the holdout sample.

The computer output shows the results for fitting a model to predict *Price* based on *Size* for the training sample. Figure 4.6 shows scatterplots for both the training and holdout samples, with the same (training) least squares line $\widehat{Price} = 56.549 + 33.611\ Size$ drawn on each.

The first house in the holdout sample is 1.200 (thousand square feet) in size and has a price of 120.5 (thousand dollars). Using the training model, the prediction is $\widehat{Price} = 56.549 + 33.611(1.2) = 96.88$. This gives an error of $120.5 - 96.88 = 23.62$. For comparison, the standard error of regression for the training sample is 34.89, so this prediction isn't too bad. Of course,

*If we fit a simple linear model to just two data cases, we always get a perfect fit with no errors. That doesn't mean we would expect no errors if using that fitted model on new data!

```
Coefficients:
            Estimate  Std. Error  t value  Pr(>|t|)
(Intercept)  56.549    16.824      3.361   0.0019
Size         33.611     9.354      3.593   0.00105 **
---
Residual standard error: 34.89 on 33 degrees of freedom
Multiple R-squared: 0.2812, Adjusted R-squared: 0.2594
F-statistic: 12.91 on 1 and 33 DF, p-value: 0.00105
```

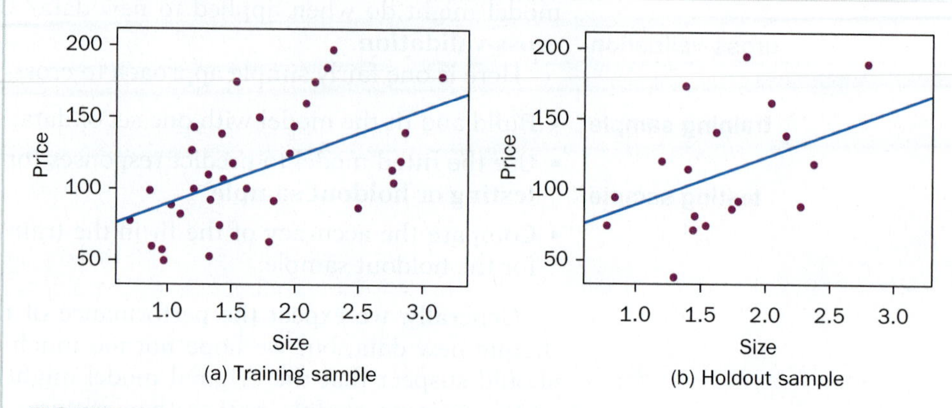

(a) Training sample (b) Holdout sample

FIGURE 4.6 *Price* versus *Size* scatterplots with training line

we need to look at predictions for more than just one point. Here are the prediction errors for all 18 houses in the holdout sample.

23.6 −33.2
−33.0 −24.2
10.2 30.0
−61.6 −23.4
9.9 36.7
−7.8 76.3
71.0 39.5
−17.4 11.4
−44.0 −28.0

The average prediction error is just 2.00 (quite close to zero) and the sum of the squared prediction errors for these 18 houses is 25,777, compared to $SSE = 40{,}162$ for the 35 houses in the training sample. Since sums of squared errors are based on different sample sizes, we prefer to compare the mean squared error from the holdout sample, $25{,}777/18 = 1432$ to the mean square error for the training sample,* $MSE = 40{,}162/33 = 1217$. Thus the predictions are slightly worse for the holdout sample (as expected), but not drastically so.

Cross-validation Correlation

Here is an alternate way to assess the effectiveness of the training model for predicting responses in the holdout sample. Recall that the coefficient of determination for a regression model can be found in two different ways: $R^2 = SSModel/SSTotal$ and by squaring the correlation between the actual and predicted values (see page 96). We use the second method to develop a numerical summary to see how much predictive power is lost when we apply a training model to the holdout data.

*For the training sample we divide by $n - 2$ df to account for the two parameters estimated from the data.

CROSS-VALIDATION CORRELATION AND SHRINKAGE

- Obtain predictions ($\hat{y}$) for the cases in a holdout sample, using the model fit for the training sample.
- The correlation between these predictions and the actual values (y) for the holdout sample is called the **cross-validation correlation**.
- The difference between the R^2 for the training sample and the square of the cross-validation correlation is known as the **shrinkage**.

The model for the training sample will (usually) fit the training sample that built it better than it fits the new holdout sample, so the cross-validation correlation is usually smaller (in magnitude) than if we had computed the correlation for the training data. Thus its square is (usually) less than the R^2 for the original model and the shrinkage is positive. How much shrinkage should cause concern? We won't give any hard and fast rules; instead we generally use shrinkage as a rough indicator of how well the model is doing at predicting the holdout sample. Shrinkage of 10% or less should not be a problem, but a shrinkage of more than 50% would be worrisome.

EXAMPLE 4.4

HseNY

Houses in NY: Cross-validation correlation and shrinkage For the model in Example 4.3 to predict house prices using size, the cross-validation correlation between actual and predicted prices for the 18 houses in the holdout sample is 0.4873, which squared is 0.2375. From the output in the previous example, we see that $R^2 = 28.12\%$ for the original training model. This means the shrinkage on cross-validation is $28.12 - 23.75 = 4.37\%$. This is a fairly small amount, so we have a good indication that a model fit to predict house prices based on sizes (at least in Canton, New York) will still predict almost as well when applied to new data.

The previous examples dealt with a single predictor regression model, but the cross-validation idea works equally well for multiple regression models (see the exercises). In that setting we should use the training sample to select predictors, determine transformations, and fit the model—ignoring the information in the holdout sample when making those decisions.

One drawback to doing cross-validation using a holdout sample is that the training model is built with a smaller sample size than would be available if the model were built for the entire original dataset. Especially for regression models with lots of terms, we would like to have plenty of data when fitting the model. If the original sample size is fairly large, we won't lose much by saving some (perhaps 25%) of the data for holdout purposes. Another suggestion is to use cross-validation to see if the model appears to translate well to new data, then recombine the data to get a bigger sample size for final estimation of parameters. Just remember that the fit optimized for one sample is likely to overestimate its effectiveness for new cases.

There are a number of variations on the idea of cross-validation that are beyond the scope of the short introduction in this section. For example, we could split the data into two pieces, use one as training and the other as holdout, then flip the roles and repeat the process. We could also take an original sample of 500 cases and randomly divide it into 5 samples of size 100 each. We let each of the smaller samples take a turn as the holdout sample, using the other 400 values for training. As an extreme example of this approach, we could let each data point be holdout sample of size one and predict it using a model fit to all of the other data points. This is the idea behind the studentized residuals that we first saw in Section 1.5 and which is discussed in more detail in the next section.

TOPIC 4.4 Identifying Unusual Points in Regression

In Section 1.5, we introduced the ideas of outliers and influential points in the simple linear regression setting. Recall that outliers in a regression setting are points that fall far from the least squares line and influential points are cases with unusual predictor values that substantially affect where the line goes. We return to those ideas in this section and examine more formal methods for identifying unusual points in a simple linear regression and extend them to models with multiple predictors.

Leverage

Generally, in a simple linear regression situation, points farther from the mean value of the predictor ($\overline{x}$) have greater potential to influence the slope of a fitted regression line. This concept is known as **leverage**. Points with high leverage have a greater capacity to pull the regression line in their direction than do low-leverage points, such as points near the predictor mean. In the case of a single predictor, we could measure leverage as just the distance from the mean; however, we will find it useful to have a somewhat more complicated statistic that can be generalized to the multiple regression setting.

leverage

LEVERAGE FOR A SINGLE PREDICTOR

For a simple linear regression on n data points, the **leverage** of any point (x_i, y_i) is defined to be

$$h_i = \frac{1}{n} + \frac{(x_i - \overline{x})^2}{\sum (x_i - \overline{x})^2}$$

The sum of the leverages for all points in a simple linear model is $\sum h_i = 2$; thus a "typical" leverage is about $2/n$. Leverages that are more than two times the typical leverage, that is, points with $h_i > 4/n$, are considered somewhat high for a single predictor model and values more than three times as large, that is, points with $h_i > 6/n$, are considered to have especially high leverage.

Note that leverage depends only on the value of the predictor and not on the response. Also, a high-leverage point does not necessarily exert large influence on the estimation of the least squares line. It is possible to place a new point exactly on an existing regression line at an extreme value of the predictor, thus giving it high leverage but not changing the fitted line at all. Finally, the formula for computing h_i in the simple linear case may seem familiar. If you look in Section 2.4 where we considered confidence and prediction intervals for the response Y in a simple linear model, you will see that leverage plays a prominent role in computing the standard errors for those intervals.

EXAMPLE 4.5

PalmBch

Butterfly ballots In Example 1.11 on page 46, we looked at a simple linear regression of votes for Pat Buchanan versus George Bush in Florida counties from the 2000 U.S. presidential election. A key issue in fitting the model was a significant departure from the trend of the rest of the data for Palm Beach County, which used a controversial butterfly ballot. The data for the 67 counties in that election are stored in **PalmBeach**.

Statistical software such as Minitab and R will generally have options to compute and display the leverage values for each data point when fitting a regression model. In fact, the default regression output in Minitab includes output showing "Unusual Observations" such as the output below from the original regression using the butterfly ballot data:

```
Unusual Observations

Obs   Bush     Buchanan   Fit      SE Fit   Residual   St Resid
  6   177279    789.0      916.9   111.1    -127.9     -0.38 X
 13   289456    561.0     1468.5   193.0    -907.5     -3.06RX
 29   176967    836.0      915.4   110.9     -79.4     -0.24 X
 50   152846   3407.0      796.8    94.2    2610.2      7.65R
 52   184312   1010.0      951.5   116.1      58.5      0.17 X

R denotes an observation with a large standardized residual.
X denotes an observation whose X value gives it large leverage.
```

Minitab uses the "three times typical" rule, that is $h_i > 6/n$, to identify large leverage points and flags them with an "X" in the output. For the $n = 67$ counties in this dataset, the cutoff for large leverage is $6/67 = 0.0896$ and the mean Bush vote is $\bar{x} = 43{,}356$. We see that there are four counties with high leverage:

- Broward (Bush = 177,279, $h_6 = 0.0986$)
- Dade (Bush = 289,456, $h_{13} = 0.2975$)
- Hillsborough (Bush = 176,967, $h_{29} = 0.0982$)
- Pinellas (Bush = 184,312, $h_{52} = 0.1076$)

By default, R uses the "two times typical" rule to identify high outliers, so with $4/67 = 0.0597$ as the cutoff, R would add two more counties to the list of high-leverage counties:

- Palm Beach (Bush = 152,846, $h_{50} = 0.07085$)
- Duval (Bush = 152,082, $h_{16} = 0.07007$)

These six "unusual" observations are highlighted in Figure 4.7. In this example, the six high-leverage counties by the $h_i > 4/n$ criterion are exactly the six counties that would be identified as outliers when building a boxplot (Figure 4.8) and looking for points more than 1.5*IQR* beyond the quartiles.

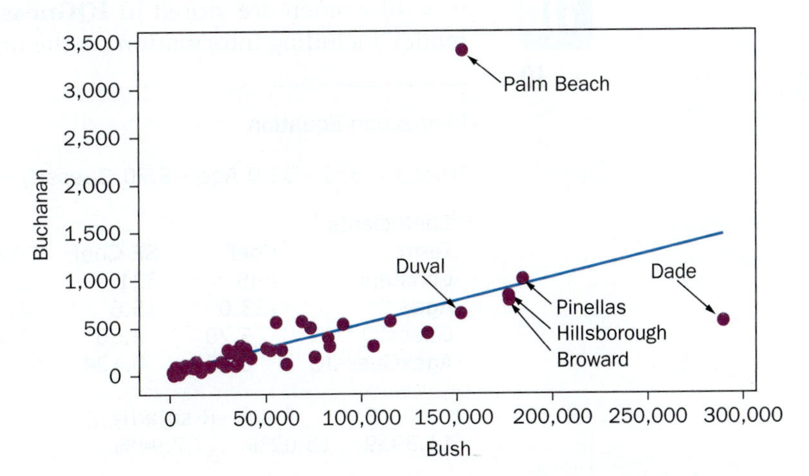

FIGURE 4.7 Unusual observations identified in the butterfly ballot model

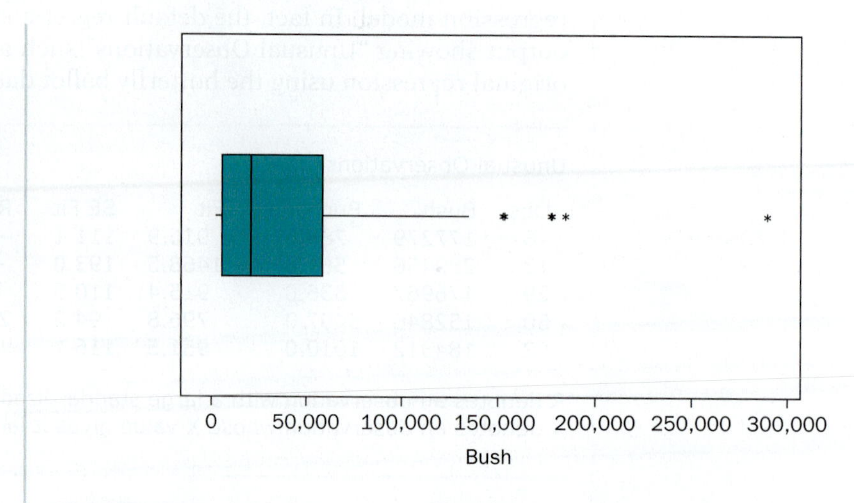

FIGURE 4.8 Boxplot of the number of Bush votes in Florida counties

In the multiple regression setting, the computation of leverage is more complicated (and beyond the scope of this book), so we will depend on statistical software. Fortunately, the interpretation is similar to the interpretation for the case of a single predictor: Points with high leverage have the potential to influence the regression fit significantly. With more than one predictor, a data case may have an unusual value for any of the predictors or exhibit an unusual combination of predictor values. For example, if a model used people's *height* and *weight* as predictors, a person who is tall and thin, say, 6 feet tall and 130 pounds, might have a lot of leverage on a regression fit even though neither the individual's height nor weight is particularly unusual.

For a multiple regression model with k predictors the sum of the leverages for all n data cases is $\sum h_i = (k+1)$. Thus, an average or typical leverage value in the multiple case is $(k+1)/n$. As with the simple linear regression case, we identify cases with somewhat high leverage when h_i is more than twice the average, $h_i > 2(k+1)/n$, and very high leverage when $h_i > 3(k+1)/n$.

EXAMPLE 4.6

IQ

IQ guessing: Leverage In Example 3.12 on page 111, we considered a multiple regression model to predict the *TrueIQ* of women using the *GuessIQ* and *Age* of the women along with an interaction term, *GuessIQ* · *Age*. Data for the $n = 40$ women are stored in **IQGuessing**. Some computer output for this model, including information on the unusual cases, is shown as follows:

```
Regression Equation

TrueIQ = 835 - 33.0 Age - 5.70 GuessIQ + 0.265 AgexGuessIQ

 Coefficients
 Term           Coef     SE Coef   T-Value   P-Value   VIF
 Constant       835      321        2.60     0.013
 Age            -33.0    15.6      -2.12     0.041     120.96
 GuessIQ        - 5.70    2.56     -2.23     0.032     211.22
 AgexGuessIQ     0.265    0.124     2.14     0.039     353.83

       S      R-sq    R-sq(adj)
 14.3939   15.02%       7.94%

 Analysis of Variance
 Source        DF    Adj SS    Adj MS   F-Value   P-Value
 Regression     3   1318.33    439.44     2.12     0.115
 Error         36   7458.64    207.18
 Total         39   8776.97
```

Fits and Diagnostics for Unusual Observations

Obs	TrueIQ	Fit	Resid	Std Resid
1	83.00	121.22	-38.22	-2.72 R
16	162.00	129.28	32.72	2.36 R
24	114.00	115.37	-1.37	-0.12 X
25	166.00	126.41	39.59	2.83 R

R Large residual
X Unusual X

The output identifies one high-leverage case. The 24th woman in the sample is 24 years old, which is tied for the highest age in the data, and has a guessed IQ of 110. The other 24-year-olds have guessed IQs between 124 and 134, so 110 is a low number within that age group. This combination of high age and low guessed IQ gives case 24 a lot of leverage. The leverage for woman #24 (as computed by software) is $h_{24} = 0.381$, while the cutoff for high leverage in this fitted model is $3(3+1)/40 = 0.3$. It is often difficult to visually spot high-leverage points for models with several predictors by just looking at the individual predictor values. Figure 4.9 shows a scatterplot of the *GuessIQ* and *Age* values for this sample of women with the unusual point identified and a dotplot of the leverage values where that point stands out more clearly as being unusual.

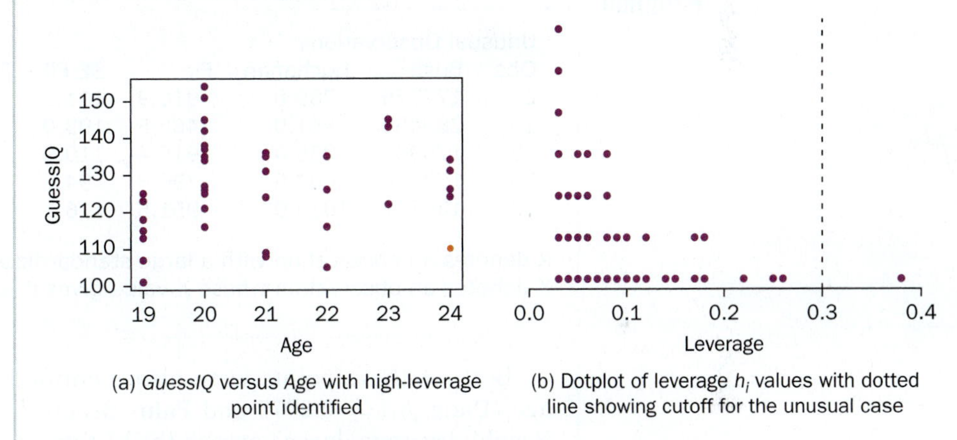

(a) *GuessIQ* versus *Age* with high-leverage point identified

(b) Dotplot of leverage h_i values with dotted line showing cutoff for the unusual case

FIGURE 4.9 High-leverage case in IQ guessing interaction model

Standardized and Studentized Residuals

In Section 1.5, we introduced the idea of standardizing the residuals of a regression model so that we could more easily identify points that were poorly predicted by the fitted model. Now that we have defined the concept of leverage, we can give a more formal definition of the adjustments used to find standardized and studentized residuals.

STANDARDIZED AND STUDENTIZED RESIDUALS

The **standardized residual** for the ith data point in a regression model can be computed using

$$\text{stdres}_i = \frac{y_i - \hat{y}_i}{\hat{\sigma}_\epsilon \sqrt{1 - h_i}}$$

where $\hat{\sigma}_\epsilon$ is the standard deviation of the regression and h_i is the leverage for the ith point.

For a **studentized** residual (also known as a **deleted-t** residual), we replace $\hat{\sigma}_\epsilon$ with the standard deviation of the regression, $\hat{\sigma}_{(i)}$, from fitting the model with the ith point omitted:

$$\text{studres}_i = \frac{y_i - \hat{y}_i}{\hat{\sigma}_{(i)}\sqrt{1 - h_i}}$$

Under the usual conditions for the regression model, the standardized or studentized residuals follow a *t*-distribution. Thus, we identify data cases with standardized or studentized residuals beyond ± 2 as moderate outliers, while values beyond ± 3 denote more serious outliers.* The adjustment in the standard deviation for the studentized residual helps avoid a situation where a very influential data case has a big impact on the regression fit, thus artificially making its residual smaller.

EXAMPLE 4.7

PalmBch

More butterfly ballots Here again is the "Unusual Observations" portion of the output for the simple linear model to predict the number of *Buchanan* votes from the number of *Bush* votes in the Florida counties. We see two counties flagged as having high standardized residuals, Dade (stdres = −3.06) and, of course, Palm Beach (stdres = 7.65).

Unusual Observations

Obs	Bush	Buchanan	Fit	SE Fit	Residual	St Resid
6	177279	789.0	916.9	111.1	-127.9	-0.38 X
13	289456	561.0	1468.5	193.0	-907.5	-3.06RX
29	176967	836.0	915.4	110.9	-79.4	-0.24 X
50	152846	3407.0	796.8	94.2	2610.2	7.65R
52	184312	1010.0	951.5	116.1	58.5	0.17 X

R denotes an observation with a large standardized residual.
X denotes an observation whose X value gives it large leverage.

Both of these points were also identified earlier as having high leverage (Dade, $h_{13} = 0.2975$ and Palm Beach, $h_{50} = 0.07085$), although Palm Beach's leverage doesn't exceed the "3 times typical" threshold to be flagged as a large leverage point in the output. Using the estimated standard deviation of the regression, $\hat{\sigma}_\epsilon = 353.92$, we can confirm the calculations of these standardized residuals:

$$\text{stdres}_{13} = \frac{561 - 1468.5}{353.92\sqrt{1 - 0.2975}} = -3.06 \qquad \text{(Dade)}$$

$$\text{stdres}_{50} = \frac{3407 - 796.8}{353.92\sqrt{1 - 0.07085}} = 7.65 \qquad \text{(Palm Beach)}$$

Although statistical software can also compute the studentized residuals, we show the explicit calculation in this situation to illustrate the effect of omitting the point and reestimating the standard deviation of the regression. For example, if we refit the model without the Dade County data point, the new standard deviation is $\hat{\sigma}_{(13)} = 330.00$, and without Palm Beach this goes down to $\hat{\sigma}_{(50)} = 112.45$. Thus, for the studentized residuals we have

$$\text{studres}_{13} = \frac{561 - 1468.5}{330.00\sqrt{1 - 0.2975}} = -3.38 \qquad \text{(Dade)}$$

*Minitab flags cases beyond the more liberal ± 2, while R uses the ± 3 bounds.

$$\text{studres}_{50} = \frac{3407 - 796.8}{112.45\sqrt{1 - 0.07085}} = 24.08 \qquad \text{(Palm Beach)}$$

When comparing the two regression models, the percentage of variability explained by the model is much higher when Palm Beach County is omitted ($R^2 = 75.2\%$ compared to $R^2 = 38.9\%$) and the standard deviation of the regression is much lower ($\hat{\sigma}_\epsilon = 112.5$ compared to $\hat{\sigma}_\epsilon = 353.9$). This is not surprising since the Palm Beach County data point added a huge amount of variability to the response variable (the Buchanan vote) that was poorly explained by the model.

Cook's Distance

The amount of influence that a particular data case has on the estimated regression equation depends both on how close the case lies to the trend of the rest of the data (as measured by its standardized or studentized residual) and on its leverage (as measured by h_i). It's useful to have a statistic that reflects both of these measurements to indicate the impact of any specific case on the fitted model.

COOK'S DISTANCE

The **Cook's distance** of a data point in a regression with k predictors is given by

$$D_i = \frac{(\text{stdres}_i)^2}{k+1}\left(\frac{h_i}{1-h_i}\right)$$

A large Cook's distance indicates a point that strongly influences the regression fit. Note that this can occur with a large standardized residual, a large leverage, or some combination of the two. As a rough rule, we say that $D_i > 0.5$ indicates a moderately influential case and $D_i > 1$ shows a case that is very influential. For example, in the previous linear regression both Palm Beach ($D_{50} = 2.23$) and Dade ($D_{13} = 1.98$) would be flagged as very influential counties in the least squares fit. The next biggest Cook's distance is for Orange County ($D_{48} = 0.016$), which is not very unusual at all.

EXAMPLE 4.8

PalmBch

Butterfly ballots again The computer output shown on page 168 for the model using *Bush* votes to predict *Buchanan* votes identified five unusual points. Four cases (Counties #6, #13, #29, and #52) were flagged due to high leverage and two cases (Counties #13 and #50) were flagged due to high standardized residuals. One of the high residual cases also had high leverage: County #13 (Palm Beach County).

So which of these counties (if any) might be considered influential using the criterion of Cook's distance? The results are summarized in Table 4.1. Two of these five cases show a Cook's D value exceeding the 1.0 threshold.

- Three of the counties (#6, #29, and #52) have high or moderately high leverage but are predicted fairly accurately so they each have a small standardized residual and small Cook's D.
- One county (#50, Palm Beach County) has a standardized residual well beyond ±3 and also has a moderately high leverage, so it has a large Cook's D.
- One county (#13, Dade County) has the most extreme leverage and also has a large (negative) standardized residual, giving it a large Cook's D.

County	Bush	Buchanan	$stdres_i$	h_i	Cook's D_i
6	177279	789	−0.38	0.0986	0.01
13	289456	561	−3.06	0.2975	1.98
29	176967	836	−0.24	0.0982	0.00
50	152846	3407	7.65	0.0807	2.23
52	184312	1010	0.17	0.1076	0.00

TABLE 4.1 Unusual cases in regression for butterfly ballot model

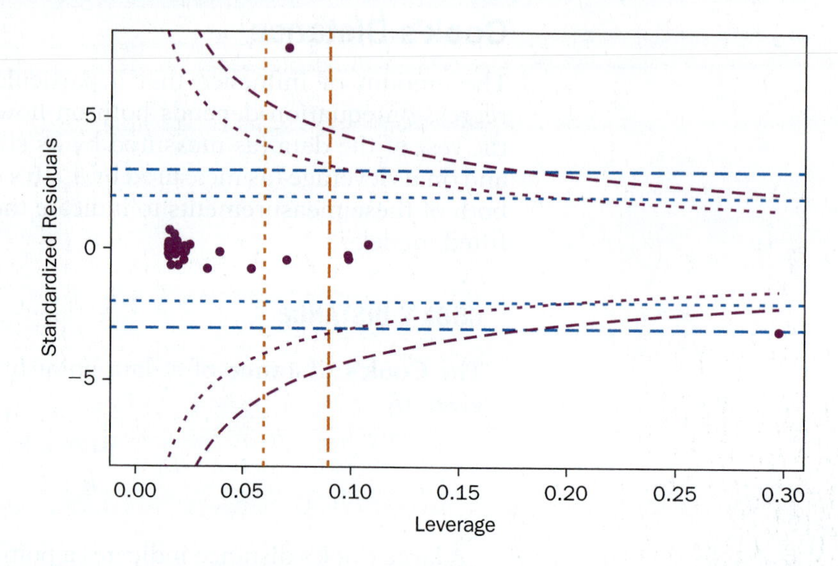

FIGURE 4.10 Unusual observations in the regression for the butterfly ballot model

One can produce a nice plot using R to summarize these relationships, as shown in Figure 4.10. Leverage (h_i) values are shown on the horizontal axis (representing unusual combinations of the predictors), and standardized residuals are plotted on the vertical axis (showing unusual responses for a given set of predictor values). Vertical and horizontal lines show the "moderately unusual" and "very unusual" boundaries for each of these quantities. The curved lines show the boundaries where the combination (as measured by Cook's D) becomes unusual, beyond 0.5 or 1.0. There's a lot going on in this plot, but if you locate each of the unusual counties from Table 4.1 in Figure 4.10, you should get a good feel for the role of the different measures for unusual points in regression.

IDENTIFYING UNUSUAL POINTS IN REGRESSION—SUMMARY

For a multiple regression model with k predictors fit with n data cases:

Statistic	Moderately unusual	Very unusual
Leverage, h_i	above $2(k+1)/n$	above $3(k+1)/n$
Standardized residual	beyond ± 2	beyond ± 3
Studentized residual	beyond ± 2	beyond ± 3
Cook's D	above 0.5	above 1.0

Some statistical software will provide additional diagnostic measures such as **DFfits** and **DFbetas** that also assess influence. Like a studentized residual, DFfits measures the difference between the fitted values for a case for models built with and without that case. Likewise, DFbetas measures the difference between coefficient estimates for models built with and without the individual case.

We conclude with a final caution about using the guidelines provided in this section to identify potential outliers or influential points while fitting a regression model. The goal of these diagnostic tools is to help us identify cases that might need further investigation. Points identified as outliers or high-leverage points might be data errors that need to be fixed or special cases that need to be studied further. Doing an analysis with and without a suspicious case is often a good strategy to see how the model is affected. *We should avoid blindly deleting all unusual cases until the data that remain are "nice." In many situations (like the butterfly ballot scenario), the most important features of the data would be lost if the unusual points were dropped from the analysis!*

TOPIC 4.5 Coding Categorical Predictors

In most of our regression examples, the predictors have been quantitative variables. However, we have seen how a *binary* categorical variable, such as the *Sex* of a butterfly in Example 1.11 or *Sex* in the regression model for children's growth rates in Example 3.10, can be incorporated into a model using an indicator variable. But what about a categorical variable that has more than two categories? If a {0,1} indicator worked for two categories, perhaps we should use {0,1,2} for three categories. As we show in the next example, a better method is to use multiple indicator variables for the different categories.

EXAMPLE 4.9

Cars17

Car prices In Example 1.1, we considered a simple linear model to predict the prices (in thousands of dollars) of used Accords offered for sale at an Internet site, based on the mileages (in thousands of miles) of the cars. Suppose now that we also have similar data on the prices of two other car models, Mazda6s and Maximas, as in the file **ThreeCars2017**. The software output below shows the descriptive statistics for each type of car.

Variable	CarType	N	Mean	StDev
Price	Accord	30	14.28	5.74
	Maxima	30	15.46	4.56
	Mazda6	30	11.50	5.68

Since the car type is categorical, we might consider coding the information numerically with a variable that assigns a 1 to each Accord, a 2 to each Mazda6, and a 3 to each Maxima. But such a method for coding the car types numerically is flawed, as it forces the fitted price for Mazda6s to be exactly halfway between the predictions for Accords and Maximas.

A better way to handle the categorical information on car type is to produce separate indicator variables for each of the car models. For example,

$$Accord = \begin{cases} 1 & \text{if Accord} \\ 0 & \text{if not} \end{cases} \qquad Mazda6 = \begin{cases} 1 & \text{if Mazda6} \\ 0 & \text{if not} \end{cases}$$

$$Maxima = \begin{cases} 1 & \text{if Maxima} \\ 0 & \text{if not} \end{cases}$$

If we try to fit all three indicator variables in the same multiple regression model

$$Price = \beta_0 + \beta_1 Accord + \beta_2 Mazda6 + \beta_3 Maxima + \epsilon$$

we should experience some difficulty since any one of the indicator variables is *exactly* a linear function of the other two. For example,

$$Accord = 1 - Mazda6 - Maxima$$

This means there is not a unique solution when we try to estimate the coefficients by minimizing the sum of squared errors. Most software packages will either give an error message if we try to include all three predictors or automatically drop one of the indicator predictors from the model. While it might seem counterintuitive at first, including all but one of the indicator predictors is exactly the right approach.

Suppose we drop the *Accord* predictor and fit the model

$$Price = \beta_0 + \beta_1 Mazda6 + \beta_2 Maxima + \epsilon$$

to the data in **ThreeCars2017** and obtain a fitted prediction equation:

$$\widehat{Price} = 14.277 - 2.773 Mazda6 + 1.187 Maxima$$

Can we still recover useful information about Accords as well as the other two types of cars? Notice that the constant coefficient in the fitted model, $\hat{\beta}_0 = 14.277$, matches the sample mean for the Accords. This is no accident since the values of the other two predictors are both zero for Accords. Thus, the predicted value for Accords from the regression is the same as the mean Accord price in the sample. The estimated coefficient of *Mazda6*, $\hat{\beta}_1 = -2.773$, indicates that the prediction should decrease by 2.773 when *Mazda6* goes from 0 to 1. Up to a round-off difference, that is how much smaller the mean Mazda6 price in this sample is compared to the mean Accord price. Similarly, the fitted coefficient of Maxima, $\hat{\beta}_2 = 1.187$, means we should add 1.187 to the Accord mean to get the Maxima mean, $14.277 + 1.187 = 15.464$.

REGRESSION USING INDICATORS FOR MULTIPLE CATEGORIES

To include a categorical variable with K categories in a regression model, use indicator variables, $I_1, I_2, \ldots, I_{K-1}$, for all but one of the categories:

$$Y = \beta_0 + \beta_1 I_1 + \beta_2 I_2 + \cdots + \beta_{K-1} I_{K-1} + \epsilon$$

We call the category that is not included as an indicator in the model the *reference* category. The constant term represents the mean for that category and the coefficient of any indicator predictor gives the difference of that category's mean from the reference category.

The idea of leaving out one of the categories when fitting a regression model with indicators may not seem so strange if you recall that, in our previous regression examples with binary categorical predictors, we used just one indicator predictor. For example, to include information on gender, we might use *Female* = 0 for males and *Female* = 1 for females, rather than using two different indicators for each gender. We can also extend the ideas of Section 3.3 to include quantitative variables in a regression model along with categorical indicators.

EXAMPLE 4.10 More car prices

CHOOSE

Cars17

Figure 4.11 shows a scatterplot of the relationship of *Price* and *Mileage* with different symbols indicating the type of car (Accord, Mazda6, or Maxima). The plot shows a decreasing trend for each of the car types, with Accords tending to be more expensive and Maximas tending to have more miles. While we could separate the data and explore separate simple linear

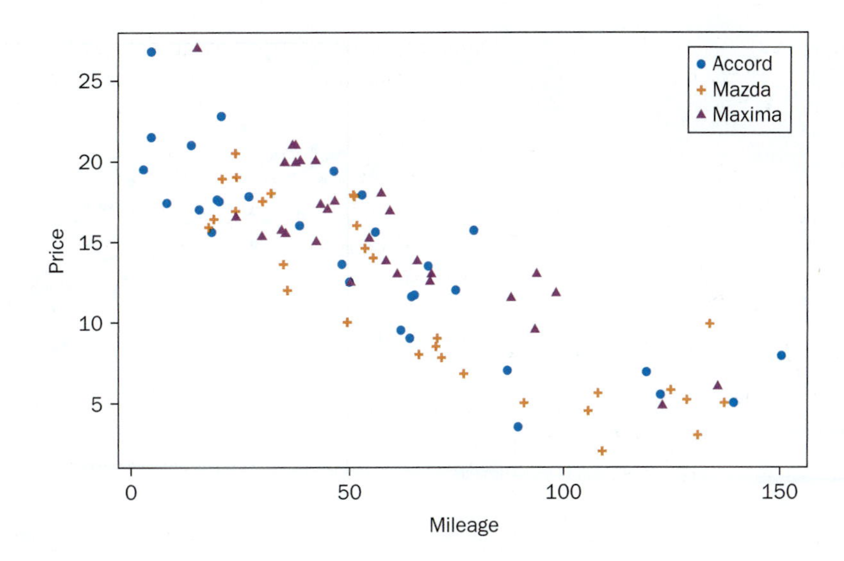

FIGURE 4.11 *Price* versus *Mileage* for three car types

regression fits for each of the car types, we can also combine all three in a common multiple regression model using indicator predictors along with the quantitative *Mileage* variable.

Using indicator predictors as defined in the previous example, consider the following model:

$$Price = \beta_0 + \beta_1 Mileage + \beta_2 Mazda6 + \beta_3 Maxima + \epsilon$$

Once again, we omit the Accord category and use it as the reference group.* This model allows for prices to change with mileage for all of the cars, but adjusts the intercept for the type of car.

FIT

Here is the R output for fitting this model to the **ThreeCars2017** data:

```
Coefficients:
(Intercept)  Mileage   Mazda6   Maxima
21.087       -0.1249   -1.2616  1.5397
```

The fitted prediction equation is

$$\widehat{Price} = 21.087 - 0.1249 Mileage - 1.2616 Mazda6 + 1.5397 Maxima$$

For an Accord ($Mazda6 = 0$ and $Maxima = 0$), we start at a base price of 21.087 (or \$21,087 when $Mileage = 0$) and see a decrease of about 0.1249 (or \$124.90) for every increase of one (thousand) miles driven. The coefficient of *Mazda6* indicates that the predicted price of a Mazda6 is 1.2616 (or \$1261.6) less than an Accord *that has the same number of miles*. Similarly, a Maxima with the same mileage is estimated to cost about 1.5397 (or \$1539.7) more than the Accord.

This model assumes that the rate of depreciation (the decrease in *Price* as *Mileage* increases) is the same for all three car models, about 0.1249 thousand dollars per thousand miles. Thus we can think of this model as specifying three different regression lines, each with the same slope but possibly different intercepts. Figure 4.12 shows a scatterplot of *Price* versus *Mileage* with the fitted model for each type of car.

*The reference group can be any of the groups. If you are interested in comparisons to a certain group, the analysis will be simplest if you use that as the reference group.

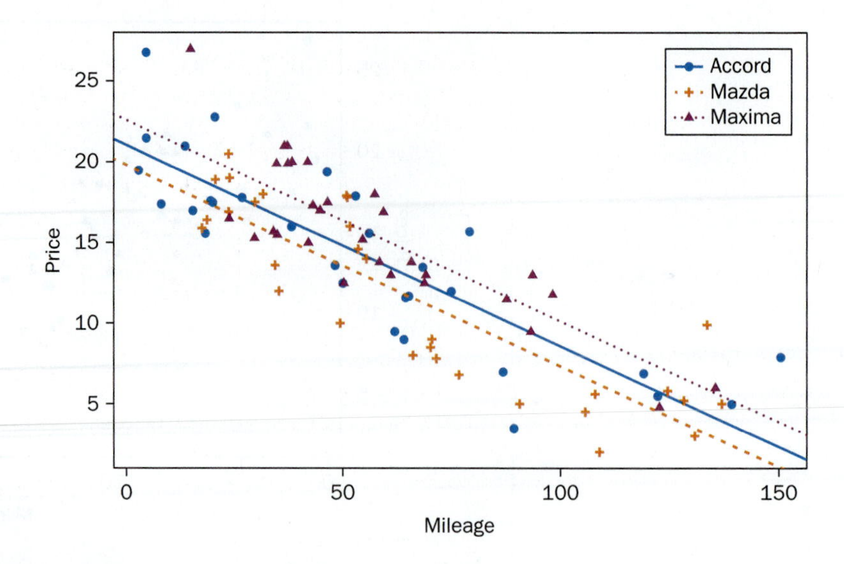

FIGURE 4.12 *Price* versus *Mileage* with equal slope fits

ASSESS

The form of the model suggests two immediate questions to consider in its assessment:

1. Are the differences between the car types statistically significant? For example, are the adjustments in the intercepts really needed, or could we do just about as well by using the same regression line for each of the car types?
2. Is the assumption of a common slope reasonable, or could we gain a significant improvement in the fit by allowing for a different slope for each car type?

The first question asks whether the model is unnecessarily complicated and a simpler model might work just as well. The second asks whether the model is complicated enough to capture the main relationships in the data. Either of these questions is often important to ask when assessing the adequacy of any model.

We can address the first question by examining the individual t-tests for the coefficients in the model:

```
Coefficients:
             Estimate  Std. Error  t value  Pr(>|t|)
(Intercept)  21.08738  0.68281      30.883  < 2e-16 ***
Mileage      -0.12491  0.008252    -15.136  < 2e-16 ***
Mazda6       -1.26155  0.733145     -1.721  0.0889
Maxima        1.53974  0.726685      2.119  0.0370 *
```

The computer output shows that most of the coefficients in the model are significantly different from zero, and even the Mazda6 coefficient shows evidence of not being zero; thus we want to keep all of the terms in the model. So, not surprisingly, we find the average price does tend to decrease as the mileage increases for all three car models. Also, the average prices at the same mileage are different for the three car models, with Maximas being the most expensive, Mazda6s the least, and Accords somewhere in between.

To consider a more complicated model that would allow for different slopes, we again extend the work of Section 3.3 to consider the following model:

$$\begin{aligned} Price = \beta_0 + \beta_1 Mileage + \beta_2 Mazda6 + \beta_3 Maxima + \beta_4 Mazda6 \cdot Mileage \\ + \beta_5 Maxima \cdot Mileage + \epsilon \end{aligned}$$

Before we estimate coefficients to fit the model, take a moment to anticipate the role of each of the terms. When the car is an Accord, we have $Mazda6 = Maxima = 0$ and the model reduces to

$$Price = \beta_0 + \beta_1 Mileage + \epsilon \qquad \text{(for Accord)}$$

so β_0 and β_1 represent the slope and intercept for predicting Accord *Price* based on *Mileage*. Going to a Mazda6 adds two more terms:

$$\begin{aligned} Price &= \beta_0 + \beta_1 Mileage + \beta_2 + \beta_4 Mileage + \epsilon \\ &= (\beta_0 + \beta_2) + (\beta_1 + \beta_4) Mileage + \epsilon \qquad \text{(for Mazda6)} \end{aligned}$$

Thus, the β_2 coefficient is the *change* in intercept for a Mazda6 compared to an Accord and β_4 is the change in slope. Similarly, β_3 and β_5 represent the difference in intercept and slope, respectively, for a Maxima compared to an Accord.

Here is some regression output for the more complicated model:

```
Coefficients:
                  Estimate   Std. Error  t value   Pr(>|t|)
(Intercept)       20.80961   0.87637     23.745    < 2e-16 ***
Mileage           -0.11981   0.01296     -9.242    1.93e-14 ***
Mazda6            -1.01679   1.35553     -0.750    0.4554
Maxima             2.46161   1.46790      1.677    0.0973
Mazda6*Mileage    -0.00460   0.01867     -0.247    0.8058
Maxima*Mileage    -0.01633   0.02254     -0.724    0.4709
```

We can determine separate regression lines for each of the car models by substituting into the fitted prediction equation:

$$\begin{aligned} \widehat{Price} = 20.81 - 0.12 Mileage - 1.02 Mazda6 + 2.46 Maxima \\ - 0.005 Mazda6 \cdot Mileage - 0.016 Maxima \cdot Mileage \end{aligned}$$

These lines are plotted on the scatterplot in Figure 4.13. Is there a statistically significant difference in the slopes of those lines? We see that neither of the new terms $Mazda6 \cdot Mileage$ and $Maxima \cdot Mileage$ has a coefficient that would be considered statistically different from zero. However, we should take care when interpreting these individual t-tests since there are obvious relationships between the predictors (such as $Mazda6$ and $Mazda6 \cdot Mileage$) in this model. For that reason we should also consider a nested F-test to assess the terms simultaneously.

To see if the different slopes are really needed, we test

$H_0 : \beta_4 = \beta_5 = 0$
$H_a :$ at least one $\beta_i \neq 0, i = 4, 5$

For the full model with all five predictors, we have $SSE_{Full} = 676.27$ with $90 - 5 - 1 = 84$ degrees of freedom, while the reduced model using just the first three terms has $SSE_{Reduced} = 680.51$ with 86 degrees of freedom. Thus, the two extra predictors account for $680.51 - 676.27 = 4.24$ in new variability explained. To compute the test statistic for the nested F-test in this case:

$$F = \frac{4.24/2}{676.27/84} = 0.26$$

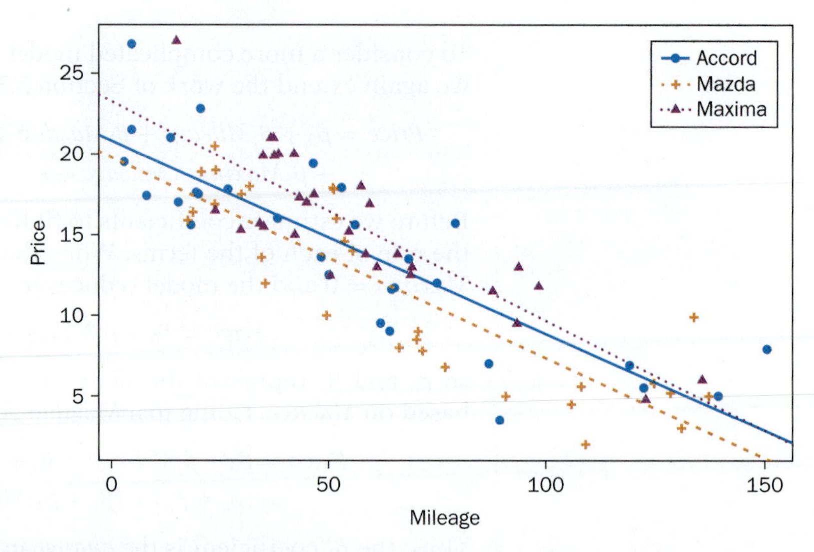

FIGURE 4.13 *Price* versus *Mileage* with different linear fits

k	SSE	R^2	adj R^2	Predictors
2	2493	9.0%	6.9%	*Mazda6, Maxima*
1	797	70.9%	70.6%	*Mileage*
3	680.5	75.2%	74.3%	*Mileage, Mazda6, Maxima*
5	676.3	75.3%	73.9%	*Mileage, Mazda6, Maxima, Mazda6 · Mileage, Maxima · Mileage*

TABLE 4.2 Several models for predicting car prices

Comparing to an *F*-distribution with 2 and 84 degrees of freedom gives a *P*-value = 0.77, which indicates that the extra terms to allow for different slopes do not produce a significantly better model.

Table 4.2 shows a summary of several possible models for predicting *Price* based on the information in *Mileage* and *CarType*. Since the interaction terms were insignificant in the nested *F*-test, the model including *Mileage* and the indicators for *Mazda*6 and *Maxima* would be a good choice for predicting prices of these types of cars.

Before we use the model based on *Mileage*, *Mazda*6, and *Maxima* to predict some car prices, we should also check the regression conditions. Figure 4.14(a) shows a plot of the studentized residuals versus fits that produce a reasonably consistent band around zero with no studentized residuals beyond ±3.* The linear pattern in the normal quantile plot of the residuals in Figure 4.14(b) indicates that the normality condition is met.

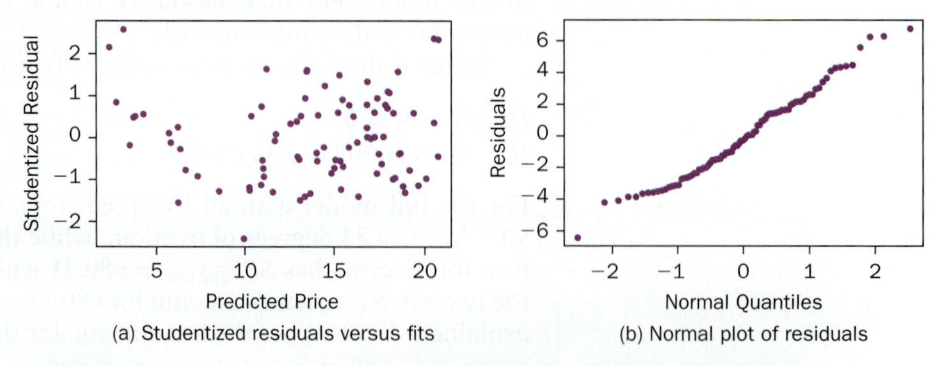

(a) Studentized residuals versus fits

(b) Normal plot of residuals

FIGURE 4.14 Residual plots for *Price* model based on *Mileage*, *Accord*, and *Mazda6*

*The residual plot shows a bit of a pattern that hints at considering a quadratic effect of mileage on price. We explore this extension to the model in the exercises.

USE

Suppose that we are interested in used cars with about 50 (thousand) miles. What prices might we expect to see if we are choosing from among Accords, Mazda6s, or Maximas? Requesting 95% confidence intervals for the mean price and 95% prediction intervals for individual prices of each car model when *Mileage* = 50 yields the information below (with prices converted to dollars):

Car	Predicted Price	95% Confidence Interval	95% Prediction Interval
Accord	$14,842	($13,818, $15,866)	($9,157, $20,527)
Mazda6	$13,581	($12,524, $14,637)	($7,890, $19,272)
Maxima	$16,381	($15,354, $17,410)	($10,696, $22,068)

TOPIC 4.6 Randomization Test for a Relationship

The inference procedures for linear regression depend to varying degrees on the conditions for the linear model being met. If the errors are not normally distributed, the relationship is not linear, or the variance is not constant over values of the predictor(s), we should be wary of conclusions drawn from tests or intervals that are based on the t-distribution. In many cases, we can find transformations (as described in Section 1.4) that produce new data for which the linear model conditions are more reasonable. Another alternative is to use different procedures for testing the significance of a relationship or constructing an interval that are less dependent on the conditions of a linear model.

EXAMPLE 4.11

SATGPA

Predicting GPAs with SAT scores In recent years, many colleges have reexamined the traditional role that scores on the Scholastic Aptitude Tests (SATs) play in making decisions on which students to admit. Do SAT scores really help predict success in college? To investigate this question, a group of 24 introductory statistics students[1] supplied the data in Table 4.3 showing their score on the Verbal portion of the SAT as well as their current grade point average (GPA) on a 0.0–4.0 scale (the data along with Math SAT scores are in **SATGPA**). Figure 4.15 shows a scatterplot with a least squares regression line to predict GPA using the Verbal SAT scores. The sample correlation between these variables is $r = 0.244$, which produces a P-value of 0.25 for a two-tailed t-test of $H_0 : \rho = 0$ versus $H_a : \rho \neq 0$ with 22 degrees of freedom. This sample provides little evidence of a linear relationship between GPA and Verbal SAT scores.

VerbalSAT	GPA	VerbalSAT	GPA	VerbalSAT	GPA
420	2.90	500	2.77	640	3.27
530	2.83	630	2.90	560	3.30
540	2.90	550	3.00	680	2.60
640	3.30	570	3.25	550	3.53
630	3.61	300	3.13	550	2.67
550	2.75	570	3.53	700	3.30
600	2.75	530	3.10	650	3.50
500	3.00	540	3.20	640	3.70

TABLE 4.3 Verbal SAT scores and GPA

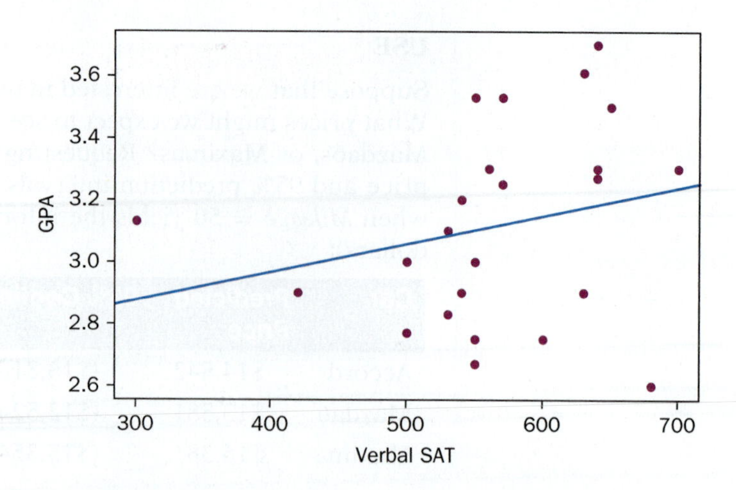

FIGURE 4.15 Linear regression for GPA based on Verbal SAT

One concern with fitting this model might be the potential influence of the high-leverage point ($h_{13} = 0.457$) for the Verbal SAT score of 300. The correlation between GPA and Verbal SAT would increase to 0.333 if this point was omitted. However, we shouldn't ignore the fact that the student with the lowest Verbal SAT score still managed to earn a GPA slightly above the average for the whole group.

If there really is no relationship between a student's Verbal SAT score and his or her GPA, we could reasonably expect to see any of the 24 GPAs in this sample associated with any of the 24 Verbal SAT scores. This key idea provides the foundation for a **randomization test** of the hypotheses:

randomization test

H_0: GPAs are unrelated to Verbal SAT scores ($\rho = 0$)
H_a: GPAs are related to Verbal SAT scores ($\rho \neq 0$).

The basic idea is to scramble the GPAs so they are randomly assigned to each of the 24 students in the sample (with no relationship to the Verbal SAT scores) and compute a measure of association, such as the sample correlation r, for the "new" sample. Table 4.4 shows the results of one such randomization of the GPA values from Table 4.3; for this randomization the sample correlation with Verbal SAT is $r = -0.188$. If we repeat this randomization process many, many times and record the sample correlations in each case, we can get a good picture of what the r-values would look like if the GPA and Verbal SAT scores were unrelated. If the correlation from the original sample ($r = 0.244$) falls in a "typical" place in this **randomization distribution**, we would conclude that the sample does not provide evidence of a relationship between GPA and Verbal SAT. On the other hand, if the original correlation falls at an extreme point in either tail of the randomization distribution, we can conclude that the sample is not consistent with the null hypothesis of "no relationship" and thus GPAs are probably related to Verbal SAT scores.

randomization distribution

Figure 4.16 shows a histogram of the sample correlations with Verbal SAT obtained from 1000 randomizations of the GPA data.* Among these permuted samples, we find 239 cases where the sample correlation was more extreme in absolute value than the $r = 0.244$ that was observed in the original sample: 116 values below -0.244 and 123 values above $+0.244$. Thus the approximate P-value from this randomization test is $239/1000 = 0.239$, and

*In general, we use more randomizations, 10,000 or even more, to estimate a P-value, but we chose a smaller number to illustrate this first example.

VerbalSAT	GPA	VerbalSAT	GPA	VerbalSAT	GPA
420	3.53	500	2.90	640	2.75
530	3.50	630	2.90	560	3.61
540	3.00	550	3.70	680	3.25
640	3.27	570	2.83	550	3.00
630	2.77	300	3.30	550	2.90
550	3.30	570	3.20	700	3.13
600	3.53	530	2.60	650	2.67
500	2.75	540	3.50	640	3.30

TABLE 4.4 Verbal SAT scores and with one set of randomized GPAs

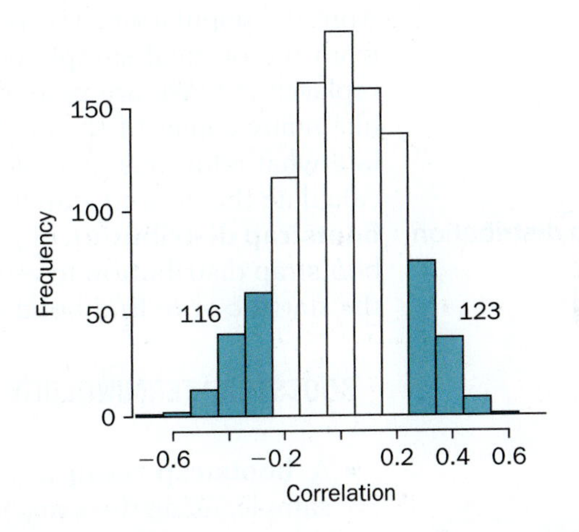

FIGURE 4.16 Randomization distribution for 1000 correlations of GPA versus Verbal SAT

we conclude that the original sample does not provide significant evidence of a relationship between Verbal SAT scores and GPAs. By random chance alone, about 24% of the randomization correlations generated under the null hypothesis of no relationship were more extreme than what we observed in our sample. Note that the randomization *P*-value would differ slightly if a new set of 1000 randomizations were generated, but shouldn't change very much. This value is also consistent with the *P*-value of 0.25 that was obtained using the standard *t*-test for correlation with the original data. However, the *t*-test is only valid if the normality condition is satisfied, whereas the randomization test does not depend on normality.

Modern computing power has made randomization approaches to testing hypotheses much more feasible. These procedures generally require much less stringent assumptions about the underlying structure of the data than classical tests based on the normal or *t*-distributions. Although we chose to use the sample correlation as the statistic of interest for each of the randomizations in the previous example, we just as easily could have used the sample slope, SSModel, the standard error of regression, or some other measure of the effectiveness of the model. Randomization techniques give us flexibility to work with different test statistics, even when a derivation of the theoretical distribution of a statistic may be unfeasible.

TOPIC 4.7 Bootstrap for Regression

Section 4.6 introduced the idea of a randomization test as an alternate procedure to doing a traditional t-test when the conditions for a linear model might not apply. In this section, we examine another technique for doing inference on a regression model that is also less dependent on conditions such as the normality of errors. The procedure is known as **bootstrapping**.* The basic idea is to use the data to generate an approximate sampling distribution for the statistic of interest, rather than relying on conditions being met to justify using some theoretical distribution.

bootstrapping

In general, a sampling distribution shows how the values of a statistic (such as a mean, standard deviation, or regression coefficient) vary when taking many samples of the same size from the same population. In practice, we generally have just our original sample and cannot generate lots of new samples from the population. The bootstrap procedure involves creating new samples from the original sample data (not the whole population) by sampling with replacement. We are essentially assuming that the population looks roughly like many copies of the original sample. Using this assumption we can simulate what additional samples might look like. For each simulated sample we calculate the desired statistic, repeating the process many times to generate a **bootstrap distribution** of possible values for the statistic. We can then use this bootstrap distribution to estimate quantities such as the standard deviation of the statistic or to find bounds on plausible values for the parameter.

bootstrap distribution

BOOTSTRAP TERMINOLOGY

- A **bootstrap sample** is chosen with replacement from an existing sample, using the same sample size.
- A **bootstrap statistic** is a statistic computed for each bootstrap sample.
- A **bootstrap distribution** collects bootstrap statistics for many bootstrap samples.

EXAMPLE 4.12

Accord

Accord prices In Section 2.1, we considered inference for a regression model to predict the price (in thousands of dollars) of used Accords at an Internet site based on mileage (in thousands of miles). The data are in **AccordPrice**. Some of the regression output for fitting this model follows; Figure 4.17 shows a scatterplot with the least squares line.

```
Coefficients:
             Estimate  Std. Error  t value  Pr(>|t|)
(Intercept)  20.8096   0.9529       21.84   < 2e-16 ***
Mileage      -0.1198   0.0141       -8.50   3.06-09 ***
```

Although we showed in Example 1.6 on page 32 that the residuals for the Accord price model are fairly well behaved, let's assume that we don't want to rely on using the t-distribution to do inference about the slope in this model and would rather construct a bootstrap distribution of sample slopes. Since the original sample size was $n = 30$ and we want to assess the accuracy

*One of the developers of this approach, Brad Efron, used the term *bootstrap* since the procedure allows the sample to help determine the distribution of the statistic on its own, without assistance from distributional assumptions, thus "pulling itself up by its own bootstraps."

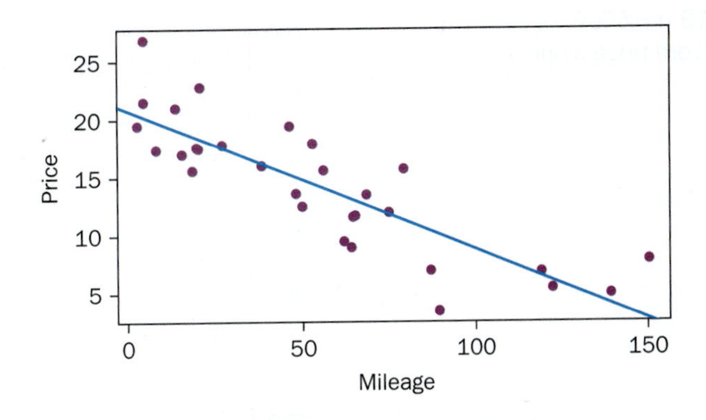

FIGURE 4.17 Original regression of Accord prices on mileage

for a sample slope based on this sample size, we select (with replacement) a random sample of 30 values from the original sample. One such sample is shown in Figure 4.18(a). The fitted regression model for this bootstrap sample is $\widehat{Price} = 22.13 - 0.121 \cdot Mileage$. Fitted regression coefficients for 9 additional bootstrap samples are shown below, with all 10 lines displayed in Figure 4.18(b):

Intercept	Slope	Intercept	Slope	Intercept	Slope
21.76	-0.149	20.25	-0.108	22.54	-0.161
21.24	-0.127	21.58	-0.131	19.35	-0.100
22.66	-0.152	21.25	-0.132	20.75	-0.116

To construct an approximate sampling distribution for the sample slope, we repeat this process many times and save all the bootstrap slope estimates. For example, Figure 4.19 shows a histogram of slopes from a bootstrap distribution based on 5000 samples from the Accord data. We can then estimate the standard deviation of the sample estimates, $SE_{\hat{\beta}_1}$, by computing the standard deviation of the slopes in the bootstrap distribution. In this case, the standard deviation of the 5000 bootstrap slopes is 0.0163 (which is similar to the standard error of the slope in the original regression output, 0.0141). In order to obtain reliable estimates of the standard error, we recommend using at least 5000 bootstrap samples. Fortunately, modern computers make this a quick task.

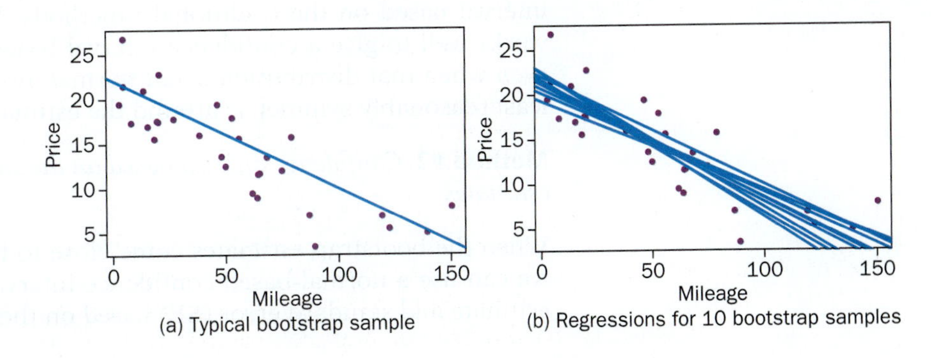

FIGURE 4.18 Accord regressions for bootstrap samples

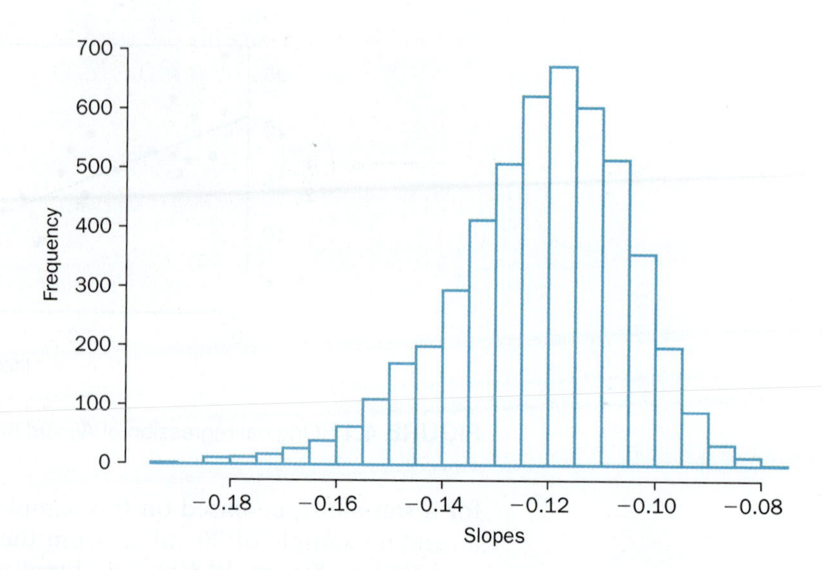

FIGURE 4.19 n = 5000 Bootstrap Accord price slopes

Confidence Intervals Based on Bootstrap Distributions

There are several methods for constructing a confidence interval for a parameter based on the information in a bootstrap distribution of sample estimates. For the Accord data the traditional t-interval is $-0.1198 \pm 0.0289 = (-0.1487, -0.0909)$. We now present three intervals based on a bootstrap distribution.

Method #1: *Confidence interval based on the quantiles from the bootstrap distribution*

One notion of a confidence interval is to find the middle 95% (or whatever confidence level you like) of the sampling distribution. Since we have a bootstrap distribution based on many simulated estimates, we can find the middle 95% of the bootstrap sample statistics. Let $q_{0.025}$ represent the point in the bootstrap distribution where only 2.5% of the estimates are smaller and $q_{0.975}$ be the point at the other end of the distribution, where just 2.5% of the sample statistics are larger. A 95% confidence interval can then be constructed as $(q_{0.025}, q_{0.975})$. For example, the quantiles using the 5000 bootstrap slopes in the previous example are $q_{0.025} = -0.1579$ and $q_{0.975} = -0.0944$.

Thus, the 95% confidence interval for the slope in the Accord price model would go from −0.1579 to −0.0944. Note that this interval is almost (but not exactly) symmetric about the estimated slope of −0.1198 and is close to the interval based on the traditional t-methods. Using the quantiles in this way works well to give a confidence interval based on the bootstrap distribution, even when that distribution is *not* normal, provided that the distribution is at least reasonably symmetric around the estimate.

Method #2: *Confidence interval based on the standard deviation of the bootstrap estimates*

When the bootstrap estimates come close to following a normal distribution, we can use a normal-based confidence interval with the original slope as the estimate and standard error (SE) based on the bootstrap distribution:

$$Estimate \pm z^* \cdot SE$$

where z^* is chosen from a standard normal distribution to reflect the desired level of confidence.

In the previous example for the Accord prices where the original slope estimate is $\hat{\beta}_1 = -0.1198$, this method gives

$$-0.1198 \pm 1.96(0.0163) = -0.1198 \pm 0.0319 = (-0.1517, -0.0879)$$

This compares to the traditional t-interval of $(-0.1487, -0.0909)$. The lower and upper confidence limits with these two different methods are relatively close.*

Method #3: *Confidence interval based on the standard deviation of the bootstrap estimates but with quantiles*

The final method we discuss for generating a confidence interval from a bootstrap distribution is, in a way, a combination of the first two methods. When using Method #2, it might be that z^* is not a great choice for a multiplier because our distribution is not symmetric. For example, suppose that in a hypothetical example such as the bootstrap distribution in Figure 4.20, the original parameter estimate is 20 and the lower quantile ($q_{0.025}$) is 4 units below the original estimate, but the upper quantile ($q_{0.975}$) is 10 units above the original estimate.

For Method #3 we replace $\pm z^*$ with quantiles from a scaled version of the bootstrap distribution. Specifically, consider the distribution of $T^* = \left(\hat{\beta}_1^* - \hat{\beta}_1\right)/SE^*$ where $\hat{\beta}_1^*$ is a bootstrap slope and SE^* is the standard error of $\hat{\beta}_1^*$, that is, the standard error of the slope for that bootstrap sample.† We find the 0.025 and 0.975 quantiles of this distribution (denoted as $qt_{0.025}$—which will be negative—and $qt_{0.975}$)‡ and use this in place of $\pm z^*$.

We construct the lower limit of the confidence interval as

$$Estimate - qt_{0.975} \cdot SE$$

and the upper limit of the confidence interval as

$$Estimate - qt_{0.025} \cdot SE$$

where the SE here is the standard error from the original regression.

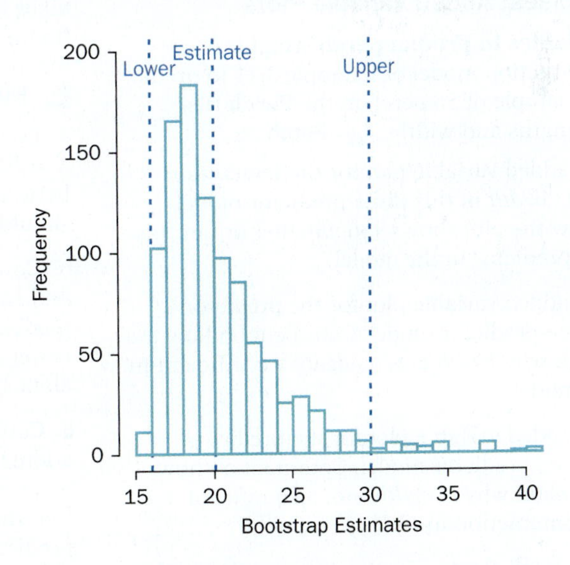

FIGURE 4.20 Hypothetical skewed bootstrap distribution

*We may have some reservations about using the normal distribution here given the slight skewness in the bootstrap distribution in Figure 4.19.

†We are following bootstrapping convention by using * here to denote a bootstrap slope. Don't confuse this with z^*, where the * donates a critical value such as $z^* = 1.96$.

‡We use the notation $qt_{0.025}$ rather than $q_{0.025}$ to distinguish between the scaled bootstrap distribution that we are using within Method #3 (and which often has close to a t-distribution) and the unscaled bootstrap distribution used within Method #1.

For the bootstrap distribution of the Accord price slopes the quantiles for the T^* values are $qt_{0.025} = -2.268$ and $qt_{0.975} = 2.221$. The SE for the slope is 0.0141, as we saw in Example 4.12, which gives the confidence interval

$$-0.1198 - 2.221(0.0141) \text{ to } -0.1198 + 2.268(0.0141) = (-0.1511, -0.0878)$$

There are a number of other, more sophisticated methods for constructing a confidence interval from a bootstrap distribution, but they are beyond the scope of this text. There are also other methods for generating bootstrap samples from a given model. For example, we could keep all of the original predictor values fixed and generate new sample responses by randomly selecting from the residuals of the original fit and adding them to the predicted values. This procedure assumes only that the errors are independent and identically distributed. The methods for constructing confidence intervals from the bootstrap distribution would remain the same. See Exercise 4.24 for an example of this approach to generating bootstrap samples.

The three methods described here for constructing bootstrap confidence intervals should help you see the reasoning behind bootstrap methods and give you tools that should work reasonably well in most situations. One of the key features of the bootstrap approach is that it can be applied relatively easily to simulate the sampling distribution of almost any estimate that is derived from the original sample. So if we want a confidence interval for a coefficient in a multiple regression model, the standard deviation of the error term, or a sample correlation, the same methods can be used: Generate bootstrap samples, compute and store the statistic for each sample, then use the resulting bootstrap distribution to gain information about the sampling distribution of the statistic.

EXERCISES

Topic 4.1 Exercises: Added Variable Plots

4.1 Adding variables to predict perch weights. Consider the interaction model of Example 3.11 to predict the weights of a sample of 56 perch in the **Perch** file based on their lengths and widths. **Perch**

a. Construct an added variable plot for the interaction predictor *Length* · *Width* in this three-predictor model. Comment on how the plot shows you whether or not this is an important predictor in the model.

b. Construct an added variable plot for the predictor *Width* in this three-predictor model. Comment on how the plot helps explain why *Width* gets a negative coefficient in the interaction model.

c. Construct an added variable plot for the predictor *Length* in this three-predictor model. Comment on how the plot helps explain why *Length* is not a significant predictor in the interaction model.

4.2 Adirondack High Peaks. Forty-six mountains in the Adirondacks of upstate New York are known as the High Peaks, with elevations near or above 4000 feet (although modern measurements show a couple of the peaks are actually slightly under 4000 feet). A goal for hikers in the region is to become a "46er" by scaling each of these peaks. The file **HighPeaks** contains information on the elevation (in feet) of each peak along with data on typical hikes including the ascent (in feet), round-trip distance (in miles), difficulty rating (on a 1 to 7 scale, with 7 being the most difficult), and expected trip time (in hours). **HiPeak**

a. Look at a scatterplot of $Y = Time$ of a hike versus $X = Elevation$ of the mountain and find the correlation between these two variables. Does it look like *Elevation* should be very helpful in predicting *Time*?

b. Consider a multiple regression model using *Elevation* and *Length* to predict *Time*. Is *Elevation* important in this model? Does this two-predictor model do substantially better at explaining *Time* than either *Elevation* or *Length* alone?

c. Construct an added variables plot to see the effect of adding *Elevation* to a model that contains just *Length* when predicting the typical trip *Time*. Does the plot show that there is information in *Elevation* that is useful for predicting *Time* after accounting for trip *Length*? Explain.

Topic 4.2 Exercises: Techniques for Choosing Predictors

4.3 Major League Baseball winning percentage. Consider the **MLBStandings2016** file described in Exercise 3.56 on page 152. In this exercise, you will consider models with four variables to predict winning percentages (*WinPct*) using any of the predictors except *Wins* and *Losses*—those would make it too easy! Don't worry if you are unfamiliar with baseball terminology and some of the acronyms for variable names have little meaning. Although knowledge of the context for data is often helpful for choosing good models, you should use software to build the models requested in this exercise rather than using specific baseball knowledge. **MLB16**

a. Use forward selection until you have a four-predictor model for *WinPct*. You may need to adjust the criteria if the procedure won't give sufficient steps initially. Write down the predictors in this model and its R^2 value.

b. Use backward elimination until you have a four-predictor model for *WinPct*. You may need to adjust the criteria if the procedure stops too soon to continue eliminating predictors. Write down the predictors in this model and its R^2 value.

c. Use a "best subsets" procedure to determine which four predictors together would explain the most variability in *WinPct*. Write down the predictors in this model and its R^2 value.

d. Find the value of Mallows's C_p for each of the models produced in (a–c).

e. Assuming that these three procedures didn't all give the same four-predictor model, which model would you prefer? Explain why.

4.4–4.6 Fertility measurements. A medical doctor[2] and her team of researchers collected a variety of data on women who were having troubling getting pregnant. Data for randomly selected patients where complete information was available are provided in the file **Fertility**, including: **Fertil**

Variable	Description
Age	Age in years
LowAFC	Smallest antral follicle count
MeanAFC	Average antral follicle count
FSH	Maximum follicle stimulating hormone level
E2	Fertility level
MaxE2	Maximum fertility level
MaxDailyGn	Maximum daily gonadotropin level
TotalGn	Total gonadotropin level
Oocytes	Number of egg cells
Embryos	Number of embryos

A key method for assessing fertility is a count of antral follicles that can be performed with noninvasive ultrasound. Researchers are interested in how the other variables are related to these counts (either *LowAFC* or *MeanAFC*).

4.4 Fertility measurements: predicting MeanAFC. Use the other variables in the **Fertility** dataset (but not *LowAFC*), to consider models for predicting average antral follicle counts (*MeanAFC*) as described below.

a. Find the correlation of *MeanAFC* with each of the other variables (except *LowAFC*) or provide a correlation matrix. Which has the strongest correlation with *MeanAFC*? Which is the weakest?

b. Test whether a model using just the weakest explanatory variable is still effective for predicting *MeanAFC*.

c. Use a "best subsets" procedure to determine which set of three predictors (not including *LowAFC*) would explain the most variability in *MeanAFC*. Write down the predictors in this model and its R^2 value.

d. Are you satisfied with the fit of the three-variable model identified in part (c)? Comment on the individual t-tests and the appropriate residual plots.

4.5 Fertility measurements: predicting LowAFC. Repeat the analysis described in parts (a–d) of Exercise 4.4 for the data in **Fertility** now using the low antral follicle count (*LowAFC*) as the response variable. Do not consider *MeanAFC* as one of the predictors.

4.6 Fertility measurements: predicting embryos. Use a stepwise regression procedure to choose a model to predict the number of embryos (*Embryos*) using the **Fertility** data.

a. Briefly describe each "step" of the stepwise model building process in this situation.

b. Are you satisfied with the model produced in part (a)? Explain why or why not.

c. Repeat part (a) if we don't consider *Oocytes* as one of the potential predictors.

4.7 Baseball game times. Consider the data introduced in Exercise 1.45 on the time (in minutes) it took to play a sample of Major League Baseball games. The datafile **BaseballTimes2017** contains four quantitative variables (*Runs*, *Margin*, *Pitchers*, and *Attendance*) that might be useful in predicting the game times (*Time*). From among these four predictors choose a model for each of the goals below. **BballT**

a. Maximize the coefficient of determination, R^2.

b. Maximize the adjusted R^2.

c. Minimize Mallows's C_p.

d. After considering the models in (a–c), what model would you choose to predict baseball game times? Explain your choice.

Topic 4.3 Exercises: Cross-validation

4.8 County health: cross validation. In Example 1.7 on page 35 we see that transforming the response variable to $\sqrt{MDs}$ was a helpful re-expression for predicting the number of physicians in counties with a linear model based on the number of hospitals. **CtyHlth**

a. Use the first 35 counties in the **CountyHealth** dataset as a training sample* to fit a model to predict *TsqrtMDs* based on *Hospitals*. Write down the equation of the least squares line.

b. Compute predictions for the holdout sample (last 18 cases in the original file) and find the cross-validation correlation.

c. Compute the shrinkage and comment on what it tells you about the effectiveness of this model.

4.9 Cross-validation of a GPA model. The data in **FirstYearGPA** contains information on 219 college students (see the start of Example 4.2 on page 156 for more details). In this exercise, you will use cross-validation to assess a model to predict first-year grade point average (*GPA*) for students at this college. Split the original **FirstYearGPA** datafile to create a training sample with the first 150 cases and holdout sample with cases #151–219. Variable selection techniques (as in Topic 4.2) show that using high school GPA (*HSGPA*), number of humanities credits (*HU*), and an indicator for race (*White* = 1 for white students, 0 otherwise) produces a reasonable model for the training sample. **Y1GPA**

a. Use the training sample to fit a multiple regression to predict *GPA* using *HSGPA*, *HU*, and *White*. Give the prediction equation along with output to analyze the effectiveness of each predictor, estimated standard deviation of the error term, and R^2 to assess the overall contribution of the model.

b. Use the prediction equation in the previous part as a formula to generate predictions of the *GPA* for each of the cases in the holdout sample. Also, compute the prediction errors by subtracting the prediction from the actual *GPA* for each case.

c. Compute the mean and standard deviation for the prediction errors. Is the mean reasonably close to zero? Is the standard deviation reasonably close to the standard deviation of the error term from the fit to the training sample?

d. Compute the cross-validation correlation between the actual and predicted *GPA* values for the cases in the holdout sample.

e. Square the cross-validation correlation and subtract from R^2 for the training sample to compute the shrinkage. Does it look like the training model works reasonably well for the holdout sample or has there been a considerable drop in the amount of variability explained?

Topic 4.4 Exercises: Identifying Unusual Points in Regression

4.10 Breakfast cereals. In Exercise 1.19, you were asked to fit a simple linear model to predict the number of calories (per serving) in breakfast cereals using the amount of sugar (grams per serving). The file **Cereal** also has a variable showing the amount of fiber (grams per serving) for each of the 36 cereals. Fit a multiple regression model to predict *Calories* based on both predictors: *Sugar* and *Fiber*. Examine each of the measures below and identify which (if any) of the cereals you might classify as possibly "unusual" in that measure. Include specific numerical values and justification for each case. **Cereal**

a. Standardized residuals
b. Studentized residuals
c. Leverage, h_i
d. Cook's D

4.11 Religiosity of countries. Does the level of religious belief in a country predict per capita gross domestic product (GDP)? The Pew Research Center's Global Attitudes Project surveyed people around the world and asked (among many other questions) whether they agreed that "belief in God is necessary for morality," whether religion is very important in their lives, and whether they pray at least once per day. The variable *Religiosity* is the sum of the percentage of positive responses on these three items, measured in each of 44 countries. This variable is part of the datafile **ReligionGDP**, which also includes the per capita GDP for each country and indicator variables that record the part of the world the country is in (East Europe, Asia, etc.). **RelGDP**

a. Transform the *GDP* variable by taking the log of each value, then make a scatterplot of *log*(*GDP*) versus *Religiosity*.

b. Regress *log*(*GDP*) on *Religiosity*. What percentage of the variability in *log*(*GDP*) is explained in this model?

c. Interpret the coefficient of *Religiosity* in the model from part (b).

d. Make a plot of the studentized residuals versus the predicted values. What is the magnitude of the studentized residual for Kuwait?

e. Add the indicator variables for the regions of the world to the model, except for Africa (which will thus serve as the reference category). What percentage of the variability in *log*(*GDP*) is explained by this model?

f. Interpret the coefficient of *Religiosity* in the model from part (e).

g. Does the inclusion of the regions' variables substantially improve the model? Conduct an appropriate test at the 0.05 level.

h. Make a plot of the studentized residuals versus the predicted values for the model with *Religiosity* and the region indicators. What is the magnitude of the studentized residual for Kuwait using the model that includes the regions' variables?

4.12 Adirondack High Peaks. Refer to the data in **HighPeaks** on the 46 Adirondack High Peaks that are described in Exercise 4.2 on page 188. **HiPeak**

* The counties are listed in alphabetical order, but we assume that is unrelated to levels of medical care.

a. What model would you use to predict the typical *Time* of a hike using any combination of the other variables as predictors? Justify your choice.

b. Examine plots using the residuals from your fitted model in (a) to assess the regression conditions of linearity, homoscedasticity, and normality in this situation. Comment on whether each of the conditions is reasonable for this model.

c. Find the studentized residuals for the model in (a), and comment on which mountains (if any) might stand out as being unusual according to this measure.

d. Are there any mountains that have high leverage or may be influential on the fit? If so, identify the mountain(s) and give values for the leverage or Cook's D, as appropriate.

Topic 4.5 Exercises: Coding Categorical Predictors

4.13 North Carolina births. The file **NCbirths** contains data on a sample of 1450 birth records that statistician John Holcomb selected from the North Carolina State Center for Health and Environmental Statistics. One of the questions of interest is how the birth weights (in ounces) of the babies might be related to the mother's race. The variable *MomRace* codes the mother's race as white, black, Hispanic, or other. We set up indicator variables for each of these categories and ran a regression model to predict birth weight using indicators for the last three categories. Here is the fitted model:

$$\widehat{BirthWeightOz} = 117.87 - 7.31 Black + 0.65 Hispanic - 0.73 Other$$

Explain what each of the coefficients in this fitted model tells us about race and birth weights for babies born in North Carolina. **NCBirth**

4.14 More North Carolina births. Refer to the model described in Exercise 4.13 in which the race of the mother is used to predict the birth weight of a baby. Some additional output for assessing this model is shown below:

Predictor	Coef	SE Coef	T	P
Constant	117.872	0.735	160.30	0.000
Black	-7.309	1.420	-5.15	0.000
Hispanic	0.646	1.878	0.34	0.731
Other	-0.726	3.278	-0.22	0.825

S = 22.1327   R-Sq = 1.9%   R-Sq(adj) = 1.7%

Analysis of Variance

Source	DF	SS	MS	F	P
Regression	3	14002.4	4667.5	9.53	0.000
Residual Error	1446	708331.7	489.9		
Total	1449	722334.1			

Assuming that the conditions for a regression model hold in this situation, interpret each of the following parts of this output. Be sure that your answers refer to the context of this problem. **NCBirth**

a. The individual t-tests for the coefficients of the indicator variables

b. The value of R^2

c. The overall F-test in the ANOVA table

4.15 Blood pressure. The dataset **Blood1** contains information on the systolic blood pressure for 500 randomly chosen adults. One of the variables recorded for each subject, *Overwt*, classifies weight as 0 = Normal, 1 = Overweight, or 2 = Obese. Fit two regression models to predict *SystolicBP*, one using *Overwt* as a single quantitative predictor and the other using indicator variables for the weight groups. Compare the results for these two models. **Blood1**

4.16 Caterpillar nitrogen assimilation and body mass. In Exercise 1.30 on page 54, we explored the relationship between nitrogen assimilation and body mass (both on log scales) for data on a sample of caterpillars in **Caterpillars**. The *Instar* variable in the data codes different stages (1 to 5) of caterpillar development. **Caterp**

a. Fit a model to predict log nitrogen assimilation (*LogNassim*) using log body mass (*LogMass*). Report the value of R^2 for this model.

b. Fit a model to predict *LogNassim* using appropriate indicators for the categories of *Instar*. Report the value of R^2 for this model and compare it to the model based on *LogMass*.

c. Give an interpretation (in context) for the first two coefficients of the fitted model in (b).

d. Fit a model to predict *LogNassim* using *LogMass* and appropriate indicators for *Instar*. Report the value of R^2 and compare it to the earlier models.

e. Is the *LogMass* variable really needed in the model of part (d)? Indicate how you make this decision with a formal test.

f. Are the indicators for *Instar* as a group really needed in the model of part (d)? Indicate how you make this decision with a formal test.

Topic 4.6 Exercises: Randomization Test for a Relationship

4.17 Baseball game times. The data in **BaseballTimes2017** contain information from 15 Major League Baseball games played on August 11, 2017. In Exercise 1.45 on page 57, we considered models to predict the time a game lasts (in minutes). One of the potential predictors is the number of runs scored in the game. Use a randomization procedure to test whether there is significant evidence to conclude that the correlation between *Runs* and *Time* is greater than zero. **BballT**

4.18 GPA by Verbal SAT slope. In Example 4.11 on page 181, we looked at a randomization test for the correlation between GPA values and Verbal SAT scores for the data in **SATGPA**. Follow a similar procedure to obtain a randomization distribution of sample slopes of a regression model to predict *GPA* based on *VerbalSAT* score, under the null hypothesis $H_0 : \beta_1 = 0$. Use this distribution to find a *P*-value for the original slope if the alternative is $H_a : \beta_1 \neq 0$. Interpret the results and compare them to both the randomization test for correlation and a traditional *t*-test for the slope. **SATGPA**

4.19 More baseball game times. Refer to the situation described in Exercise 4.17 for predicting baseball game times using the data in **BaseballTimes2017**. We can use the randomization procedure described in Section 4.6 to assess the effectiveness of a multiple regression model. For example, we might be interested in seeing how well the number of *Pitchers* and *Attendance* can do together to predict game *Time*. **BballT**

a. Fit a multiple regression model to predict *Time* based on *Pitchers* and *Attendance* using the original data in **BaseballTimes2017**.

b. Choose some value, such as R^2, *SSE*, *SSModel*, *ANOVA F-statistic*, or S_ϵ, to measure the effectiveness of the original model.

c. Use technology to randomly scramble the values in the response column *Time* to create a sample in which *Time* has no consistent association with either predictor. Fit the model in (a) for this new randomization sample and record the value of the statistic you chose in (b).

d. Use technology to repeat (c) until you have values of the statistic for 10,000 randomizations (or use just 1000 randomizations if your technology is slow). Produce a plot of the randomization distribution.

e. Explain how to use the randomization distribution in (d) to compute a *P*-value for your original data, under a null hypothesis of no relationship. Interpret this *P*-value in the context of this problem.

f. We can also test the overall effectiveness of a multiple regression using an ANOVA table. Compare the *P*-value of the ANOVA for the original model to the findings from your randomization test.

Topic 4.7 Exercises: Bootstrap for Regression

4.20 Bootstrapping Adirondack hikes. Consider a simple linear regression model to predict the *Length* (in miles) of an Adirondack hike using the typical *Time* (in hours) it takes to complete the hike. Fitting the model using the data in **HighPeaks** produces the prediction equation **HiPeak**

$$\widehat{Length} = 1.10 + 1.077\ Time$$

One rough interpretation of the slope, 1.077, is the average hiking speed (in miles per hour). In this exercise you will examine some bootstrap estimates for this slope.

a. Fit the simple linear regression model and use the estimate and standard error of the slope from the output to construct a 90% confidence interval for the slope. Give an interpretation of the interval in terms of hiking speed.

b. Construct a bootstrap distribution with slopes for 5000 bootstrap samples (each of size 46 using replacement) from the High Peaks data. Produce a histogram of these slopes and comment on the distribution.

c. Find the mean and standard deviation of the bootstrap slopes. How do these compare to the estimated coefficient and standard error of the slope in the original model?

d. Use the standard deviation from the bootstrap distribution to construct a 90% confidence interval for the slope.

e. Find the 5th and 95th quantiles from the bootstrap distribution of slopes (i.e., points that have 5% of the slopes more extreme) to construct a 90% percentile confidence interval for the slope of the Adirondack hike model.

f. See how far each of the endpoints for the percentile interval is from the original slope estimate. Subtract the distance to the upper bound from the original slope to get a new lower bound, then add the distance from the lower estimate to get a new upper bound.

g. Do you see much difference between the intervals of parts (a), (d), (e), and (f)?

4.21 Bootstrap standard error of regression. Consider the simple linear regression to predict the *Length* of Adirondack hikes using the typical hike *Time* in the **HighPeaks** data file described in Exercise 4.2. The bootstrap method can be applied to any quantity that is estimated for a regression model. Use the bootstrap procedure to find a 90% confidence interval for the standard deviation of the error term in this model, based on each of the three methods described in Section 4.7. *Hint*: In R, you can get the standard deviation of the error estimate for any model with `summary(model)$sigma`. **HiPeak**

4.22 Bootstrap for a multiple regression coefficient. Consider the multiple regression model for *Weight* of perch based on *Length*, *Width*, and an interaction term *Length* · *Width* that was used in Example 3.11 (page 107). Use the bootstrap procedure to generate an approximate sampling distribution for the coefficient of the interaction term in this model. Construct 95% confidence intervals for the interaction coefficient using each of the three methods discussed in Section 4.7 and compare the results to a *t*-interval obtained from the original regression output. The data are stored in **Perch**. **Perch**

4.23 Bootstrap confidence interval for correlation. In Example 4.11, we considered a randomization test for the correlation between *VerbalSAT* scores and *GPA* for data on 24 students in **SATGPA** Rather than doing a test, suppose that we want to construct a 95% confidence interval for the population correlation. **SATGPA**.

a. Generate a bootstrap distribution of correlations between *VerbalSAT* and *GPA* for samples of size 24 from the data in the original sample. Produce a histogram and normality plot of the bootstrap distribution. Comment on whether assumptions of normality or symmetry appear reasonable for the bootstrap distribution.

b. Use at least two of the methods from Section 4.7 to construct confidence intervals for the correlation between *VerbalSAT* and *GPA*. Are the results similar?

c. Do any of your intervals for the correlation include zero? Explain what this tells you about the relationship between Verbal SAT scores and GPA.

4.24 Bootstrap regression based on residuals. In Section 4.7, we generated bootstrap samples by sampling with replacement from the original sample of Honda Accord prices. An alternate approach is to leave the predictor values fixed and generate new values for the response variable by randomly selecting values from the residuals of the original fit and adding them to the fitted values. **Accord**

a. Run a regression to predict the *Price* of cars based on *Mileage* using the **AccordPrice** data. Record the coefficients for the fitted model and save both the fits and residuals.

b. Construct a new set of random errors by sampling with replacement from the original residuals. Use the same sample size as the original sample and add the errors to the original fitted values to obtain a new price for each of the cars. Produce a scatterplot of *NewPrice* versus *Mileage*.

c. Run a regression model to predict the new prices based on mileages. Compare the slope and intercept coefficients for this bootstrap sample to the original fitted model.

d. Repeat the process 1000 times, saving the slope from each bootstrap fit. Find the mean and standard deviation of the distribution of bootstrap slopes.

e. Use each of the three methods discussed in Section 4.7 to construct confidence intervals for the slope in this regression model. Compare your results to each other and the intervals constructed in Section 4.7.

© Gary M. Hart, Gary Hart Photography, www.EloquentImages.com

UNIT B

Unit B: Analysis of Variance

Response: Quantitative
Predictor(s): Categorical

For this unit, we assume that you are familiar with material from Chapters 1 and 2 on fitting and doing inference for a simple linear regression model.

CHAPTER 5: ONE-WAY ANOVA AND RANDOMIZED EXPERIMENTS

Identify a one-way ANOVA model for a quantitative response based on a categorical predictor and examine the principles behind a randomized experimental design. Recognize how the process of data collection affects the scope of conclusions. Use transformations to help choose and assess an ANOVA model. Perform inference to assess when groups have different means and to measure the possible size of group differences.

CHAPTER 6: BLOCKING AND TWO-WAY ANOVA

Extend the ideas of the previous chapter to consider main effects ANOVA models with two explanatory factors arising from a block design.

CHAPTER 7: ANOVA WITH INTERACTION AND FACTORIAL DESIGNS

Add interaction to the two-way ANOVA model for factorial designs. Create and interpret interaction graphs.

CHAPTER 8: ADDITIONAL TOPICS IN ANALYSIS OF VARIANCE

Apply Levene's Test for Homogeneity of Variances. Develop methods for performing multiple inference procedures. Apply inference procedures for comparing special combinations of two or more groups. Apply nonparametric versions of the two-sample t-test and analysis of variance when normality conditions do not hold. Use randomization methods to perform inference in ANOVA settings. Use regression models with indicator variables to fit and assess ANOVA models. Use ANOVA models for designs that have repeated measures. Apply analysis of covariance when there are both categorical and quantitative predictors and the quantitative predictor is a nuisance variable.

Bill Kennedy/Shutterstock

CHAPTER 5

One-way ANOVA and Randomized Experiments

In this chapter you will learn to:

- Identify a one-way ANOVA model and examine the principles behind a randomized experimental design.
- Recognize how the process of data collection affects the scope of conclusions.
- Use transformations to help choose and assess an ANOVA model.
- Perform inference to assess when groups have different means and to measure the possible size of group differences.

Does the source of protein in a diet (beef, cereal, or pork) affect weight gain in rats? How much does the average commute time differ among four cities (Boston, Houston, Minneapolis, Washington)? In both of these examples, we have a quantitative response variable (weight gain, commute time), and a categorical explanatory variable (protein source, city). The data for the first situation (see Example 5.1) are the result of an experiment where rats were randomly assigned to different diets. The data for the second situation (see Example 5.19) come from an observational study of commute times for randomly sampled commuters in the four cities.

In both examples we want to explore the differences and similarities in the response variable between the explanatory variable groups. The first question, about diets for rats, is asking about *whether or not* there are differences, and the second question, about commute times, presumes that there are differences and asks *how big* the differences might be. These are the two main questions we typically want to address using analysis of variance (ANOVA) to compare group means while taking into consideration the variability in the response values within the groups.

5.1 Overview of ANOVA

EXAMPLE 5.1

FatRat

Fat rats: A randomized experiment Are humans like rats? (Bite your tongue.) In many ways the physiology of the rat mirrors that of the human. One study, which has become a "statistical classic," was designed to compare the effect of three different high-protein diets on weight gain in baby rats.[1] The data are stored in **FatRats** (which also has cases from three additional low-protein diets). The three high-protein diets had different sources of protein: beef, cereal, or pork. The subjects for the study were 30 baby rats. All 30 were fed a high-protein diet and their weight gains were recorded. Ten rats got their protein from beef, 10 from cereal, and 10 from pork. Note that this was an experiment because the scientists assigned the rats to the diets (at random).

The response variable is the weight gain in grams (*Gain*), which is quantitative. The predictor is type of protein (*Source*), which is categorical. The side-by-side dotplots and boxplots in Figure 5.1 show the results. (There are 30 dots, one per rat.)

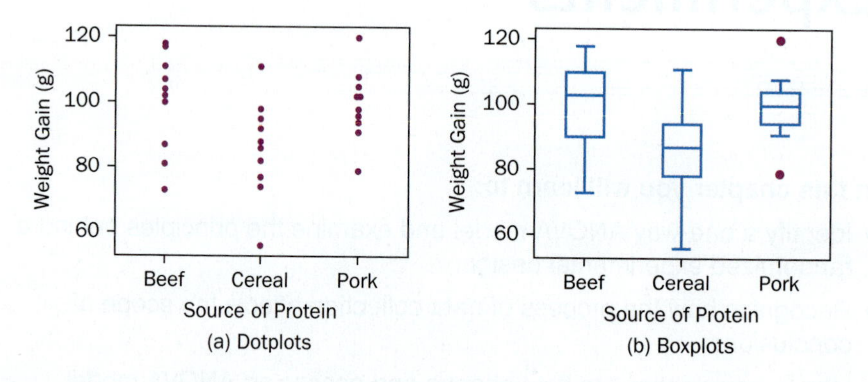

FIGURE 5.1 Weight gain versus protein source for baby rats on a high protein diet

On average, the 30 rats gained 95.13 grams each, or 95 to the nearest gram. The 10 rats on the beef diet gained an average of 100 grams, 5 grams above the overall average. The 10 on the cereal diet gained 85.9 grams, about 9 below the overall average. The 10 who got pork gained 99.5 grams, roughly 4 above the grand average. We are also interested in the variability of the weight *Gain*, which can be measured by the standard deviation for the overall sample as well as the standard deviation within each protein *Source*. For all 30 rats taken as one large group, the standard deviation of weight gains is 14.9 grams, and for the different groups it is 15.1 (beef), 15.0 (cereal), and 10.9 (pork).

In the context of the previous example, the key questions for this chapter are the following:

- Question 1: How strong is the evidence that protein source makes a difference in weight gain?
- Question 2: If there is a difference due to protein source, how big is it (and which sources differ)?

We can often address these questions informally using side-by-side plots (as in Figure 5.1) and by comparing summary statistics, such as the group means. For this particular random assignment of rats, the mean gain is smallest for the cereal group and largest for beef group, with a difference of about 14 grams. But for *any* set of three means, we'll generally find one that is the smallest and one that is the largest, even if the response is totally unrelated to the groups. How do we know when the differences in group means are large enough to conclude they cannot be attributed to random chance and instead are due to the explanatory variable? Note that this is the same question that we considered in Example 0.2 (page 7, on financial incentive for weight loss), only now we have more than two groups. Although the computational methodology differs, we can view the ANOVA models in this chapter as a generalization of the models for doing a two-sample t-test that we saw in Chapter 0.

In assessing the differences in the group means (100, 85.9, and 99.5), we should factor in the variability within the groups (standard deviations of 15.1, 15.0, and 10.9), as well as the sample sizes (10 rats each). If there is little variability within each group (think of Figure 5.1 with the dots or boxplots all clustered very close to the current centers of each group), we can be more confident of a "real" difference in the groups. On the other hand, if each group's responses were much more spread out (think of weight gains ranging from 10 to 200 in each group), those group means might look very similar. But are the actual group weight gains from *this* experiment different enough to conclude that it is unlikely that the differences are due to random chance alone? The answer to that question will have to wait until we develop some more formal inference techniques for ANOVA models later in this chapter.

EXAMPLE 5.2

Leaf

Leafhopper diets If you eat nothing but sugar, how long will you live? Would it depend on the type of sugar? Needless to say, if your study is experimental, and your response is length of life, the choice of what animal to use as your subject is an ethical issue. Clearly, humans are out. Probably, all mammals are out. The investigators in one study[2] chose to go with the potato leafhopper.

What sorts of sugar could they use? "Simple" sugars, like glucose and fructose, have 6 carbon atoms per molecule. A "compound" sugar like sucrose has 12. Among simple sugars, do glucose (blood sugar) and fructose (fruit sugar) differ in their capacity to sustain life? How do these simple sugars compare with the 12-carbon sucrose? How does a sugar diet compare with distilled water? The experimenters prepared eight petri dishes, two for each diet: control, sucrose, glucose, and fructose. Eight leafhoppers were put into each dish. Diets were randomly assigned to dishes. The response variable was the time (in days) until half the leafhoppers in a dish had died. The data are stored in **Leafhoppers** and displayed in Figure 5.2.

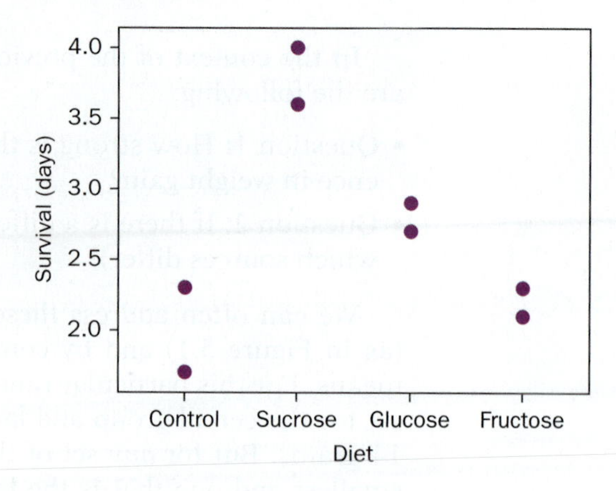

FIGURE 5.2 Survival time in days versus diet

Note that, even though there were 64 leafhoppers used in the experiment (8 in each of 8 petri dishes), the units here are the petri dishes—not the leafhoppers. Thus there are only 8 dots (two for each of the four groups) in the dotplot (and making boxplots comparing the groups would be silly). We see some fairly dramatic differences in the "lifetimes" shown in the dotplots: the sucrose group had the two longest values, both of the glucose cases came next, while fructose and control lagged near the bottom. But there are only two cases in each group, so we are again faced with the question of whether these differences could plausibly be due to random chance alone.

The leafhopper data are a good example of a designed experiment. The sugar diets in each petri dish were controlled and assigned by the researchers. Much of the work to develop the ANOVA methods of this unit was motivated by the need to analyze such experiments. However, the techniques can also be used to compare group means that arise from observational data, such as in the next example. When doing so, we must be careful about the type of conclusions we draw, which are more limited with nonexperimental data.

EXAMPLE 5.3

Teen

Teen pregnancy and the Civil War The data in **TeenPregnancy** give the teen pregnancy rate (number of pregnancies per 1000 teenage girls) for each of the 50 U.S. states in 2010. One of the other variables in that dataset records the role of the state in the U.S. Civil War, as described in Table 5.1, which also shows the mean teen pregnancy rate for each group of states.

Group	Number of states	Mean Teen Pregnancy
C = Confederate states that seceded	11	64.64
B = Border states, sympathized but did not secede	3	61.67
U = Union states that remained	21	48.24
O = Other states, admitted later	15	55.07

TABLE 5.1 Mean teen pregnancy rate by Civil War status of states

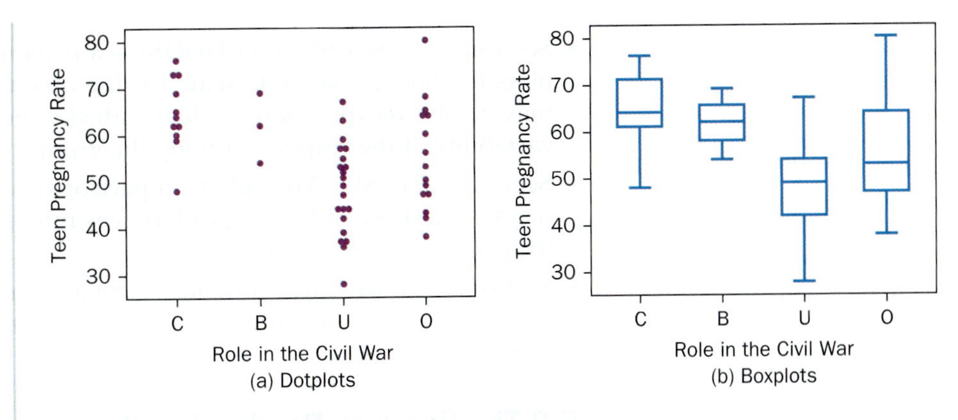

FIGURE 5.3 Plots of teen pregnancy rate versus role in the U.S. Civil War

Figure 5.3 shows dotplots and boxplots to graphically compare the teen pregnancy rates between states based on Civil War status. The plots and group means make clear that there are substantial differences among the four groups. The rates are generally higher in former Confederate and Border states; and generally lower in Union states.

Although this is strong evidence of a relationship between the two variables (Civil War status and teen pregnancy rate), the *pattern* is one thing, the *reason* for it is quite another. Whatever the reason for this relationship, we can be pretty sure it is not a matter of direct cause-and-effect. It would be hard to imagine any way a state's role in the Civil War in 1860 could "cause" the state's teen pregnancy rate 150 years later. Even more preposterous would be a way, even with a time machine, that a recent rate of teen pregnancy could go back through the decades to cause states to take sides in the Civil War. The ANOVA techniques of this chapter might be effective for establishing an association for these observational data, but we hope this extreme example—relating teen pregnancy and the Civil War—will serve to remind you of the important cliche: "association is not causation."

Preview of the Sections That Follow

Our usual workflow follows the steps CHOOSE, FIT, ASSESS, and USE. The next four sections in this chapter follow that order. The last sections provide more detail on these steps and then summarize the whole process with a case study.

- Section 5.2 (CHOOSE—data collection) has two parts. The first subsection describes the simplest experimental structure for an ANOVA model, the completely randomized experiment. This has a single categorical predictor, and we create the data for this choice of model by randomly assigning units to the categories of that variable. Another subsection deals with the situation for one-way observational data.
- Section 5.3 (FIT) shows how, for ANOVA data, you can split the response values into a set of numerical layers and then compare the variability between the group means to the variability within the groups.
- Section 5.4 (ASSESS and USE) introduces a formal inference method, the ANOVA *F*-test, for addressing the question, "Is there really a difference in mean response between the groups?" and then deals with issues of the scope of inference.
- Section 5.5 (USE) introduces more inference techniques, confidence intervals and effect sizes, for addressing the question, "How big are the differences?"

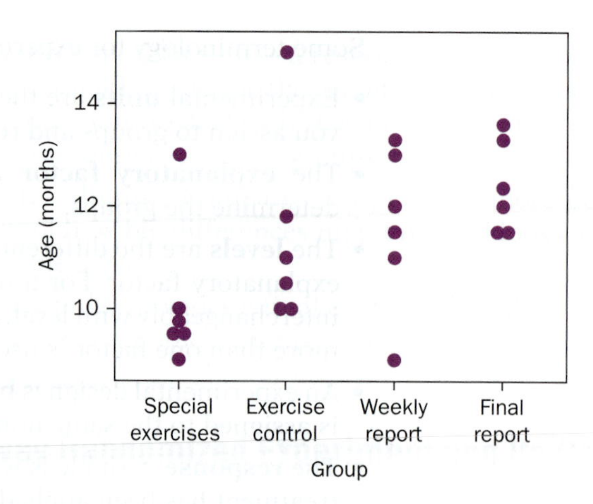

FIGURE 5.4 Walking babies: Time (in months) to walk unaided related to instructions to parents

The leafhopper and walking babies examples are both experimental, because the levels of the explanatory variable were randomly assigned by the scientists. Comparing category means can also be applied to observational data.

ANOVA for Observational Studies

ANOVA can be applied in several different settings, united by the shared abstract structure of their models, but divided by how they get their data and by what conclusions are supported by the numbers. Sections 5.3 and 5.6 rely on the common abstract structure and treat experimental and observational data as one. To see that the structure is the same, regardless of source, note that data sets of both kinds can be represented using side-by-side dotplots or boxplots. The algebraic model is the same for data of both kinds, as is the fitting of the model, and the numerical analysis. The difference comes when we get to formal inference (Section 5.4) where the scope and type of conclusions we can draw depend on the way randomization was used (or not used) in producing the data.

REASONS TO RANDOMIZE:

- to justify the use of a probability model
- to protect against bias and confounding
- to support inference about cause-and-effect

To make these three points concrete, compare the data situations for Example 5.3 (teen pregnancy and state Civil War status), an example where we compare commute times for samples of residents from four cities, and Example 5.1 (weight gain due to rat diets).

1. *What is the justification for using a probability model?*
   Teen pregnancy: Was randomization used to select states or assign them to different Civil War statuses? No. This isn't a random sampling of states—the 50 states make up the whole population. We only observed the status of each state in the Civil War. Thus we don't use a probability model here.

Metropolitan commutes: Do average commute times in various U.S. cities differ? If so, by how much? To investigate these questions, we used data from the 2007 American Housing Survey,[4] which involved representative samples of American households in various metropolitan areas. We selected independent random samples of 500 commuters from Boston, Houston, Minneapolis, and Washington, D.C. The data on one-way daily commuting distance (in miles) and time (in minutes) appear in **MetroCommutes**.

The probability model here is based on the random sampling of commuters in each of the four cities. This is an observational study, because the researchers didn't assign commuters to live in different cities; they only asked for the commute times of the people who were selected for each sample.

Fat rats: The probability model is based on the random assignment of the diets to the rats. This is an experiment, because the researches controlled and used a random mechanism to choose the diet for each rat.

2. *Could there be bias or confounding variables?*

   Teen pregnancy: Yes. Since the states in different Civil War groups are located together in different regions of the country, there could be many cultural, demographic, or lifestyle differences in those regions that contribute to differences in teen pregnancy rates.

   Metropolitan commutes: The use of random samples from each city eliminates the concern about bias. This allows us to generalize the results from the sample to the larger populations of commuters in each city. On the other hand, if, for example, the researchers had only contacted commuters in a few convenient neighborhoods in each city, they might easily have gotten misleading estimates of the mean commute times.

   Fat rats: Confounding is extremely unlikely. Randomizing the diets should ensure that possible confounders are spread more or less evenly among the four sugar diet groups. If this study had not been randomized and controlled, for example, by letting the rats choose on their own from four different feeding tubes, we would not be able to tell if a factor that influenced which diet was chosen might also influence weight gain.

3. *Is there justification to support a conclusion about cause-and-effect?*

   Teen pregnancy: No, inference about cause is not justified. However, there are lots of possible geographic and demographic variables that might influence both teen pregnancy and the role a state played in the Civil War.

   Metropolitan commutes: No, inference about cause is not justified. There *might* be infrastructure conditions in each city that cause differences in mean commute times, but there could also be lifestyle choices that attract different types of commuters to each city. To get a causal conclusion, we could randomly assign people to move to each of the four cities and then measure commute times, although that would be an expensive and cumbersome experiment to run for 500 people going to each city!

   Fat rats: Yes, inference about cause is justified. Random assignment of diets has also randomized the assignment of any and all unmeasured confounders, so that confounding is not a plausible explanation for the pattern. By elimination (unless something really unlikely has happened) the differences must be due to the diets.

We return to the distinctions between experimental and observational data and the sorts of conclusions they allow in Section 5.4. Before that, we look at specifying and fitting the ANOVA model using the general data format (quantitative response, categorical predictor) that is common to both settings.

5.3 Fitting the Model

For ANOVA, once you have chosen a design, you have your initial model in hand. We haven't given explicit mathematical notation for the model yet, in the form of $Y = f(\text{Explanatory variable}) + \epsilon$ that we introduced in Chapter 0 and saw frequently for regression models in Unit A. However, that is coming soon, after we consider an overview for how the ANOVA model is organized.

There are two main ways to think about the process of fitting an ANOVA model, two ways that correspond to the two main ways to think about ANOVA itself: (1) You can regard ANOVA as a special case of regression with categorical predictors, create 0,1 indicator variables for the categories, and fit the resulting regression model. Section 8.10 shows how to do this. Alternatively, you can think of ANOVA as a generalization of the two-sample t-test (as in Example 0.2 on page 7) and think about fitting the model by computing and comparing averages.

The big advantage of the first approach is that you can fit ANOVA models without having to learn a new way to think about ANOVA as different from regression. The big disadvantage of the first approach is that you can fit ANOVA models without having to learn a new way to think about ANOVA as different from regression. If you want to be able to understand ANOVA beyond one-way ANOVA, the simple model of this chapter, it is worth learning the approach based on averages, as in this section.

The aim is to show how to see ANOVA as much more than just a special case of regression. This aim is captured by the metaphor of a CAT scan. Just as a CAT scan lets a radiologist "see" your body as a set of layers, fitting an ANOVA model using averages lets you "see" your data as a set of numerical overlays.[5] Imagine that you want to study human organ systems, and that a CAT scan shows the inside of your body as a set of three overlaid transparencies:

$$\text{Body} = \text{Skeleton} + \text{Organ System} + \text{Everything Else}$$

Now think of ANOVA as a numerical CAT scan that splits your response values into a sum of three numerical overlays:

$$\text{Response} = \text{Grand Average} + \text{Treatment Effect} + \text{Residual}$$

In important ways, this way of thinking parallels simple linear regression:

$$\text{Response} = \text{Grand Average} + (\text{slope}) \cdot (\text{Change in predictor}) + \text{Residual}$$

Down the road, however, ANOVA models get more complicated and the regression approach gets a lot harder. Looking ahead, the metaphor of layers extends more easily:

$$\text{Body} = \text{Skeleton} + \text{Digestive System} + \text{Lymphatic System} + \text{Everything Else}$$

$$\text{Response} = \text{Grand Average} + \text{Main Effect A} + \text{Main Effect B} + \text{Interaction} + \text{Residual}$$

One-way ANOVA Model

In Example 0.2 on page 7 we introduced a model for the two-sample t-test as

$$Y = \mu_i + \epsilon$$

where μ_1 and μ_2 represent the means of the two populations (or treatments) being compared. We can easily extend this model to the more general one-way ANOVA situation where we are comparing more than two groups, for example, by using μ_1, μ_2, μ_3, and μ_4 if we have four groups (as in the leafhopper example).

However, given the overlay analogy to a CAT scan, we can also think of the model as specifying an overall mean (Grand average), departures from that mean for each group (Treatment Effect), and additional deviations for individual cases within each group (Residual).

ONE-WAY ANOVA MODEL

For a quantitative response variable and single categorical explanatory variable, the **one-way ANOVA model** is

$$Y = \mu + \alpha_i + \epsilon$$

where μ is the grand mean, α_i is the treatment effect (difference from the grand mean for the ith group mean), and ϵ is the usual random error term.

Notice that, under this model, the mean for the ith group is $\mu_i = \mu + \alpha_i$ so the model is equivalent to $Y = \mu_i + \epsilon$. In what follows, we first show how to split data from a one-way ANOVA situation into overlays to estimate the terms in this model. Then we show how to use the overlays to compare the two kinds of variation we care about, group-to-group variation and unit-to-unit variation within groups. Because averages are familiar, we show all this without formulas, but if at any point you find yourself wanting algebra, you can find a boxed summary at the end of the section.

Fitting the Data as a Sum of Overlays

Recall from Section 5.1 that *the goal of ANOVA is to compare category averages*: How strong is the evidence that they differ from each other? What can we say about the size of the differences? These two questions direct our numerical strategy: compute overlays, then compare.

Compute overlays: split each observed value into three pieces:

Response		
=	Grand average	
+	Treatment effect	(the differences we care about)
+	Residual	(the differences we use for comparison)

Then compare: How big are the treatment differences—the differences we care about—compared to the residuals—the unit-to-unit variation?

The overlays come from three questions:

- Question 0: What is the overall average?
- Question 1: How far is each group average from the overall average?
- Question 2: How far is each response from its group average?

EXAMPLE 5.5

Leaf

Seeing the leafhopper data as a sum of overlays

Question 0: What is the overall or "grand" average?

Answer: This is the anchor point, or the benchmark we will use to compute group differences.

For the leafhopper data, the grand average ($\bar{y}$) is the sum of the eight response values divided by 8: $\bar{y} = (2.3 + 1.7 + 3.6 + 4.0 + 2.9 + 2.7 + 2.1 + 2.3)/8 = 2.7$.

Question 1: How far is each group average from the grand average?

Answer: Compute group averages, then subtract the grand average.

$$\text{Treatment effect} = \text{group average} - \text{grand average} = \bar{y}_i - \bar{y}$$

For the leafhoppers, the control average is $(1.7 + 2.3)/2 = 2.0$, which is 0.7 day below the grand average of 2.7, or −0.7 unit away as shown in Figure 5.5.

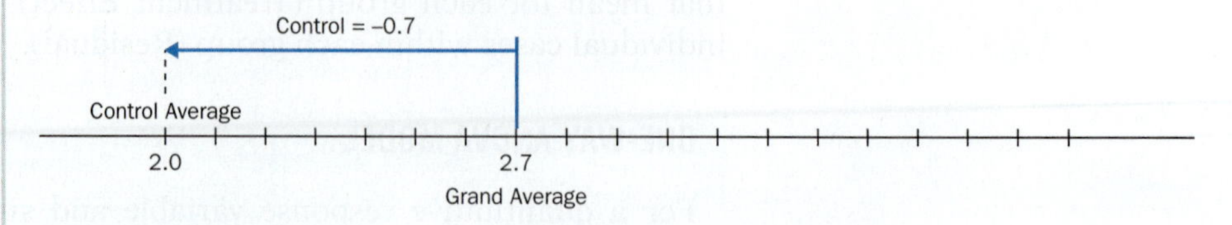

FIGURE 5.5 Control effect = Control Average − Grand Average

Doing the same for each of the other diets is easy to organize in a table:

	Diet				
	Control	**Sucrose**	**Glucose**	**Fructose**	
Data (2 cases per group)	2.3	3.6	2.9	2.1	
	1.7	4.0	2.7	2.3	
Average	2.0	3.8	2.8	2.2	2.7
Group effect	−0.7	1.1	0.1	−0.5	

You can help build your intuition if you visualize the results as we have done in Figure 5.6.

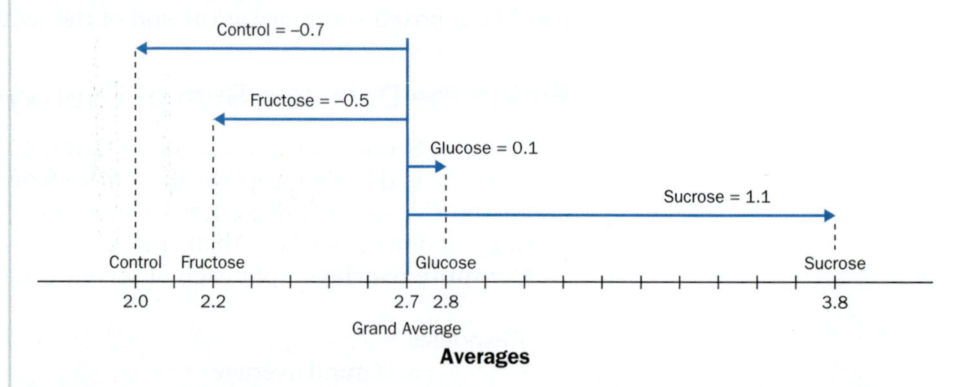

FIGURE 5.6 Group effects address Question 1: "How far is each group average from the grand average?"

Here, as with every one-way ANOVA, we want to compare the sizes of these group-to-group differences with the sizes of the unit-to-unit differences, and so we ask Question 2.

Question 2: How far is each response from its group average?
Answer: Compute each residual = response − group average = $y - \bar{y}_i$

For the Control diet, the two response values are 2.3 and 1.7, with an average of 2.0, so the residuals are +0.3 and −0.3.*

If we do the arithmetic for the rest of the data, we can summarize the results with both a table and a picture. (See Figures 5.7 and 5.8.)

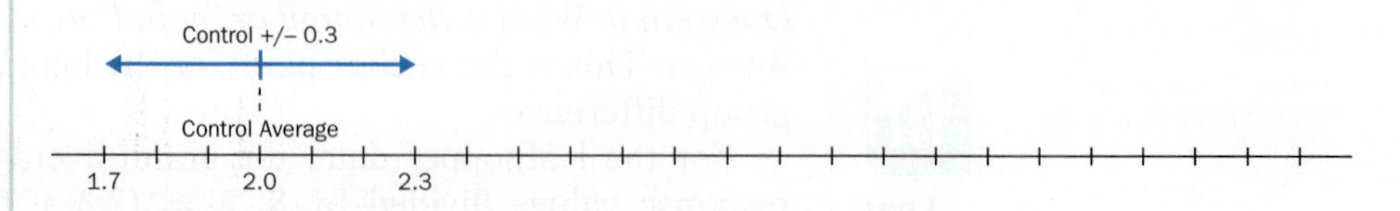

FIGURE 5.7 Residual = Response − Group Average

*Note that, if we think of the group average as a "predicted" value, this is the same idea as the residuals we saw in Unit A.

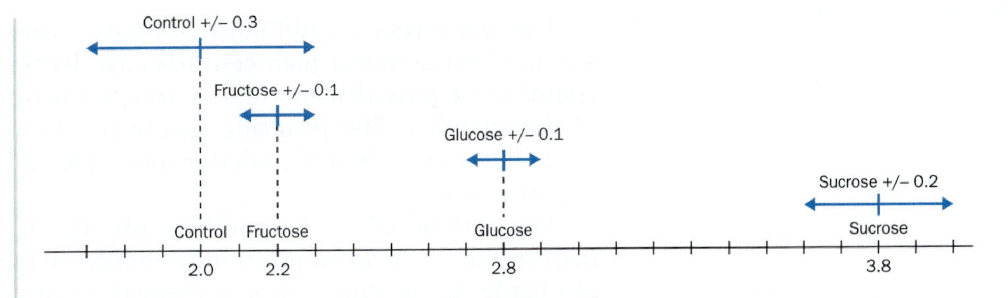

FIGURE 5.8 Residuals tell the answer to Question 2: "How far is each response value from its group average?"

0.3	−0.2	0.1	−0.1
−0.3	0.2	−0.1	0.1

At this point, our CAT scan is done: we have split the data into a sum of three overlays.

Response value	2.3	3.6	2.9	2.1
	1.7	4.0	2.7	2.3

=

Grand Average	2.7	2.7	2.7	2.7
	2.7	2.7	2.7	2.7

+

Treatment effects	−0.7	1.1	0.1	−0.5
	−0.7	1.1	0.1	−0.5

+

Residuals	0.3	−0.2	0.1	−0.1
	−0.3	0.2	−0.1	0.1

FIGURE 5.9 Leafhopper data as a sum of overlays

Figure 5.9 is most useful if you can learn to read it two different ways: (1) one response at a time, and (2) one box at a time.

1. *One response at a time.* Focus on the upper left corner of the top box of Figure 5.9. The response value of 2.3 tells us that for the first dish of leafhoppers on the control diet, it took 2.3 days until half the leafhoppers had died. Now move your eyes down the set of four boxes, keeping your focus on the upper left corners of the four boxes. You should see that

$$\begin{array}{ccccccc} 2.3 & = & 2.7 & + & (-0.7) & + & 0.3 \\ \text{Response} & = & \text{Grand Average} & + & \text{Diet effect} & + & \text{Residual} \end{array}$$

   You can do the same with each of the other seven response values. Taken together, they offer one way to think about the overlays.

2. *One box at a time.* The first, top, box is the data. Our arithmetic has split the data into grand average + diet effects + residuals. As a rule, we don't much care about the grand average—it's just a convenient benchmark. What we do care about is the bottom pair of boxes, one for the treatment effects (group-to-group differences) and the last one for the residuals (unit-to-unit differences). For the leafhoppers it is clear that the treatment effects are quite a bit larger than the residuals.

The two ways to read Figure 5.9 echo a very deep and powerful idea from science, first demonstrated centuries ago by Sir Isaac Newton. In a darkened chamber, he passed white light through a prism onto a wall at the other end of the chamber. The resulting spectrum of colors convinced the world that white light, until then regarded as pure and indivisible, was in fact a mixture of components.

The idea of splitting something observable into its fundamental components is one of the most profound concepts in the history of science. Molecules are made up of atoms. Atoms themselves are made of subatomic particles. Proteins are built from amino acids. Your genome is a sequence of letters chosen from an alphabet of four, A, G, C, and T. Another deep example, from physics, is Fourier analysis, which enables scientists to think about sound waves in two ways, which they call the "time domain" and the "frequency domain." Mathematicians have showed that ANOVA is in fact a kind of Fourier analysis.[6] Informally, we can borrow, from the physics of sound and from information theory, the idea of signal and noise. The unit-to-unit differences are the noise. Our question, "Are there group-to-group differences?" is a question about whether the group differences we observe are signal, or just more noise.

For one-way designs, whether experimental or observational, the same methods for decomposing the observed values work for unbalanced data also.

EXAMPLE 5.6

Undo

Evaluating psychotherapy Insurance companies are reluctant to pay for long-term, talk-based psychotherapy. This reluctance, together with the growing emphasis on evidence-based medicine, has led researchers to find ways to evaluate the effectiveness of long-term therapy. Here is one pioneering study.* What matters for the example: What is the response, what are the groups, and what are the units?

- The response: The patient had been diagnosed with severe OCD (obsessive/compulsive disorder). The response is a measure of the patient's freedom from a particular OCD symptom called the "defense of undoing," rated on a four-point scale. On this scale 1 is bad, 4 is good.
- The groups: Detailed notes for each of 108 hour-long psychotherapy sessions were grouped into six sets of 18 sessions each, from Group I (sessions 1–18) to Group VI (sessions 91–108).
- The units: In some of the 108 sessions no OCD symptoms were observed, but sometimes an OCD symptom was observed (e.g., this happened in 4 of the 18 Group I sessions). A unit of observation was an OCD symptom that was observed (and then rated on the 1–4 scale).

The 108 sets of session notes were presented to the raters (experienced therapists) one at a time in a random order, randomized so that the raters could not tell which notes were from an early therapy session and which were from a later session. The results, shown in Figure 5.10 and stored in **Undoing**, show a steady improvement over time.

Rather than use the entire data set, for simplicity, this example will work with the data from three representative groups, I, III, and IV:

Group I	1	1	1	2		
Group III	2	2				
Group IV	2	2	2	2	3	4

*This study is a marvel of ingenious design. The data are derived from observations of a single person. You might guess that the study must be observational, that $n = 1$, and that there can be no randomized experiment. Not so! Because the details are somewhat complicated, we omit them here.

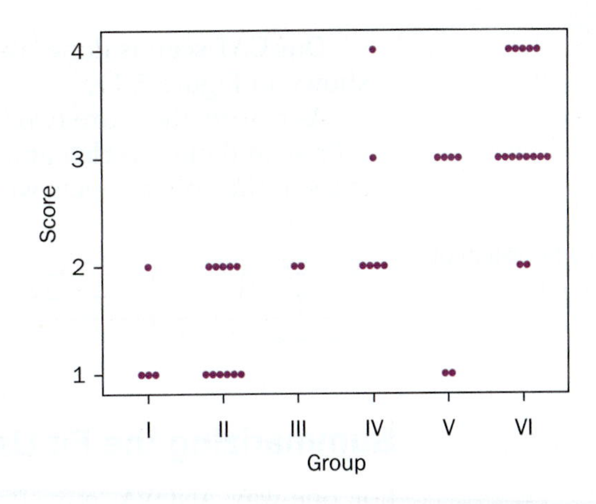

FIGURE 5.10 The Undoing data. The pattern is clear: The scores increase steadily and markedly over time

Note that for this example, we use rows instead of columns to represent the groups. It doesn't matter which we use, rows or columns; the logic is the same either way. What *does* matter is the groups, which are decided by meaning.

The decomposition answers the same three questions as in Example 5.5.

Question 0: What is the grand average?
Answer: $(1 + 1 + 1 + 2 + 2 + 2 + 2 + 2 + 2 + 2 + 3 + 4)/12 = 24/12 = 2$. This value, $\bar{y} = 2.0$, is our benchmark for measuring group-to-group differences.

Question 1: How far is each group average from the grand average?
Answer: Compute group averages, and subtract the grand average.*

Group effect = group average − grand average

							Ave	Effect
Group I	1	1	1	2			1.25	−0.75
Group III	2	2					2.00	0.00
Group IV	2	2	2	2	3	4	2.50	0.50
All three							2.00	

Aligning the effects with their groups, we have:

Effects						
	−0.75	−0.75	−0.75	−0.75		
	0.00	0.00				
	0.50	0.50	0.50	0.50	0.50	0.50

For Group I, the typical score is three-fourths of a point below the grand average; for Group III, the typical score is the same as the grand average; and for Group IV, the typical score is half a point above the grand average. These are the group-to-group differences we want to compare with the unit-to-unit differences.

Question 2: How far is each response from its group average?
Answer: Compute each residual = response − group average

For the first group, the response values are 1, 1, 1, and 2. Subtracting their group mean of 1.25 gives residuals of −0.25, −0.25, −0.25, and 0.75.

Residuals						
	−0.25	−0.25	−0.25	0.75		
	0.00	0.00				
	−0.50	−0.50	−0.50	−0.50	0.50	1.50

*For unbalanced data, as in this example, there are two ways to estimate group effects. We have chosen the way that gives an additive decomposition whose sums of squares add.

Our CAT scan is done. We can now write the data as a sum of overlays as shown in Figure 5.11.

As before, there are two ways to read the decomposition, one response at a time, and one overlay at a time. Also as before, the goal is to compare the sizes of the group effects with the sizes of the residuals.

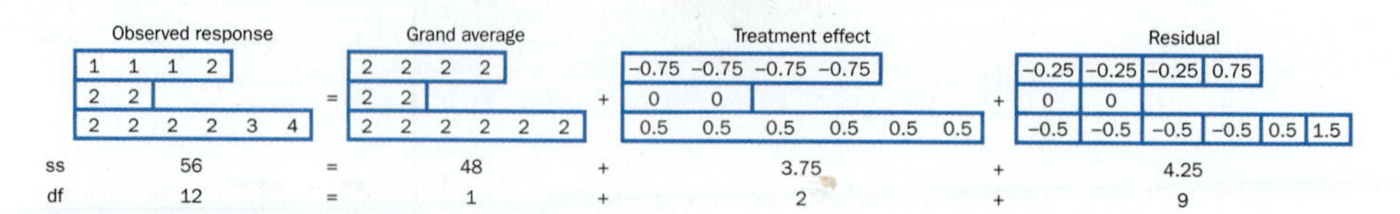

FIGURE 5.11 Decomposition of the psychotherapy data

Summarizing the Fit Using the ANOVA Table

For one-way ANOVA, our strategy will be to compute a single statistic, the *F*-ratio, that compares the size of the treatment effects (the group-to-group differences we care about) with the size of the residuals (the differences we use to measure unit-to-unit variation). The *F*-ratio is computed in four steps. Although the formulas are different, the general ideas and terminology are similar to the way we use ANOVA in regression (see Section 2.2). Here is an overview/review, followed by two examples.

COMPUTING THE *F*-RATIO

SS: **sums of squares**, a measure of variation due to a source
df: **degrees of freedom**, an adjusted sample size
MS: **means square**, a measure of average variation for a source
F: **a ratio**, mean square for the model divided by mean square error

Before we explain the mechanics, we start with a preview.

EXAMPLE 5.7

Leaf

ANOVA table for the leafhoppers Here is the ANOVA table* for the leafhopper data of Example 5.5.

Source	df	SS	MS	F
Diets	3	3.92	1.307	17.42
Residual	4	0.30	0.075	
Total	7	4.22		

In the ANOVA table, we see that the sum of squares (*SS*), which measures overall variation, is much larger for *Diets* than for *Residuals*. The residuals estimate the error term (ϵ) in the model, so we can refer to that variability as the sum of squared errors (*SSE*). Thus we have $SSDiets = 3.92$ and $SSE = 0.30$. As with the regression ANOVA, the degrees of freedom (*df*) is an adjusted sample size (more soon), and the mean square (*MS*) is a measure of average variation, equal to the measure of variation (*SS*) divided by the adjusted sample size (*df*). The *MS* column of the table shows that the average variation due to diets ($MSDiets = 1.307$) is a lot larger than the average variation for the errors ($MSE = 0.075$). The *F*-ratio ($17.42 = 1.307/0.075$) tells us that the variation due to *Diets* is 17.42 times larger than the variation due to errors. The *F*-ratio is the statistic that we will use to compare the two sources of variation.

*To fully match the decomposition we would want to add a row for "Grand average," but because we never test the grand average, the software does not include such a row.

WHAT THE *F*-RATIO TELLS

- A value of F near 1 indicates that the two sizes of variation, group-to-group and unit-to-unit, are roughly the same size.
- A large value of F indicates that group-to-group variation is large compared with unit-to-unit variation.

EXAMPLE 5.8

Undo

ANOVA table for the psychotherapy data Here's the ANOVA table for the subset of the **Undoing** data of Example 5.6.

Source	df	SS	MS	F
Groups	2	3.75	1.875	3.97
Residual	9	4.25	0.472	
Total	11	8.00		

For this psychotherapy dataset, the F-ratio indicates that the average variation in symptom scores between groups is about 4 times larger than the average variation within groups. This is not nearly as large as the F-ratio (more than 17) for the leafhoppers.

We will return in the next section to the important question of how to put the F-ratios for the two examples, leafhoppers and psychotherapy, on the same scale by computing a P-value. First, however, we summarize the computations for one-way ANOVA.

- *SS*. To compute the sum of squares for a box of the decomposition, just square each number in the box and add them up. Many books on ANOVA include messy formulas for sums of squares, but those are vestiges of the age before computers and have little conceptual content. You don't need formulas for the key idea: When the overall variation in a box is large, at least some of the numbers will be far from zero, their squares will be large, and the *SS* will be large.

  It will always be true that the sums of squares add:

  $$SS\text{Observed values} = SS\text{Grand average} + SS\text{Group effects} + SS\text{Residuals}$$

- *df*. The *SS* tells, "How big?" for the variation in a box of the decomposition. The degrees of freedom tells "How many chances to vary?"

EXAMPLE 5.9

Leaf

Degrees of freedom for the leafhopper data For the leafhoppers there are 8 response values in all, so 8 chances to vary, or 8 df. In the box for the grand average, there are also 8 numbers, but all 8 are the same, so there is only one chance to vary, and $df = 1$. In the box for diets, there are also 8 numbers, but they come in pairs, and there are only 4 pairs, so you might guess $df = 4$, but in fact $df = 3$. The reason is that the 4 diet effects have to add to 0, so as soon as you know any three of them, the value of the last one is no longer free to vary. In the box for residuals, the 8 numbers also come in pairs, this time one + and one − in each pair. There is one df for each pair, so 4 in all.

The diagram in Figure 5.12 shows how to count the df. In each box, an R stands for a repeated value, and a + indicates a value that is fixed by the requirement of adding to 0.

FIGURE 5.12 Degrees of freedom for the leafhoppers

Observed values	1 2	3 4	5 6	7 8	df = 8
Grand Average	1 R	R R	R R	R R	df = 1
Diet effects	1 R	2 R	3 R	+ R	df = 3
Residuals	1 +	2 +	3 +	4 +	df = 4

If you prefer, you can leave finding the *df* to a computer, or just use a formula in the next box:

COMPUTE DEGREES OF FREEDOM

df(Observed values) = # observed values
df(Grand average) = 1
df(Group effects) = # groups − 1
df(Residuals) = # observed values − # groups

You can use the formulas to check that the degrees of freedom will always add:

$$df(\text{Observed values}) = df(\text{Grand average}) + df(\text{Group effects}) + df(\text{Residuals})$$

Note that the usual ANOVA table (as shown in Examples 5.7 and 5.8) moves the "Grand average" term to the other side of the equation, so that the "Total" reflects overall variation in the sample as the sum of squared deviations from the grand mean. In this form the term has *df*(Total) = # observed values − 1. We also have *SSTotal* = *SS*Observed values − *SS*Grand average.

Recall from Section 2.2 that the regression ANOVA partitions the total variability in the quantitative response variable into two parts: some explained by the model and the remaining unexplained in the error term. We can think of the sum of squares of the group effects (call it *SSGroups*) as what's explained by the model and the sum of squared residuals (call it *SSE*) as the unexplained error variability. We can then measure total variability in the same way as in regression, with the sum of squared deviations for all response values from their grand mean. This gives the partition

$$SSTotal = SSGroups + SSE$$

that we see in the ANOVA table (and most computer output). Also note that the degrees of freedom for this "total" ($n - 1$) matches that of regression and is the sum of the degrees of freedom for the group-to-group variation and the unit-to-unit variation.

- *MS* and *F*. The *SS* tells "How big?" for the variation in a box of the decomposition. The degrees of freedom tells "How many chances to vary?" So the mean square tells the average variation per chance to vary. The *F*-ratio compares the average variation from the two sources, the group effects and the residual errors.

In practice, we generally rely on computer technology to compute the various sums of squares, mean squares, and *F*-ratios for an ANOVA table. The important idea is the comparison of the average group-to-group variation to the average unit-to-unit variation. If the *MSGroups* is large relative to the *MSE*, we have evidence suggesting there is a difference in means between the groups. How large? We leave that question to the formal inference in Section 5.4. We conclude this section with a quick look at some computational details behind constructing an ANOVA table.

Algebraic Notation and Formulas

So far everything we have done in this chapter has been purely descriptive: looking for patterns and for ways to show and summarize those patterns. The next section is about inference, trying to use the patterns we can see in order to learn about things we can't see, either possible cause-and-effect relationships or differences in some larger population, or both. This goal of inference requires that we review a critical distinction between two kinds of quantities: statistics (quantities we can observe because we can compute their values from the data) and parameters (quantities whose values we can never know because they only exist in theory).

We follow the usual convention and use Greek letters α, β, γ, μ, σ, etc., for parameters; we use x's and y's and SS, df, t, p, etc., for statistics. Thus in the model

$$Y = \mu + \alpha_i + \epsilon$$

μ and the α_i's are parameters that need to be estimated and quantities such as $\bar{y}$ and $\bar{y}_i - \bar{y}$ are statistics measured from the data that provide numerical estimates.

While we may need to introduce additional subscripts for fancier models in later chapters, at this point we'll provide formulas with minimal use of subscripts. So y represents an individual response value (just as in regression), $\bar{y}_i$ is the mean of its group, and $\bar{y}$ is the grand mean for all the response values.

SUMMARY OF FORMULAS FOR ONE-WAY ANOVA

Meaning	Parameter	Statistic
Overall mean	μ	$\bar{y}$
Group mean	μ_i	$\bar{y}_i$
Group effect	$\alpha_i = \mu_i - \mu$	$\bar{y}_i - \bar{y}$
Observed response	$\mu_i + \epsilon$	y
Random error, residual	ϵ	$y - \bar{y}_i$

Decomposition

Question 0: The grand average	$\bar{y}$
Question 1: How far from the group average to the grand average?	$\bar{y}_i - \bar{y}$
Question 2: How far from the response to the group average?	$y - \bar{y}_i$

All sums below are over **all** observed data values:

	Sums of squares	**Degrees of freedom**
Observed values	$\sum y^2$	# observed values
Grand average	$\sum \bar{y}^2$	1
Group effects	$\sum (\bar{y}_i - \bar{y})^2 = SSGroups$	# groups − 1
Residuals	$\sum (y - \bar{y}_i)^2 = SSE$	# observed values − # groups

Mean Square: $MSGroups = SSGroups/df(Groups)$ $MSE = SSE/df(Error)$
F-ratio: $F = MSGroups/MSE$

Algebraically armed, we now turn to inference.

5.4 Formal Inference: Assessing and Using the Model

For the leafhoppers of Example 5.7 we have a large F-ratio: the size of the average variation (MS) due to diets is more than 17 times as big as the size of the average variation due to unit differences. However, the adjusted sample size (residual df) for the units is only 4. Contrast those results with the psychotherapy data of Example 5.8. We have a much smaller F-ratio (3.97 instead of 17.42). At the same time our adjusted sample size for the residuals is roughly twice as large, $df = 9$ instead of $df = 4$. How can we evaluate the difference between the strength of evidence in leafhopper data (large F, small df) and the psychotherapy data (smaller F, larger df)?

Recall the two key questions that motivate our work in this chapter: "Is there a difference?" and "If so, how big might it be?" We start with formal inference for addressing the first question, using the F-ratio from an ANOVA table.

Is There Really a Difference? ANOVA *F*-test

For the leafhoppers the first key question is, "Does the type of sugar in a diet affect leafhopper lifetimes?"; for the psychotherapy it's, "Do symptom scores differ over time during psychotherapy?" We know that a larger F-ratio provides stronger evidence for a difference in means between the groups, but that "evidence" may be hard to interpret since it depends on adjusted sample sizes as measured by the degrees of freedom. Fortunately, as with ANOVA for regression, we can quantify this strength of evidence with a P-value computed from an F-distribution, if certain conditions on the model are reasonably met. This gives us a *hypothesis test* to address the "Is there a difference?" question, based on the ANOVA table.

In regression ANOVA, the null hypothesis is that the response variable has no linear relationship with the predictor(s). In the one-way ANOVA setting, the equivalent null hypothesis is that the means do not differ between the groups. We can state this a couple of ways. Extending the hypotheses for a two sample t-test we have

$$H_0: \mu_1 = \mu_2 = \mu_3 = \cdots$$
$$H_a: \text{Some } \mu_i \neq \mu_j$$

where the null says that all group means are the same and the alternative says that at least two groups have different means.

We can also use the one-way ANOVA model

$$Y = \mu + \alpha_i + \epsilon$$

to express the hypotheses as

$$H_0: \alpha_1 = \alpha_2 = \alpha_3 = \cdots = 0$$
$$H_a: \text{Some } \alpha_i \neq 0$$

Shortly we will give a box with conditions under which we can use an F-distribution, which has both numerator and denominator degrees of freedom, to compute a P-value for the F-ratio in an ANOVA table. For now, let's assume the conditions hold and see what we get for P-values for the leafhopper and psychotherapy datasets.

EXAMPLE 5.10

Leaf

Undo

Leafhoppers and psychotherapy: *P*-values For the leafhopper experiment the F-ratio in the ANOVA table is 17.42. The degrees of freedom for the F-distribution are 3 (groups df) for the numerator and 4 (error df) for the denominator (as shown in the ANOVA table). The P-value is the area in the tail of an $F_{3,4}$ distribution beyond the observed F-ratio of 17.42, which turns out to be 0.009. This completes the ANOVA table that was introduced in Example 5.7.

Source	df	SS	MS	F	P-value
Diets	3	3.92	1.307	17.42	0.009
Residual	4	0.30	0.075		
Total	7	4.22			

This is a very small *P*-value so we have strong evidence against the null claim that the type of sugar diet does not make a difference in longevity for leafhoppers. Since the data arose from an experiment, we have strong evidence to support the claim that the type of sugar diet affects the average lifespan of leafhoppers. The answer to key question number 1 is yes.

For the subset of the psychotherapy data, the area beyond $F = 3.97$ in an $F_{2,9}$ distribution gives a *P*-value of 0.058. If the patient's scores were purely random, we would get a value of F as large as 3.97 only 1 time out of about 17. The null hypothesis should be embarrassed,* but not mortified. This evidence is not as convincing as with the leafhoppers; there may be some differences in mean symptom scores between different points in the treatment, or the differences might be just due to random chance. (In an exercise you will have an opportunity to consider the full psychotherapy dataset [with six groups, rather than just three] where the evidence is more than strong enough to mortify the null hypothesis.)

The following box summarizes the details for performing the hypothesis test for a one-way ANOVA model.

ANOVA TEST

To test for differences between I group means, the hypotheses are

$H_0: \alpha_1 = \alpha_2 ... = \alpha_I = 0$
H_a: At least one $\alpha_i \neq 0$

and the **ANOVA table** is

Source	Degrees of freedom	Sum of Squares	Mean Square	*F* statistic
Groups	$I - 1$	*SSGroups*	*MSGroups*	$F = \frac{MSGroups}{MSE}$
Error	$n - I$	*SSE*	*MSE*	
Total	$n - 1$	*SSTotal*		

If the proper conditions hold, the *P*-value is calculated using the upper tail of an *F*-distribution with $I - 1$ and $n - I$ degrees of freedom.

As a concept and a construct for data analysis, the *P*-value is controversial. You may want to look back at some cautions (page 62) from the section on inference for regression models that still apply in this setting. Thanks to computers, the number itself has become as easy to get as shoving coins into a vending machine. Just insert your data and model, push the button, and your *P*-value comes rolling down the chute. Getting the bottle is easy. The hard question, the one that matters, is this: "What is it you are about to swallow?"

Conditions for ANOVA Inference

The main goal of one-way ANOVA is to investigate whether (and then how much) the mean response differs among the explanatory groups. A key question to address the goal is: *How big are the group-to-group differences compared with the unit-to-unit differences?* This question is easier to answer when datasets are "well behaved." Just as with the regression models of Unit A, we have conditions under which inference based on ANOVA is more valid.

*Recall the discussion in Section 2.1 on how to think about *P*-values.

An important part of the previous key question depends on the unit-to-unit differences within each of the explanatory groups. This is reflected in the separate distributions of the response variable for each of the groups. The comparison needed for that key question tends to be easier to see and reliably quantify when the distributions have the following properties:*

- The distributions of the response within each group are relatively symmetric (and preferably reasonably normal), with no (or few) outliers.
- The variability in the response variable is similar within each of the groups.

We summarize these conditions in terms of the ANOVA model in the following box.

CONDITIONS FOR A ONE-WAY ANOVA MODEL

For the one-way ANOVA model

$$Y = \mu + \alpha_i + \epsilon$$

we have conditions on the effects (α_i):

- The effects are constant and additive

and on the residuals (ϵ):

- The residuals are random and independent
- The residuals have the same variability in each group
- The residuals are normally distributed

We have already seen the main tool for assessing these properties in the ANOVA setting: side-by-side dotplots or boxplots that show the distribution of the sample response values within each of the explanatory groups. Examples include Figure 5.1 (weight gain by protein source for rats), Figure 5.2 (lifetimes by sugar type for leafhoppers), and Figure 5.3 (teen pregnancy rates by Civil War status for states). We need to check that the data are relatively symmetric with few big outliers (*normality*) and the group standard deviations are not too different (*equal variability*). These both depend on the distribution of the responses *within* the groups which is controlled by the random error term (ϵ). Both of these conditions are met in the plots for those examples (although it's hard to tell much for the small samples in the leafhopper data), so each would be classified as relatively well behaved.

As always, the mathematical theory works only in a perfect world, which differs from the messier world of reality. All the same, imperfections acknowledged, we can and should use mathematical theory to guide our thinking about what a small *P*-value can tell us. The answer depends on how closely the real world that gave us our data behaves like the perfect world of the mathematician.†

The next example shows two additional plots for checking ANOVA conditions that should be familiar from your work with regression models.

*Note that these are analogous to the equal variance and normality conditions that you saw in regression models with Unit A.

†"Real" and "perfect" are not meant as sarcasm. The progress of science depends on how well we relate the two worlds.

EXAMPLE 5.11

FFlies

Fruit fly lifetimes An experiment investigated the research question: Does male sexual activity shorten length of life?* The subjects were 125 male fruit flies. Each fly was randomly assigned to one of five treatment groups (shown below), based on who they got for roommates in their individually assigned test tube. The response variable (*Longevity*) is the lifetime for each fruit fly (in days). The data are stored in **FruitFlies**.

None:	no tube-mates
Pregnant1:	one pregnant female
Pregnant8:	eight pregnant females
Virgin1:	one virgin female
Virgin8:	eight virgin females

Note that no mating occurs in the pregnant (or control) groups, but the number of females in the pregnant groups will duplicate conditions, like competition for resources, that might also affect lifetimes.

The lifetimes are displayed graphically with side-by-side dotplots and boxplots in Figure 5.13. The dotplots allow us to see the distribution of individual lifetimes in each treatment group, showing relatively symmetric distributions with similar spreads. The boxplots are also relatively symmetric, show no outliers, and allow us to visually compare statistics such as the medians. These show that the distributions of lifetimes appear to be a bit lower in the *Virgin1* groups and even shorter for the male fruit flies living with 8 virgin females. Note that the sample sizes in this case are much larger than they are in the earlier example comparing leafhopper lifetimes. This may give some support for the idea that more sexual activity tends to shorten average lifetimes (at least for male fruit flies). We'll test this more formally in the next example.

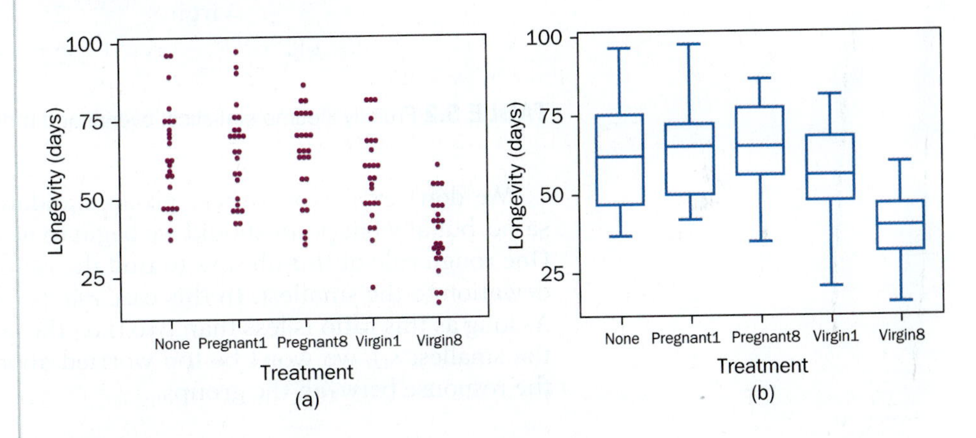

FIGURE 5.13 Fruit fly longevity by treatment groups

So far, we don't see any serious concerns with the conditions for the ANOVA model, but borrowing from our regression work, we can also examine residual plots. The residuals in this case are the differences between each lifetime and the mean for its treatment group ($y - \bar{y}_i$). Figure 5.14(a) shows boxplots of the residuals for each group. These are very similar to the boxplots in Figure 5.13, except the residuals are always centered at zero for each group. This makes it a bit easier to compare the spreads. The normal quantile plot of residuals in Figure 5.14(b) shows that the distribution of all 125 residuals taken together appears to be reasonably normal.

*There's a joke about old statisticians: They never die, they just get broken down by age and sex.

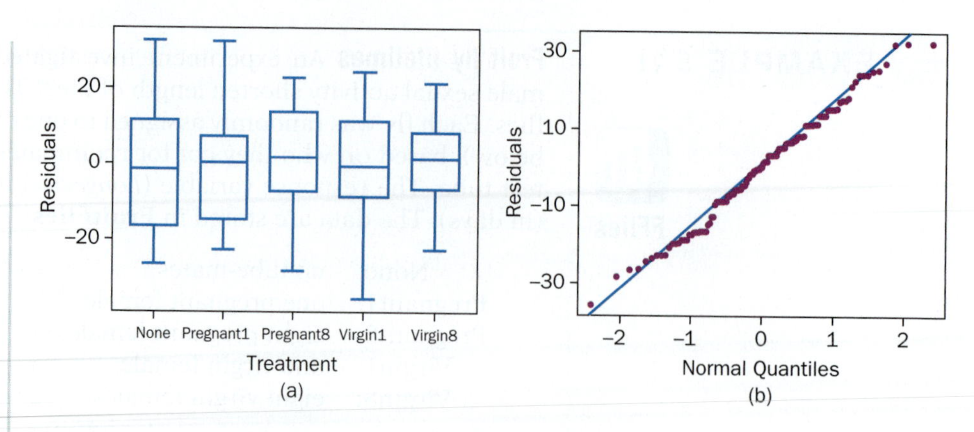

FIGURE 5.14 Residual plots for fruit fly lifetime ANOVA

While we can often compare variabilities within the groups by eyeing the plot, we can also compare numerical summaries such as the standard deviations in the response variable for each of the groups. For the fruit fly longevities these are shown in Table 5.2.

Group	Treatment	n	$\bar{y}_i$	s_i
1	None	25	63.56	16.45
2	Pregnant1	25	64.80	15.65
3	Pregnant8	25	63.36	14.54
4	Virgin1	25	56.76	14.93
5	Virgin8	25	38.72	12.10
All		125	57.44	17.56

TABLE 5.2 Fruit fly lifetime statistics; overall and broken down by groups

We don't ever expect these group standard deviations to be exactly the same, but at what point should we begin to worry that they are too different? One rough rule of thumb says to find the ratio of the largest group standard deviation to the smallest. In this case $max(s_i)/min(s_i) = 16.45/12.10 = 1.36$. As long as this ratio is less than two (i.e., the largest s_i is not more than twice the smallest s_i), we won't be too worried about differences in variability of the response between the groups.

Now that it looks like the ANOVA conditions we can check for the fruit fly lifetimes are reasonably satisfied, we can proceed with the formal inference to see if the experimental living conditions really make a difference in average male fruit fly lifetimes.

EXAMPLE 5.12

FFlies

Fruit fly lifetimes: Is there a difference? For the ANOVA model

$$Longevity = \mu + \alpha_i + \epsilon$$

where α_i is the effect of being in the ith treatment group. The hypotheses to test are:

H_0: $\alpha_1 = \alpha_2 = \alpha_3 = \alpha_4 = \alpha_5 = 0$
H_a: Some $\alpha_i \neq 0$

Equivalently, if we let μ_i represent the mean lifetime in each treatment situation, we have

$H_0: \mu_1 = \mu_2 = \mu_3 = \mu_4 = \mu_5$
H_a: Some $\mu_i \neq \mu_j$

Using the information from Table 5.2 we can estimate the effect, $\hat{\alpha}_i$, for each treatment by seeing how far the mean for that group is from the overall mean lifetime.

None	$\hat{\alpha}_1 = 63.56 - 57.44 = 6.12$
Pregnant1	$\hat{\alpha}_2 = 64.80 - 57.44 = 7.36$
Pregnant8	$\hat{\alpha}_3 = 63.36 - 57.44 = 5.92$
Virgin1	$\hat{\alpha}_4 = 56.76 - 57.44 = -0.68$
Virgin8	$\hat{\alpha}_5 = 38.72 - 57.44 = -18.72$

After partitioning the variability into the amount due to the differences in the group means and the natural variability within each group we obtain computer output* for the ANOVA table.

Response: Longevity

	Df	Sum Sq	Mean Sq	F value	Pr(>F)
Treatment	4	11939	2984.82	13.612	3.516e-09
Residuals	120	26314	219.28		

The large F-statistic and very small P-value give us strong evidence to reject the null hypothesis. Thus we have convincing evidence from this experiment that the number and type of tube-mates can affect the mean lifetime of male fruit flies. Note that the ANOVA test does not tell us *which* groups are different or by *how much*. From the plots and estimated group effects, we would certainly be suspicious that living with 8 virgin females is probably not conducive to a long life for a male fruit fly. We come back to this example in Section 5.5 where we consider confidence intervals for group effects and ways to formally estimate and test for differences between specific pairs of groups.

Cautions: Scope and Limits of ANOVA Inference

We can reduce the cautions to a slogan by borrowing from one you may have heard in kindergarten: "*Stop, look, and think, before you trust your* P*-value. Use your eyes, use your brain, and only then . . .*"

For a randomized, controlled experiment or independent random samples, a small P-value (< 0.05 or < 0.01) tells us there is strong evidence that the group-to-group differences we wanted to compare are not due just to the chancelike variation between units. We may be on the track of something important. For other studies, the message from a small P-value is less clear.

The checks so far have been visual (*use your eyes*). There are two harder checks that cannot be made visual: constant effects and randomness. Time to *use your brain*.

Constant effects. Whether effects are constant depends mainly on lurking variables. Typically, if there is a lurking variable, its presence will make the residuals artificially large, will make the residual mean square artificially large, will make the F-ratio artificially small, and so will make it harder to detect the group-to-group differences you care about. (Chapter 6 has more on this.)

*Note that the output shown here omits the row showing the total variability.

Randomness. Many old-style statistics books (and sadly, too many newer ones also) ask you to pretend that your ANOVA dataset behaves like a set of random samples. At best this pretense is a useful metaphor, at worst a deceptive fraud. In practice, an ANOVA dataset almost never comes from random sampling. In practice, almost all ANOVA datasets come from one of three sources: (a) randomized, controlled experiments, (b) independent random samples, (c) observational data that can plausibly be regarded as behaving "like" random samples from populations, or (d) observational data with no basis to claim a probability model. The *P*-value can be computed for all three situations and can be useful for each of the three, but the meaning of the *P*-value is not at all the same.

a. *Data from randomized controlled experiments*. A significant *P*-value ($p < 0.05$ or $p < 0.01$) confirms that there are detectable differences among the means of the treatments you chose to study. Examples: Fat rats, leafhoppers, and fruit flies. The data support a conclusion about cause and effect.

b. *Data from independent random samples*. Here, also, a small *P*-value leads to a conclusion that the means differ between the groups, and that conclusion can be generalized to the populations from which the samples were drawn. Example: The metropolitan commute data. In Section 5.6 we will see that comparing the mean commute times between the cities yields a small *P*-value. Since we have random samples from each city, we can make inferences about the differences in mean commute times for the larger populations of commuters in those cities. However, since the explanatory factor (city) was not assigned at random to the observational units, the data do not support a conclusion about cause and effect.

c. *Data from convenience samples that can be plausibly regarded as random*. To the extent that your samples can be regarded as "like random" you can justify extrapolating from samples to populations. Example: Data collected on hawks (Exercise 5.70) strongly support a conclusion that different hawk species have different beak lengths.

d. *Observational data with absolutely no basis for a probability model*. You can still use the *P*-value as a way to measure the strength of evidence. Example: Teen pregnancy and the Civil War. There is no random sampling, because we have the entire population, all 50 states. There is no random assignment: Both the predictor values and response values are simply observed. In no way can we argue that the data were created by a probability model. All the same, we can still use the *P*-value. Suppose you have 50 chips, one for each state, with the state's teen pregnancy rate written on its chip. You randomly toss the chips into four baskets, 11 chips in the basket for the Confederate states, 3 chips in the basket for the Border states, 21 chips in the basket for the Union states, and the rest in the basket for the Other states. How likely is it that the resulting random *F*-ratio will be as large as the one from the actual data? A small *P*-value suggests that the pattern of association in the data may be worth serious thought. A moderate-to-large *P*-value tells you that the pattern could easily have happened just by chance, so don't waste your time.

Haven't We Already Seen ANOVA?

We have two different ways to address this question.

- In Section 2.2 we look at the use of an ANOVA table to help assess the effectiveness of a simple linear regression model. We also generalize this to multiple regression in Section 3.2 and test subsets of predictors in

Section 3.6. Each of these applications deals with measuring variability explained by a model and comparing to unexplained variability. That is still true in this chapter (and the rest of Unit B), but now we are interested in comparing the variability "explained" by differences in response means to the "unexplained" variability within the categorical groups.

- It is possible to create a model for a quantitative response based on a categorical predictor using multiple regression with indicator variables for the categories (see Section 4.5 and Section 8.10). However, a more common approach (for many people) to handling a categorical predictor is to use the ANOVA procedures of this chapter.

In the next example, we start with a traditional regression setting involving two quantitative variables and then shift to a grouped regression where we treat the explanatory variable as more of a categorical factor.

EXAMPLE 5.13

Teen

Church attendance and teen pregnancy In Example 5.3 on page 200 we looked at a possible relationship between a state's status in the Civil War and its teenage pregnancy rate. Another variable in the **TeenPregnancy** dataset is percentage of state residents in a Gallup survey who said they attend church at least once per week (*Church*). Are these variables related?

Regression. Figure 5.15 is a scatterplot of *Teen* pregnancy rate versus rate of regular *Church* attendance. The pattern is roughly linear and moderately strong, although we might have some concern about the data point for Utah that is likely to be an influential outlier. We have clear evidence of a relationship, although—as always!—no mere pattern by itself can explain the reason for it.

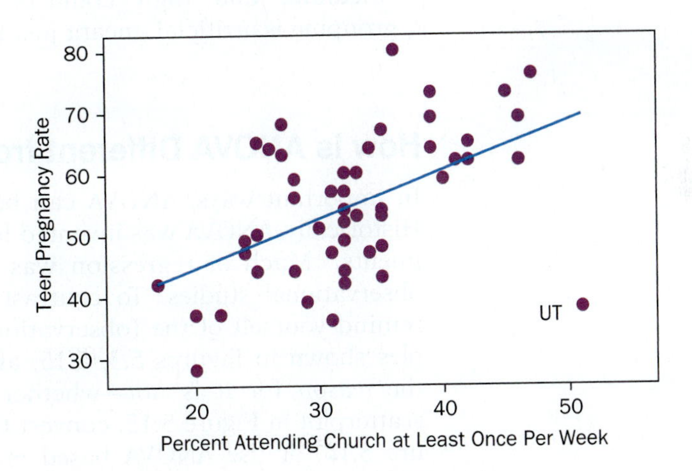

FIGURE 5.15 Teen pregnancy rate versus church attendance for the 50 U.S. states

Regression for grouped data. Figure 5.16 shows side-by-side dotplots and boxplots for the same data (similar to the plots for Civil War status in Figure 5.3). For these plots, the states have been sorted into three groups based on the predictor (weekly church attendance) classed as low (< 28%), medium, or high (> 35%). Even though compacting church attendance to just three categories gives up some information, the pattern is still clear. States with higher rates of church attendance also have higher rates of teen pregnancy. *But*: As with the scatterplot in Figure 5.15, these plots show only a pattern, and not the reason for it. (There is no reason to think going to church causes teen pregnancy, or that getting pregnant as a teen increases church attendance.)

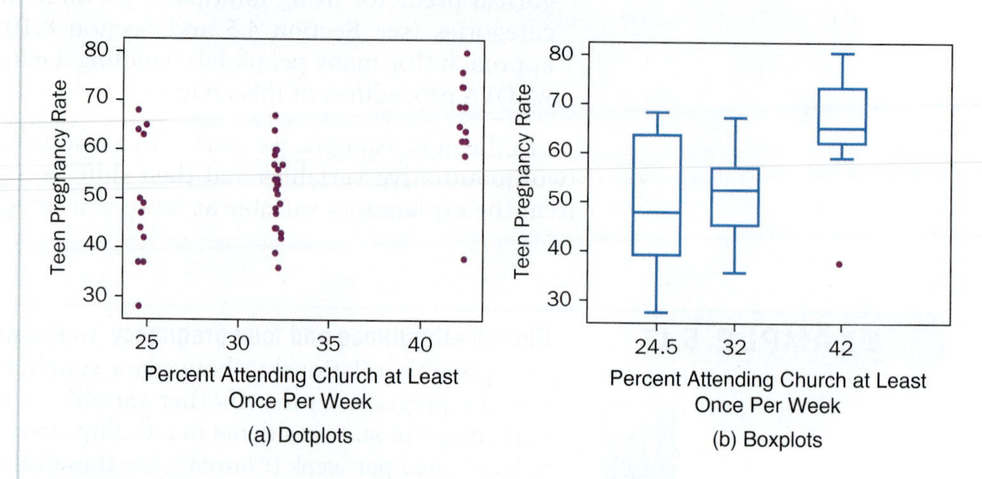

FIGURE 5.16 Teen pregnancy rate versus church attendance

The plots in Figure 5.16 hint at how regression and ANOVA are similar in the abstract. The response, teen pregnancy rate, is quantitative. The predictor, church attendance, is quantitative, but for the grouped data, "low," "medium," and "high" could be regarded as categorical, even though the grouping is artificial, meant just to create a bridge to ANOVA.

How Is ANOVA Different from Regression?

In important ways, ANOVA can be viewed quite distinctly from regression. Historically, ANOVA was invented for data from randomized controlled experiments.* Much of regression was developed for and applied to data from observational studies. To contrast observational and experimental studies, remind yourself of the (observational) data in the teenage pregnancy examples shown in Figures 5.3, 5.15, and 5.16. Even though the *pattern* is clear, the *reason* for it is not—whether we analyze the data with a regression scatterplot in Figure 5.15, convert the quantitative predictor to groups in Figure 5.16, or use ANOVA based on the boxplots of Civil War categories in Figure 5.3.

Experimental studies are planned in advance to allow you to understand the *reason* for any differences you find. Consider Example 5.1 where experimenters randomly assigned baby rats to be fed high-protein diets from one of three different sources. If it turns out that the differences in mean weight gains between the groups are larger than we should reasonably expect to see by random chance, we would be able to conclude that the association *is* due to causation; the choice of protein sources causes differences in mean weight gain.

*Nonetheless, some experimental situations such as response surfaces in chemical engineering have traditionally been handled with regression models.

Random assignment in the rat experiment provides the potential to conclude that the differences in protein source could not plausibly be due to confounding factors, because the randomization that assigned diets to rats has also automatically, simultaneously, and randomly assigned values of all possible confounding variables. Unless the randomization has done something truly unlikely, values of confounders are roughly evenly distributed among the three diet groups.

Compare the rat study (Figure 5.1) with the dataset relating the teen pregnancy rate of a state and its role in the Civil War (Figure 5.3). For the observational study there is no way to go back in time and use random numbers to decide which states would secede. True, that dataset has the same abstract structure as for the rats—response is quantitative, predictor is categorical. However, despite that structural similarity, random assignment makes a gigantic difference in what you can conclude from a tiny *P*-value.

Although ANOVA was invented for randomized experiments, the method is often applied to observational data also, like the data in Figure 5.3. The two key questions are the same: (1) How strong is the evidence that the category averages differ? (2) What can we say about the size of the differences? The models and mechanisms for addressing these questions with ANOVA are the same for both experimental and observational data, but the way we can interpret the results depends on the type of study. This also applies equally well to regression. Generally, we should draw only causal conclusions for data generated from an experiment, regardless of whether we use ANOVA or regression models to produce the results.

The difference between experimental and observational data, like the distinction between ANOVA and regression models, has important consequences. We summarize three of them here, in order of importance, starting at the top and working down.

1. **Use the model: Scope of inference.** Most important, experimental data have the potential to justify conclusions about cause and effect.

2. **Choose a model.** For observational data, you typically gather the data first, then look for a model that describes the patterns (see Unit A). For experimental data, you often choose a model first (e.g., picking a categorical explanatory factor), then obtain the data according to the model you chose (see Section 5.2).

3. **Fit the model.** For regression data, the formulas for the fitted slopes are not intuitive, unless you are good at algebra and know some calculus. (Go back and look at the formulas for Unit A.) For ANOVA data, you can fit the model just by computing and comparing averages (see Section 5.3). The formulas for ANOVA are more straightforward and intuitive.

5.5 How Big Is the Effect?: Confidence Intervals and Effect Sizes

Testing a null hypothesis using a *P*-value is the traditional way to answer the IS question, "Is there an effect?" At best, the *P*-value tells only whether there is detectable evidence of differences anywhere among the group means. For a well-designed, randomized, controlled experiment, this can be what we most want to know. *But*: The *P*-value doesn't tell how precisely we can estimate the means. The *P*-value doesn't tell whether the detected differences matter in real life. To address these two other important issues, the IT questions ("How big is it?"), we need interval estimates and effect sizes.

Interval Estimates

For the leafhoppers on the control diet (no sugar) we know that the observed average length of life is 2.0 days, based on observing two dishes of leafhoppers. For the fruit flies in the control condition (no roommates) we observed that the average length of life is 63.56 days based on observing 25 flies each living alone. We also know that dish-to-dish survival for the leafhoppers has a standard deviation of 0.27 day, and that the individual fly-to-fly variation in length of life has a standard deviation of 14.8 days. What can we say about how precisely we can estimate the two average lengths of life? The answer is given by confidence intervals.

CONFIDENCE INTERVAL FOR A GROUP MEAN

The confidence interval for a group mean μ_i is:

$$\bar{y}_i \pm t^* \cdot SD\sqrt{1/n_i}$$

where $SD = \sqrt{MSE}$ and t^* is from a t-distribution with the error degrees of freedom.

The margin of error is a product of three terms:

- t^* depends on $df(Error)$ and on the level of confidence, e.g., 95%, 99%. The larger the df, the narrower the interval; the higher the confidence you want, the wider the interval.
- $SD = \sqrt{MSE}$ tells the size of the unit-to-unit variation. The larger the SD, the wider the interval.
- *Sample size factor* $= \sqrt{1/n_i}$, depends on the sample size used to find that $\bar{y}_i$. The larger the group's size, the narrower the interval.

EXAMPLE 5.14

Confidence interval for a group mean This example shows the arithmetic that leads to 95% confidence intervals for the mean lifetime for the leafhoppers who get no sugar, and the mean lifetime for the male fruit flies who get no companions.

Leaf

a. Leafhoppers.

- The mean leafhopper lifetime for the no sugar group is $\bar{y}_1 = 2.0$.
- *t-value.* For a 95% level of confidence and $df(Error) = 4$, the t-value is 2.776.
- $SD = \sqrt{MSE} = \sqrt{0.075} = 0.274$
- *Sample size factor* $= \sqrt{1/n_1} = \sqrt{1/2} = 0.707$
- The margin of error is $(2.776)(0.274)(0.707) = 0.5378$
- The interval is mean ± margin of error, which is 2.0 ± 0.5378

For the leafhoppers on the control diet, we can be 95% confident that the mean length of life is within ±0.54 day of the observed mean of 2.0 days, that is, between 1.46 and 2.54 days.

FFlies

b. Fruit flies.

- The mean fruit fly lifetime for the control group is $\bar{y}_1 = 63.56$.
- *t-value.* For a 95% level of confidence and $df(Error) = 120$, the t-value is 1.980.
- $SD = \sqrt{MSE} = \sqrt{219} = 14.80$

- *Sample size factor* $= \sqrt{1/n_1} = \sqrt{1/25} = 0.200$
- The margin of error is (1.98)(14.80)(0.200) = 5.86
- The interval is mean ± margin of error, which is 63.56 ± 5.86

For the male fruit flies living alone, we can be 95% confident that the mean length of life is within ±5.86 days of the observed mean of 63.56 days, or between 57.70 and 69.42 days.

The logic of the confidence interval for a single group mean extends easily to the difference of two group means. The only change is to the factor for sample size. Instead of one size n_i for a single group mean, we have two sizes, n_1 and n_2, for the two groups.

CONFIDENCE INTERVAL FOR THE DIFFERENCE OF TWO GROUP MEANS

The confidence interval for the difference of two group means $\mu_1 - \mu_2$ is:

$$(\bar{y}_1 - \bar{y}_2) \pm t^* \cdot SD\sqrt{1/n_1 + 1/n_2}$$

where $SD = \sqrt{MSE}$

The last two boxed summaries, intervals for a single mean and intervals for the difference of two means, are closely parallel. The only change in the arithmetic is in the factor for sample size. However, the intervals for differences have greater reach. Each interval for a difference between group means ($\mu_1 - \mu_2$) is also an automatic test of the null hypothesis that the group means are equal (H_0: $\mu_1 = \mu_2$).

EQUIVALENCE OF CONFIDENCE INTERVALS AND TESTS

Two group means are significantly different at the 5% (1%) level if and only if a 95% (99%) confidence interval for their difference does not include 0.

EXAMPLE 5.15

FFlies

Confidence interval for the difference of two group means To illustrate, return to the fruit flies, and consider 95% confidence intervals for a pair of group differences.

a. Control versus 1 pregnant

- The difference in sample mean lifetimes between the Control and Pregnant1 groups is $\bar{y}_1 - \bar{y}_2 = 63.56 - 64.80 = -1.24$.
- *t-value.* For a 95% level of confidence and $df(Error) = 120$, the t-value is 1.98.
- $SD = \sqrt{MSE} = \sqrt{219} = 14.80$
- *Sample size factor* $= \sqrt{1/n_1 + 1/n_2} = \sqrt{1/25 + 1/25} = 0.2828$
- The margin of error is (1.98)(14.80)(0.2828) = 8.287
- The interval is difference in means ± margin of error, which is −1.24 ± 8.29, or −9.53 to 7.05.

Based on this interval we are 95% sure that the mean lifetime for male fruit flies with no tube-mates is between 9.53 days *less* and 7.05 days *more* than the mean for male fruit flies with 1 pregnant female. This interval contains zero, so a t-test at the 5% level would not reject the null hypothesis $H_0 : \mu_C = \mu_{P1}$. Our interval shows the difference might be positive, negative, or not exist at all, so there is no detectable difference between the mean lengths of life for male fruit flies living alone or with 1 pregnant female.

b. Control versus 8 virgins

- The difference in sample mean lifetimes between the Control and Virgin8 groups is $\bar{y}_1 - \bar{y}_5 = 63.56 - 38.72 = 24.84$.
- The computation of the margin of error (8.287) is exactly the same as for the previous interval.
- The interval is difference in means ± margin of error, which is 24.84 ± 08.29, or 16.55 to 33.13.

Based on this interval we are 95% sure that the mean lifetime for male fruit flies with no tube-mates is between 16.55 and 33.13 days *more* than the mean for male fruit flies with 8 virgin females. This interval is nowhere near zero, so a t-test at the 5% level would reject the null hypothesis $H_0: \mu_C = \mu_{V8}$. Our interval clearly shows that the difference in means should be positive and thus male fruit flies living alone will tend, on average, to have longer lives than male fruit flies living with 8 virgin females.

The margin of error and confidence interval tell us how well we can estimate the theoretical parameters or their differences. *Neither the* P*-value nor the margin of error tells us anything at all about whether the differences we have detected and estimated matter in actual practice.* For that, we need effect sizes.

Effect Sizes: How Much Do the Differences Matter?

The idea of effect size comes from common sense: Size matters. The *P*-value and the margin of error can never tell the whole story. The phrase "statistically significant" is often misleading. The words suggest, wrongly, that a "significant" result is important. In fact, "statistically significant" means "too big to be due just to unit-to-unit variation," or, more briefly, "statistically detectable." In this sense, the difference "signifies."

To see what *P*-values and interval estimates miss, consider two examples, boy babies and lung cancer:

EXAMPLE 5.16 **Why effect size matters**

a. (a) Boy babies and girl babies: The effect is highly significant but tiny. The chance that a baby born in the United States is a boy is about 51.2%, and the chance of a girl is 48.8%. We have so much data on sex at birth (as defined by chromosomes) that these ratios are rock solid. The margin of error is essentially zero. The *P*-value is essentially zero. The difference is "real." But the effect size? For most purposes, the difference between 51/49 and 50/50 makes no difference at all. Bottom line: We are sure that the difference is "real," we can estimate the size of the difference with near certainty, but, most of the time, the size of the difference is so small that 50/50 is as good as the true ratio of 51/49. Who cares?

b. (b) Smoking and lung cancer: There is a large effect that at one time had not been shown to be statistically significant. Back in the 1940s the possibility of a link between smoking and lung cancer was in the air. The available data were observational, and there was no clear way to compute a P-value to compare smokers with nonsmokers. As we know now, the effect of smoking on lung cancer is substantial, and for all too many individuals, catastrophic. Back in the 1940s, science had no tiny P-value for smoking data, but—if only we'd known—there was a large effect size.

The examples capture the main idea, and we have already looked at effect size in the setting of a two-sample t-test in Chapter 0. Effect size is about the IT question, "How big is IT?" More specifically, the effect size answers the question, "How large is the group-to-group difference, compared with ordinary unit-to-unit variation?"

EFFECT SIZE

The **effect size** is one commonly used way to measure how much practical importance a numerical difference might make in real life. In the current one-way ANOVA setting we can measure effect size by finding the ratio of a difference to the standard deviation within all groups. This can be done for a single group

$$D_i = \hat{\alpha}_i/SD = (\bar{y}_i - \bar{y})/SD$$

or a difference between any pair of groups

$$D_{ij} = (\bar{y}_i - \bar{y}_j)/\text{SD}$$

where $SD = \sqrt{MSE}$. Note: When the context is clear, we may drop the subscripts and just refer to an effect size as D.

EXAMPLE 5.17

FFlies

Fruit flies: Effect size The "typical" size of the fly-to-fly differences in length of life can be measured by the estimated standard deviation $SD = \sqrt{MSE} = 14.8$ days.* We can use this "typical" size as a yardstick to measure the relative effect on length of life. "If I'm a fruit fly, and I know the actuarial data, I know that my projected length of life is squishy, with a big margin of error, give or take almost 15 days. If I get to choose how to rent out the space in my test tube, does my decision matter?"

a. *Virgin8 group by itself.* The estimated effect for mean lifetime of a male fruit fly living with 8 virgins is $\hat{\alpha}_5 = 38.72 - 57.44 = -18.72$. Scaling this by $\sqrt{MSE} = 14.8$ gives an effect size of $D_5 = -18.72/14.8 = -1.26$. This is a substantial amount. We estimate that the drop in mean lifetime for a male fruit fly when living with 8 virgin female fruit flies is about 126% of the SD in lifetimes.

b. *Alone versus one pregnant female.* The 25 male flies raised alone had an average length of life of 63.56 days. The average male fly who shared his tube with one pregnant female fly had an average length of life of 64.80 days. Does this difference of 1.24 days really matter? One way to judge is to express the difference of 1.24 days as a percentage of the standard

*We put quotation marks around "typical" to warn that the word is meant informally, to guide your intuition. There is no formal definition of "typical" and there are many ways to describe it.

deviation of 14.8 days. This fraction, $(63.56 - 64.80)/14.8 = -0.084$, is the effect size. For this particular comparison, the effect size of 8% is tiny. The difference between the two environments hardly matters.

c. *Alone versus eight virgins.* Suppose, for a contrasting example, that we had compared the average length of life for the bachelor flies living solo (63.56 days) with the length of life for those who share their little glass pads with eight virgins (38.72 days). The mean difference in length of life $(63.56 - 38.72)$ is 24.84 days, and the effect size is $D =$ (mean difference/SD) $= 24.84/14.80 = 1.68$, or 168% of the SD. If you are a male fruit fly, the effect size says to curb your enthusiasm. If you are in the business of underwriting life insurance for these two groups of males, you really need to know their living environment.

To summarize, this section has offered a mix of methods and cautions. The methods are meant to help you answer three questions, one about existence and two about size: (1) The *P*-value: IS there an effect? (2) The interval estimate: If yes, there is an effect, how precisely can we measure IT? (3) The effect size: If yes, there is an effect, is IT big enough to matter?

To review these questions, and to build your intuition, consider how the answers depend on sample size. Example 5.18 will offer some data, but first, by way of introduction, a small thought experiment may help. Compare the effect of two exercise regimens on body weight: no exercise, and running stadium steps for 10 minutes each day. Abstractly, we want to answer the same three questions as above: (1) Is there a difference in size of effect? (2) If so, how precisely can we estimate the sizes? (3) Does the size difference matter in real life? In this concrete context, the three questions become: (1) Does running steps make a difference in weight loss? (2) If so, how precisely can we measure the difference? (3) Is the weight loss big enough to be worth the effort?

Refer these questions to Jack and Jill. Dutiful Jill climbs up the stadium hill every day. Lazy Jack just sits and watches. Each day Jack and Jill record their weight. How much do they learn from one day to the next? After just one day, they know almost nothing. (1) After 6 months they know, almost surely, that running steps makes a difference. (2) After 6 years, they can be very confident about how much daily difference Jill will get for her pain. (3) What about effect size? With more and more data, they can estimate the effect size better and better, but the size of the daily effect is pretty much the same from one day to the next. The calories you burn, and their effect on your weight, won't change much from day to day.

More abstractly, as your sample size increases, your *P*-value decreases (you can be more confident that there is a difference), your margin of error decreases (you can be more precise about the size of the difference), but the effect size doesn't change much. The benefit from running steps stays the same. A calorie is a calorie.

The moral: Take this example as a lesson about why no single summary, whether *P*-value, margin of error, or effect size, offers you more than just one summary of your data.

EXAMPLE 5.18

FFlies

What you get from larger samples To see what happens as sample sizes increase, consider the two mean differences of the previous example, Control versus Pregnant 1 and Control versus Virgin 8. For both fruit fly differences, the standard deviation σ is the same, and does not depend on n, the number of test tubes with a male fly. For the sake of the example, we pretend that the value of σ is 14.8, the same as the observed value. For both comparisons, the standard error for the difference is the same, and depends only on σ and n. The same is true of the 95% margin of error (M95 in Figure 5.17): the value depends only on σ and n. For the first comparison, C versus P1, the observed difference of the two means is 1.24 days, and the effect size D is tiny, about 8%. For the second comparison, the observed difference between the two means is 24.84 days, and the effect size is much larger, 168%. These effect sizes depend on the SD and the mean difference, but not on the sample size n.

What *does* depend on n is the value of the t-statistic, and the corresponding P-value. As the sample size increases, the *effect size* based on the group means μ_i and standard deviation σ remains constant. However, even though the effect size remains constant, the margin of error gets smaller and smaller, and the evidence that the effect is nonzero gets stronger and stronger, the t-statistic gets larger and larger, and the P-value gets closer and closer to zero.

			C versus P1		C versus V8	
			SD =	14.8	*SD* =	14.8
			mean diff =	1.24	mean diff =	24.84
			D =	0.08	*D* =	1.68
n	SE	M95	t	p	t	p
10	6.619	13.906	0.187	0.4267	3.753	0.0007
100	2.093	4.128	0.592	0.2771	11.868	0.0000
1000	0.662	1.298	1.873	0.0306	37.530	0.0000
10000	0.209	0.410	5.924	0.0000	118.679	0.0000

FIGURE 5.17 What happens as the sample size increases?

Notice that for the first comparison, C versus P1, for which the mean difference is small, it takes 1000 tubes of each kind to achieve a P-value of 0.03. With 10,000 tubes of each kind the P-value is zero to four decimal places and the evidence of a difference is rock solid, even though the effect size is still a tiny 8%. For the other comparison, C versus V8, the effect size is large, and with as few as 10 tubes of each kind, the difference registers with a P-value of less than 0.001.

In conclusion: The P-value, the margin of error, and the effect size are all important. Also the cautions are important: *The methods work only to the extent that the required probability model applies*.

5.6 Using Plots to Help Choose a Scale for the Response

This section is based on a cardinal rule of statistics: *Always plot your data*. To quote from one of the great statistics educators of our time, David Moore, "No one has ever invented a data analysis tool that beats the combination of the human eye and brain."[7]

Using a Reexpression to Choose a Scale

In Section 1.4 we saw how transformations or reexpressions of the variables can help address issues of skewness or changing variability in a linear regression setting. The same ideas apply here and are a bit easier to implement since we have only one quantitative variable to deal with (and we don't have to worry about issues of linearity). The next few examples illustrate these ideas.

EXAMPLE 5.19

Metropolitan commutes We return to the metropolitan commutes scenario and start looking at the actual data.

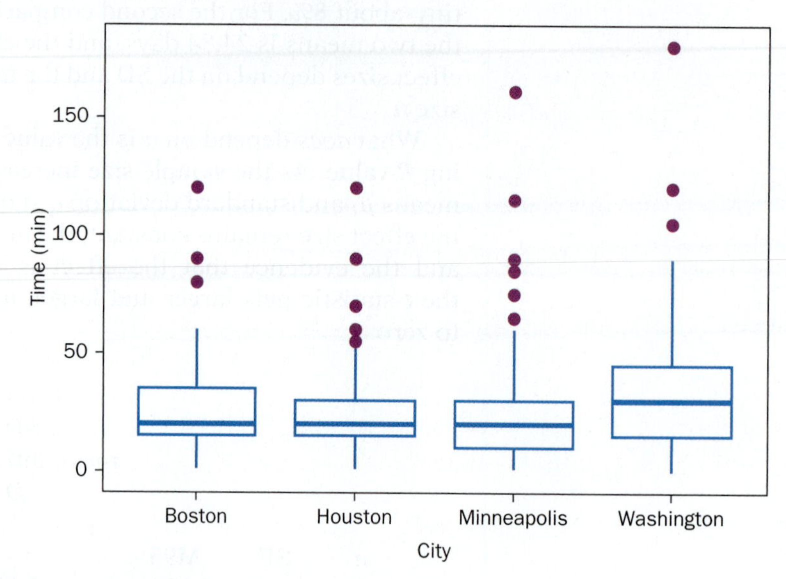

FIGURE 5.18 Commute times in four cities

Comparative boxplots* of the commuting times in Figure 5.18 indicate that Washington tends to have somewhat longer commuting times than the other cities. The boxplots also show lots of variability in commuting times within each city and not surprisingly, lots of overlap in commuting times across the four cities. They also reveal that the distributions of commuting times show considerable skewness to the high end and many outliers.

Summary statistics shown in the following computer output indicate that Washington commuters have an average commute time ($\bar{y}_4 = 31.69$) of more than six minutes longer than the other cities, with Minneapolis having the smallest average commuting time ($\bar{y}_3 = 22.74$). When comparing variabilities we see that the largest standard deviation in commute times ($s_4 = 21.16$ for Washington) is only 1.25 times greater than the smallest ($s_3 = 16.88$ for Minneapolis). The boxplots also show similar variability in commute times for all four cities, so the main issue with the ANOVA conditions is the heavy skewness in each distribution.

Summary for Time:

	City	mean	sd	n
1	Boston	25.75	17.59	500
2	Houston	25.10	17.46	500
3	Minneapolis	22.74	16.88	500
4	Washington	31.69	21.16	500

Recall in Section 1.4 we considered some conditions (on page 43) for when a log transformation might be helpful. One of those is to deal with distributions that are skewed to the high end. Perhaps a log transformation will work with the commute times?

*Note that comparative dotplots are not so useful when we have large sample sizes in each group, so we show only the boxplots here.

We need to make a small adjustment before calculating logs because there are a few commute times of zero (likely people who work at home) and the logarithm of zero is undefined. For this reason the transformed response variable will be $LogTime = log(Time + 1)$. The summary statistics for the groups are given below and we see that $max(s_i)/min(s_i) = 0.786/0.699 = 1.12$, so we still have similar variabilities in the reexpressed response variable.

```
Summary for LogTime:

  City          mean    sd      n
1 Boston        3.031   0.786   500
2 Houston       3.042   0.700   500
3 Minneapolis   2.941   0.699   500
4 Washington    3.262   0.737   500
```

Figure 5.19 shows the distribution within the four cities after we reexpress the response variable as the logarithm of commute times (plus one). Although some outliers remain, there are fewer and they are now distributed at both the high and low ends of the boxplots. The boxplots look more symmetric and still have similar spreads.

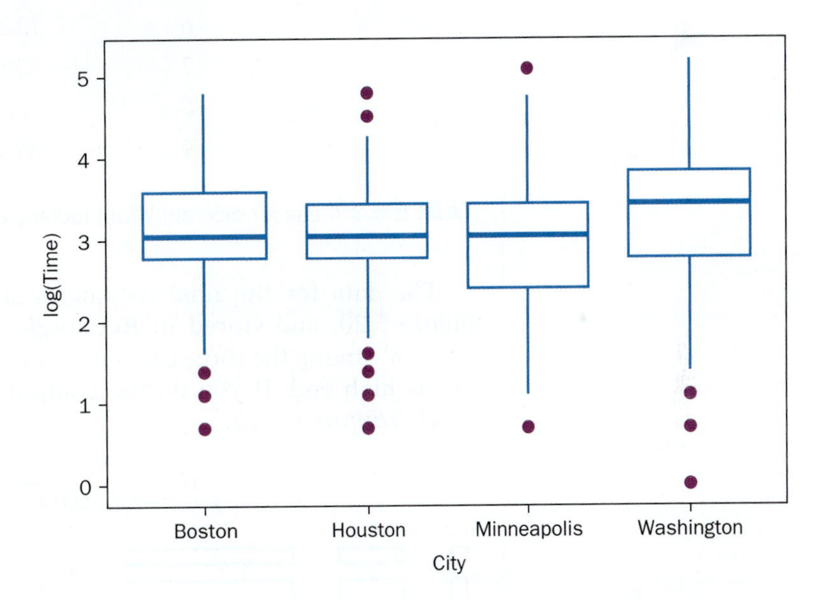

FIGURE 5.19 Logarithm transform of commute times in four cities

Notice that Washington still has slightly larger values in the log scale, but there is still considerable overlap in the boxplots. Are the sample differences large enough to be confident that there really are differences among the cities, or could this still just reflect random chance in getting four samples? We return to this question in Exercise 5.62.

In Section 1.4 when we look at transformations for a regression model, Example 1.9 on page 42 presents a hypothetical example dealing with "circular science" to show how a log transformation can help with showing the relationship between area and radius of circles. The next example follows a similar theme to illustrate how a log reexpression can help with the response variable in an ANOVA setting.

EXAMPLE 5.20

Rectang

Areas of rectangles For the sake of this example, imagine that you are a pioneer in the emerging field of rectangle science. Little is known, and your goal is to understand the relationship between the area of a rectangle (quantitative response) and its width (categorical predictor). Your lab produces the following data, which you plan to analyze using one-way ANOVA. You have three categories of rectangle widths, thin (width = 1″), medium (width = 4″), and thick (width = 10″). There are three rectangles in each group, with lengths of 1″, 4″, and 10″, but due to the primitive state of rectangle science, the relevance of length has not yet been recognized and so was not recorded. The key idea, of course, but not yet known to rectangle science, is that the effect of width on area is proportional to length, or as we learned in school, $A = L \cdot W$.

Rectangle	Width	Area
1	Thin	1
2	Thin	4
3	Thin	10
4	Medium	4
5	Medium	16
6	Medium	40
7	Thick	10
8	Thick	40
9	Thick	100

TABLE 5.3 *Areas* for different *Width* rectangles

The data for the nine rectangles are shown in Table 5.3, displayed in Figure 5.20, and stored in **Rectangles**. Clearly, spreads of areas are very unequal among the three categories and the distributions of areas are skewed to the high end. If you do the computations, you can verify that the ratio $max(s_i)/min(s_i) = 10$.

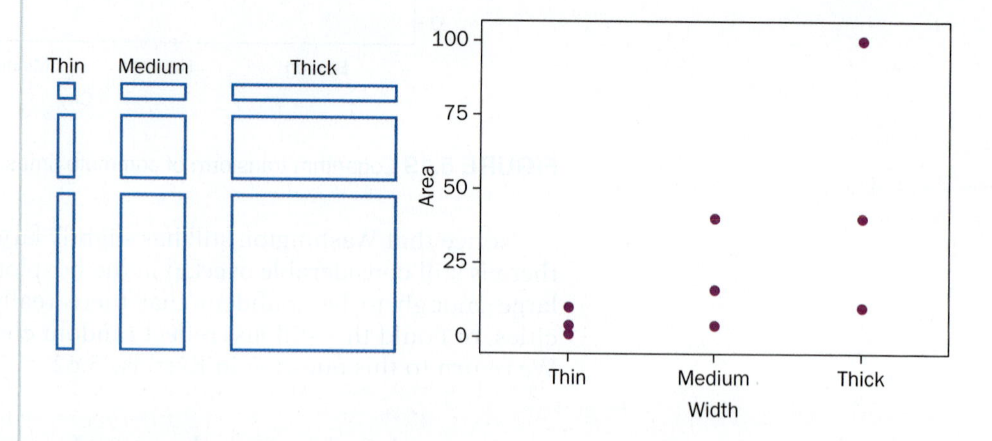

FIGURE 5.20 Area versus width for three groups of rectangles

With benefit of hindsight, we know that logs convert multiplication into addition: If $A = L \cdot W$, then $\log(A) = \log(L) + \log(W)$. If we reexpress area in a

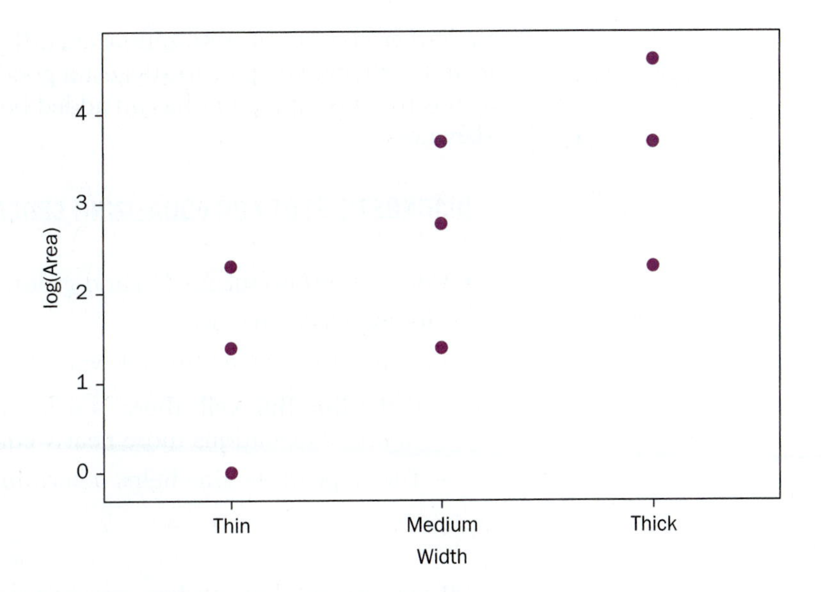

FIGURE 5.21 log(Area) versus width for three groups of rectangles

log scale, the standard deviations of the *log*(*Area*) values for the three groups are exactly equal, as shown in Figure 5.21.

Although the previous example is artificially simplistic, it demonstrates the utility of using a log transformation to address imbalances in variability and skewness in the response variable in an ANOVA setting (just as the log transformation helped produce a linear relationship in the earlier circle example). When dealing with reexpressions in the regression setting, we noted several signs for when a log transformation might be appropriate (page 43). Some of these, such as curvature in the scatterplot, are not relevant in the ANOVA setting. Signs that a log transformation might be helpful in an ANOVA setting include:

- The distribution of the response is skewed toward the high end in one or more groups.
- The responses in one or more groups range over two or more orders of magnitude.
- The variability (e.g., standard deviation) varies considerably between groups, especially when the changes in spread track with changes in group means.

Just as with the regression setting, we can consider other transformations, such as reciprocals, squares, or square roots, to help stabilize variability and reduce skewness. In Section 1.4 we looked at ways to use concavity of scatterplots to help choose an appropriate power transformation. While a scatterplot is not feasible for an ANOVA model, is there some other graphical construction that might be helpful?

The examples that come next will show you an important technique, a recipe you can follow to use your data to help CHOOSE a scale for reexpression of the response variable. As with any recipe, you have to pay attention to the details, but as with any recipe also, it is important to remember that the details are there to serve a larger goal. In that spirit, keep in mind that a major goal of science is to find simple patterns that relate cause and effect. A good choice of scale can make those relationships easier to see, easier to summarize, and easier to think about. In that sense, the recipe for finding a good scale serves the larger goal of understanding scientific relationships. For ANOVA, there is a simple diagnostic plot to suggest a good scale to make group standard deviations roughly equal that has an added bonus of also helping address issues of skewness.[8]

DIAGNOSTIC PLOT FOR EQUALIZING SPREADS

- Compute group means ($\bar{y}_i$) and group standard deviations (s_i).
- Plot $\log(s_i)$ versus $\log(\bar{y}_i)$
- Fit a line by eye (or with software).
  - If the line fits well, there is a reexpression that will make the group standard deviations more nearly equal. (If not, not.)
  - The slope of the line helps determine the reexpression.

If you are good at algebra, you can use a formula to find the transformation: reexpress the response y as y^p where the exponent for the transformation $p = 1 - \text{slope}$. ($p = 0$ means take logs)*

If you prefer to avoid formulas, here are the most common transformations:

Fitted slope	Exponent = 1 − slope	Transformation	Replace y with
0	1	(no change)	y
0.5	0.5	square roots	$\sqrt{y}$
1	0	logarithms	$\log(y)$
2	−1	reciprocals	$1/y$

The next two examples show the diagnostic plot in action. For Example 5.21, the plot tells us to take logs. For Example 5.22, we investigate what the diagnostic plot says about the **MetroCommutes** data.

EXAMPLE 5.21

VVerb

Left brain, right brain: Just take logs According to the theory of hemispheric specialization, the two sides of your brain carry out different kinds of cognitive tasks. To oversimplify, your right brain is visual, your left brain is verbal. To test this hypothesis, a classic experiment asks subjects to carry out two kinds of tasks, one visual, one verbal, and to report the results in either of two ways, one visual, one verbal. According to theory, if task and report are of the same kind, both visual or both verbal, it will take longer than if you can use both sides of your brain at once, one side for the task, the other side to report. In this experiment the *Task* was to identify parts of letters (Visual) or sentences (Verbal) and *Respond* by pointing (Visual) or speaking (Verbal). The data for this example come from an experiment with 20 psychology students at Mount Holyoke College and are stored in **VisualVerbal**. More details for this experiment are given in Example 6.16 on page 00. The results (response *Time* in seconds) for each combination of task and report method are shown as side-by-side plots in Figure 5.22. As you look at the plots, use them to think about whether reexpression is indicated.

The group standard deviations of *Time* are not drastically different ($max(s_i)/min(s_i) = 1.8$), but within-group distributions are markedly skewed. Figure 5.22(a) shows that dots are dense at the bottom of each column,

*If you like calculus, you can use l'Höpital's rule to explain why $p = 0$ corresponds to logs. Take limits of $(y^p - 1)/(p - 1)$ as $p \to 0$.

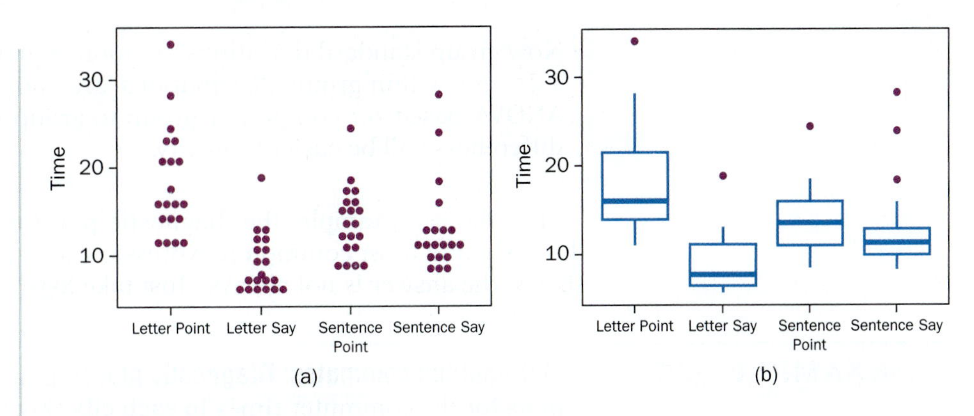

FIGURE 5.22 Completion time for combinations of task (letter or sentence) and report method (point or say)

spread out at the high end. In Figure 5.22(b), all the outliers are at the high end. Reexpression is indicated. To find a scale, Figure 5.23 plots $log(s_i)$ versus $log(\bar{y}_i)$ for the four groups.

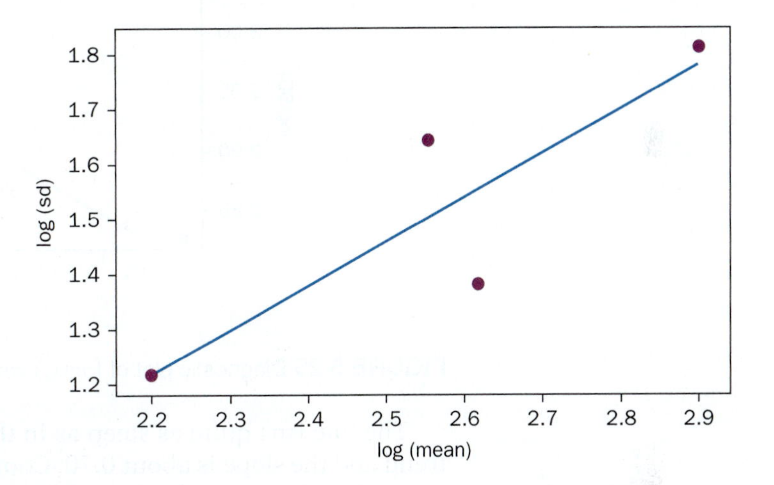

FIGURE 5.23 Diagnostic plot of $\log(s_i)$ versus $\log(\bar{y}_i)$ for the four groups

The slope of the fitted line is a little below 1, so the exponent for the reexpression (1 − slope) is close to zero, suggesting a transformation to log of *Time*. Figure 5.24 shows the same data in the new scale.

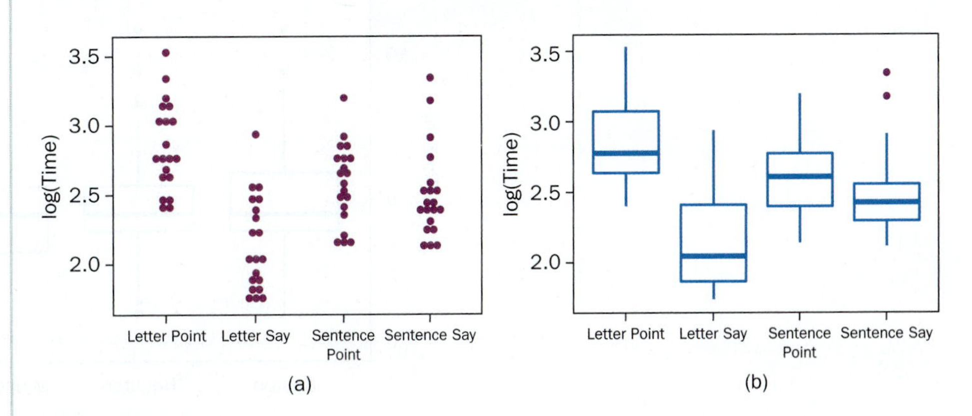

FIGURE 5.24 Completion time for combinations of task (letter or sentence) and report method (point or say)

Now group standard deviations are more equal ($max(s_i)/min(s_i)$ is less than 1.2) and within-group distributions are more nearly symmetric. Formal ANOVA based on comparing group-to-group differences with unit-to-unit differences will be easier to justify.

For the last example, the diagnostic plot for equalizing spreads led us to the usual and most common reexpression: Just take logs. As the next example shows, the answer is not always "Just take logs."

EXAMPLE 5.22

MetCom

Metropolitan commutes: Diagnostic plot In Example 5.19 we see very skewed plots for the commuter times in each city (Figure 5.18) and a more symmetric distribution (Figure 5.19) after doing a log transformation. What would the diagnostic plot say about a possible transformation? Figure 5.25 plots the log of the group standard deviations (as shown in the first table on page 232) versus the log of the mean commute *Time* for each city.

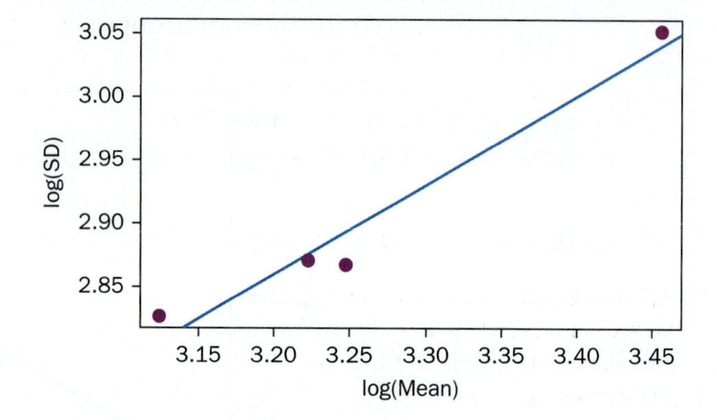

FIGURE 5.25 Diagnostic plot of $\log(s_i)$ versus $\log(\bar{y}_i)$ for commute times in four cities

The line isn't quite as steep as in the previous example, but there is some trend and the slope is about 0.70. Computing $1 - slope$ we get $1 - 0.7 = 0.3$, so somewhere between a log transformation ($p = 0$) and a square root ($p = 1/2$). We saw the log transformation in Example 5.19. Figure 5.26 shows boxplots using the square root of commute times as the response variable.

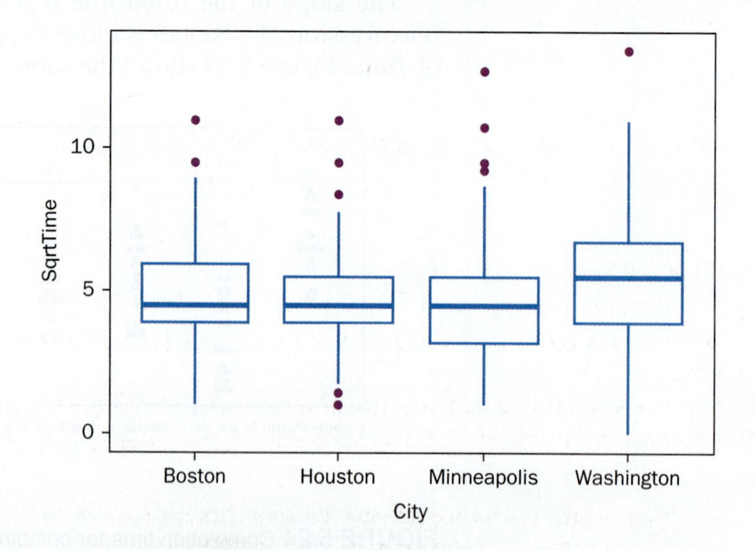

FIGURE 5.26 Square root transformation for commute times

Similar to the log transformation in Figure 5.19, we see better symmetry and fewer outliers. Comparing variances in the following table, we see that $max(s_i)/min(s_i) = 1.840/1.613 = 1.14$, well below the warning level of two and, again, similar to what we saw for the log transformation for these data. For the commute times, either the log or the square root transformation would be reasonable choices. We consider the ANOVA tables for each of these situations in Exercises 5.62 and 5.63.

```
Summary for sqrt(Time):

  City          mean     sd     n
1 Boston        4.759  1.762  500
2 Houston       4.730  1.652  500
3 Minneapolis   4.489  1.613  500
4 Washington    5.321  1.840  500
```

This section provided a deeper look at issues arising from checking the conditions of an ANOVA model that were introduced in Section 5.4. The next section looks more carefully at issues arising from comparing specific groups to check for differences after doing an ANOVA *F*-test.

5.7 Multiple Comparisons and Fisher's Least Significant Difference

The ideas of this section can be summarized by borrowing from four extraordinary scientists, I. J. Good, Linus Pauling, John Tukey, and R. A. Fisher.

Irving Jack Good, back in the 1940s, was one of the crackerjack young data scientists who helped Alan Turing design and build one of the ancestors of today's electronic computers and use that computer to crack the Enigma code of Hitler's Germany during World War II. Jack Good is relevant here because he once said, famously, "If you torture your data long enough, they will confess." "Multiple comparisons" is a technical term for one way to torture your data: Keep probing until, in desperation, your data coughs up something "significant." This section tells how to stay honest in your search for truth.

Linus Pauling, another extraordinary scientist, remains the only person ever to win two unassisted Nobel Prizes. Pauling is relevant here because of something he once said about how he won his two Nobels: "You have to have lots of ideas, and throw away the bad ones." Pauling's aphorism juxtaposes two contending goals of statisics, to have lots of ideas, and to throw away the bad ones; to explore, and to test; to look for patterns in the data we have, and to design studies to make sure that our idea—our research hypothesis—is not one of the bad ones.

Because of uncertainty, these two important goals are at odds. You already know their trade-off. The more ideas you have, the higher the chance that some of them are bad ones. To make the trade-off official, statistics uses formal language. A Type I error is a false alarm, a bad hunch that deserves to be thrown away. A Type II error is a miss, a failure to detect a difference, a failure to catch an important clue that slid past unnoticed.

John Tukey, a research scientist who worked at Bell Labs, was a pioneer in what he called exploratory data analysis, a system of methods and a way of thinking he invented as an approach to looking for patterns in data. Tukey used the words "exploratory" and "confirmatory" to label the two competing goals and two corresponding types of error in statistics. As Tukey and many others have said, you should be extremely careful about using the same dataset for both exploration and confirmation.

A question that often gets too little attention is this, "Which error matters more?" To see this, contrast two situations, (a) getting a medical screening test, and (b) submitting a scientific paper for publication. (a) Medical screening. If you are being screened for a serious disease, a miss (Type II) is much worse than a false alarm (Type I). If a medical screening test is positive, the news can be scary, but if the positive test is truly a false-positive, the follow-up test will let you know, and the scare has been only temporary. On the other hand, if the screening test gives a false-negative, a miss, your life may be at risk even though the test gives no alert. This same logic applies to exploration in science—looking for new drugs, or looking for genes related to cancer. You can always use follow-up studies to sort false-positives from the real thing, but a miss can never tell you that you ought to try again. A miss could be a major oversight. A false alarm is not so bad. (b) Preparing to publish. Suppose that you have completed your analysis, and you have found some differences that register as "statistically significant" (i.e., $p < 0.05$). Should you publish? If the *P*-value is a false alarm, not only do you risk damage to your own reputation as a scientist, but also, and worse, you may mislead other scientists.[9]

The two official categories of error, with those clever names of Type I and Type II, are attempts to reduce the fuzzy ideas of reality to sharp, abstract ideas we can study using the power of mathematics. R. A. Fisher, one of the greatest geneticists of all time, is relevant here because he used mathematics to offer a way to control the rate of false alarms while you explore for differences.* His LSD (least significant difference) method is the focus of this section. Section 8.2 will offer more options.

To make all this concrete, return to the fruit flies once again. Suppose you are allowed to use two-sample *t*-tests over and over, as many *t*-tests as you want, and you win a gold-plated Mercedes if you can find two groups of observed values whose mean difference gives a *P*-value less than 0.01. The next example shows two strategies for choosing the groups to compare.

EXAMPLE 5.23

FFlies

Fruit fly lifetimes: Two strategies for choosing which means to compare For the fruit flies, there were five treatment groups: control (C), one pregnant (P1), eight pregnant (P8), one virgin (V1), and eight virgins (V8). Here are two different ways to create two groups for a *t*-test.

a. **Pairwise comparisons.**The most basic comparison is between one treatment group and another, known as a *pairwise comparison*. For the fruit flies, there are 5 treatment groups and 10 possible pairwise differences: C/P1, C/P8, C/V1, C/V8, P1/P8, P1/V1, P1/V8, P8/V1, P8/V8, V1/V8. (If you have studied combinations and permutations, you know "5 choose 2 = 10.")

b. **Contrasts**. A *contrast* comes from comparing two sets of treatments that don't overlap. For example, we might contrast virgin and pregnant (V1&V8/P1&P8). We might contrast control versus pregnant (C/P1&P2). We might contrast control with the rest (C/P1&P8&V1&V8). Such contrasts offer another way to use a two-sample *t*-test to make comparisons. Each contrast defines two sets of treatments, a mean for each set, two sample sizes, and a *t*-test. For an experiment with five groups, there are many dozens of possible contrasts. We discuss contrasts in more detail in Section 8.3.

*Other methods that control the overall experiment-wise error rate, including the popular Tukey and Bonferroni procedures, are considered in Section 8.2.

With this example in mind, consider the two strategies in the context of your goal, to advance science or to win the Mercedes. For trying to win the Mercedes, $p < 0.01$ wins the prize. False alarms don't matter. Strategy (b), looking at all possible contrasts, is the best choice. For science, your goal is to learn from data, and false alarms can be very costly. Holding down the false alarm rate is important. In the rest of this section, we show how to control the false alarm rate for strategy (a), all pairwise comparisons, using the method known as Fisher's LSD. Section 8.2 shows ways to control the false alarm rate for strategy (b).

All methods that go by the name "multiple comparisons" aim to control the false alarm rate. (If your main concern is to avoid Type II errors, if you are exploring and want to cast a wide net, if a Type I error is not your main concern, these methods for multiple comparisons are not what you want.)

All the methods for controlling Type I errors are based on the mathematics of two different kinds of Type I error rate, individual and familywise. If there actually is no difference between two groups (the null is true) and we use a test at a 5% significance level, the chance our sample data provides enough evidence to conclude there is a difference (reject the null) is about 5%. To borrow the words of Linus Pauling, 5% is the chance that you think you've found a good idea, but it is actually a bad one. Common sense tells us that the more ideas we have (the more tests we do), the more likely it is that at least one of them is a bad one that we don't recognize (a Type I error). Just as when spinning a spinner with only a small 5% sliver of a bad outcome (the individual Type I error rate), the chance of hitting that outcome at least once when you spin (test) a lot (the familywise Type I error rate) can be much larger than 5%.

INDIVIDUAL FALSE ALARM RATE AND FAMILYWISE FALSE ALARM RATE

As always, the Type I error (false alarm) rate is the chance that a test is statistically significant when in fact the null hypothesis (no difference) is true.

- The **individual** false alarm rate is the chance of a Type I error for a single test.
- The **familywise** false alarm rate is the chance of at least one Type I error among a family of multiple tests.

Consider rolling a fair die, and pretend that rolling a six is a false alarm. For a single roll, the chance of a false alarm is 1/6 or about 16%. For two rolls, the chance of at least one six is 11/36, about 31%. For ten rolls, the chance of at least one six is about 84%. Die rolls are independent, and pairwise tests are not, but the basic idea still applies: The more tests you do, the more likely it is that at least one "significant" result in your family of tests is a false alarm.

FISHER'S LSD: CONTROLLING THE FAMILYWISE TYPE I ERROR RATE FOR ALL PAIRWISE DIFFERENCES

Step 1: Is the ANOVA F-test statistically significant at your chosen significance level (10%, 5%, 1%)?
No: Stop. There are no "significant" pairwise differences.
Yes: Go to Step 2.

Step 2: Find the least significant difference (LSD):
Find the value of t^* for your $df(Error)$ and your level of confidence (90%, 95%, 99%).

For each pair of groups, compute the standard error $SE = \sqrt{MSE} \cdot \sqrt{1/n_1 + 1/n_2}$ Compute $LSD = t^* \cdot SE$.

Step 3: Compare: Declare each difference of group means "significant" if the difference is as large as the LSD.

EXAMPLE 5.24

FFlies

Fisher's LSD for the fruit flies For the fruit flies there are five groups and therefore 10 pairwise differences. The F-test from a one-way ANOVA is highly significant, with $p < 0.0001$. The evidence of group-to-group differences is extremely strong. Accordingly, we compute Fisher's LSD, which works out to be 8.29 for each pair (because each group had the same sample size), and we ask, "Which pairwise differences are greater than 8.29?" Most of the differences are clear (see Figure 5.27).

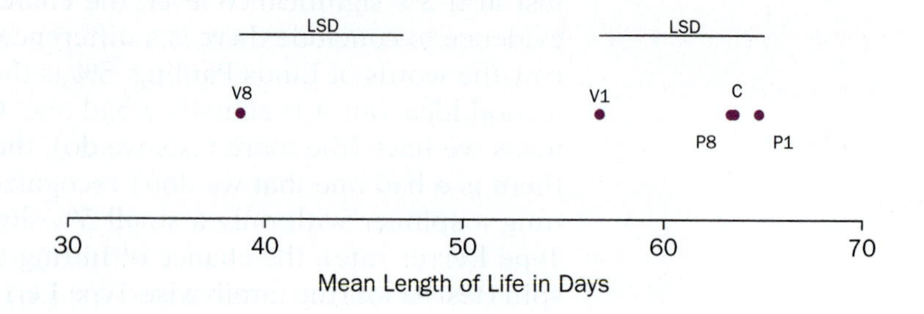

FIGURE 5.27 Differences among the means of the five groups of fruit flies

The means for C (63.56), P1 (64.80), and P8(63.36) are close together. Their differences are not even close to 8.29. At the other extreme, the mean for V8 (38.72) is far away from the other four means, and we declare those differences significant. The only pairwise difference that is not clear-cut is $P1 - V1 = 64.80 - 56.76 = 8.04$, which is not quite as large as the least significant difference of 8.29 and thus we do not declare a difference for those two groups.

Here is a summary of the process:

Step 1: F-test
P-value < 0.0001 — OK to check for differences

Step 2: Compute LSD
$SE = 4.19$ — $SE = \sqrt{MSE} \cdot \sqrt{(1/n_1) + (1/n_2)}$
$t^* = 1.98$ — $df(Error) = 120$, confidence level $= 95\%$
$LSD = t^* \cdot SE = 8.29$

Step 3: Compare groups — Which differences are $\geq LSD$? (see Figure 5.27)

5.8 Case Study: Words with Friends

Words

The game Words with Friends is an online version of Scrabble®. Each player has seven letters (at a time) and they take turns creating words using those letters, with some letters being worth more points than others. In addition to the letters of the alphabet, there are two blank tiles in the collection of 107 total tiles that are used in the game. A blank tile can be used as any letter, A – Z, and thus is valuable despite being worth zero points on its own.

One of your authors likes to play Words with Friends, but sadly he doesn't have many friends. Thus he plays the "solo" version of the game, in which his opponent is the Words with Friends software, which uses an algorithm to find the best words to play.*

EXAMPLE 5.25

The value of a blank tile The file **WordsWithFriends** has data on 444 games and includes the variables *Points* (the number of points the author scored in the game) and *Blanks* (how many of the two blank tiles the author was lucky enough to get during the game).

The idea that "blank tiles are valuable" is a nice theory, but does it hold up in practice? To explore, we start with a plot. The sample size is large, so parallel boxplots are useful. Figure 5.28 shows that getting more blank tiles is associated with scoring more points. There are a few outliers, but that is to be expected whenever we have hundreds of observations, as we do here: there are 111 observations for 0blanks, 217 observations for 1blank, and 116 observations for 2blanks. Each of the three boxplots is symmetric and the spreads are similar. Indeed, the three SDs are 33.5, 34.3, and 34.3, which are remarkably close.

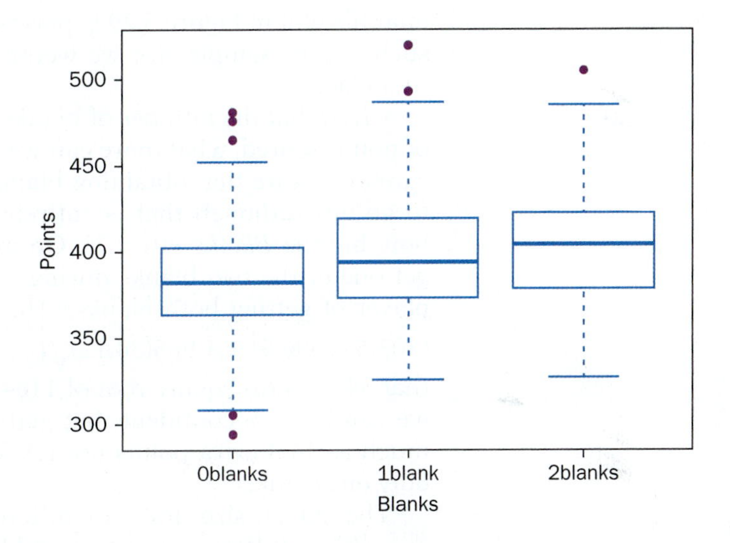

FIGURE 5.28 Parallel boxplots of the words with friends data

Based on the boxplots, we can feel comfortable fitting an ANOVA model to the raw data; there is no need to transform. The overall average is 395.6 and the group averages are 384.7, 396.9, and 403.5. Thus the fitted group effects are $384.7 - 395.6 = -10.9$, $396.9 - 395.6 = 1.3$, and $403.5 - 395.6 = 7.9$. Here is the ANOVA table:

	Df	Sum Sq	Mean Sq	F value	Pr(>F)
Blanks	2	20786	10393	8.93	0.00016
Residuals	441	513427	1164		

The *P*-value is tiny, so there is strong evidence that the number of blanks obtained is associated with the number of points scored. The normal

*Fortunately, the algorithm isn't very good, so the author almost always wins. You can analyze the winning margin in Exercise 5.52.

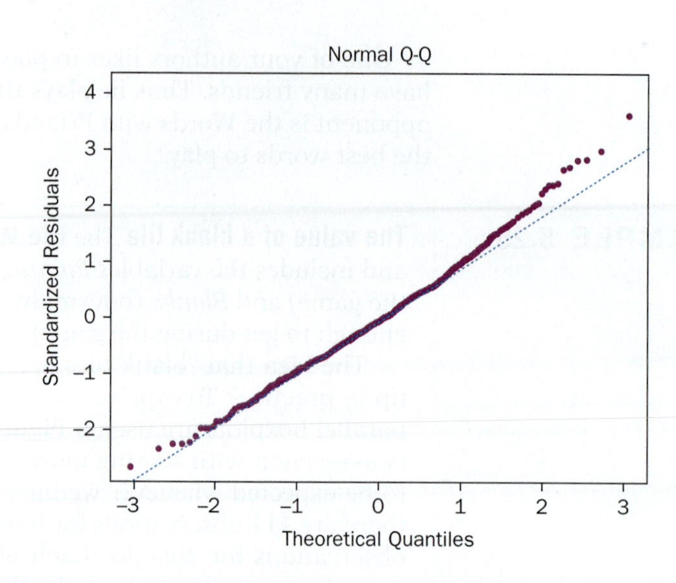

FIGURE 5.29 Normal quantile plot of residuals from ANOVA fit to words with friends data

quantile plot in Figure 5.29 supports the normality condition (although with such a large sample size we weren't really worried about normality in the first place).

Given that the number of blanks obtained is associated with the number of points scored, what more can we say? These are observational data, so we cannot be sure that obtaining blanks "causes" the number of *Points* scored to go up—although that is entirely plausible. What we can do is quantify how big the *Blanks* effect is. On average we would expect each player to get one of the two blanks during a game. How large is the advantage to a player of getting both blanks? The 95% confidence interval for $\mu_2 - \mu_1$ is $(403.5 - 396.9) \pm 1.965(34.12)\sqrt{\left(\frac{1}{217} + \frac{1}{116}\right)}$ or 6.6 ± 7.7 or $(-1.1, 14.3)$. (Note that 34.12 is the square root of 1164, the MSE from the ANOVA table.) Thus we can be 95% confident that getting two blanks is worth, on average, as much as 14.3 extra points (or 1.1 fewer points) when compared to getting only one blank.

The effect size for the difference between 2blanks and 1blank is $\frac{403.5 - 396.9}{34.12} = 0.193$, so getting two blanks is worth about 20% of a standard deviation, which is a modest but noticeable benefit.

CHAPTER SUMMARY

As you conclude this chapter, look back and notice how much ANOVA and regression have in common. Both have a quantitative response, and in their simplest versions, both have a single predictor. Both rely on *xy*-plots to explore and visualize patterns, with y = response and x = predictor: scatterplots for regression and side-by-side plots for ANOVA. Both methods, regression and ANOVA, often use fitted models simply to describe and summarize patterns, but both can also use probability models, with caution, for formal inference. For both regression and ANOVA, probability-based inference should be trusted only to the extent that the dataset satisfies a set of essential conditions, conditions that are basically the same for both methods.

ANOVA differs from regression in its reliance on group averages. For ANOVA, this reliance makes it possible to "see" the fitted model as a set of overlays. ANOVA also differs in more often being connected with randomized, controlled experiments. Both of these two differences will prove to be important in the remainder of Unit B.

EXERCISES

Conceptual Exercises

Exercises 5.1–5.4 are True/False.

5.1 Random selection. True or False: Randomly selecting units from populations and performing an ANOVA allow you to generalize from the samples to the populations.

5.2 No random selection. True or False: In datasets in which there was no random selection, you can still use ANOVA results to generalize from samples to populations.

5.3 Independence transformation? True or False: If the dataset does not meet the independence condition for the ANOVA model, a transformation might improve the situation.

5.4 Comparing groups, two at a time. True or False: It is appropriate to use Fisher's LSD only when the *P*-value in the ANOVA table is small enough to be considered significant.

Exercises 5.5–5.8 are multiple choice. Choose the answer that best fits.

5.5 Why ANOVA? The main purpose of an ANOVA model in the setting of this chapter is to learn about:

a. the variances of several groups.

b. the mean of one group.

c. the means of several groups.

d. the variance of one group.

5.6 Not a condition? Which of the following statements is *not* a condition of the ANOVA model?

a. Error terms have the same standard deviation.

b. Error terms are all positive.

c. Error terms are independent.

d. Error terms follow a normal distribution.

5.7 ANOVA plots. The two best plots to assess the conditions of an ANOVA model are:

a. normal probability plot of residuals and dotplot of residuals.

b. scatterplot of response versus predictor and dotplot residuals.

c. bar chart of explanatory groups and normal probability plot of residuals.

d. normal probability plot of residuals and scatterplot of residual versus fits.

5.8 When to transform. Which of the following is a reason to try transforming the data before fitting an ANOVA model?

a. The group means are not similar.

b. The residuals appear to have approximately a normal distribution.

c. The group standard deviations do not appear to be similar.

d. The data are independent.

5.9 List the three main reasons to randomize the assignment of treatments to units.

5.10 Diagnostic plot. One diagnostic tool introduced in this chapter is a plot of log(SD) versus log(ave).

a. What characteristic of your dataset would lead you to consider making such a plot?

b. If the plot exhibits a straight line, what does the slope suggest for your ANOVA analysis?

5.11 North Carolina births. The file **NCbirths** contains data on a random sample of 1450 birth records in the state of North Carolina in the year 2001. This sample was selected by John Holcomb, based on data from the North Carolina State Center for Health and Environmental Statistics. In the study, the response was the birth weight of a newborn infant, and the factor of interest was the racial/ethnic group of the mother. **NCBirth**

a. Explain why random assignment is impossible, and how this impossibility limits the scope of inference: Why is inference about cause not possible?

b. Under what circumstances would inference from samples to populations be justified? Were the requirements for this kind of inference satisfied?

5.12 Heavy metal, lead foot? Is there a relationship between the type of music teenagers like and their tendency to exceed the speed limit? A study[10] was designed to answer this question (among several). For this question, the response is the number of times a person reported driving over 80 mph in the last year. The study was done using random samples taken from four groups of students at a large high school: (1) those who described their favorite music as acoustic/pop; (2) those who preferred mainstream rock; (3) those who preferred hard rock; and (4) those who preferred heavy metal.

a. Results showed that, on average, those who preferred heavy metal reported more frequent speeding. Give at least two different reasons why it would be wrong to conclude that listening to heavy metal causes teens to drive fast.

b. What kinds of generalizations from this study *are* justified?

5.13 Student survey. You gather data on the following variables from a sample of 75 undergraduate students on your campus:

- Major
- Sex
- Class year (first year, second year, third year, fourth year, other)
- Political inclination (liberal, moderate, conservative)

- Sleep time last night
- Study time last week
- Body mass index
- Total amount of money spent on textbooks this year

Assume that you have a quantitative response variable and that all of the above are possible explanatory variables. Classify each variable as quantitative or categorical. For each categorical variable, assume that it is the explanatory variable in the analysis and determine whether you could use a two-sample t-test or whether you would have to use an ANOVA.

5.14 Car ages. Suppose that you want to compare the ages of cars among faculty, students, administrators, and staff at your college or university. You take a random sample of 200 people who have a parking permit for your college or university, and then you ask them how old their primary car is. You ask several friends if it's appropriate to conduct ANOVA on the data, and you obtain the following responses. Indicate how you would respond to each.

a. "You can't use ANOVA on the data, because there are four groups to compare."

b. "You can't use ANOVA on the data, because the response variable is not quantitative."

c. "You can't use ANOVA on the data, because the sample sizes for the four groups will probably be different."

d. "You can do the calculations for ANOVA on the data, even though the sample sizes for the four groups will probably be different, but you can't generalize the results to the populations of all people with a parking permit at your college/university."

5.15 Comparing fonts. Suppose that an instructor wants to investigate whether the font used on an exam affects student performance as measured by the final exam score. She uses four different fonts (Times, Courier, Helvetica, Comic Sans) and randomly assigns her 40 students to one of those four fonts.

a. Identify the explanatory and response variables in this study.

b. Is this an observational study or a randomized experiment? Explain how you know.

c. Even though the subjects in this study were not randomly selected from a population, it's still appropriate to conduct an analysis of variance. Explain why.

5.16 Dog food. Suppose a researcher wants to learn about a possible relationship between the main ingredient in dog food (salmon, chicken, beef) and the amount of energy that dogs have. The researcher obtains 45 Border Collies and randomly assigns each dog to one of the three diets. She then measures the average amount of time each dog sleeps per 24 hours for the next week as an indication of the amount of energy each dog has.

a. Identify the explanatory and response variables in this study.

b. Is this an observational study or a randomized experiment? Explain how you know.

c. Even though the subjects in this study were not randomly selected from a population, it's still appropriate to conduct an analysis of variance. Explain why.

5.17 Comparing fonts: study details. Refer to Exercise 5.15.

a. What are the units in this experiment?

b. What are the treatments?

c. What would the researcher have to do to ensure that this experiment is balanced?

5.18 Dog food: study details. Refer to Exercise 5.16.

a. What are the units in this experiment?

b. What are the treatments?

c. What would the researcher have to do to ensure that this experiment is balanced?

5.19 Comparing fonts: *df*. Refer to Exercise 5.15. Determine each of the entries that would appear in the "degrees of freedom" column of the ANOVA table.

5.20 Dog food: *df*. Refer to Exercise 5.16. Determine each of the entries that would appear in the "degrees of freedom" column of the ANOVA table.

5.21 All false. Reconsider Exercise 5.19. Now suppose that the P-value from the ANOVA F-test turns out to be 0.003. All of the following statements are *false*. Explain why each one is false.

a. The probability is 0.003 that the four groups have the same mean score.

b. The data provide very strong evidence that all four fonts produce different mean scores.

c. The data provide very strong evidence that the Comic Sans font produces a different mean score than the other fonts.

d. The data provide very little evidence that at least one of these fonts produces a different mean score than the others.

e. The data do not allow for drawing a cause-and-effect conclusion between font and exam score.

f. Conclusions from this analysis can be generalized to the population of all students at the instructor's school.

5.22 Can this happen? You conduct an ANOVA to compare three groups.

a. Is it possible that all of the residuals in one group are positive? Explain how this could happen, or why it cannot happen.

b. Is it possible that all but one of the residuals in one group is positive? Explain how this could happen, or why it cannot happen.

c. If you and I are two subjects in the study, and if we are in the same group, and if your score is higher than mine, is it possible that the magnitude of your residual is smaller than mine? Explain how this could happen, or why it cannot happen.

d. If you and I are two subjects in the study, and if we are in different groups, and if your score is higher than mine, is it possible that the magnitude of your residual is smaller than mine? Explain how this could happen, or why it cannot happen.

5.23 Schooling and political views. You want to compare years of schooling among American adults who describe their political viewpoint as liberal, moderate, or conservative. You gather a random sample of American adults in each of these three categories of political viewpoint.

a. State the appropriate null hypothesis, both in symbols and in words.

b. What additional information do you need about these three samples in order to conduct ANOVA to determine if there is a statistically significant difference among these three means?

c. What additional information do you need in order to assess whether the conditions for ANOVA are satisfied?

5.24 Turtles and habitat. A researcher is interested in studying the size of hatchling Ornate Box Turtles based on the state in which they are found. He gathers a random sample of turtles in each of three different states: Nebraska, Oklahoma, and Texas, wondering if the size changes from North to South.

a. State the appropriate null hypothesis, both in symbols and in words.

b. What additional information do you need about these three samples in order to conduct ANOVA to determine if there is a statistically significant difference among these three means?

c. What additional information do you need in order to assess whether the conditions for ANOVA are satisfied?

5.25 Schooling and political views: variation. Now suppose that the three sample sizes in Exercise 5.23 are 25 for each of the three groups, and also suppose that the standard deviations of years of schooling are very similar in the three groups. Assume that all three populations do, in fact, have the same standard deviation. Suppose that the three sample means turn out to be 11.6, 12.3, and 13.0 years.

a. Without doing any ANOVA calculations, state a value for the standard deviation that would lead you to reject H_0. Explain your answer, as if to a peer who has not taken a statistics course, without resorting to formulas or calculations.

b. Repeat part (a), but state a value for the standard deviation that would lead you to fail to reject H_0.

5.26 Turtles and habitat: variation. Now suppose that the three sample sizes in Exercise 5.24 are 15 for each of the three groups, and also suppose that the standard deviations of the turtle sizes are very similar in the three groups. Assume that all three populations do, in fact, have the same standard deviation. Suppose that the three sample means turn out to be 18.3, 20.1, and 22.4 mm.

a. Without doing any ANOVA calculations, state a value for the standard deviation that would lead you to reject H_0. Explain your answer, as if to a peer who has not taken a statistics course, without resorting to formulas or calculations.

b. Repeat part (a), but state a value for the standard deviation that would lead you to fail to reject H_0.

Guided Exercises

5.27 Mouse serotonin: exploratory. Serotonin is a chemical that influences mood balance in humans. But how does it affect mice? Scientists genetically altered mice by "knocking out" the expression of a gene, tryptophan hydroxylase 2 (Tph2), that regulates serotonin production. With careful breeding, the scientists produced three types of mice that we label as "Minus" for Tph2−/−, "Plus" for Tph2+/+, "Mixed" for Tph2+/−. The variable Genotype records Minus/Plus/Mixed. The variable Contacts is the number of social contacts that a mouse had with other mice during an experiment and the variable Sex is "M" for males and "F" for females. These variables are collected in the file **MouseBrain**.[11] **Mouse**

a. Produce boxplots and calculate descriptive statistics to compare contacts across the three genotypes. Comment on what these reveal.

b. Verify that the conditions required for ANOVA are met for these data.

5.28 Discrimination: exploratory. The city of New Haven, Connecticut, administered exams (both written and oral) in November and December of 2003 to firefighters hoping to qualify for promotion to either Lieutenant or Captain in the city fire department. A final score consisting of a 60% weight for the written exam and a 40% weight for the oral exam was computed for each person who took the exam. Those people receiving a total score of at least 70% were deemed to be eligible for promotion. In a situation where t openings were available, the people with the top $t + 2$ scores would be considered for those openings. A concern was raised, however, that the exams were discriminatory with respect to race and a lawsuit was filed. The data are given in the data file **Ricci**.[12] For each person who took the exams, there are measurements on their race (black, white, or Hispanic), which position they were trying for (Lieutenant, Captain), scores on the oral and written exams, and the combined score.

The concern over the exams administered by the city was that they were discriminatory based on race. Here we concentrate on the overall, combined score on the two tests for these people seeking promotion and we analyze the average score for the three different races. **Ricci**

a. Use a graphical approach to answer the question of whether the average combined score is different for the three races. What do the graphs suggest about any further analysis that could be done? Explain.

b. Check the conditions necessary for conducting an ANOVA to determine if the combined score is significantly different for at least one race.

5.29 Child poverty. A stratified random sample of 12 counties in Iowa (stratified by size of county–small, medium, or large) includes information about the child poverty rate in the counties. We are interested in seeing if child poverty rates in Iowa are associated with the size of the county. Figure 5.30 is the dotplot of the child poverty rates by type of county. Does this dotplot raise any concerns for you with respect to the use of ANOVA? Explain.

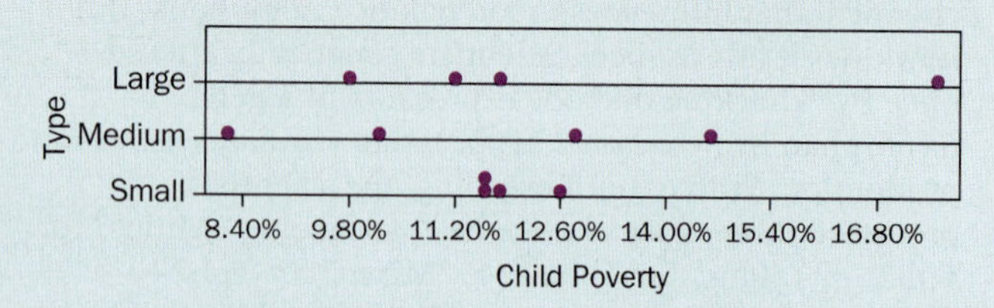

FIGURE 5.30 Child poverty rates in Iowa by size of county

5.30 Palatability. A food company was interested in how texture might affect the palatability of a particular food. They set up an experiment in which they looked at two different aspects of the texture of the food: the concentration of a liquid component (low or high) and the coarseness of the final product (coarse or fine). The experimenters randomly assigned each of 16 groups of 50 people to one of the four treatment combinations. The response variable was a total palatability score for the group. Figure 5.31 is the dotplot of the palatability score for each of the two concentration levels. Does this dotplot raise any concerns for you with respect to the ANOVA? Explain.

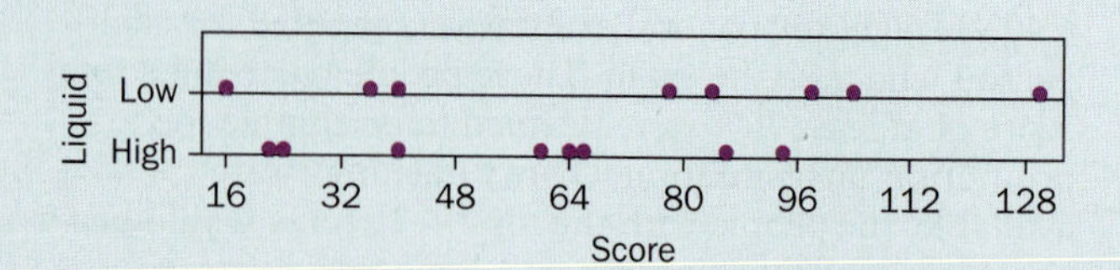

FIGURE 5.31 Palatability scores by level of concentration of liquid component

5.31 Amyloid. Amyloid - β (Abeta) is a protein fragment that has been linked to Alzheimer's disease. Autopsies from a sample of Catholic priests included measurements of Abeta (pmol/g tissue from the posterior cingulate cortex) from three groups: subjects who had exhibited no cognitive impairment before death, subjects who had exhibited mild cognitive impairment, and subjects who had mild to moderate Alzheimer's disease. The data are in the file **Amyloid**.[13] **Amyloid**

a. Report the sample sizes, sample means, and sample standard deviations for each group.

b. Make parallel boxplots of the data by group. What do these plots indicate about whether an analysis of variance model would be appropriate?

5.32 Hawks. Students and faculty at Cornell College in Mount Vernon, Iowa, collected data over many years at the hawk blind at Lake MacBride, near Iowa City, Iowa.[14] They captured over a thousand birds, recording many different measurements for each bird before releasing it. The dataset (**Hawks**) that we will analyze here is a subset of the original dataset, using only those species for which there were more than 10 observations. Data were collected on random samples of three different species of hawks: red-tailed, sharp-shinned, and Cooper's hawks. In this exercise we consider the *weight* of the hawks as our response variable. **Hawk**

a. Create dotplots to compare the values of *weight* between the three species. Do you have any concerns about the use of ANOVA based on the dotplots? Explain.

b. Compute the standard deviation of the weights for each of the three species groups. Do you have any concerns about the use of ANOVA based on the standard deviations? Explain.

5.33 Life spans. The *World Almanac* and *Book of Facts* lists notable people of the past in various occupational categories; it also reports how many years each person lived. Do these sample data provide evidence that notable people in different occupations have different average lifetimes? To investigate this question, we recorded the lifetimes for 973 people in various occupation categories. Consider the following ANOVA output:

Source	DF	SS	MS	F	P
Occupation			2749		0.000
Error	968	195149	202		
Total	972	206147			

a. Fill in the three missing values in this ANOVA table. Also show how you calculate them.

b. How many different occupations were considered in this analysis? Explain how you know.

c. Summarize the conclusion from this ANOVA (in context).

5.34 Aphid honeydew. Aphids (a type of small insect) produce a form of liquid waste, called honeydew, when they eat plant sap. An experiment was conducted to see whether the amounts of honeydew produced by aphids differ for different combinations of type of aphid and type of host plant. The following ANOVA table was produced with the data from this experiment.[15]

Analysis of Variance

Source	DF	SS	MS	F-Value	P-Value
aphid race - host plant combina			4.9807		0.000
Error	46	39.87	0.8667		
Total	51	64.77			

a. Fill in the three missing values in this ANOVA table. Also show how you calculate them.

b. How many different aphid/plant combinations were considered in this analysis? Explain how you know.

c. Summarize the conclusion from this ANOVA (in context).

5.35 Meth labs. Nationally, the abuse of methamphetamine has become a concern, not only because of the effects of drug abuse, but also because of the dangers associated with the labs that produce them. A stratified random sample of a total of 12 counties in Iowa (stratified by size of county—small, medium, or large) produced the following ANOVA table relating the number of methamphetamine labs to the size of the county. Use this table to answer the following questions:[16]

One-way ANOVA: meth labs versus type

Source	DF	SS	MS	F
type		37.51		
Error				
Total		70.60		

a. Fill in the values missing from the table.

b. What does the MS for county type tell you?

c. Find the *P*-value for the *F*-test in the table.

d. Describe the hypotheses tested by the *F*-test in the table, and using the *P*-value from part (c), give an appropriate conclusion.

5.36 Palatability: ANOVA table.[17] Exercise 5.30 introduced a study about the palatability of food. In that exercise, the explanatory variable was concentration of a liquid component. For this analysis you will focus on how the coarseness of the product affected the total palatability score. The data collected resulted in the following ANOVA table. Use this table to answer the following questions:

One-way ANOVA: Score versus Coarseness

Source	DF	SS	MS	F
Coarseness				
Error		6113		
Total		16722		

a. Fill in the values missing from the table.

b. What does the MS for the coarseness level tell you?

c. Find the *P*-value for the *F*-test in the table.

d. Describe the hypotheses tested by the *F*-test in the table, and using the *P*-value from part (c), give an appropriate conclusion.

5.37 Mouse serotonin: inference. Refer to Exercise 5.27. Conduct a one-way ANOVA. Report the ANOVA table and interpret the results. Do these data provide strong evidence that the mean number of contacts differs based on the type of mouse? Explain. **Mouse**

5.38 Meniscus: stiffness. An experiment was conducted to compare three different methods of repairing a meniscus (cartilage in the knee). Eighteen lightly embalmed cadaveric specimens were used, with each being randomly assigned to one of the three treatments: vertical suture, meniscus arrow, FasT-Fix. Each knee was evaluated on three different response variables: load at failure, stiffness, and displacement. The data are located in the file **Meniscus**.[18] For this exercise we will concentrate on the stiffness response variable (variable name *stiffness*). **Menisc**

a. Give the hypotheses that would be tested in an ANOVA procedure for this dataset.

b. Show that the conditions for ANOVA are met for these data.

c. Conduct an ANOVA. Report the ANOVA table and interpret the results. Do the data provide strong evidence that the mean value of *stiffness* differs based on the type of meniscus repair? Explain.

5.39 Discrimination: inference. In Exercise 5.28 you showed that the conditions were met to perform an ANOVA to answer the question about potential differences in test scores based on the race of the test taker. Here we continue the analysis of that dataset. **Ricci**

a. Write out in words and in symbols the hypotheses that would be tested by an ANOVA model in this setting.

b. What conclusions do you reach from the ANOVA analysis? Write a paragraph discussing your conclusions. Include the relevant statistics (*F*-value, *P*-value, etc.) as part of your discussion.

5.40 Meniscus: displacement Refer to the data discussed in Exercise 5.38. A second response variable that the researchers were interested in was the amount of displacement (variable name *displacement*). **Menisc**

a. Give the hypotheses that would be tested in an ANOVA procedure for this dataset.

b. Show that the conditions for the ANOVA procedure and *F*-test are satisfied with these data.

c. Conduct an ANOVA. Report the ANOVA table and interpret the results. Do the data prove strong evidence that the mean value of *displacement* differs based on the type of meniscus repair? Explain.

5.41 Simple illustration. This exercise was invented to show the diagnostic plot at work using simple numbers. Consider a dataset with four groups and three observations per group:

Group	Observed Values
A	0.9, 1.0, 1.1
B	9, 10, 11
C	90, 100, 110
D	900, 1000, 1100

Notice that you can think of each set of observed values as $m - s$, m, and $m + s$. Find the ratio s/m for each group, and notice that the "errors" are constant in *percentage* terms. For such data, a transformation to logarithms will equalize the standard deviations, so applying Steps 1–4 should show that transforming is needed, and that the right transformation is $p = 0$.

a. Compute the means and standard deviations for the four groups. (Don't use a calculator. Use a shortcut instead: Check that if the response values are $m - s$, m, and $m + s$, then the mean is m and the standard deviation is s.)

b. Compute S_{max}/S_{min}. Is a transformation called for? Plot $\log_{10}(s)$ versus $\log_{10}(m)$, and fit a line by eye. (Note that the fit is perfect: The right transformation will make $S_{max}/S_{min} = 1$ in the new scale.)

c. What is $p = 1 -$ slope? What transformation is called for?

d. Use a calculator to transform the data and compute new group means and standard deviations.

e. Check S_{max}/S_{min}. Has changing scales made the standard deviations more nearly equal?

5.42 Diamonds. Diamonds have several different characteristics that people consider before buying them. Most think, first and foremost, about the number of carats in a particular diamond, and probably also price. But there are other attributes that a more discerning buyer will look for, such as cut, color, and clarity. The file **Diamonds2** contains several variables measured on 307 randomly selected diamonds. (This is a subset of the 351 diamonds listed in the dataset **Diamonds** used in Chapter 3. This dataset contains all of those diamonds with color D, E, F, and G—those colors for which there are many observations.) Among the measurements made on these diamonds are the color and the number of carats of that particular diamond. A prospective buyer who is interested in diamonds with more carats might want to know if a particular color of diamond is associated with more or fewer carats. Here are the means, standard deviations, and their logs for the diamond data. **Diam2**

Color	Mean	St. dev.	log(Mean)	log(st. dev.)
D	0.8225	0.3916	−0.0849	−0.4072
E	0.7748	0.2867	−0.1108	−0.5426
F	1.0569	0.5945	0.0240	−0.2258
G	1.1685	0.5028	0.0676	−0.2986

a. Plot log(s) versus log(ave) for the four groups. Do the points suggest a line?

b. Fit a line by eye and estimate its slope.

c. Compute $p = 1 -$ slope. What transformation, if any, is suggested?

5.43 Hawks: diagnostic plot. Return to the **Hawks** data from Exercise 5.32. **Hawk**

a. Plot log(s) versus log(ave) for the three species. Do the points suggest a line?

b. Fit a line by eye and estimate its slope.

c. Compute $p = 1 -$ slope. What transformation, if any, is suggested?

5.44 Fenthion: diagnostic plot. Fenthion is a pesticide used against the olive fruit fly in olive groves. It is toxic to humans, so it is important that there be no residue left on the fruit or in olive oil that will be consumed. One theory was that if there is residue of the pesticide left in the olive oil, it would dissipate over time. Chemists set out to test that theory by taking a random sample of small amounts of olive oil with fenthion residue and measuring the amount of fenthion in the oil at 3 different times over the year—day 0, day 281 and day 365.[19] **Olive**

a. Two variables given in the dataset **Olives** are *fenthion* and *time*. Which variable is the response variable and which variable is the explanatory variable? Explain.

b. Check the conditions necessary for conducting an ANOVA to analyze the amount of fenthion present in the samples.

c. Produce a diagnostic plot. What transformation is suggested?

5.45 Mouse serotonin: Fisher's LSD. In Exercise 5.37 we discovered that there was a significant difference between the mean number of contacts for the three different mouse types. Use Fisher's LSD to find the differences that are significant. **Mouse**

5.46 Meniscus stiffness: Fisher's LSD. In Exercise 5.38 we discovered that there was a significant difference between the treatments with respect to the response variable *stiffness*. For this variable, larger values are better (less stiffness to the specimen). The researchers were comparing a potential new treatment (FasT-Fix) to two commonly used treatments (vertical suture and meniscus arrow). Use Fisher's LSD to determine which differences exist between the treatments and discuss the ramifications of your conclusions for doctors. Menisc

5.47 Discrimination: Fisher's LSD. In Exercise 5.39 you discovered that there was at least one significant difference in mean combined score for the three different groups of firefighters who took the promotion exam. Use Fisher's LSD to find the differences that are significant. Ricci

5.48 Meniscus displacement: Fisher's LSD. In Exercise 5.40 we discovered that there was a significant difference between the treatments with respect to the response variable *displacement*. For this variable, smaller values are better, but the values for the meniscus arrow were artificially low because these measures are taken during the load at failure testing and none of the meniscus arrow knees completed this test. All of the knees using the other two methods did complete the test. Once again, the researchers were comparing a potential new treatment (FasT-Fix) to two commonly used treatments (vertical suture and meniscus arrow). Use Fisher's LSD to determine which differences exist between the treatments and discuss the ramifications of your conclusions for doctors. Menisc

5.49 Hawks: transformation. Return to the **Hawks** data from Exercise 5.43. In that exercise you discovered that a transformation might help. Use that transformation for this exercise. Hawk

a. Transform the data by using the transformation suggested by Exercise 5.43. Report the sample means and sample standard deviations for each group in the transformed scale.

b. Make parallel boxplots of the transformed data by group. What do these plots indicate about the relationship between the transformed mean weights in the three groups?

c. What do the parallel boxplots indicate about whether an analysis of variance model would be appropriate?

5.50 Fenthion: transformation Return to the Fenthion data of Exercise 5.44. There was no clear "simple" transformation suggested by the diagnostic plot as $p \approx 2.4$. The closest "obvious" transformations would be a square or a cube transformation. Olive

a. Transform the amount of fenthion using the square transformation as potentially suggested by the diagnostic plot. Check the conditions necessary for conducting an ANOVA to analyze the square of the amount of fenthion present in the samples. If the conditions are met, report the results of the analysis.

b. Now try a second transformation. Transform the original amount of fenthion using the exponential. Check the conditions necessary for conducting an ANOVA to analyze the exponential of the amount of fenthion present in the samples. If the conditions are met, report the results of the analysis.

c. Which transformation did you prefer for this analysis? Explain.

5.51 Amyloid: transformation. Return to the **Amyloid** data from Exercise 5.31. Amyloid

a. Transform the data by taking the square root of each observation. Report the sample means and sample standard deviations for each group in the transformed scale.

b. Make parallel boxplots of the transformed data by group. What do these plots indicate about the amount of Abeta levels in the three groups?

c. What do the parallel boxplots indicate about whether an analysis of variance model would be appropriate?

d. Conduct an ANOVA using the transformed data. Interpret the results. What do you conclude about Abeta and cognitive impairment?

5.52 Words with Friends. Revisit the dataset **WordsWithFriends** that was analyzed in Section 5.8. In the analysis in the Case Study, we questioned whether there number of blank tiles that a player receives was related to the final score. In that analysis, our conclusion was that there is a noticeable difference between the final scores of the games, depending on how many blank tiles the player receives. In this exercise we ask the same question, but with respect to the winning margin rather than the final score. Words

a. Show that the conditions for ANOVA are met for these data.

b. Conduct an ANOVA. Report the ANOVA table and interpret the results.

5.53 Oil. This dataset is the result of a science fair experiment run by a high school student. The basic question was whether exposing sand with oil in it (think oil spill) to ultrasound could help the oil deapsorb from it better than sand that was not exposed to ultrasound. There were two levels of ultrasound tested (5 minutes and 10 minutes) and two levels of oil (5 ml and 10 ml). There was also a question of whether exposure to saltwater or freshwater made a difference, so half the samples had saltwater, the others distilled water. Each combination of factor levels was replicated 5 times. There were also an equivalent number of control observations run, all factors being the same but without any exposure to ultrasound. Each experimental run was paired with an appropriate control run, and the response variable is the difference in the amount of oil left in the experimental run and the control run. The data are in the file **OilDeapsorbtion**.[20]

The researcher who ran this experiment had reason to believe that exposing sand contaminated with oil to ultrasound would affect the amount of oil deapsorbed. He ran the experiment with two different levels of oil in the sand, 5 ml and 10 ml. Does the amount of oil in the sample affect how well ultrasound works? OilD

a. Create dotplots and compute descriptive statistics to compare the values of *diff* between the two different oil levels (*Oil*). Comment on what the plots and descriptive statistics reveal. Do you think there might be a statistical difference between the two oil levels?

b. Since there are only two groups, we can analyze the data using a t-test. Do so and report the results.

c. Now conduct an ANOVA to analyze the same data (be sure to check the necessary conditions). Report the results.

d. Compare your answers to parts (b) and (c). Are your conclusions the same? Explain.

5.54 Blood pressure. A person's systolic blood pressure can be a signal of serious issues in their cardiovascular system. Are there differences between average systolic blood pressure based on smoking habits? The dataset **Blood1** has the systolic blood pressure and the smoking status of 500 randomly chosen adults.[21] **Blood1**

a. Create dotplots and compute descriptive statistics to compare values of the systolic blood pressure between smokers and nonsmokers. Comment on what the plots and descriptive statistics reveal. Do you think there might be a statistical difference between the systolic blood pressures of smokers and nonsmokers?

b. Perform a two-sample t-test, using the assumption of equal variances, to determine if there is a significant difference in systolic blood pressure between smokers and nonsmokers.

c. Compute an ANOVA table to test for differences in systolic blood pressure between smokers and nonsmokers. What do you conclude? Explain.

d. Compare your answers to parts (a) and (b). Discuss the similarities and differences between the two methods.

5.55 Oil: length of ultrasound. The same researcher as in Exercise 5.53 wondered whether the length of time that the sand was exposed to ultrasound affected the amount of oil deapsorbed. He ran some observations with sand exposed to ultrasound for 5 minutes and some with sand exposed to ultrasound for 10 minutes. Does the length of ultrasound (*Ultra*) exposure affect the amount of oil deapsorbed (*diff*)? (In Chapter 7 we will consider both amount of oil and length of ultrasound together in the same model.) **OilD**

a. Create dotplots and compute descriptive statistics to compare the values of *diff* between the two different ultrasound levels (*Ultra*). Comment on what the plots and descriptive statistics reveal. Do you think there might be a statistical difference between the two oil levels?

b. Since there are only two groups, we can analyze the data using a t-test. Do so and report the results.

c. Now conduct an ANOVA to analyze the same data (be sure to check the necessary conditions). Report the results.

d. Compare your answers to parts (b) and (c). Are your conclusions the same? Explain.

5.56 Blood pressure: size. The dataset used in Exercise 5.54 also measured the size of people using the variable *Overwt*. This is a categorical variable that takes on the values 0 = Normal, 1 = Overweight, and 2 = Obese. Is the mean systolic blood pressure different for these three groups of people? **Blood1**

a. Why should we not use two-sample t-tests to see what differences there are between the means of these three groups?

b. Compute an ANOVA table to test for differences in systolic blood pressure between normal, overweight, and obese people. What do you conclude? Explain.

c. Use Fisher's LSD to find any differences that exist between these three groups' mean systolic blood pressures. Comment on your findings.

5.57 Meniscus failure load. Refer to the data discussed Exercise 5.38. A third response variable that the researchers were interested in was the pressure the repair could handle (variable name *FailureLoad*). **Menisc**

a. Give the hypotheses that would be tested in an ANOVA procedure for this dataset.

b. Check to see if the conditions for the ANOVA procedure and F-test are satisfied with these data. Discuss why some statisticians might choose not to complete the ANOVA procedure for this response variable.

5.58 Salary. A researcher wanted to know if the mean salaries of men and women are different. She chose a stratified random sample of 280 people from the 2000 U.S. Census consisting of men and women from New York State, Oregon, Arizona, and Iowa. The researcher, not understanding much about statistics, had Minitab compute an ANOVA table for her. It is shown below:

```
One-way ANOVA: salary versus sex

Source  DF          SS          MS      F      P
sex      1  8190848743  8190848743  12.45  0.000
Error  278  1.82913E+11  657958980
Total  279  1.91103E+11

S = 25651    R-Sq = 4.29    R-Sq(adj) = 3.94
```

a. Is a person's sex significant in predicting their salary? Explain your conclusions.

b. What value of R^2 value does the ANOVA model have? Is this good? Explain.

c. The researcher did not look at residual plots. They are shown in Figure 5.32. What conclusions do you reach about the ANOVA after examining these plots? Explain.

5.59 North Carolina births: summary statistics. We return to the data described in Exercise 5.11. For the purposes of this analysis, we will consider four racial

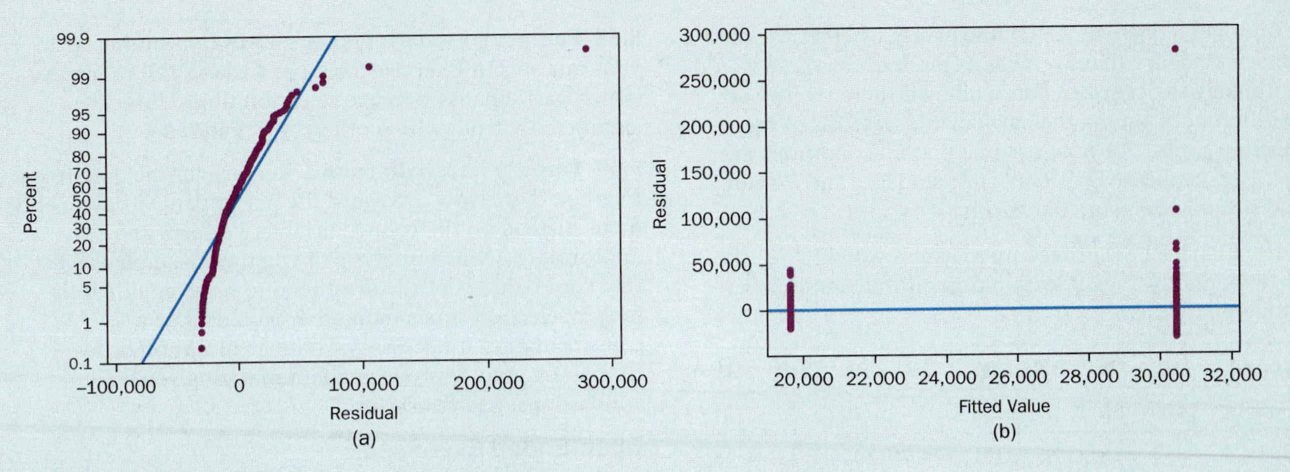

FIGURE 5.32 Normal probability plot and plot of residuals versus fitted values for salaries data

groups: white, black, Hispanic, and other (including Asian, Hawaiian, and Native American). Use the variable *MomRace*, which gives the races with descriptive categories. (The variable *RaceMom* uses only numbers to describe the races.) **NCBirth**

a. Produce graphical displays of the birth weights (in ounces) separated by mothers' racial group. Comment on both similarities and differences that you observe in the distributions of birth weight among the races.

b. Report the sample sizes, sample means, and sample standard deviations of birth weights for each racial group.

c. Explain why it's not sufficient to examine the four sample means, note that they all differ, and conclude that all races do have birth weight distributions that differ from each other.

5.60 North Carolina births: ANOVA. Return to the data discussed in the previous exercise. **NCBirth**

a. Comment on whether the conditions of the ANOVA procedure and *F*-test are satisfied with these data.

b. Conduct an ANOVA. Report the ANOVA table, and interpret the results. Do the data provide strong evidence that mean birth weights differ based on the mothers' racial group? Explain.

5.61 North Carolina births: Fisher's LSD. We return to the birth weights of babies in North Carolina one more time. **NCBirth**

a. Apply Fisher's LSD to investigate which racial groups differ significantly from which others. Summarize your conclusions, and explain how they follow from your analysis.

b. This is a fairly large sample so even relatively small differences in group means might yield significant results. Do you think that the differences in mean birth weight among these racial groups are important in a practical sense?

5.62 Metropolitan commutes: log transformation. In Example 5.19 on page 231 we looked at data in **MetroCommutes** on commute times for a sample of four cities. We saw that the distribution of commute times was quite skewed in each city, but a transformation to the natural logarithm, *log*(*Time*), did a better job as satisfying the ANOVA conditions. Find the ANOVA table for comparing mean log commute times and make a decision about whether there is convincing evidence of a difference in means among the four cities. **MetCom**

5.63 Metropolitan commutes: square root transformation. In Example 5.22 on page 238 we looked at data in **MetroCommutes** on commute times for a sample of four cities and found that reexpressing the response variable (*Time*) with a square root transformation helped in satisfying the ANOVA conditions. Find the ANOVA table for comparing mean square root commute times and make a decision about whether there is convincing evidence of a difference in means among the four cities. **MetCom**

5.64 Metropolitan commutes: pairwise differences for logs. In Exercise 5.62 you are asked to produce an ANOVA table for comparing the natural logs of commute times for the cities in **MetroCommutes**. Now determine which (if any) pairs of cities show evidence of a difference in means. **MetCom**

5.65 Metropolitan commutes: pairwise differences for square roots. In Exercise 5.63 you are asked to produce an ANOVA table for comparing the square roots of commute times for the cities in **MetroCommutes**. Now determine which (if any) pairs of cities show evidence of a difference in means. **MetCom**

5.66 Fantasy baseball. A group of friends who participate in a "fantasy baseball" league became curious about whether some of them take significantly more or less time to make their selections in the "fantasy draft" through which they select players.[22] The table at the end of this exercise reports the times (in seconds) that each of the eight friends (identified by their initials) took to make their 24 selections in 2008 (the data are also available in the datafile **FantasyBaseball**): **FanBase**

a. Produce boxplots and calculate descriptive statistics to compare the selection times for each participant. Comment on what they reveal. Also identify (by initials) which participant took the longest and which took the shortest time to make their selections.

b. Conduct a one-way ANOVA to assess whether the data provide evidence that averages as far apart as these would be unlikely to occur by chance alone if there really were no differences among the participants in terms of their selection times. For now, assume that all conditions are met. Report the ANOVA table, test statistic, and *P*-value. Also summarize your conclusion.

c. Use Fisher's LSD procedure to assess which participants' average selection times differ significantly from which others.

Round	DJ	AR	BK	JW	TS	RL	DR	MF
1	42	35	49	104	15	40	26	101
2	84	26	65	101	17	143	43	16
3	21	95	115	53	66	103	113	88
4	99	41	66	123	6	144	16	79
5	25	129	53	144	6	162	113	48
6	89	62	80	247	17	55	369	2
7	53	168	32	210	7	37	184	50
8	174	47	161	164	5	36	138	84
9	105	74	25	135	14	118	102	163
10	99	46	60	66	13	112	21	144
11	30	7	25	399	107	17	55	27
12	91	210	69	219	7	65	62	1
13	11	266	34	436	75	27	108	76
14	93	7	21	235	5	53	23	187
15	20	35	26	244	19	120	94	19
16	108	61	13	133	25	13	90	40
17	95	124	9	68	5	35	95	171
18	43	27	9	230	5	52	72	3
19	123	26	13	105	6	41	32	18
20	75	58	50	103	13	38	57	86
21	18	11	10	40	8	88	20	27
22	40	10	119	39	6	51	46	59
23	33	56	20	244	6	38	13	41
24	100	18	27	91	11	23	31	2

5.67 Fantasy baseball: transformation.

a. In Exercise 5.66, part (a), you produced boxplots and descriptive statistics to assess whether an ANOVA model was appropriate for the fantasy baseball selection times of the various members of the league. Now produce the normal probability plot of the residuals for the ANOVA model in Exercise 5.66 and comment on the appropriateness of the ANOVA model for these data. **FanBase**

b. Transform the selection times using the natural log. Repeat your analysis of the data and report your findings.

5.68 Fantasy baseball: Fisher's LSD. Continuing with your analysis in Exercise 5.67, use Fisher's LSD to assess which participants' average selection times differ significantly from which others. **FanBase**

5.69 Fantasy baseball: round. Reconsider the data from Exercise 5.66. Now disregard the participant variable, and focus instead on the round variable. Perform an appropriate ANOVA analysis of whether the data suggest that some rounds of the draft tend to have significantly longer selection times than other rounds. Use a transformation if necessary. Write a paragraph or two describing your analysis and summarizing your conclusions. **FanBase**

Open-Ended Exercises

5.70 Hawks: culmen. The dataset on hawks was used in Exercise 5.32 to analyze the weight based on species. Other response variables were also measured in the **Hawks** data. Analyze the length of the *culmen* (a measurement of beak length) for the three different species represented. Report your findings. **Hawk**

5.71 Sea slugs.[23] Sea slugs, common on the coast of Southern California, live on vaucherian seaweed. The larvae from these sea slugs need to locate this type of seaweed to survive. A study was done to try to determine whether chemicals that leach out of the seaweed attract the larvae. Seawater was collected over a patch of this kind of seaweed at 5-minute intervals as the tide was coming in and, presumably, mixing with the chemicals. The idea was that as more seawater came in, the concentration of the chemicals was reduced. Each sample of water was divided into six parts. Larvae were then introduced to this seawater to see what percentage metamorphosed. Is there a difference in this percentage over the six time periods? Open the dataset **SeaSlugs**, analyze it, and report your findings. **Slug**

5.72 Auto pollution.[24] In 1973 testimony before the Air and Water Pollution Subcommittee of the Senate Public Works Committee, John McKinley, president of Texaco, discussed a new filter that had been developed to reduce pollution. Questions were raised about the effects of this filter on other measures of vehicle performance. The dataset **AutoPollution** gives the results of an experiment on 36 different cars. The cars were randomly assigned to receive either this new filter or a standard filter and the noise level for each car was measured. Is the new filter better or worse than the standard? The variable *Type* takes the value 1 for the standard filter and 2 for the new filter. Analyze the data and report your findings. **AutPol**

5.73 Auto pollution: size of car. The experiment described in Exercise 5.72 actually used 12 cars, each of three different sizes (small = 1, medium = 2, and large = 3). These cars were, presumably, chosen at random from many cars of these sizes. **AutPol**

a. Is there a difference in noise level among these three sizes of cars? Analyze the data and report your findings.

b. Regardless of significance (or lack thereof), how must your conclusions for Exercise 5.72 and this exercise be different and why?

5.74 Psychotherapy. In Example 5.6 we considered data used to evaluate the effectiveness of long-term therapy. For simplicity, we chose to analyze only three of the six groups of data. In this exercise we ask you to do a more complete analysis of the dataset. The data are found in the file **Undoing**. **Undo**

a. Now consider all time periods except the one labeled III and perform a complete analysis.

b. We excluded time period III from the analysis in part (a) because there are only two observations and they are identical, leading to a group standard deviation of 0. This does not overly concern us because the sample size is so small and there are only five possible values for the response variable. Rerun the analysis using all six groups and discuss whether there are any differences between the analyses in parts (a) and (b).

Susla/Shutterstock

CHAPTER 6

Blocking and Two-way ANOVA

In this chapter you will learn to:

- Identify a randomized complete block design.
- Fit a main effects ANOVA model with two explanatory factors.
- Assess and use a main effects ANOVA model with two explanatory factors.

If you know about the paired *t*-test, you already know about the simplest block design: Each pair in a paired *t*-test is a block of size two. The word *block* comes from agriculture, and refers to a block of farmland, but abstractly, a block can also be a group of matched subjects or a single subject used multiple times. These designs come from one of the most basic of all principles for research design: *Isolate the effects of interest*. Compare "like with like." Block designs and their analysis of variance generalize the paired *t*-test.

6.1 Choose: RCB Design and Its Observational Relatives

In the language of statisticians, this section defines a "block" and introduces the randomized complete block design. To psychologists, this section introduces the concept of repeated measures and introduces the "within-subjects" design. The statisticians' block design was first developed in the first half of the 1900s, and became one of the major innovations in scientific agriculture. The same abstract idea became every bit as influential in psychology as the within-subjects design.

This section introduces three ways to create block designs and offers examples of observational studies with the same structure. As a goal, we urge you to read this chapter with the aim of understanding how, when it comes to design, a plot of farmland, a rat, and a human subject can be abstractly equivalent. (Not obvious, but worth the effort.)

The Randomized Complete Block (RCB) Design

To introduce the idea of a block design, remind yourself of the example of financial incentives for weight loss from Chapter 0. The goal was to see whether subjects who were offered a financial incentive would lose more weight than subjects in the "control condition." Imagine that you sit on a scientific panel to decide whether to fund the research, and your response to the researchers is:

> Your design doesn't have much power to detect group-to-group differences. Your unit-to-unit differences will be large, because your experimental unit is a person, and people differ. They differ in their commitment to losing weight, and they differ in how they respond to financial incentives. These differences will give you a large standard deviation of the residuals, a small effect size, and a wide interval estimate.

Now that you have crushed the investigators and saved taxpayers thousands of dollars, you can sit back smugly... but wait... back comes a revised grant proposal:

> We recognize the wisdom of the review panel, and we have revised our design accordingly. Instead of a two-sample *t*-test, we now plan a paired *t*-test. We have devised a questionnaire to measure the two things you were concerned about: (1) a subject's commitment to weight loss, and (2) a subject's responsiveness to financial incentives. Based on this questionnaire, we plan to create matched pairs of subjects, with the two subjects in each pair similar to each other on both (1) and (2). Using a coin toss, one subject in each pair will be offered the incentive; the other will serve as the control. We hope you recognize that our new design responds to your concerns. The two subjects in a pair will be similar in both of the two ways you identify. Our new design lets us compare incentive/control without the added "noise" of the differences you pointed out. The standard deviation of the residuals will be smaller, our power to detect a difference will be greater, our effect size will be larger, and our confidence interval will be narrower. Hand over the grant money.

EXAMPLE 6.1

Weight loss Here are two ways to conduct a weight-loss experiment.

1. *Two-sample* t*-test (a completely randomized design)*. Each subject was randomly assigned to one or the other of two treatments, Control or Financial Incentive.
2. *Paired* t*-test (a randomized complete block design)*. Subjects were sorted into pairs (blocks) based on a pre-test. Within each pair, one subject was chosen at random to be offered the financial incentive; the other subject was assigned to the control condition.

The principle here is profound:

IDENTIFYING SOURCES OF VARIATION

A good design isolates the effects of interest.
Where you have control, arrange to compare like with like, that is, arrange to compare different treatments using similar units.

- The main goal is to reduce the size of residual error.
- The key idea is to compare "like with like."
- The strategy is to create blocks: groups of experimental units that are as much alike as possible.

The number of units in a block should equal the number of treatments you want to compare (the number of categories in your predictor). For each block, you assign one treatment (one category) to each unit, using a chance device. Each unit gets a treatment; each block gets all of the treatments, one per unit.

The idea of a block design goes back to Sir Ronald Fisher, knighted for his work in statistics and genetics. In his 1935 book, *The Design of Experiments*, he described a randomized study to compare the yield from five varieties of wheat.

EXAMPLE 6.2

Fisher's randomized field trial of wheat varieties Figure 6.1 illustrates Fisher's block design for comparing varieties of wheat.

The two panels show the same information in different ways. The left panel shows varieties as columns. Each cell tells which plot the variety was assigned to. The right panel shows the same experimental plan, but in a different format. Each column corresponds to a plot, numbered from 1 to 5, and the cell tells which variety gets assigned to that plot. For example, in the left panel, we see that in block I, variety A was assigned to plot 1. In the right panel, we see that in block I, plot 1 was assigned to be planted with variety A.

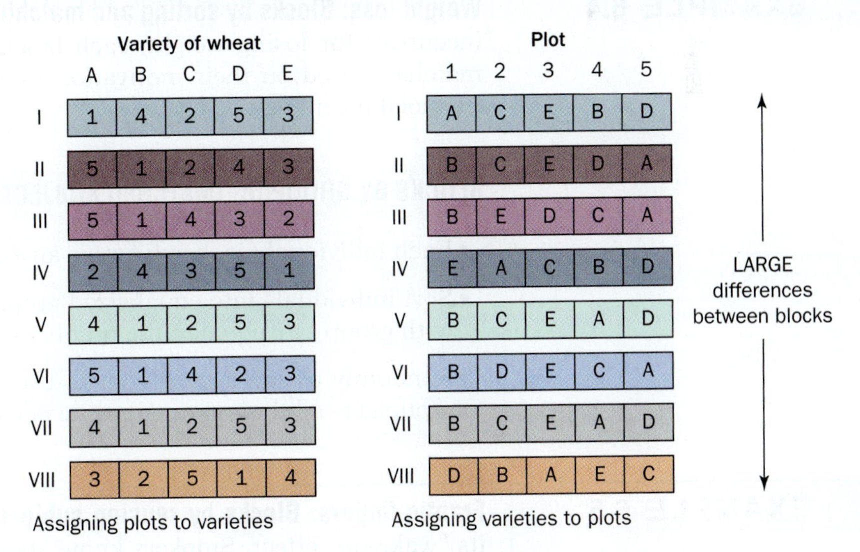

FIGURE 6.1 Schematic of Fisher's experimental design

Although the earliest block designs came from agriculture, the idea of comparing similar units spread quickly to other areas. Fisher's original blocks came from **subdividing** a plot of farmland, but blocks can also come from **matching** units or from **reusing** subjects.

subdivide
match
reuse

Three Kinds of Randomized Complete Block Designs

RANDOMIZED COMPLETE BLOCK DESIGNS

There are three main ways to create blocks of similar units: subdivide, sort, or reuse.

- **Subdivide** a large plot of land (the block) into subplots (the units).
- **Sort** and match subjects or other units into groups of similar units (the blocks).
- **Reuse** each subject (the block) multiple times. Each time slot is a unit.

EXAMPLE 6.3

Fisher's wheat: Blocks by subdividing For Fisher's field trial of wheat varieties, blocks came from subdividing farm fields. It is common in agricultural experiments to divide large fields into smaller plots or strips of land. The blocks are the large fields and the units are the smaller plots or strips of land. In this case, each farm was a block, which was subdivided into smaller plots or strips of land to provide the units.

BLOCKS BY SUBDIVIDING

- Divide each large block of land into equal-sized smaller plots, with the same number of plots per block, equal to the number of treatments.
- Randomly assign a treatment to each plot so that each block gets all the treatments, one per block, and each treatment goes to some plot in each and every block.

EXAMPLE 6.4

Weight loss: Blocks by sorting and matching For the revised study of financial incentives for losing weight, each block was a pair of subjects, who were matched based on their motivation to lose weight and their receptivity to financial incentives.

BLOCKS BY GROUPING (MATCHED SUBJECTS DESIGN)

- Each individual (each subject) is an experimental unit.
- Sort individuals into equal-sized groups (blocks) of similar individuals, with group size equal to the number of treatments.
- Randomly assign a treatment to each subject in a group so that each group gets all the treatments, one per subject.

EXAMPLE 6.5

Frantic

Frantic fingers: Blocks by reusing subjects We all know about caffeine and its "wake-up" effect. Smokers know about the effect of nicotine. Chocolate "addicts" may not know that the cocoa bean contains theobromine, a chemical in the same alkaloid family as caffeine and nicotine. Suppose you want to compare the effects of caffeine and theobromine with the effects of a placebo. One strategy would be to use the ideas of Chapter 5 to create a completely randomized design: For example, find 12 subjects and randomly assign 4 subjects to the placebo, to caffeine, or to theobromine. Two hours later, time the subjects while they tap a finger in a specified manner (that they had practiced earlier, to control for learning effects).

Of course, people differ in their response to these stimulants, so you might prefer a block design with each of four subjects getting each of the three treatments, one at a time. Each subject provides three units. Each unit is a "time slot." For each subject, treatments are assigned to time slots using random numbers. This is how the actual experiment was carried out.

Reusing subjects in a design introduces a number of concerns. One concern is the possibility of a carry-over effect. If you give me theobromine now and caffeine an hour later, the effects will overlap and muddy the results. The experimenters dealt with this concern by allowing an entire two days as a "wash out" period. If I get caffeine today, I don't get my next drug until the day after tomorrow. This gives me almost 48 hours to wash out the effect of the caffeine.

A second concern is the possibility of a time trend from Day 1 to Day 3 to Day 5. For example, we should probably not give all subjects the placebo first, caffeine second, theobromine last. One way to account for this is to randomize the order for each subject, with the hope that any effects will "average out."* That was done in this study for the data that appear in Table 6.1 and are stored in **FranticFingers** (with a different format for the columns).

Subject	Placebo	Caffeine	Theobromine
A	11	26	20
B	56	83	71
C	15	34	41
D	6	13	32

TABLE 6.1 Tap rates for four subjects

BLOCKS BY REUSING (WITHIN-SUBJECTS DESIGN, REPEATED MEASURES DESIGN)

- Each subject is a block; each time slot is an experimental unit.
- Each subject gets every treatment, each treatment in a different time slot = unit.
- Each treatment is given once to each subject.
- Treatments are assigned to time slots (units) by chance, with a separate randomization for each subject.

EXAMPLE 6.6

VVerb

Visual/Verbal: A classic experiment in cognitive psychology Recall from Example 5.21 on page 236 the classic cognitive psychology experiment comparing reaction times for "left brain" (verbal) and "right brain" (visual) tasks and reporting. Now that we have the two-way ANOVA model, we can analyze these data coming from a within-subjects design.

*A more sophisticated design, called a Latin square, can sometimes be used in this sort of situation to ensure equal representation in the days. For more, see Section 8.8.

For this study, as for many studies in cognitive psychology, the response is the time it takes a subject (you or me) to complete a task. The design challenge is to invent a set of tasks that let you, the investigator, use response time to answer questions about how the human brain works. In this case, find tasks that could indicate whether the two sides of your brain work more efficiently if they divide and share the work. This classic design separates the task into two parts, the decision (visual or verbal) and the report (visual or verbal). If the sides of your brain do different tasks, the theory says you should be faster when the task and report use different sides of your brain. If both task and report call on the same side of the brain, it should take longer. Table 6.2 shows a summary of the results of a version of the experiment from a psychology class at Mount Holyoke College. The data are stored in **VisualVerbal** in a somewhat different format (all response times in a single column with other columns identifying the *Subject*, *Task*, and *Report* method).

Decision	Visual Task (Letter)		Verbal Task (Sentence)	
Report Subject	Visual (Point)	Verbal (Say)	Visual (Point)	Verbal (Say)
1	11.60	13.06	16.05	11.51
2	22.71	6.27	13.16	23.86
⋮	⋮	⋮	⋮	⋮
20	33.98	12.64	15.48	28.18
Average	**18.17**	**9.01**	**13.69**	**12.85**

TABLE 6.2 Some data from **VisualVerbal**

For this experiment there are two tasks, "Letter" and "Sentence." The Letter Task (visual) is to look at a block letter F (Figure 6.2) and go clockwise from corner to corner, deciding Yes/No for each corner, is it an inside corner or not? The Sentence Task (verbal) was to read the sentence "The pencil is on the desk." one word at a time, deciding for each word Yes/No, is this a noun?

The two ways to report are Point (visual) and Say (verbal). To report visually, you point to one or the other of the words "Yes" and "No." To report verbally, you say the word, "Yes" or "No."

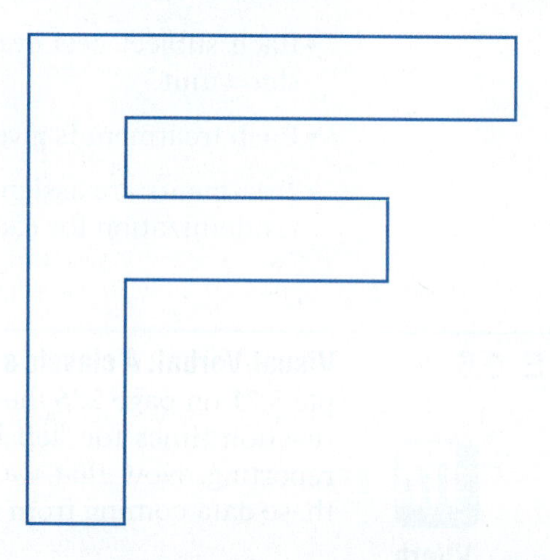

FIGURE 6.2 The Letter task. Start with the upper left corner and go clockwise. For each corner, decide Yes/No, is it an inside corner?

What matters here is not which kind of decision takes longer on average (letter or sentence), or which kind of report takes longer (point or say). One task may take longer on average than the other, and one way to report may take longer on average. The theory of "left brain, right brain" predicts that the time to respond will be shorter *if the decision and report are different*.

In fact, on average, the time for the decision is about the same for both kinds of tasks: $(18.17 + 9.01)/2 = 13.59$ for visual, $(13.69 + 12.85)/2 = 13.27$ for verbal. The time for verbal report (saying) is much shorter than the time for visual reporting (pointing): $(9.01 + 12.85)/2 = 10.93$ for saying, $(18.17 + 13.69)/2 = 15.93$ for pointing. For this study these differences are secondary. What matters is how the decision and report go together. Here, when the decision is visual (letter task), subjects are twice as fast to report by saying (9 seconds) compared with pointing (18 seconds). The result tends to support the left/right hypothesis. When the decision is verbal, there is not much difference in the two kinds of reports. The concept raised here is interaction, which is the subject of Chapter 7.

All three versions of the randomized block design share the same abstract structure:

THE RANDOMIZED COMPLETE BLOCK DESIGN

A block is a set of similar experimental units. For a randomized balanced complete block (RCB) design, each block has the same number of units, equal to the number of treatments. Within each block, each unit is assigned a different treatment; each block receives a complete set of treatments.* The assignment of treatments to units is randomized separately for each block.

All three designs in this subsection rely on randomization to assign treatments to units, with a separate (independent) randomization for each block. Even when randomization is not possible, the abstract structure of the block design can be useful.

Observational Versions of the Complete Block Design

The next set of three examples illustrates observational versions of the block design. To practice your understanding of how blocks are created, you might try to decide, for each example, whether the blocks come from subdividing, as for Fisher's wheat; from matching, as for the Weight losers; or from reusing, as for the Fingers.

EXAMPLE 6.7

Three observational versions of the block design: Radioactive twins, sleeping shrews, and river iron

Twin

a. Radioactive twins.

This example uses a block design with blocks of size 2. Data from designs like this one are often analyzed using a paired t-test instead of ANOVA. In that sense, ANOVA for block designs simply generalizes the paired t-test to matched sets of size 2 or more.

*We only cover RCB designs with one observation per treatment within each block. There are more complicated block designs that include replications and incomplete assignments of the treatments to the blocks, but these are beyond the scope of this book.

The research hypothesis for this study was that people who live in the country have healthier lungs than those who live in a big city. The plan was to compare two groups of subjects, those living in a city and those living in the country. To assess lung health, scientists measured tracheobronchial clearance rate, that is, in English, "How fast do your lungs get rid of nasty stuff?" Each subject agreed to inhale an aerosol of Teflon particles tagged with a weakly radioactive isotope with a fast half-life of six hours. The response was the percent of radioactivity remaining in the lungs after one hour.*

You might guess that to create comparison groups, urban and rural, the scientists would choose people at random, some from cities, some from the country, using a one-way design. But no! The scientists managed to find 15 pairs of identical twins with one twin in each pair living in the country and one living in a major city (Stockholm). Moreover, both twins in each pair agreed to inhale an aerosol of radioactive Teflon particles. Each twin pair was a block, and the two subjects in a block were genetically identical. The main difference between them was their environment, urban or rural.

This study can only be observational. You can't force people to decide where they live on the basis of your coin toss. The data are stored (in a different format) in **RadioactiveTwins** and displayed in Table 6.3. Note that to save space, we have reversed the format: Blocks are columns.

Twin pair	Rural	Urban
1	10.1	28.1
2	51.8	36.2
3	33.5	40.7
4	32.8	38.8
5	69.0	71.0
6	38.8	47.0
7	54.6	57.0
8	42.8	65.9
9	15.2	20.1
10	72.2	61.6
11	23.7	16.4
12	38.0	52.3
13	84.8	78.6
14	12.5	22.3
15	52.2	39.7

TABLE 6.3 Data for the Radioactive Twins

Shrew

b. The sleeping of a shrew.
Science has established that REM (rapid eye movement) sleep is associated with dreaming, and has divided non-REM sleep into DSW (deep slow wave) and LSW (light slow wave). Scientists who designed

*This number was estimated by fitting a line to multiple readings taken over the course of two hours.

this study wanted to know whether heart rate was different for the three different sleep phases. For subjects, they recruited six tree shrews, who served as blocks.

Each shrew was attached to a heart monitor and a brain wave monitor. The data in Table 6.4 (and stored in **SleepingShrews**) shows the average heart rate for each shrew (A–F) during each of its three phases of sleep (REM, DSW, and LSW).

The study can only be observational. You can't command a shrew to sleep according to a random number. All the same, it makes sense to regard each shrew as a block, and to regard sleep phases as a factor of interest. In short, it may not be randomized, but it has the same abstract structure as an RCB design.

SHREW	LSW	DSW	REM
A	14.0	11.7	15.7
B	25.8	21.1	21.5
C	20.8	19.7	18.3
D	19.0	18.2	17.1
E	26.0	23.2	22.5
F	20.4	20.7	18.9

TABLE 6.4 Data for the Sleeping Shrews

c. River iron

RivElt

RivIron

The chemical composition of river water changes as the river flows from its source to its mouth. Geoscientists in upstate New York gathered data on the four major rivers of their area: the Grasse, Oswegatchie, Racquette, and St. Regis. For each river, they chose three sites: upstream (near the source), midstream, and downstream (near the mouth), and measured the chemical composition at each site. See Figure 6.3 for a map of the area. Some of the data from this study are saved in the file **RiverElements**.[1] The data in Table 6.5 give the iron concentration in parts per million (ppm), but because chemists typically work with concentrations in the log scale (think pH), the file **RiverIron** also stores the log base 10 of the concentrations.

River	Up	Mid	Down
Grasse	944	525	327
Oswegatchie	860	229	130
Racquette	108	36	30
St. Regis	751	568	350

TABLE 6.5 Iron concentration in ppm

The **RiverIron** dataset illustrates blocking by subdividing, the twins are sorted and matched, and the shrews reuse subjects.

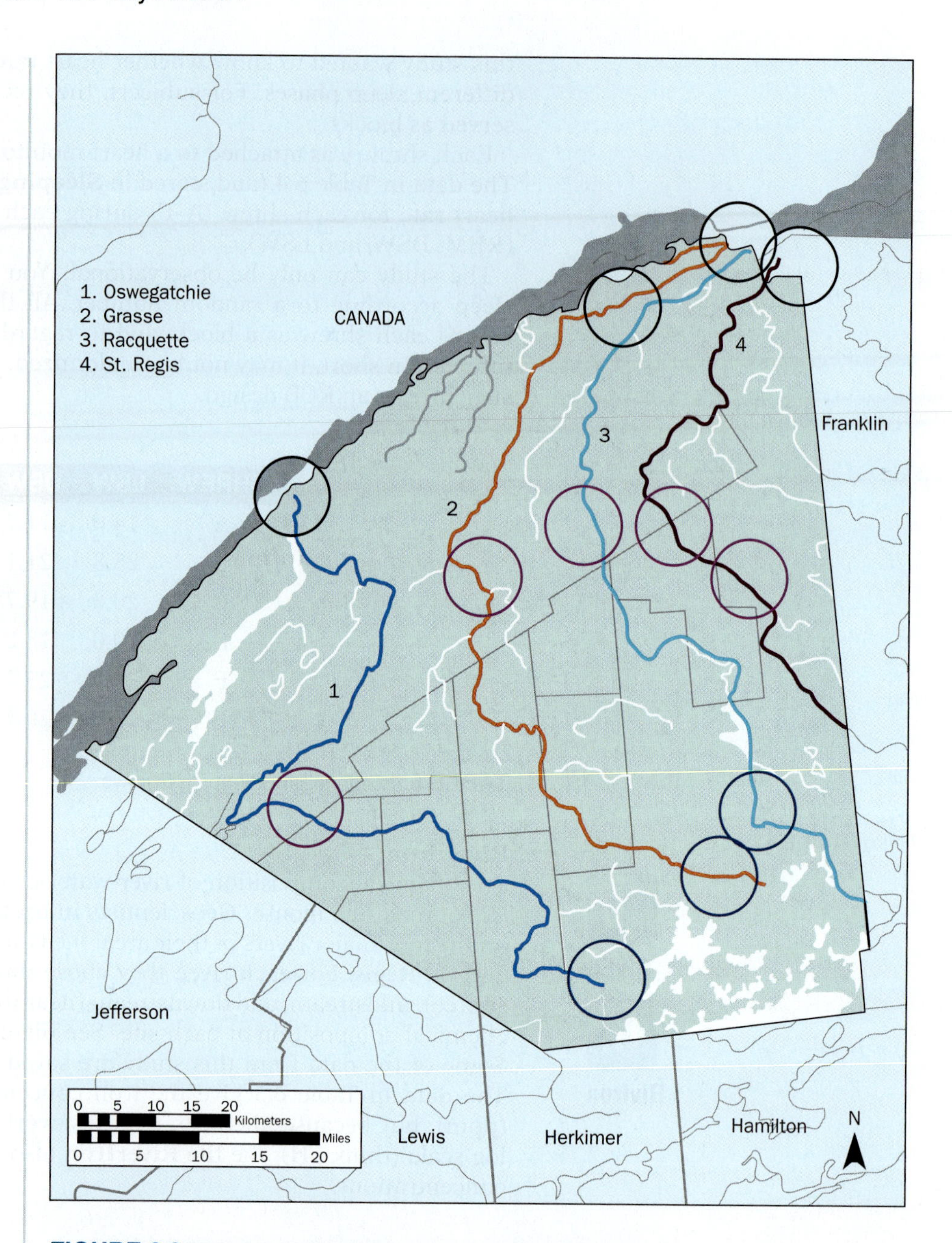

FIGURE 6.3 Sites for the **RiverElements** and **RiverIron** data

There is a standard terminology for describing study designs. Rather than give formal, abstract definitions, we illustrate the usual vocabulary using the frantic fingers data.

DESIGN TERMINOLOGY, USING THE FRANTIC FINGERS DATA

Factors and levels. *Subjects* and *Drugs* are **factors**, that is, *factors are categorical predictor variables*. The factor *Subjects* has four *levels*: I, II, III, IV. The factor *Drugs* has three levels: placebo, caffeine, theobromine. (Each *category* is a *level* of its factor.)

Crossed factors and cells. The two factors *Subjects* and *Drugs* are **crossed**. This means that we collect data on all combinations of *Subject* and *Drugs*. This is easiest to see in Table 6.1: Each subject is a row, each drug is a column, a **cell** is where a row and column cross.

Nuisance factors and factors of interest. A **factor of interest** is *Drugs*, directly related to the goal of the study. A **nuisance factor** is *Subjects*, included as a factor only because it is a source of variation. For the sleeping shrew data, sleep *Phase* is a factor of interest; *Shrew* is a nuisance factor.

Fixed and random factors. Typically, *factors of interest* are ***fixed***, which means they are an unknown constant. *Nuisance factors* are ***random***, which means they behave according to chance. *Nuisance* refers to a factor's role in the study. *Random* refers its role in the statistical model.

Main effects. Just as in Chapter 5, an **effect** tells the difference between a category average and the grand average. **Main effect** refers to the effect of a factor. For a block design, there are two sets of main effects, for blocks and for treatments (e.g, for *Subjects* and for *Drugs*).

Additive main effects. The two sets of main effects, for subjects and for drugs, are called additive if each drug adds or subtracts the same amount to the tap rate of all four subjects. The model for a complete block design assumes (requires) that main effects are additive. (Chapter 7 treats models with nonadditive main effects.)

6.2 Exploring Data from Block Designs

This section, on exploration, is about what you can learn from data *before* you fit a model. The plots of this section are based on response values. Later, in Section 6.4 on assessing the fit of a model, you will see plots based on residuals from the fitted values.

For plotting and exploring for patterns, you can think of data from a block design as much like data from a completely randomized design, but with the additional structure that comes from the blocks. For data from a block design, the side-by-side dot plots of Chapter 5 don't show the blocks. We can modify the plot, using line segments to connect points from the same blocks, as in the next example.

Side-by-side Plots for Block Designs

EXAMPLE 6.8

Frantic

Side-by-side dot plots for the frantic fingers data Figure 6.4 shows two different side-by-side plots for the **FranticFingers** data. Figure 6.4(a) is a side-by-side plot of response values (tap rates) with one important addition: Points from the same block are joined by line segments. That plot makes it easy to see three important patterns in the data. (1) Subjects vary. Subject 2 is hyper, at roughly 70 taps per minute; the other three subjects are similar to each other and much slower. (2) All four subjects tap faster after swallowing a drug than after a placebo. (3) Subjects 1 and 4 appear to react differently to the two drugs: Subject 1 is faster after caffeine; Subject 4 is faster after theobromine. The patterns for Subjects 1 and 2 are similar; the patterns for Subjects 3 and 4 are similar.

Figure 6.4(b) is a side-by-side plot for response values after adjusting for (subtracting) block effects. The idea of adjusting for blocks gets a little ahead of the story, because the adjusted values come from a partial fit of the model. Section 6.3 will show how to fit the model, but for now focus only on the general idea of "taking out" the effects of the blocks. Suppose we could compute and remove the differences from one subject to the next and plot the adjusted values that remain. We would end up with a plot like Figure 6.4(b).

The two plots of this section, side-by-side dot plots and Anscombe's block plots, are plots of response values. These plots can alert you to the possibility that a model with additive main effects may offer a bad fit to the data, or at least that you may need to reexpress the response to get a good fit. Later, in Section 6.4 on ways to assess the fitted model, we show how three different kinds of residual plots can be useful. To use these plots, however, you first have to fit the model.

6.3 Fitting the Model for a Block Design

There are two very different ways to fit the model for a complete block design. These two ways correspond to two different formats for listing the data, known in statistics as tall and wide.

TWO FORMATS FOR TWO-WAY ANOVA DATA: TALL AND WIDE

- Tall: One row for each case, one column for each variable. (Best for regression with indicator variables as in Topic 8.10.)
- Wide: One row for each block, one column for each treatment. (Best for ANOVA based on a sum of overlays.)

The tall format lends itself to fitting the model using regression with indicator variables, as set out in Topic 8.10. The wide format invites you to compute averages: a row average for each block and a column average for each treatment. Table 6.7 shows both the tall and wide formats for the frantic fingers data. Once you have these averages, it is easy to write the data as a sum of overlays. Whichever way you fit the model, you end up with the same numbers: the same estimated effects, fitted values, residuals, sums of squares, and degrees of freedom.

So what is the model in this case? The two-way ANOVA model can be written as:

$$Y = \mu + \tau_i + \beta_j + \epsilon$$

Tall format

Case	Subj	Drug	Rate
1	A	Pl	11
2	A	Ca	26
3	A	Th	20
4	B	Pl	56
5	B	Ca	83
6	B	Th	71
7	C	Pl	15
8	C	Ca	34
9	C	Th	41
10	D	Pl	6
11	D	Ca	13
12	D	Th	32

Wide format

	Drug			
Subj	Pl	Ca	Th	Ave
A	11	26	20	19
B	56	83	71	70
C	15	34	41	30
D	6	13	32	17
Ave	22	39	41	34

TABLE 6.7 Tall and wide formats for the frantic fingers data

where μ is the population mean, τ_i is the treatment main effect, β_j is the block effect, and ϵ is the random error term.*

The Goal of the Fit: To Represent the Data as a Sum of Overlays

As preparation, remind yourself of the metaphor of a CAT scan for the data. The overlays display the data as a set of layers added together. Just as in Section 5.3, the decomposition of the data into overlays is based on a set of questions. We urge you to focus on these questions as a way to understand a dataset. Thinking about how the questions and group averages are related can be useful in a way that just memorizing computing rules or just using computer software will not be. If you find algebraic formulas helpful, you can find them in a boxed summary at the end of the section.

PARTITIONING THE SOURCES OF VARIATION

Question 0: What is the grand (overall) average (μ)?
Questions 1a and 1b: Estimated main effects.

a. Block effects (β_j): How far is each block average from the grand average?

b. Treatment effects (τ_i): How far is each treatment average from the grand average?

Question 2: Residuals (ϵ). How far is each response value from its fitted value?†

$$\text{Fitted Value} = \text{Grand Average} + \text{Block Effect} + \text{Treatment Effect}$$

EXAMPLE 6.12

Frantic

Overlays for the frantic fingers data Finding answers to questions 0 and 1 is easy to organize in a two-way table, as in Table 6.8.

0. Compute row and column averages. The overall average will be the average of either the row averages or the column averages.
1. Compute the two sets of main effects by subtracting the grand average from the row and column averages.

Subject	Pl	Ca	Th	Ave	Eff
I	11	26	20	19	−15
II	56	83	71	70	36
III	15	34	41	30	−4
IV	6	13	32	17	−17
Ave	22	39	41	34	
Eff	−12	5	7		

TABLE 6.8 Averages and effects for the frantic fingers data

*We used α in Chapter 5 to refer to the factor. In this chapter, we use τ for the factor of interest (think "treatment") and β for the blocking factor.

†We wrote the model as $Y = \mu + \tau_i + \beta_j + \epsilon$ but when analyzing the model we find it helpful to reorder the treatment and block terms to that we have Fitted Value = Grand Average + Block Effect + Treatment Effect.

Question 2 requires two steps, as in Tables 6.9 and 6.10.

2a. Compute fitted values: Grand Average + Row Effect + Column Effect

Fitted Values					Grand Average				Subject Effects				Drug Effects		
Subj	Pl	Ca	Th												
I	7	24	26		34	34	34		−15	−15	−15		−12	5	7
II	58	75	77	=	34	34	34	+	36	36	36	+	−12	5	7
III	18	35	37		34	34	34		−4	−4	−4		−12	5	7
IV	5	22	24		34	34	34		−17	−17	−17		−12	5	7

TABLE 6.9 Fitted values for the frantic fingers data

2b. Compute residuals: Response − Fitted value

Residuals					Observed				Fitted		
Subj	Pl	Ca	Th								
I	4	2	−6		11	26	20		7	24	26
II	−2	8	−6	=	56	83	71	−	58	75	77
III	−3	−1	4		15	34	41		18	35	37
IV	1	−9	8		6	13	32		−5	22	24

TABLE 6.10 Residuals for the frantic fingers data

Combining these steps gives the overlays in Table 6.11.

Observed Values					Grand Average				Subject Effects				Drug Effects				Residuals		
Subj	Pl	Ca	Th																
I	11	26	20		34	34	34		−15	−15	−15		−12	5	7		4	2	−6
II	56	83	71	=	34	34	34	+	36	36	36	+	−12	5	7	+	−2	8	−6
III	15	34	41		34	34	34		−4	−4	−4		−12	5	7		−3	−1	4
IV	6	13	32		34	34	34		−17	−17	−17		−12	5	7		1	−9	8

TABLE 6.11 Overlays for the frantic fingers data

Just as in Chapter 5, the entire set of overlays can be displayed compactly in a bordered residual table, as in Table 6.12, which shows the residuals (from Table 6.10) bordered by the subject effects and drug effects, from Table 6.11, along with the grand average, from Table 6.11. This one table shows the entire decomposition.

Subject	Pl	Ca	Th	Effect
I	4	2	−6	−15
II	−2	8	−6	36
III	−3	−1	4	−4
IV	1	−9	8	−17
Effect	−12	5	7	34

TABLE 6.12 Bordered residual table for the frantic fingers data

Reading the Overlays

Here again, as in Section 5.4, there are two ways to read the overlays, one observation (position) at a time, and one box at a time.

EXAMPLE 6.13 Reading the overlays for the frantic fingers data

Frantic

1. One observation at a time.
Consider Subject I, Placebo, in the upper left corners of the boxes in Table 6.11. The decomposition shows, as you read across:

11	=	34	+	(−15)	+	(−12)	+	4
Observed Value	=	Grand Average	+	Subject effect	+	Drug effect	+	Residual

2. One box at a time.
Reading the overlays one box at a time lets you compare the sizes of variation from the different sources: blocks, treatments, and residuals. From looking at the boxes for the frantic fingers overlays, it is clear that variation in the box for subjects is by far the largest, variation for drugs is substantially smaller, and the residual variation is much smaller still. Squaring the numbers in a box and adding the squares confirms this. The sums of squares are 5478 for subjects, 872 for treatments, and 332 for residuals.

The mean squares adjust for the degrees of freedom, or "chances to vary." For the grand average, there is only one number, one chance to vary, so 1 df. For subjects, there are four values, but they add to zero, so there are 3 df. For drugs, there are three numbers, required to add to zero, so 2 df. Finally, for residuals, the values are required to add to zero two ways: across rows and down columns. There are 12 numbers, but only 6 df. See Figure 6.8.

Subj	Pl	Ca	Th
I	1	2	3
II	4	5	6
III	7	8	9
IV	10	11	12

= Grand Average

1	R	R
R	R	R
R	R	R
R	R	R

\+ Subject Effects

1	R	R
2	R	R
3	R	R
+	R	R

\+ Drug Effects

1	2	+
R	R	R
R	R	R
R	R	R

\+ Residuals

1	2	+
3	4	+
5	6	+
+	+	+

df 12 = 1 + 3 + 2 + 6

FIGURE 6.8 Degrees of freedom for the **frantic fingers** data. In each box, "R" denotes a repeated value, and "+" denotes a value fixed by the requirement of adding to zero.

DEGREES OF FREEDOM FOR THE COMPLETE BLOCK DESIGN

Source	*df*
Grand Average	1
Blocks	(# blocks − 1)
Treatments	(# treatments − 1)
Residual	(# blocks − 1) · (# treatments − 1)
Total	(# observations) = (# blocks) · (# treatments)

The Triple Decomposition and the ANOVA Table

For the complete block design, the observed values add, the sums of squares add, and the degrees of freedom add, as shown in Figure 6.9 for the frantic fingers data.

FIGURE 6.9 Triple decomposition for the frantic fingers data

Subj	Pl	Ca	Th		Grand Average				Subject Effects				Drug Effects				Residuals		
I	11	26	20		34	34	34		−15	−15	−15		−12	5	7		4	2	−6
II	56	83	71	=	34	34	34	+	36	36	36	+	−12	5	7	+	−2	8	−6
III	15	34	41		34	34	34		−4	−4	−4		−12	5	7		−3	−1	4
IV	6	13	32		34	34	34		−17	−17	−17		−12	5	7		1	−9	8
ss		20,554		=		13,872		+		5478		+		872		+		332	
df		12		=		1		+		3		+		2		+		6	

The analysis of variance table summarizes this triple decomposition, omitting the Grand Average, which does not vary. Here is some output showing the ANOVA table for the frantic fingers data.

Response: Rate

	Df	Sum Sq	Mean Sq	F value	Pr(>F)
Subj	3	5478	1826.00	33.0000	0.0003993
Drug	2	872	436.00	7.8795	0.0209669
Residuals	6	332	55.33		

The mean squares tell average variation (SS/df): 1826 for subjects, 436 for drugs, and 55.333 for residuals. The F-ratios compare mean squares. The *MS* for subjects is 33 times as large as the mean square for residuals. The *MS* for drugs is 7.88 times as big as the mean square for residuals. From our work with one-way ANOVA in Chapter 5, you probably sense that the last two values in the table are P-values that help judge whether those F-statistics should be considered unusually large. Such intuition would be correct as you will see when we consider formal inference for this model in Section 6.5.

Finally, we conclude this section with an example to show numerically why blocking can be so effective.

EXAMPLE 6.14

Frantic

The effect of using a design with blocks To see why using blocks can make such a difference, compare two models for the frantic fingers data, one with blocks and one without.

1. No blocks (one-way model).

Tap rate	=	Grand Average	+	Drug effect	+	Residual

If we fit the one-way model, the subject differences get included in the residuals. The residual sum of squares for the one-way model equals the sum of the subject and residual sums of squares for the block model. Computer output for the ANOVA table for the one-way model is shown as follows.

Response: Rate

	Df	Sum Sq	Mean Sq	F value	Pr(>F)
Drug	2	872	436.00	0.6754	0.533
Residuals	9	5810	645.56		

2. Complete block model.

For the earlier ANOVA table with both *Subj* and *Drug* in the model, the subject-to-subject differences are removed from the residuals. This produces a smaller MSE (55.33 compared to 645.6). Thus we see a larger F-ratio (7.88) and a smaller P-value (0.021) for *Drug* in the two-way ANOVA table.

As in Chapter 5, we follow the usual convention and use Greek letters α, β, γ, μ, σ, etc., for parameters; we use x's and y's and SS, df, t, p, etc., for statistics. We also follow a standard convention for subscripts, sums, and averages. For two-way ANOVA, each response value y_{ij} has two subscripts, i and j. The first subscript, i, tells the treatment number and the second subscript, j, tells the block number. For the frantic fingers data, there are three treatments, $i = 1, 2, 3$, and four blocks (subjects), $j = 1, 2, 3, 4$.

If you replace a subscript with a dot and put a bar over the y, the notation means to take an average by adding over the subscript that's been replaced. For example, $\bar{y}_{i.}$ represents the mean of all responses with treatment level i.

SUMMARY OF FORMULAS

Meaning	Parameter	Statistic
Overall mean	μ	$\bar{y}_{..}$
Treatment mean	$\mu_{i.}$	$\bar{y}_{i.}$
Treatment effect	$\tau_i = \mu_{i.} - \mu$	$\bar{y}_{i.} - \bar{y}_{..}$
Block mean	$\mu_{.j}$	$\bar{y}_{.j}$
Block effect	$\beta_j = \mu_{.j} - \mu$	$\bar{y}_{.j} - \bar{y}_{..}$
Observed response	$\mu + \tau_i + \beta_j + \epsilon_{ij}$	y_{ij}
Random error, residual	ϵ_{ij}	$y_{ij} - \bar{y}_{i.} - \bar{y}_{.j} + \bar{y}_{..}$

Decomposition

Question 0: The grand average	$\bar{y}_{..}$
Question 1a: How far is each block average from the grand average?	$\bar{y}_{.j} - \bar{y}_{..}$
Question 1b: How far is each treatment average from the grand average?	$\bar{y}_{i.} - \bar{y}_{..}$
Question 2: How far is the response value from its fitted value?	$y_{ij} - \bar{y}_{i.} - \bar{y}_{.j} + \bar{y}_{..}$

Let T = the total number of treatments.
Let B = the total number of blocks.

	Sums of squares	Degrees of freedom
Observed values	$\sum_{i,j} y_{ij}^2$	$B \cdot T$
Grand average	$\sum_{i,j} \bar{y}_{..}^2$	1
Treatment effects	$\sum_{i,j} (\bar{y}_{i.} - \bar{y}_{..})^2$	$T - 1$
Block effects	$\sum_{i,j} (\bar{y}_{.j} - \bar{y}_{..})^2$	$B - 1$
Residuals	$\sum_{i,j} (y_{ij} - \bar{y}_{i.} - \bar{y}_{.j} + \bar{y}_{..})^2$	$(B - 1) \cdot (T - 1)$

$$MS = \frac{SS}{df} \qquad F_{Treatment} = \frac{MS_{Treatment}}{MS_{Residuals}}$$

6.4 Assessing the Model for a Block Design

As you saw in Section 6.2, there are two plots (side-by-side dot plots and Anscombe block plots) that do not rely on residuals. You can use these not just to reveal patterns in your data, but also to help choose a scale for the response, and as a preliminary assessment of how well an additive complete block model may fit. The methods of this section are useful once you have a fitted model in hand, because the three diagnostic plots of this section rely on

EXAMPLE 6.20

VVerb

Tukey nonadditivity plots for the visual/verbal data Figure 6.14 shows two Tukey nonadditivity plots for the data from Example 6.16.

For these two plots the points are widely scattered, and no one line is an obvious best fit. All the same, there are four conditions of interest and 20 subjects, so 80 points for each plot. With so many points, we can have some confidence in the fitted slopes. For Figure 6.14(a), with time as response, the fitted slope is 1.26, about 1, suggesting a transformation to logs. (Exponent = 1 − slope ≈ 0.) For Figure 6.14(b), with log time as response, the fitted slope of −0.27 is near 0, suggesting to keep the scale as is. (Exponent ≈ 1.)

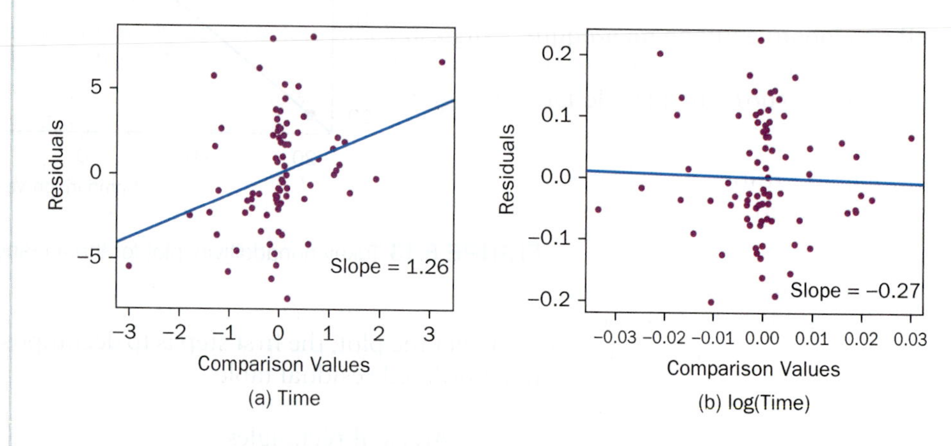

FIGURE 6.14 Tukey nonadditivity plots for the visual/verbal data

For the visual/verbal data, even though points were scattered and there was no obvious best-fitting line, there were so many data points that the fitted slopes sent a consistent message. Transforming to logs was indicated. For the next example, there are fewer points and no clear message.

EXAMPLE 6.21

RivIron

Tukey nonadditivity plots for the river iron data Figure 6.15 shows two Tukey nonadditivity plots for the data from Example 6.17.

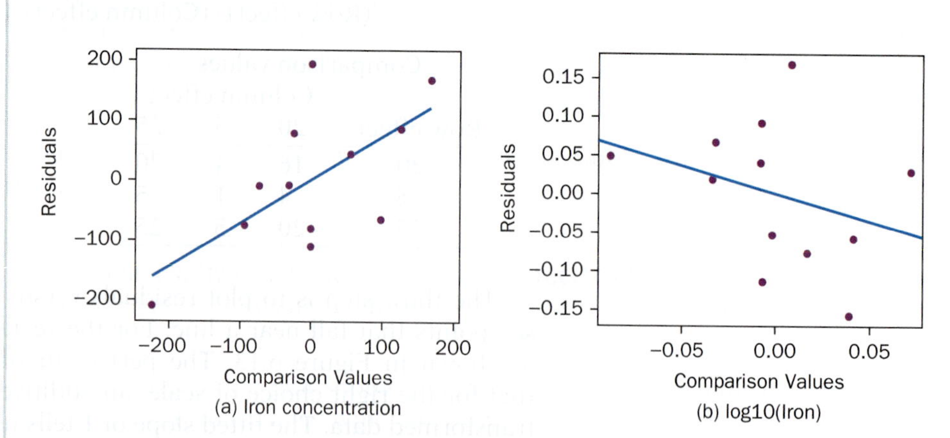

FIGURE 6.15 Tukey nonadditivity plots for the river iron data

Figure 6.15(a) shows residuals versus comparison values with iron concentration (ppm) as the response. Figure 6.15(b) shows log10(Iron). Both plots show wide scatter about their fitted lines. Neither scale, neither plot, has a story to tell. Either way, it looks like mush. For some kinds of plots, notably plots of residuals versus fitted values, a "mushy" plot can be a happy sign that your model has captured what matters, but not here. Sometimes, "mush" signals a dead end. The relationships you want to understand are too complicated to give up their secrets to a block model with additive main effects. (More in Chapter 7.)

At this point we assume that you have found a model that offers a good fit to the data, or decided to learn what you can from a less satisfactory model. With model in hand, the next step is to use the model for inference.

6.5 Using the Model for a Block Design

As in Chapter 5, there are three complementary tools for formal (probability-based) inference: tests (P-values), intervals, and effect sizes. The mechanical aspects are straightforward and can be summarized using formulas. If you are friends with software and know how to ask politely, a computer will give you the numbers. As with all of inference, however, the mere numbers alone are worthless. The value of the tools comes from how well you use them.

This section starts with the mechanical aspects first, before turning to the harder part, the scope of inference. The ideas are pretty much the same as in Section 5.4. We look at F-tests based on the ANOVA table to test hypotheses about which factors have differences (effects) that are larger than we would expect to see by random chance. We also consider confidence intervals for single group means or differences in means for pairs of groups. Finally, as in one-way ANOVA, we look at effect sizes for helping judge when differences might really matter.

Testing a Null Hypothesis

We start with three facts:

- An F-ratio compares two mean squares, each of which measures average variation. A large F-ratio indicates that the variation due to the treatments or blocks is large compared to the unit-to-unit variation (due to the error term).
- Also, the P-value for an F-ratio hopes to answer the question "If
  (1) the null hypothesis is true, and
  (2) the only patterns in the data are due purely to chance variation,
  then how likely is it that we would get an F-ratio as large as shown in the ANOVA table?" Informally, if (1) and (2) are true, how embarrassed should the null hypothesis be? Note that in a two-way ANOVA table we have distinct tests for *two* different questions: Are there differences in means due to the treatments? Are there differences in means due to the blocks?
- Condition (2) is critical. The P-value is just one piece of the puzzle. A small P-value by itself does not tell you whether the difference is, in fact, important.

The details for testing hypotheses for this two-way ANOVA model are summarized in the next box. Since this procedure can be applied in more general settings, we identify the two explanatory factors with a generic Factor A and Factor B.* In the current examples, Factor A corresponds to the treatments and Factor B is the blocking variable.

*Since we are talking about generic factors, we use α and β rather than τ and β.

2. An effect size measures the importance of an effect: "How big is the difference, compared with the unit-to-unit variation?"
3. The effect size can be useful even if a probability model is not appropriate, because the standard deviation of the difference can be taken as a descriptive measure of spread.

EXAMPLE 6.24

Frantic

Frantic fingers: Effect sizes For frantic fingers, the effect sizes take the form "Difference over standard deviation of the difference." These are:

Caffeine versus Placebo:	$17/7.439 = 2.29$, or 229%
Theobromine versus Placebo:	$19/7.439 = 2.55$, or 255%
Theobromine versus Caffeine:	$2/7.439 = 0.27$, or 27%

Again we see evidence of sizable effects when either drug is compared to a placebo, but not nearly so large an effect when the drugs are compared to each other.

Examples 6.22–6.24 illustrated the mechanics only. For the mere numbers you can go to a computer. The meaning is what really matters.

Scope of Inference

The basic issues about scope of inference are the same as in Chapter 5, except that for the datasets of this chapter, we have two factors, not just one. The questions are the same:

1. If the levels of a factor are meant to be typical of some larger population, were they chosen by random sampling from that population? If not, how plausible is it that the levels represent some population anyway? Which population?
2. If we care about cause and effect, were the treatments assigned to units using random numbers? If not, how plausible is it that a chance model applies anyway?

In what follows, we compare two extreme examples, frantic fingers and river iron.

EXAMPLE 6.25

Frantic

Scope of inference for the frantic fingers data

- For the frantic fingers data, "Drugs" is a fixed factor of interest; "Subjects" is a nuisance factor. We don't care about the subjects as individuals, and so we hope we can regard them as typical of some larger population, as if they had been chosen at random, even though we know they were not. Extrapolation to a larger population is possible, but far from guaranteed.
- The treatments (drugs) were randomly assigned to units (time slots), with a separate randomization for each subject. Inference about cause-and-effect is justified.

The various plots show that a model with additive main effects offers a reasonable fit to the data in the original scale.

For this study, the random assignment of drugs to subjects justifies the use of a probability model for inference. *P*-values and interval estimates make sense. Moreover, the *P*-value for *Drugs* is significant, so if we are concerned about multiple comparisons, we can invoke the logic of Fisher's LSD to help support our conclusions. *Subjects* and *Drugs* are highly significant—the effects cannot be attributed to chance alone. We can take the confidence intervals more or less at face value: the two drugs, caffeine and theobromine, have a substantial, causal effect on tap rate, but the difference in the effects of the two drugs might well be zero. The effect sizes for the two drugs, at 229% and 255%, are very large; their difference has a small effect size of only 27%.

EXAMPLE 6.26

RivIron

Scope of inference for the river iron data For the river iron data, both factors, "River" and "Site," are fixed factors of interest. There is no nuisance factor, because the four rivers are of interest in their own right. As the only four major rivers of Upstate New York, they constitute the entire population of interest. They cannot be regarded as chosen at random to represent a larger population of the rivers of Upstate New York. Generalization to a larger population makes little sense.

Both factors are observational; there was no random assignment. Inference about cause-and-effect makes little sense.

The various plots give a very mixed message. Theory suggests that we should measure concentration in a logarithmic scale, but the plots are far from clear about choice of scale for the response.

No scale gives a good fit for an additive model. Rather, the plots suggest that there is an interaction between river and site. If interaction is present, there is no way to use the available data to separate the interaction effects from the unit-to-unit variation. (Chapter 7 will address interaction in more detail.)

For the reasons just listed, *P*-values are largely worthless for this study. Does that mean that the study is worthless? In no way. You don't need *P*-values to tell you that the Racquette River has a lot less iron than the other three. You don't need *P*-values to tell you that iron concentration decreases as river water flows downhill toward its mouth, and you don't need *P*-values to recognize the implied underlying question: "Why, and what are the consequences?"

When it comes to scope of inference, the last two examples can serve as extremes. For the frantic fingers, formal inference is justified. For the river iron, probability-based inference is not appropriate, but the study is valuable and informative nonetheless.

We rely on one last example to tie together the sections of this chapter.

EXAMPLE 6.27

Twin

Radioactive twins

Research question and design. The goal of this study was to see whether and to what extent environment affects lung health. In particular, do those who live in a city have a slower rate of lung clearance than those who live in the country? The data come from a (nonrandomized) block design with pairs of identical twins as blocks. One twin in each pair had been living in a large city (Stockholm) for at least the last 10 years; the other twin had been living in the surrounding suburbs and countryside for at least the last 10 years.

Data and exploration. The data from Table 6.3 are repeated in Table 6.13, this time with block and environment averages, and pairwise differences (Urban − Rural). Notice that the averages for the twin pairs vary considerably, from a low near 17% to a high above 80%. By comparison, the environment averages are quite close: 45.05% for Urban, 42.13% for Rural.

Pair	Rural	Urban	Avg.	Diff.
1	10.1	28.1	19.1	18.0
2	51.8	36.2	44.0	−15.6
3	33.5	40.7	37.1	7.2
4	32.8	38.8	35.8	6.0
5	69.0	71.0	70.0	2.0
6	38.8	47.0	42.9	8.2
7	54.6	57.0	55.8	2.4
8	42.8	65.9	54.4	23.1
9	15.2	20.1	17.7	4.9
10	72.2	61.6	66.9	−10.6
11	23.7	16.4	20.1	−7.3
12	38.0	52.3	45.2	14.3
13	84.8	78.6	81.7	−6.2
14	12.5	22.3	17.4	9.8
15	52.2	39.7	46.0	−12.5
Avg.	42.13	45.05	43.59	2.91

TABLE 6.13 Data, averages, and differences for the radioactive twins. Percent of radioactive tracer remaining in the lungs after 60 minutes.

For data in blocks of size two it can be useful to look at pairwise differences, as shown in the bottom row of Table 6.13. The mean difference is just 2.91%, which is comparatively small, especially when compared to the standard deviation (11.4) of the differences.

Fitting the model, summarizing the fit. Our complete block model is

$$\text{Response} = \text{Grand Average} + \text{Twin effect} + \text{Environment effect} + \text{Residual}$$

Looking ahead to formal inference, we are particularly interested in the environment effects: are they detectably different from zero? How wide is a 95% confidence interval? What is the effect size?

The decomposition, based on the model, is shown in Figure 6.16, then summarized in the ANOVA table of Figure 6.17.

Three features of the decomposition stand out. (1) Variation from one twin pair to the next is quite large. Given that all readings are percentages between 0 and 100, it is striking that 7 of 15 twin effects are greater than 20% in absolute value. (2) Residual variation (between individual twins within a pair) is substantially less, but still large. In comparison, the estimated effect of environment is tiny, at ±1.5%.

	Clearance rate%			Grand Mean			Twin Effect			Env. Effect			Residual	
	10.1	28.1		43.6	43.6		−24.5	−24.5		−1.5	1.5		−7.5	7.5
	51.8	36.2		43.6	43.6		0.4	0.4		−1.5	1.5		9.3	−9.3
	33.5	40.7		43.6	43.6		−6.5	−6.5		−1.5	1.5		−2.1	2.1
	32.8	38.8		43.6	43.6		−7.8	−7.8		−1.5	1.5		−1.5	1.5
	69.0	71.0		43.6	43.6		26.4	26.4		−1.5	1.5		0.5	−0.5
	38.8	47.0		43.6	43.6		−0.7	−0.7		−1.5	1.5		−2.6	2.6
	54.6	57.0		43.6	43.6		12.2	12.2		−1.5	1.5		0.3	−0.3
	42.8	65.9	=	43.6	43.6	+	10.8	10.8	+	−1.5	1.5	+	−10.1	10.1
	15.2	20.1		43.6	43.6		−25.9	−25.9		−1.5	1.5		−1.0	1.0
	72.2	61.6		43.6	43.6		23.3	23.3		−1.5	1.5		6.8	−6.8
	23.7	16.4		43.6	43.6		−23.5	−23.5		−1.5	1.5		5.1	−5.1
	38.0	52.3		43.6	43.6		1.6	1.6		−1.5	1.5		−5.7	5.7
	84.8	78.6		43.6	43.6		38.1	38.1		−1.5	1.5		4.6	−4.6
	12.5	22.3		43.6	43.6		−26.2	−26.2		−1.5	1.5		−3.4	3.4
	52.2	39.7		43.6	43.6		2.4	2.4		−1.5	1.5		7.7	−7.7
SS	69,142.43		=	57,002.64		+	11,164.44		+	63.66		+	911.69	
df	30		=	1		+	14		+	1		+	14	

FIGURE 6.16 Decomposition of the radioactive twins

Source	df	SS	MS	F
Pairs	14	11164.44	797.46	12.246
Env	1	63.66	63.66	0.978
Res	14	911.69	65.12	
Total	29	12139.79		

FIGURE 6.17 ANOVA for the radioactive twins

Mean squares in the ANOVA table summary confirm these observations. The average variation within twin pairs is (65.12), which is about the same size as the variation due to environment (63.66), with an F-ratio of 0.978. (Compare this with the huge mean square of nearly 800 for blocks, more than 12 times the mean square for residuals.)

Our research hypothesis is that compared with a rural environment, an urban environment is associated with a lower rate of lung clearance. The corresponding null hypothesis is that the two environments have the same mean clearance rate (i.e, $H_0 : \mu_{Urban} = \mu_{Rural}$). Before testing this hypothesis, we first check the fit of the model.

Assessing the fit of the model. Figure 6.18 shows four diagnostic plots: (a) residuals versus fitted values, (b) a normal quantile plot of residuals, (c) a Tukey plot for nonadditivity, and (d) a scatterplot of Urban versus Rural with blocks as points.

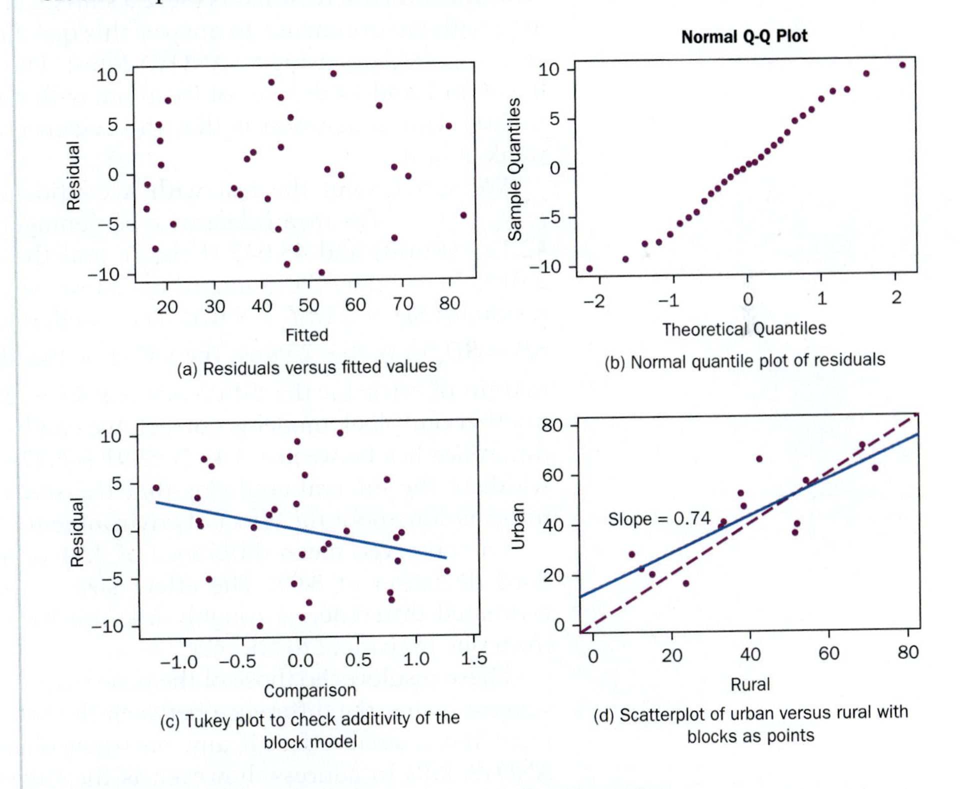

FIGURE 6.18 Four diagnostic plots for the radioactive twins

(a) Residuals versus fitted values. The plot looks good. There is no suggestion of curvature, and no suggestion that spread is related to the size of the fitted values.

(b) Normal quantile plot. Here, too, the plot looks good. The points lie near a line, with no clumps, gaps, or outliers.

(c) Tukey plot. There is not much of a linear pattern in this plot, so we are unlikely to find a transformation that gives us additivity. The slope of the line is −2.46 which suggests trying the scale $\text{Rate}^{3.46}$. Changing to that scale has no qualitative effect on the analysis.

The first two plots raise no serious concerns about conditions for a complete block model. However, (d) the scatterplot of Urban versus Rural, with slope 0.74, suggests an alternative analysis based on the fitted line, as described in Example 6.10.

Scope of inference. The data are observational. (It was not possible to use coin flips to tell the subjects where they were supposed to have been living for the previous 10 years.) The twin pairs were not chosen by random sampling from some larger population. All subjects lived either in Stockholm or in the surrounding rural and suburban areas. The authors do not try to generalize their results to other cities.

Without either random assignment or random sampling, there is no direct justification for a probability model based on how the data were produced. However, it is reasonable to think, based on the diagnostic plots, that a *P*-value computed from the usual normal distribution can serve as a rough guide.

Results and conclusions. Is there a statistically detectable difference associated with environment? To answer this question, we test the null hypothesis $H_0 : \mu_{Urban} = \mu_{Rural}$ using an ANOVA *F*-test. The *F*-statistic for environments is 0.978 on 1 and 14 degrees of freedom, with a *P*-value of 0.34, nowhere close to 0.05. Our conclusion is that the evidence against the null hypothesis is weak at best.

We supplement the test with a confidence interval for the difference $\mu_{Urban} - \mu_{Rural}$. The mean clearance percentages for the two environments are 42.133 (Rural) and 45.047 (Urban), and the difference (*Urban* − *Rural*) is 2.913. To estimate the standard deviation we use the mean square error for residuals: $SD = \sqrt{MSE} = 8.070$. The standard error of the mean difference is $SE = SD\sqrt{\frac{1}{15} + \frac{1}{15}} = 2.9466$; the 95% t^* value on 14 *df* is 2.1448, and the 95% margin of error for the difference is $t^*SE = 2.1448(2.9466) \approx 6.32$. Assuming that our block model is correct, we can be 95% confident that the mean difference lies between −3.41 (= 2.91 − 6.32) and 9.23 (= 2.91 + 6.32). The width of the interval confirms that the study gives us comparatively little information about the effect of environment.

The observed mean difference of 2.91 is small compared with the standard deviation of 8.07. The effect size is the ratio 2.91/8.07 = 36%: the estimated difference is roughly one-third the typical size of the variation from one individual to another.

These results echo those of the investigators: “What the investigation does suggest is that the difference between the Stockholm and the rural environment has a small effect if any, on tracheobronchial clearance.”[2] What the ANOVA fails to address, however, is the suggestion from the scatterplot of Figure 6.18(d) that the size of the effect of environment may be related to other factors that affect lung health.

CHAPTER SUMMARY

This chapter introduced you to the **randomized complete block (RCB)** design and the ANOVA model for data that come from such a design. Most often RCB designs are associated with experiments, but we showed you that they can also arise in observational studies. Typically RCB designs are formed using one of three methods: **subdivide** large plots of land (blocks) into smaller subplots (units), **sort** and match subjects or other units into groups of similar units, or **reuse** each subject multiple times.

We introduced you to the two-way additive model:

$$Y = \mu + \tau_i + \beta_j + \epsilon$$

We used several familiar plots (side-by-side dot plots, residuals versus fit plots, and normal quantile plots), and introduced two new plots **Anscombe block plots**, and **Tukey nonadditivity plots** to assess the fit of this model to the data. Then we used overlays to help understand where the values in the ANOVA table come from and what is being tested in the hypotheses.

We described which hypothesis tests can be conducted in this setting, as well as the appropriate interval estimates and effect sizes. Finally, we noted that the scope of inference is dependent on how the data were produced. If the data came from a randomized experiment, then a cause and effect conclusion is warranted. If the data came from an observational study, such a conclusion is not appropriate.

Finally, all the models in this chapter assume that "effects are additive and constant"; that is, there is no interaction between blocks and treatments, and there are no lurking confounders. For many two-way datasets the additive model fits well, but there are also many two-way datasets that require an additional set of terms for interaction. For a simple example, consider again the rectangle data. The effect of width on area is different for rectangles of different lengths. Length and width interact in their effects on area. In this chapter we noted that sometimes we can find a scale to make the effects additive. In particular, taking the natural log of the response variable will sometimes work. But for many other datasets, like the river iron, no change of scale will make the interaction go away. These datasets require an additional set of terms for interaction. In the next chapter we will discuss adding interaction effects to the two-way model.

EXERCISES

Conceptual Exercises

***6.1–6.6* Factors and levels.** For each of the following studies, (a) give the response; (b) the name of the two factors; (c) for each factor, tell whether it is observational or experimental and the number of levels it has; and (d) tell whether the study is a complete block design.

6.1 Bird calcium. Ten male and 10 female robins were randomly divided into groups of five. Five birds of each sex were given a hormone in their diet; the other 5 of each sex were given a control diet. At the end of the study, the researchers measured the calcium concentration in the plasma of each bird.

6.2 Crabgrass competition. In a study of plant competition,[3] two species of crabgrass (*Digitaria sanguinalis [D.s.]* and *Digitatria ischaemum [D.i.]*) were planted together in a cup. In all, there were 20 cups, each with 20 plants. Four cups held 20 D.s. each, another four held 15 D.s. and 5 D.i., another four held 10 of each species, still another four held 5 D.s. and 15 D.i., and the last four held 20 D.i. each. Within each set of four cups, two were chosen at random to receive nutrients at normal levels; the other two cups in the set received nutrients at low levels. At the end of the study, the plants in each cup were dried and the total weight was recorded.

6.3 Behavior therapy for stuttering. Thirty-five years ago the journal *Behavior Research and Therapy*[4] reported a study that compared two mild shock therapies for stuttering. There were 18 subjects, all of them stutterers. Each subject was given a total of three treatment sessions, with the order randomized separately for each subject. One treatment administered a mild shock *during* each moment of stuttering, another gave the shock *after* each stuttered word, and the third treatment was a control, with no shock. The response was a score that measured a subject's adaptation.

6.4 Noise and ADHD. It is now generally accepted that children with attention deficit and hyperactivity disorder tend to be particularly distracted by background noise. About 20 years ago a study was done to test this hypothesis.[5] The subjects were all second-graders.

Some had been diagnosed as hyperactive; the other subjects served as a control group. All the children were given sets of math problems to solve, and the response was their score on the set of problems. All the children solved problems under two sets of conditions, high noise and low noise. (Results showed that the controls did better with the higher noise level, whereas the opposite was true for the hyperactive children.)

6.5 Running dogs. In a study conducted at the University of Florida,[6] investigators compared the effects of three different diets on the speed of racing greyhounds. (The investigators wanted to test the common belief among owners of racing greyhounds that giving their dogs large doses of vitamin C will cause them to run faster.) In the University of Florida study, each of 5 greyhounds got all three diets, one at a time, in an order that was determined using a separate randomization for each dog. (To the surprise of the scientists, the results showed that when the dogs ate the diet high in vitamin C, they ran slower, not faster.)

6.6 Fat rats. Is there a magic shot that makes dieting easy? Researchers investigating appetite control measured the effect of two hormone injections, leptin and insulin, on the amount eaten by rats.[7] Male rats and female rats were randomly assigned to get one hormone shot or the other. (The results showed that for female rats, leptin lowered the amount eaten, compared to insulin; for male rats, insulin lowered the amount eaten, compared to leptin.)

6.7 *F*-ratios. Suppose you fit the two-way main effects ANOVA model for the river data of Example 6.7, this time using the log concentration of copper (instead of iron) as your response, and then do an *F*-test for differences between rivers.

a. If the *F*-ratio is near 1, what does that tell you about the differences between rivers?

b. If the *F*-ratio is near 10, what does that tell you?

6.8 *F*-ratios. Suppose you fit the two-way main effects ANOVA model for the shrew data of Example 6.7, and then do an *F*-test for differences between sleep phases.

a. If the *F*-ratio is near 0.5, what does that tell you about the differences between sleep phases?

b. If the *F*-ratio is near 1, what does that tell you?

6.9 Degrees of freedom. If you carry out a two-factor ANOVA (main effects model) on a dataset with Factor A at four levels and Factor B at five levels, with one observation per cell, how many degrees of freedom will there be for:

a. Factor A?

b. Factor B?

c. Error?

6.10 Degrees of freedom. If you carry out a two-factor ANOVA (main effects model) on a dataset with Factor A at three levels and Factor B at three levels, with one observation per cell, how many degrees of freedom will there be for:

a. Factor A?

b. Factor B?

c. Error?

6.11 Fill in the blank. If your dataset has two factors and you carry out a one-way ANOVA, ignoring the second factor, your SSE will be too ___________ (small, large) and you will be ___________ (more, less) likely to detect real differences than would a two-way ANOVA.

6.12 Fill in the blank. If your dataset has one factor of interest and one blocking factor, and you carry out a two-way ANOVA, your SSE will be ___________ (smaller, larger) than if you ignored the blocks and did a one-way ANOVA, and you will be ___________ (more, less) likely to detect real differences than would a one-way ANOVA.

Exercises 6.13–6.16 are True or False exercises. If the statement is false, explain why it is false.

6.13 In a randomized complete block study, at least one of the factors must be experimental.

6.14 The conditions for the errors for the two-way additive ANOVA model are the same as the conditions for the errors of the one-way ANOVA model.

6.15 In a complete block design, blocks and treatments are crossed. Explain.

6.16 Block designs result only from observing subjects several times, each time with a different treatment.

6.17 Fenthion. In the study of fenthion in olive oil (Exercise 5.50), the response was the concentration of the toxic chemical fenthion in samples of olive oil and the factor of interest was the time when the concentration was measured, with three levels: day 0, day 281, and day 365. Consider two versions of the study: In Version A, 18 samples of olive oil are randomly divided into three groups of six samples each. Six samples are measured at time 1, six at time 2, and six at time 3. In Version B, the olive oil is divided into six larger samples. At each of times 1–3, a subsample is taken from each, and the fenthion concentration is measured.

a. One version is a one-way, completely randomized design; the other is a complete block design. Which is which? Explain.

b. What is the advantage of the block design?

6.18 Give one example each (from the examples in the chapter) of three kinds of block designs: one that creates blocks by reusing subjects, one that creates blocks by matching subjects, and one that creates blocks by subdividing experimental material. For each, identify the blocks and the experimental units.

6.19 Give the two most common reasons why creating blocks by reusing subjects may not make sense, and give an example from the chapter to illustrate each reason.

6.20 Recall the two versions of the finger tapping study in Example 6.5. Version 1, the actual study, used 4 subjects three times each, with subjects as blocks and time slots as units. Version 2, a completely randomized design, used 12 subjects as units and had no blocks.

a. List two advantages of the block design for this study.

b. List two advantages of the completely randomized design for this study.

6.21 Crossing. Give two examples of one-way designs that can be improved by crossing the factor of interest with a second factor of interest. For each of your designs, tell why it makes sense to include the second factor.

6.22 Why is it that in a randomized complete block design, the factor of interest is nearly always experimental rather than observational?

6.23 Behavior therapy for stuttering: design decisions. Exercise 6.3 on page 293 described a study[8] that compared two mild shock therapies for stuttering. Each of the 18 subjects, all of them stutterers, was given a total of three treatment sessions, with the order randomized separately for each subject. One treatment administered a mild shock during each moment of stuttering, another gave the shock after each stuttered word, and the third "treatment" was a control, with no shock. The response was a score that measured a subject's adaptation.

a. Explain why the order of the treatments was randomized.

b. Explain why this study was run as a block design with subjects as blocks of time slots instead of a completely randomized design with subjects as experimental units.

6.24 Sweet smell of success. Chicago's Smell and Taste Treatment and Research Foundation funded a study, "Odors and Learning," by A. R. Hirsch and L. H. Johnson. One goal of the study was to see whether a pleasant odor could improve learning. Twenty subjects participated.* If you had been one of the subjects, you would have been timed while you completed a paper-and-pencil maze two times: once under control conditions, and once in the presence of a "floral fragrance."

a. Suppose that for all 20 subjects, the control attempt at the maze came first, the scented attempt came second, and the results showed that average times for the second attempt were shorter and the difference in times was statistically significant. Explain why it would be wrong to conclude that the floral scent caused subjects to go through the maze more quickly.

b. Consider a modified study. You have 20 subjects, as above, but you have two different mazes, and your subjects are willing to do both of them. Your main goal is to see whether subjects solve the mazes more quickly in the presence of the floral fragrance or under control conditions, so the fragrance factor has only two levels, control and fragrance. Tell what design you would use. Be specific and detailed: What are the units? Nuisance factors? How many levels? What is the pattern for assigning treatments and levels of nuisance factors to units?

6.25 Rainfall. According to theory, if you release crystals of silver iodide into a cloud (from an airplane), water vapor in the cloud will condense and fall to the ground as rain. To test this theory, scientists randomly chose 26 of 52 clouds and seeded them with silver iodide.[9] They measured the total rainfall, in acre-feet, from all 52 clouds.

Explain why it was not practical to run this experiment using a complete block design. (Your answer should address the three main ways to create blocks.)

6.26 Hearing. Audiologists use standard lists of 50 words to test hearing; the words are calibrated, using subjects with normal hearing, to make all 50 words on the list equally hard to hear. The goal of the study[10] described here was to see how four such lists, denoted by L1–L4 in Table 6.14, compared when played at low volume with a noisy background. The response is the percentage of words identified correctly. The data are stored in **HearingTest**. **Hearing**

Table 6.14 Percentage of words identified for each of four lists

Sub	L1	L2	L3	L4	Mean
1	28	20	24	26	24.5
2	24	16	32	24	24.0
3	32	38	20	22	28.0
4	30	20	14	18	20.5
5	34	34	32	24	31.0
6	30	30	22	30	28.0
7	36	30	20	22	27.0
8	32	28	26	28	28.5
9	48	42	26	30	36.5
10	32	36	38	16	30.5
11	32	32	30	18	28.0
12	38	36	16	34	31.0
13	32	28	36	32	32.0
14	40	38	32	34	36.0
15	28	36	38	32	33.5
16	48	28	14	18	27.0
17	34	34	26	20	28.5
18	28	16	14	20	19.5
19	40	34	38	40	38.0
20	18	22	20	26	21.5
21	20	20	14	14	17.0
22	26	30	18	14	22.0
23	36	20	22	30	27.0
24	40	44	34	42	40.0
Mean	33	30	25	26	28.3

a. Is this an observational or experimental study? Give a reason for your answer. (Give the investigators the benefit of any doubts: If it was possible to randomize, assume they did.)

* Actually, there were 21 subjects. (We've simplified reality for the sake of this exercise.)

b. List any factors of interest and any nuisance factors.

c. What are the experimental (or observational) units?

d. Are there blocks in this design? If so, identify them.

6.27 Running dogs: *df*. To test the common belief that racing greyhounds run faster if they are fed a diet containing vitamin C, scientists at the University of Florida used a randomized complete block design (*Science News*, July 20, 2002) with 5 dogs serving as blocks. See Exercise 6.5 on page 294 for more detail. Over the course of the study, each dog got three different diets, in an order that was randomized separately for each dog. Suppose the scientists had been concerned about the carry-over effects of the diets, and so had decided to use a completely randomized design with 15 greyhounds as experimental units.

a. Complete the following table to compare degrees of freedom for the two designs:

BLOCKS		NO BLOCKS	
Source	**df**	**Source**	**df**
Diets		Diets	
Blocks			
Error		Error	
Total	14	Total	14

b. Compare the advantages and disadvantages of the two designs.

6.28 Migraines. People who suffer from migraine headaches know that once you get one, it's hard to make the pain go away. A comparatively new medication, Imatrex, is supposed to be more effective than older remedies. Suppose you want to compare Imatrex (I) with three other medications: Fiorinol (F), Acetaminophen (A), and Placebo (P).

a. Suppose you have available four volunteers who suffer from frequent migraines. Tell how to use a randomized complete block design to test the four drugs: What are the units? How will you assign treatments (I, F, A, and P) to units?

b. Notice that you have exactly the same number of subjects as treatments and that it is possible to reuse subjects, which makes it possible to have the same number of time slots as treatments. Thus you have four subjects (I, II, III, and IV), four time slots (1, 2, 3, 4), and four treatments (A, F, I, P). Rather than randomize the order of the four treatments for each subject, which might by chance assign the placebo always to the first or second time slot, it is possible to balance the assignment of treatments to time slots so that each treatment appears in each time slot exactly once. Such a design is called a **Latin square**. Create such a design yourself, by filling in the squares below with the treatment letters A, F, I, and P, using each letter four times, in such a way that each letter appears exactly once in each row and column.

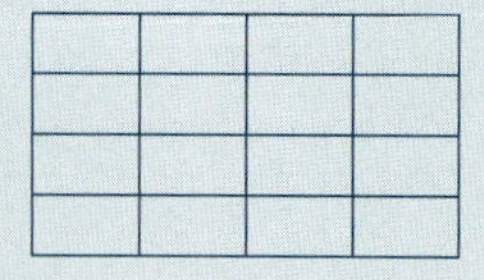

6.29 Fat rats: block design. Researchers investigating appetite control measured the effect of two hormone injections, leptin and insulin, on the amount eaten by rats (*Science News*, July 20, 2002). Male rats and female rats were randomly assigned to get one hormone shot or the other. See Exercise 6.6 on page 294 for more detail.

a. Tell how this study could have been run using each rat as a block of two time slots.

b. Why do you think the investigators decided not to use a design with blocks?

Guided Exercises

6.30 River iron. This is an exercise with a moral: Sometimes, the way to tell that your model is wrong requires you to ask, "Do the numbers make sense in the context of the problem?" Consider the New York river data of Example 6.7, with iron concentrations in the original scale of parts per million: **RivIron**

	Grasse	Oswegatchie	Raquette	St. Regis	Mean
Upstream	944	860	108	751	665.75
Midstream	525	229	36	568	339.50
Downstream	327	130	30	350	209.25
Mean	598.7	406.3	58.0	556.3	404.83

a. Fit the two-way additive model $FE = River + Site + Error$.

b. Obtain a normal probability plot of residuals. Is there any indication from this plot that the normality condition is violated?

c. Obtain a plot of residuals versus fitted values. Is there any indication from the shape of this plot that the variation is not constant? Are there pronounced clusters? Is there an unmistakable curvature to the plot?

d. Finally, look at the leftmost point, and estimate the fitted value from the graph. Explain why this one fitted value strongly suggests that the model is not appropriate.

6.31 Sleeping shrews. Consider the shrew data of Example 6.7 where we are interested in whether and how heart rates differ between the three different phases of sleep. **Shrew**

	LSW	DSW	REM	Mean
A	14.0	11.7	15.7	13.8
B	25.8	21.1	21.5	22.8
C	20.8	19.7	18.3	19.6
D	19.0	18.2	17.1	18.1
E	26.0	23.2	22.5	23.9
F	20.4	20.7	28.9	20.0
Mean	21.0	19.1	19.0	19.7

a. Fit the two-way additive model $HeartRate = Shrew + Phase + Error$.

b. Obtain a normal probability plot of residuals. Is there any indication from this plot that the normality condition is violated?

c. Obtain a plot of residuals versus fitted values. Is there any indication from the shape of this plot that the variation is not constant? Are there pronounced clusters? Is there an unmistakable curvature to the plot? Are there other issues?

6.32 Oral contraceptives: Anscombe plots.[11] Researchers were interested in how a new drug might interact with a standard oral contraceptive. In particular, they wondered if the new drug might affect the amount of a component of the contraceptive that the body would be able to use. They recruited 22 female subjects. Each subject took an oral contraceptive throughout the study. They each, also, spent some time taking the new drug and some time taking a placebo (in random order). The response variable is the bioavailability of one of the components of the contraceptive, ethinyl estradiol (*EE*) measured in pg*hr/ml. The data are in **Contraceptives**. **Contracp**

a. Create the Anscombe plot for this dataset. What is the slope?

b. What does the slope tell you about the need for transformation?

6.33 Frantic fingers: Anscombe plots. Return to the **FranticFingers** data of Example 6.5. **Frantic**

a. Because there are three treatments in this study (placebo, caffeine, and theobromine), there will be three different Anscombe plots for these data. Create these three plots.

b. What are the slopes of the lines for each of these three plots? What do they tell you?

6.34 River iron: Anscombe plots. Return to the **RiverIron** data of Exercise 6.30. In that exercise we saw that there might be a problem with the fit of the ANOVA model. Here we use a graphical approach to address the same question. **RivIron**

a. Because there are three sites in this study (effectively three treatments), there will be three different Anscombe plots for this data. Create these three plots.

b. What are the slopes of the lines for each of these three plots? What do they tell you?

6.35 Iron deficiency. In developing countries, roughly one-fourth of all men and half of all women and children suffer from anemia due to iron deficiency. Researchers[12] wanted to know whether the trend away from traditional iron pots in favor of lighter, cheaper aluminum could be involved in this most common form of malnutrition. They compared the iron content of 12 samples of three Ethiopian dishes: one beef, one chicken, and one vegetable casserole. Four samples of each dish were cooked in aluminum pots, four in clay pots, and four in iron pots. A parallel dotplot of the data follows.

Describe what you consider to be the main patterns in the plot. Cover the usual features, keeping in mind that in any given plot, some features deserve more attention than others: How are the group averages related (to each other and to the researchers' question)? Are there gross outliers? Are the spreads roughly equal? If not, is there evidence that a change of scale would tend to equalize spreads?

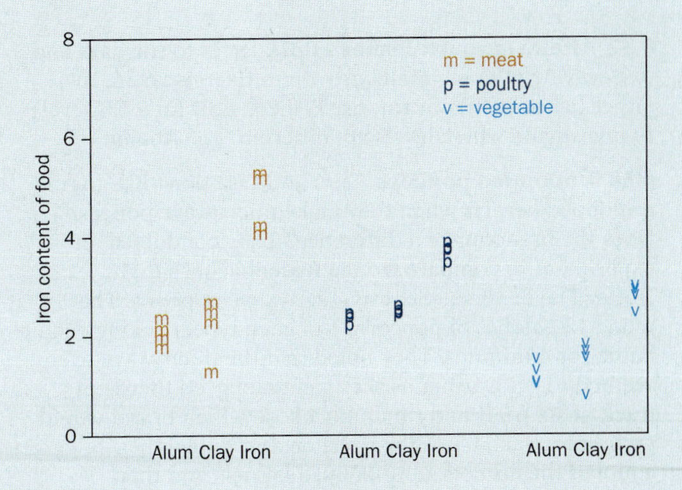

6.36 Alfalfa sprouts. Some students were interested in how an acidic environment might affect the growth of plants. They planted alfalfa seeds in 15 cups and randomly chose five to get plain water, five to get a moderate amount of acid (1.5M HCl), and five to get a stronger acid solution (3.0M HCl). The plants were grown in an indoor room, so the students assumed that the distance from the main source of daylight (a window) might have an effect on growth rates. For this reason, they arranged the cups in five rows of three, with one cup from each *Acid* level in each row. These are labeled in the dataset as *Row*: *a* = farthest from the window through *e* = nearest to the window. Each cup was an experimental unit, and the response variable was the average height of the alfalfa sprouts in each cup after four days (*Ht*4). The data are shown in the table below and stored in the **Alfalfa** file: **Alfalfa**

Treatment/Cup	a	b	c	d	e
water	1.45	2.79	1.93	2.33	4.85
1.5 HCl	1.00	0.70	1.37	2.80	1.46
3.0 HCl	1.03	1.22	0.45	1.65	1.07

a. Find the means for each row of cups (a, b, ..., e) and each treatment (water, 1.5 HCl, 3.0 HCl). Also find the average and standard deviation for the growth in all 15 cups.

b. Construct a two-way main effects ANOVA table for testing for differences in average growth due to the acid treatments using the rows as a blocking variable.

c. Check the conditions required for the ANOVA model.

d. Based on the ANOVA, would you conclude that there is a significant difference in average growth due to the treatments? Explain why or why not.

e. Based on the ANOVA, would you conclude that there is a significant difference in average growth due to the distance from the window? Explain why or why not.

6.37 Alfalfa sprouts: Fisher's LSD. Refer to the data and two-way ANOVA on alfalfa growth in Exercise 6.36. If either factor is significant, use Fisher's LSD (at a 5% level) to investigate which levels are different. **Alfalfa**

6.38 Unpopped popcorn. Lara and Lisa don't like to find unpopped kernels when they make microwave popcorn. Does the brand make a difference? They conducted an experiment to compare Orville Redenbacher's Light Butter Flavor versus Seaway microwave popcorn. They made 12 batches of popcorn, 6 of each type, cooking each batch for 4 minutes. They noted that the microwave oven seemed to get warmer as they went along, so they kept track of six trials and randomly chose which brand would go first for each trial. For a response variable, they counted the number of unpopped kernels and then adjusted the count for Seaway for having more ounces per bag of popcorn (3.5 vs. 3.0). The data are shown in Table 6.15 and stored in **Popcorn**. **Popcorn**

Table 6.15 Unpopped popcorn by *Brand* and *Trial*

Brand/Trial	1	2	3	4	5	6
Orville Redenbacher	26	35	18	14	8	6
Seaway	47	47	14	34	21	37

a. Find the mean number of unpopped kernels for the entire sample, and estimate the effects (α_1 and α_2) for each brand of popcorn.

b. Run a two-way ANOVA model for this randomized block design. (Remember to check the required conditions.)

c. Does the brand of popcorn appear to make a difference in the mean number of unpopped kernels? What about the trial?

6.39 River iron: Tukey additivity plot. Here are the **RiverIron** data in the original scale: **RivIron**

Site	Grasse	Oswegatchie	Raquette	St. Regis
Upstream	944	860	108	751
Midstream	525	229	36	568
Downstream	327	130	30	350

The bordered residual table is as follows:

Site	Grasse	Oswegatchie	Raquette	St. Regis	Eff
Upstream	84.42	192.75	−210.92	−66.25	260.92
Midstream	−8.33	−112.00	43.33	77.00	−65.33
Downstream	−76.08	−80.75	167.58	−10.75	−195.58
Eff	193.83	1.50	−346.83	151.51	404.83

a. Use a calculator or spreadsheet to compute the comparison values.

b. Plot the residuals versus the comparison values.

c. Fit a line by eye and estimate its slope. What transformation is suggested?

6.40 Oral contraceptives: Tukey additivity plot. In Exercise 6.32 you determined, by using an Anscombe block plot, that a transformation was probably not necessary. Let's see what Tukey's additivity plot has to say about that here. The dataset **Contraceptives** has columns for the residuals and the comparison values. Use these columns to create Tukey's additivity plot. Discuss what you learn from this plot. **Contracp**

Open-Ended Exercises

6.41 Mental health and the moon. For centuries, people looked at the full moon with some trepidation. From stories of werewolves coming out, to more crime sprees, the full moon has gotten the blame. Some researchers[13] in the early 1970s set out to actually study whether there is a "full-moon" effect on the mental health of people. The researchers collected admissions data for the emergency room at a mental health hospital for 12 months. They separated the data into rates before the full moon (mean number of patients seen 4–13 days before the full moon), during the full moon (the number of patients seen on the full moon day), and after the full moon (mean number of patients seen 4–13 days after the full moon). They also kept track of which month the data came from since there was likely to be a relationship between admissions and the season of the year. The data can be found in the file **MentalHealth**. Analyze the data to answer the researcher's question. **MentHlt**

MARK GARLICK/SCIENCE PHOTO LIBRARY/Getty Images

CHAPTER 7

ANOVA with Interaction and Factorial Designs

In this chapter you will learn to:

- Create and interpret interaction graphs.
- Assess a two-way interaction model.
- Fit and use a two-way interaction model.

This chapter has just one new idea, interaction. Everything else is a matter of putting together the ideas of Chapters 5 and 6. Our plan for the chapter is to start first with the one new idea, then rely as much as possible on ideas you have already seen. We have three main goals:

1. to introduce and incorporate the new idea, interaction,
2. to review and reinforce the ideas from the last two chapters, and
3. to show how these basic ideas can lead to new ways to think about data.

We hope you will work through the chapter with these three goals in mind.

interaction

We follow the same general structure as in the last two chapters, apart from our one extra section at the start, introducing examples of **interaction**. The sections after Section 7.1 cover (2) choosing a model (and the two-way design), (3) choosing a scale (and exploring two-way data), (4) fitting the model, (5) assessing the fit, and (6) using the model.

Just because there is only one new idea does not mean the chapter is easy. Learning to rethink and extend old ideas can be a challenge.

7.1 Interaction

You have already seen two-factor models in the last chapter, with one factor for blocks and one for a factor of interest. Those two-factor models were assumed (required) to be additive: The rate of finger tapping in Example 6.5 was assumed (required) to be the sum of the grand average plus a subject effect plus a drug effect. The concentration of river iron in Example 6.7 was assumed to be the sum of the grand average plus a river effect plus a site effect. For the river iron, as for many studies, the simple additive model was highly suspect.

Most of the time, as for the river iron, reality is not as simple as the additive model asks you to pretend. Typically, factors interact. Deprive a cornfield of nitrogen and the crop will die, even if you fertilize with phosphorus. Deprive the field of phosphorus and the crop will die, even if you offer nitrogen. Nitrogen and phosphorus are both factors that affect growth, but their effects interact. If you are a cornstalk, you need both. No amount of nitrogen can make up for being short of phosphorus.

This section offers three ways to recognize interaction: contextual, numerical, and visual.

Contextual: Interaction in a Concrete Setting

Sometimes you can tell just from the applied context that two factors will interact. We start with one such example. As you read through it, notice that reasoning about interaction involves comparing pairs (or sets) of differences.

EXAMPLE 7.1

Cats and claws, kids and grown-ups[1] First ask yourself two questions:

1. Does a cat have claws?
2. Does a cat have a head?

Now reflect on how you arrived at your answers. Although adults can answer these questions very quickly, researchers are interested in response time differences that are measured in tiny fractions of a second. Most adults are quick to answer, “Yes,” a cat does have claws, because we have learned to associate the two words “cat” and “claws.” We tend to be a bit slower to answer the question “Does a cat have a head?” because we can’t rely on a learned word association. Instead, we have to create a mental image of a cat, and use that image to recognize that, “Yes,” a cat does have a head. For adults, it takes longer to answer the cat/head question.

Now go back in time and imagine that you are a first-grader. You don’t yet read much, if at all. You don’t have the experience to know that the words “cats” and “claws” go together. When you get to the cat/claws question you have to create a mental image, just as for the cat/head question. You have no shortcut based on word association. The time it takes you to answer will be pretty much the same for both questions.

For adults, it takes longer to answer the head question than the claws question. For first-graders it takes about the same time for both questions. In short, there is an interaction between age (adult versus first grade) and type of question (claws versus head).

For the cats and claws (Example 7.1) it was possible to use the context to predict that interaction would be present. The next pair of examples (7.2 a and b) is not like that. Unless you know a lot of biophysiology, context alone can't tell you whether the two factors will interact. You need data. However, if you do have the data you need, mere arithmetic can do the rest.

EXAMPLE 7.2 Birds and hormones, Pigs and vitamins

BirdCa

a. **Birds and hormones.**[2] Do male and female birds respond in the same way to a hormone supplement added to their food? The hormone in question was meant to raise the concentration of calcium in the blood. To investigate this question, 10 male robins were divided (at random) into two groups of 5. One group was treated with the hormone; the other wasn't. A similar procedure was applied to 10 female robins, 5 getting the hormone and 5 not. The response variable of interest is the blood calcium level (measured in mg per 100 ml). The data are shown in Table 7.1 and are stored in **BirdCalcium**.

	No Hormone	Hormone
Male	14.5 11.0 10.8 14.3 10.0	32.0 23.8 28.8 25.0 29.3
Female	16.5 18.4 12.7 14.0 12.8	31.9 26.2 21.3 35.8 40.2

TABLE 7.1 Birds and blood calcium (mg per 100 ml)

It is clear in the data table that the hormone supplement raises the calcium level, but is the size of this effect much different for male and female birds? That's the question that looks at a possible interaction.

Pig

b. **Pigs and vitamins.**[3] In theory, adding vitamin B12 to the diet of baby pigs should cause them to gain weight faster. In practice, however, the presence of bacteria in a pig's gut can prevent the use of vitamin B12. Would taking antibiotics make the use of vitamin B12 more effective?

You want to compare 4 diets based on crossing two factors, *B12* (0 mg or 5 mg) added to the diet, and *Antibiotics* (0 mg or 40 mg): no/no, no/yes, yes/no, and yes/yes. With 12 pigs and 4 diets, you set up a lottery to assign diets to pigs, with three tickets for each diet. The tickets tell the pigs what they get to eat—no choice. Table 7.2 shows the data* (also stored in **PigFeed**) from one such experiment.

	No Antibiotics			Antibiotics		
No B12	30	19	8	5	0	4
B12	26	21	19	52	56	54

TABLE 7.2 Weekly weight gain, in hundredths of a pound above 1 pound

The question of whether the presence/absence of antibiotics affects the utility of the vitamin B12 supplements is a question about the interaction of these two factors.

*The response is weekly weight gain, measured in hundredths of a pound above 1 pound. Thus the 30 in the upper left corner records 30/100 of a pound above 1.00, or 1.30 pounds per week. This recoding makes the patterns easier to see and doesn't change what matters for inference: *F*-ratios, df, and *P*-values. Also note that the value of 19 in the bottom row was actually equal to 21. We changed it by 0.02, from 1.21 to 1.19 to make the arithmetic give whole numbers.

We will rely on the birds and pigs throughout the chapter, first to show numerical and graphical ways to tell when interaction is present, and later to show how the presence or absence of interaction helps determine which contrasts, confidence intervals, and effect sizes you should look at.

Before we get to the numerical and graphical explorations for interactions with the **BirdCalcium** and **PigFeed** data, think back over the three examples so far, keeping in mind this informal definition: *"Interaction is a difference of differences."*

EXAMPLE 7.3 Cats, birds, and pigs: The concept of interaction as a difference of differences

- For the cats and claws (Example 7.1), a sample of adults took more time with the cat/head question than with the cat/claws question. In a sample of first-graders, the two questions took about the same time. Their difference was about zero. There was a difference in the averages for adults, but the difference was near zero for first-graders: In short, the two differences were different. Interaction is present.
- For the birds and hormones (Example 7.2a), you will soon see that the difference between hormone and no hormone was about the same for male and female birds. The differences (hormone versus control) were not different for males and females. No interaction.
- For the pigs and vitamins (Example 7.2b), the experimental data will show the effect of B12 was different depending on whether the pigs got antibiotics. With no antibiotics, B12 made almost no difference. With antibiotics in the diet, adding B12 made a huge difference, giving a much higher average weight gain than when used without antibiotics. The differences were different. Interaction is present.

INTERACTION

Interaction is a "difference of differences."

To investigate an interaction numerically, we can look at the difference in cell means across one factor and see how they compare as we move between levels of the other factor.

EXAMPLE 7.4 Birds and pigs: Interaction as a numeric difference of differences

BirdCa

a. **Birds: no interaction.** Figure 7.1 shows a table of means for each of the four treatment cells from the birds experiment, along with the difference in cell means for each row and column. If you look at the rightmost **Diff** column, and round to the nearest whole number, you see that for male birds, the hormone supplement raised the concentration of blood calcium by about 16 mg/100 ml. For female birds the supplement raised the calcium level also by about 16. The differences are roughly the same. Conclusion: No interaction.

Now look at the bottom **Diff** row, which shows the male/female mean differences for control groups (left) and hormone groups (right). Again, rounding to the nearest whole number, the differences are both roughly equal to 3. No difference, no interaction.

	Hormone? No	Yes	Diff
Male	12.12	27.78	15.66
Female	14.88	31.08	16.20
Diff	2.76	3.30	

Differences roughly equal: no interaction (rows)

Differences roughly equal: no interaction (columns)

FIGURE 7.1 Difference of differences for the birds

Whether you look at row differences or column differences, the message is the same: No interaction.

For the birds, the "difference of differences" shows no interaction.

Pig

b. **Pigs: interaction present.** Figure 7.2 shows shows a table of means for each of the four treatment cells from the pigs experiment. The interaction is easiest to see by looking at the rows and comparing the differences due to antibiotics. For pigs who got no B12, adding antibiotics to the diet actually slowed their weight gain from a mean of 19 down to 3. The difference was −16 (hundredths of a pound per week). For the pigs who got B12, antibiotics increased their weekly weight gain by an average of 32/100 of a pound, 54 − 22. The differences are not only a two-to-one ratio but in opposite directions. Interaction is present.

The message from comparing column differences—the effect of B12—is the same. With 0 mg antibiotics (left column), adding B12 makes hardly any difference (3/100 pounds per week). With 40 mg (right column), adding B12 to the diet adds half a pound (51/100) to the average weekly weight gain. The differences are different, and the message is the same: Interaction is present.

		Antibiotics 0 mg	40 mg	Diff
B12	**0 mg**	19	3	−16
	5 mg	22	54	32
	Diff	3	51	

Differences are different: interaction is present (rows)

Differences are different: interaction is present (columns)

FIGURE 7.2 Difference of differences for the pigs

Visual: Interaction Plots

interaction plot

This same, important, recurring idea, "interaction is a difference of differences," is easy to show visually in an **interaction plot**. An interaction plot is drawn using the cell means plotted on the vertical axis versus groups defined by one of the factors along the horizontal axis. Lines connect the cell means for each group determined by the second factor. When there is no interaction, these lines will be roughly parallel, but when there is a "difference of differences," the lines will have very different slopes. *Interaction means lines are not parallel*. The next example illustrates the use of interaction plots for the birds and pigs examples.

EXAMPLE 7.5 Birds and pigs: Interaction plots

BirdCa

a. **Birds: no interaction.** Figure 7.3 shows an interaction plot for the **Bird-Calcium** data. Each line shows the effect of the hormone: control on the left (no), hormone on the right (yes), with response (calcium concentration) on the vertical axis. The solid line is for male birds, the dashed line is for female birds. The lines go up: adding hormone increases the level of blood calcium. The vertical distance between lines shows the sex difference. The lines are almost exactly parallel: the male/female differences are almost exactly the same regardless of whether birds had the hormone in their diet. *Lines are parallel; differences are equal; there is no interaction*.

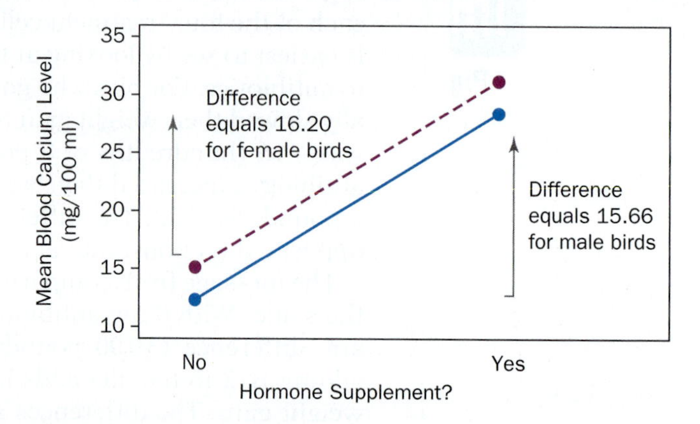

FIGURE 7.3 Lines are nearly parallel: No interaction in the bird data

BirdCa

b. **Pigs: interaction present.** Figure 7.4 shows an interaction plot for the pig data. Each line shows the effect of antibiotics: 0 mg on the left, 40 mg on the right, with response (weight gain) on the vertical axis. The lines are far from parallel. With B12 present (dashed line), giving antibiotics raises weight gain by about 32/100 of a pound per week. With no B12 (solid line), giving antibiotics actually reduces weight gain, by 16, roughly one-sixth of a pound per week. *The differences are different; the lines are not even close to parallel; interaction is present*.

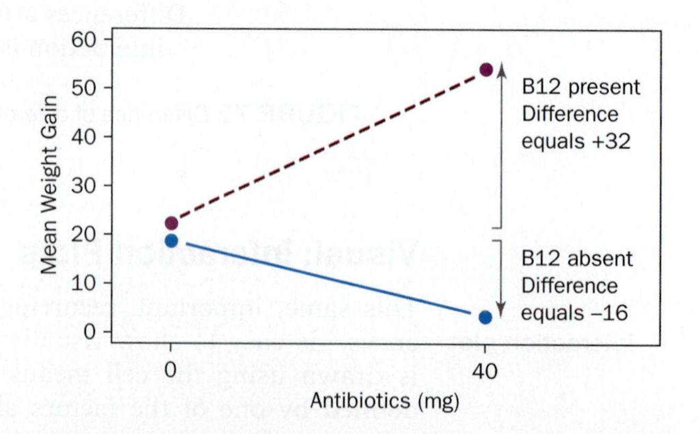

FIGURE 7.4 Lines are far from parallel: The pig data show striking interaction

INTERACTION PLOT

An interaction plot shows:

- Levels for one factor identified along the horizontal axis.
- Cell means displayed above each level according to a numeric scale on the vertical axis.
- Lines connecting the cell means for cells within the same level of the second factor.

When looking at an interaction plot:

- Lines that are roughly parallel indicate a lack of interaction.
- Lines that have very different slopes indicate interaction and show the nature of that interaction.

Having introduced the new topic of interaction, we now turn to the same five topics as in the last two chapters: design, exploration, fit, assessment, and use.

7.2 Design: The Two-way Factorial Experiment

You have already seen the five most important principles for designing an experiment: comparison (Chapter 5), replication (Chapter 5), blocking (Chapter 6), crossing (Chapter 6), and randomization (Chapter 5).

FIVE BASIC PRINCIPLES FOR EXPERIMENTAL DESIGN:

1. **Comparison**: Don't study just one condition in isolation.
2. **Replication**: The only way to measure unit-to-unit variation is to replicate: measure more than one unit for each condition.
3. **Blocking**: Isolate the effects of interest. Arrange to compare like with like.
4. **Crossing**: If you have two or more factors of interest, make sure you record data for all possible combinations of levels.
5. **Randomization**: Never accept "just what happens" when you can use random numbers.

The introduction to this chapter listed three goals: (a) to incorporate the one new idea, interaction, into what you have already seen, (b) to review and reinforce the ideas from the last two chapters, and (c) to show how these basic ideas can lead to new ways to think about data. This section addresses the first two of those goals. The first and more immediate of these two is to illustrate the two-way factorial design as a way to learn about interaction.

EXAMPLE 7.6

RivIron

River iron: A difficulty when you have just one observation per cell Keep the five design principles in mind as you review the study of iron in the rivers of upstate New York. There were two crossed factors: three sites for each of the four major rivers. The study was observational because randomization was not possible.

The main obstacle to the analysis of the data is this: There is only one observation per cell, and so it is not possible to separate interaction effects from unit-to-unit differences. *In order to measure interaction, you need to replicate, that is arrange to get more than one observation per cell.*

The Two-way Factorial Design

You have already seen an example of a two-way factorial design in Chapter 5 (see Example 5.21). We didn't label it as a two-way design, but it was. In a sense, a two-way design is just a one-way design with extra structure that comes from crossing two factors.

EXAMPLE 7.7 **The two-way factorial design: Birds and pigs**

a. **Birds: a two-way factorial design with one experimental factor.** Table 7.1 on page 301 shows the **BirdCalcium** data from the study of hormones and blood calcium, grouped to show the two-way structure: male versus female (rows) and hormone versus control (columns). There are four groups in all, with five birds per group. Notice that the design combines ideas from Chapters 5 and 6.

   Crossing: As in Chapter 6, there are two factors: *Sex* (male/female) and *Hormone* (no/yes). As in a complete block design, the factors are crossed: All possible combinations are present: male control, female control, male hormone, and female hormone.

   Replication and randomization: These are ideas from Chapter 5. There are 5 birds in each of the four conditions. Each one of the 10 male birds, and each of the 10 female birds, was randomly assigned to either the control group or the hormone group. (Note that it is not possible to randomly assign a sex to each bird.)

b. **Pigs: a two-way factorial design with both factors experimental.** Look back at the **PigFeed** data as displayed in Table 7.2 on page 301. As with the birds, the design combines ideas from Chapters 5 and 6.

   Crossing: As in Chapter 6, there are two factors, *B12* and *Antibiotics*. As in a complete block design, the factors are crossed: All possible combinations are present.

   Replication and randomization: As in Chapter 5, the design uses both. There are 3 pigs in each of the 4 conditions. Each of the 12 pigs was randomly assigned to one of the four diets.

For both of the birds and pigs examples, at least one factor was experimental, using randomization to assign treatments to units. Often you get a two-way structure from observational data.

EXAMPLE 7.8 **Dinosaurs: An unbalanced two-way observational study**[4]

Question: When, in the course of cataclysmic events, does it become necessary for evolutionary biology, archeology, geology, data science, and science fiction to come together?

Answer: When mass extinction of dinosaurs is at issue. Why did they die?

The data for this example are shown in Table 7.3 and stored in **Dinosaurs**. They were gathered in part to help resolve a controversy about the cause of a mass extinction. Was it an asteroid in the Yucatan peninsula or a giant volcano eruption in southern China?[5] According to the "Alvarez hypothesis," named for a father-son team, if an asteroid (the size of Manhattan!) had hit the earth, the impact would raise a dust cloud that would block the sun for years, killing plants that depend on photosynthesis, killing herbivores that depend on plants for food, and killing carnivores that depend on herbivores for food. Of course, a gigantic volcanic eruption would also raise a huge dust cloud, with similar consequences. Can iridium tell which one was the villain?

The goal of this study was to resolve the controversy using the patterns in iridium concentration in samples of rock far below the earth's surface. Why iridium? This metal is extremely rare in the earth's soil, but more common in certain kinds of asteroids. An asteroid impact would deposit more iridium than a volcano eruption. The settling of atmospheric dust would deposit layers that could be detected in rock samples hundreds of meters below the earth's current surface. What was the pattern of iridium concentration over time? Does the iridium data indict the asteroid or the volcano?

Table 7.3 shows the concentration of iridium (parts per billion) in samples taken in Gubbio, Italy. As in the two previous examples, there are two crossed factors of interest, this time the type of rock (limestone or shale) and the depth of the sample in meters (345, 346, 347, 348, 349, and 350).* The depth is an indicator of time: The deeper the sample, the older the deposited layer.

Setting aside the context and viewing the data in terms of its abstract structure, several features stand out:

- As noted, both factors are observational. There is no randomization.
- Unlike the last two examples, whose levels were "2 by 2," this one is "2 by 6." The column factor, *Depth*, has six levels.
- The six levels of the column factor come from a quantitative measurement, but will be viewed as categories in this example.
- The dataset is **unbalanced**, with very unequal cell counts, which range from a minimum of one (shale, 346 and 348) to a maximum of five (shale, 349).

	Depth (meters)					
	345	346	347	348	349	350
Limestone	75	120	290	170	120	5
	20	210	450	205	135	90
			620	260		105
Shale	110	315	710	400	120	145
	501		875		130	215
					135	
					10	
					290	

TABLE 7.3 Dying dinosaurs: Iridium concentration in parts per billion

*"345" means "$345 \leq$ Depth < 346."

With Examples 7.7 and 7.8 as context, here is a summary of the essential concepts of design related to the two-way factorial experiment:

factor
levels
crossed
cell
balanced
two-way factorial design
main effects
interaction effects

A **factor** is a predictor variable that can be regarded as a set of categories, or **levels**. Two or more factors are **crossed** if every possible combination of factor levels is present in the design, that is, has at least one response value. Each combination of factor levels is called a **cell**. A factorial design is **balanced** if it has the same number of observations per cell.

A **two-way factorial design** crosses two factors of interest. *In order to measure interaction, you need more than just one observation per cell.* The effects of the factors that get crossed are called **main effects**, to distinguish them from **interaction effects**.

7.3 Exploring Two-way Data

The main thing that is new here is to recognize that nothing here is really new. There are three main plots. You've seen all three before. Because the kinds of plots themselves are not new, we focus here not on how to make the plots but on what they can tell you, using a variety of datasets for each kind of plot. The examples that come next start with side-by-side dotplots (Examples 7.9, 7.10, and 7.11), look briefly at plots of log standard deviation versus log average (Example 7.12), and then look at interaction plots (Examples 7.13 to 7.15).

Side-by-Side Dotplots for Two-way Data

Just as in Chapter 5, these plots show individual observed values as points stacked in columns, with one column per group. For a one-factor design, each group is a level of the factor; for a two-way design, each group is a combination of factor levels. For either design, the plots let you estimate centers and spreads by eye. What is the pattern of the group centers? Are there outliers? Are spreads roughly equal? If not, how different are they? Is there a pattern relating center and spread? Are points in a column distributed symmetrically, or are they dense at one end, sparse at the other?

For two-way data, there are two standard versions of the dotplot, illustrated in Examples 7.9–7.11.

EXAMPLE 7.9

Side-by-side dotplots: Birds, pigs, and rats The plots of Figure 7.5 show each combination of factor levels in a column of its own, with columns in groups sorted by levels of the first factor. (For the birds, in Figure 7.5(a), the first two columns of dots are for male birds, the next two are for female birds.)

In reading these plots, you should focus your attention on two attributes: center and spread. For centers, look at the plots one at a time, and estimate the centers by eye. You may find it helpful to visualize line segments joining the centers, and try to imagine an interaction plot. Are there main effects? Is there any evidence of interaction? When comparing spreads, compare the distributions of the points within each column.

In each of the three experiments presented here, we first discuss the centers and then the spreads seen in the plots.

BirdCa

a. **Birds:** A line segment joining the centers of the two columns for the males slants upward with a steep slope. (Note: line segments are not shown on the plots, but we encourage you to draw them by eye as you look at the plots.) A segment for the female birds is similar, also upward, with a nearly equal slope. The segment for females is slightly higher, in short, a large main effect for hormone, a much smaller effect for sex, and no evidence of interaction.

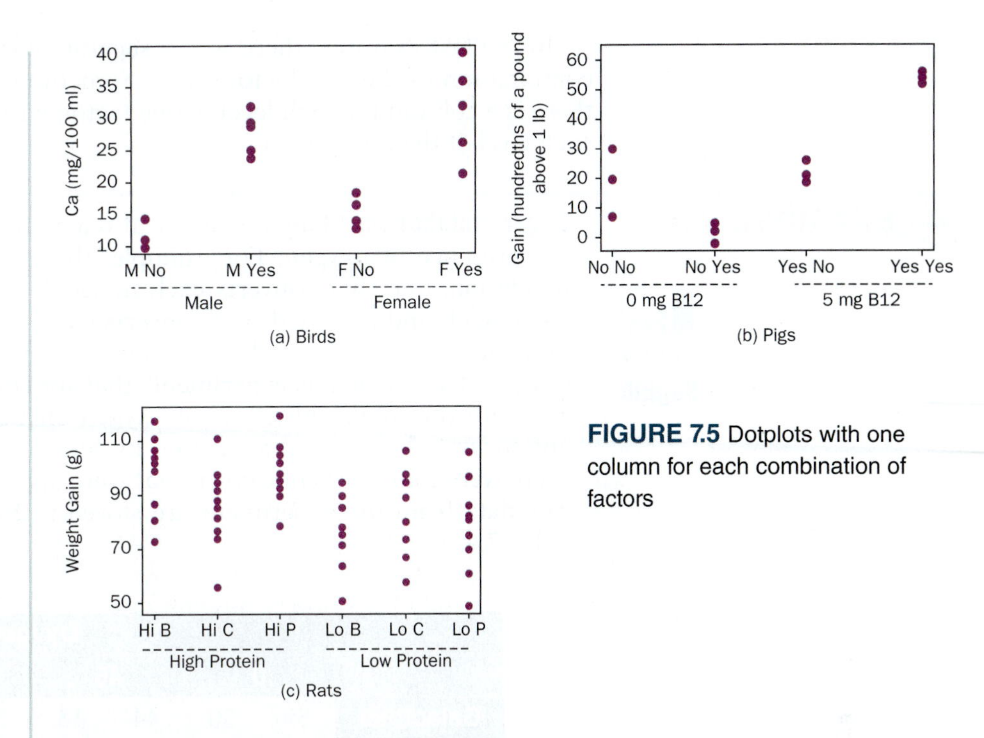

FIGURE 7.5 Dotplots with one column for each combination of factors

The standard deviations of the responses within each treatment combination are unequal, smallest for male birds with no hormone, largest for female birds with hormone present. The difference between the two spreads is quite substantial, and the spreads are related to centers, large with large, small with small. A transformation is suggested.

b. **Pigs:** A line segment joining the centers of the first two columns slants sharply down: with no B12, adding antibiotics decreases weight gain. For the two columns on the right, a line segment joining centers slopes steeply up: with B12 present, adding antibiotics increases weight gain. There is a pronounced interaction.

The spreads are also unequal, but even though the standard deviations of the gains for no, no (0 mg B12, 0 mg antibiotics) is much larger than for yes, yes (5 mg B12, 40 mg antibiotics), there is no suggestion to transform. For one thing, there are only three points per group, so estimated standard deviations are very approximate. Moreover, three of the groups have similar spreads. Finally, there is no pattern relating spread and center.

c. **Rats:** Recall the **FatRats** dataset that was introduced in Example 5.1 on page 198. The experiment included giving baby rats diets that had beef, cereal, or pork as a source of protein. In the earlier example we only considered a subset of the data consisting of 30 rats that were given a high-protein diet from the assigned source (10 rats for each source). There were also another 30 rats that got low-protein versions of the same diets. Thus we have two factors: amount of *Protein* (*Lo* or *Hi*) and *Source* of protein (*Beef*, *Cereal*, *Pork*).

For the left three columns in the rats dotplots (high protein) line segments joining the column centers form a V shape. Average weight gains for high-protein beef and high-protein pork are almost the same, and much greater than for high-protein cereal. For the three columns on the right (low protein), the segments form an inverted V. There is evidence of an interaction. Again, beef and pork are about the same, but this time the average for cereal is higher than the average for meat.

In this case, all six spreads are roughly the same. There is no need to transform.

In the last example, there was a separate column in the dotplot for each combination of the two factor levels. Sometimes patterns are easier to see if there is a column for each level of one factor, with a different print symbol for each level of the other factor.

EXAMPLE 7.10

SugEth

Sugar metabolism Many biochemical reactions are slowed or prevented by the presence of oxygen. For example, there are two simple forms of fermentation, one that converts each molecule of sugar to two molecules of lactic acid, and a second that converts each molecule of sugar to one each of lactic acid, ethanol, and carbon dioxide. The second form is inhibited by oxygen. The particular experiment[6] that we consider here was designed to compare the inhibiting effect of oxygen on the metabolism of two different sugars, glucose and galactose, by *Streptococcus* bacteria. In this case, there were four levels of oxygen that were applied to the two kinds of sugar. The data from the experiment are stored in **SugarEthanol** and appear in Table 7.4.

	Oxygen							
	0		46		92		138	
Galactose	59	30	44	18	22	23	12	13
Glucose	25	3	13	2	7	0	0	1

TABLE 7.4 Ethanol concentration at various levels of oxygen for two sugars

Figure 7.6 shows a dotplot of the *Ethanol* concentrations for each level of *Oxygen*. Solid circles show data points for glucose and open circles are for galactose.

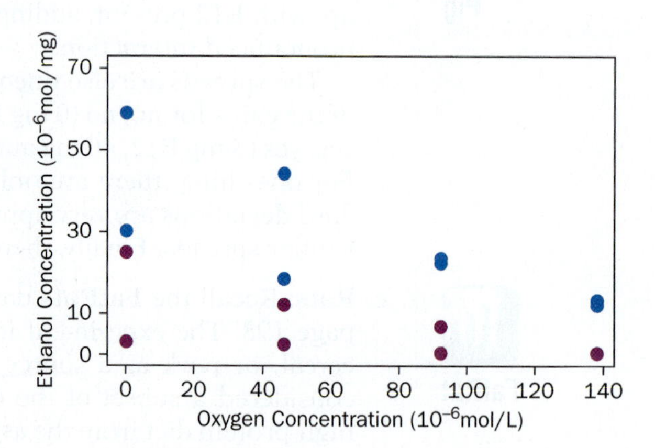

FIGURE 7.6 Dotplot for the **SugarEthanol** data

Three main patterns stand out: (1) For both sugars, the ethanol concentration drops off as oxygen concentration increases, tending to an asymptote of zero, and the rate of decrease is related to the type of sugar. (2) Spreads are very unequal, and the size of the spread is large when the response is large, small when the response is small. Both patterns suggest transforming to logs. (3) Concentrations for galactose are consistently higher than concentrations for glucose, and the downward trend is steeper for galactose.

EXAMPLE 7.11

Dino

Dinosaurs: Dotplots with two symbols Figure 7.7 shows a dotplot for the **Dinosaur** dataset.

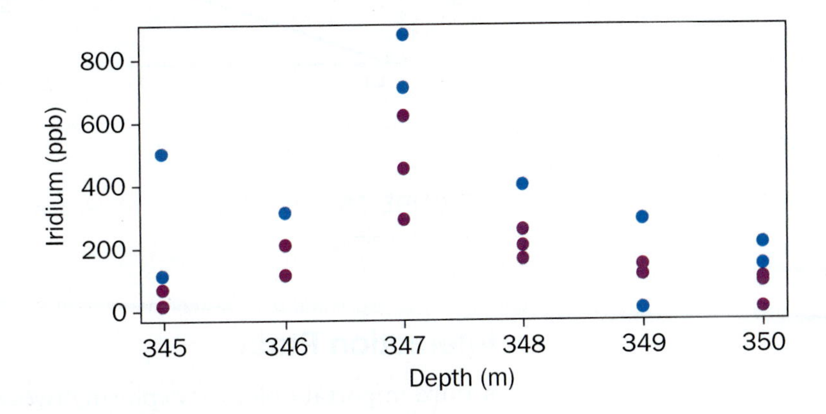

FIGURE 7.7 Dotplot for Iridium by depth and type of rock

The *Depth* factor (in meters) is on the horizontal axis. Solid circles are for limestone samples; open circles are for shale. The pattern is not as easy to see as in sugar metabolism, but here, also, we can look at the pattern of centers and the pattern of spreads. (1) Centers: Shale has higher concentrations of iridium than does limestone. For both kinds of rock, iridium concentration increases at first, reaches a peak at 347 meters, then declines. (2) Spreads: For both kinds of rock, points are clustered more densely for lower concentrations of iridium, more sparsely for larger values. This relationship, together with the fact that we are measuring concentrations, suggests transforming to logs.

For the birds, sugar, and dinosaurs, the dotplots suggest transforming "down" to roots ($p = 1/2$), logs ($p = 0$), or reciprocals ($p = -1$). For all three datasets, the original response was measured as a concentration, which often lends itself to analysis in the log scale. One way to choose the transformation is to plot log(standard deviation) versus log(average) with groups as points, fit a line, and compute "Power = 1 − slope."

Plots of Log(Standard Deviation) versus Log(Average)

EXAMPLE 7.12

BirdCa

Birds: Plots of Log(Standard Deviation) versus Log(Average) The message from the two plots of Figure 7.8 is clear. Figure 7.8(a) shows log(standard deviation) versus log(average) for calcium concentrations in the original scale (mg / 100 ml). A line fits well, and the fitted slope of 1.05 suggests transforming to log concentrations. (Power = 1 − slope = −0.05, essentially 0.) Figure 7.8(b) shows log(standard deviation) versus log(average) for the transformed data. No fitted line can come close to both of the two right-most points, suggesting that no additional transformation can do much better. The line that is shown has a fitted slope of 0.27, with power 1 − 0.27 = 0.73, close to 1: no change.

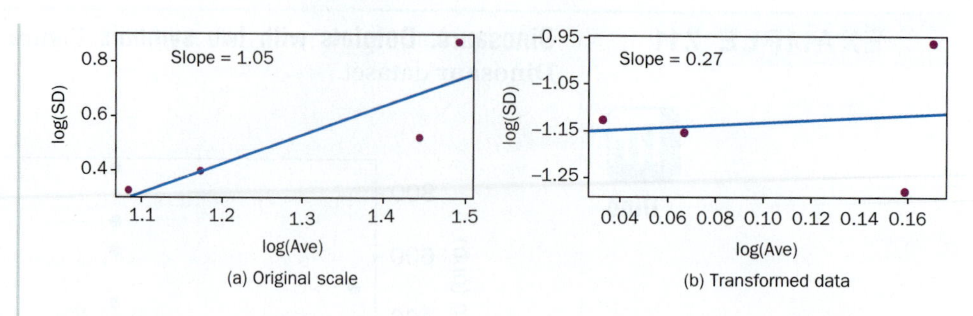

FIGURE 7.8 Plots of log(standard deviation) versus log(average) for response in two scales

Interaction Plots

A third important plot for exploring two-way data is the interaction plot.

EXAMPLE 7.13

FatRat

The two versions of the interaction plot: Rats For two-way data you have two choices for your interaction plot, as in Figure 7.9. The two plots show the same six numbers, the cell means for the 2 × 3 design. Although the six cell means are the same for both plots, the two plots look very different.

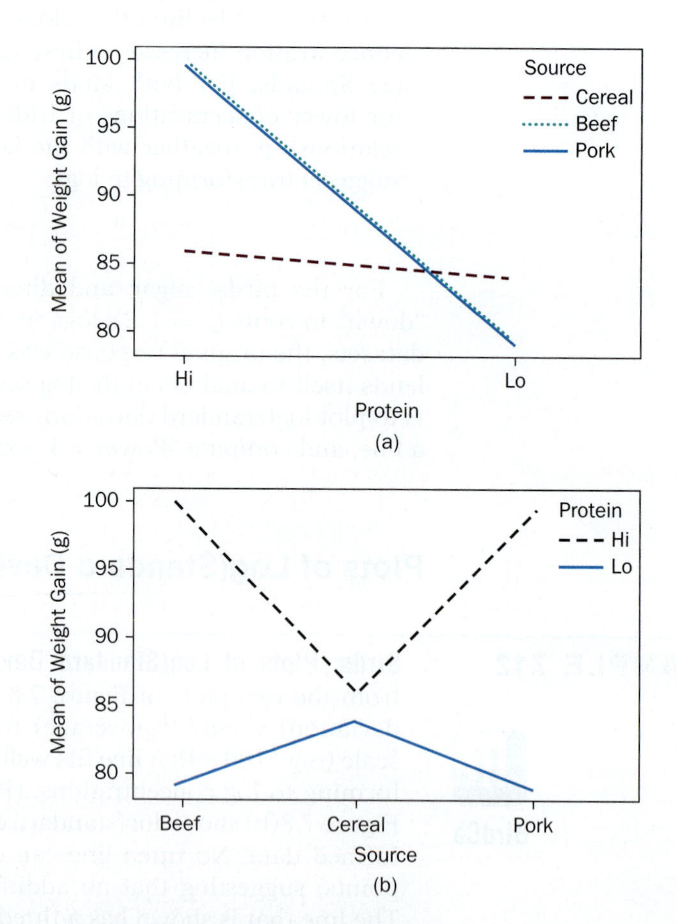

FIGURE 7.9 The two interaction plots for the rat data

In Figure 7.9(a), there is a separate line segment for each source (beef, cereal, pork). The other factor, high/low, is shown on the x-axis. The plot in figure 7.9(b) reverses these choices. The source, beef/cereal/pork, is shown on the x-axis, with a separate set of line segments for each of high and low protein.

For the rat data, Figure 7.9(a) shows three clear patterns:

- Most striking, beef and pork are essentially equal when your goal is to fatten rats. For understanding patterns in the data, we might as well lump beef and pork together in a single category, "meat."
- All three lines slope down. If you are a rat hoping to gain weight fast, high protein is better.
- The slope for meat is steep. The slope for grain is shallow. High-protein meat is a lot better than high-protein grain. Low-protein grain is better than low-protein meat.

Figure 7.9(b) shows the same set of six cell means in a different way. Visually the pattern is quite different, but the message from the data is the same.

- The two line segments for high protein lie above the two segments for low protein. If weight gain is your goal, high protein is better.
- The corners of the bow-tie shape come close to forming a perfect rectangle with sides parallel to the axes. Beef and pork look like twins, nutritionally speaking.
- The V-shape (high) and inverted V (low) show the interaction. The difference between high and low is large for meat, small for cereal. The differences are different.

For the rats, both plots are useful, and there is no logic to tell how to choose one over the other in advance. However, for two-way data with a quantitative factor, it is nearly always best to put the quantitative factor on the x-axis, as in the next example, because it is generally easier to see any patterns that might exist.

EXAMPLE 7.14

Dino

Interaction plots with a quantitative factor: Dinosaurs two ways For the dinosaurs, one factor, depth, is quantitative; the other, source, is categorical. Figure 7.10 shows the two choices for interaction plots. Notice how Figure 7.10(a) is much more useful.

With the quantitative factor on the x-axis, the key patterns are much easier to see:

- Time trend: Iridium concentrations start low at the deepest, earliest levels, spike at 347 meters, then fall off sharply.
- Source effect: Iridium concentrations are consistently greater in shale (metamorphic and compressed) than in limestone (sedimentary, less dense.)
- For data measured as concentrations, there is a suggestion of interaction. Line segments are steeper for shale than for limestone. It is worth considering changing to log concentrations.

For Figure 7.10(b), none of these patterns are easy to see.

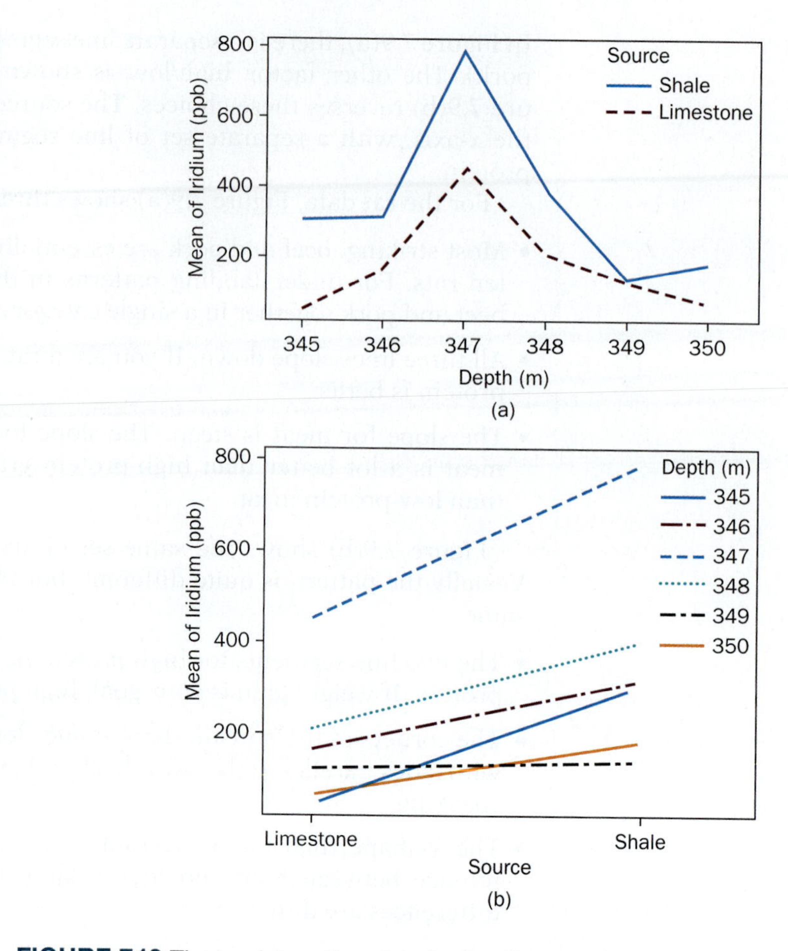

FIGURE 7.10 The two interaction plots for the dinosaur data

As the last example suggests, and the next example shows more clearly, putting a quantitative factor on the x-axis can help us think about choice of scale.

EXAMPLE 7.15

SugEth

Linear relationships with a quantitative factor: Sugar three ways Figure 7.11 shows interaction plots for the sugar data, using three possible scales for the response: concentration ($p = 1$) in Figure 7.11(a), square root of the concentration ($p = 1/2$) in Figure 7.11(b), and log concentration ($p = 0$) in Figure 7.11(c). What stands out most for these plots is that for all three scales, and for both sugars, there is a near-linear relationship between oxygen concentration and ethanol concentration. The more O_2, the slower the sugar gets converted to alcohol. (Clear proof of an anaerobic process.)

What changes most, as you go from left to right, is the pattern of the slopes.

- Figure 7.11(a), which measures the response as concentration, is useful because it shows the curved relationship, with zero as asymptote, that is, the drop in ethanol concentration slows as the oxygen concentration increases. (The "rate of change is proportional to how much is there.") Both features are characteristic of exponential decay, a recurring pattern in science.
- In Figure 7.11(b), using square roots of the concentration ($p = 1/2$ instead of $p = 1$ or $p = 0$) the linear fit is still clear, and now the slopes of the two fitted lines are almost identical, suggesting a very simple model of two parallel lines fitted to the eight values. This model requires only 3 degrees of

freedom, leaving 5 degrees of freedom for residuals. However, extrapolating from the fitted lines leads to impossible values in the form of negative concentrations.

- Figure 7.11(c), plotting log concentrations, the suggested linear fit is still clear, but now the slopes are reversed. A fitted line for glucose is steeper than a fitted line for galactose. Does a log transform overshoot?

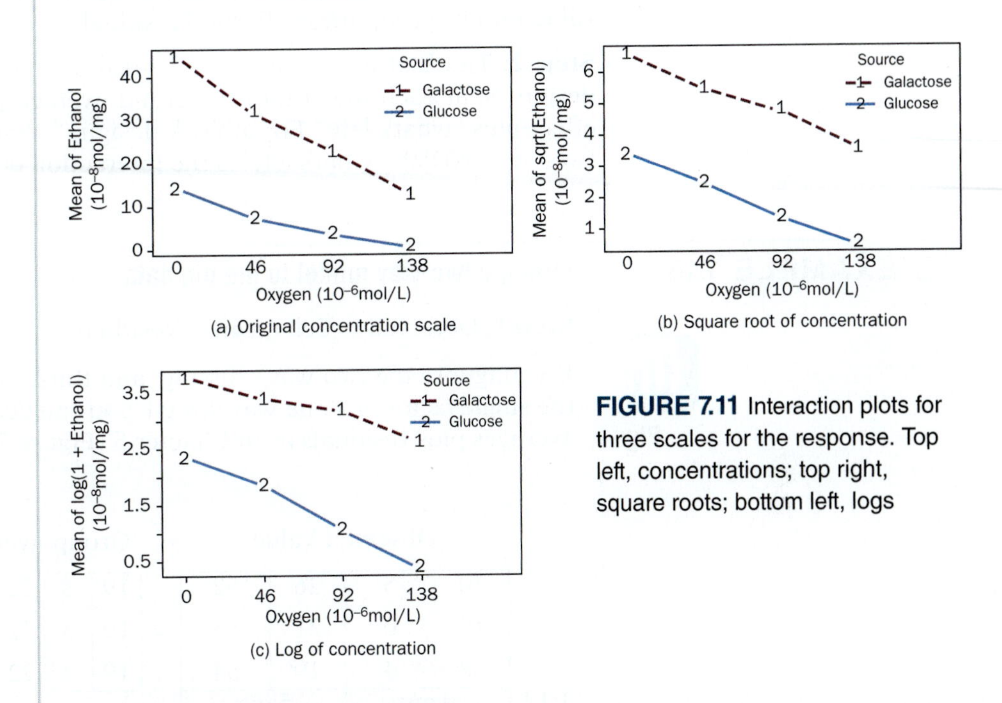

FIGURE 7.11 Interaction plots for three scales for the response. Top left, concentrations; top right, square roots; bottom left, logs

Bottom line: Even though fitting a model to log concentrations requires two slopes instead of just one, the theory of chemical reactions fits better for the log model. It is worth doing all three analyses, one for each of the scales. If your conclusions are about the same for all three, it won't matter much which scale you choose.

As usual for ANOVA, once you have chosen a design and a scale, the next step is to fit the model.

7.4 Fitting a Two-way Balanced ANOVA Model

For balanced two-way data, as for all balanced datasets, fitting the model is a simple matter of computing and subtracting averages. The reason we spend space on the process is because we want you to see the logic applied in a concrete setting. This is especially true for balanced two-way models, where fitting is just a matter of combining two steps, a first step from Chapter 5 and a second step from Chapter 6.

Even so, why bother? Why not just leave it to the computer? That's a reasonable question, one that deserves a serious answer. We have already agreed that fitting ANOVA models by hand should not be a priority. All the same, we urge you to use the example that follows to solidify your understanding of ANOVA organized as a set of questions based on meaning. ANOVA compares averages, using the size of unit-to-unit differences as its yardstick.

If your dataset is unbalanced, like the dinosaur data, you can still use ANOVA software to fit the model, but for understanding some aspects of the data, you may find it useful to adopt the regression approach as a different way to think about the data (see online Section 8.8).

Here, as in the last section, *there are no new methods. The only new idea is to put together the ideas you have already seen.* You can do this in two steps:

Step 1: Think of the data as coming from a one-way design. Split each observed value into its group mean (fit) and residual.

Step 2: Think of each cell (group) mean as coming from a complete block design. Split each mean into the grand average plus row effect plus column effect plus "what's left." For a block design, "what's left" is the residual. For a two-way ANOVA, "what's left" is the interaction effect.

EXAMPLE 7.16

Fitting a two-way model to the pig data

Step 1: Observed = Cell mean + Residual

If you ignore the two-way crossing, and think of the four groups as levels of the single factor of a one-way design, you can decompose the data into group averages plus residuals as in Chapter 5 (Figure 7.12).

Observed Value

	30	5	26	52
	19	0	21	56
	8	4	19	54
B12	0 mg		5 mg	
Anti	0 mg	40 mg	0 mg	40 mg

= Group Average

19	3	22	54
19	3	22	54
19	3	22	54

+ Residual

11	2	4	−2
0	−3	−1	2
−11	1	−3	0

FIGURE 7.12 Think of the data as a one-way design: Observed = Cell mean + Residual

Now restore the two-way structure and you get Figure 7.13.

Observed

30	5
19	0
8	4
26	52
21	56
19	54

= Cell Mean

19	3
19	3
19	3
22	54
22	54
22	54

+ Residual

11	2
0	−3
−11	1
4	−2
−1	2
−3	0

	Observed		Cell Mean		Residual
SS	11,600	=	11,310	+	290
df	12	=	4	+	8

FIGURE 7.13 Restore the two-way structure. Observed = Cell mean + Residual

This first step of the decomposition shows that differences among the cell means are large in comparison with the residuals. The next step is to decompose the cell means.

Step 2:
Cell mean = Grand average + Row effect + Column effect + Interaction

Our goal is to split the set of cell means into a sum of four overlays. To do this we first compute row averages, column averages, and the grand average as shown in Figure 7.14. To get the row and column effects, we subtract the grand average from the row and column averages. (Notice that arithmetic here is the same as for decomposing a block design with blocks as rows and treatments as columns.)

		Antibiotics			
		0 mg	40 mg	Average	Effect
B12	0 mg	19	3	11	−13.5
	5 mg	22	54	38	13.5
Average		20.5	28.5	24.5	
Effect		−4	4		

FIGURE 7.14 Averages and main effects for the pig data. First compute row, column, and grand averages. Then subtract the grand average to get row and column effects.

To get the interaction terms, we add together the grand average plus row and column effects, then subtract the result from the cell averages, as shown in Figure 7.15.

Grand Average			B12 Effect			Anti Effect			Additive Fit	
24.5	24.5	+	−13.5	−13.5	+	−4	4	=	7	15
24.5	24.5		13.5	13.5		−4	4		34	42

Cell Average			Additive Fit			Interaction	
19	3	−	7	15	=	12	−12
22	54		34	42		−12	12

FIGURE 7.15 Interaction effects for the pig data: Interaction = Cell average − (Grand average + Row effect + Column effect)

Cell Mean			Grand Average			B12 Effect			Anti Effect			Interaction	
19	3	=	24.5	24.5	+	−13.5	−13.5	+	−4	4	+	12	−12
22	54		24.5	24.5		13.5	13.5		−4	4		−12	12

FIGURE 7.16 Think of the cell means as a complete block design: Cell mean = Grand average + Row effect + Column effect + Interaction

	Cell Mean			Grand Average			B12 Effect			Anti Effect			Interaction	
	19	3		24.5	24.5		−13.5	−13.5		−4	4		12	−12
	19	3		24.5	24.5		−13.5	−13.5		−4	4		12	−12
	19	3		24.5	24.5		−13.5	−13.5		−4	4		12	−12
	22	54	=	24.5	24.5	+	13.5	13.5	+	−4	4	+	−12	12
	22	54		24.5	24.5		13.5	13.5		−4	4		−12	12
	22	54		24.5	24.5		13.5	13.5		−4	4		−12	12
SS	11,310		=	7203		+	2187		+	192			1728	
df	4		=	1		+	1		+	1		+	1	

FIGURE 7.17 Restore the repeated values: Cell mean = Grand average + Row effect + Column effect + Interaction

Combining Steps 1 and 2 gives the complete decomposition, as shown in Figures 7.16 and 7.17. We can see that

- There is a large main effect for *B12* (±13.5, *SS* = 2187).
- There is a modest main effect for *Antibiotics* (±4, *SS* = 192).
- There is an interaction effect almost as large as the *B12* main effect (±12, *SS* = 1728).

Putting the two steps together gives the full decomposition in Figure 7.18.

	Observed			Grand Average			B12 Effect			Anti Effect			Interaction			Residuals	
	30	5		24.5	24.5		−13.5	−13.5		−4	4		12	−12		11	2
	19	0		24.5	24.5		−13.5	−13.5		−4	4		12	−12		0	−3
	8	4		24.5	24.5		−13.5	−13.5		−4	4		12	−12		−11	1
	26	52	=	24.5	24.5	+	13.5	13.5	+	−4	4	+	−12	12	+	4	−2
	21	56		24.5	24.5		13.5	13.5		−4	4		−12	12		−1	2
	19	54		24.5	24.5		13.5	13.5		−4	4		−12	12		−3	0
SS	11,600		=	7203		+	2187		+	192		+	1728		+	290	
df	12		=	1		+	1		+	1		+	1		+	8	

FIGURE 7.18 The two steps combined: Response = Grand average + Row effect + Column effect + Interaction + Residual

Compare Figure 7.18 with 7.13 and 7.17, and check that the complete set of overlays in 7.18 is just a matter of combining 7.13 from Step 1 and 7.17 from Step 2. The overlays in 7.18, along with the sums of squares and degrees of freedom, show all the patterns in a single display.

The decomposition shows clearly that *B12* has an effect (±13.5) that is large compared with the typical size of the residuals. However, the interaction is almost as large (±12) as the marginal effect of *B12*. Because of the large interaction, we learn more from the cell means than from comparing row and column averages. The overall effect of antibiotics is small compared with the effects of *B12* and the interaction. These comparisons are summarized by the ANOVA table.

The boxed summary that follows shows algebraic notation for the decomposition. As in Chapters 5 and 6, we follow the usual convention and use Greek letters α, β, γ, μ, σ, etc., for unknown parameters in the model. Latin letters (x, y, df, SS, t, p, etc.) refer to observed values and summaries computed from them. We also follow a standard convention for subscripts. For two-way ANOVA, the first subscript i tells the level of the first (row) main effect α_i. The second subscript j tells the level for the second (column) main effect β_j. Each (i,j) pair refers to the cell for row effect i and column effect j, and also the corresponding interaction term γ_{ij}. The third subscript k counts units within each (i, j) cell. The notation $\sum_{i,j,k}$ indicates summing across all three subscripts (one term in the sum for each data point).

NOTATION AND FORMULAS

Meaning	Parameter	Statistic
Overall mean	μ	$\bar{y}$
Factor A mean	$\mu_{i.}$	$\bar{y}_{i.}$
Factor A effect	$\alpha_i = \mu_i - \mu$	$\bar{y}_{i.} - \bar{y}$
Factor B mean	$\mu_{.j}$	$\bar{y}_{.j}$
Factor B effect	$\beta_j = \mu_{.j} - \mu$	$\bar{y}_{.j} - \bar{y}$
Interaction effect	$\gamma_{ij} = \mu_{ij} - \mu$	$\bar{y}_{ij} - \bar{y}_{i.} - \bar{y}_{.j} + \bar{y}$
Random error, residual	ϵ_{ijk}	$y_{ijk} - \bar{y}_{ij}$
Observed response	$\mu + \alpha_i + \beta_j + \gamma_{ij} + \epsilon_{ijk}$	y_{ijk}

Decomposition

Step 1: Observed value = Cell mean + Residual

$$y_{ijk} = \bar{y}_{ij} + (y_{ijk} - \bar{y}_{ij})$$

Step 2: Cell mean = Grand average + Factor A effect + Factor B effect + Interaction

$$\bar{y}_{ij} = \bar{y} + (\bar{y}_{i.} - \bar{y}) + (\bar{y}_{.j} - \bar{y}) + (\bar{y}_{ij} - \bar{y}_{i.} - \bar{y}_{.j} + \bar{y})$$

	Sums of squares	**Degrees of freedom**
Observed values	$\sum_{i,j,k} y_{ijk}^2$	IJK = # of observed values
Grand average	$\sum_{i,j,k} \bar{y}^2$	1
Factor A	$\sum_{i,j,k} (\bar{y}_{i.} - \bar{y})^2$	$(I - 1)$ = # of A levels – 1
Factor B	$\sum_{i,j,k} (\bar{y}_{.j} - \bar{y})^2$	$(J - 1)$ = # of B levels – 1
Interaction	$\sum_{i,j,k} (\bar{y}_{.j} - \bar{y})^2$	$(I - 1)(J - 1)$
Residuals	$\sum_{i,j,k} (y_{ijk} - \bar{y}_{ij})^2$	$IJ(K - 1)$

$$MS = \frac{SS}{df} \qquad F_{Treatment} = \frac{MS_{Treatment}}{MS_{Residuals}}$$

Once you have the overlays in hand, you can summarize them, as in the last two chapters, using an ANOVA table. For now, we regard these ANOVA tables as nothing more than a formal way to summarize the decomposition of a dataset into overlays. Of course, the ultimate goal of the summaries is to use them for inference.

We return to the pigs and the birds to compare ANOVA tables for datasets with and without interactions.

EXAMPLE 7.17

Pig

ANOVA table for the pigs: A large interaction effect The F-ratios in Table 7.5 tell us that

- The average variation ($MSB12$) for the main effect of $B12$ is more than 60 times as large as the average variation from pig to pig (MSE).
- The average variation due to antibiotics is much smaller, a little more than 5 times as large as the average residual variation.
- The average variation due to interaction is large, almost 48 times the residual variation.

Source	SS	df	MS	F
B12	2187	1	2187.00	60.331
Antibiotics	192	1	192.00	5.297
Interaction	1728	1	1728.00	47.669
Residual	290	8	36.25	
Total	4397	11		

TABLE 7.5 ANOVA for the pigs

The very large *F*-ratio for interaction indicates that the effect of *B12* depends on whether or not a pig got antibiotics. We will have more to say about this when we look at confidence intervals and effect sizes in the section on inference.

No doubt you have noticed that there is no column for *P*-values in the ANOVA table. The reason is that we have not yet finished our assessment of the conditions required for formal inference. We pick up the story in the next two sections.

For the pigs, changing the scale is not an issue. For the birds, it is, so we look at two ANOVA tables.

EXAMPLE 7.18

BirdCa

Two ANOVA tables for the birds: Two scales Side-by-side dotplots and the plots of log(standard deviation) versus log(average) both suggest transforming to logs. All the same, for some purposes, raw (untransformed) concentrations might be better. Does it matter?

Table 7.6 shows ANOVA tables in both scales, for concentrations in the upper panel and log concentrations in the lower panel. The two ANOVA tables tell the same story. Regardless of scale, whether concentrations or log concentrations, there is strong evidence that hormone supplements in the diet raise calcium concentration, a suggestion that female birds may have slightly higher blood calcium levels than males, and good reason to think that the increase is the same for birds of both sexes.

Source	df	SS	MS	F
Sex	1	45.90	45.90	2.346
Hormone	1	1268.82	1268.82	64.841
Interaction	1	0.36	0.36	0.0186
Residual	16	313.09	19.57	

Source	df	SS	MS	F
Sex	1	0.021	0.021	3.325
Hormone	1	0.572	0.572	89.947
Interaction	1	0.003	0.003	0.473
Residual	16	0.102	0.0064	

TABLE 7.6 ANOVA tables for the bird calcium, concentration (top) and $\log_{10}$ concentration (bottom)

This is a good place to remind yourself about the limitations of every ANOVA table. You can take any design, make up arbitrary numbers, decompose them into overlays, and use the overlays to compute the ANOVA table.

ANOVA tables are just summaries of a bunch of numbers. In particular, whether the P-values mean anything depends on how we assess the fit of the model and how we justify using that model for inference.

The two ANOVA tables for the bird calcium data show similar results for concentration and log(concentration).

- The variation due to interaction is tiny, less than half the variation from bird to bird.
- The average difference between male and female birds is modest, two or three times as large as the average variation from bird to bird.
- The average variation due to the hormone supplement is huge, either 66 or 90 times the "error variation."

For the birds we can focus just on the main effects, because there is no evidence of interaction. Just as for the pig data, we hope to use our fitted model for formal inference in the form of hypothesis tests and confidence intervals. First, however, we need to assess the required conditions.

7.5 Assessing Fit: Do We Need a Transformation?

Because the plots in this chapter are the same as in the last two, we focus here not on the types of plots, but rather on how to use them together. We illustrate this process with the **Dinosaur** data where the conclusions are fairly consistent from plot to plot. In other examples the messages may be more mixed.

EXAMPLE 7.19

Dino

Diagnosing dinosaurs: Four plots, one message—transform to logs Figure 7.19 shows the four diagnostic plots for the dinosaur data in the original scale, iridium concentration in parts per billion.

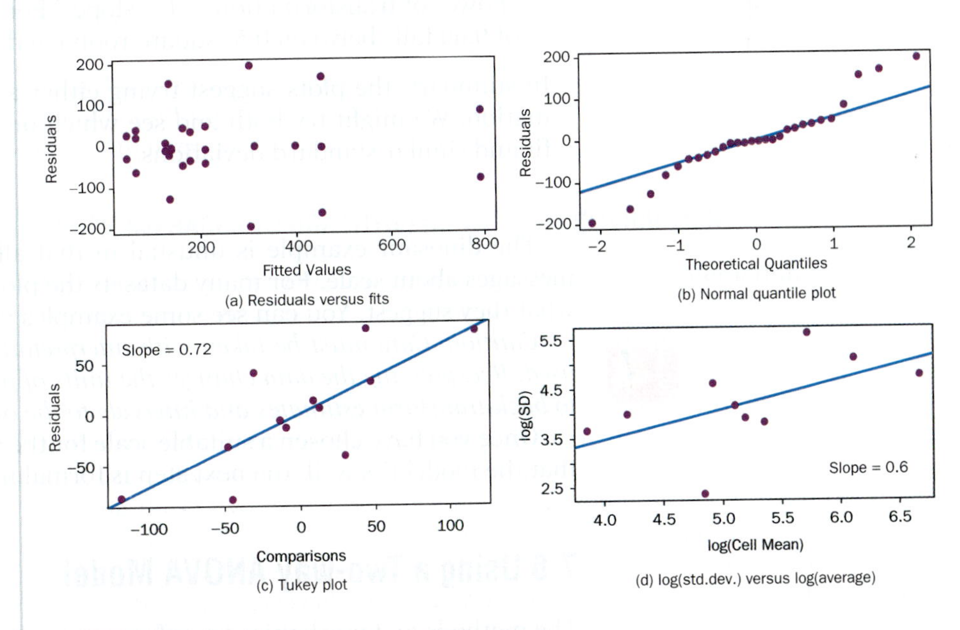

FIGURE 7.19 Diagnostic plots for the Dinosaurs

As you have seen, there is no suggestion of interaction in the bird calcium data, but a huge interaction in the pig data. This difference has important implications for tests, comparisons, confidence intervals, and effect sizes.

EXAMPLE 7.20

BirdCa

Birds: Inference when there is no interaction present When there is no interaction it makes sense to focus on inference about main effects:

- Is there evidence of a male/female difference? (H_0: $\mu_{Male} = \mu_{Female}$)
- Is there evidence of an effect due to the hormone supplement? (H_0: $\mu_{Hormone} = \mu_{Control}$)

What can the data tell us about the sizes of the mean differences, ($\mu_{Male} - \mu_{Female}$) and ($\mu_{Hormone} - \mu_{Control}$)? We answer these questions two ways, with confidence intervals and effect sizes.

As suggested earlier in this chapter, we begin by transforming the data to logs. Since there is no interaction, we focus on the two sets of "marginal" averages, to compare male versus female, and hormone versus control.

Scope and scale. The model fits. It has to fit, because the data were created to follow the design. For the male birds, and separately for the female birds, the supplement (hormone or control) was assigned using random numbers. This randomization justifies the use of probability-based methods for inference. The units (birds) are distinct entities, so it is reasonable to regard the observed values as independent.

The birds were not chosen at random from a larger population. To generalize about male/female differences to a larger population of birds depends on how, whether, and to what extent the 20 birds in the sample represent the larger population you care about.

We show here the results for log concentrations, leaving the parallel results for concentrations as an exercise.

Test. Regardless of which scale you look at, concentration or log concentration, there is not even a hint of interaction. The absence of interaction means that it makes sense to ask separately about main effects (marginal differences).

- If you ignore (i.e., average over) the male/female differences, what is the average effect of the hormone supplement?
- If you ignore (average over) whether hormone is present, what is the average male/female difference?

Here as always, *P*-values can tempt the user to stop with a misleadingly simple yes or no. Confidence intervals and effect sizes give important ways to learn more.

Table 7.7 repeats the bottom ANOVA table (for log(concentration)) from Table 7.6, this time with the *P*-values included. We use these *P*-values to test three different null hypotheses.

Source	df	SS	MS	F	p
Sex	1	0.021	0.021	3.325	0.0870
Hormone	1	0.572	0.572	89.947	<0.0001
Interaction	1	0.003	0.003	0.473	0.5016
Residual	16	0.102	0.0064		

TABLE 7.7 ANOVA tables for the bird calcium, using $\log_{10}$ concentration

- Interaction.
  H_0 : There are no interaction effects, or in symbols $H_0 : \gamma_{ij} = 0$ for all (i, j) pairs

  The P-value of 0.50 is very large.

- Sex differences.
  H_0: There are no male/female differences, or equivalently
  H_0: The mean calcium concentrations for male and female birds are equal

  In symbols, we use $H_0 : \mu_{Male} = \mu_{Female}$ or $H_0 : \alpha_i = 0$ for all i.

  The P-value of 0.087 is not significant at the level of 0.05, but it would be a major mistake just to say "not significant" and stop with that. We should also look at a confidence interval and effect size.

- Hormones.
  H_0: There is no effect due to the hormone supplement, or equivalently
  H_0: The means for supplement and no supplement are equal

  In symbols, we use $H_0 : \mu_{Hormone} = \mu_{Control}$ or $H_0 : \beta_j = 0$ for all j.

  The P-value is unambiguous, less than 0.0001.

Intervals. For marginal comparisons, we compute differences of row averages (sex, male/female) or differences of column averages (hormone, yes/no). For the birds, to get such comparisons, we compare averages of two groups (two row averages or two column averages) with 10 birds per average. For confidence intervals, we need the standard deviation of the residuals, the standard error of the differences, and the 95% margin of error.

- Standard deviation (SD). As always for regression and ANOVA, the standard deviation of the errors—the typical size of unit-to-unit differences—is estimated by the square root of the mean square for error.

$$SD = \sqrt{MSE} = \sqrt{0.0064} = 0.08$$

- Standard error (SE). The sample sizes for the two groups are 10 and 10.

$$SE = SD\sqrt{\frac{1}{n_1} + \frac{1}{n_2}} = (0.08)\sqrt{\frac{1}{10} + \frac{1}{10}} = 0.036$$

- Margin of error (ME). The 95% t-value with 16 df for error is 2.120. The 95% margin of error is

$$ME = t \cdot SE = (2.120)(0.036) = 0.076$$

The intervals: difference ± margin of error

- Female versus male: $(1.325 - 1.260) \pm 0.076$, or -0.011 to 0.141
- Hormone versus control: $(1.462 - 1.123) \pm 0.076$, or 0.263 to 0.415

In words: We can be 95% confident that the average male/female difference in log concentration lies somewhere between -0.011 and 0.141.* At one extreme (-0.011) there is essentially no effect; at the other extreme (0.141) there is a moderate effect.

*More formally, the interval $(-0.011, 0.141)$ was constructed using a method that captures the parameter value with probability 0.95.

antibiotics, adding B12 might lower the average weight gain by as much as 0.08 pound per week, or raise the gain by as much as 0.14 pound (again with 95% confidence).

Effect sizes. Standard deviations and differences are the same as above. The conditional effect size = difference/typical error.

- Antibiotics (yes versus no) with B12 present: (54 – 22)/6.02 = 5.32 or 532%
- Antibiotics (yes versus no) with B12 absent: (3 – 19)/6.02 = –2.66 or –266%
- B12 (yes versus no) with antibiotics present: (54 – 3)/6.02 = 8.47 or 847%
- B12 (yes versus no) with antibiotics absent: (22 – 19)/6.02 = 0.498 or 49.8%

The four effect sizes make it clear: there is a huge benefit to adding the combination of antibiotics together with B12. The conditional effect sizes are 532% (adding antibiotics when B12 is present) and 847% (adding B12 when antibiotics are present). Giving antibiotics alone cuts the weekly weight gain by 266%, almost three times the typical size of pig-to-pig variation. Giving B12 alone has a much smaller effect, roughly half the size of pig-to-pig variation.

Summary: If you are a pig farmer in Iowa, what you should know is that your pigs need both B12 and antibiotics. If you are a scientist studying the physiology of piglets, you have a solid basis for a new grant proposal: What is the numerical relationship between weight gain and the quantities of B12 and antibiotics in the diet?

There are many ways a model can be more complicated than it needs to be. The two examples that follow show just two ways.

EXAMPLE 7.22

FatRat

Rats: Two categories that are essentially the same As with the pigs of the previous example (7.21), there is no need to transform, and there is a very pronounced interaction. Inference should focus on conditional effects. For this study, one of the factors has three levels (beef, cereal, pork), but the differences between beef and pork are so tiny that it makes sense to combine them and compare meat (beef and pork together) versus cereal.

Scope and scale. Here again, the model must fit.* The random assignment of diets to rats justifies probability-based inference about cause and effect. It is less clear whether the rats represent a larger population of all baby rats, but most likely they do. As to scale, there is no evidence against using the original scale, weekly weight gain.

Tests. A pair of ANOVA tables tells an interesting story, one that illustrates some of the limitations of significance tests. Table 7.9 shows the two ANOVAs. For the first one (top panel), the variable "Source" has three categories, beef, cereal, and pork. Protein quality is very highly significant ($p = 0.0034$), and the interaction between quality and source is not quite significant ($p = 0.075$). For the second ANOVA (bottom panel), the categories "beef" and "pork" have been combined into a single category, "meat."

**A big advantage of ANOVA compared with regression is that you get to choose the model first, then obtain the data to fit the model. The model has to fit. A big disadvantage of ANOVA compared with regression is that observational data rarely let you choose the model in advance. Regression offers more options for learning from "found" data.

As before, protein quality is significant ($p = 0.0026$), but for this analysis, the interaction is also significant ($p = 0.021$, as opposed to 0.075 for the first analysis). Why the difference in the *P*-values? What are we to make of the apparent contradiction?

	Df	Sum Sq	Mean Sq	*F*-value	Pr(>F)
Quality (high/low)	1	3,139	3,139.3	14.656	0.00034
Source (beef/cereal/pork)	2	272	135.8	0.634	0.53
Interaction	2	1,166	582.9	2.721	0.075
Residuals	54	11,567	214.2		

	Df	Sum Sq	Mean Sq	*F*-value	Pr(>F)
Quality (high/low)	1	3,139	3,139.3	15.196	0.00026
Source (meat/cereal)	1	270	270.0	1.307	0.26
Interaction	1	1,166	1,165.6	5.643	0.021
Residuals	56	11,568	206.6		

TABLE 7.9 Two ANOVAs for the rat data. (The bottom ANOVA combines beef and pork into a single category, meat. Note the difference in the *P*-value for interaction.)

Compare the two ANOVA tables, with an eye to what is the same and what has changed. Look first at sums of squares. Those for quality and interaction are unchanged, and the other two barely differ. Now look at the column for degrees of freedom. Degrees of freedom for source and interaction have changed from 2 down to 1, because we changed the number of categories for protein source from 3 down to 2. The consequence of these reductions in degrees of freedom shows up in the column for mean squares. Notice that the means squares for protein quality and for residuals are either exactly or almost the same, but the mean squares for protein source and interaction are roughly twice as large as in the first ANOVA. This doubling is due, of course, to reducing the corresponding degrees of freedom from 2 to 1.

The pattern in the changes offers a deeper message. For the first ANOVA, the analysis created what we can now see as an artifact, an arbitrary division of what should be a single category, meat, into two categories that ought to be regarded as one. Such divisions dilute the effects of the categories and blur the focus on what matters. (Imagine how the analysis would change if the scientists had chosen to have separate categories for Iowa pork, North Carolina pork, Texas beef, and Kansas beef, Nebraska wheat, and North Dakota wheat. They would have six categories and 5 degrees of freedom for protein source, and 5 degrees of freedom for interaction. Unless state-to-state differences mattered, the sums of squares would be roughly the same as for the actual study, but the mean squares would be much, much smaller because of the increase in degrees of freedom.)

To assure yourself that for the purpose of the analysis, the difference between beef and pork does not matter, compare the two sets of cell means in Table 7.10. There is hardly any difference between beef and pork averages, whether high protein or low.

	Source		
Quality	**Cereal**	**Beef**	**Pork**
High	85.90	100.00	99.50
Low	83.90	79.20	78.90

	Source	
Quality	**Cereal**	**Meat**
High	85.90	99.75
Low	83.90	79.05

TABLE 7.10 Two sets of cell means for the rat data. Note how the means for beef and pork are nearly identical

The second ANOVA shows strong evidence of an interaction between the two main effects ($p = 0.021$). Because of this interaction, it makes little sense to look at marginal main effects, that is, the overall average effect of protein

include a product of *Oxygen* and the *Sugar* indicator to allow for different slopes—although the lack of interaction would suggest that the latter term is not needed. See Exercise 8.63.

Bottom line: The chemists turn out to be right, of course. Oxygen concentration does have an effect on sugar metabolism. ANOVA fails to show this only because we have represented a quantitative variable (O_2 concentration) as a set of four (unordered) categories. Models matter. Data analysis is not automatic. It takes a brain, and it takes experience.

If you look back over this section, we hope you will see that the formulas and computations are the same every time, that the logic is the same every time, but that the ways the formulas and logic go together will vary from one study to the next, based partly on what you want to learn from your data, and mostly on what your data have to tell you.

CHAPTER SUMMARY

As promised, we end with a summary example that follows the pig example from beginning to end.

EXAMPLE 7.24

Pig

Pigs one last time Just as for the final example of the last chapter, we follow an analysis through a sequence of steps: design and data; exploration and scale; fitting the model and summarizing the fit; assessing the fit; scope of inference, results, and conclusions.

Design and Data

The research focus of this study was a particular interaction, between a B12 supplement (present or absent) and antibiotics (present or absent). Both factors of interest, *B12* and *Antibiotics*, were experimental. Crossing the two factors gave four diets, which were randomly assigned to 12 baby pigs, one diet per pig, 3 pigs per diet. (See Table 7.2, which is repeated here.)

	Antibiotics	
B12	0 mg	40 mg
	30	5
0 mg	19	0
	8	4
	26	52
5 mg	21	56
	19	54

Exploration and Scale

As expected, the data show a large interaction (Figure 7.20). A quartet of diagnostic plots (see Figure 7.21) confirms that there is no need to transform the data, and no transformation that will give a satisfactory fit to an additive model with no interaction term.

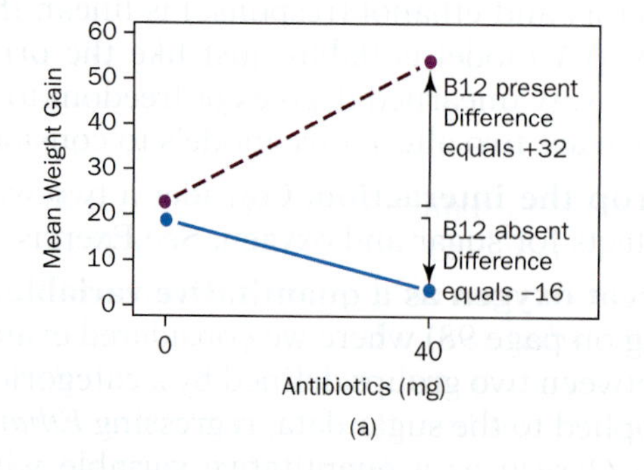

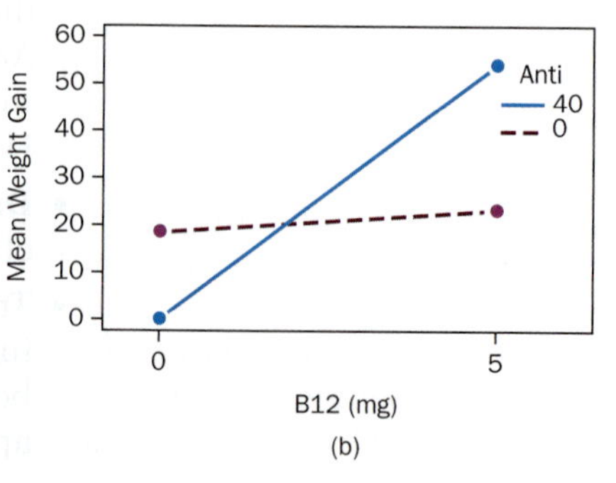

FIGURE 7.20 Interaction plots for the pig data

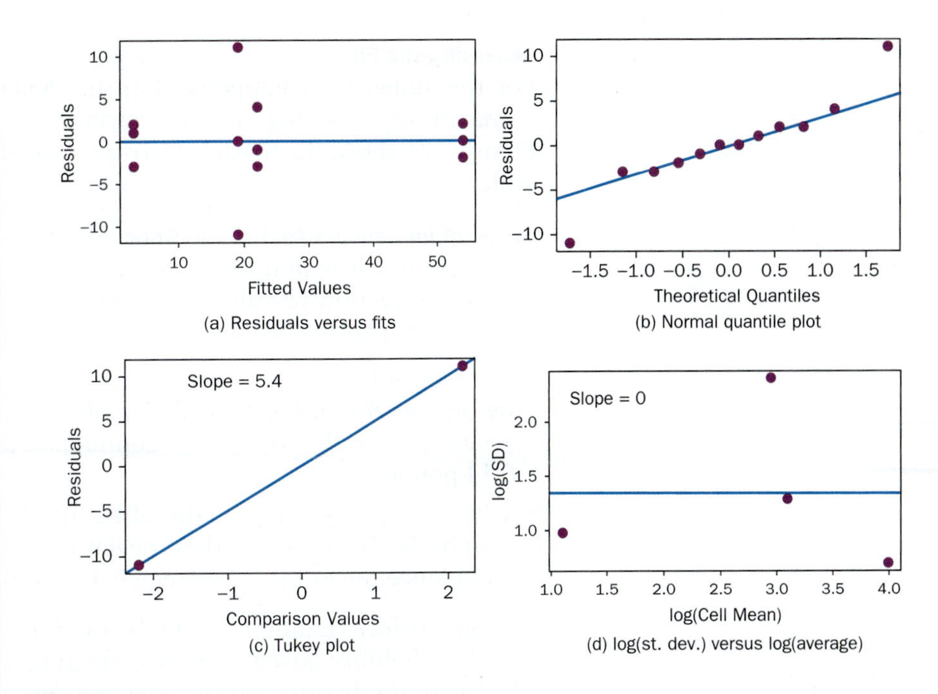

FIGURE 7.21 Diagnostic plots for the pig data

Fitting the Model and Summarizing the Fit

Figure 7.22 shows the triple decomposition of the pig data.

	Observed			Grand Average			B12 Effect			Anti Effect			Interaction			Residuals	
	30	5		24.5	24.5		−13.5	−13.5		−4	4		12	−12		11	2
	19	0		24.5	24.5		−13.5	−13.5		−4	4		12	−12		0	−3
	8	4		24.5	24.5		−13.5	−13.5		−4	4		12	−12		−11	1
	26	52	=	24.5	24.5	+	13.5	13.5	+	−4	4	+	−12	12	+	4	−2
	21	56		24.5	24.5		13.5	13.5		−4	4		−12	12		−1	2
	19	54		24.5	24.5		13.5	13.5		−4	4		−12	12		−3	0
SS	11,600		=	7203		+	2187		+	192		+	1728		+	290	
df	12		=	1		+	1		+	1		+	1		+	8	

FIGURE 7.22 Decomposition of the pig data

The decomposition shows that the interaction effects (±12) are almost as large as the marginal *B12* effects (±13.5), and that these two sets of effects are large compared with the overall effect of the antibiotics, and the residual. These comparisons are summarized by the ANOVA table.

Source	SS	df	MS	F	p
B12	2187	1	2187.00	60.331	<0.0001
Antibiotics	192	1	192.00	5.297	0.0504
Interaction	1728	1	1728.00	47.669	0.0001
Residual	290	8	36.25		
Total	4397	11			

Although we have included *P*-values as part of the ANOVA table, it would be a mistake to rush ahead to formal inference before first assessing the fit of the model.

Assessing the Fit

For this dataset, as happens often for ANOVA, assessing the fit is mainly a matter of choosing an appropriate scale. As the diagnostic plots of Figure 7.21 show, the model offers a reasonable fit to the data in the original scale.

Scope of Inference, Results, and Conclusions

The main result is that the evidence strongly supports the research hypothesis of interaction (*P*-value < 0.0001). The effect of B12 thus depends on whether the pigs are fed antibiotics.

- Without antibiotics in the diet, the effect of B12 is to raise weight gain by only 49.8% of the typical size of the pig-to-pig variation (roughly 0.06 pound per week). With 95% confidence the effect is between −0.08 and 0.14 pound per week.
- With antibiotics present, the effect of B12 is to raise the weekly weight gain by 847% of the standard deviation of 0.06 pound per week. With 95% confidence the effect is between 0.40 and 0.62 pound per week.

As noted before (Example 7.21) the random assignment of diets justifies the use of probability-based inference about cause and effect. Because the units (the pigs) are distinct entities, we can reasonably regard the observed values as independent. Although the pigs were not chosen by random sampling from a particular population, it is reasonable to think the results generalize to similar pigs.

EXERCISES

Conceptual Exercises

7.1 Degrees of freedom. If you carry out a two-factor ANOVA (with interaction) on a dataset with Factor A at four levels and Factor B at five levels, with three observations per cell, how many degrees of freedom will there be for:

a. Factor A?

b. Factor B?

c. interaction?

d. error?

7.2 More degrees of freedom. If you carry out a two-factor ANOVA (with interaction) on a dataset with Factor A at three levels and Factor B at three levels, with three observations per cell, how many degrees of freedom will there be for:

a. Factor A?

b. Factor B?

c. interaction?

d. error?

7.3 Interaction plot: when no interaction. If there is no interaction present,

a. the lines on the interaction plot will be nonparallel.

b. the lines on the interaction plot will be approximately the same.

c. the lines on the interaction plot will be approximately parallel.

d. it won't be obvious on the interaction plot.

7.4 Interaction plot: when interaction. If there is an interaction present,

a. the lines on the interaction plot will be nonparallel.

b. the lines on the interaction plot will all have positive slope.

c. the lines on the interaction plot will be approximately parallel.

d. it won't be obvious on the interaction plot.

7.5 Fill in the blank. If you have two-way data with one observation per cell, the only model you can fit is a main effects model and there is no way to tell whether interaction is present. If, in fact, there is interaction present, your SSE will be too ____________ (small, large) and you will be ____________ (more, less) likely to detect real differences due to each of the factors.

7.6 Main effects when interaction is present. Suppose you have two-way data with Factor A at four levels and Factor B at three levels, with two observations per cell.

If there is interaction present, then how will the main effect for A compare with the main effect of B? Will it be smaller, larger, or can we not tell?

7.7 Heart and soul. Is interaction present in the following data? How can you tell?

	Heart	Soul
Democrats	2, 3	10, 12
Republicans	8, 4	11, 8

7.8 Blood, sweat, and tears. Is interaction present in the following data? How can you tell?

	Blood	Sweat	Tears
Males	5, 10	10, 20	15, 15
Females	15, 20	20, 30	25, 25

Exercises 7.9–7.10 are True or False exercises. If the statement is false, explain why it is false.

7.9 If interaction is present, it is not possible to describe the effects of a factor using just one set of estimated main effects.

7.10 The conditions for the errors of the two-way ANOVA model with interaction are the same as the conditions for the errors of the one-way ANOVA model.

7.11 Difference of differences. Here is a table of means from an experiment. Does the "difference of differences" indicate that there is an interaction present?

		Factor A	
		No	Yes
Factor B	No	23	33
	Yes	18	28

7.12 More difference of differences. Here is a table of means from an experiment. Does the "difference of differences" indicate that there is an interaction present?

		Factor A	
		No	Yes
Factor B	No	35	47
	Yes	23	27

7.13 ANOVA with a log transformation: what changes? Consider two ANOVA tables for data analyzed in the original scale and the log scale. You may assume that all of the original response values are greater than one.

a. *Sum of squares.* Which of the following is true?

i. All SSs will be larger for the transformed data.

ii. All SSs will be smaller for the transformed data.

iii. All SSs will remain the same as for the original data.

iv. Some SSs will be smaller for the transformed data, and the others will either remain the same or become larger.

v. No way to tell without doing both ANOVAs.

b. *Degrees of freedom.* Which of the following is true?

i. All dfs will be larger for the transformed data.

ii. All dfs will be smaller for the transformed data.

iii. All dfs will remain the same as for the original data.

iv. Some dfs will be smaller for the transformed data, and the others will either remain the same or become larger.

v. No way to tell without doing both ANOVAs.

c. *Mean squares.* Which of the following is true?

i. All MSs will be larger for the transformed data.

ii. All MSs will be smaller for the transformed data.

iii. All MSs will remain the same as for the original data.

iv. Some MSs will be smaller for the transformed data, and the others will either remain the same or become larger.

v. No way to tell without doing both ANOVAs.

d. F-*ratios.* Which of the following is true?

i. All F-ratios will be larger for the transformed data.

ii. All F-ratios will be smaller for the transformed data.

iii. All F-ratios will remain the same as for the original data.

iv. Some F-ratios will be smaller for the transformed data, and the others will either remain the same or become larger.

v. No way to tell without doing both ANOVAs.

e. P-*values.* Which of the following is true?

i. All P-values will be larger for the transformed data.

ii. All P-values will be smaller for the transformed data.

iii. All P-values will remain the same as for the original data.

iv. Some P-values will be smaller for the transformed data, and the others will either remain the same or become larger.

v. No way to tell without doing both ANOVAs.

7.14 Hypothetical decomposition #1. Table 7.12 show a decomposition for a hypothetical dataset using a two-way ANOVA model with interaction. Without explicitly computing the ANOVA table and doing as little arithmetic as feasible, put the means squares (MSA, MSB, MSAB, and MSE) in order from smallest to largest. The sum of squared residuals (SSE) is 60.

Table 7.12 Decomposition #1 for Exercise 7.14

Grand Average			Factor A			Factor B			Interaction AB			Residuals	
30	30		10	10		5	−5		4	−4		2	−3
30	30		10	10		5	−5		4	−4		−2	3
30	30	+	−10	−10	+	5	−5	+	−4	4	+	4	−1
30	30		−10	−10		5	−5		−4	4		−4	1

7.15 Hypothetical decomposition #2. Table 7.12 shows a decomposition for a hypothetical dataset using a two-way ANOVA model with interaction. Without explicitly computing the ANOVA table and doing as little arithmetic as feasible, put the means squares (MSA, MSB, MSAB, and MSE) in order from smallest to largest. The sum of squared residuals (SSE) is 180.

Table 7.13 Decomposition #2 for Exercise 7.15

Grand Average		Factor A		Factor B		Interaction AB		Residuals
30 30		4 4		2 −2		5 −5		8 −3
30 30		4 4		2 −2		5 −5		−8 3
30 30	+	−4 −4	+	2 −2	+	−5 5	+	4 −1
30 30		−4 −4		2 −2		−5 5		−4 1

7.16 Hypothetical decomposition #3. Table 7.14 shows a decomposition for a hypothetical dataset using a two-way ANOVA model with interaction. Without explicitly computing the ANOVA table and doing as little arithmetic as feasible, put the means squares (MSA, MSB, MSAB, and MSE) in order from smallest to largest. The sum of squared residuals (SSE) is 60.

Table 7.14 Decomposition #3 for Exercise 7.16

Grand Average		Factor A		Factor B		Interaction AB		Residuals
30 30		2 2		3 −3		1 −1		2 −3
30 30		2 2		3 −3		1 −1		−2 3
30 30	+	−2 −2	+	3 −3	+	−1 1	+	4 −1
30 30		−2 −2		3 −3		−1 1		−4 1

7.17 Comparing hypothetical decompositions. Tables 7.12–7.14 of the previous three exercises show two-way ANOVA decompositions for three hypothetical datasets. For this exercise you should compare the three datasets and try to determine the relative order for the *F*-ratios of each of the three terms (main effect for A, main effect for B, and interaction) between the three datasets. For example, looking at Factor A which of the hypothetical decompositions would give the smallest *F*-ratio for the main effect for A, which would be in the middle, and which would give the largest *F*-ratio for Factor A. Repeat for Factor B and again for the $A \times B$ interaction. Try to make these decisions with minimal arithmetic computations. Note that the respective residual sums are $SSE_1 = 60, SSE_2 = 180$, and $SSE_3 = 60$.

7.18 Five basic principles. A garden was divided into three parts: Eastern section, Middle section, Western section. In each section four radish seedlings were planted in a 2 × 2 grid. Based on a coin toss (with separate coin tosses for each of the three sections of the garden) either the top two seedlings were given fertilizer while the bottom two were not, or vice versa. Likewise, based on a coin toss (with separate coin tosses for each of the three sections of the garden) either the left two seedlings were given extra water each day while the right two were not, or vice versa. After several weeks of growth, the heights of all of the seedlings were measured and appropriate averages were compared.

Which of the five basic principles for experimental design were met? Which were not met?

7.19 Five basic principles, again. Forty tomato plants were randomly divided into four groups of 10 each. Random numbers from 1 to 4 were drawn to decide which group would get fertilizer and extra water, which group would get fertilizer but no extra water, which group would get extra water but no fertilizer, and which group would get no fertilizer and no extra water. After several weeks of growth, the weights of all of the tomato plants were measured and appropriate averages were compared.

Which of the five basic principles for experimental design were met? Which were not met?

Guided Exercises

7.20 Reading dotplots. A scientist measured concentration of adenosine triphosphate (ATP) in the roots of two tree species under each of two conditions. The dotplot below shows mean ATP for each of the four combinations of condition and species. What does the plot say about the data?

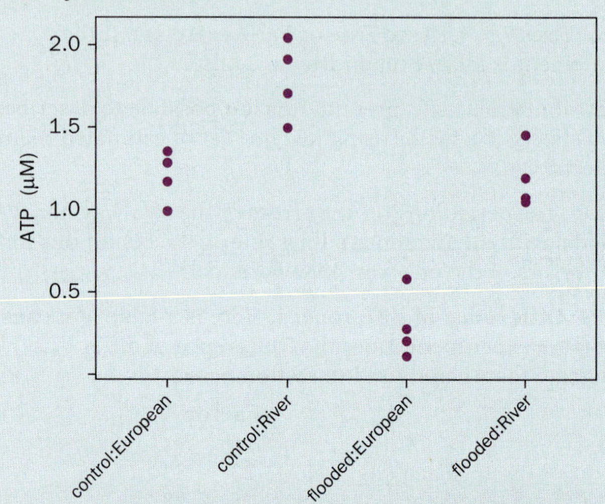

7.21 Reading dotplots, again. Researchers studied iron retention in people who drank fruit drinks that were fortified with zinc (Zn), or with iron (Fe), or both, or neither. The following dotplot shows mean ATP for each of the four combinations of condition and species. What does the plot say about the data?

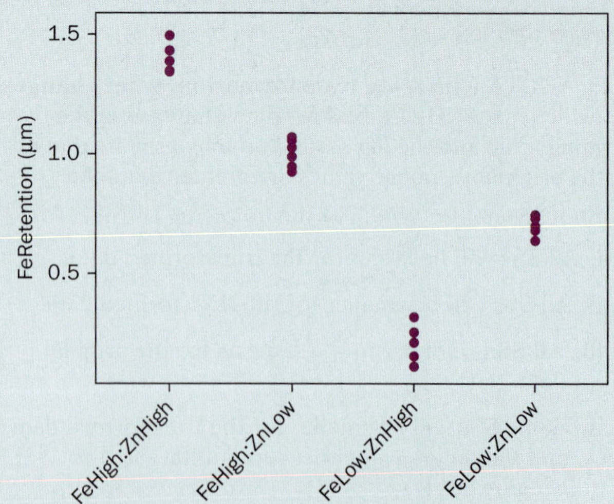

7.22 Reading interaction plots. A scientist measured concentration of adenosine triphosphate (ATP) in the roots of two tree species under each of two conditions.

Does the interaction plot that follows indicate that the conditions and the species interact in their effects on ATP concentration?

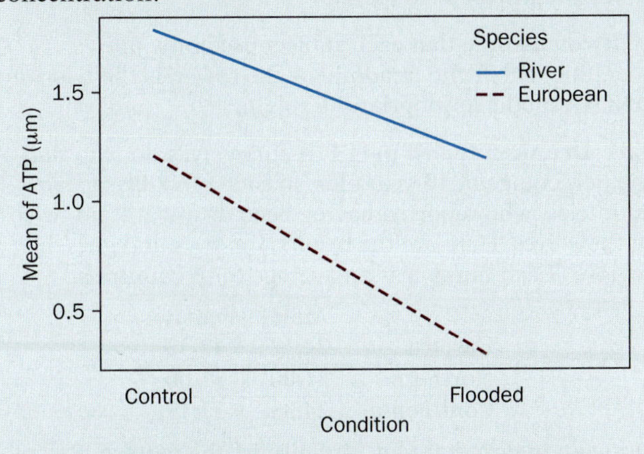

7.23 Reading interaction plots, again. Researchers studied iron retention in people who drank fruit drinks that were fortified with zinc (Zn), or with iron (Fe), or both, or neither. Does the interaction plot that follows indicate that the zinc and iron fortification interact in their effects on iron retention (FeRetention)?

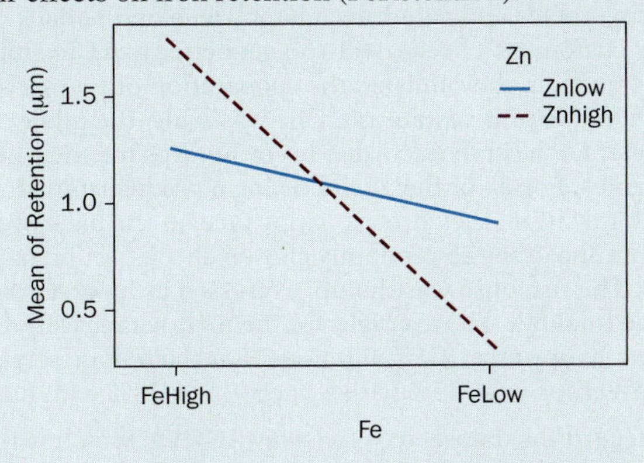

7.24 Reading diagnostic plots. A scientist measured concentration of adenosine triphosphate (ATP) in the roots of two tree species under each of two conditions. The plot that follows shows log(standard deviation) versus log(average) for the groups. Does the plot indicate that a transformation is needed? (Hint: The slope of the line is 0.014.)

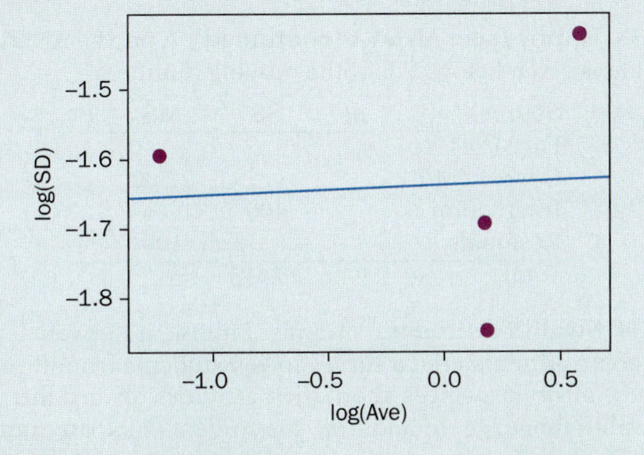

7.25 Reading diagnostic plots, again. Researchers studied iron retention in people who drank fruit drinks that were fortified with zinc (Zn), or with iron (Fe), or both, or neither. The plot that follows shows log(standard deviation) versus log(average) for the groups. Does the plot indicate that a transformation is needed? (Hint: The slope of the line is −0.099.)

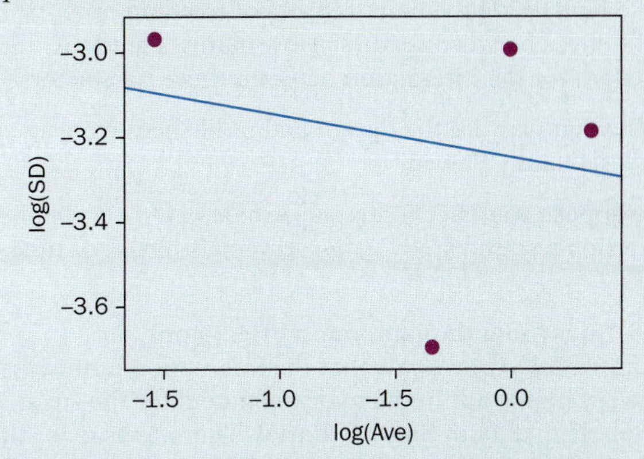

7.26 Burning calories. If you really work at it, how long does it take to burn 200 calories on an exercise machine? Does it matter whether you use a treadmill or a rowing machine? An article[7] in *Medicine and Science in Sports and Exercise* reported average times to burn 200 calories for men and for women using a treadmill and a rowing machine for heavy exercise. Suppose you have 20 subjects, 10 male and 10 female.

a. Tell how to run the experiment with each subject as a unit. How many factors are there? For each factor, tell whether it is experimental or observational.

b. Now tell how to run the experiment with each subject as a block of two time slots. How many factors are there this time?

c. Your design in (b) has units of two sizes, one size for each factor of interest. Subjects are the larger units. Each subject is a block of smaller units, the time slots. Which factor of interest goes with the larger units? Which with the smaller units?*

7.27 Happy face, sad design. Researchers at Temple University[8] wanted to know whether drawing a happy face on the back of restaurant customers' checks would lead to higher tips. They enlisted the cooperation of 2 servers at a Philadelphia restaurant, 1 male server, 1 female. Each server recorded tips for 50 consecutive tables. For 25 of the 50, following a predetermined randomization, they drew a happy face on the back of the check. The other 25 randomly chosen checks got no happy face. The response was the tip, expressed as a percentage of the total bill. See Exercise 7.32 on page 338 for the results of the study.

a. Although the researchers who reported this study analyzed it as a two-way completely randomized design, with the *Sex* of server (Male/Female) and *Happy Face*

*Statisticians often call the design in (b) a split plot/repeated measures design. Psychologists call it a mixed design, "mixed" because there is both a between-subjects factor and a within-subjects factor.

(Yes/No) as crossed factors, and with tables serving as units, that analysis is deeply flawed. Explain how to tell from the description that the design is not completely randomized and why the two-way analysis is wrong. (*Hint*: Is a server a unit or a block? Both? Neither?)

b. How many degrees of freedom are there for Male/Female? How many degrees of freedom for differences between servers? How many degrees of freedom for the interaction between those two factors?

c. In what way are the design and published two-way analysis fatally flawed?

d. Suppose you have 6 servers: 3 male, 3 female. Tell how to design a sound study using servers as blocks of time slots.

7.28 Sugar metabolism: main effects only In Example 7.21 (b) we saw that the interaction term was not very important in the two-way model for the sugar metabolism data in **SugarEthanol**. This was true for the original Ethanol concentration scale and when we considered either a log or square root transformation (see Table 7.11 on page 331). Try a two-way ANOVA model *without* the interaction term (i.e., main effects only). Don't worry about doing any transformations. How do the results for this model compare to what we saw in the previous example with interaction? **SugEth**

7.29 Noise and ADHD. Exercise 6.4 on page 293 described a study to test the hypothesis that children with attention deficit and hyperactivity disorder tend to be particularly distracted by background noise.[9] The subjects were all second-graders, some of whom had been diagnosed as hyperactive, and others who served as controls. All were given sets of math problems under two conditions, high noise and low noise. The response was their score. (Results showed an interaction effect: The controls did better with the higher noise level; the opposite was true for the hyperactive children.)

a. Describe how the study could have been done without blocks.

b. If the study had used your design in (a), would it still be possible to detect an interaction effect?

c. Explain why the interaction is easier to detect using the design with blocks.

7.30 Burning calories. If you really work at it, how long does it take to burn 200 calories on an exercise machine? Refer back to Exercise 7.26 for a study in which the researchers reported average times to burn 200 calories for men and for women using a treadmill and a rowing machine for heavy exercise.
The results:

Average Minutes to Burn 200 Calories

	Men	Women
Treadmill	12	17
Rowing machine	14	16

a. Draw both interaction graphs.

b. If you assume that each person in the study used both machines, in a random order, is the two-way ANOVA model appropriate? Explain.

c. If you assume that each subject used only one machine, either the treadmill or the rower, is the two-way ANOVA model appropriate? Explain.

7.31 Drunken teens: part 1. A survey was done to find the percentage of 15-year-olds, in each of 18 European countries, who reported having been drunk at least twice in their lives. Here are the results, for boys and girls, by region. (Each number is an average for 6 countries.)

	Male	Female
Eastern	24.17	42.33
Northern	51.00	51.00
Continental	24.33	33.17

Draw an interaction plot, and discuss the pattern. Relate the pattern to the context. (Don't just say "The lines are parallel, so there is no interaction," or "The lines are not parallel, so interaction is present.")

7.32 Happy face: interaction. Researchers at Temple University[10] wanted to know the following: If you work waiting tables and you draw a happy face on the back of your customers' checks, will you get better tips? To study this question, they enlisted the cooperation of two servers at a Philadelphia restaurant. One was male, the other female. Each server recorded his or her tips for their next 50 tables. For 25 of the 50, following a predetermined randomization, they drew a happy face on the back of the check. The other 25 randomly chosen checks got no happy face. The response was the tip, expressed as a percentage of the total bill. The averages for the male server were 18% with a happy face, 21% with none. For the female server, the averages were 33% with a happy face, 28% with none.

a. Regard the dataset as a two-way ANOVA, which is the way it was analyzed in the article. Name the two factors of interest, tell whether each is observational or experimental, and identify the number of levels.

b. Draw an interaction graph. Is there evidence of interaction? Describe the pattern in words, using the fact that an interaction, if present, is a difference of differences.

7.33 Happy face: ANOVA (continued). A partial ANOVA table is given below. Fill in the missing numbers.

Source	*df*	SS	MS	*F*
Face (Yes/No)				
Gender (M/F)			2,500	
Interaction		400		
Residuals			100	
Total		25,415		

7.34 Swahili attitudes.[11] Hamisi Babusa, a Kenyan scholar, administered a survey to 480 students from Pwani and Nairobi provinces about their attitudes toward the Swahili language. In addition, the students took an exam on Swahili. From each province, the students were from 6 schools (3 girls' schools and 3 boys' schools), with 40 students sampled at each school, so half of the students from each province were males and the other half females. The survey instrument contained 40 statements about attitudes toward Swahili and students rated their level of agreement on each. Of these questions, 30 were "positive" questions and the remaining 10 were "negative" questions. On an individual question, the most positive response would be assigned a value of 5, while the most negative response would be assigned a value of 1. By summing (adding) the responses to each question, we can find an overall *Attitude Score* for each student. The highest possible score would be 200 (an individual who gave the most positive possible response to every question). The lowest possible score would be 40 (an individual who gave the most negative response to every question). The data are stored in **Swahili**. Swahili

a. Investigate these data using *Province* (Nairobi or Pwani) and *Sex* to see if attitudes toward Swahili are related to either factor or an interaction between them. For any effects that are significant, give an interpretation that explains the direction of the effect(s) in the context of this data situation.

b. Do the normality and equal variance conditions look reasonable for the model you chose in (a)? Produce a graph (or graphs) and summary statistics to justify your answers.

c. The **Swahili** data also contain a variable coding the school for each student. There are 12 schools in all (labeled A, B, ... , L). Despite the fact that we have an equal sample size from each school, explain why an analysis using *School* and *Province* as factors in a two-way ANOVA would *not* be a balanced complete factorial design.

7.35 Sugar metabolism: log(SD) versus log(average). The following plot shows a scatterplot of the log(SD) versus log(average) for the **SugarEthanol** data from Example 7.10 on page 310. SugEth

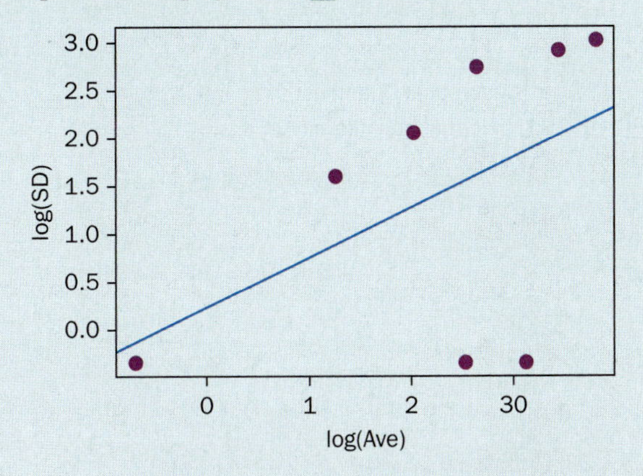

a. Fit a line by eye to all eight points, estimate the slope, and compute $p = 1 - \text{slope}$. What transformation is suggested?

b. Let's see what happens if the two points are ignored. The following plot shows that the remaining six points lie very close to a line (see the figure below). Estimate its slope and compute $p = 1 - \text{slope}$. What transformation is suggested?

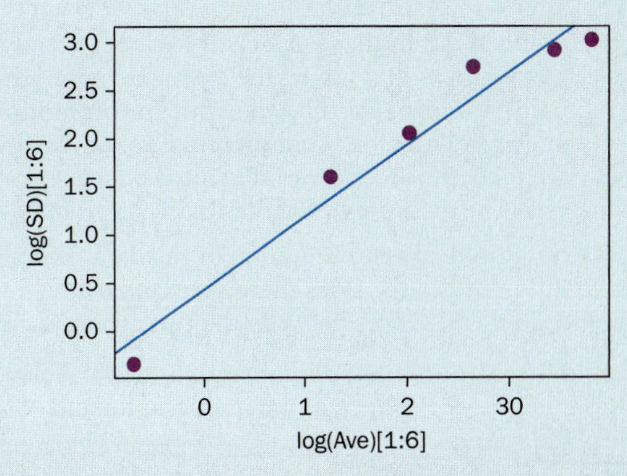

7.36 Sugar metabolism: residuals versus comparison plot. Figure 7.23 shows the plot of residuals versus comparison values for the **SugarEthanol** data from Example 7.10 on page 310. Fit a line by eye and estimate its slope. What transformation is suggested? SufEth

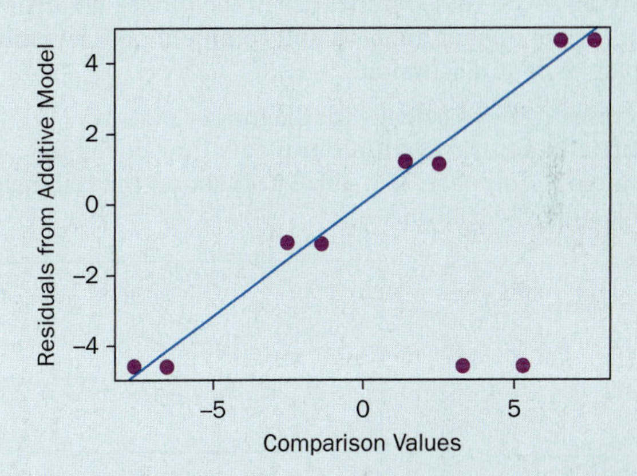

FIGURE 7.23 Graph of the residuals versus the comparison values for the sugar metabolism study

7.37 Oil. In Chapter 5, Exercises 5.53 and 5.55, we considered two different explanatory variables to see whether ultrasound had an effect on the amount of oil deapsorbed from sand. The variables we considered were the amount of oil in the sand and the amount of time the sand was exposed to ultrasound. We now use those variables together to model the difference in oil deapsorbed between treated samples (exposed to ultrasound) and control samples (not exposed to ultrasound). OilD

a. Create a set of dotplots to investigate whether these two explanatory variables might have an effect on the amount of deapsorbtion. What do the plots reveal? Do you think that an ANOVA analysis will show some significance?

b. Create an interaction plot. Does it appear that we should include an interaction term in the model? Explain.

c. Check the conditions for a two-way ANOVA with interaction model. Are they met? If so, run the model and discuss the results. If not, explain why not.

7.38 Discrimination. In Exercises 5.39 and 5.47 you analyzed promotion exam data for firefighters in New

Haven, Connecticut. In those exercises we focused only on the race of the individual taking the exam. However, there were actually two forms of the exam: one for promotion to Lieutenant, and one for promotion to Captain. In this exercise we take this second variable into consideration in our analysis. The data are in the file **Ricci**. Ricci

a. Create a graphical representation of the data to explore the two different exams. Does there seem to be a difference in mean scores between the two groups?

b. Could there be an interaction between the variables of position and race? Produce an interaction plot and discuss the results.

c. Check the conditions for running a two-way ANOVA model with interaction. Discuss whether the model is appropriate to fit to this data.

d. Discuss your findings. If the two-way ANOVA model with interaction is appropriate, give the results of this analysis. If you find significance in the model, do any appropriate further analysis and comment on the results as part of your discussion.

Exercises 7.39–7.41 deal with diagnostic plots in Figure 7.24 for the **BirdCalcium** data introduced in Example 7.2 on page 301. Table 7.15 shows the cell means and standard deviations.

Table 7.15 Cell means and standard deviations for **BirdCalcium** data

	Hormone?			
	No		Yes	
	Mean	(SD)	Mean	(SD)
Male	12.12	(2.12)	27.78	(3.34)
Female	14.88	(2.49)	31.08	(7.51)

7.39 Bird calcium: residual versus fitted values plot. The following questions deal with the residual versus fitted values plot, Figure 7.24(a), for the **BirdCalcium** data. BirdCa

a. The column of five points at the far right of the plot all come from the same cell. Which cell is that?

b. What feature of the residual versus fitted values plot most clearly indicates that a transformation might be needed for the data?

c. What information from Table 7.15 gives numerical support for the issue identified in (b)?

d. Which transformation of the response values would be more likely to address the issue identified in (b): square or square root?

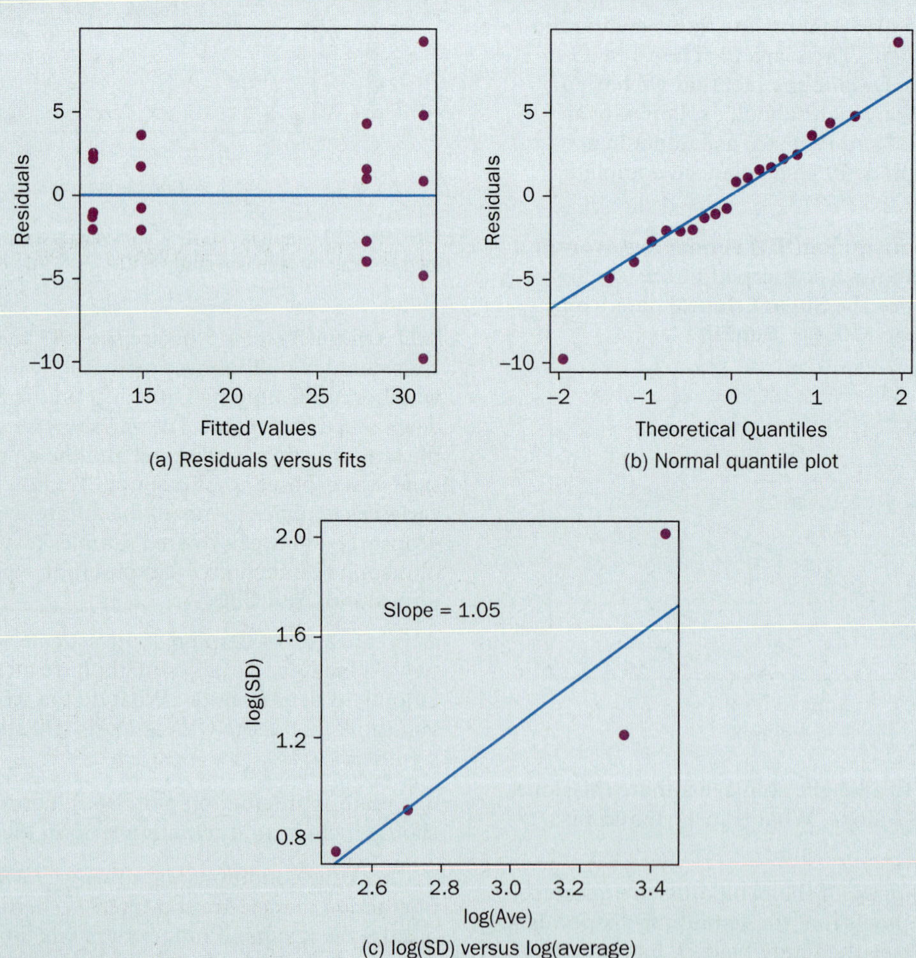

FIGURE 7.24 Diagnostic plots for the **BirdCalcium** data

7.40 Bird calcium: normal quantile plot. The following questions deal with the normal quantile plot, Figure 7.24(b), for the **BirdCalcium** data. **BirdCa**

a. All but two points appear to fall close to the line of the normal quantile plot, but there are single points at both ends that deviate a bit from the pattern. Those two points both come from the same cell of the data. Which cell is that?

b. Does the normal quantile plot indicate a strong need for a transformation to address a clear problem with normality of the residuals?

7.41 Bird calcium: log(SD) versus log(average) plot. The following questions deal with the log(SD) versus log(average), Figure 7.24(c), for the **BirdCalcium** data. **BirdCa**

a. One point on the plot falls almost exactly on the fitted line. Which cell corresponds to that point?

b. What transformation (if any) would be suggested by this plot?

c. *Mystery challenge (optional)*. The plot shown in Figure 7.24(c) looks a lot like Figure 7.8(a) on page 312, but the scales are different. Can you guess the reason for the difference?

7.42 Fat rats: conditional inference. We found evidence for an interaction in the **FatRats** data after combining beef and pork into a single "meat" group; see the second ANOVA table in Table 7.9 on page 329. Later in that example we showed inferences for the effect of protein source (meat, cereal) conditional on protein quality (low, high). There is second set of conditional inferences, for the effect of protein quality conditional on protein source. Parts (a) and (b) ask you to find and interpret these conditional intervals and effect sizes. The following summary numbers might be useful in doing so.

For *meat* the cell means are 85.90 (high quality) and 83.90 (low quality) with sample sizes of 20 in each group. For *cereal* the cell means are 99.75 (high) and 79.05 (low) with sample sizes of 10 in each group. The mean square error from the ANOVA table is 206.6 with 56 degrees of freedom. **FatRat**

a. Construct and interpret the two confidence intervals for the effect of protein quality conditional on protein source.

b. Construct and interpret effect sizes for protein quality conditional on the protein source.

Open-ended Exercises

7.43 Fruit flies. Researchers were interested in male fruit flies and set up the following experiment. They randomly assigned virgin male fruit flies to one of two treatments: live alone or live in an environment where they can sense one other male fly (*Alone*). Then the flies were randomly allocated to either have mating opportunities with female flies or to not have such opportunities (Mated). Two response variables were measured: *Lifespan* (hours after the 11th day of life), and *Activity* (number of times the fly triggered a movement sensor in the 12th day of life). The data are in the file **FruitFlies2**. The size of the thorax was also measured and reported in the variable *Size*. Analyze the response variable *Lifespan*. Consider both of the treatment variables as part of your analysis. Write a complete report as your response. **FFiles2**

7.44 Fruit flies (continued). Refer back to Exercise 7.43. For this exercise, analyze the response variable *Activity*. Consider both of the treatment variables as part of your analysis. Write a complete report as your response. **FFiles2**

7.45 Oil (continued). In Exercise 7.37 we considered two explanatory variables (length of time exposed to ultrasound, and amount of oil in the sample) to predict the difference in oil deapsorbed between treated samples (exposed to ultrasound) and control samples (not exposed to ultrasound). There was actually a third potential explanatory variable, designed to mimic seawater and freshwater. Some samples were exposed to salt water, while others were exposed to distilled water. We now have three potential explanatory variables: Ultra, Oil, Salt. Use what you have learned about ANOVA in this unit to fully analyze this data and write a report on your findings. **OilD**

7.46 Discrimination (continued). In Exercises 5.39, 5.47, and 7.38 we examined the combined scores on promotion tests taken by firefighters in New Haven, Connecticut. For this exercise analyze the oral portion of the exam. Your solution should be written as a report to the head of the fire department. **Ricci**

7.47 Fat rats: diagnostic plots. Perform an analysis along the lines of Example 7.19 for **Dinosaurs**, Example 7.24 for **PigFeed**, and Exercises 7.39–7.41 for **BirdCalcium**, using the **FatRats** data to investigate what transformation (if any) might be helpful. **FatRat**

Monkey Business Images/Shutterstock

CHAPTER 8

Additional Topics in Analysis of Variance

In this chapter you will learn to:

- Apply Levene's Test for Homogeneity of Variances.
- Develop methods for performing multiple inference procedures.
- Apply inference procedures for comparing special combinations of two or more groups.
- Apply nonparametric versions of the two-sample *t*-test and analysis of variance when normality conditions do not hold.
- Use randomization methods to perform inference in ANOVA settings.
- Use ANOVA models for designs that have repeated measures.
- Use regression models with indicator variables to fit and assess ANOVA models.
- Apply analysis of covariance when there are both categorical and quantitative predictors and the quantitative predictor is a nuisance variable.

TOPICS

In Chapters 5, 6, and 7, we introduced you to what we consider to be the basics of analysis of variance. Those chapters contain the topics we think you must understand to be equipped to follow someone else's analysis and to be able to adequately perform an ANOVA on your own dataset. This chapter takes a more in-depth look at the subject of ANOVA and introduces you to ideas that, while not strictly necessary for a beginning analysis, will substantially strengthen an analysis of data.

We prefer to think of the sections of this chapter as topics rather than subdivisions of a single theme. Each topic stands alone and the topics can be read in any order. Because of this, the exercises at the end of the chapter have been organized by topic.

TOPIC 8.1 Levene's Test for Homogeneity of Variances

When we introduced the ANOVA procedure in Chapter 5, we discussed the conditions that are required in order for the model to be appropriate, and how to check them. In particular, we discussed the fact that all error terms need to come from a distribution with the same variance. That is, if we group the error terms by the levels of the factor, all groups of errors need to be from distributions that have a common variance. In Chapter 5, we gave you two ways to check this: making a residual plot and comparing the ratio $max(s_i)/min(s_j)$ to 2. Neither of these guidelines is very satisfying. The value 2 is just a "rule of thumb" and the usefulness of this rule depends on the sample sizes (when the sample sizes are equal, the value 2 can be replaced by 3 or more). Moreover, the residual plot also carries a lot of subjectivity with it: No two groups will have exactly the same spread. So how much difference in spread is tolerable and how much is too much?

In this section, we introduce you to another way to check this condition: **Levene's Test for Homogeneity of Variances**. You may be wondering why we didn't introduce you to this test in Chapter 5 when we first discussed checking conditions. The answer is simple: Levene's test is a form of an ANOVA itself, which means that we couldn't introduce it until we had finished introducing the ANOVA model.

Levene's test

Levene's test is designed to test the hypotheses

$H_0 : \sigma_1^2 = \sigma_2^2 = \sigma_3^2 = \cdots = \sigma_I^2$ (no differences in variances)
H_a : Not all variances are equal

Notice that for our purposes, what we are hoping to find is that the data are consistent *with H_0, rather than the usual hope of rejecting H_0.* This means we hope to find a fairly large *P*-value rather than a small one.

Levene's test first divides the data points into groups based on the level of the factor they are associated with. Then the median value of each group is computed. Finally, the absolute deviation between each point and the median of its group is calculated. In other words, we calculate $|\text{observed} - \text{Median}_k|$ for each point using the median of the group that the observation belongs to. This gives us a set of estimated error measurements based on the grouping variable. These absolute deviations are not what we typically call residuals (observed – predicted), but they are similar in that they give an idea of the amount of variability within each group. Now we want to see if the average absolute deviation is the same for all groups or if at least one group differs from the others. This leads us to employ an ANOVA model.

LEVENE'S TEST FOR HOMOGENEITY OF VARIANCES

Levene's Test for Homogeneity of Variances tests the null hypothesis of equality of variances by computing the absolute deviations between the observations and their group medians, and then applying an ANOVA model to those absolute deviations. If the null hypothesis is *not* rejected, then the condition of equality of variances required for the original ANOVA analysis can be considered met.

The procedure, as described previously, sounds somewhat complicated if we need to compute every step, but most software will perform this test for you from the original data, without you having to make the intermediate calculations. We now illustrate this test with several datasets.

EXAMPLE 8.1

FFlies

Checking equality of variances in the fruit fly data The data for this example are in the file **FruitFlies**. Recall from Chapter 5 that we applied both of our previous methods for checking the equality of the variances in the various groups of fruit flies. Figure 8.1 is the dotplot of the residuals for this model. We observed in Chapter 5 that the spreads seem to be rather similar in this graph.

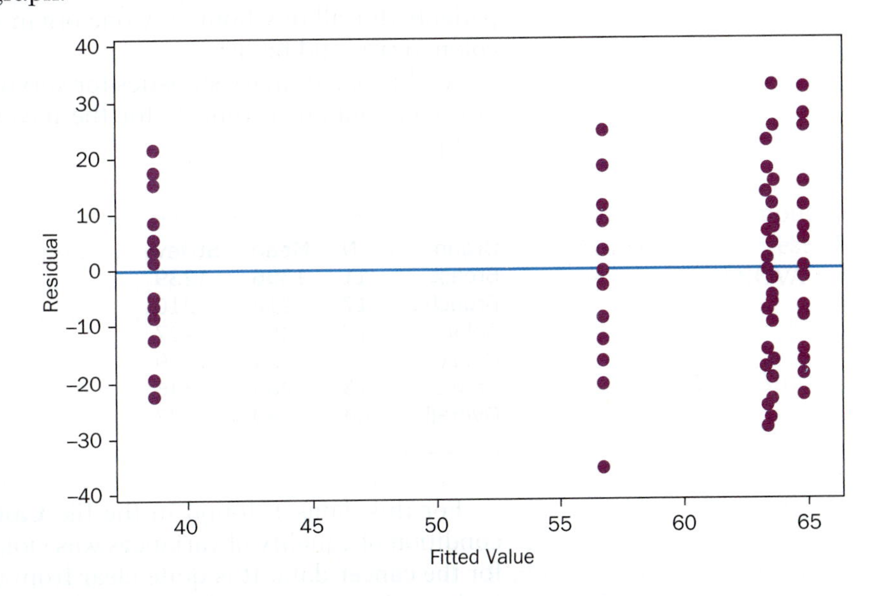

FIGURE 8.1 Residuals versus fitted values for fruit flies

We also computed the ratio of the extreme standard deviations to be $16.45/12.10 = 1.36$, which we observed was closer to 1 than to 2 and so seemed to be perfectly acceptable.

Now we apply Levene's test to these data. Computer output is given below:

```
Levene's Test
Test statistic = 0.49, p-value = 0.742
```

TOPIC 8.2 Multiple Tests

What happens to our analysis when we have concluded that there is at least one difference among the groups in our data? One question to ask would be: Which groups are different from which other groups? But this may involve doing many comparisons. We begin by summarizing what we presented in Section 5.7, and then we expand on the options for dealing with this situation.

Why Worry about Multiple Tests?

Remember that when we compute a 95% confidence interval, we use a method that, if used many times, will result in a confidence interval that contains the appropriate parameter in 95% of cases. If we only compute one confidence interval, we feel fairly good about our interval as an estimate for the parameter because we know that most of the time such intervals contain the parameter.

But what happens if we compute lots of such intervals? Although we may feel good about any one interval individually, we have to recognize that among those many intervals there is likely to be at least one interval that doesn't contain its parameter. And the more intervals we compute, the more likely it is that we will have at least one interval that doesn't capture its intended parameter.

Familywise Error Rate

What we alluded to in the previous discussion is the difference between individual error rates and family-wise error rates. If we compute 95% confidence intervals, then the individual error rate is 5% for each of them, but the family-wise error rate (the likelihood that at least one interval among the group does not contain its parameter) increases as the number of intervals increases.

There are quite a few approaches to dealing with multiple comparisons. Each one deals with the two different kinds of error rates in different ways. In Chapter 5, we introduced you to one such method: Fisher's LSD. Here, we review Fisher's LSD and introduce you to two more methods: the Bonferroni adjustment and Tukey's HSD. We also discuss the relative merits of all three methods.

We start by noting that all three methods produce confidence intervals of the form

$$\bar{y}_i - \bar{y}_j \pm \text{margin of error}$$

What differs from method to method is the definition of the margin of error. In fact, for all three methods, the margin of error will be

$$cv\sqrt{MSE\left(\frac{1}{n_i} + \frac{1}{n_j}\right)}$$

where cv stands for the critical value. What differs between the methods is the critical value.

Fisher's LSD: A Liberal Approach

Fisher's LSD

Recall that, in Section 5.7, we introduced you to **Fisher's Least Significant Difference (LSD)**. This method is the most liberal of the three we will discuss, producing intervals that are often much narrower than those produced by the other two methods and thus is more likely to identify differences (either real or false). The reason that its intervals are narrower is that it focuses only on the individual error rate. We employ this method only when the F-test from the ANOVA is significant. Since we have already determined that there are differences (the ANOVA is statistically significant), we feel comfortable identifying differences with the liberal LSD method. Because LSD only controls the individual error rate, this method has a larger family-wise error rate. But in its favor, it has a small chance of missing actual differences that exist.

The method starts with verifying that the F-test is significant. If it is, then we compute the usual two-sample confidence intervals using the MSE as our estimate of the sample variance and using as the critical value a t^* with the MSE degrees of freedom, $n - I$, and the individual α-level of choice.

FISHER'S LSD

To compute multiple confidence intervals comparing pairs of means using Fisher's LSD:

1. Verify that the F-test is significant.
2. Compute the intervals

$$\bar{y}_i - \bar{y}_j \pm t^* \sqrt{MSE\left(\frac{1}{n_i} + \frac{1}{n_j}\right)}$$

Use an individual α-level (e.g., 0.05) and the MSE degrees of freedom, $n-I$, for finding t^*.

Bonferroni Adjustment: A Conservative Approach

Bonferroni method

On the other end of the spectrum is the **Bonferroni method**. Whereas Fisher's LSD places its emphasis on the individual error rate (and only controls the family-wise error rate through the ANOVA F-test), the Bonferroni method places its emphasis solely on the family-wise error rate. It does so by using a smaller individual error rate for each of the intervals.

In general, if we want to make m comparisons, we replace the usual α with α/m and make the corresponding adjustment in the confidence levels of the intervals. For example, suppose that we are comparing means for $K = 5$ groups, which require $m = 10$ pairwise comparisons. To ensure a family-wise error rate of at most 5%, we use $0.05/10 = 0.005$ as the significance level for each individual test, or equivalently, we construct the confidence intervals using the previous formula with a t^* value to give 99.5% confidence. Using this procedure, in at least 95% of datasets drawn from populations for which the means are all equal, our *entire set* of conclusions will be correct (no differences will be found significant). In fact, the family-wise error rate is often smaller than 5%, but we cannot easily compute exactly what it is. That is why we say the Bonferroni method is *conservative*. The actual family-wise confidence level is at least 95%. And since we have accounted for the family-wise error rate, it is less likely to incorrectly signal a difference between two groups that actually have the same population mean.

The bottom line is that the Bonferroni method is easy to put in place and provides an upper bound on the family-wise error rate. These two facts make it an attractive option.

Group Difference	Confidence Interval	Contains 0?
1 preg – none	$(-7.05, 9.53)$	yes
8 preg – none	$(-8.49, 8.09)$	yes
1 virgin – none	$(-15.09, 1.49)$	yes
8 virgin – none	$(-33.13, -16.55)$	no
8 preg – 1 preg	$(-9.73, 6.85)$	yes
1 virgin – 1 preg	$(-16.33, 0.25)$	yes
8 virgin – 1 preg	$(-34.37, -17.79)$	no
1 virgin – 8 preg	$(-14.89, 1.69)$	yes
8 virgin – 8 preg	$(-32.93, -16.35)$	no
8 virgin – 1 virgin	$(-26.33, -9.75)$	no

TABLE 8.1 Fisher's LSD confidence intervals

Note that the conclusion here is that living with 8 virgins does significantly reduce the life span of male fruit flies in comparison to all other living conditions tested, but none of the other conditions are significantly different from each other.

Bonferroni

Computer output is given below for the 95% Bonferroni intervals. For this example, we see that the conclusions are the same as for those found using the Fisher's LSD intervals. That is, the life span of male fruit flies living with 8 virgins is significantly shorter than that of all other groups, but none of the other groups are significantly different from each other.

```
Bonferroni 95.0% Simultaneous Confidence Intervals
Response Variable Longevity
All Pairwise Comparisons among Levels of Treatment
Treatment = 1 pregnant subtracted from:

 Treatment    Lower   Center   Upper   -----+---------+---------+---------+-
 1 virgin    -20.02    -8.04    3.94          (----*----)
 8 pregnant  -13.42    -1.44   10.54              (---*----)
 8 virgin    -38.06   -26.08  -14.10   (----*---)
 none        -13.22    -1.24   10.74              (----*---)
                                       -----+---------+---------+---------+-
                                          -25         0        25        50
Treatment = 1 virgin subtracted from:

 Treatment    Lower   Center   Upper   -----+---------+---------+---------+-
 8 pregnant   -5.38     6.60  18.578               (----*---)
 8 virgin    -30.02   -18.04  -6.062     (----*----)
 none         -5.18     6.80  18.778               (----*----)
                                       -----+---------+---------+---------+-
                                          -25         0        25        50
Treatment = 8 pregnant subtracted from:

 Treatment    Lower   Center   Upper   -----+---------+---------+---------+-
 8 virgin    -36.62   -24.64  -12.66   (----*----)
 none        -11.78     0.20   12.18             (----*----)
                                       -----+---------+---------+---------+-
                                          -25         0        25        50

Treatment = 8 virgin subtracted from:

 Treatment    Lower   Center   Upper   -----+---------+---------+---------+-
 none         12.86    24.84   36.82                       (----*----)
                                       -----+---------+---------+---------+-
                                          -25         0        25        50
```

Tukey's HSD

Finally, we present the 95% intervals computed using Tukey's HSD method. Again, we see that the conclusions are the same for this method as they were for the other two:

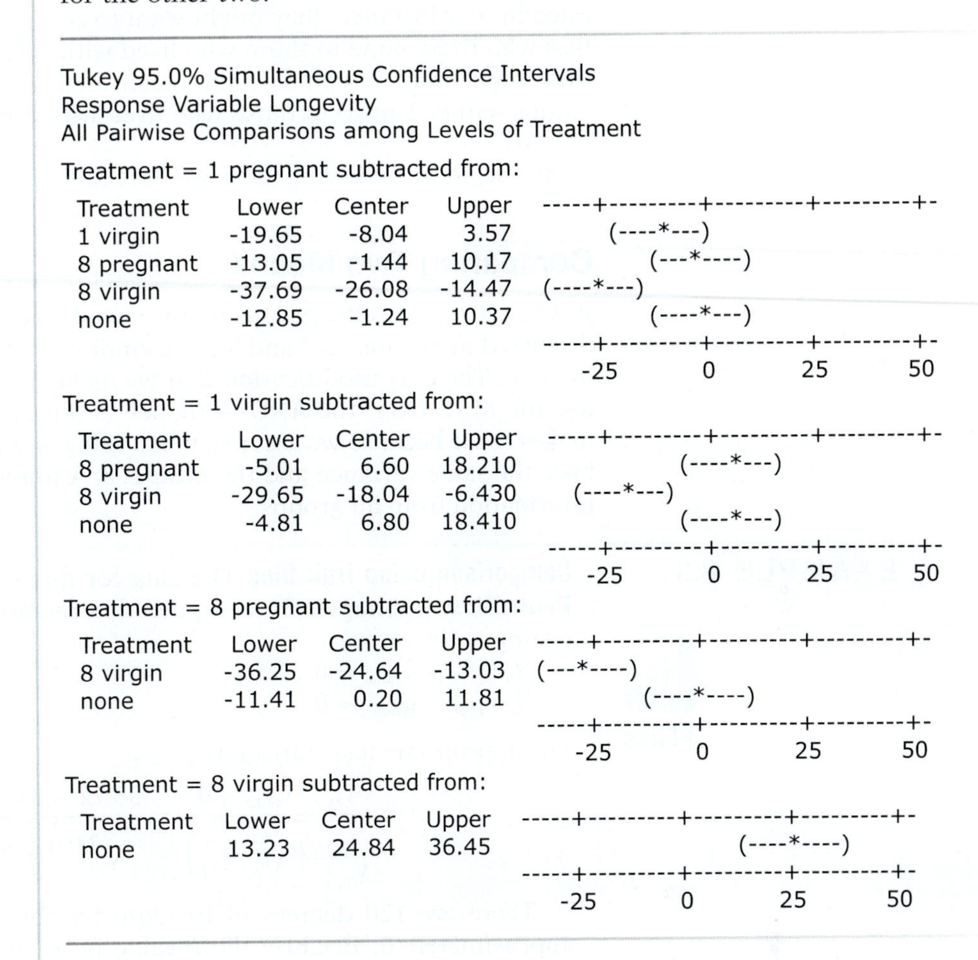

```
Tukey 95.0% Simultaneous Confidence Intervals
Response Variable Longevity
All Pairwise Comparisons among Levels of Treatment
Treatment = 1 pregnant subtracted from:

Treatment     Lower   Center   Upper  -----+---------+---------+---------+-
1 virgin     -19.65    -8.04    3.57          (----*---)
8 pregnant   -13.05    -1.44   10.17             (---*----)
8 virgin     -37.69   -26.08  -14.47  (----*---)
none         -12.85    -1.24   10.37             (----*---)
                                      -----+---------+---------+---------+-
                                         -25         0        25        50

Treatment = 1 virgin subtracted from:

Treatment     Lower   Center   Upper   -----+---------+---------+---------+-
8 pregnant    -5.01     6.60  18.210              (----*---)
8 virgin     -29.65   -18.04  -6.430      (----*---)
none          -4.81     6.80  18.410              (----*---)
                                       -----+---------+---------+---------+-
                                          -25         0        25        50

Treatment = 8 pregnant subtracted from:

Treatment     Lower   Center   Upper  -----+---------+---------+---------+-
8 virgin     -36.25   -24.64  -13.03  (---*----)
none         -11.41     0.20   11.81            (----*----)
                                      -----+---------+---------+---------+-
                                         -25         0        25        50

Treatment = 8 virgin subtracted from:

Treatment     Lower   Center   Upper  -----+---------+---------+---------+-
none          13.23    24.84   36.45                     (----*----)
                                      -----+---------+---------+---------+-
                                         -25         0        25        50
```

We also draw your attention to the relative lengths of the intervals. As an example, consider the lengths of the 8 virgins – none intervals. For Fisher's LSD, the interval is $(-33.13, -16.55)$, which has a length of $|-33.13 + 16.55| = 16.58$ days. The Bonferroni interval is $(-36.82, -12.86)$, which has a length of $|-36.82 + 12.86| = 23.96$. Finally, the Tukey's HSD interval is $(-36.45, -13.23)$, which has a length of $|-36.45 + 13.23| = 23.22$. As predicted, Fisher's LSD results in the shortest intervals. This allows it to find more differences to be significant.* Bonferroni's intervals are the longest, reflecting the fact that it is the most conservative of the three methods. Tukey's HSD fits in between, though for this example, since we have a relatively small number of intervals and small sample sizes, it is similar to Bonferroni.

TOPIC 8.3 Comparisons and Contrasts

We start this topic by returning to the fruit fly example from Chapter 5. Remember that the researchers were interested in the life spans of male fruit flies and how they were affected by the number and type of females that were living with each male. In Chapter 5, we performed a basic ANOVA and learned

*In this example, all three methods give the same answers about which means differ and which do not; but this will not always be the case.

that there are significant differences in mean lifetime between at least two of the treatment groups. We also introduced the idea of comparisons, where we compared the treatments to each other in pairs.

But there are other types of analyses that the researchers might be interested in. For instance, they might want to compare the mean lifetimes of fruit flies who lived alone to those who lived with virgin females (either 1 or 8). Or they might want to compare the mean lifetimes of fruit flies who lived with virgins (either 1 or 8) to those who lived with pregnant females (again, either 1 or 8).

We begin by reviewing ideas we have already seen.

Comparing Two Means

If we have two specific groups that we would like to compare to each other, as discussed in Sections 5.7 and 8.2, we employ a two-sample t-test or confidence interval. The only modification that we make in the ANOVA setting is that we use the MSE from the ANOVA model as our estimate of the variance. This makes sense because we assume, when using an ANOVA model, that all groups have the same variance and the MSE is an estimate of that variance that uses information from all groups.

EXAMPLE 8.5

FFlies

Comparison using fruit flies The data for this example are found in the file **FruitFlies**. In Chapter 5, we argued that researchers would be interested in testing

$$H_0 : \mu_{8v} - \mu_{none} = 0$$
$$H_a : \mu_{8v} - \mu_{none} \neq 0$$

The appropriate test statistic is

$$t = \frac{\bar{y}_{8v} - \bar{y}_{none} - 0}{\sqrt{MSE\left(\frac{1}{n_i} + \frac{1}{n_j}\right)}} = \frac{38.72 - 63.56 - 0}{\sqrt{219\left(\frac{1}{25} + \frac{1}{25}\right)}} = -5.93$$

There are 120 degrees of freedom for this statistic and the P-value is approximately 0. Because the P-value is so small, we conclude that there is a difference in the mean lifetime of fruit flies who live with 8 virgins and fruit flies who live alone.

We can also construct a 95% confidence interval for the (population) difference. The estimated difference is $38.72 - 63.56 = -24.84$. The standard error, which we used in the denominator of the t-statistic, is 4.19. The t-multiplier, with 120 degrees of freedom, is $t^* = 1.98$, so the confidence interval is

$$-24.84 \pm 1.98(4.19)$$

which works out to be $(-33.1, -16.5)$. We are 95% confident that living with 8 virgins shortens the life of a male fruit fly by between 33.1 and 16.5 days, on average, when compared to living alone.

But how do we proceed if our question of interest involves more than two groups? For example, in the fruit flies example there are three questions we might like to answer that fit this situation. We have already identified one of them: Is the life span for males living with pregnant females different from that of males living with virgins? The researchers thought so. Their assumption was that the pregnant females would not welcome advances from the males and so living with pregnant females would result in a different mean life span than living with virgins.

In fact, the researchers thought that living with pregnant females would be like living alone. This would lead us to test the null hypothesis that the fruit flies living with pregnant females would have the same lifetime as those living alone. In this case, we would be working with three groups: those living alone, those with 1 pregnant female, and those with 8 pregnant females.

Combining the previous two ideas could lead us to ask a final question in which we ask if living with virgins (either 1 or 8, it doesn't matter) is different from living alone. Again, we would focus on three groups, using the two groups living with virgins together to compare to the group living alone.

To tackle these complicated, and interesting, questions we need a more general idea, that of a linear combination of means.

Linear Combinations of Means

A linear combination of means is a quantity of the form

$$c_1\mu_1 + c_2\mu_2 + \cdots + c_I\mu_I$$

for some coefficients $c_1, c_2, \cdots, c_I$. The sample (observed) value of the linear combination is the same thing, but using sample means in place of the unknown population means:

$$c_1\bar{y}_1 + c_2\bar{y}_2 + \cdots + c_I\bar{y}_I$$

For example, if we want to compare group 1 to group 2 (and ignore all other groups), then we can choose $c_1 = 1$, $c_2 = -1$, and $c_3 = c_4 = \cdots = c_I = 0$. The sample value of the linear combination would then be $\bar{y}_1 - \bar{y}_2$. So any pairwise comparison that we have seen before is a special case of a linear combination.

The Standard Error for a Linear Combination

To find the standard error of a linear combination of means, we build on something we already know. If we compare just two groups to each other, as we did in Chapter 5, we estimate the variability of our statistic, $\bar{y}_1 - \bar{y}_2$, with $\sqrt{MSE(\frac{1}{n_1} + \frac{1}{n_2})}$. Recall that the actual standard deviation of $\bar{y}_1 - \bar{y}_2$ is $\sqrt{\frac{\sigma_1^2}{n_1} + \frac{\sigma_2^2}{n_2}}$, but we only use the ANOVA model when all groups have the same variance. So we call that common variance σ^2 and factor it out, getting $\sqrt{\sigma^2(\frac{1}{n_1} + \frac{1}{n_2})}$. Of course, we don't know σ^2 so we estimate it using the MSE.

Now we need to consider the more general case where our linear combination is $c_1\bar{y}_1 + c_2\bar{y}_2 + \cdots + c_k\bar{y}_k$. Note that, in the preceding discussion, when we were comparing just two groups, $c_1 = 1$ and $c_2 = -1$. As we consider the more general case, we will continue to use the MSE as our estimate for the common variance, but now we need to take into consideration the sample sizes of all groups involved and the coefficients used in the linear combination.

STANDARD ERROR OF A LINEAR COMBINATION

The general formula of the standard error for contrasts is

$$\sqrt{MSE \sum_{i=1}^{I} \frac{c_i^2}{n_i}}$$

This formula might look intimidating, but when we compare two means, the formula gives us a standard error of $\sqrt{MSE(\frac{1}{n_i} + \frac{1}{n_j})}$, which we have seen before, in Example 8.5 and elsewhere.

contrasts

Linear combinations of means can take many forms, but usually the most interesting linear combinations are **contrasts**. A contrast is a linear combination for which the coefficients add up to zero.

CONTRASTS

In general, we will write contrasts in the form

$$c_1\mu_1 + c_2\mu_2 + \cdots + c_I\mu_I$$

where

$$c_1 + c_2 + \cdots + c_I = 0$$

and some c_i might be 0. The contrast is estimated by substituting sample means for the population means:

$$c_1\bar{y}_1 + c_2\bar{y}_2 + \cdots + c_I\bar{y}_I$$

Let's return to the question of how to compare living with virgin females, either 1 or 8, to living alone.

EXAMPLE 8.6

FFlies

Fruit flies (virgins vs. alone) In this case, we want to compare the life spans of those living with either 1 or 8 virgin females to those living alone. This leads us to the null hypothesis

$$H_0 : \frac{1}{2}(\mu_{1v} + \mu_{8v}) = \mu_{\text{none}}$$

or

$$H_0 : \frac{1}{2}(\mu_{1v} + \mu_{8v}) - \mu_{\text{none}} = 0$$

The estimate of the difference is

$$\frac{1}{2}\bar{y}_{1v} + \frac{1}{2}\bar{y}_{8v} - \bar{y}_{\text{none}}$$

Notice that this is a contrast: the coefficients are 1/2, 1/2, and −1.* The sample value of the contrast is

$$\left(\frac{1}{2}\right)(56.76) + \left(\frac{1}{2}\right)(38.72) - 1(63.56) = -15.82.$$

The standard error for the contrast is

$$\sqrt{219\left(\frac{\left(\frac{1}{2}\right)^2}{25} + \frac{\left(\frac{1}{2}\right)^2}{25} + \frac{(-1)^2}{25}\right)} = \sqrt{219\left(\frac{1.5}{25}\right)} = 3.6249$$

The t-test statistic we need has the familiar form of

$$\text{test statistic} = \frac{\text{estimate} - \text{hypothesized value}}{\text{standard error of the estimate}}$$

In this case we have a t-test. The test statistic is

$$t = \frac{-15.82}{3.6249} = -4.36$$

which yields a small P-value.† We have strong evidence that average lifespan for male fruit flies living with virgin fruit flies is different than for those living alone.

*There are five groups in the FruitFlies data and we are only considering three of them here. Technically, we are using coefficients of 0 for the "1 pregnant" and "8 pregnant" groups.

†If you care to know the exact P-value, it is 0.00003.

EXAMPLE 8.7

FFlies

Fruit flies (virgins vs. pregnant) Consider the question of whether the mean life span of the male fruit flies living with virgins is different from that of those living with pregnant females. The contrast that we need is

$$\frac{1}{2}\bar{y}_{1v} + \frac{1}{2}\bar{y}_{8v} - \frac{1}{2}\bar{y}_{1p} - \frac{1}{2}\bar{y}_{8p}.$$

From Chapter 5, we know that $\bar{y}_{1v} = 56.76$, $\bar{y}_{8v} = 38.72$, $\bar{y}_{1p} = 64.80$, and $\bar{y}_{8p} = 63.36$, so that the sample value of the contrast is -16.34.

All groups consist of 25 fruit flies and all coefficients are $\pm\frac{1}{2}$ so the standard error is

$$\sqrt{219\left(\frac{\left(\frac{1}{2}\right)^2}{25} + \frac{\left(\frac{1}{2}\right)^2}{25} + \frac{\left(-\frac{1}{2}\right)^2}{25} + \frac{\left(-\frac{1}{2}\right)^2}{25}\right)} = \sqrt{219\left(\frac{\frac{1}{4}}{25} + \frac{\frac{1}{4}}{25} + \frac{\frac{1}{4}}{25} + \frac{\frac{1}{4}}{25}\right)}$$

$$= \sqrt{219\left(\frac{1}{25}\right)} = 2.9597$$

The test statistic is

$$t = \frac{-16.34}{2.9597} = -5.52$$

Comparing this to a t-distribution with 120 degrees of freedom, we find that the P-value is approximately 0 and we conclude that there is a difference between the life spans of those fruit flies that live with virgins compared to those fruit flies that live with pregnant females.

A 95% confidence interval is $-16.34 \pm 1.98(2.9597)$ or $(-22.2, -10.5)$. We are 95% confident that living with virgins (1 or 8) reduces average lifetime by between 22.2 and 10.5 days, compared to living with pregnant flies (1 or 8).

We leave the remaining tests for the fruit flies to the exercises, and we conclude this topic with one final example that puts together all of the pieces.

EXAMPLE 8.8

Babies

Walking babies We return to the walking babies study, first seen in Example 5.4. Recall that we are wondering if the age at which babies begin walking will be different depending on the exercises they do earlier.

To start the analysis, we ask the simple question: Is there a difference in the mean time to walking for the four groups of babies? So we start by evaluating whether ANOVA is an appropriate analysis tool for this dataset by checking the necessary conditions.

The method of fitting the model guarantees that the residuals always add to zero, so there's no way to use residuals to check the condition that the mean error is zero. Essentially, the condition says that the equation for the structure of the response hasn't left out any terms. Since babies are randomly assigned to the four treatment groups, we hope that all other variables that might affect walking age have been randomized out.

The independence of the errors condition says, in effect, that the value of one error is unrelated to the others. For the walking babies, this is almost surely the case because the time it takes one baby to learn to walk doesn't depend on the times it takes other babies to learn to walk.

The next condition says that the amount of variability is the same from one group to the next. One way to check this is to plot residuals versus fitted values and compare columns of points. Figure 8.3 shows that the amount of variability is similar from one group to the next. We can also compare the largest and smallest group standard deviations by computing the ratio max_{s_i}/min_{s_i}.

$$\frac{max_s}{min_s} = \frac{1.898}{0.871} = 2.18$$

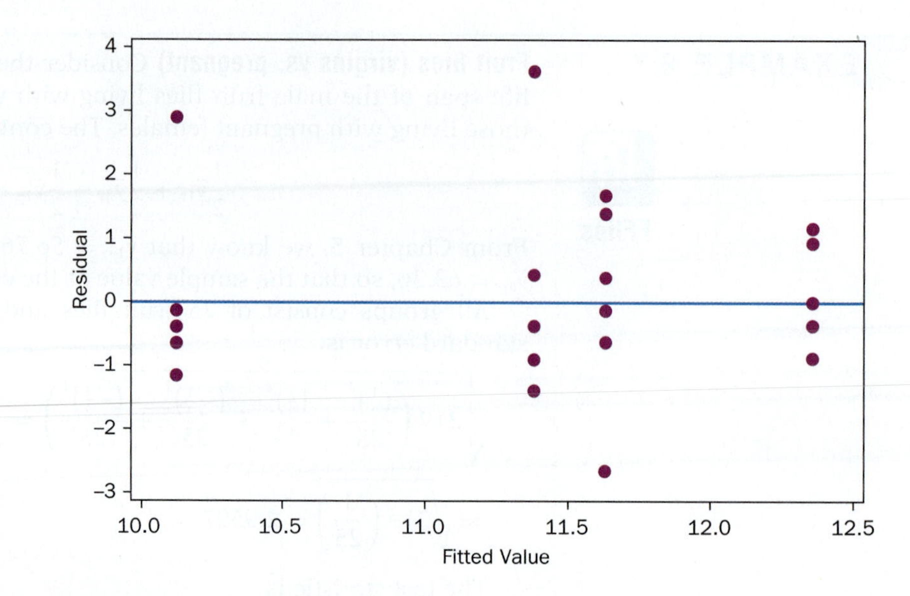

FIGURE 8.3 Residual plot for walking babies ANOVA model

group	StDev
Control_E	1.898
Weekly	1.558
Final	0.871
Special	1.447

Although this ratio is slightly larger than 2, we are talking about 4 groups with only 6 observations in each group so we are willing to accept this ratio as being small enough to be consistent with the condition that the population variances are all the same.

Finally, the last condition says that the error terms should be normally distributed. To check this condition, we look at a normal probability plot, which should suggest a line. Figure 8.4 looks reasonably straight so we are willing to accept the normality of the errors.

The end result for this example is that the conditions seem to be met and we feel comfortable proceeding with our analysis using ANOVA.

The ANOVA table is given below:

One-way ANOVA: age versus group

Source	DF	SS	MS	F	P
group	3	15.60	5.20	2.34	0.104
Error	20	44.40	2.22		
Total	23	60.00			

From the table, we see that the *F*-statistic is 2.34 with 3 and 20 degrees of freedom and the *P*-value is 0.104. We also compute $R^2 = 1 - \frac{SSE}{SSTotal} = 1 - \frac{44.40}{60.00} = 1 - 0.74 = 0.26$. This tells us that the group differences account for 26% of the total variability in the data. And the test of the null hypothesis that the mean time to walking is the same for all groups of children is not significant. In other words, we do not have significant evidence that any group is different from the others.

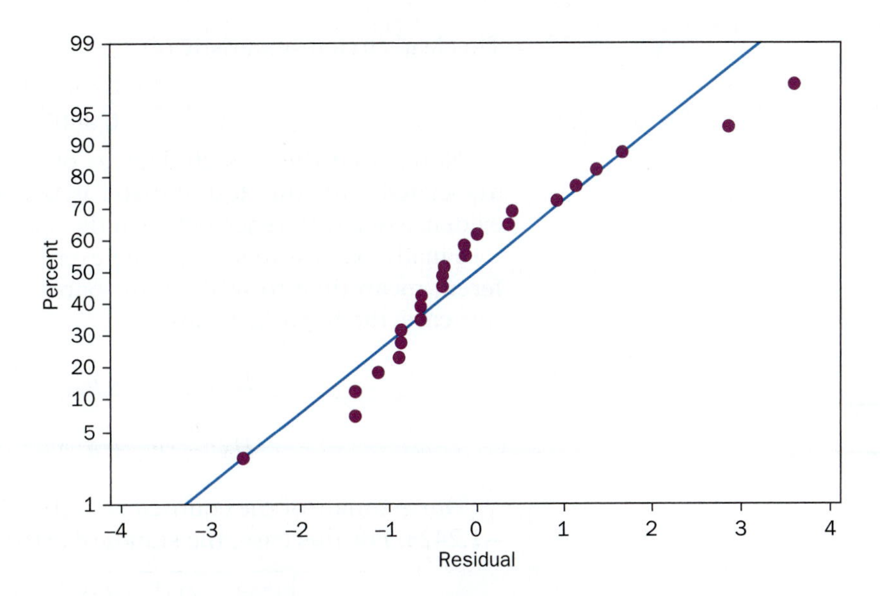

FIGURE 8.4 Normal probability plot of residuals

However, notice in this case that we really have one treatment group and three kinds of control groups. Each control group is meant to ferret out whether some particular aspect of the treatment is, in fact, important. Is it the special exercise itself, or is it any exercise at all that will help? The first control group could help us decide this. Is it extra attention, not the exercise itself? This is where the second control group comes in. Finally, is it all of the above? This is what the third control group adds to the mix.

While the F-test for this dataset was not significant, the researchers had two specific questions from the outset that they were interested in. First was the question of whether there was something special to the exercises labeled "special." That is, when compared to the control group with exercises (treated in every way the same as the treatment group except for the type of exercise), is there a difference? As a secondary question, they also wondered whether there was a difference between children who used some type of exercise (any type) versus those who did not use exercise. The first question calls for a comparison; the second requires a more complicated contrast.

We start with the first question: Is there a difference between the group that received the special exercises and the group that were just told to exercise? The hypotheses here are

$$H_0 : \mu_{se} - \mu_{ce} = 0$$
$$H_a : \mu_{se} - \mu_{ce} \neq 0$$

The group means are given in the table below:

	Group Mean
Control_E	11.383
Weekly	11.633
Final	12.360
Exercise	10.125

Our estimate of the comparison is $\bar{y}_{se} - \bar{y}_{ce} = 10.125 - 11.383 = -1.258$. Since there are 6 babies in each group, the standard error of the comparison is

$$\sqrt{MSE\left(\frac{1^2}{6} + \frac{(-1)^2}{6}\right)} = \sqrt{2.22\left(\frac{1}{6} + \frac{1}{6}\right)} = 0.8602$$

This leads to a test statistic of

$$t = \frac{-1.258 - 0}{0.8602} = -1.46$$

Notice that this has 20 degrees of freedom, from the MSE. The *P*-value associated with this test statistic is 0.1598, so we do not have significant evidence of a difference between the special exercises and any old exercises.

Finally, we test to see if using exercises (special or otherwise) gives a different mean time to walking for babies in comparison to no exercises. For this case, the hypotheses are

$$H_0 : \frac{1}{2}(\mu_{\text{se}} + \mu_{\text{ce}}) - \frac{1}{2}(\mu_{\text{w}} + \mu_{\text{f}}) = 0$$

$$H_0 : \frac{1}{2}(\mu_{\text{se}} + \mu_{\text{ce}}) - \frac{1}{2}(\mu_{\text{w}} + \mu_{\text{f}}) \neq 0$$

Our estimate of the contrast is $\frac{1}{2}(10.125 + 11.383) - \frac{1}{2}(11.633 + 12.360) = -1.2425$. For this case, the standard error is

$$\sqrt{2.22\left(\frac{\left(\frac{1}{2}\right)^2}{6} + \frac{\left(\frac{1}{2}\right)^2}{6} + \frac{\left(\frac{-1}{2}\right)^2}{6} + \frac{\left(\frac{-1}{2}\right)^2}{6}\right)} = \sqrt{\frac{2.22}{6}} = 0.6083$$

This leads us to the test statistic

$$t = \frac{-1.2425 - 0}{0.6083} = -2.043$$

with 20 degrees of freedom and a *P*-value of 0.0544. Here, we conclude that we have moderate evidence against the null hypothesis. That is, it appears that having babies take part in exercises may lead to earlier walking.

We end this example with the following note. You will have noticed that we did the analysis for the comparison and contrast of interest even though the *F*-test in the ANOVA table was not significant. This is because these two comparisons were planned at the outset of the experiment. These were the questions that the researchers designed the study to ask. Planned comparisons can be undertaken even if the overall *F*-test is not significant.

TOPIC 8.4 Nonparametric Statistics

Here, we consider alternative statistical methods that do not rely on having data from a normal distribution. These procedures are referred to as nonparametric statistical methods or distribution-free procedures, because the data are not required to follow a particular distribution. As you will see, we still have conditions for these procedures, but these conditions tend to be more general in nature. For example, rather than specifying that a variable must follow the normal distribution, we may require that the distribution be symmetric. We introduce competing procedures for two-sample and multiple-sample inferences for means that we have considered earlier in the text.

Two Sample Nonparametric Procedures

Wilcoxon-Mann-Whitney test

Rather than testing hypotheses about two means, as we did in Chapter 0 with the two-sample *t*-test, we will now consider a procedure for making inferences about two medians. The **Wilcoxon-Mann-Whitney*** test procedure is used to

*Equivalent tests, one developed by Wilcoxon and another developed by Mann and Whitney, are available in this setting. The Wilcoxon procedure is based on the sum of the joint ranks for one of the two samples. The Mann-Whitney procedure is based on the number of all possible pairs where the observation in the second group is larger than an observation in the first group.

make inferences for two medians. Recall from Chapter 0 that data for two independent samples can be written as $DATA = MODEL + ERROR$. In that chapter we concentrated on the two-sample t-test, and we used the following model:

$$Y = \mu_i + \epsilon$$

where $i = 1, 2$.

When considering the Wilcoxon-Mann-Whitney scenario, the model could be written as

$$Y = \theta_i + \epsilon$$

where θ_i is the population median for the ith group and ϵ is the random error term that is symmetrically distributed about zero. The conditions for our new model are similar to the conditions that we have seen with regression and ANOVA models for means. In fact, the conditions can be relaxed even more because the error distribution does not need to be symmetric.

WILCOXON-MANN-WHITNEY MODEL CONDITIONS

The error terms must meet the following conditions for the Wilcoxon-Mann-Whitney model to be applicable:

- Have median zero
- Follow a continuous, but not necessarily normal, distribution*
- Be independent

Since we have only two groups, this model says that

$$Y = \theta_1 + \epsilon \quad \text{for individuals in Group 1}$$
$$Y = \theta_2 + \epsilon \quad \text{for individuals in Group 2}$$

We consider whether an alternative (simpler) model might fit the data as well as our model with different medians for each group. This is analogous to testing the hypotheses:

$$H_0 : \theta_1 = \theta_2$$
$$H_a : \theta_1 \neq \theta_2$$

If the null hypothesis (H_0) is true, then the simpler model $Y = \theta + \epsilon$ uses the same median for both groups. The alternative (H_a) reflects the model we have considered here that allows each group to have a different median.

Nonparametric tests are usually based on ranks. Consider taking the observations from the two groups (say the sample sizes are m and n) and making one big collection. The entire collection of data is then ranked from smallest, 1, to largest, $m + n$. If H_0 is true, then the largest overall observation is equally likely to have come from either group; likewise for the second largest observation, and so on. Thus the average rank of data that came from the first group should equal the average rank of the data that came from the second group. On the other hand, if H_0 is false, then we expect the large ranks to come mostly from one group and the small ranks from the other group. We will use statistical software to help us decide which model is better for a particular set of data.

*The least restrictive condition for the Wilcoxon-Mann-Whitney test requires that the two distributions be stochastically ordered, which is the technical way of saying that one of the variables has a consistent tendency to be larger than the other.

One issue we need to consider is ties. Nonparametric procedures must be adjusted for ties, if they exist. Ties take place when two observations, one from each group, have exactly the same value of the response variable. Most software packages will provide options for adjusting the *P*-value of the test when ties are present, and we recommend the use of such adjusted *P*-values.

EXAMPLE 8.9

HwkTail

HwkTail2

Hawk tail length[2] A researcher is interested in comparing the tail lengths for red-tailed and sharp-shinned hawks near Iowa City, Iowa. The data are provided in two different formats because some software commands require that the data be unstacked, with each group in separate columns, and other commands require that the data be stacked, identified with one response variable and another variable to identify the group. The stacked data are in the file **HawkTail** and the unstacked data are in the file **HawkTail2**.

Figure 8.5 shows dotplots for the tail lengths for the two different types of hawks. Both distributions contain outliers and the distribution of the tail lengths for the sharp-shinned hawks is skewed to the right, so we would prefer inference methods that do not rely on the normality condition. The visual evidence is overwhelming that the two distributions (or two medians) are not the same, but we'll proceed with a test to illustrate how it is applied.

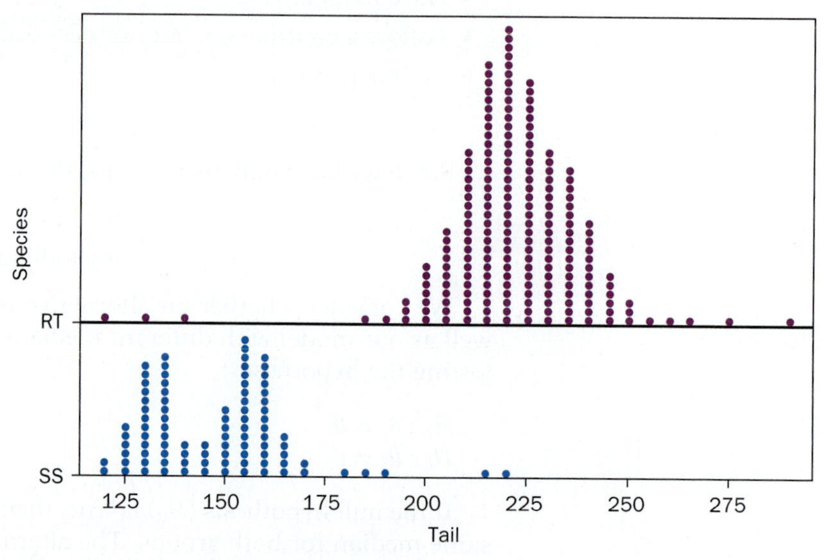

FIGURE 8.5 Dotplot of hawk tail lengths for red-tailed and sharp-shinned hawks

One way to formally examine the tail lengths for the two groups of hawks is with the results of the Wilcoxon-Mann-Whitney test (as shown in the Minitab output that follows). The Minitab output provides the two sample sizes, estimated medians, a confidence interval for the difference in the two medians (red-tailed minus sharp-shinned), the Wilcoxon rank-sum statistic, and the corresponding *P*-values. The 95% confidence interval (74, 78) is far above zero, which indicates that there is strong evidence of a difference in the median tail lengths for these two types of hawks. Minitab provides the Wilcoxon rank-sum statistic, which is identified as $W = 316{,}058$ in the output. The software provides the approximate *P*-value of 0.000.

```
Mann-Whitney Test and CI: Tail_RT, Tail_SS

            N   Median
 Tail_RT  577   221.00
 Tail_SS  261   150.00

Point estimate for ETA1-ETA2 is 76.00
95.0 Percent CI for ETA1-ETA2 is (74.00,78.00)
W = 316058.0
Test of ETA1 = ETA2 vs ETA1 not = ETA2 is significant at 0.0000
The test is significant at 0.0000 (adjusted for ties)
```

Nonparametric ANOVA: Kruskal-Wallis Test

In this subsection, we consider an alternative to the one-way ANOVA model. Rather than testing hypotheses about means, as we have done in Chapters 5, 6, and 7, we will now test hypotheses about medians. Our one-way ANOVA model has the same overall form, but we no longer require the error terms to follow a normal distribution. Thus our one-way ANOVA model for medians is

$$Y = \theta_i + \epsilon$$

where θ_i is the median for the ith population and ϵ has a distribution with median zero. The form of this model is identical to the model we discussed for the Wilcoxon-Mann-Whitney procedure for two medians and the conditions for the error terms are the same as those provided in the box.

KRUSKAL-WALLIS MODEL CONDITIONS

The error terms must meet the following conditions for the Kruskal-Wallis model to be applicable:

- Have median zero
- Follow a continuous, but not necessarily normal, distribution*
- Be independent.

As usual, the hypothesis of no differences in the populations is the null hypothesis, and the alternative hypothesis is a general alternative suggesting that at least two of the groups have different medians. In symbols, the hypotheses are

$H_0 : \theta_1 = \theta_2 = \theta_3 = \cdots = \theta_I$
H_a: At least two θ_is are different

The Kruskal-Wallis test statistic is a standardized version of the rank sum for each of the treatment groups. Without getting into the technical details, the overall form of the statistic is a standardized comparison of observed and expected rank sums for each group. Note that H_0 says that the average ranks should be the same across the I groups. Thus the P-value is obtained by using exact tables or approximated using the chi-square distribution with degrees of freedom equal to the number of groups minus 1.

*The I different treatment distributions should only differ in their locations (medians). If the variability is substantially different from treatment to treatment, then the Kruskal-Wallis procedure is not distribution-free.

EXAMPLE 8.10

CanSurv

Cancer survival In Example 8.2, we analyzed survival for cancer patients who received an ascorbate supplement (see the file **CancerSurvival**). Patients were grouped according to the organ that was cancerous. The organs considered in this study are the stomach, bronchus, colon, ovary, and breast. The researchers were interested in looking for differences in survival among patients in these five different groups.

Figure 8.6 shows boxplots for these five different groups. The boxplots clearly show that the five distributions have different shapes and do not have the same variability. Figure 8.7 shows the modified distributions after making the log transformation. Now the distributions have roughly the same shape and variability.

As you will soon see, using the log transformation is not necessary, because the Kruskal-Wallis procedure uses the rank transformation. Thus we can proceed with our analysis.

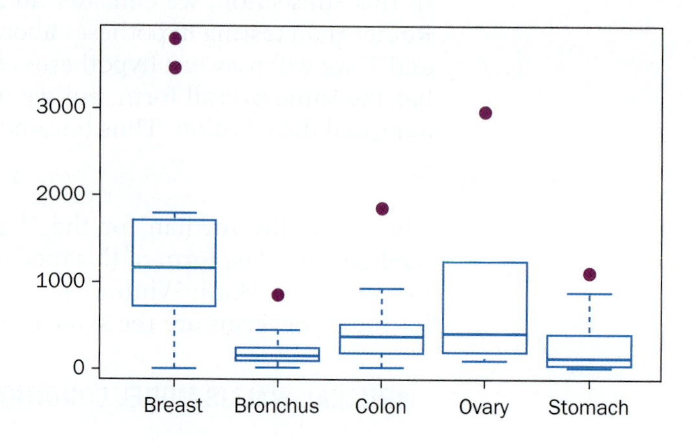

FIGURE 8.6 Boxplot of survival for different groups of cancer patients

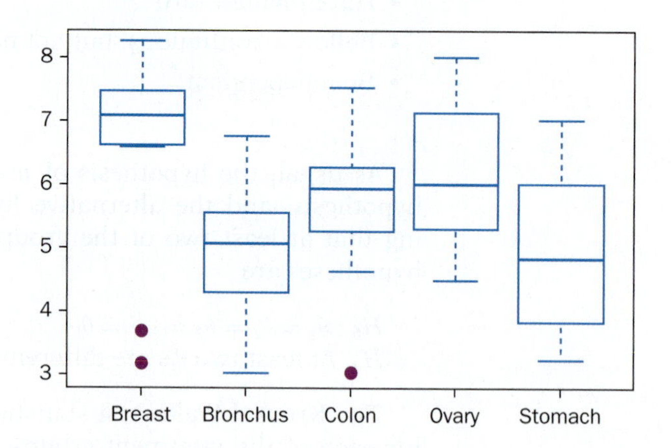

FIGURE 8.7 Boxplot of transformed survival time, log(Survival), for different groups of cancer patients

Computer output is shown as follows. The value of the Kruskal-Wallis statistic is 14.9539. Since the *P*-value of 0.0048 is small, we reject the null hypothesis of no difference in the survival rates and conclude that there are different survival rates, depending on what organ is affected with cancer.

```
Kruskal-Wallis rank sum test
data: lSurvival by Organ
Kruskal-Wallis chi-squared = 14.9539, df = 4, p-value = 0.004798
```

The Minitab output below provides the default output for applying the Kruskal-Wallis test on the original survival rates. Notice that the test statistic, $H = 14.95$, is the same as the test statistic on the transformed survival times provided by R. In addition to the test statistic and *P*-value, the Minitab output includes the sample sizes, estimated medians, average rank, and standardized *Z*-statistic for each group. The average ranks can be used to identify the differences using multiple comparison procedures, but this topic is beyond the scope of this text. However, as we can see in Figure 8.7, the two groups with the lowest average ranks are Bronchus and Stomach.

Kruskal-Wallis Test on Survival

Organ	N	Median	Ave Rank	Z
Breast	11	1166.0	47.0	2.84
Bronchus	17	155.0	23.3	-2.37
Colon	17	372.0	35.9	0.88
Ovary	6	406.0	40.2	1.06
Stomach	13	124.0	24.2	-1.79
Overall	64		32.5	

H = 14.95   DF = 4   P = 0.005

One of the interesting features of nonparametric procedures is that they are often based on counts or ranks. Thus if we apply the Kruskal-Wallis test to the survival rates, rather than the transformed rates, we will get exactly the same value of the test statistic (and *P*-value). You can verify this feature for yourself with one of the exercises.

TOPIC 8.5 Randomization *F*-Test

As you saw in the Chapter 5, there are many reasons to "randomize the assignment of treatments to units," that is, to use chance to decide which units get which treatments. The three most important reasons are:

1. To justify using a probability model
2. To protect against bias
3. To permit inference about cause and effect

This section is about the first reason and its consequences. If you use probability methods to assign treatments, you automatically get a built-in probability model to use for inference. Errors don't have to be normal, independence is automatic, and the size of the variation doesn't matter. The model depends *only* on the random assignment. If you randomize, there are no other conditions to check!* This is a very different approach from the inference methods of Chapters 5–7; those methods do require that your data satisfy the usual four conditions.

*In what follows, the randomization test uses the same *F*-statistic as before. Depending on conditions, a different statistic might work somewhat better for some datasets.

The rest of this section has three main parts: the mechanics of a randomization test, the logic of the randomization test, and a comparison of the randomization test to the usual F-test.

Mechanics of the Randomization F-Test

A concrete way to think about the mechanics of the randomization F-test is to imagine putting each response value on a card, shuffling the cards, dealing them into treatment groups, and computing the F-statistic for this random set of groups. The P-value is the probability that this randomly created F-statistic will be at least as large as the one from the actual groups. To estimate the P-value, shuffle, deal, and compute F, over and over, until you have accumulated a large collection of F-statistics, say, 10,000. The proportion of times you get an F at least as large as the one from the actual data is an estimate of the P-value. (The more shuffles you do, the better your estimate. Usually, 10,000 is good enough.)

RANDOMIZATION F-TEST FOR ONE-WAY ANOVA

Step 1: Observed F: Compute the value F_{obs} of the F-statistic in the usual way.
Step 2: Randomization distribution.

- Step 2a: Rerandomize: Use a computer to create a random reordering of the data.
- Step 2b: Compute F_{Rand}: Compute the F-statistic for the rerandomized data.
- Step 2c: Repeat and record: Repeat Steps 2a and 2b a large number of times (e.g., 10,000) and record the set of values of F_{Rand}.

Step 3: P-value: Find the proportion of F_{Rand} values from Step 2c that are greater than or equal to the value of F_{obs} from Step 1. This proportion is an estimate of the P-value.

EXAMPLE 8.11

CaBP

Calcium and blood pressure[3] The purpose of this study was to see whether daily calcium supplements can lower blood pressure. The subjects were 21 men; each was randomly assigned either to a treatment group or to a control group. Those in the treatment group took a daily pill containing calcium. Those in the control group took a daily pill with no active ingredients. Each subject's blood pressure was measured at the beginning of the study and again at the end. The response values below (and stored in **CalciumBP**) show the decrease in systolic blood pressure. Thus a negative value means that the blood pressure went up over the course of the study.

												Mean
Calcium	7	−4	18	17	−3	−5	1	10	11	−2		5.000
Placebo	−1	12	−1	−3	3	−5	5	2	−11	−1	−3	−0.273

Step 1: Value from the data F_{obs}.
A one-way ANOVA gives an observed F of 2.6703. This value of F is the usual one, computed using the methods of Chapter 5. The P-value in the computer output is also based on the same methods. For the alternative method based on randomization, we use a different approach to compute the P-value. Ordinarily, the two methods give very similar P-values, but the logic is different, and the required conditions for the data are different.

```
           Df  Sum Sq  Mean Sq  F value  Pr(>F)
Groups      1  145.63  145.628   2.6703  0.1187
Residuals  19 1036.18   54.536
```

Step 2: Create the randomization distribution.

- *Step 2a: Rerandomize.* The computer randomly reassigns the 21 response values to groups, with the result shown below:

												Mean
Calcium	18	−5	−1	17	−3	7	−2	3	−1	10		4.300
Placebo	11	−3	−1	−3	1	5	−5	12	−4	2	−11	−0.364

- *Step 2b: Compute F_{Rand}.* The F-statistic for the rerandomized data turns out to be 1.4011:

```
           Df  Sum Sq  Mean Sq  F value  Pr(>F)
Groups      1   81.16   81.164   1.4011  0.2511
Residuals  19 1100.65   57.929
```

- *Step 2c: Repeat and record.* Here are the values of F for the first 10 rerandomized datasets:

1.4011	0.8609	6.4542	0.0657	0.8872
0.5017	0.0892	0.04028	0.0974	0.2307

As you can see, only 1 of the 10 randomly generated F-values is larger than the observed value $F_{obs} = 2.6703$. So based on these 10 repetitions, we would estimate the P-value to be 1 out of 10, or 0.10. For a better estimate, we repeat Steps 2a and 2b a total of 10,000 times, and summarize the **randomization distribution** with a histogram (Figure 8.8).

randomization distribution

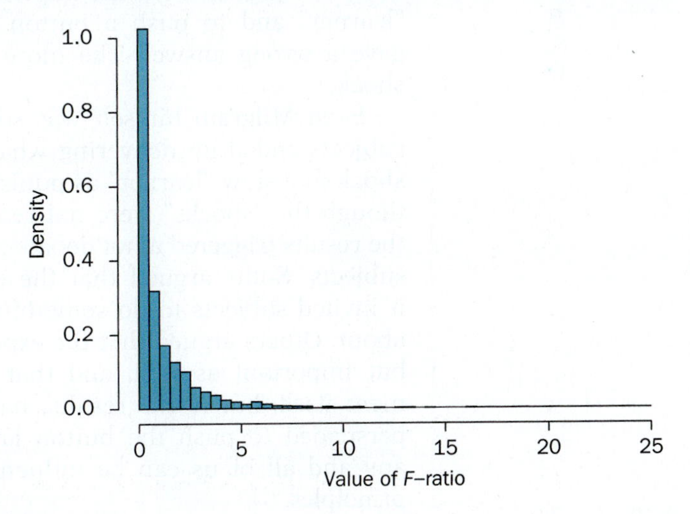

FIGURE 8.8 Distribution of F-statistics for 10,000 rerandomizations of the calcium data

Step 3: P*-value.*
The estimated P-value is the proportion of F-values greater than or equal to $F_{obs} = 2.6703$ (Figure 8.9). For the calcium data, this proportion is 0.119. The P-value of 0.119 is not small enough to rule out chance variation as an explanation for the observed difference.

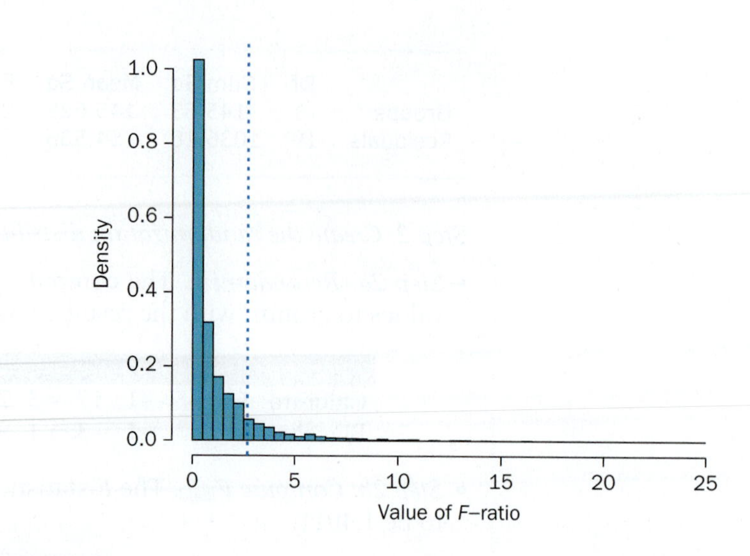

FIGURE 8.9 The *P*-value equals the area of the histogram to the right of the dashed line

EXAMPLE 8.12

Milgrm

Rating Milgram One of the most famous and most disturbing psychological studies of the twentieth century took place in the laboratory of Stanley Milgram at Yale University.[4] The unsuspecting subjects were ordinary citizens recruited by Milgram's advertisement in the local newspaper. The ad simply asked for volunteers for a psychology study, and offered a modest fee in return. Milgram's study of "obedience to authority" was motivated by the atrocities in Nazi Germany and was designed to see whether people off the street would obey when "ordered" to do bad things by a person in a white lab coat. If you had been one of Milgram's subjects, you would have been told that he was studying ways to help people learn faster, and that your job in the experiment was to monitor the answers of a "learner" and to push a button to deliver shocks whenever the learner gave a wrong answer. The more wrong answers, the more powerful the shock.

Even Milgram himself was surprised by the results: Every one of his subjects ended up delivering what they *thought* was a dangerous 300-volt shock to a slow "learner" as punishment for repeated wrong answers. Even though the "shocks" were not real and the "learner" was in on the secret, the results triggered a hot debate about ethics and experiments with human subjects. Some argued that the experiment itself was unethical, because it invited subjects to do something they would later regret and feel guilty about. Others argued that the experiment was not only ethically acceptable but important as well, and that the uproar was due not to the experiment itself, but to the results, namely that every one of the subjects was persuaded to push the button labeled "300 volts: XXX," suggesting that any and all of us can be influenced by authority to abandon our moral principles.

To study this aspect of the debate, Harvard graduate student Mary Ann DiMatteo conducted a randomized comparative experiment. Her subjects were 37 high school teachers who did not know about the Milgram study. Using chance, Mary Ann assigned each teacher to one of three treatment groups:

- Group 1: Actual results. Each subject in this group read a description of Milgram's study, including the actual results that every subject delivered the highest possible "shock."
- Group 2: Many complied. Each subject read the same description given to the subjects in Group 1, except that the actual results were replaced by fake results, that many but not all subjects complied.
- Group 3. Most refused. For subjects in this group, the fake results said that most subjects refused to comply.

After reading the description, each subject was asked to rate the study according to how ethical they thought it was, from 1 (not at all ethical) to 9 (completely ethical). Here are the results (which are also stored in **Milgram**):

														Mean
Actual results	6	1	7	2	1	7	3	4	1	1	1	6	3	3.308
Many complied	1	3	7	6	7	4	3	1	1	2	5	5	5	3.846
Many refused	5	7	7	6	6	6	7	2	6	3	6			5.545

For the **Milgram** data, the high school teachers who served as subjects were randomly assigned to one of three groups. On average, those who read the actual results gave the study the lowest ethical rating (3.3 on a scale from 1 to 7); those who read that most subjects refused to give the highest shocks gave the study a substantially higher rating of 5.5.

To decide whether results like these could be due just to chance alone, we carry out a randomization F-test.

Step 1: F_{obs} value from the data.
For this dataset, software tells us that the value of the F-statistic is $F_{obs} = 3.488$.

Step 2: Create the randomization distribution.

- *Steps 2a and 2b: Rerandomize and compute F_{Rand}.* We used the computer to randomly reassign the 37 response values to groups, with the result shown below. Notice that for this rerandomization, the group means are very close together. The value of F_{Rand} is 0.001.

														Mean
Actual results	6	3	1	6	2	1	3	5	1	7	7	7	4	4.077
Many complied	3	2	7	1	7	5	6	7	6	5	1	1	6	4.385
Many refused	6	3	1	4	7	6	1	6	1	7	3			4.091

- *Step 2c: Repeat and record.* Figure 8.10 shows the randomization distribution of F-values based on 10,000 repetitions.

Step 3: P-*value.*
The estimated P-value is the proportion of F-values greater than or equal to F_{obs} (see Figure 8.11). For the **Milgram** data, this proportion is 0.044.

The P-value of 0.044 is below 0.05, and so we reject the null hypothesis and conclude that the ethical rating does depend on which version of the Milgram study the rater was given to read.

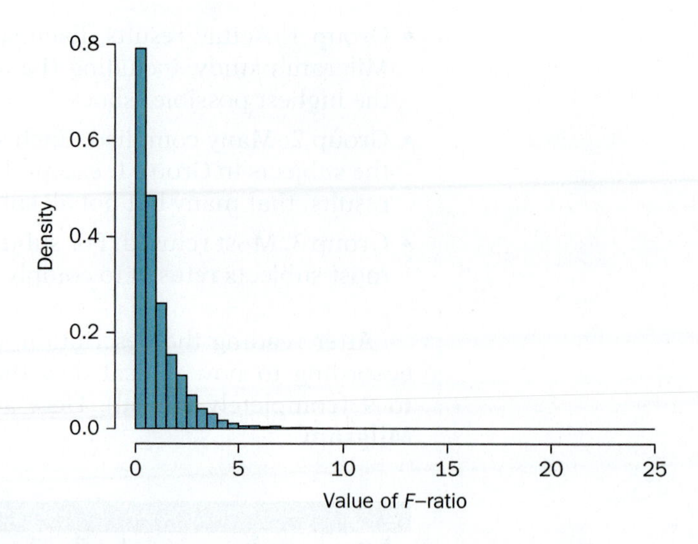

FIGURE 8.10 Distribution of *F*-statistics for 10,000 rerandomizations of the Milgram data

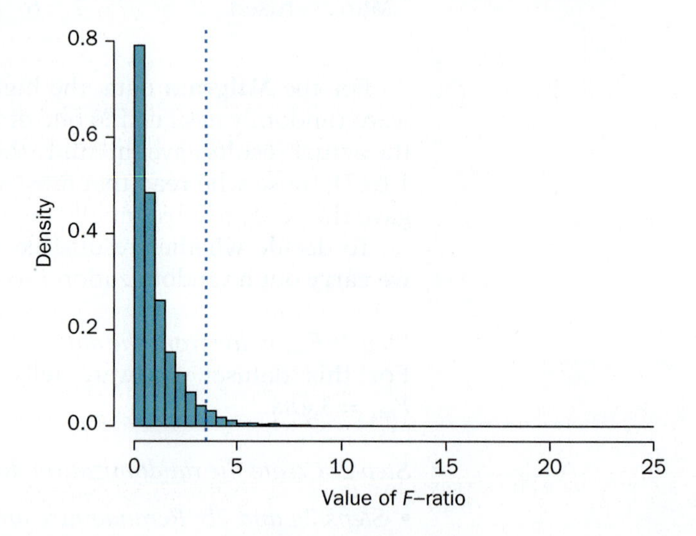

FIGURE 8.11 The *P*-value equals the area of the histogram to the right of the dashed line

We now turn from the mechanics of the randomization test to the logic of the test: What is the relationship between the randomization *F*-test and the *F*-test that comes from an ANOVA table? What is the justification for the randomization test, and how does it compare with the justification for the ordinary *F*-test?

The Logic of the Randomization *F*-test

The key to the logic of the randomization *F*-test is the question, *"Suppose there are no group differences. What kinds of results can I expect to get?"* The answer: If there are no group differences (no treatment effects), then the *only* thing that can make one group different from another is the random assignment. If that's true, then the data I got should look much like other datasets I *might* have gotten by random assignment, that is, datasets I get by rerandomizing the assignment of response values to groups. On the other hand, if my actual dataset looks quite different from the typical datasets that I get by rerandomizing, then that difference is evidence that it is the treatments, and not just the random assignment, causing the differences to be big. To make this reasoning concrete, consider a simple artificial example based on the calcium study.

EXAMPLE 8.13

CaBP

Calcium (continued) Consider the 11 subjects in the calcium study who were given the placebo. Here are the decreases in blood pressure for these 11 men:

$$-1 \quad 12 \quad -1 \quad -3 \quad 3 \quad -5 \quad 5 \quad 2 \quad -11 \quad -1 \quad -3$$

The 11 men were treated exactly the same, so these values tell how these men would respond when there is no treatment effect. Suppose you use these 11 subjects for a pseudo-experiment, by randomly assigning each man to one of two groups, which you call "A" and "B," with 6 men assigned to Group A and 5 men assigned to B.

							Mean
Group A	−1	3	2	−11	−1	−3	−1.8
Group B	12	−1	−3	−5	5		1.6

Although you have created two groups, there is no treatment and no control. In fact, the men aren't even told which group they belong to, so there is absolutely no difference between the two groups in terms of how they are treated. True, the group means are different, but *the difference is due entirely to the random assignment*. This fact is a major reason for randomizing: If there are no treatment differences, then any observed differences can only be caused by the randomization. So if the observed differences are roughly the size you would expect from the random assignment, there is no evidence of a treatment effect. On the other hand, if the observed differences are too big to be explained by the randomization alone, then we have evidence that some other cause is at work. Moreover, because of the randomization, the only other possible cause is the treatment.

We now carry out the randomization F-test to compare Groups A and B:

Step 1: Value from the data F_{obs}.
For this dataset, the value of the F-statistic is $F_{obs} = 0.9155$.

Step 2: Create the randomization distribution.
Figure 8.12 shows the distribution of F-statistics based on 10,000 repetitions.

Step 3: P*-value.*
Because 3994 of 10,000 rerandomizations gave a value of F_{Rand} greater than or equal to $F_{obs} = 0.9155$ (Figure 8.12), the estimated P-value is 0.40.

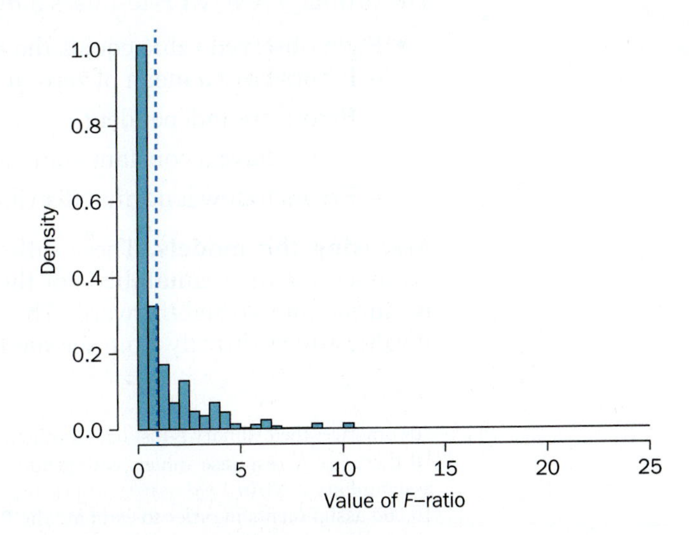

FIGURE 8.12 3994 of 10,000 values of F_{Rand} were at least as large as $F_{obs} = 0.9155$

The *P*-value tells us that close to half (40%) of random assignments will produce an *F*-value of 0.9155 or greater. In other words, the observed *F*-value looks much like what we can expect to get from random assignment alone. There is no evidence of a treatment effect, which is reassuring, given that we know there was none.

Isn't it possible to get a large value of *F* purely by chance, due just to the randomization? Yes, that outcome (a very large value of *F*) is always possible, but unlikely. The *P*-value tells us how unlikely. Whenever we reject a null hypothesis based on a small *P*-value, the force of our conclusion is based on "either/or": Either the null hypothesis is false or an unlikely event has occurred. Either there are real treatment differences or the differences are due to a very unlikely and atypical random assignment.

The Randomization *F*-test and the Ordinary *F*-test Compared

Many similarities. In this subsection, we compare the randomization *F*-test with the "ordinary" *F*-test for one-way ANOVA. The two varieties of *F*-tests are quite similar in most ways:

- Both *F*-tests are intended for datasets with a quantitative response and one categorical predictor that sorts response values into groups.
- Both *F*-tests are designed to test whether observed differences in group means are too big to be due just to chance.
- Both *F*-tests use the same data summary, the *F*-statistic, as the test statistic.
- Both *F*-tests use the same definition of a *P*-value as a probability, namely the probability that a dataset chosen at random will give an *F*-statistic at least as large as the one from the actual data: *P*-value $= Pr(F_{Rand} \geq F_{obs})$.
- Both *F*-tests use a probability model to compute the *P*-value.*

Different probability models. The last of these points of similarity is also the key to the main difference between the two tests: They use very different probability models.

The randomization *F*-test uses a model of equally likely assignments:

- All possible assignments of treatments to units are equally likely.†

The ordinary ANOVA *F*-test uses a model of independent normal errors:

- Each observed value equals the mean plus error: $y = \mu + \epsilon$.
  - Errors have a mean of zero: $\mu_\epsilon = 0$.
  - Errors are independent.
  - Errors have a constant standard deviation: σ_ϵ.
  - Errors follow a normal distribution: $\epsilon \sim N(0, \sigma_\epsilon)$.

Assessing the models. These differences in the models have consequences for assessing their suitability. For the randomization test, assessing conditions is simple and straightforward: The probability model you use to compute a *P*-value comes directly from the method you use to produce the data.

*In practice, the ordinary *F*-test uses mathematical theory as a shortcut for finding the *P*-value.

†If there are *N* response values, with group sizes $n_1, n_2, ..., n_J$, then the number of equally likely assignments is $N!/(n_1!\, n_2! \cdots n_J!)$. In practice, this number is so large that we randomly generate 10,000 assignments in order to estimate the *P*-value, but in theory, we could calculate the *P*-value exactly by considering every one of these possible assignments.

Checking conditions for the randomization model:

- Were treatments assigned to units using chance? If yes, we can rely on the randomization *F*-test.

For the normal errors model, the justification for the model is not nearly so direct or clear. There is no way to know with certainty whether errors are normal or whether standard deviations are equal. Checks are empirical, and may mislead us.

Checking conditions for the normal errors model:

- Ideally, we should check all four conditions for the errors. In practice, this means at the very least looking at a residual plot and a normal probability plot or normal quantile plot.

With the normal errors model, if you are wrong about whether the necessary conditions are satisfied, you risk being wrong about your interpretation of small *P*-values.

Scope of inference. The randomization *F*-test is designed for datasets for which the randomization is clear from the way the data were produced. As you know, there are two main ways to randomize, and two corresponding inferences. If you randomize the assignment of treatments to units, your experimental design eliminates all but two possible causes for group differences, the treatments and the randomization. A tiny *P*-value lets you rule out the randomization, leaving the treatments as the only plausible cause. Thus when you have randomized the assignment of treatments to units, you are justified in regarding significant differences as evidence of cause and effect. The treatments caused the differences. If you could not randomize, or did not, then to justify an inference about cause, you must be able to eliminate other possible causes that might have been responsible for observed differences. Eliminating these alternative causes is often tricky, because they remain hidden, and because the data alone can't tell you about them.

The second of the two main ways to randomize is by random sampling from a population; like random assignment, random sampling eliminates all but two possible causes of group differences: Either the observed differences reflect differences in the populations sampled or else they are caused by the randomization. Here, too, a tiny *P*-value lets you rule out the randomization, leaving the population differences as the only plausible cause of observed differences in group means.

Although the exposition in this section has been based on random assignment, you can use the same randomization *F*-test for samples chosen using chance.[5] As with random assignment, when you have used a probability-based method to choose your observational units, you have an automatic probability model that comes directly from your sampling method. Your use of random sampling guarantees that the model is suitable, and there is no need to check the conditions of the normal errors model. On the other hand, if you have not used random sampling, then you cannot automatically rule out selection bias as a possible cause of observed differences, and you should be more cautious about your conclusions from the data.

A surprising fact. As you have seen, the probability models for the two kinds of *F*-tests are quite different. Each of the models tells how to create random pseudo-datasets, and so each model determines a distribution of *F*-statistics computed from those random datasets. It would be reasonable to expect that if you use both models for the same dataset, you would get quite different distributions for the *F*-statistic and different *P*-values. Surprisingly, that doesn't often happen. Most of the time, the two methods give very similar distributions

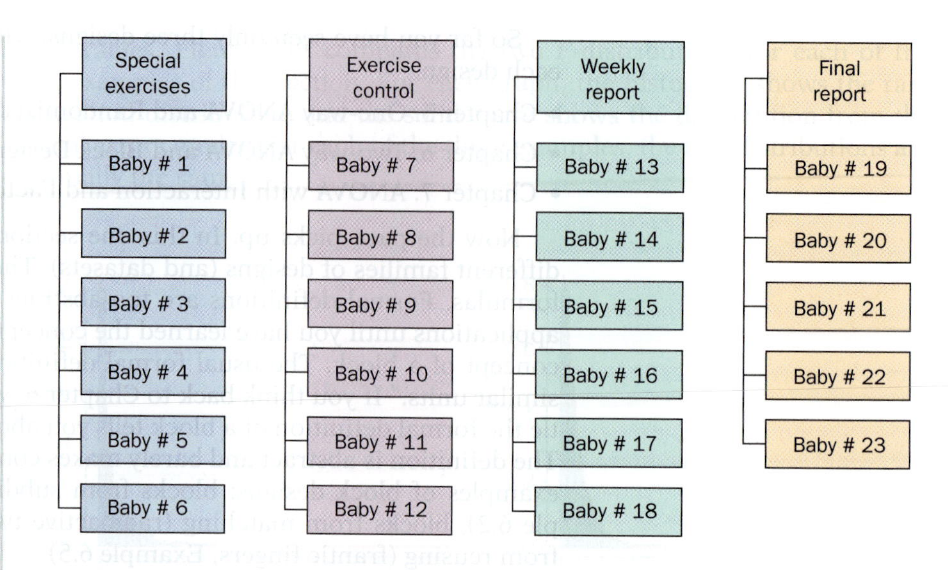

FIGURE 8.14 The walking babies design, with babies (subjects) as units. The babies here are numbered from 1 to 23, but in the actual study, they were assigned at random.

For a between-subjects design, each subject is a unit and gets a single treatment. For a within-subjects design, each subject is a block of units (time slots) and receives a complete set of treatments.

WITHIN-SUBJECTS DESIGN

A **within-subjects design** is a complete block design with each subject as a block of units. The experimental unit is a "time slot" for a subject. Each subject provides a block of time slots and receives a complete set of treatments, one at a time, in sequence. Each treatment is assigned to a time slot.

EXAMPLE 8.15

Within-subjects designs

a. **The study of caffeine and theobromine (Example 6.5)** is a within-subjects design. Each subject provided a block of three time slots. Each subject received all three drugs, one at a time, in a random order, as shown in Figure 8.15.

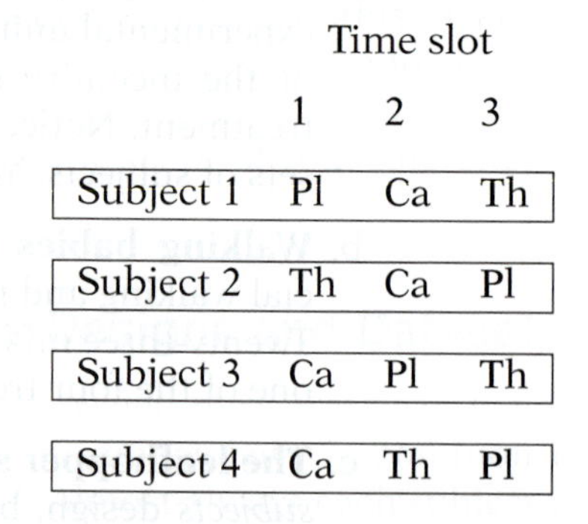

	Time slot 1	2	3
Subject 1	Pl	Ca	Th
Subject 2	Th	Ca	Pl
Subject 3	Ca	Pl	Th
Subject 4	Ca	Th	Pl

FIGURE 8.15 The frantic fingers design, with subjects as blocks of time slots

b. **Shrews.** The study of sleeping shrews (Example 6.7) is also a within-subjects design. There were six shrews (subjects) and each was measured during three phases of sleep (three time slots). The order could not be randomly assigned, because you can't tell a shrew what sleep phase you want to observe. All the same, each shrew was observed under all three phases.

c. **Wheat (Example 6.2)** is not, strictly speaking, a within-*subjects* design, because the block is not a subject, and the experimental unit is not a time slot, but a plot of land. Abstractly, the design is the same. Either way, each block of units receives a complete set of treatments. For agriculture, it is space rather than time that gets subdivided.

CROSSING FACTORS

Basic designs can be extended by **factorial crossing**. You can have two (or more) crossed between-subjects factors. You can have two (or more) crossed within-subjects factors. The possibilities are endless.

EXAMPLE 8.16

Two crossed between-subjects factors Many of the examples in Chapter 7 had two crossed between-subjects factors.

a. **Pigs.** For the study of *B12* and *Antibiotics* (Example 7.2), the 12 subjects were baby pigs, and each piglet was assigned to receive one of four diet combinations, obtained by crossing, *B12* (Yes/No) and *Antibiotics* (Yes/No).

b. **Rats.** For the study of rat diets (Example 7.9) there were 60 subjects (baby rats). Each rat received one of six diets, obtained by crossing *Source* (Beef, Cereal, or Pork) with protein *Quality* (High or Low).

EXAMPLE 8.17

Two crossed within-subjects factors

a. **Visual/verbal**
Task (letter/sentence) crossed with *Report* (point/say) within subjects. Remind yourself of Example 6.6. The study was designed to use a subject's response time to test the theory that the left and right sides of the brain specialize, with verbal on the left and visual on the right. Each subject provided a block of four time slots. The four treatments assigned to those time slots came from crossing two factors, the task (visual or verbal), and the report. This study had two crossed within-subjects factors. We'll sometimes use notation like *Task* × *Report* as a shorthand to denote crossed factors.

b. **Remembering words**
Abstraction crossed with *Frequency* within subjects. Imagine yourself as a subject in this version of a standard within-subjects experiment in cognitive psychology. The goal is to learn which of four kinds of words are easier to remember. As a subject, you hear four different lists of 25 nouns and are then asked to list all the nouns you remember. The response is the percent of words you recall from each list. The words are of four kinds, defined by crossing two within-subjects factors, *Abstraction* (abstract/concrete) and *Frequency* (frequent/infrequent).

Concrete	Frequent:	dog, bathtub, grandmother
Concrete	Infrequent:	vat, snorkel, catapult
Abstract	Frequent:	size, fitness, quality
Abstract	Infrequent:	sloth, ardor, entropy

Here are the average numbers of words remembered (out of 25) by 10 subjects in one version of the experiment:

	Frequent	Infrequent
Abstract	10.3	9.7
Concrete	11.0	10.6

Now that we have looked at many ideas and examples of experimental design, we conclude this section by revisiting the **VisualVerbal** data.

EXAMPLE 8.19

FFlies

Fruit flies (two categories) Consider the **FruitFlies** data from Chapter 5, where we examined the life span of male fruit flies. Two of the groups in that study were *8 virgins* and *none*, which we compare again here (ignoring, for now, the other three groups in the study). The 25 fruit flies in the *8 virgins* group had an average life span of 38.72 days, whereas the average for the 25 fruit flies living alone (in the *none* group) was 63.56 days.

Is the difference, 38.72 versus 63.56, statistically significant? Or could the two sample means differ by 38.72 − 63.56 = −24.84 just by chance? Let's approach this question from three distinct directions: the pooled two-sample *t*-test of Chapter 0, one-way ANOVA for a difference in means as covered in Chapter 5, and regression with an indicator predictor as described in Chapter 3.

Pooled Two-sample *t*-Test

Parallel dotplots, shown in Figure 8.17, give a visual summary of the data. The life spans for most of the fruit flies living alone are greater than those of most of the fruit flies living with 8 virgins. In the computer output for a pooled two-sample *t*-test, the value of the test statistic is −6.08, with a *P*-value of approximately 0 based on a *t*-distribution with 48 degrees of freedom. This small *P*-value gives strong evidence that the average life span for fruit flies living with 8 virgins is smaller than the average life span for those living alone.

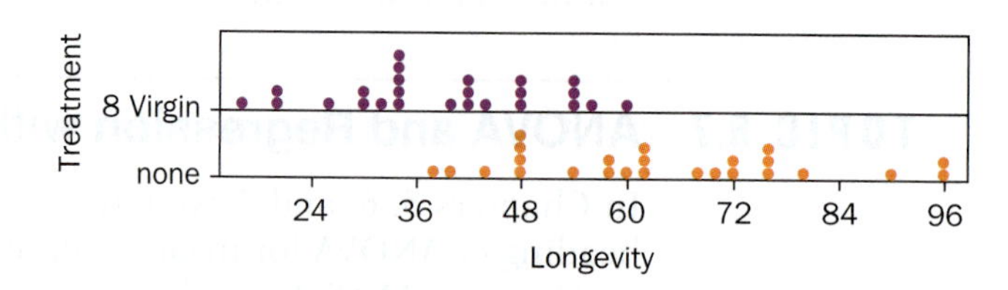

FIGURE 8.17 Life spans for 8 virgins and living alone groups

```
Two-sample T for Longevity

Treatment   N   Mean   StDev   SE Mean
8 virgin   25   38.7    12.1       2.4
none       25   63.6    16.5       3.3

Difference = mu (8 virgin) - mu (none)
Estimate for difference: -24.84
95% CI for difference: (-33.05, -16.63)
T-Test of difference = 0 (vs not =): T-Value = -6.08 P-Value = 0.000 DF = 48
Both use Pooled StDev = 14.4418
```

ANOVA with Two Groups

As we showed in Chapter 5, these data can also be analyzed using an ANOVA model. The output for the ANOVA table follows:

```
Source      DF     SS    MS      F      P
Treatment    1   7713  7713  36.98  0.000
Error       48  10011   209
Total       49  17724

S = 14.44  R-Sq = 43.52%  R-Sq(adj) = 42.34%
```

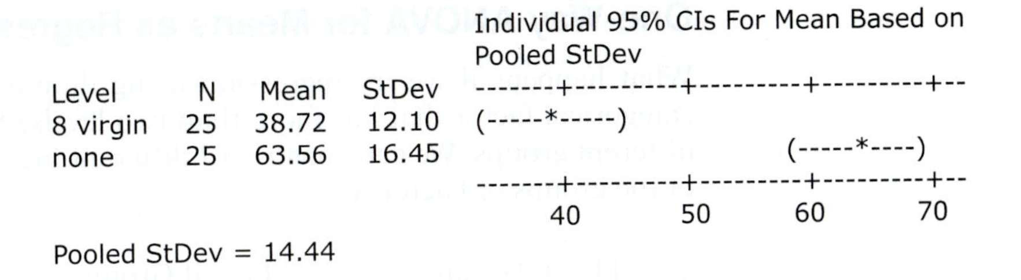

```
                                Individual 95% CIs For Mean Based on
                                Pooled StDev
Level      N   Mean   StDev   -------+---------+---------+---------+--
8 virgin  25  38.72   12.10   (-----*-----)
none      25  63.56   16.45                            (-----*----)
                              -------+---------+---------+---------+--
                                    40        50        60        70

Pooled StDev = 14.44
```

Notice that the P-value is approximately 0, the same that we found when doing the two-sample test. Notice also that the F-value is 36.98, which is $(-6.08)^2$ approximately; the only difference between the $(-6.08)^2$ and 36.98 is due to rounding error. In fact, the F-statistic will always be the square of the t-statistic and the P-value will be the same no matter which test is run.

Regression with an Indicator

In Chapter 3, we introduced the idea of using an indicator variable to code a binary categorical variable as 0 or 1 in order to use it as a predictor in a regression model. What if that is the *only* predictor in the model? For the fruit flies example, we can create an indicator variable, *V8*, to be 1 for the fruit flies living with 8 virgins and 0 for the fruit flies living alone. The results for fitting a regression model to predict life span using *V8* are shown as follows. Figure 8.18 shows a scatterplot of *Lifespan* versus *V8* with the least-squares line. Note that the intercept for the regression line (63.56) is the mean life span for the sample of 25 *none* fruit flies ($V8 = 0$). The slope $\hat{\beta}_1 = -24.84$ shows how much the mean decreases when we move to the *8 virgin* fruit flies (mean = 38.72). We also see that the t-test statistic, degrees of freedom, and P-value for the slope in the regression output are identical to the corresponding values in the pooled two-sample t-test.

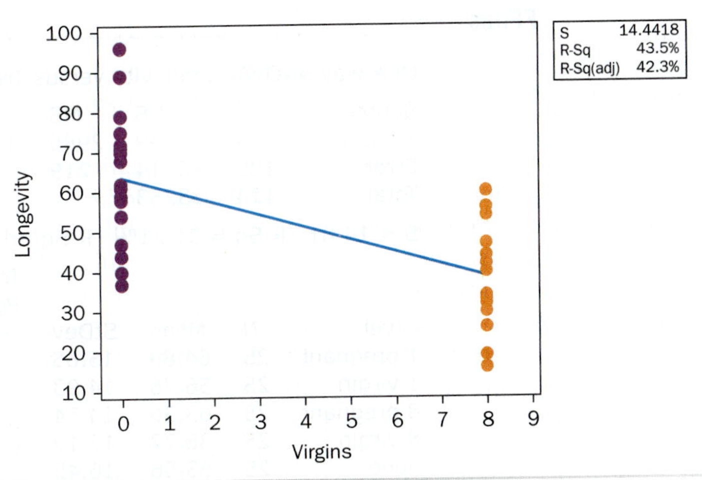

FIGURE 8.18 Life spans for fruit flies living alone and with 8 virgins

```
The regression equation is
Longevity = 63.6 - 24.8 v8

Predictor      Coef   SE Coef       T      P
Constant     63.560     2.888   22.01  0.000
v8          -24.840     4.085   -6.08  0.000

S = 14.4418  R-Sq = 43.5%  R-Sq(adj) = 42.3
Analysis of Variance

Source           DF       SS       MS      F      P
Regression        1   7712.8   7712.8  36.98  0.000
Residual Error   48  10011.2    208.6
Total            49  17724.0
```

EXAMPLE 8.21

Pig

Feeding pigs: Dummy regression In Example 7.2, we looked at a two-factor model to see how the presence or absence of antibiotics and vitamin B12 might affect the weight gain of pigs. Recall that the data in **PigFeed** had two levels (yes or no) for each the two factors and three replications in each cell of the 2×2 factorial design, for an overall sample size of $n = 12$. Since $I = 2$ and $J = 2$, we need just one indicator variable per factor to create a regression model that is equivalent to the two-way ANOVA with main effects:

$$WgtGain = \beta_0 + \beta_1 A + \beta_2 B + \epsilon$$

where A is one for the pigs who received antibiotics in their feed (zero if not) and B is one for the pigs who received vitamin B12 (zero if not).

The two-way ANOVA (main effects only) follows:

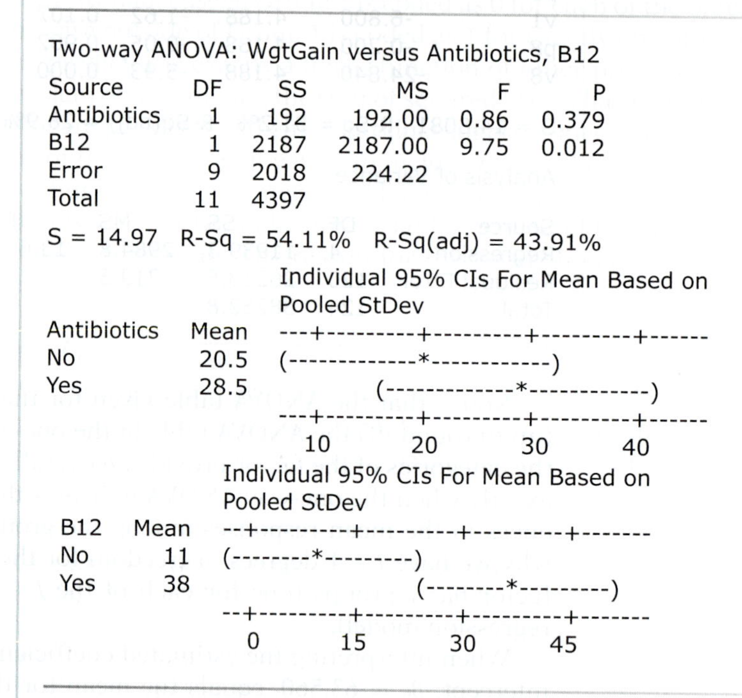

```
Two-way ANOVA: WgtGain versus Antibiotics, B12
Source       DF    SS       MS     F      P
Antibiotics   1   192   192.00  0.86  0.379
B12           1  2187  2187.00  9.75  0.012
Error         9  2018   224.22
Total        11  4397

S = 14.97  R-Sq = 54.11%  R-Sq(adj) = 43.91%
                     Individual 95% CIs For Mean Based on
                     Pooled StDev
Antibiotics  Mean   ---+---------+---------+---------+------
No           20.5   (-------------*------------)
Yes          28.5            (-------------*------------)
                    ---+---------+---------+---------+------
                      10        20        30        40
                 Individual 95% CIs For Mean Based on
                 Pooled StDev
B12  Mean        --+---------+---------+---------+-------
No     11        (--------*---------)
Yes    38                          (--------*---------)
                 --+---------+---------+---------+-------
                   0        15        30        45
```

And here is the multiple regression output:

```
Regression Analysis: WgtGain versus A, B
The regression equation is
WgtGain = 7.00 + 8.0 A + 27.00 B

Predictor    Coef  SE Coef     T      P
Constant    7.000    7.487  0.93  0.374
A           8.000    8.645  0.93  0.379
B          27.000    8.645  3.12  0.012

S = 14.9741  R-Sq = 54.1%  R-Sq(adj) = 43.9%
Analysis of Variance

Source          DF      SS      MS     F      P
Regression       2  2379.0  1189.5  5.31  0.030
Residual Error   9  2018.0   224.2
Total           11  4397.0

Source  DF  Seq SS
A        1   192.0
B        1  2187.0
```

Now we have to dig a little deeper to make the connections between the two-way ANOVA (main effects) output and the multiple regression using the two indicators. Obviously, we see a difference in the ANOVA tables themselves since the multiple regression combines the effects due to both A and B into a single component while they are treated separately in the two-way ANOVA. However, the regression degrees of freedom and SSModel terms are just the sums of the individual factors shown in the two-way ANOVA model; for example, $2 = 1 + 1$ for the degrees of freedom and $2379 = 192 + 2187$ for the sum of squares. Furthermore, the "Seq SS" numbers given by software regression output show the contribution to the variability explained by the model for each factor. These sums of squares match those in the two-way ANOVA output. Furthermore, note that the P-values for the individual t-tests of the coefficients of the two indicators in the multiple regression match the P-values for each factor as main effects in the two-way model; this is a bonus that comes when there are only two levels in a factor.

Comparing group means in the two-way ANOVA output shows that the difference for *Antibiotic* is $28.5 - 20.5 = 8$ and for *B12* is $38 - 11 = 27$, exactly matching the coefficients of the respective indicators in fitted regression. What about the estimated constant term of $\hat{\beta}_0 = 7.0$? Our experience tells us that this should have something to do with the no antibiotic, no B12 case ($A = B = 0$), but the mean of the data in that cell is $\bar{y}_{11} = 19$. Remember that the main effects model also had some difficulty predicting the individual cells accurately. In fact, you can check that the predicted means for each cell using the two-indicator regression match the values generated from the estimated effects in the main effects only ANOVA. In Chapter 7 we included an interaction term in the model to explain the effects of antibiotics and vitamin B12 on pig weight growth. Let's see how to translate the interaction model into a multiple regression setting.

Recall that, in earlier regression examples (such as comparing two regression lines in Section 3.3 or the interaction model for perch weights in Example 3.11), we handled interaction by including a term that was a product of the two interacting variables. The same reasoning works for indicator variables. For the **PigFeed** data, the appropriate model is

$$WgtGain = \beta_0 + \beta_1 A + \beta_2 B + \beta_3 AB + \epsilon$$

Output for the two-way ANOVA with interaction is shown as follows:

Source	DF	SS	MS	F	P
Antibiotic	1	192.0	192.0	5.30	0.050
B12	1	2187.0	2187.0	60.33	0.000
Antibiotic*B12	1	1728.0	1728.0	47.67	0.000
Error	8	290.0	36.3		
Total	11	4397.0			

S = 6.02080 R-Sq = 93.40% R-Sq(adj) = 90.93%

Means

Antibiotic	B12	N	WgtGain
No	No	3	19.000
No	Yes	3	22.000
Yes	No	3	3.000
Yes	Yes	3	54.000

Note that there are 2 degrees of freedom for *Filling*, 3 degrees of freedom for *Bread*, and 6 degrees of freedom for the interaction. The output from running this regression follows:

```
The regression equation is
Ants = 58.5 - 21.5 A1 - 25.0 A2 - 1.00 B1 - 2.00 B2 - 9.00 B3 + 4.3 A1B1
       + 3.3 A1B2 + 18.0 A1B3 - 3.5 A2B1 + 7.2 A2B2 + 12.8 A2B3

Predictor      Coef  SE Coef      T      P
Constant     58.500    6.692   8.74  0.000
A1          -21.500    9.463  -2.27  0.029
A2          -25.000    9.463  -2.64  0.012
B1           -1.000    9.463  -0.11  0.916
B2           -2.000    9.463  -0.21  0.834
B3           -9.000    9.463  -0.95  0.348
A1B1           4.25    13.38   0.32  0.753
A1B2           3.25    13.38   0.24  0.810
A1B3          18.00    13.38   1.34  0.187
A2B1          -3.50    13.38  -0.26  0.795
A2B2           7.25    13.38   0.54  0.591
A2B3          12.75    13.38   0.95  0.347

S = 13.3832  R-Sq = 40.2%  R-Sq(adj) = 22.0%

Analysis of Variance
Source          DF       SS     MS     F      P
Regression      11   4338.0  394.4  2.20  0.037
Residual Error  36   6448.0  179.1
Total           47  10786.0

Source  DF  Seq SS
A1       1   234.4
A2       1  3486.1
B1       1    25.0
B2       1     6.1
B3       1     9.4
A1B1     1    10.1
A1B2     1    68.1
A1B3     1   180.2
A2B1     1   155.0
A2B2     1     1.0
A2B3     1   162.6
```

As we saw in the previous example, the ANOVA *F*-test for the multiple regression model combines both factors and the interaction into a single test for the overall model. Check that the degrees of freedom ($11 = 2+3+6$) and sum of squares ($4338 = 3720.5 + 40.5 + 577.0$) are sums of the three model components in the two-way ANOVA table. In the multiple regression setting, the tests for the main effects for Factor A (coefficients of A_1 and A_2), Factor B (coefficients of B_1, B_2 and B_3), and the interaction effect (coefficients of the six indicator products) can be viewed as nested *F*-tests. For example, to test the interaction, we sum the sequential SS values from the Minitab output, $10.1 + 68.1 + 180.2 + 155.0 + 1.0 + 162.6 = 576.9$, matching (up to roundoff) the interaction sum of squares in the two-way ANOVA table and having 6 degrees of freedom. In this way, we can also find the sum of squares for Factor A ($234.4 + 3486.1 = 3720.5$) with 2 degrees of freedom and for Factor B ($25.0 + 6.1 + 6.4 = 4.5$) with 3 degrees of freedom.

Let's see if we can recover the cell means in Table 8.2 from the information in the multiple regression output. The easy case is the constant term, which represents the cell mean for the indicators that weren't included in the model (*HamPickles* on *MultiGrain*). If we keep the bread fixed at *MultiGrain*, we see that the cell means drop by −21.5 ants and −25.0 ants as we move to *PeanutButter*, and *Vegemite*, respectively. On the other hand, if we keep *HamPickles* as the filling, the coefficients of B_1, B_2, and B_3 indicate what happens to the cell means as we change the type of bread to *Rye*, *White*, and *WholeWheat*, respectively. For any of the other cell means, we first need to adjust for the filling, then the bread, and finally the interaction. For example, to recover the cell mean for *Vegemite* (A_2) and *Rye* (B_1), we have $58.5 - 25.0 - 1.0 - 3.5 = 29.0$, the sample mean number of ants for a vegemite on rye sandwich.

With most statistical software, we can run the multiple regression version of the ANOVA model without explicitly creating the individual indicator variables, as the software will do this for us. Once we recognize that we can include any categorical factor in a regression model by using indicators, the next natural extension is to allow for a mixture of both categorical and quantitative terms in the same model. Although we have already done this on a limited scale with single binary categorical variables, the next topic explores these sorts of models more fully.

TOPIC 8.8 Analysis of Covariance

We now turn our attention to the setting in which we would like to model the relationship between a continuous response variable and a categorical explanatory variable, but we suspect that there may be another quantitative variable affecting the outcome of the analysis. Were it not for the additional quantitative variable, we would use an ANOVA model, as discussed in Chapter 5 (assuming that conditions are met). But we may discover, often after the fact, that the experimental or observational units were different at the onset of the study, with the differences being measured by the additional continuous variable. For example, it may be that the treatment groups had differences between them even before the experimental treatments were applied to them. In this case, any results from an ANOVA analysis become suspect. If we find a significant group effect, is it due to the treatment, the additional variable, or both? If we do not find a significant group effect, is there really one there that has been masked by the extra variable?

Setting the Stage

One method for dealing with both categorical and quantitative explanatory variables is to use a multiple regression model, as discussed in Section 3.3. But in that model, both types of explanatory variables have equal importance with respect to the response variable. Here we discuss what to do when the quantitative variable is more or less a nuisance variable: We know it is there, we have measured it on the observations, but we really don't care about its relationship with Y other than how it interferes with the relationship between Y and the factor of interest. This type of variable is called a *covariate*.

COVARIATE

A continuous variable X_c not of direct interest, but that may have an effect on the relationship between the response variable Y and the factor of interest, is called a **covariate**.

The type of model that takes covariates into consideration is called an *analysis of covariance model*.

Of course, the ANCOVA model is only appropriate under certain conditions:

CONDITIONS

The conditions necessary for the ANCOVA model are:

- All of the ANOVA conditions are met for Y with the factor of interest.
- All of the linear regression conditions are met for Y with X_c.
- There is no interaction between the factor and X_c.

Note that the last condition is equivalent to requiring that the linear relationship between Y and X_c have the same slope for each of the levels of the factor.

We saw in Example 8.23 that the discount did not seem to have an effect on the sales of the products. This finding is consistent with Figure 8.19, which shows considerable overlap between the three groups.

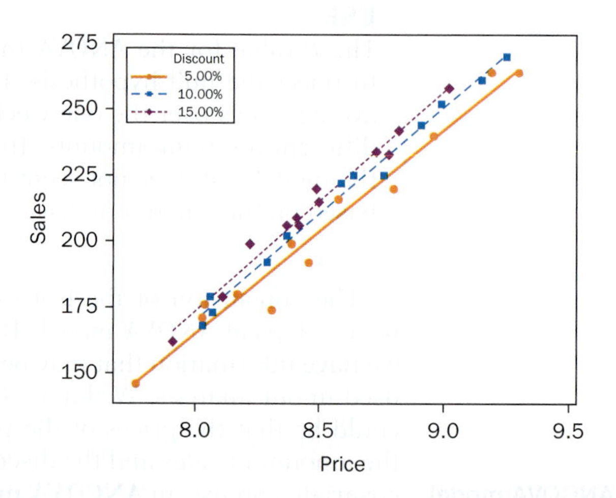

FIGURE 8.21 Scatterplot of *Sales* versus *Price* by *Discount*

Now consider Figure 8.21. This is a scatterplot of the product sales versus the product price with different symbols for the three different discount amounts. We have also included the regression lines for each of the three groups. This graph shows a more obvious difference between the three groups. If we collapse the y values across the x-axis (as the dotplot does), there is not much difference between the sales at the different discount levels. But if we concentrate on specific values for the product price, there is a difference. The scatterplot suggests that the higher the discount, the more the product will sell. This suggests that an ANCOVA model may be appropriate.

EXAMPLE 8.24

Groc

Grocery store data: Checking conditions In Example 8.23, we already discovered that the conditions for an ANOVA model with discount predicting the product sales were met, so we move on to checking the conditions for the linear regression of the sales on the price of the product. Figure 8.22 is a scatterplot of the response versus the covariate. There is clearly a strong linear relationship between these two variables. Next, we consider Figure 8.23. Figure 8.23(a) gives the normal probability plot of the residuals and Figure 8.23(b) shows the residuals versus fits plot. Both of these graphs are consistent with the conditions for linear regression. We also have no reason to believe that one product's sales will have affected any other product's sales, so we can consider the error terms to be independent.

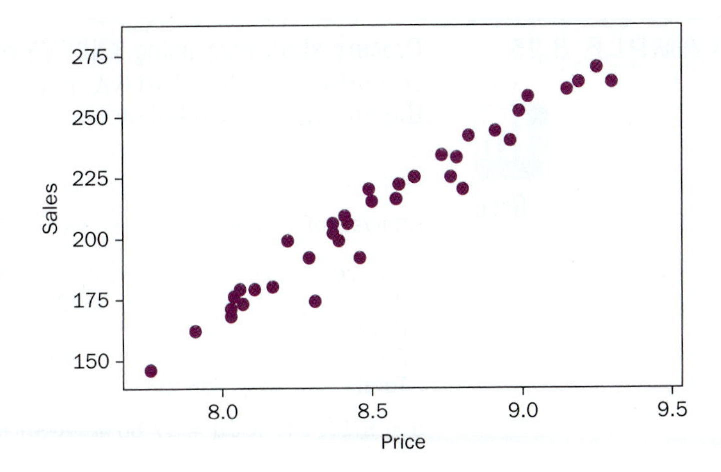

FIGURE 8.22 Scatterplot of sales versus price

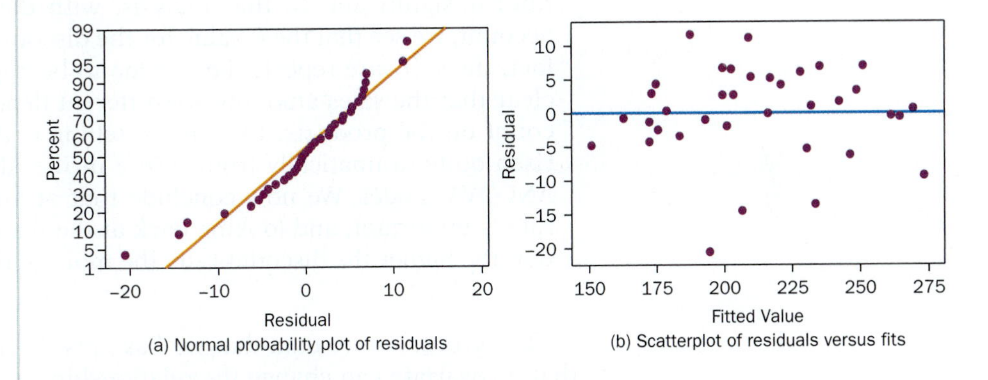

FIGURE 8.23 Plots to assess the fit of the linear regression model

Finally, we need to consider whether the slopes for the regression of sales on price are the same for all three groups. Figure 8.21 shows that while they are not identical, they are quite similar. In fact, because of variability in data, it would be very surprising if a dataset resulted in slopes that were exactly the same.

It appears that all of the conditions for the ANCOVA model have been met and we can proceed with the analysis.

At this point, we have discussed the general reason for using an ANCOVA model, what the equation for the model looks like, and what the conditions are for the model to be appropriate. What's left to discuss is how to actually fit the model and how to interpret the results. As with any regression model, we rely on software to find least-squares estimates to fit the model. When interpreting the results in the ANOVA table, the basic ideas behind the sums of squares are the same as when we looked at regression and ANOVA separately in Chapters 3 and 5. In fact, the output looks very similar to that of the ANOVA model but with one more line in the ANOVA table corresponding to the covariate. When we interpret this output, we typically continue to concentrate on the *P*-value associated with the factor, as we did in the ANOVA model. Specifically, we are interested in whether the effect of the factor is significant now that we have controlled for the covariate, and whether the significance has changed in moving from the ANOVA model to the ANCOVA model. If the answer to the latter question is yes, then the covariate was indeed important to the analysis.

EXAMPLE 8.25

Groc

Grocery store data using ANCOVA model In Example 8.24, we saw that the conditions for the ANCOVA model were satisfied. The computer output for the model is shown below:

Analysis of Variance for Sales, using Adjusted SS for Tests

Source	DF	Seq SS	Adj SS	Adj MS	F	P
Price	1	36718	36230	36230	1372.84	0.000
Discount	2	800	800	400	15.15	0.000
Error	32	844	844	26		
Total	35	38363				

S = 5.13714 R-Sq = 97.80% R-Sq(adj) = 97.59%

Once again, we check to see whether the F-test for the factor (discount rate) is significant. In this analysis, with the covariate (*Price*) taken into account, we see that the P-value for the discount rate is indeed significant. In fact, the software reports the P-value to be approximately 0. It seems quite clear that the sales amounts are different depending on the amount of discount on the products. Even more telling is the fact that the R^2 value has risen quite dramatically from 3.36% in the ANOVA model to 97.8% in the ANCOVA model. We now conclude that at any given price level, discount rate is important, and looking back at the scatterplot in Figure 8.21, we see that the higher the discount rate, the more of the product that is sold.

The grocery store data that we have just been considering shows one way that a covariate can change the relationship between a factor and a response variable. In that case, there did not seem to be a relationship between the factor and the response variable until the covariate was taken into consideration. *It might also be that the factor and the covariate interact*. If so, then we cannot isolate the effect of the factor, since "controlling for the covariate" will change across values of the covariate. In the **Grocery** setting, if *Discount* and *Price* interacted, then we might have a conclusion that would be something like "When *Price* is low, the effect of discounting is xxx but when *Price* is high the effect of discounting is yyy" where xxx and yyy are two different effects. Or maybe the conclusion would be "When *Price* is low, the effect of discounting is xxx but when *Price* is high discounting has no effect."

The next example illustrates another way that a covariate can change the conclusions of an analysis.

EXAMPLE 8.26

Pulse

Exercise and heart rate[8] Does how much exercise you get on a regular basis affect how high your heart rate is when you are active? This was the question that a Stat 2 instructor set out to examine. He had his students rate themselves on how active they were generally (1 = not active, 2 = moderately active, 3 = very active). This variable was recorded with the name *Exercise*. He also measured several other variables that might be related to active pulse rate, including the student's sex and whether he or she smoked or not. The last explanatory variable he had them measure was their resting pulse rate. Finally, he assigned the students to one of two treatments (walk or run up and down a flight of stairs three times) and measured their pulse when they were done. This last pulse rate was their "active pulse rate" (*Active*) and the response variable for the study. The data are found in the file **Pulse**.

CHOOSE

We start with the simplest model. That is, we start by looking to see what the ANOVA model tells us since we have a factor with three levels and a quantitative response.

FIT

The ANOVA table is given as follows:

```
One-way ANOVA: Active versus Exercise

Source     DF     SS     MS      F      P
Exercise    2  10523   5261  16.90  0.000
Error     229  71298    311
Total     231  81820

S = 17.64  R-Sq = 12.86%  R-Sq(adj) = 12.10%
```

ASSESS

Figure 8.24(a) shows the residual versus fits plot and Figure 8.24(b) shows the normal probability plot of the residuals for this model. Both indicate that the conditions of normality and equal variances are met. Levene's test (see Section 8.1) also confirms the consistency of the data with an equal variances model.

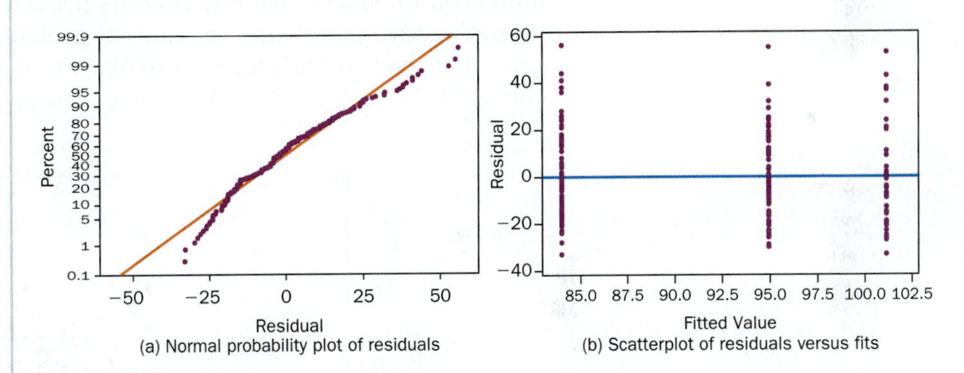

FIGURE 8.24 Plots to assess the fit of the ANOVA model

```
Levene's Test

Test statistic = 2.11, p-value = 0.124
```

USE

In this case, the *P*-value is quite small in the ANOVA table and therefore it seems that the amount of exercise a student gets in general does affect how high his or her pulse is after exercise. But we also notice that the R^2 value is still fairly low at 12.86%, confirming that this model does not explain everything about the after-exercise pulse rate.

Now we use the fact that we have a covariate that might affect the analysis: the resting pulse rate.

CHOOSE

We redo the analysis, this time using analysis of covariance.

FIT

The ANCOVA table is given below:

Analysis of Variance for Active, using Adjusted SS for Tests

Source	DF	Seq SS	Adj SS	Adj MS	F	P
Rest	1	29868	19622	19622	86.57	0.000
Exercise	2	276	276	138	0.61	0.544
Error	228	51676	51676	227		
Total	231	81820				

S = 15.0549 R-Sq = 36.84% R-Sq(adj) = 36.01%

ASSESS

We already assessed the conditions of the ANOVA model. Now we move on to assessing the conditions of the linear regression model for predicting the active pulse rate from the resting pulse rate. Figure 8.25 is a scatterplot of the response versus the covariate. There is clearly a strong linear relationship between these two variables. Next, we consider Figure 8.26. Figure 8.26(a) gives the normal probability plot of the residuals and Figure 8.26(b) gives the residuals versus fits plot. Although the residuals versus fits plot is consistent with the conditions for linear regression of equal variance and we have no reason to believe that one person's active pulse rate will affect any other person's active pulse rate, the normal probability plot of the residuals suggests that the residuals have a distribution that is right-skewed. So we have concerns about the ANCOVA model at this point.

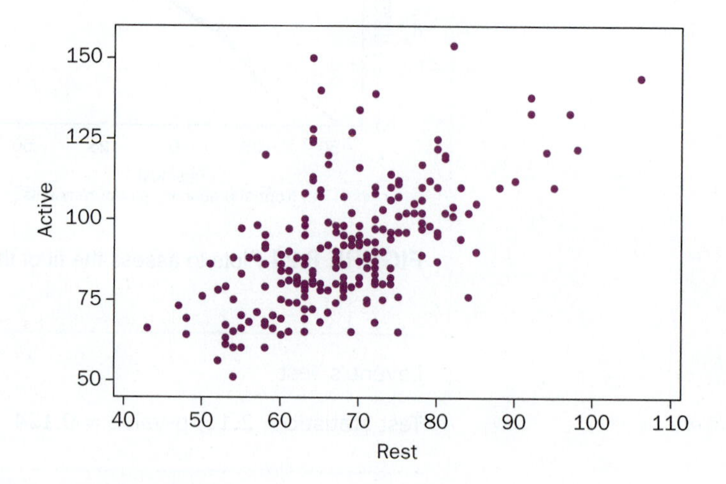

FIGURE 8.25 Scatterplot of active pulse rate versus resting pulse rate

For completeness, we check whether the slopes for the regression of active pulse rate on resting pulse rate are the same for all three groups. Figure 8.27 shows that, while they are not identical, they are reasonably similar.

CHOOSE (again)

At this point, we cannot proceed with the ANCOVA model, because the normality condition of the residuals for the linear model has not been met. This leads us to try using a log transformation on both the active pulse rate and resting pulse rate.

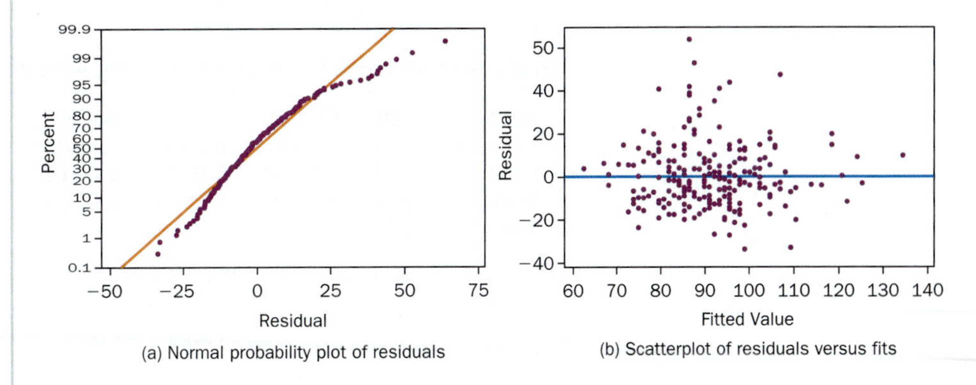

(a) Normal probability plot of residuals (b) Scatterplot of residuals versus fits

FIGURE 8.26 Plots to assess the fit of the linear regression model

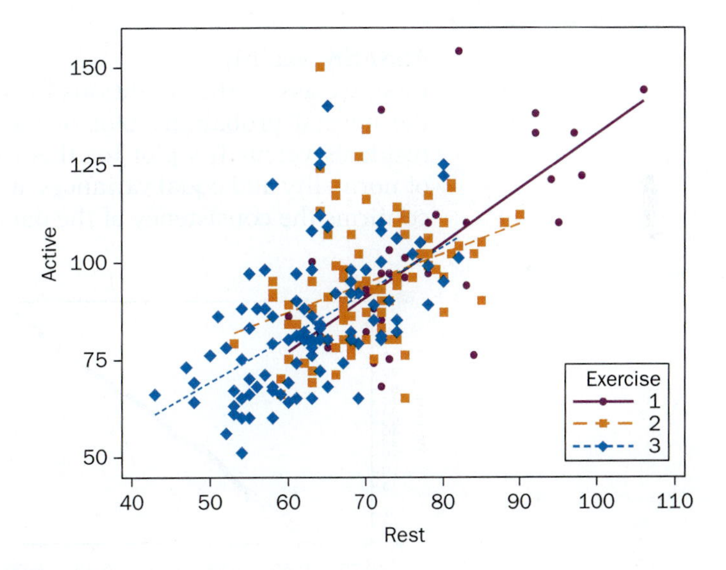

FIGURE 8.27 Scatterplot of active pulse rate versus resting pulse rate for each exercise level

FIT (again)
Since we are transforming both our response variable and our covariate, we rerun both the ANOVA table and ANCOVA table.

First, we display the ANOVA table:

One-way ANOVA: log(active) versus Exercise

Source	DF	SS	MS	F	P
Exercise	2	1.3070	0.6535	18.40	0.000
Error	229	8.1316	0.0355		
Total	231	9.4386			

S = 0.1884 R-Sq = 13.85% R-Sq(adj) = 13.10%

Here we note that the factor is still significant (though we still need to recheck the conditions for this ANOVA) and the R^2 is a similar value at 13.85%.

Next, we display the ANCOVA table:

Analysis of Variance for log(active), using Adjusted SS for Tests

Source	DF	Seq SS	Adj SS	Adj MS	F	P
log(rest)	1	3.5951	2.3305	2.3305	91.59	0.000
Exercise	2	0.0424	0.0424	0.0212	0.83	0.436
Error	228	5.8011	5.8011	0.0254		
Total	231	9.4386				

S = 0.159510 R-Sq = 38.54% R-Sq(adj) = 37.73%

And now we note that the exercise factor is not significant, though again, we still have to recheck the conditions.

ASSESS (again)
First, we assess the conditions for the ANOVA model. Figure 8.28(a) shows the normal probability plot of the residuals and Figure 8.28(b) gives the residuals versus fits plot for this model. Both indicate that the conditions of normality and equal variances are met. Levene's test (see Section 8.1) also confirms the consistency of the data with an equal variances model.

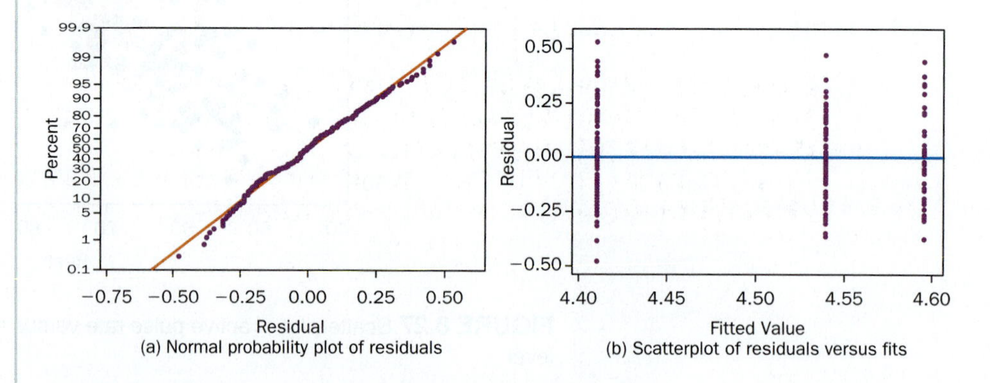

FIGURE 8.28 Plots to assess the fit of the ANOVA model

Levene's Test

Test statistic = 1.47, p-value = 0.231

We have already decided that the independence condition holds for that dataset, so we do not need to check for that again.

Now we check the conditions for the linear regression model. Figure 8.29 is a scatterplot of the response versus the covariate. There is clearly a strong linear relationship between these two variables. Next, we consider Figure 8.30. Figure 8.30(a) gives the normal probability plot of the residuals and Figure 8.30(b) shows the residuals versus fits plot. This time we do not see anything that concerns us about the conditions.

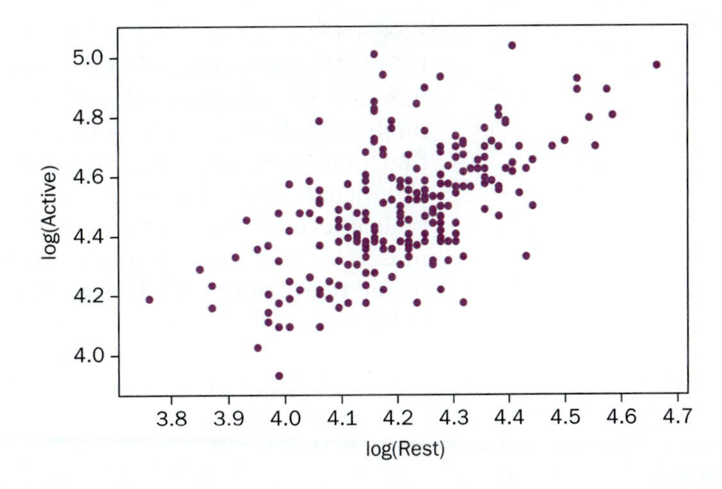

FIGURE 8.29 Scatterplot of the log of active pulse rate versus the log of resting pulse rate

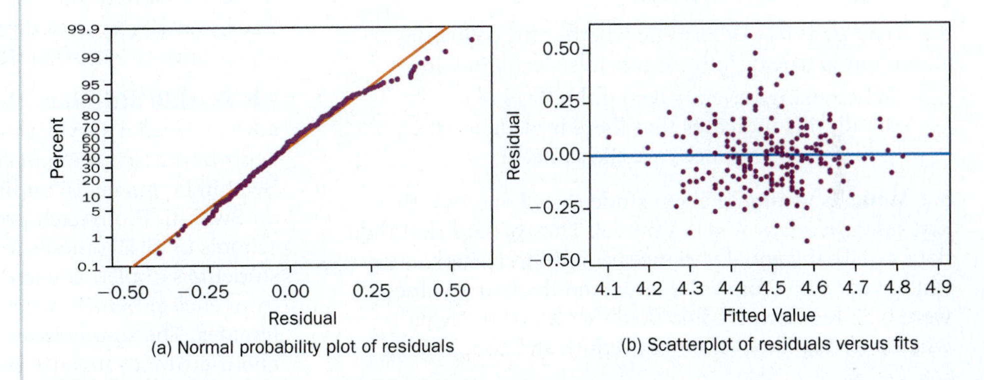

FIGURE 8.30 Plots to assess the fit of the linear regression model

Finally, we again check to make sure the slopes are approximately the same, and Figure 8.31 shows us that they are.

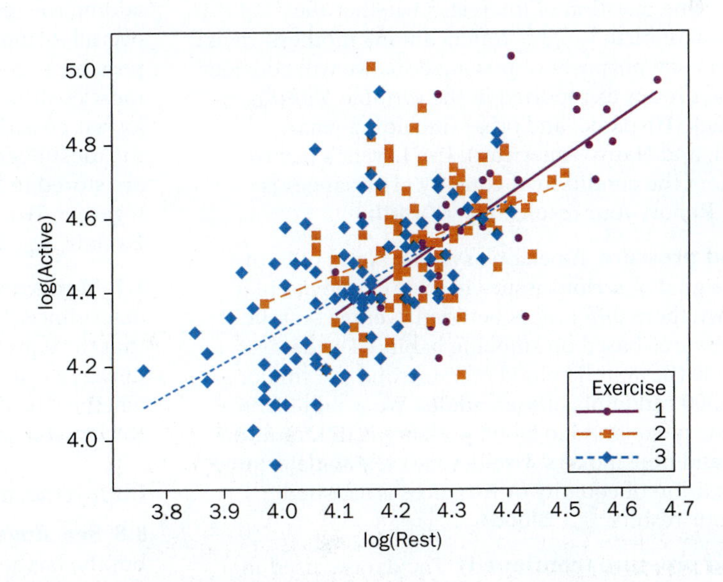

FIGURE 8.31 Scatterplot of the log of active pulse rate versus the log of resting pulse rate for each exercise level

USE

It finally appears that all of the conditions for the ANCOVA model have been met and we can proceed with the analysis. Looking back at the ANCOVA table with the log of the active pulse rate for the response variable, we see that now the factor measuring the average level of exercise is no longer significant, even though it was in the ANOVA table. This means that the typical level of exercise was really just another way to measure the resting pulse rate and that we would probably be better off just running a simple linear regression model using the resting pulse rate to predict the active pulse rate.

EXERCISES

Topic 8.1 Exercises: Levene's Test for Homogeneity of Variances

8.1 True or false. Determine whether the following statement is true or false. If it is false, explain why.

> When using Levene's test, if the *P*-value is small, this indicates that there is evidence the population variances are the same.

8.2 What is wrong? Fellow students tell you that they just ran a two-way ANOVA model. They believe that their data satisfy the equal variances condition because they ran Levene's test for each factor, and the two *P*-values were 0.52 for Factor A and 0.38 for Factor B. Explain what is wrong about their procedure and conclusions.

8.3 North Carolina births. The file **NCbirths** contains data on a random sample of 1450 birth records in the state of North Carolina in the year 2001. This sample was selected by John Holcomb, based on data from the North Carolina State Center for Health and Environmental Statistics. One question of interest is whether the distribution of birth weights differs among mothers' racial groups. For the purposes of this analysis, we will consider four racial groups as reported in the variable *MomRace*: white, black, Hispanic, and other (including Asian, Hawaiian, and Native American). Use Levene's test to determine if the condition of equality of variances is satisfied. Report your results. **NCBirth**

8.4 Blood pressure. A person's systolic blood pressure can be a signal of serious issues in their cardiovascular system. Are there differences between average systolic blood pressures based on smoking habits? The dataset **Blood1** has the systolic blood pressure and the smoking status of 500 randomly chosen adults. We would like to know if the mean systolic blood pressure is different for smokers and nonsmokers. Use Levene's test to determine if the condition of equality of variances is satisfied. Report your results. **Blood1**

8.5 Blood pressure (continued). The dataset used in Exercise 8.4 also measured the sizes of people using the variable *Overwt*. This is a categorical variable that takes on the values 0 = Normal, 1 = Overweight, and 2 = Obese. We would like to know if the mean systolic blood pressure differs for these three groups of people. Use Levene's test to determine if the condition of equality of variances is satisfied. Report your results. **Blood1**

8.6 Swahili attitudes. Hamisi Babusa, a Kenyan scholar, administered a survey to 480 students from Pwani and Nairobi provinces about their attitudes toward the Swahili language. In addition, the students took an exam on Swahili. From each province, the students were from 6 schools (3 girls' schools and 3 boys' schools) with 40 students sampled at each school, so half of the students from each province were males and the other half females. The survey instrument contained 40 statements about attitudes toward Swahili and students rated their level of agreement to each. Of these questions, 30 were positive questions and the remaining 10 were negative questions. On an individual question, the most positive response would be assigned a value of 5, while the most negative response would be assigned a value of 1. By adding the responses to each question, we can find an overall *Attitude Score* for each student. The highest possible score would be 200 (an individual who gave the most positive possible response to every question). The lowest possible score would be 40 (an individual who gave the most negative response to every question). The data are stored in **Swahili**. Use Levene's test with both factors together. Would a two-way ANOVA model be appropriate? Explain. **Swahili**

8.7 Meniscus In Chapter 5, Exercise 5.57, you determined, based on graphs and descriptive statistics, that the equal variances condition was not met. Use Levene's test to make a more formal decision about whether the condition of equality of variances is satisfied. Report your results. **Menisc**

Supplemental Exercise

8.8 Sea slugs. Sea slug larvae need vaucherian seaweed, but the larvae from these sea slugs must locate this type of seaweed to survive. A study was done to try to determine whether chemicals that leach out of the seaweed attract the larvae. Seawater was collected over a patch of this kind of seaweed at 5-minute intervals as the tide was coming in and presumably, mixing with the chemicals. The idea was that as more seawater came in, the concentration of the chemicals was reduced. Each sample of water was divided into six parts. Larvae were then introduced to this seawater to see what percentage metamorphosed. The question of interest is whether or not there is a difference in this percentage over the 5 time periods. Open the dataset **SeaSlugs**. We will use this dataset to illustrate the way that Levene's test is calculated. **Slug**

a. Find the median percent of metamorphosed larvae for each of the six time periods. Find the absolute deviation between those medians and the actual observations. Plot those deviations on one dotplot, grouped by time period. Does it look as if there is a difference in the average absolute deviation for the six different time periods? Explain.

b. Compute an ANOVA table for the absolute deviations. What is the test statistic? What is the *P*-value?

c. Run Levene's test on the original data. What is the test statistic? What is the *P*-value? How do these values compare to the values that you computed in part (b)?

Topic 8.2 Exercises: Multiple Tests

8.9 Method comparison. The formulas for the three types of confidence intervals considered in this section (Bonferroni, Tukey, and Fisher) are all very similar. There is one piece of the formula that is different in all three cases. What is it?

8.10 Bonferroni intervals. Why is the Bonferroni method considered to be conservative? That is, what does it mean to say that "Bonferroni is conservative"?

8.11 Fantasy baseball. Recall the data from Exercise 5.66 (page 253). The data recorded the amount of time each of 8 "fantasy baseball" participants took in each of 24 rounds to make their selection. The data are listed in Exercise 5.66 and are available in the datafile **FantasyBaseball**. In Exercise 5.67, we asked you to transform the selection times using the natural log before continuing with your analysis because the residuals were not normally distributed. We ask you to do the same again here. **FanBase**

a. Use Tukey's HSD to compute confidence intervals to identify the differences in average selection times for the different participants. Report your results.

b. Which multiple comparisons method would you use to compute confidence intervals to assess which rounds' average selection times differ significantly from which other? Explain.

8.12 North Carolina births. The file **NCbirths** contains data on a random sample of 1450 birth records in the state of North Carolina in the year 2001. In Exercises 5.59–5.61 (pages 252–253), we conducted an analysis to determine whether there was a race-based difference in birth weights. In Exercise 5.60, we found that the test in the ANOVA table was significant. Use the Bonferroni method to compute confidence intervals to identify differences in birth weight for babies of moms of different races. Report your results. **NCBirth**

8.13 Blood pressure (continued). The dataset used in Exercise 8.4 also measured the sizes of people using the variable *Overwt*. This is a categorical variable that takes on the values 0 = Normal, 1 = Overweight, and 2 = Obese. Are the mean systolic blood pressures different for these three groups of people? **Blood1**

a. Use Bonferroni intervals to find any differences that exist between these three group mean systolic blood pressures. Report your results.

b. Use Tukey's HSD intervals to find any differences that exist between these three group mean systolic blood pressures. Report your results.

c. Were your conclusions in (a) and (b) different? Explain. If so, which would you prefer to use in this case and why?

8.14 Sea slugs. Sea slugs, common on the coast of Southern California, live on vaucherian seaweed. But the larvae from these sea slugs need to locate this type of seaweed to survive. A study was done to try to determine whether chemicals that leach out of the seaweed attract the larvae. Seawater was collected over a patch of this kind of seaweed at 5-minute intervals as the tide was coming in and, presumably, mixing with the chemicals. The idea was that as more seawater came in, the concentration of the chemicals was reduced. Each sample of water was divided into six parts. Larvae were then introduced to this seawater to see what percentage metamorphosed. Is there a difference in this percentage over the five time periods? Open the dataset **SeaSlugs**. **Slug**

a. Use Fisher's LSD intervals to find any differences that exist between the percent of larvae that metamorphosed in the different water conditions.

b. Use Tukey's HSD intervals to find any differences that exist between the percent of larvae that metamorphosed in the different water conditions.

c. Were your conclusions to (a) and (b) different? Explain. If so, which would you prefer to use in this case and why?

Topic 8.3 Exercises: Comparisons and Contrasts

8.15 Diamonds. In Exercise 3.37 on page 148, we considered four different color diamonds and wondered if they were associated with differing numbers of carats. In that example, after using a log transformation on the number of carats, we discovered that there was, indeed, a significant difference. Suppose we have reason to believe that diamonds of color D and E have roughly the same number of carats and diamonds of color F and G have the same number of carats (although different from D and E). **Diam2**

a. What hypotheses would we be interested in testing?

b. How would the contrast be written in symbols?

8.16 Fantasy baseball. In Exercises 5.66–5.69 (pages 253–254), you considered the dataset **FantasyBaseball**. These data consist of the amount of time (in seconds) that each of 8 friends took to make their selections in a "fantasy draft." In Exercise 5.67, you took the log of the times so that the conditions for ANOVA would be met, and you discovered that the *F*-statistic in the ANOVA table was significant. The friends wonder whether this is because TS makes his choices significantly faster than the others. Should this question be addressed with a comparison or a contrast? Why? **FanBase**

8.17 Fruit flies. Use the fruit fly data in **FruitFlies** to test the second question the researchers had. That is, is there a difference between the life spans of males living with pregnant females and males living alone? **FFlies**

a. What are the hypotheses that describe the test we would like to perform?

b. Write the contrast of interest in symbols and compute its estimated value.

c. What is the standard error of the contrast?

d. Perform the hypothesis test to test the alternative hypothesis that the mean life span of fruit flies living with pregnant females is different from that of fruit flies living alone. Be sure to give your conclusions.

8.18 Blood pressure. A person's systolic blood pressure can be a signal of serious issues in their cardiovascular system. Are there differences between average systolic blood pressure based on weight? The dataset **Blood1** includes the systolic blood pressure and the weight status of 300 randomly chosen adults. The categorical variable *Overwt* records the values 0 = Normal, 1 = Overweight, and 2 = Obese for each individual. We would like to compare those who are of normal weight to those who are either overweight or obese. **Blood1**

a. What are the hypotheses that describe the test we would like to perform?

b. Write the contrast of interest in symbols and compute its estimated value.

c. What is the standard error of the contrast?

d. Perform the hypothesis test to test the alternative hypothesis that the mean systolic blood pressure is different for people of normal weight, as compared to people who are overweight or obese. Be sure to give your conclusions.

8.19 Auto pollution. The dataset **AutoPollution** gives the results of an experiment on 36 different cars. The cars were randomly assigned to get either a new filter or a standard filter, and the noise level for each car was measured. For this problem, we are going to ignore the treatment itself and just look at the sizes of the cars (also given in this dataset). The 36 cars in this dataset consisted of 12 randomly selected cars in each of three sizes (1 = Small, 2 = Medium, and 3 = Large). The researchers wondered if the large cars just generally produced a different amount of noise than the other two categories combined. **AutPol**

a. What are the hypotheses that describe the test we would like to perform?

b. Write the contrast of interest in symbols and compute its estimated value.

c. What is the standard error of the contrast?

d. Perform the hypothesis test to see if mean noise level is different for large cars, as compared to small- or medium-sized cars. Be sure to give your conclusions.

8.20 Cancer survivability. Example 8.2 discusses the dataset from the following example. In the 1970s, doctors wondered if giving terminal cancer patients a supplement of ascorbate would prolong their lives. They designed an experiment to compare cancer patients who received ascorbate to cancer patients who did not receive the supplement. The result of that experiment was that, in fact, ascorbate did seem to prolong the lives of these patients. But then a second question arose. Was the effect of the ascorbate different when different organs were affected by the cancer? The researchers took a second look at the data. This time they concentrated only on those patients who received the ascorbate and divided the data up by which organ was affected by the cancer. Five different organs were represented among the patients (for all of whom only one organ was affected): stomach, bronchus, colon, ovary, and breast. In this case, since the patients were not randomly assigned to which type of cancer they had, but were instead a random sample of those who suffered from such cancers, we are dealing with an observational study. The data are available in the file **CancerSurvival**. In Example 8.10, we discovered that we needed to take the natural log of the survival times for the ANOVA model to be appropriate. **CanSurv**

a. We would like to see if the survival times for breast and ovary cancer are different from the survival times of the other three types. What are the hypotheses that describe the test we would like to perform?

b. Write the contrast of interest in symbols and compute its estimated value.

c. What is the standard error of the contrast?

d. Perform the hypothesis test to test the alternative hypothesis that the mean survival time is different for breast and ovary cancers, as compared to bronchus, colon, or stomach cancers. Be sure to give your conclusions.

Topic 8.4 Exercises: Nonparametric Tests

8.21 Wilcoxon-Mann-Whitney versus *t*-test. Is the following statement true or false? If it is false, explain why it is false.

> The Wilcoxon-Mann-Whitney test can be used with any dataset for comparing two means, while the two-sample *t*-test can only be used when the residuals from the model are normal, random, and independent.

8.22 Nonparametric tests versus parametric tests. Is the following statement true or false? If it is false, explain why it is false.

The null hypothesis for a nonparametric test is the same as the null hypothesis in its counterpart parametric test (Wilcoxon-Mann-Whitney compared to two-sample *t*-test, Kruskal-Wallis compared to ANOVA).

8.23 Nursing homes[9] Data collected by the Department of Health and Social Services of the State of New Mexico in 1988 on nursing homes in the state are located in the file **Nursing**. Several operating statistics were measured on each nursing home as well as whether the home was located in a rural or urban setting. We would like to see if there is a difference in the size of a nursing home, as measured by the number of beds in the facility, between rural and urban settings. **Nurs**

a. Create one or more graphs to assess the conditions required by two-sample *t*-test. Comment on why the Wilcoxon-Mann-Whitney test might be more appropriate than a two-sample *t*-test.

b. Run the Wilcoxon-Mann-Whitney test and report the results and conclusions.

8.24 Cloud seeding. Researchers were interested in whether seeded clouds would produce more rainfall. An experiment was conducted in Tasmania between 1964 and 1971, and rainfall amounts were measured in inches per rainfall period. The researchers measured the amount of rainfall in two target areas: East (TE) and West (TW). They also measured the amount of rainfall in four control locations. Clouds were coded as being either seeded or unseeded. For this exercise, we will concentrate on the winter results in the east target area. The data can be found in **CloudSeeding**.[10] **Cloud**

a. Create one or more graphs to assess the conditions required by a two-sample *t*-test. Comment on why the Wilcoxon-Mann-Whitney test might be more appropriate than a two-sample *t*-test.

b. Run the Wilcoxon-Mann-Whitney test and report the results and conclusions.

8.25 Daily walking.[11] A statistics professor regularly keeps a pedometer in his pocket. It records not only the number of steps taken each day, but also the number of steps taken at a moderate pace, the number of minutes walked at a moderate pace, and the number of miles total that he walked. He also added to the dataset the day of the week; whether it was rainy, sunny, or cold (on sunny days he often biked, but on rainy or cold days he did not); and whether it was a weekday or weekend. For this exercise, we will focus on the number of steps taken at a moderate pace and whether the day was a weekday or a weekend day. The data are in the file **Pedometer**. **Ped**

a. Create one or more graphs to assess the conditions required by a two-sample *t*-test. Comment on whether such a test would be appropriate, or whether it would be better to use a Wilcoxon-Mann-Whitney test.

b. Run the two-sample *t*-test and report the results and conclusions.

c. Run the Wilcoxon-Mann-Whitney test and report the results and conclusions.

d. Compare your answers to parts (b) and (c).

8.26 Pulse rates. A Stat 2 instructor collected data on his students; they may be found in the file **Pulse**. He first had the students rate themselves on how active they were generally (1 = Not active, 2 = Moderately active, 3 = Very active). This variable has the name *Exercise*. Then he randomly assigned the students to one of two treatments (walk or run up and down a flight of stairs three times) and measured their pulse when they were done. This last pulse rate is recorded in the variable *Active*. One question that we might ask is whether there is a difference in the active heart rate between people with different regular levels of activity. **Pulse**

a. Create one or more graphs to assess the conditions required by ANOVA. Comment on why the Kruskal-Wallis test might be more appropriate than an ANOVA.

b. Run the Kruskal-Wallis test and report the results and conclusions.

8.27 Cuckoo eggs.[12] Cuckoos lay their eggs in the nests of host species. O. M. Latter collected data on the lengths of the eggs laid and the type of bird's nest. He observed eggs laid in the nests of six different species and measured their lengths. Are the egg lengths different in the bird nests of the different species? The data are available in the file **Cuckoo**. **Cuckoo**

a. Create one or more graphs to assess the conditions required by ANOVA. Comment on why the Kruskal-Wallis test might be more appropriate than an ANOVA.

b. Run the Kruskal-Wallis test and report the results and conclusions.

8.28 Cloud seeding: season. In Exercise 8.24, you analyzed data from clouds that were both seeded and unseeded, during winter months, to see if there was a difference in rainfall. Now, rather than looking at whether clouds are seeded or not, we will focus on the season of the year. Again, we will limit our analysis to the east target area (TE). Do the clouds in different seasons produce different amounts of rainfall? Or is the amount of rainfall per cloud consistent across the year? The data found in **CloudSeeding2** contain information on all clouds observed during the years in question. **Cloud2**

a. Create one or more graphs to assess the conditions required by ANOVA. Comment on why the Kruskal-Wallis test might be more appropriate than an ANOVA.

b. Run the Kruskal-Wallis test and report the results and conclusions.

8.29 Daily walking: each day of the week. In Exercise 8.25, you analyzed data collected by a statistics professor using his pedometer. In that exercise, we compared the median number of moderate steps for weekdays and weekend days. But of course, there are 5 different weekdays and 2 different weekend days. In this case, we are going to treat each of the 7 days as a separate category and analyze whether the median number of steps is different across the days. The data are located in the file **Pedometer**. **Ped**

a. Create one or more graphs to assess the conditions required by ANOVA. Comment on whether ANOVA would be appropriate, or whether it would be better to use a Kruskal-Wallis test.

b. Run the Kruskal-Wallis test and report the results and conclusions.

c. The results of the Kruskal-Wallis test seem to be in contradiction to the results of the Mann-Whitney-Wilcoxen test in Exercise 8.25. Comment on why this occurred.

8.30 Cancer survival. In Example 8.10, we analyzed the transformed survival rates, after the log transformation was applied, to test the hypothesis that the median survival rate was the same for all five organs. Use the Kruskal-Wallis test on the untransformed survival rates to verify that the test statistic and *P*-value remain the same. That is, the value of the test statistic is 14.95 and the *P*-value is 0.0048. (This exercise illustrates that nonparametric procedures are not very sensitive to slight departures from the conditions.) **CanSurv**

Topic 8.5 Exercises: Randomization *F*-Test

8.31 Recovery times. A study[13] of two surgical methods compared recovery times, in days, for two treatments, the standard and a new method. Four randomly chosen patients got the new treatment; the remaining three patients got the standard. Here are the results:

Recovery times, in days, for seven patients		Average
New procedure	19, 22, 25, 26	23
Standard	23, 33, 40	32

There are 35 ways to choose three patients from seven. If you do this purely at random, the 35 ways are equally likely. What is the probability that a set of three randomly chosen from 19, 22, 23, 25, 26, 33, 40 will have an average of 32 or more?

8.32 Challenger: Sometimes statistics *is* rocket science.[14] Bad data analysis can have fatal consequences. After the Challenger exploded in 1987, killing all five astronauts aboard, an investigation uncovered the faulty data analysis that had led Mission Control to OK the launch despite cold weather. The fatal explosion was caused by the failure of an O-ring seal, which allowed liquid hydrogen and oxygen to mix and explode. **ORing**

a. For the faulty analysis, engineers looked at the seven previous launches with O-ring failures. The unit is a launch; the response is the number of failed O-rings. Here are the numbers:

Number of failed O-rings for launches with failures:
Above 65°: 1 1 2
Below 65°: 1 1 1 3

The value of the *F*-statistic for the actual data is 0.0649. Consider a randomization *F*-test for whether the two groups are different. We want to know how likely it is to get a value at least as large as 0.0649 purely by chance. Take the seven numbers 1, 1, 1, 1, 1, 2, 3 and randomly choose four to correspond to the four launches below 65°. There are four possible samples, which occur with the percentages shown below. Use the table to find the *P*-value. What do you conclude about temperature and O-ring failure?

A = Above			B = Below 65°				% of samples	Means: A	B	MSE	F
3	2	1	1	1	1	1	14.3%	2.00	1.00	0.400	4.2857
3	1	1	2	1	1	1	28.6%	1.67	1.25	0.683	0.4355
2	1	1	3	1	1	1	28.6%	1.33	1.50	0.733	0.0649
1	1	1	3	2	1	1	28.6%	1.00	1.75	0.550	1.7532

b. The flaw in the analysis done by the engineers is this: They ignored the launches with zero failures. There were 17 such zero-failure launches, and for all 17, the temperature was above 65°. Here (and stored in **Orings**) is the complete dataset that the engineers should have chosen to analyze:

Number of failed O-rings for all launches:
Above 65°: 0 0 0 0 0 0 0 0 0 0 0 0 0 0 0 0 0 1 1 2
Below 65°: 1 1 1 3

The value of the *F*-statistic for these two groups is 14.427. To carry out a randomization *F*-test, imagine putting the 24 numbers on cards and randomly choosing four to represent the four launches with temperatures below 65°. The table in Figure 8.32 summarizes the results of repeating this process more than 10,000 times. Use the table to find the *P*-value, that is, the chance of an *F*-statistic greater than or equal to 14.427. What do you conclude about temperature and O-ring failure?

B = Below 65°				% of samples	Means: A	B	MSE	F
0	0	0	0	22.4%	0.50	0.00	0.619	1.346
1	0	0	0	32.0%	0.45	0.25	0.652	0.204
1	1	0	0	12.8%	0.40	0.50	0.657	0.051
2	0	0	0	6.4%	0.40	0.50	0.657	0.051
1	1	1	0	1.6%	0.35	0.75	0.633	0.842
2	1	0	0	6.4%	0.35	0.75	0.633	0.842
3	0	0	0	6.4%	0.35	0.75	0.633	0.842
1	1	1	1	0.0%	0.30	1.00	0.581	2.811
2	1	1	0	1.6%	0.30	1.00	0.581	2.811
3	1	0	0	6.4%	0.30	1.00	0.581	2.811
2	1	1	1	0.1%	0.25	1.25	0.500	6.667
3	1	1	0	1.6%	0.25	1.25	0.500	6.667
3	2	0	0	1.3%	0.25	1.25	0.500	6.667
3	1	1	1	0.1%	0.20	1.50	0.390	14.427
3	2	1	0	0.8%	0.20	1.50	0.390	14.427
3	2	1	1	0.1%	0.15	1.75	0.252	33.811

FIGURE 8.32 Randomization distribution of *F*-statistics for full Challenger data

8.33 Social learning in monkeys.[15] Some behaviors are "hardwired"—we engage in them without having to be taught. Other behaviors we learn from our parents. Still other behaviors we acquire through "social learning" by watching others and copying what they do. To study whether chimpanzees are capable of social learning, two scientists at the University of St. Andrews randomly divided 7 chimpanzees into a treatment group of 4 and control group of 3. Just as in Exercise 8.31 (recovery times), there are 35 equally likely ways to do this. During the "learning phase," each chimp in the treatment group was exposed to another "demo" chimp who knew how to use stones to crack hard nuts; each chimp in the control group was exposed to a "control" chimp who did not know how to do this. During the test phase, each chimp was provided with stones and nuts, and the frequency of nut-cracking bouts was observed.

The results for the 7 chimpanzees in the experiment:

Frequency of nut cracking bouts		Average
Treatment group	5, 12, 22, 25	16
Control group	2, 3, 4	3

Assume that the treatment has no effect, and that the only difference between the two groups resulted from randomization.

a. What is the probability that just by chance, the average for the treatment group would be greater than or equal to 16?

b. The F-value for the observed data (when all the larger values are in the treatment group) is 5.65. That is one of the 35 equally likely random assignments of the observed values to a treatment group of 4 and control group of 3. The only other assignment that has an F-value as high or higher is to put the three largest values in the control group:

Control			Treatment				Group Means			
							Ctrl	Tr	MSE	F
12	22	25	2	3	4	5	19.67	3.50	19.53	22.94

What is the P-value from the randomization F-test?

c. Why is the probability in (a) different from the P-value in (b)?

d. If the question of interest to the researchers is whether or not the chimpanzees exposed to "demos" (treatment) tend to crack *more* nuts than those who aren't (control), would it be more reasonable to use the proportion from (a) or from (b) in assessing the results? Explain.

e. The correct answer in (d) is related to the difference between a one-tailed t-test to compare two groups, and the F-test for ANOVA. Explain.

8.34 Fruit fly lifetimes: randomization test. In Example 5.12 on page 220 we see the ANOVA table for testing for a difference in mean lifetime of fruit flies living with different numbers of potential mates. For the data in **FruitFlies** the usual ANOVA procedure gives an F-statistic of 13.61 with 4 and 120 degrees of freedom. This gives a very small P-value, 3.5×10^{-9}. This would indicate that randomization test for this same set of hypotheses should yield very few (if any) F-statistics beyond 13.61 among the randomization samples. Try this for 10,000 randomizations. How many of your randomization F-statistics were beyond 13.62? What was the biggest value you got for an F-statistic? **FFlies**

8.35 Pigs: with randomization. In Example 7.2 starting on page 301 we introduce the **PigFeed** data, looking at possible effects of vitamin B12 (yes/no) and *Antibiotics* (yes/no) on weight gain for baby pigs. The ANOVA table for the two-way model with interaction from Example 7.20 on page 324 is repeated below.

Source	SS	df	MS	F	p
B12	2187	1	2187.00	60.331	< 0.0001
Antibiotics	192	1	192.00	5.297	0.0504
Interaction	1728	1	1728.00	47.669	0.0001
Residual	290	8	36.25		
Total	4397	11			

We can use randomization in this situation by scrambling the column of weight gains in **PigFeed** and then recomputing the ANOVA table. This shows what sort of F-statistics we should expect to see for both main effects and the interaction, under a null assumption that all effects are zero. Create 10,000 simulations in this manner and use them to estimate the P-values for the F-statistics ($F_{B12} = 60.331$, $F_{Antibiotics} = 5.297$, and $F_{Interaction} = 47.669$). Compare these to the P-values obtained from F-distributions in the original ANOVA. **Pig**

Topic 8.6 Exercises: Repeated Measures Designs and Datasets

8.36 Identify the design. Each of 18 patients had their blood pressure measured three times: once after taking a placebo, once after taking drug A, and once after taking drug B. We think that "medicine" might matter (i.e., that the placebo, drug A, and drug B could have different effects on blood pressure).

What kind of repeated measures design is present here? That is, how many within-subjects variables, if any, are there and what are they? How many between-subjects variables, if any, are there and what are they?

8.37 Identify another design. Each of 24 subjects was given lists of words to memorize under four conditions. Sometimes the words were typed, but sometimes they were handwritten. Sometimes the memorization was done while listening to music, but sometimes it was done in silence. Each subject was tested under each of the four possible combinations of conditions: typed/music, typed/silence, handwritten/music, and handwritten/silence. Researchers recorded how many words the subject could remember under each of the four conditions (with the order of the four conditions being randomized for each subject).

What kind of repeated measures design is present here? That is, how many within-subjects variables, if any, are

a. Exercise 5.50 asked for an ANOVA to determine if there was a difference in fenthion residue over these three testing periods. For comparison sake, we ask you to run that ANOVA again here (using an exponential transformation of *fenthion* so that conditions are met).

b. Now analyze these data using regression, but treat the time period as a categorical variable. That is, create indicator variables for the time periods and use those in your regression analysis. Interpret the coefficients in the regression model.

c. Discuss how the results are the same between the regression analysis and the analysis in part (a).

d. In this case, the variable that we used as categorical in the ANOVA analysis really constitutes measurements at three different values of a continuous variable (*Time*). Treat *Time* as a continuous variable and use it in your regression analysis. Are your conclusions any different? Explain.

e. Which form of the analysis do you think is better in this case? Explain.

8.53 Oil. In Exercise 7.37 on page 339, we used ANOVA to model the amount of oil deapsorbed from samples of sand based on two factors (amount of oil and length of time exposed to ultrasound). The data can be found in **OilDeapsorbtion**. **OilD**

a. Repeat the analysis you did in Exercise 7.37 using regression with indicator variables. Interpret the coefficients in the regression model.

b. What are your conclusions from the regression analysis? Explain.

c. Discuss how the results are the same between the regression analysis and the ANOVA procedure.

8.54 Sea slugs. Sea slugs, common on the coast of Southern California, live on vaucherian seaweed, but the larvae from these sea slugs need to locate this type of seaweed to survive. A study was done to try to determine whether chemicals that leach out of the seaweed attract the larvae. Seawater was collected over a patch of this kind of seaweed at 5-minute intervals as the tide was coming in and, presumably, mixing with the chemicals. The idea was that as more seawater came in, the concentration of the chemicals was reduced. Each sample of water was divided into six parts. Larvae were then introduced to this seawater to see what percentage metamorphosed. Is there a difference in this percentage over the five time periods? Exercise 5.71 (page 254) asked that the dataset in **SeaSlugs** be analyzed using ANOVA. **Slug**

a. Repeat the analysis you did in Exercise 5.71 using regression with indicator variables. Interpret the coefficients in the regression model.

b. What are your conclusions from the regression analysis? Explain.

c. Now notice that the grouping variable is actually a time variable that we could consider to be continuous. Use regression again, but this time use *Time* as a continuous explanatory variable instead of using indicator variables. What are your conclusions? Explain.

d. Which form of the analysis do you think is better in this case? Explain.

8.55 Auto pollution. The dataset **AutoPollution** gives the results of an experiment on 36 different cars. This experiment used 12 randomly selected cars, each of three different sizes (small, medium, and large), and assigned half in each group randomly to receive a new air filter or a standard air filter. The response of interest was the noise level for each of the 3 cars. **AutPol**

a. Run an appropriate two-way ANOVA on this dataset. What are your conclusions? Explain.

b. Now run the same analysis except using indicator variables and regression. Interpret the coefficients in the regression model.

c. What are your conclusions from the regression analysis? Explain.

d. In what ways are your analyses (ANOVA and regression) the same and how are they different? Explain.

Topic 8.8 Exercises: Analysis of Covariance

8.56 Conditions.

a. What are the conditions required for ANOVA?

b. What are the conditions required for linear regression?

c. How are the conditions for ANCOVA related to the conditions for ANOVA and linear regression?

8.57 Transformations and ANCOVA: Multiple choice. You have a dataset where fitting an ANCOVA model seems like the best approach. Assume that all of the conditions for linear regression with the covariate have been met. Also, all conditions for ANOVA with the factor of interest have been met except for the equal variances condition. You transform the response variable and see that now the equal variances condition has been met. How do you proceed? Explain.

a. Run the ANCOVA procedure and make your conclusions based on the significance tests using the transformed data.

b. Recheck the ANOVA normality condition, as it might have changed with the transformation. If it is OK, proceed with the analysis as in (a).

c. Recheck the normality and equal variances conditions for both ANOVA and linear regression. Also recheck the interaction condition. If all are now OK, proceed with the analysis as in (a).

8.58 Weight loss. Losing weight is an important goal for many individuals. An article in the *Journal of the American Medical Association* describes a study[20] in which researchers investigated whether financial incentives would help people lose weight more successfully. Some participants in the study were randomly assigned to a treatment group that was offered financial incentives for achieving weight-loss goals, while others were assigned to a control group that did not use financial incentives. All participants were monitored over a 4-month period, and the net weight change (*Before* – *After* in pounds) at the end of this period was recorded for each individual. Then the individuals were left alone for 3 months, with a follow-up weight check at the 7-month mark to see whether weight losses persisted after the original 4 months of treatment. For both weight loss variables, the data are the change in weight (in pounds) from the beginning of the study to the measurement time and positive values correspond to weight losses. The data are stored in the file **WeightLossIncentive**. Ultimately, we don't care as much about whether incentives work during the original weight loss period, but rather, whether people who were given the incentive to begin with continue to maintain (or even increase) weight losses after the incentive has stopped. **WtLossIn**

a. Use an ANOVA model to first see if the weight loss during the initial study period was different for those with the incentive than for those without it. Be sure to check conditions and transform if necessary.

b. Use an ANOVA model to see if the final weight loss was different for those with the incentive than for those without it. Be sure to check conditions and transform if necessary.

c. It is possible that how much people were able to continue to lose (or how much weight loss they could maintain) in the second, unsupervised time period might be related to how much they lost in the first place. Use the initial weight loss as the covariate and perform an ANCOVA. Be sure to check conditions and transform if necessary.

8.59 Fruit flies. Return to the fruit fly data used as one of the main examples in Chapter 5. Recall that the analysis that was performed in that chapter was to see if the mean life spans of male fruit flies were different depending on how many and what kind of females they were housed with. The result of our analysis was that there was, indeed, a difference in the mean life spans of the different groups of male fruit flies.

There was another variable measured, however, on these male fruit flies—the length of the thorax (variable name *Thorax*). This is a measure of the size of the fruit fly and may have an effect on its longevity. **FFlies**

a. Use an ANCOVA model on this data, taking the covariate length of thorax into account. Be sure to check conditions and transform if necessary.

b. Compare your ANCOVA model to the ANOVA model fit in Chapter 5. Which model would you use for the final analysis? Explain.

8.60 Fruit flies again. In Exercise 7.44 on page 341 we analyzed the response variable *Activity* using the explanatory variables *Alone* and *Mated*. We also have information on the size of the thorax for each male fruit fly (*Size*). One might surmise that the size of the thorax of the fly would be a marker for how healthy the fly is and how long he might last. In Exercise 7.44 we found that the only significant variable was *Mated*. For this exercise we will only consider the one-way ANOVA model. **FFlies**

a. Use an ANCOVA model on this data, taking the covariate *Size* into account. Be sure to check conditions and transform if necessary.

b. Compare your ANCOVA model to the ANOVA model fit in Chapter 7. Which model would you use for the final analysis? Explain.

8.61 Horse prices.[21] Undergraduate students at Cal Poly collected data on the prices of 50 horses advertised for sale on the Internet. The response variable of interest is the price of the horse and the explanatory variable is the sex of the horse and the data are in **HorsePrices**. **Horse**

a. Perform an ANOVA to answer the question of interest. Be sure to check the conditions and transform if necessary.

b. Perform an ANCOVA to answer the question of interest using the height of the horse as the covariate.

c. Which analysis do you prefer for this dataset? Explain.

8.62 Three car manufacturers. Do the prices of three different brands of cars differ? We have a dataset of cars offered for sale at an Internet site. All of the cars in this dataset are Mazda6s, Accords, or Maximas. We have data on their price, their age, and their mileage as well as what brand car they are in **ThreeCars2017**. **Cars17**

a. Start with an ANOVA model to see if there are different mean prices for the different brands of car. Be sure to check conditions. Report your findings.

b. Now move to an ANCOVA model with age as the covariate. Be sure to check conditions and report your results.

c. Finally use an ANCOVA model with mileage as the covariate. Be sure to check your conditions and report your results.

d. Which model would you use in a final report as your analysis and why?

8.63 Sugar metabolism: Using ANCOVA. In Example 7.23 we saw that the interaction term was not very important in the two-way model for the sugar metabolism data in **SugarEthanol**. One of the factors in that model (*Oxygen*) could also be treated as a quantitative variable, leaving the other explanatory variable (*Sugar*) to be handled as categorical. Fit the regression model for *Ethanol* using these two predictor variables and compare the resulting ANOVA table to the results in Table 7.11 on page 331 for the two-way interaction model. In particular, comment on the degrees of freedom for the two models.

We see that the *P*-values for both *Sugar* and *Oxygen* are much smaller for the regression model than in the two-way ANOVA model. Although the Oxygen predictor in regression explains somewhat less new variability than

OtmarW/Shutterstock

CHAPTER 9

Logistic Regression

In this chapter you will learn to:

- Understand the logistic transformation, the idea of odds, and the odds ratio.
- Identify and fit the logistic regression model.
- Assess the conditions required for the logistic regression model.
- Use formal inference in the setting of logistic regression.

Are students with higher GPAs more likely to get into medical school? If you carry a heavier backpack, are you more likely to have back problems? Is a single mom more likely to get married to the birth father if their child is a boy rather than a girl?

You can think about these three questions in terms of explanatory and response variables, as in previous chapters:

	Explanatory	Response
1.	GPA	accepted into medical school? (Y/N)
2.	weight of backpack	back problems? (Y/N)
3.	sex of baby (boy/girl)	marry birth father? (Y/N)

For the first two questions, the explanatory variable is quantitative, as in Unit A. For the third question, the explanatory variable is categorical, as in

binary

logistic regression

Unit B. What's new here is that for all three questions, the response is **binary**: There are only two possible values, Yes and No.

Statistical modeling when your response is binary uses **logistic regression**. For logistic regression, we regard the response as an indicator variable, with 1 = Yes, 0 = No. Although logistic regression has many parallels with ordinary regression, there are important differences due to the Yes/No response. As you read through this chapter, you should keep an eye out for these parallels and differences.

In what follows, Sections 9.1 and 9.2 describe the logistic regression model. Section 9.3 shows how to assess the fit of the model, and Section 9.4 how to use the model for formal inference. Because the method for fitting logistic models is something of a detour, we have put that in Chapter 11 as an optional topic.

9.1 Choosing a Logistic Regression Model

We start with an example chosen to illustrate how a logistic regression model differs from the ordinary regression model and why you need to use a transformation when you have a binary response. It is this need to transform that makes logistic regression more complicated than ordinary regression. Because this transformation is so important, almost all of this section focuses on it. Here is a preview of the main topics:

- Example: The need to transform
- The logistic transformation
- Odds
- Log(odds)
- Two versions of the model: transforming back
- Randomness in the logistic regression model

EXAMPLE 9.1

Losing

Losing sleep As teenagers grow older, their sleep habits change. Table 9.1 shows the first few cases in a dataset called **LosingSleep** that comes from a random sample of 446 teens, aged 14 to 18, who answer the question,"On an average school night, how many hours of sleep do you get?" The responses are classified as "Yes" (coded as 1 in the data) for teens who reported getting 7 (or more) hours of sleep (on average) and "No" (coded as 0) for those that averaged less than 7 hours of sleep. As you read through the rest of this example, keep in mind that the only possible response values in the raw data of **LosingSleep** are 0's and 1's, not the actual hours of sleep. This fact will be important later on.

Person no.	Age	Outcome
1	14	1 = Yes
2	18	0 = No
3	17	0 = No
⋮	⋮	⋮
446	17	1 = Yes

TABLE 9.1 Some of the data in **LosingSleep** in cases-by-variables format

The data in Table 9.2 summarize the responses to show the number of teens in the "Yes" (7 or more hours of sleep) and "No" (fewer than 7 hours of sleep) groups for each age. Notice that the observational units in this study are the 446 teens, not the ages, so the second table is just giving a convenient summary of the raw data.

Age	14	15	16	17	18
Fewer than seven hours	12	35	37	39	27
Seven hours or more	34	79	77	65	41
Total responses by age	46	114	114	104	68
Proportion getting less than 7 hours	0.74	0.69	0.68	0.63	0.60

TABLE 9.2 Summary of the number in each *Sleep* group for each *Age*

CHOOSE

Figure 9.1 shows a plot of the proportion of Yes answers against age. *Notice that the plot looks roughly linear. You may be tempted to fit an ordinary regression line. Don't!* There are many reasons not to use ordinary regression. Here is just one. For ordinary regression, we model the response y with a linear predictor of the form $\beta_0 + \beta_1 X$. If the error terms are not large, you can often see the x–y relationship clearly enough to get rough estimates of the slope and intercept by eye, as in Figure 9.1.

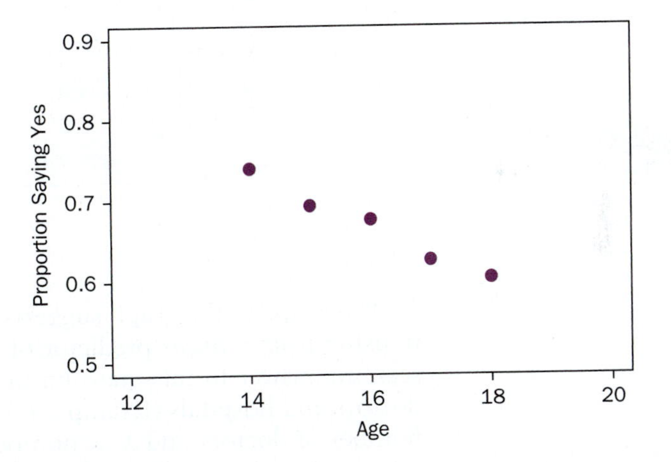

FIGURE 9.1 Scatterplot of *Age* and proportion of teenagers who say they get at least 7 hours of sleep at night

For logistic regression, the response is Yes or No, and we want to model $p = P(Success)$. We still use a linear predictor of the form $\beta_0 + \beta_1 X$, but not in the usual way. If we were to use the model $\hat{p} = \hat{\beta}_0 + \hat{\beta}_1 X$, we would run into the problem of impossible values for $\hat{p}$. We need, instead, a model that takes values of $\beta_0 + \beta_1 X$ and gives back probabilities between 0 and 1.

Suppose you ignore what we just said and fit a line anyway as in Figure 9.2. The line gives a good fit for ages between 14 and 18, but if you extend the fitted line, you run into big trouble. (Check what the model says for 1-year-olds and for 40-year-olds.) Notice also that unless your fitted slope is exactly 0, *any* regression line will give fitted probabilities less than 0 and greater than 1. The way to avoid impossible values is to transform.

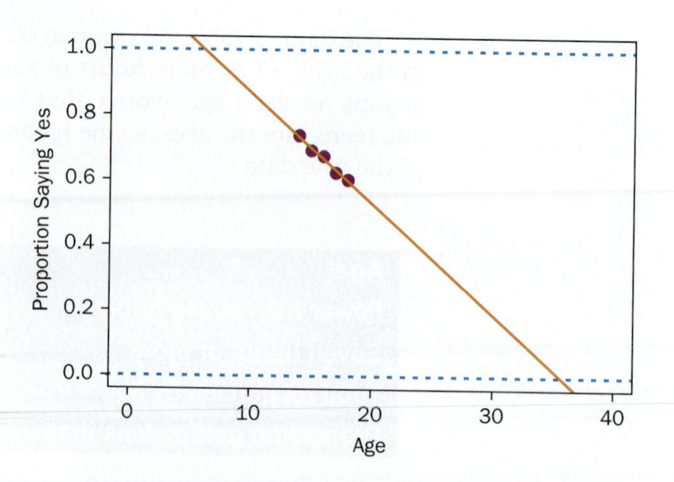

FIGURE 9.2 Scatterplot of *Age* and proportion of teenagers who say they get at least 7 hours of sleep at night with the regression line drawn

The Logistic Transformation

The logistic regression model, which is the topic of this chapter, always gives fitted values between 0 and 1. Figure 9.3 shows the fitted logistic model for the sleep data. Note the elongated "S" shape—a backward S here—that is typical of logistic regression models.

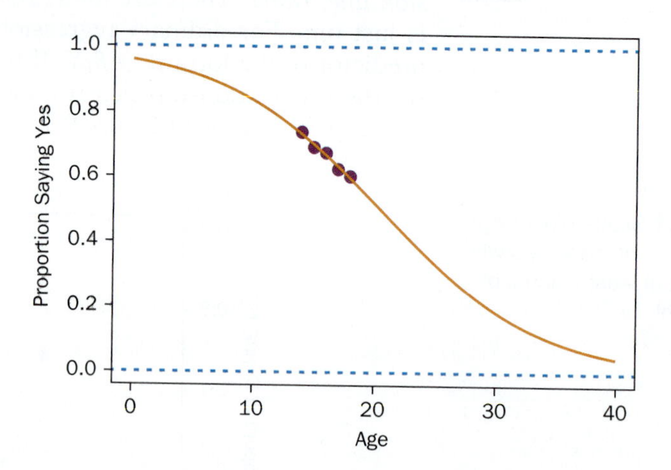

FIGURE 9.3 Scatterplot of *Age* and proportion of teenagers who say they get at least 7 hours of sleep at night with logistic regression model

The shape of the graph suggests a question: How can we get a *curved* relationship from a *linear* predictor of the form $\beta_0 + \beta_1 X$? In fact, you've already seen an answer to this question in Section 1.4. For example, for the data on doctors and hospitals (Example 1.7 on page 35), the relationship between $y =$ number of doctors and $X =$ number of hospitals was curved. To get a linear relationship, we transformed the response (to square roots), fit a line, and then transformed back to get the fitted curve.

That's what we'll do here for logistic regression, although the transformation is more complicated, and there are other complications as well. Here's a schematic summary so far:

$$\begin{aligned} &\text{Ordinary regression:} \\ &\text{Response} \approx \text{Intercept} + \text{Slope} \cdot X \\ &\text{Doctors and hospitals (Example 1.7):} \\ &(\text{Number of doctors})^{1/2} \approx \text{Intercept} + \text{Slope} \cdot X \\ &\text{Logistic regression:} \\ &??? \approx \text{Intercept} + \text{Slope} \cdot X \end{aligned}$$

The ??? on the left-hand side of the logistic equation will be replaced by a new transformation called the *log(odds)*.

ODDS AND LOG(ODDS)

Let $\pi = P(Y = 1)$ be a probability with $0 < \pi < 1$. Then the **odds** that $Y = 1$ is the ratio odds $= \frac{\pi}{1-\pi}$ and the **log(odds)** is given by log(odds) $= \log\left(\frac{\pi}{1-\pi}\right)$. Here, as usual, log is the natural log.

The transformation from π to log(odds) is called the **logistic** or **logit** transformation (pronounced "low-JIS-tic" and "LOW-jit").

EXAMPLE 9.2

Losing sleep: Logit fit Figure 9.4 shows a plot of logit-transformed probabilities versus age, along with the fitted line log(odds) $= 3.12 - 0.15Age$. To get the fitted curve in Figure 9.3, we "transform back," starting with the fitted line and reversing the transformation. Details for how to do this will come soon, but first as preparation, we spend some time with the transformation itself.

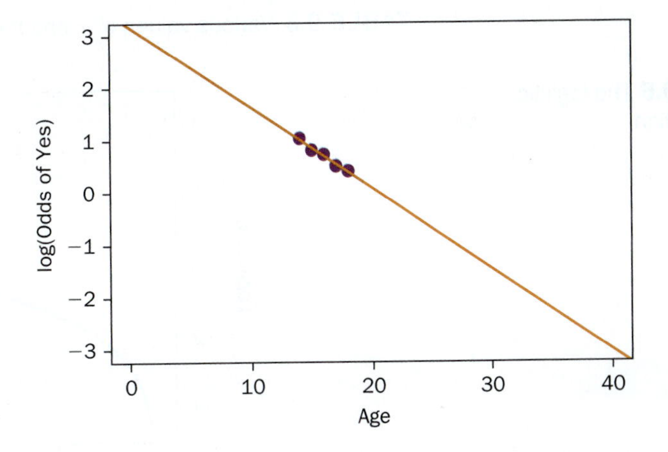

FIGURE 9.4 Logit versus *Age*

You already know about logs from Section 1.4, so we focus first on odds.

Odds

You may have run into odds before. Often, they are expressed using two numbers, for example, "4 to 1" or "2 to 1" or "3 to 2." Mathematically, all that matters is the ratio, so, for example, 4 to 2 is the same as 2 to 1, and both are equal to 2: $4/2 = 2/1 = 2$.

If you think about a probability by visualizing a spinner, you can use that same spinner to visualize the odds. (See Figure 9.5.) According to the data of Example 9.1, the chance that a 14-year-old gets at least 7 hours of sleep is

FIGURE 9.5 A spinner for $p = 3/4$, odds $= 3$

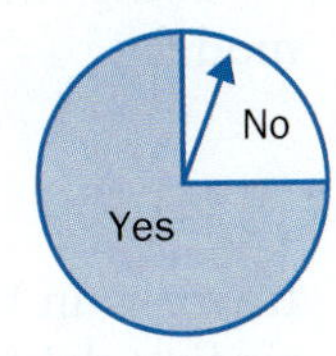

almost exactly 3/4. That is, out of every 4 randomly chosen 14-year-olds, there will be 3 Yes and 1 No. Thus, the odds of a 14-year-old getting at least 7 hours of sleep are 3 to 1 or $3/1 = 3$.

Log(odds)

To go from odds to log(odds), you do just what the words suggest: Take the logarithm. That raises the question of *which* logarithm: base 10? base *e*? Does it matter? The answer is that either will work, but natural logs (base *e*) are somewhat simpler, and so that's what statisticians use.

Table 9.3 shows several values of π, the corresponding odds $= \pi/(1-\pi)$, and the log(odds) $= \log(\pi/(1-\pi))$. Figure 9.6 is the corresponding graph.

π (fraction)	1/20	1/10	1/5	1/4	1/2	3/4	4/5	9/10	19/20
π (decimal)	0.05	0.10	0.20	0.25	0.50	0.75	0.80	0.90	0.95
odds	1/19	1/9	1/4	1/3	1/1	3/1	4/1	9/1	19/1
log(odds)	−2.94	−2.20	−1.39	−1.10	0	1.10	1.39	2.20	2.94

TABLE 9.3 Various values of π and their corresponding odds and log(odds)

FIGURE 9.6 The logistic transformation

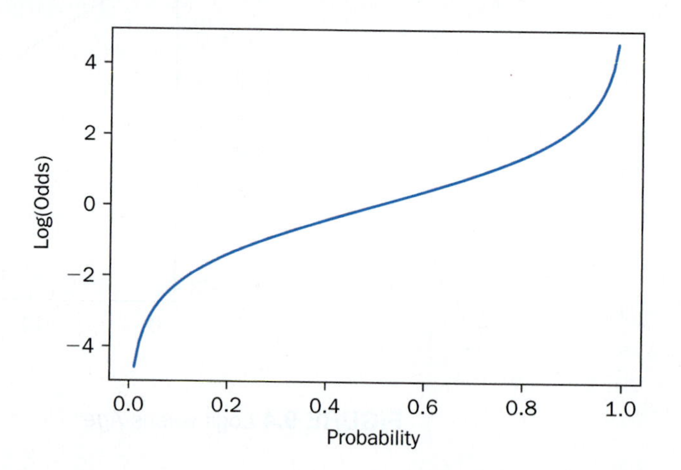

Two features of this transformation are important for our purposes:

(1) The relationship is one-to-one: for every value of π—with the two important exceptions of 0 and 1—there is one, and only one, value of $\log(\pi/(1-\pi))$. This means that *the logit transform is reversible*. If someone gives us the value of π, we can give them a unique value of $\log(\pi/(1-\pi))$, and vice versa: They give us a value of the log(odds), and we can give back the unique value of π that corresponds to the log(odds).

(2) The log(odds) can take on any value from $-\infty$ to ∞. This means that we can use a linear predictor of the form $\beta_0 + \beta_1 X$. If the predictor takes negative values, no problem: log(odds) can be negative. If the predictor is greater than 1, again no problem: log(odds) can be greater than 1.

Using "linear on the right, log(odds) on the left" gives the linear logistic model:

$$\text{log(odds)} = \log\left(\frac{\pi}{1-\pi}\right) = \beta_0 + \beta_1 X$$

It works in theory. Does it work in practice? Before we get to an example, we first define the notation that we will use throughout the unit.

FOUR PROBABILITIES

For any fixed value of the predictor x, there are four probabilities:

	True value	Fitted value
Actual	p = true P(Yes) for this x	$\hat{p}$ = #Yes/(#Yes + #No) for this x
Model	π = true P(Yes) from model	$\hat{\pi}$ = fitted P(Yes) from model

Note that $\hat{p}$ is a proportion based only on the sample cases where the predictor equals x, while $\hat{\pi}$ is an estimate based on the fitted model using all of the cases. If the model is exactly correct, then $p = \pi$ and the two fitted values estimate the same number.

EXAMPLE 9.3

Losing

Losing sleep: Observed and fitted proportions Table 9.4 shows the values of x = age, and $\hat{p}$ = (#Yes)/(#Yes + #No) from Table 9.2, along with the values of $\log(\hat{p}/(1-\hat{p}))$, and the fitted values of $\log(\hat{\pi}/(1-\hat{\pi}))$, and $\hat{\pi}$ from the logistic model. Figure 9.7 plots observed values of $\log(\hat{p}/(1-\hat{p}))$ versus x, along with the fitted line and fitted values from Table 9.4.

Age	**x**	2	12	**14**	**15**	**16**	**17**	**18**	20	30
Observed	$\hat{p}$			**0.74**	**0.69**	**0.68**	**0.63**	**0.60**		
	$\log(\hat{p}/(1-\hat{p}))$			**1.05**	**0.80**	**0.75**	**0.53**	**0.41**		
Fitted	$\log(\hat{\pi}/(1-\hat{\pi}))$	2.81	1.30	**0.99**	**0.84**	**0.69**	**0.54**	**0.39**	0.08	−1.43
	$\hat{\pi}$	0.94	0.79	**0.73**	**0.70**	**0.67**	**0.63**	**0.60**	0.52	0.19

TABLE 9.4 Fitted values of $\log(\pi/(1-\pi))$

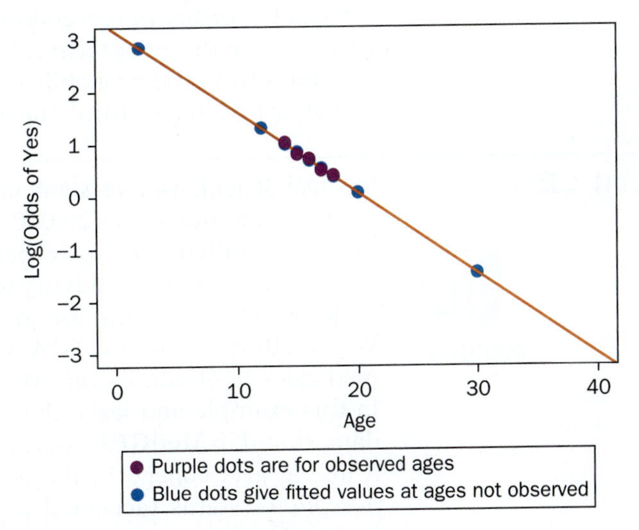

FIGURE 9.7 Observed log(odds) versus *Age*

Of course, log(odds) is just a stand-in. What we really care about are the corresponding probabilities. The graph in Figure 9.6 shows that the logistic transformation is reversible. How do we go backward from log(odds)?

Two Versions of the Logistic Model: Transforming Back

The logistic model approximates the log(odds) using a linear predictor $\beta_0+\beta_1X$. If we know $\log(\pi/(1-\pi)) = \beta_0 + \beta_1X$, what is the formula for π? Because the emphasis in this book is on applications, we'll simply give a short answer and an example here.

1. To go from log(odds) to odds, we use the exponential function e^x:

$$\text{odds} = e^{\log(\text{odds})}$$

2. You can check that if odds $= \frac{\pi}{1-\pi}$, then solving for π gives $\pi = \frac{\text{odds}}{1+\text{odds}}$.

Putting 1 and 2 together, gives

$$\pi = \frac{e^{\log(\text{odds})}}{1 + e^{\log(\text{odds})}}$$

Finally, if log(odds) $= \beta_0 + \beta_1 x$, we have

$$\pi = \frac{e^{\beta_0+\beta_1 X}}{1 + e^{\beta_0+\beta_1 X}}$$

We now have two equivalent forms of the logistic model.

LOGISTIC REGRESSION MODEL FOR A SINGLE PREDICTOR

The **logistic regression** model for the probability of success π of a binary response variable based on a single predictor X has either of two equivalent forms:

Logit form: $$\log\left(\frac{\pi}{1-\pi}\right) = \beta_0 + \beta_1 X$$

or

Probability form: $$\pi = \frac{e^{\beta_0+\beta_1 X}}{1 + e^{\beta_0+\beta_1 X}}$$

Recall the ordinary regression model $Y = \beta_0 + \beta_1 X + \epsilon$. The randomness in the model is in the error term ϵ. For the logistic regression model, the randomness comes from the probability of success at each x. We discuss this in more detail after looking at fitting the logistic regression model.

EXAMPLE 9.4

MedGPA

Medical school: Two versions of the logistic regression model Every year in the United States, over 120,000 undergraduates submit applications in hopes of realizing their dreams to become physicians. Medical school applicants invest endless hours studying to boost their GPAs. They also invest considerable time in studying for the Medical College Admission Test, or MCAT. Which effort, improving GPA or increasing MCAT scores, is more helpful in medical school admission? We look at a model relating admission to GPA in this example and leave the model based on MCAT to Exercise 9.19. The data, stored in **MedGPA**, was recorded for 55 medical school applicants from a liberal arts college in the Midwest. For each applicant, medical school *Acceptance* status (accepted or denied), *GPA*, and *MCAT* scores were collected. The response variable *Acceptance* status is a binary response, where 1 is a *success* and 0 a *failure*.

FIT

Figure 9.8(a) plots *MCAT* score versus *GPA*. This addresses the question, "How are MCAT scores related to GPA?," both quantitative variables, using an ordinary linear regression model as in Unit A. Figure 9.8(b) plots *Acceptance* versus *GPA*, where *Acceptance* has two categories, coded as 0 (not accepted) or 1 (accepted). In Figure 9.8(b), the 0/1 response values have been "jittered"* slightly in order to show all cases. In this figure, the y-axis shows $P(\textit{Accept})$ and the curve is the probability version of the fitted model.

*Jittering adds a small random amount to each number.

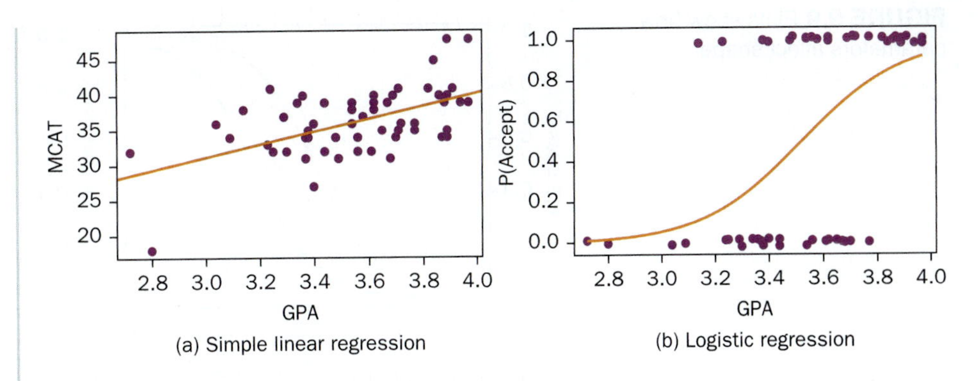

FIGURE 9.8 Scatterplots for regression models.

For the plot and data on the left (ordinary regression), the fitted model is linear: $\widehat{MCAT} = 3.92 + 9.10GPA$. For the plot and data on the right, the fitted model is linear in the log(odds) scale: $logit(P(Accept)) = -19.21 + 5.45GPA$. To find the fitted equation for $P(Accept)$, we "transform back." This takes two steps:

- Step 1. Exponentiate to go from log(odds) to odds: odds $= e^{\log(\text{odds})}$, so

$$\text{odds}(Acceptance) = e^{-19.21+5.45GPA}$$

- Step 2. Add 1 and form the ratio to go from odds to the probability:

$$P(Accept) = \frac{\text{odds}}{1+\text{odds}} = \frac{e^{-19.21+5.45GPA}}{1+e^{-19.21+5.45GPA}}$$

This is the equation of the fitted curve for the logistic regression in Figure 9.8.

For a concrete numerical example, consider a student with a *GPA* of 3.6. From the graph in Figure 9.8, we can estimate that the chance of acceptance is about 0.6. To get the actual fitted value, we transform back with $x = 3.6$:

- Step 1. Exponentiate odds $= e^{\log(\text{odds})}$.

$$\log(\text{odds}) = -19.21 + 5.45(3.6) = 0.41$$
$$\text{odds} = e^{0.41} = 1.51$$

- Step 2. Add 1 and form the ratio: $\hat{\pi} = \frac{\text{odds}}{1+\text{odds}}$:

$$\hat{\pi} = \frac{1.51}{1+1.51} = 0.60$$

So according to this model, the fitted chance of acceptance for a student with a 3.6 GPA is 60%.

How Parameter Values Affect Shape

The slope and constant terms of the linear expression in the logit form of the model affect the curve in the probability form. To get a feel for how, compare the curves in Figure 9.9. The value of the slope β_1 determines the slope of the curve at the point where $\pi = 1/2$. If $\beta_1 > 0$, the curve has a positive slope; see Figure 9.9(a). If $\beta_1 < 0$, the curve has a negative slope; see Figure 9.9(b). Regardless of sign, the larger $|\beta_1|$, the steeper the curve. As for β_0, the value of the constant term determines the right-to-left position of the curve. More precisely, the "midpoint" where $\pi = 1/2$ corresponds to $x = -\beta_0/\beta_1$.

FIGURE 9.9 Plots show how parameters affect shape

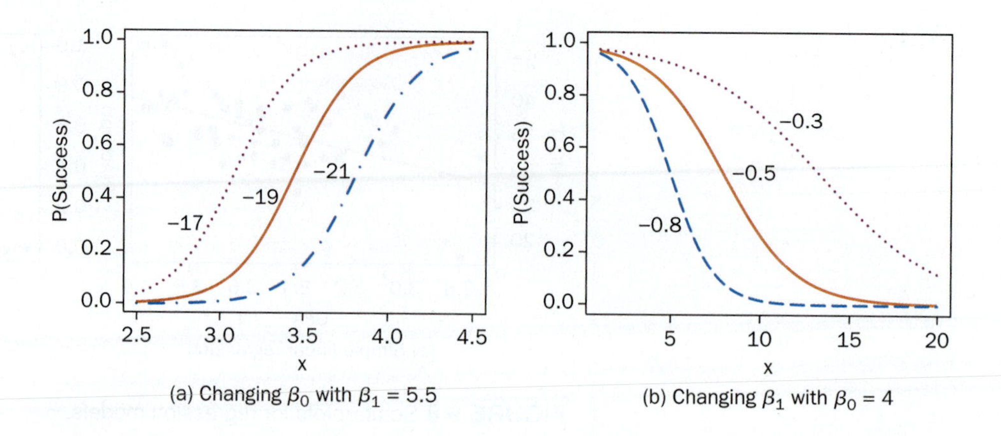

THE SHAPE OF THE LOGISTIC REGRESSION MODEL

The "midpoint" on the y-axis,"the 50% point," where $\pi = 1/2$, occurs at $x = -\beta_0/\beta_1$.

The slope of the curve at the midpoint is $\beta_1/4$.

Summary So Far

The logistic regression model is for data with a binary (Yes or No) response. The model fits the log(odds) of Yes as a linear predictor of the form $\beta_0 + \beta_1 X$, where just as in ordinary regression, β_0 and β_1 are unknown parameters.

The logistic model relies on a transformation back and forth between probabilities $\pi = P(\text{Yes})$ and $\text{log(odds)} = \log\left(\frac{\pi}{1-\pi}\right)$:

probability		odds		log(odds)
π	$\rightarrow$	$\frac{\pi}{1-\pi}$	$\rightarrow$	$\log\left(\frac{\pi}{1-\pi}\right) = l$

The backward transformation goes from $\text{log(odds)} = l$ to probabilities:

log(odds)		odds		probability
l	$\rightarrow$	e^l	$\rightarrow$	$\frac{e^l}{1+e^l} = \pi$

The logistic model has two versions:

$$\begin{aligned} \text{log(odds)} &= l \approx \beta_0 + \beta_1 X \\ P(\text{Yes}) &= \pi \approx \frac{e^{\beta_0+\beta_1 X}}{1+e^{\beta_0+\beta_1 X}} \end{aligned}$$

Randomness in the Logistic Model: Where Did the Error Term Go?

Randomness in the logistic regression model is different from randomness in the linear regression model in Chapters 1 to 3. There is no explicit error term in the model for π, and we don't have response values varying according to some normal distribution around a mean value. The responses now are binary categories (Yes/No) that don't fit so well into the *Data* = *Model* + *Error* framework of earlier chapters. The next example helps illustrate the role of randomness in logistic regression in a concrete setting.

EXAMPLE 9.5

MedGPA

Randomness in the models for acceptance to medical school Look again at the two plots of Figure 9.8, with ordinary regression (*MCAT* versus *GPA*) on the left and logistic regression (*Acceptance* versus *GPA*) on the right. To compare the two ways the two models deal with randomness, focus on a particular *GPA*, say, 3.6. For students with $GPA = 3.6$, the regression model says that the *MCAT* scores follow a normal distribution with a mean equal to the value of the linear predictor when $x = 3.6$. For those same students with $GPA = 3.6$, the logistic model says that the distribution of 1s and 0s behaves as if generated by a spinner, with $\text{logit}(\pi) =$ the value of the linear (logistic) predictor when $GPA = 3.6$. A random 0, 1 outcome that behaves like a spinner is said to follow a **Bernoulli distribution**. Figure 9.10 shows the two models.

Bernoulli distribution

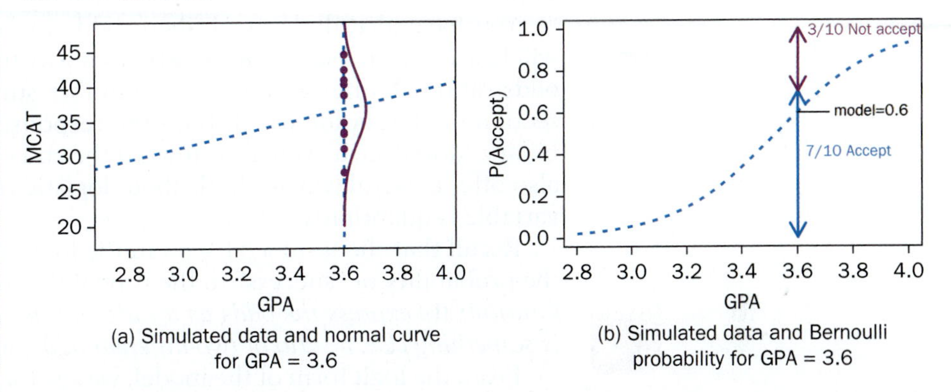

(a) Simulated data and normal curve for GPA = 3.6

(b) Simulated data and Bernoulli probability for GPA = 3.6

FIGURE 9.10 The normal model of randomness for ordinary regression (a) and the Bernoulli model of randomness for the logistic regression model (b), for $GPA = 3.6$

Figure 9.10(a) shows the ordinary regression model for MCAT versus GPA. The fitted equation $\widehat{MCAT} = 3.92 + 9.10GPA$ is shown as a dotted line. At each fixed $x = GPA$, according to the model, the values of $y = MCAT$ follow a normal curve. The line of vertical dots at $x = 3.6$ represents simulated y-values. The curve to the right of the dots represents the theoretical model—the normal distribution.

Figure 9.10(b) shows part of the logistic model for *Acceptance* versus *GPA*. The curve shows the probability version of the fitted equation whose linear form is $\log(\text{odds}) = -19.21 + 5.45GPA$. At each fixed $x = GPA$, according to the model, the values of $y = Acceptance$ follow a Bernoulli distribution with $P(Y = 1) = \pi$ and $P(Y = 0) = 1 - \pi$. For example, when $GPA = 3.6$, the fitted proportion shown on the curve is about 0.60. In the simulated data there are 10 cases with $GPA = 3.6$; 7 of which are accepted, the other 3 rejected. Thus the simulated data would have 7 "dots" at 1 and 3 "dots" at zero in the plot. The vertical arrows represent the sample proportions at $GPA = 3.6$, $\hat{p} = 0.7$ for the seven successes and $1 - \hat{p} = 0.3$ for the three not accepted. Note that the proportion in the sample is slightly above the proportion predicted from the fitted model.

To summarize:

Ordinary Regression	Logistic Regression
Response is quantitative.	Response is binary.
Outcomes are normal with mean $= \mu$ and SD $= \sigma$.	Outcomes are Bernoulli with $P(\text{success}) = \pi$.
Outcomes are independent.	Outcomes are independent.
μ depends only on X.	π depends only on X.
μ is a linear function of X, with parameters β_0 and β_1.	$\text{logit}(\pi)$ is a linear function of X, with parameters β_0 and β_1.

So far, you have seen two versions of the logistic regression model, the linear version for log(odds), and the curved version for probabilities. As a connection of these two versions, there is a way to understand the slope of the fitted line in terms of ratios of odds. That's what the next section is about.

9.2 Logistic Regression and Odds Ratios

So far in this chapter the explanatory variable has been quantitative. In this section we consider binary explanatory variables. (In Chapter 10, you will see examples with both kinds of explanatory variables in the same model.) Working with binary predictors leads naturally to an interpretation for the fitted slope.

odds ratio

You have already learned about odds in Section 9.1. In this section, you will learn how to use the **odds ratio** as a way to compare two sets of odds. The odds ratio will offer a natural one-number summary for the examples of this section—datasets for which both the response and explanatory variables are binary. Moreover, as you'll see toward the end of the section, the odds ratio will also offer a useful way to think about logistic regression when the explanatory variable is quantitative.

Recall that the term $\pi/(1-\pi)$ in the logit form of the model is the ratio of the probability of "success" to the probability of "failure," the odds of success. *Caution: We express the odds as a ratio, but as you will soon see, the odds ratio is something else, a ratio of two different odds, which makes it a ratio of ratios.*

From the logit form of the model, we see that the log of the odds is assumed to be a linear function of the predictor: $\log(\text{odds}) = \beta_0 + \beta_1 X$. Thus the odds in a logistic regression setting with a single predictor can be computed as odds $= e^{\beta_0+\beta_1 X}$. To gain some intuition for what odds mean and how we use them to compare groups using an *odds ratio*, we consider some examples in which both the response and predictor are binary.

Odds and Odds Ratios

EXAMPLE 9.6

Migrain

Zapping migraines A study investigated whether a handheld device that sends a magnetic pulse into a person's head might be an effective treatment for migraine headaches.[1] Researchers recruited 200 subjects who suffered from migraines and randomly assigned them to receive either the TMS (transcranial magnetic stimulation) treatment or a sham (placebo) treatment from a device that did not deliver any stimulation. Subjects were instructed to apply the device at the onset of migraine symptoms and then assess how they felt two hours later.

The explanatory variable here is which treatment the subject received (a binary variable). The response variable is whether the subject was pain-free two hours later (also a binary variable). The results are stored in **Migraines** and summarized in the following 2 × 2 table:

	TMS	Placebo	Total
Pain-free two hours later	39	22	61
Not pain-free two hours later	61	78	139
Total	100	100	200

Notice that 39% of the TMS subjects were pain-free after two hours compared to 22% of the placebo subjects, so TMS subjects were more likely to be pain-free. Although comparing percentages (or proportions) is a natural way to compare success rates, another measure—the one preferred for some situations, including the logistic regression setting—is through the use of odds. Although we defined the odds above as the probability of success divided by the probability of failure, we can calculate the odds for a sample by dividing the number of successes by the number of failures. Thus, the odds of being pain-free for the TMS group is (39/100)/(61/100), which simplifies to 39/61 = 0.639, and the odds of being pain-free for the placebo group is 22/78 = 0.282. Comparing odds gives the same basic message as does comparing probabilities: TMS increases the likelihood of success. The important statistic we use to summarize this comparison is called the *odds ratio* and is defined as the ratio of the two odds. In this case, the odds ratio (OR) is given by

$$OR = \frac{39/61}{22/78} = \frac{0.639}{0.282} = 2.27$$

We interpret the odds ratio by saying "the odds of being pain free were 2.27 times higher with TMS than with the placebo."

Suppose that we had focused not on being pain-free but on still having some pain. Let's calculate the odds ratio of still having pain between these two treatments. We calculate the odds of still having some pain as 61/39 = 1.564 in the TMS group and 78/22 = 3.545 in the placebo group. If we form the odds ratio of still having pain for the placebo group compared to the TMS group, we get 3.545/1.564 = 2.27, exactly the same as before. One could also take either ratio in reciprocal fashion. For example, 1.564/3.545 = 1/2.27 = 0.441 tells us that the odds for still being in pain for the TMS group are 0.441 times the odds of being in pain for the placebo group. When we have a binary predictor, it is usually more natural to interpret a statement with an OR greater than 1, so we prefer to use the OR of 2.27 when talking about how TMS compares to a placebo.

EXAMPLE 9.7

Letrozole therapy The November 6, 2003, issue of the *New England Journal of Medicine* reported on a study of the effectiveness of letrozole in postmenopausal women with breast cancer who had completed five years of tamoxifen therapy. Over 5000 women were enrolled in the study; they were randomly assigned to receive either letrozole or a placebo. The primary response variable of interest was disease-free survival. The article reports that 7.2% of the 2575 women who received letrozole suffered death or disease, compared to 13.2% of the 2582 women in the placebo group.

These may seem like very similar results, as 13.2% – 7.2% is a difference of only 6 percentage points. But the odds ratio is 1.97 (check this for yourself by first creating the 2 × 2 table; you should get the same answer as a ratio of the two proportions). This indicates that the odds of experiencing death or disease were almost twice as high in the placebo group as in the group of women who received letrozole.

These first two examples are both randomized experiments, but odds ratios also apply to observational studies, as the next example shows.

EXAMPLE 9.8

Transition to marriage Does the chance of a single mother getting married depend on the sex of their child? Researchers investigated this question by examining data from the Panel Study of Income Dynamics. For mothers who gave birth before marriage, they considered the child's sex as the explanatory variable, *Baby's Sex*, and whether or not the mother eventually married as the (binary) response, *Mother Married*. The data are summarized in the following table:

	Boy child	Girl child	Total
Mother eventually married	176	148	324
Mother did not marry	134	142	276
Total	310	290	600

We see that $176/310 = 0.568$ of the mothers with a boy eventually married, compared with $148/290 = 0.510$ of mothers with a girl. The odds ratio of marrying between mothers with a boy versus a girl is $(176/134)/(148/142) = 1.26$. So the odds of marrying are slightly higher (1.26 times higher) if the mother has a boy than if she has a girl.

This example differs from the previous one in two ways. First, the relationship between the variables is much weaker, as evidenced by the odds ratio being much closer to 1 (and the success proportions being closer to each other). Is that a statistically significant difference or could a difference of this magnitude be attributed to random chance alone? We will investigate this sort of inference question shortly. Second, these data come from an observational study rather than a randomized experiment. One implication of this is that even if the difference between the groups is determined to be statistically significant, we can only conclude that there is an association between the variables, not necessarily that a cause-and-effect relationship exists between them.

Odds Ratio and Slope

For ordinary regression, the fitted slope is easy to visualize: The larger the slope, the steeper the line. Moreover, the slope has a natural quantitative interpretation based on "rise over run" or "change-in-y over change-in-x." The slope tells you how much the fitted response changes when the explanatory variable increases by one unit.

For logistic regression, there is a similar interpretation of the fitted slope, but working with log(odds) makes things a bit more complicated, since it's the log(odds) that is linearly related to the predictor variable. To help visualize this relationship, we can use the log of the sample odds at different values of the predictor.

EMPIRICAL LOGIT

The **empirical logit** for any value of the predictor equals the log of the observed odds in the sample:

$$\text{Empirical logit} = \text{logit}(\hat{p}) = \log\left(\frac{\hat{p}}{1-\hat{p}}\right) = \log\left(\frac{\#\text{Yes}}{\#\text{No}}\right)$$

For ordinary regression, a scatterplot of "response versus explanatory" showing the fitted line is a useful summary. For logistic regression with a binary explanatory variable, plotting the empirical logit versus the 0, 1 predictor gives a useful way to summarize a 2×2 table of counts with a picture.

EXAMPLE 9.9

Migraines: Empirical logits As you saw earlier, the odds ratio for the migraine data is 2.27 (Example 9.6). The odds of being pain free after two hours are 2.27 times larger for TMS therapy than for the placebo.

The following table gives a summary:

MIGRAINES			
Pain-free?	TMS = 1		Placebo = 0
Yes	39		22
No	61		78
Total	100		100
Odds	0.64		0.28
OR		2.27	
log(odds)	−0.45		−1.27
slope		0.82	

Figure 9.11 shows a plot of empirical logits (logs of observed odds) versus the binary explanatory variable. Because there are only two points in the plot, fitting a line is easy, and because the change in x is $1 - 0 = 1$, the fitted slope is equal to the change in log(odds).

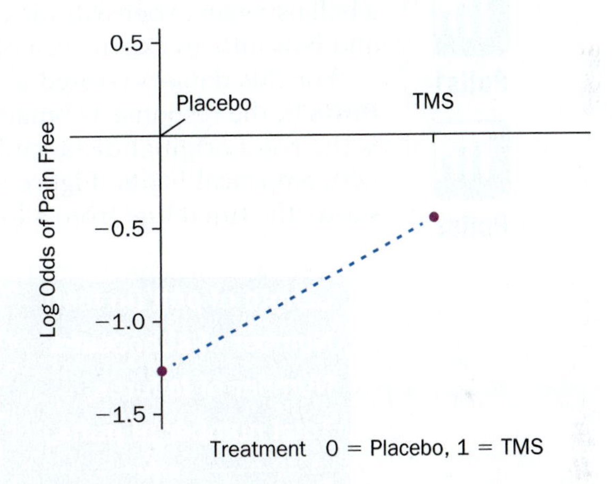

FIGURE 9.11 Change in log(odds) of relief equals fitted slope

Use the example to check that the statements in the following box are correct:

FITTED SLOPE AND ODDS RATIO

The fitted slope is

$$\text{rise/run} = \frac{\text{change in log odds}}{1 - 0}$$
$$= \text{log(odds ratio)}$$

Thus,

$$\text{odds ratio} = e^{\text{fitted slope}}$$

The bottom line, in words, is "*e to the slope equals the odds ratio.*" What this means in practice is that when the output from a logistic regression analysis tells you the fitted slope, you can translate that number to an odds ratio using "*e* to the slope."

Slope and Odds Ratios When the Predictor Is Quantitative

As promised in the section introduction, there is a way to use the odds ratio to interpret the fitted slope even when the predictor is quantitative. The challenge is this: In the original (probability) scale of the response, the fitted slope keeps changing as X changes. In the log(odds) scale, the slope is constant, but its meaning keeps changing. This makes interpretation more complicated than for ordinary regression. To anticipate: *The constant slope in the log(odds) scale corresponds to a constant odds ratio.* The probabilities may change as X changes, but the odds ratio does not change. An example will help make this clearer.

EXAMPLE 9.10

Author with a club One of your authors is an avid golfer. In a vain attempt to salvage something of value from all the hours he has wasted on the links, he has gathered data on his putting prowess. (In golf, a putt is an attempt to hit a ball using an expensive stick—called a club—so that the ball rolls a few feet and falls into an expensive hole in the ground—called a cup.)

Putts1

Putts2

For this dataset (stored as raw data in **Putts1** or as a table of counts in **Putts2**), the response is binary (1 = Success, 0 = Failure), and the predictor is the ball's original distance from the cup. Table 9.5 shows the data, along with empirical logits. Figure 9.12 plots empirical logits versus distance and shows the fitted line from a logistic regression.

Length of putt (in feet)	3	4	5	6	7
Number of successes	84	88	61	61	44
Number of failures	17	31	47	64	90
Total number of putts	101	119	108	125	134
Proportion of successes	0.832	0.739	0.565	0.488	0.328
Odds of success	4.941	2.839	1.298	0.953	0.489
Empirical logit	1.60	1.04	0.26	−0.05	−0.72

TABLE 9.5 Putting prowess

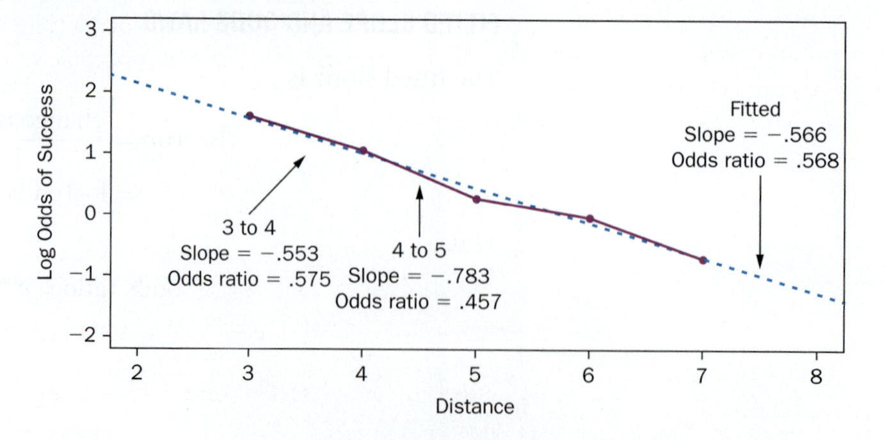

FIGURE 9.12 Slopes and odds ratios for the putting data

We can use the odds ratio to compare the odds for any two lengths of putts. For example, the odds ratio of making a 4-foot putt to a 3-foot putt is calculated as $2.839/4.941 = 0.57$. So the odds of making a 4-foot putt are about 57% of the odds of making a 3-foot putt. Comparing 5-foot putts to 4-foot putts gives an odds ratio of 0.46, comparing 6-foot to 5-foot putts gives an odds ratio of 0.73, and comparing 7-foot to 6-foot putts gives an odds ratio of 0.51. Each time we increase the putt length by 1 foot, the odds of making a putt are reduced by a factor somewhere between 0.46 and 0.73. These empirical odds ratios are the second row of the following table, the third row of which gives the corresponding odds ratios computed from the fitted model:

Length of putt (in feet)	4 to 3	5 to 4	6 to 5	7 to 6
Empirical data: Odds ratio	0.575	0.457	0.734	0.513
Fitted logistic model: Odds ratio	0.568	0.568	0.568	0.568

A consequence of the logistic regression model is that the model constrains these odds ratios (as we increase the predictor by 1) to be constant. The third row of the table—the odds ratios from the fitted model—illustrates this principle with all being 0.568. In assessing whether a logistic model makes sense for our data, we should ask ourselves if the empirical odds ratios appear to be roughly constant. That is, are the empirical odds ratios very different from one another? In this example, the empirical odds ratios reflect some variability, but seemingly not an undue amount. This is analogous to the situation in ordinary simple linear regression where the predicted mean changes at a constant rate (the slope) for every increase of 1 in the predictor—even though the sample means at successive predictor values do not follow this pattern exactly.

EXAMPLE 9.11

MedGPA

Medical school admissions: Interpreting slope In our medical school acceptance example, we said that the estimated slope coefficient is 5.45. Here is Minitab output that shows how *GPA* is related to acceptance.

```
                                                Odds      95% CI
Predictor      Coef       SE       Z      P     Ratio   Lower    Upper
Constant   -19.2065  5.62922   -3.41  0.001
GPA         5.45417  1.57931    3.45  0.001   233.73   10.58  5164.74
```

In ordinary simple linear regression, we interpret the slope as the change in the mean response for every increase of 1 in the predictor. Since the logit form of the logistic regression model relates the log of the odds to a linear function of the predictor, we can interpret the sample slope $\hat{\beta}_1 = 5.45$, as the typical change in log(odds) for each one-unit increase in *GPA*.

However, log(odds) is not as easily interpretable as odds itself; thus we exponentiate both sides of the logit form of the model for a particular *GPA*, to get

$$\frac{\hat{\pi}_{GPA}}{1-\hat{\pi}_{GPA}} = \text{odds}_{GPA} = e^{-19.21+5.45GPA}$$

If we increase the *GPA* by one unit, we get

$$\frac{\hat{\pi}_{GPA+1}}{1-\hat{\pi}_{GPA+1}} = \text{odds}_{GPA+1} = e^{-19.21+5.45(GPA+1)}$$

So an increase of one *GPA* unit can be described in terms of the odds ratio:

$$\frac{\text{odds}_{GPA+1}}{\text{odds}_{GPA}} = \frac{e^{-19.21+5.45(GPA+1)}}{e^{-19.21+5.45GPA}} = e^{5.45}$$

Therefore, a one-unit increase in *GPA* is associated with an $e^{5.45}$, or 233.7-fold, increase in the odds of acceptance! We see here a fairly direct interpretation of the estimated slope, $\hat{\beta}_1$: Increasing the predictor by one unit gives an odds ratio of $e^{\hat{\beta}_1}$; that is, the odds of success is multiplied by $e^{\hat{\beta}_1}$. In addition to the slope coefficient, Minitab gives the estimated odds ratio (233.73) in its logistic regression output.

The magnitude of this increase appears to be extraordinary, but in fact it serves as a warning that the magnitude of the odds ratio depends on the units we use for measuring the predictor (just as the slope in ordinary regression depends on the units). Increasing your GPA from 3.0 to 4.0 is dramatic and you would certainly expect remarkable consequences. It might be more meaningful to think about a tenth of a unit change in grade point as opposed to an entire unit change. We can compute the odds ratio for a tenth of a unit increase by $e^{5.454(0.1)} = 1.73$.

An alternative would have been to redefine the *X* units into "tenths of GPA points." If we multiply the GPAs by 10 (call it *GPA*10) and refit the model, we can see from the Minitab output below that the odds of acceptance nearly doubles (1.73) for each tenth unit increase in *GPA*, corroborating the result of the previous paragraph. The model assumes this is true no matter what your GPA is (e.g., increasing from 2.0 to 2.1 or from 3.8 to 3.9); your odds of acceptance go up by a factor of $e^{0.5454} = 1.73$. Note that this refers to odds and not probability.

Logistic Regression Table

					Odds	95% CI	
Predictor	Coef	SE Coef	Z	P	Ratio	Lower	Upper
Constant	-19.2065	5.62922	-3.41	0.001			
GPA10	0.545417	0.157931	3.45	0.001	1.73	1.27	2.35

Once you have chosen a logistic regression model, the next step is to find fitted values for the intercept β_0 and the slope β_1. Because the method not only relies on some new concepts but also is hard to visualize, and because our emphasis in this book is on applications rather than theory, we have chosen to put an explanation for how the method works into Chapter 11 as a topic. Topic 11.1 explains what it is that computer software does to get fitted values. For applied work, you can simply trust the computer to crunch the numbers. Your main job is to choose models, assess how well they fit, and put them to appropriate use. The next section tells how to assess the fit of a logistic regression model.

9.3 Assessing the Logistic Regression Model

This section will deal with three issues related to the logistic model: linearity, randomness, and independence. Randomness and independence are essential for formal inference—the tests and intervals of Section 9.4. Linearity is about how close the fitted curve comes to the data. If your data points are "logit-linear," the fitted logistic curve can be useful even if you can't justify formal inference.

- *Linearity is about pattern, something you can check with a plot.* You don't have to worry about how the data were produced.
- *Randomness and independence boil down to whether a spinner model is reasonable.* As a rule, graphs can't help you check this. You need to think instead about how the data were produced.

Linearity

The logistic regression model says that the log(odds)—that is, $\log(\pi/(1-\pi))$—are a linear function of x. (What amounts to the same thing: The odds ratio for a change in x of 1 is constant, regardless of x.) In what follows, we check linearity for three kinds of datasets, sorted by type of predictor and how many y values there are for each x:

1. Datasets with *binary* predictors. Examples include migraines (Example 9.6) and marriage (Example 9.8) from the last section. *When your predictor is binary, linearity is automatic,* because your empirical logit plot has only two points. See Figure 9.11.

2. Datasets with a *quantitative* predictor and *many* response values y for each value of x. As an example, recall the author with a club example (Example 9.10). For such datasets, the empirical logit plot gives a visual check of linearity, as in Figure 9.12.

CHECKING LINEARITY FOR DATA WITH MANY *y*-VALUES FOR EACH *x*-VALUE

Use an empirical logit plot of log(odds) versus x to check linearity.

3. Other datasets: *quantitative* predictor with many x-values but *few* response values for each x. For an example, recall the medical school admissions data (Example 9.4) from Section 9.1. For such datasets, we (1) slice the x-axis into intervals, (2) compute the average x-value and find the empirical logit for each slice, then (3) plot logits as in the example that follows.

EXAMPLE 9.12

MedGPA

Medical school admissions: Empirical logit plot by slicing

ASSESS

For the dataset of Example 9.4, there are many values of *GPA*, with few response values for each x. To compute empirical logits, we slice the range of *GPA* values into intervals. The intervals are somewhat arbitrary, and you should be prepared to try more than one choice. For the admissions data there are 55 cases, and it seems natural to compare two choices: 5 intervals of 11 cases each, and 11 intervals of 5 cases each. Within each slice, we compute the average $x = GPA$, and compute the number Yes, percent Yes, and empirical logit, as shown in Table 9.6 for the set of five slices.

Finally, we construct the empirical logit plot, and use it to assess how well the data points follow a line, as required by the linear logistic model.

		GPA		Admitted		Proportion		
Group	# Cases	Range[a]	Mean	Yes	No	TRUE	ADJ[b]	logit (ADJ)
1	11	2.72 – 3.34	3.13	2	9	0.18	0.21	−1.34
2	11	3.36 – 3.49	3.41	4	7	0.36	0.38	−0.51
3	11	3.54 – 3.65	3.59	6	5	0.55	0.54	0.17
4	11	3.67 – 3.84	3.73	7	4	0.64	0.63	0.51
5	11	3.86 – 3.97	3.95	11	0	1.00	0.96	3.14

[a] Ranges are based on choosing equal-sized groups, 5 groups of 11 each.
[b] Because $p = 0$ and $p = 1$ cannot be converted to logits, we use the standard "fudge": Instead of (# Yes)/(# Yes + # No), we include a fictitious observation, split half-and-half between Yes and No, so that $p = (1/2 + \#\text{Yes})/(1 + \#\text{Yes} + \#\text{No})$.

TABLE 9.6 Empirical logits by slicing: Medical school admissions data

Figure 9.13(a) is based on 5 groups of 11 cases each. Figure 9.13(b) is based on 11 groups of 5 cases each, and shows more scatter about a fitted line. In that panel, the four left-most points lie close to the line, with the right-most point as an outlier above. Linearity seems reasonable, except possibly for very high GPAs.*

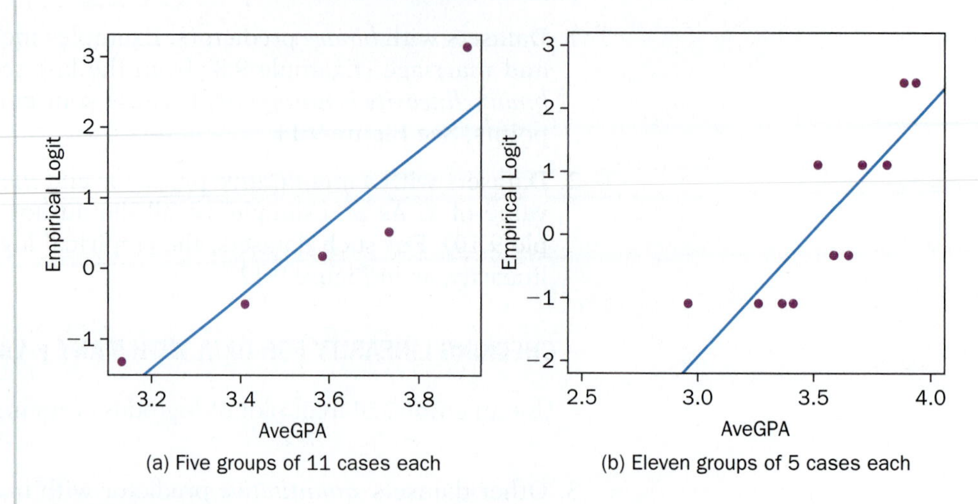

FIGURE 9.13 Empirical logit plots for the medical school admissions data

We now summarize the process of creating an empirical logit plot.

EMPIRICAL LOGIT PLOTS FOR QUANTITATIVE PREDICTORS

1. Divide the range of the predictor into intervals with roughly equal numbers of cases.[a]
2. Compute the mean value of the predictor for each interval.
3. Compute the observed proportion $\hat{p}$ for each interval.[b]
4. Compute $\text{logit}(\hat{p}) = \log(\hat{p}/(1 - \hat{p}))$.
5. Plot $\text{logit}(\hat{p})$ versus the mean value of the predictor, with one point for each interval.[c]

[a] *How many intervals?* Two intervals will give you a sense of the direction and size of the relationship. Three will also give you an indication of departures from linearity, but the plot alone can't tell how much of the departure is systematic, and how much is chance-like. If you have enough cases, four or five intervals is better.
[b] *If group sizes are small*, use $\hat{p} = (0.5 + \#\text{Yes})/(1 + n)$.
[c] *If you have enough cases*, you can use plots to explore two predictors at once, as shown in the examples.

Question: What if your plot is not linear?
Answer: You have already seen one useful strategy, in connection with ordinary regression. Look for a way to transform the x-variable to make the relationship more nearly linear.

*This pattern suggests that there is an "end-effect" due to the fact that GPA cannot be greater than 4.0, in much the same way that probabilities cannot be greater than 1.0. There are transformations that can sometimes remove such end-effects.

EXAMPLE 9.13

Transforming for linearity: Dose-response curves In some areas of science (e.g., drug development or environmental safety), it is standard practice to measure the relationship between a drug dose and whether it is effective, or between a person's exposure to an environmental hazard and whether they experience a possible effect such as thyroid cancer. Experience in these areas of science has shown that typically the relationship between logit(p) and the dose or level of exposure is *not* linear, but that transforming the dose or exposure to logs *does* make the relationship linear. There are graphical ways to check whether to try a log transformation and other methods for fitting nonlinear logistic relationships.

CHOOSE

Figures 9.14(a) and (b) show what a typical dose-response relationship would look like if you were to plot p against the dose or exposure level (a) or the log of the dose (b). Figures 9.14(c) and (d) show the corresponding plots for logits. As you can see, the relationship in Figure 9.14(c) is curved. Figure 9.14(d) shows the same relationship with logit(p) plotted against the log concentration. This plot is linear.

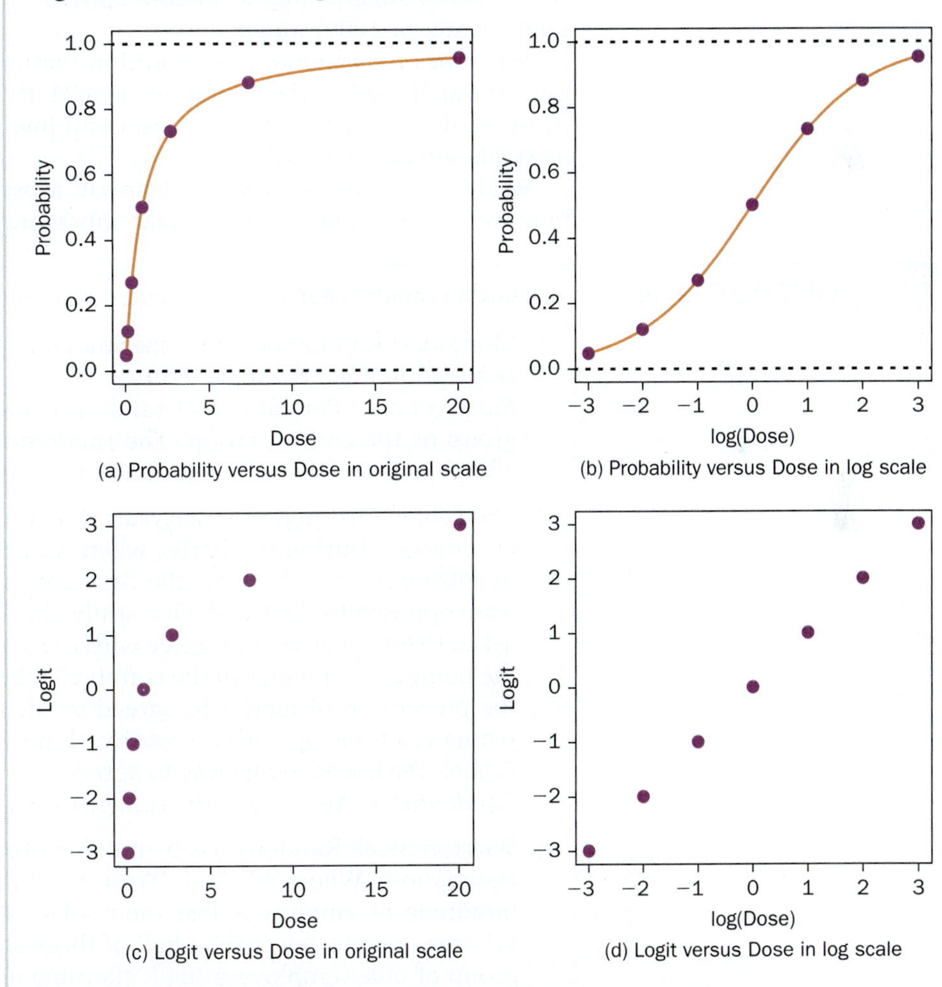

FIGURE 9.14 Typical (but hypothetical) dose-response relationship with dose in the original (left) and log (right) scales. The top panels plot probabilities on the vertical axis; the bottom panels plot logits.

The linearity condition relates observed proportions to values of the explanatory variable x. Notice that you can assess linearity without needing to think about the other two conditions: randomness and independence.

Linearity is about the *shape* of the relationship, but not about how the data were *produced. This means that if the relationship is linear, logistic regression can be useful for describing patterns even if randomness or independence fail.* For formal inference, however—tests and intervals—you *do* need randomness and independence.

The linearity condition tells how the Yes/No proportions are related to the x-values. The next two conditions, randomness and independence, are about whether and in what way the proportions are based on probabilities. Is it reasonable to think of each response Y coming from an independent spin of a spinner, with a different spinner for each x-value?

Randomness

Some proportions come from probabilities; others don't. For example, 50% of all coin flips land Heads. The 50% is based on a probability model. It is reasonable to model the outcome of a coin flip using a spinner divided 50/50 into regions marked Heads and Tails. For contrast, your body is about 60% water. This proportion is not random in the Yes/No spinner sense. It would be ridiculous to suggest that a random spinner model decides whether you end up 0% water or 100% water.

Why does the randomness condition matter? Because statistical tests and intervals are based on the probability model. If the spinner model offers a good fit to the data, you can trust the tests and intervals. If not, then the tests and intervals should be avoided.

Reality, as usual, is rarely so clear-cut. Most applications of the theory fall in between. Examples will help make this more concrete.

EXAMPLE 9.14

Checking randomness Here are seven different scenarios to consider.

1. Migraines: Randomness by experimental design.
   Description. (See Example 9.6.)
   Randomness? Patients were randomly assigned either to the treatment group or the control group. The randomness in the assignment method allows us to justify using a probability model.

2. Male chauvinist pigs of yesteryear: Randomness by sampling plan.
   Description. During the 1970s, when women were entering the workforce in substantial numbers for the first time since World War II, many men were opposed to the trend. One study chose a random sample of men and asked them to agree or disagree with the statement "Women should stay in the home and let men run the country." A linear logistic regression relating the proportion of men who agreed to their years of education showed a strong relationship with a negative slope: The more time a man spent in school, the less likely he was to agree.
   Randomness? Yes, due to the random sampling.

3. Wierton Steel: Randomness by null hypothesis.
   Description. When Wierton (West Virginia) Steel declared bankruptcy, hundreds of employees lost their jobs. After another company bought Wierton's assets, fewer than half of those same employees were rehired. A group of older employees sued, claiming age discrimination in the hiring decisions, and a logistic regression showed a strong relationship between age and whether a person was rehired.
   Randomness? For this situation, it is clear that the hiring decisions were *not* based on spinners. However, that is not the legal issue. What matters is this: Is it plausible that an age-blind spinner model could have produced results as extreme as the observed results? Here, the spinner model derives from the null hypothesis to be tested, even though we know that the model is entirely hypothetical.

4. Medical school: Randomness is plausible as an approximation.
Description. (See Example 9.4.)
Randomness? We know that medical schools do not choose students at random regardless of GPA, so random assignment to Yes/No is not a basis for a probability model. We also know that students who apply to medical school are not a random sample from the population of U.S. college students; in this sense, random sampling is not a basis for a probability model.

However, consider a different way to look at the question: (1) Is there reason to think that the students in the dataset, on balance, differ in a systematic way from some population of interest? This population might be the set of applicants from the same college in recent past and near future years, or applicants from similar colleges. (2) Is there reason to think that for these students there is a systematic pattern that makes GPA misleading as a predictor of admission? For example, at some colleges your GPA might tend to be higher or lower depending on what courses you choose to take. However, students applying to medical school tend to take much the same courses, so this possible pattern is probably not an issue.

On balance, there is a plausible argument for thinking that randomness is a reasonable approximation to the selection process.

5. Putting prowess: Randomness is plausible as an approximation.
Description. (See Example 9.10.)
Randomness? The outcome of a putt is not random. Even so, whether the ball goes in the cup is the result of physical forces so numerous and so subtle that we can apply a probability model. (Think about flipping a coin. Whether the coin lands Heads or Tails is also the result of numerous physical forces, but we have no trouble regarding the outcome as random.)

6. Moldy bread: Randomness fails.
Description. Back in the early days of "hands-on learning," one of your authors introduced the logistic curve in his calculus classes using as an example the growth of bread mold: Put a slice of nonpreservative bread in a plastic bag with a moistened tissue, and wait for the black mold to appear. Each day, put a grid of small squares over the bread, and count the number of squares that show mold. Plot the logit of the proportion versus time (number of days).
Randomness? No. But even so, the logistic fit is definitely useful as a description of the growth of mold over time, a model that ecologists use to describe biological growth in the presence of limited resources.

7. Bluegrass banjo: Randomness is ridiculous.
Description. One of your authors, not athletic enough to be any good at golf, is an avid banjo player. In a vain attempt to salvage something of value from the hours he has wasted plucking, he has foolishly tried to apply logistic regression to a bluegrass banjo "roll," an eight-note sequence with a fixed pattern. According to the logistic model, the predictor is the time in the sequence when the note is played (1 to 8) and the response is whether the note is picked with the thumb:

Forward roll:	1	2	3	4	5	6	7	8
Thumb? (1 = Yes)	1	0	0	1	0	0	1	0

Randomness? Not unless the banjo player is totally incompetent. This example is deliberately extreme. There is no randomness, because the sequence is fixed. Notice that it is possible to compute *P*-values and interval estimates. Notice also that a (brainless) computer will do it for you if you ask. But most important, notice that the output would be completely meaningless. (Commit "brainless banjo" to memory as a caution.)

Independence

Even if outcomes are random, they may not be independent. For example, if you put tickets numbered 1 to 10 in a box, mix them up, and take them out one at a time, the sequence you get is random, but the individual outcomes are not independent. If your first ticket is #9, your second cannot be. However, if you put #9 back and mix again before you grab the next ticket, your outcomes are both random *and* independent.

If you decide that randomness fails, you don't need to check independence, because you already know you don't have the probability model you need to justify formal inference. Suppose, though, you have decided that it is reasonable to regard the outcomes as random. How can you check independence? It may help to think about time, space, and the Yes/No decision. (1) Time: The ticket example suggests one thing to check: Are the results from a time-ordered process? If so, is it reasonable to think that one outcome does not influence the next outcome in the sequence? (2) Space: If your observational units have a spatial relationship, you should ask whether it is reasonable to think that the outcome for one unit is independent of the nearby units. In this context, space may be only implied, as with children in the same grade school class. (3) The Yes/No decision: Some decisions are clear—was the ticket #9? Other decisions may depend on subjective judgment—is this Medicare claim justified? When Yes/No decisions are not objective, there is a possibility that the decision process introduces dependence.

Here, as with randomness, many judgments about independence are less clear-cut than we would want, and examples can help.

EXAMPLE 9.15

Assessing independence See Example 9.14 for descriptions.

1. Migraines and assignment.
   Independence is reasonable because of the random assignment.
2. Male chauvinist pigs and sampling.
   Independence is reasonable because of the random sampling.
3. Wierton Steel and the Yes/No decision.
   If we assume that hiring decisions were made through a single coordinated process, then decisions cannot possibly be independent. (If you and a friend apply for the same job and you get hired, your friend is out of luck.) So if you and your friend are the only ones in the job pool, independence fails, big time. But if you and the friend are just two among thousands of applicants applying for hundreds of similar jobs, it's a different story. If you get a job offer, your friend's chances are not much reduced. Bottom line: Independence is wrong but may be a reasonable approximation.
4. Medical school.
   The situation is like Wierton, only more so. For any one school, the Wierton argument applies: Many applicants, many similar positions. But there are hundreds of medical schools. If Georgetown offers you a place, that won't have much effect on a friend's chances at Washington University, St. Louis. Bottom line: Independence is reasonable.
5. Putting prowess and time.
   Independence may be reasonable. Suppose you argue that my success or failure on the last putt affects the outcome of the putt I am about to attempt. Even so, the distance for this putt is likely to be different from the distance for the last one. What matters here is whether consecutive

putts at the same distance are independent. Typically, these will not be consecutive in time.*

6. Moldy bread and space.
   Independence fails. A little square with no mold today is more likely to have mold tomorrow if it is next to a square with mold, but less likely if not.

7. Bluegrass banjo.
   Independence is moot because the sequence of notes is fixed, not random.

9.4 Formal Inference: Tests and Intervals

Even if your check of conditions convinces you that the Bernoulli (spinner) model is not appropriate, you can still use logistic regression for description, and sometimes for prediction as well. If, in addition, the probability model is reasonable (outcomes are random and independent), you can also test hypotheses and construct confidence intervals.

This section reveals how to do tests and intervals in the context of three examples: medical school admissions, putting, and migraines. We assume you already know how to do tests and intervals for simple linear regression. The exposition here relies on parallels between that context and logistic regression.

Quantitative Predictors

Recall that in simple linear regression we had several ways to assess whether the relationship between the predictor and response was stronger than one would expect to see by random chance alone and to assess the strength of the relationship. These included:

- A t-test to see if the slope differs from zero
- An ANOVA F-test to see if a significant amount of variability is explained by the linear fit
- Quantities such as the variability explained by the model (*SSModel*), the variability left unexplained (*SSE*), the percent of variability explained (R^2), or the estimated standard deviation of the error term ($\hat{\sigma}_\epsilon$), which can be used to compare the effectiveness of competing models

Do we have analogous tests and measurements to help assess the effectiveness of a model in the logistic setting? The answer is Yes, although some of the analogous procedures will look more similar to the simple regression setting than others: The procedure to test to see if a slope is different from zero will look very similar, but for many other procedures, we need to discuss a substitute for the sums of squared deviations that forms the basis of most of the procedures in ordinary least squares regression.

EXAMPLE 9.16

MedGPA

Medical school admissions: Inference for slope Here is some computer output for fitting a logistic regression model to predict medical school acceptance with GPA, but using *GPA*10—which is 10*GPA* so that GPA is measured in tenths of points rather than full points—as the predictor (Example 9.11 on page 429).

```
Coefficients:
             Estimate  Std. Error  z value   Pr(>|z|)
(Intercept)  -19.2065    5.6287    -3.412    0.000644
GPA10          0.5454    0.1579     3.454    0.000553
```

*Note that for this dataset, only one putt was recorded for a given hole. Otherwise, independence would fail because a golfer who misses a putt can learn from watching the ball roll.

ASSESS

The meaning is much the same as for ordinary regression. The small *P*-value shown for the test of the slope gives strong evidence that there is a log-linear relationship between medical school acceptance and *GPA10*. Software typically supplies the estimated coefficient, $\hat{\beta}_1$, and its standard error, $SE_{\hat{\beta}_1}$, as well as the other details.

If we use the output for the medical school data using the *GPA*10 predictor, we can compute a 95% confidence interval for the slope of that model with

$$0.5454 \pm 1.96(0.1579) = 0.5454 \pm 0.3095 = (0.2359, 0.8549)$$

However the slope, measuring the change in log(odds) for every unit change in the predictor, is often difficult to interpret in a practical sense. As you saw in the last section, you can convert the slope to an estimated odds ratio using $e^{0.5454} = 1.73$ ("*e* to the slope = odds ratio"). Applying the same process to the confidence bounds for the slope produces a confidence interval for the odds ratio. With R the confidence interval is not given as part of the logistic regression output. You must enter an additional command. But Minitab output reports this interval directly.

Minitab output for medical school acceptance data:

Predictor	Coef	SE Coef	Z	P	Odds Ratio	95% CI Lower	Upper
Constant	-19.2065	5.62922	-3.41	0.001			
GPA10	0.545417	0.157931	3.45	0.001	1.73	1.27	2.35

The following box summarizes inference procedures for the slope in simple logistic regression:

TEST AND CONFIDENCE INTERVAL FOR THE SLOPE OF A SIMPLE LOGISTIC REGRESSION MODEL

To test whether the slope in a model with a single predictor is different from zero, the hypotheses are

$H_0 : \beta_1 = 0$
$H_a : \beta_1 \neq 0$

and the test statistic is

$$z = \frac{\hat{\beta}_1}{SE_{\hat{\beta}_1}}$$

Assuming we have a reasonably large sample (with independent, random outcomes), the *P*-value is determined from a normal distribution. This *z*-statistic is also called the *Wald statistic*.

We can also compute a confidence interval for the slope using

$$\hat{\beta}_1 \pm z^* \cdot SE_{\hat{\beta}_1}$$

where z^* is found using the normal distribution and the desired level of confidence.

The second important tool we used in assessing the fit of a simple linear regression model was the *F*-test based on the ANOVA table. The main principle in deriving that test was to see how much improvement was gained by using the linear model instead of just a constant model. We measured "improvement" by how much the sum of squared residuals changed between the two models. Recall that estimates in the ordinary regression setting were obtained by minimizing that sum of squared residuals. In logistic regression, we choose coefficients to minimize a different quantity ($-2\log L$) that behaves in much the same way. Comparing values of $-2\log L$ for models with and without the linear predictor serves as the basis for a test of $H_0 : \beta_1 = 0$.

What is the quantity $-2\log L$? A detailed explanation would take us on a detour into theory, away from applications, and so we have put the details in Topic 11.1. Here is a bare-bones definition.

LIKELIHOOD, MAXIMUM LIKELIHOOD, AND DEVIANCE

The **likelihood** of the data, denoted L, is the probability of the data, regarded as a function of the unknown parameters with the data values fixed. (This is parallel to sums of squares being regarded as a function of the parameters with data values held fixed for regular regression.)

The **method of maximum likelihood** chooses parameter values to maximize L, or, equivalently, to minimize $-2\log L$, which is called the **deviance**. The deviance behaves similarly to residual sum of squares in regular regression.

We compare nested models by observing the change in deviance, much as we observed changes in residual sums of squares for nested regression models.

Logistic Regression Table

					Odds	95% CI	
Predictor	Coef	SE Coef	Z	P	Ratio	Lower	Upper
Constant	-19.2065	5.62922	-3.41	0.001			
GPA10	0.545417	0.157931	3.45	0.001	1.73	1.27	2.35

Log-Likelihood = -28.420

Test that all slopes are zero: G = 18.952, DF = 1, p-value = 0.000

To test $H_0 : \beta_1 = 0$ versus $H_a : \beta_1 \neq 0$, we need to compare the models

$$\log\left(\frac{\pi}{1-\pi}\right) = \beta_0 \qquad \text{and} \qquad \log\left(\frac{\pi}{1-\pi}\right) = \beta_0 + \beta_1 X$$

The null model (the model on the left) assumes the *same* probability of success for every data case. We shouldn't be surprised that the best estimate for this common probability is the proportion of successes in the entire sample; that is, $constant = \hat{\pi}_0 = \frac{\#successes}{\#trials}$. For the medical school acceptance data, there were 30 students accepted to medical school out of 55 in the sample, so $\hat{\pi}_0 = 30/55 = 0.5454$ (and it's only a coincidence that these digits match the first four digits in the estimate of the logistic slope!).

For ordinary regression, there are two common tests, one based on the fitted slope, and one based on sums of squares. For logistic regression, there are also two common tests. The one you have just seen is based on the fitted slope. The other follows the logic of the nested F-test, using $-2\log(L)$ in place of the sum of squares. This test rests on the fact that if outcomes are random and independent and the sample size is large, then $-2\log(L)$ follows an approximate *chi-square* distribution (denoted by χ^2). The degrees of freedom equals the sample size minus the number of coefficients estimated. We start with the arithmetic, then give an explanation and a summary of the test:

$$-2\log(L_0) = 75.79 = \chi^2 \text{ value for constant model, } df = 55 - 1 = 54$$
$$-2\log(L) = 56.84 = \chi^2 \text{ value for logistic model, } df = 55 - 2 = 53$$
$$75.79 - 56.84 = 18.95 = \text{difference in } \chi^2,\ df = 54 - 53 = 1$$

Computer software will give you this information for each model you fit.

For the logistic model using GPA, we regard the value of $-2\log(L) = 56.84$ as a value from a chi-square distribution with 53 degrees of freedom. For the null or constant model, we regard $-2\log(L_0) = 75.79$ as the value of a chi-square distribution with 54 degrees of freedom. Informally, we can think of $-2\log(L)$ as assuming the role *SSE* plays in ordinary regression. The test rests on a difference in these $-2\log(L)$ values, $75.79 - 56.84 = 18.95$, which we view as the *improvement gained* by adding the GPA term to the model. This difference also follows a chi-square distribution, with degrees of freedom obtained by subtraction, and the difference in $-2\log(L)$ is our test statistic.

The *P*-value in this example is, therefore, the area to the right of 18.95 in a chi-square distribution with 1 degree of freedom, which gives a *P*-value of 0.000013. There is compelling evidence of a relationship between GPA and medical school acceptance.

Note that the value of our test statistic, the 18.95, is the same $G = 18.952$ in the previous output and the *P*-value of 0.000013 is essentially the 0.000 that the output gives.

The test described above is called the "likelihood ratio test (LRT) for utility of a simple regression model." You will also see it called the "*G*-test" or the "drop-in-deviance test." In Chapter 10, we will generalize this test to a "full-versus-reduced model test" analogous to what appeared earlier in our discussions of multiple regression using an *F*-distribution. The *P*-value we get from the drop-in-deviance test will not always agree precisely with the *P*-value for the Wald test for the slope, but they are testing essentially the same thing. In fact, there are instances where the *z*-test for slope and the likelihood ratio test will disagree substantially. In such instances, you should trust the likelihood ratio test to be the more accurate test.

LIKELIHOOD RATIO TEST FOR UTILITY OF A SIMPLE LOGISTIC REGRESSION MODEL

To test the overall effectiveness of a logistic regression model with a single predictor

$H_0 : \beta_1 = 0$ versus $H_a : \beta_1 \neq 0$

we use the test statistic $G = -2\log(L_0) - (-2\log(L))$, where L_0 is the likelihood for a constant model and L is the likelihood using the logistic model. We compare this improvement in $-2\log(L)$ to a chi-square distribution with 1 degree of freedom.

EXAMPLE 9.17

Putts1

Putting prowess: Does length matter? Can the length of a putt predict its outcome? The complete summary of the logistic model, based on the data in **Putts1**, is shown here.

```
Deviance Residuals:
    Min       1Q    Median      3Q      Max
-1.8705  -1.1186   0.6181  1.0026   1.4882

Coefficients:
             Estimate Std. Error z value Pr(>|z|)
(Intercept)   3.25684    0.36893   8.828   <2e-16 ***
Length       -0.56614    0.06747  -8.391   <2e-16 ***
---
Signif. codes: 0 '***' 0.001 '**' 0.01 '*' 0.05 '.' 0.1 ' ' 1

(Dispersion parameter for binomial family taken to be 1)

    Null deviance: 800.21  on 586  degrees of freedom
Residual deviance: 719.89  on 585  degrees of freedom
AIC: 723.89
Number of Fisher Scoring iterations: 4
```

ASSESS

Note that the output shown does not explicitly calculate the statistic G, but it does give the $-2\log(L)$ values for both the null (constant) model and the logistic model based on *Length*, labeled as "Null deviance" and "Residual deviance," respectively. Thus the G-statistic is

$$G = 800.21 - 719.89 = 80.32$$

When compared to a chi-square distribution with 1 degree of freedom (note $586 - 585 = 1$), we find the P-value to be essentially zero. Thus we have very strong evidence that the success of making a putt depends on the length of the putt.

Binary Predictors

What if both the explanatory and response variables are binary? We can still use logistic regression to model the relationship between the variables and test whether there is a statistically significant association between them.

EXAMPLE 9.18

Migrain

Zapping migraines Recall the earlier example of a study that investigated whether a handheld device that sends a magnetic pulse into the head might be an effective treatment for migraine headaches. Both the explanatory and response variables are binary. The results are stored in **Migraines** and summarized in the following 2×2 table:

	TMS	Placebo	Total
Pain-free two hours later	39	22	61
Not pain-free two hours later	61	78	139
Total	100	100	200

As we saw before, the success proportions are 0.39 for the TMS group and 0.22 for the placebo group, a difference of 0.17. The odds ratio of 2.27

	Balanced ANOVA	Ordinary Regression	Logistic Regression
Response	Quantitative	Quantitative	Binary
Predictor	Categorical	Either	Either
Estimation method	Minimize SSE	Minimize SSE	Minimize $-2\log(L)$
SEs, intervals and tests	Exact, if conditions satisfied	Exact, if conditions satisfied	Approximate, if samples large enough
Test or interval for slope	Contrasts, not slopes	Based on t	Based on z
Comparing nested models	Nested $F = \dfrac{\Delta SSE}{SSE_{\text{Full}}}$	Nested $F = \dfrac{\Delta SSE}{SSE_{\text{Full}}}$	Drop-in-deviance = $G = \Delta[-2\log(L)]$
Distributions	$\Delta SSE \sim \chi^2$ $SSE_{\text{Full}} \sim \chi^2$	$\Delta SSE \sim \chi^2$ $SSE_{\text{Full}} \sim \chi^2$	$-2\log(L) \sim \chi^2$ $\Delta[-2\log(L)] \sim \chi^2$
Error terms	Additive Normal Independent SD is constant, but unknown	Additive Normal Independent SD is constant, but unknown	Built-in 0 or 1 Independent Not constant, but known from π

EXERCISES

Conceptual Exercises

9.1 Ordinary regression. Why does simple linear regression used in previous chapters not work well when the response is binary?

9.2 Residuals. Why do we not rely on residual plots when checking conditions for logistic regression?

9.3 Probability to odds.

a. If the probability of an event occurring is 0.5, what are the odds?

b. If the probability of an event occurring is 0.9, what are the odds?

c. If the probability of an event occurring is 0.1, what are the odds?

9.4 Probability to odds.

a. If the probability of an event occurring is 0.8, what are the odds?

b. If the probability of an event occurring is 0.25, what are the odds?

c. If the probability of an event occurring is 0.6, what are the odds?

9.5 Odds to probabilities.

a. If the odds of an event occurring are 2:1, what is the probability?

b. If the odds of an event occurring are 10:1, what is the probability?

c. If the odds of an event occurring are 1:4, what is the probability?

9.6 Odds to probabilities.

a. If the odds of an event occurring are 1:3, what is the probability?

b. If the odds of an event occurring are 5:2, what is the probability?

c. If the odds of an event occurring are 1:9, what is the probability?

9.7 Odds ratio for recovery. If the probability of a recovery with treatment is 0.3 and without is 0.1, what is the odds ratio for recovery when treated versus not treated?

9.8 Odds ratio for birth defects. If the probability of a birth defect with exposure to a potential teratogen is 0.6 and without exposure the probability is 0.01, what is the odds ratio for a birth defect when exposed versus not exposed?

9.9 Effects of slope and intercept. Suppose that we have a logistic model with intercept $\beta_0 = 5$ and slope $\beta_1 = 2$. Explain what happens to a plot of the probability form of the model in each of the following circumstances:

a. The slope β_1 decreases to 1.

b. The intercept β_0 increases to 8.

c. The slope changes sign to become $\beta_1 = -2$.

9.10 Effects of slope and intercept. Suppose that we have a logistic model with intercept $\beta_0 = 2$ and slope $\beta_1 = -3$. Explain what happens to a plot of the probability form of the model in each of the following circumstances:

a. The slope β_1 increases to -1.

b. The intercept β_0 increases to 5.

c. The slope changes sign to become $\beta_1 = 3$.

9.11 Must, might, cannot: Different slopes. Suppose that two logistic models have the same intercepts but different slopes. For each of (a–e), state whether the statement *must* be true, *might* be true, or *cannot* be true.

a. The graphs of log(*odds*) versus x for both models cross the x-axis at the same value of x.

b. The graphs of log(*odds*) versus x for both models cross the y-axis at the same value of y.

c. The graphs of $P(Y = 1)$ versus x for both models have the same horizontal asymptotes.

d. The graphs of $P(Y = 1)$ versus x for both models cross the line $P(Y = 1) = 0.5$ at the same value of x.

e. The graphs of $P(Y = 1)$ versus x for both models cross the line $x = 0.5$ at the same value of y.

9.12 Must, might, cannot: Different intercepts. Suppose that two logistic models have the same slopes but different intercepts. For each of (a–e), state whether the statement *must* be true, *might* be true, or *cannot* be true.

a. The graphs of log(*odds*) versus x for both models cross the x-axis at the same value of x.

b. The graphs of log(*odds*) versus x for both models cross the y-axis at the same value of y.

c. The graphs of $P(Y = 1)$ versus x for both models have the same horizontal asymptotes.

d. The graphs of $P(Y = 1)$ versus x for both models cross the line $P(Y = 1) = 0.5$ at the same value of x.

e. The graphs of $P(Y = 1)$ versus x for both models cross the line $x = 0.5$ at the same value of y.

9.13 Must, might, cannot: Different curves. Suppose that two logistic models have different slopes and different intercepts. For each of (a–e), state whether the statement *must* be true, *might* be true, or *cannot* be true.

a. The graphs of log(*odds*) versus x for both models cross the x-axis at the same value of x.

b. The graphs of log(*odds*) versus x for both models cross the y-axis at the same value of y.

c. The graphs of $P(Y = 1)$ versus x for both models have the same horizontal asymptotes.

d. The graphs of $P(Y = 1)$ versus x for both models cross the line $P(Y = 1) = 0.5$ at the same value of x.

e. The graphs of $P(Y = 1)$ versus x for both models cross the line $x = 0.5$ at the same value of y.

9.14 Putting prowess. If you fit the same logistic model in two scales, Model A with distance as a predictor in feet and Model B with distance as a predictor in yards, how will the fitted coefficients for the two predictors differ? Consider a one-unit change of the predictor in each of the two models: How will the effects of the one-unit change on the odds ratios compare for the two models?

9.15 Fitted slope and odds ratio. If you have a 0, 1 response and 0, 1 predictor, what value of the fitted slope will give each of the following odds ratios: 0.01, 1, 2, e?

9.16 Fitted slope and odds ratio. If you have a 0, 1 response and 0, 1 predictor, what value of the odds ratio will give each of the following fitted slopes: 1, −0.5, 5, −4?

9.17 Matching. Here are nine pairs of fitted values (intercept, slope): $(-1, -1)$, $(-1, 0)$, $(-1, 1)$, $(0, -1)$, $(0, 0)$, $(0, 1)$, $(1, -1)$, $(1, 0)$, $(1, 1)$. Match each with its corresponding logistic curve.

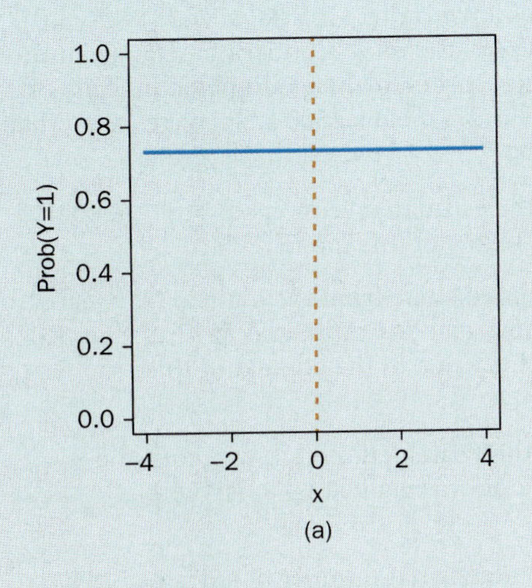

(a)

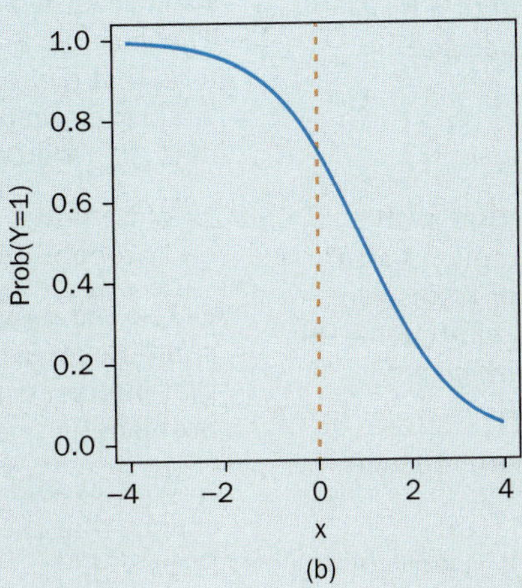

(b)

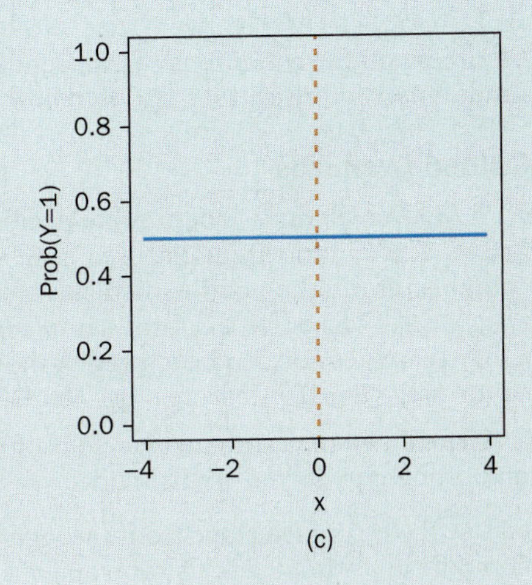

(c)

(continued on next page)

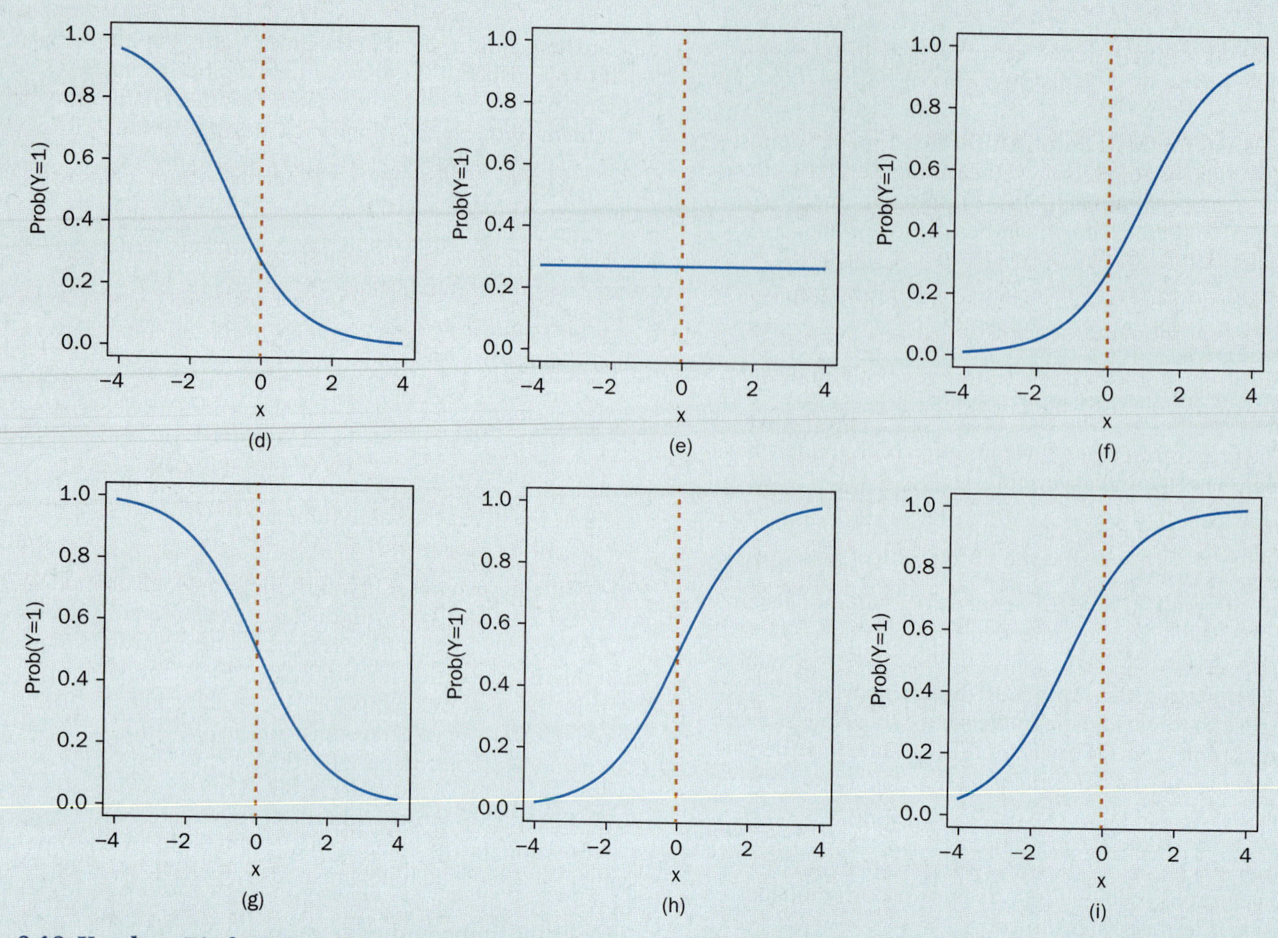

9.18 Kershaw Pitches: Conditions. Clayton Kershaw is a baseball player who won the 2013 Cy Young Award as the best pitcher in the National League while pitching for the Los Angeles Dodgers. The file **Kershaw** contains information on each of 3204 pitches he threw during that season.[2] One of the variables (*StartSpeed*) is the speed of the ball (in mph) when it leaves the pitcher's hand. Will faster pitches tend to be more successful? The variable *Result* broadly characterizes the outcome of each pitch from the pitcher's perspective as either positive ("Pos" = a strike or an out) or negative ("Neg" = a ball or a hit). Comment on the conditions of independence and randomness in this setting. **Kershaw**

Guided Exercises

9.19 Medical school acceptance: predicting with MCAT. The datafile **MedGPA** used in Example 9.4 also contains information on the medical school admission test (MCAT) scores for the same sample of 55 students. Fit a logistic regression model to predict the *Acceptance* status using the *MCAT* scores. **MedGPA**

a. Write down the estimated versions of both the logit and probability forms of this model.

b. Use the estimated slope from the logistic regression to compute an estimated odds ratio and write a sentence that interprets this value in the context of this problem.

c. What would the estimated model say about the chance that a student with $MCAT = 40$ is accepted to medical school?

d. For approximately what *MCAT* score would a student have roughly a 50-50 chance of being accepted to medical school? (*Hint:* You might look at a graph or solve one of the equations algebraically.)

9.20 Walking the dog. "Are you going to walk the dogs this afternoon?" That is a question often asked in the house of one of the authors. The file **WalkTheDogs** has data from 223 days in 2016. For each of those days the author either did walk the dogs (68 days, with the variable Walk recorded as 1) or didn't (155 days, with Walk = 0). The variable Steps is the total number of steps the author took that day, as recorded by his pedometer, in units of 1,000 (i.e., Steps = 2.5 means 2500 steps were taken that day). **WlkDog**

a. Write down the estimated versions of both the logit and probability forms of this model.

b. Use the estimated slope from the logistic regression to compute an estimated odds ratio, and write a sentence that interprets this value in the context of this problem.

c. What would the estimated model say about the chance that the dog was walked on a day with *Steps* = 4?

d. For approximately what number of steps would we have roughly a 50-50 chance of the dog being walked? (*Hint:* You might look at a graph or solve one of the equations algebraically.)

9.21 Metastasizing cancer: Odds and probability. In a study of 31 patients with esophageal cancer, it was found that in 18 of the patients the cancer had metastasized to the lymph nodes. Thus an overall estimate of the probability of metastasis is $18/31 = 0.58$. A predictor variable measured on each patient is *Size* of the tumor (in cm). A fitted logistic regression model is

$$log\left(\frac{\hat{\pi}}{1-\hat{\pi}}\right) = -2.086 + 0.5117Size$$

a. Use this model to estimate the odds of metastasis, $\pi/(1-\pi)$, if a patient's tumor size is 6 cm.

b. Use the model to predict the probability of metastasis if a patient's tumor size is 6 cm.

c. How much do the estimated odds change if the tumor size changes from 6 cm to 7 cm? Provide and interpret an odds ratio.

d. How much does the estimate of π change if the tumor size changes from 6 cm to 7 cm?

9.22 Dementia: Odds and probability. Two types of dementia are Dementia with Lewy Bodies and Alzheimer's disease. Some people are afflicted with both of these. The file LewyBody2Groups includes the variable *Type*, which has two levels: "DLB/AD" for the 20 subjects with both types of dementia and "DLB" for the 19 subjects with only Lewy Body dementia. The variable *MMSE* measures change in functional performance on the Mini Mental State Examination. We are interested in using *MMSE* to predict whether or not Alzheimer's disease is present. A fitted logistic model is

$$\log\left(\frac{\hat{\pi}}{1-\hat{\pi}}\right) = -0.742 - 0.294MMSE$$

a. Use this model to estimate the odds of Alzheimer's disease, $\pi/(1-\pi)$, if a patient's *MMSE* is -4.

b. Use this model to estimate the probability of Alzheimer's disease if a patient's *MMSE* is -4.

c. How much do the estimated odds change if the *MMSE* changes from -4 to -3?

d. How much does the estimate of π change if the *MMSE* changes from -4 to -3?

9.23 Metastasizing cancer: Estimating *X* from probability. Consider the fitted logit model from Exercise 9.21. How large does a tumor need to be for the estimated probability of metastasis to be at least 0.80?

9.24 Dementia (continued): Estimating *X* from probability. Consider the fitted logit model from Exercise 9.22. What value of *MMSE* is necessary to have a probability of Alzheimer's of at least 0.75?

9.25 Transition to marriage: Empirical logit plot. Example 9.8 on page 425 shows a two-way table looking at whether a mother who gives birth before marriage eventually marries versus the sex of the child. In that sample, 176 of the 310 mothers of boys eventually married, as did 148 out of 290 mothers of girls. Find the log odds of eventual marriage for the samples of sexes of children and and create a plot of the empirical logits (similar the one for migraines in Figure 9.11 on page 427). Assume the boy children are coded as 0 and the girls as 1. Find the slope of the line.

9.26 Logit and probability forms of the model. The box on page 420 shows two forms of the logistic regression model for a single predictor.

$$\text{Logit form: } \log\left(\frac{\pi}{1-\pi}\right) = \beta_0 + \beta_1 X$$

$$\text{Probability form: } \pi = \frac{e^{\beta_0+\beta_1 X}}{1+e^{\beta_0+\beta_1 X}}$$

Show algebraically that these forms are equivalent. Start with one version and use algebraic manipulations to arrive at the other version.

9.27–9.32 Titanic. The *Titanic* was a British luxury oceanliner that sank famously in the icy North Atlantic Ocean on its maiden voyage in April 1912. Of the approximately 2200 passengers on board, 1500 died. The high death rate was blamed largely on the inadequate supply of lifeboats, a result of the manufacturer's claim that the ship was "unsinkable." A partial dataset of the passenger list was compiled by Philip Hinde in his *Encyclopedia Titanica* and is given in the datafile **Titanic**.

Two questions of interest are the relationship between survival and age and the relationship between survival and sex. The following variables will be useful for your work on the following questions: **Titanic**

Age	which gives the passenger's age in years
Sex	which gives the passenger's sex (male or female)
Survived	a binary variable, where 1 indicates the passenger survived and 0 indicates death
SexCode	which numerically codes male as 0 and female as 1

9.27 Titanic: Survival and age, CHOOSE and FIT.

a. Use a plot to explore whether there is a relationship between survival and the passenger's age. What do you conclude from this graph alone?

b. Use software to fit a logistic model to the survival and age variables to decide whether there is a statistically significant relationship between age and survival, and if there is, what its direction and magnitude are. Write the estimated logistic model using the output and interpret the output in light of the question.

9.28 Titanic: Survival and sex, CHOOSE and FIT.

a. Use a two-way table to explore whether survival is related to the sex of the passenger. What do you conclude from this table alone? Write a summary statement that interprets *Sex* as the explanatory variable and *Survived* as the response variable, and that uses simple comparisons of conditional proportions or percentages.

b. Use software to fit a logistic model to the survival and sex variables to decide whether there is a statistically significant relationship between sex and survival. If there is, what are the nature and magnitude of the relationship? Does the relationship found by the logistic model confirm the descriptive analysis? (*Note:* You will actually use *SexCode* as the predictor in the logistic model.)

9.29 Titanic: Survival and sex, ASSESS and USE. Using the information from Exercise 9.28, answer the following questions:

a. Present a calculation that compares the estimated slope coefficient of the model with *SexCode* as a predictor to the estimated odds ratio. Then give a sentence that interprets the odds ratio in the context of the *Titanic* tragedy.

b. Write a sentence that interprets a 95% CI for the odds ratio discussed in (a).

c. Present a calculation from the two-way table that leads to an estimated coefficient from the output for the model found in (b).

d. Use the model to estimate the probability that a female would have survived the sinking of the *Titanic*.

e. Assess the model conditions for the model relating *Survival* to *SexCode*. Write a short summary of the assessment.

9.30 Titanic: Survival and age, ASSESS and USE. Using the information from Exercise 9.27, answer the following questions:

a. Present a calculation that compares the estimated slope coefficient of the model with *Age* as a predictor to the estimated odds ratio. Then give a sentence that interprets the odds ratio in the context of the *Titanic* tragedy.

b. Write a sentence that interprets a 95% CI for the odds ratio discussed in (a).

c. Use the model to estimate the probability that a 40-year-old would have survived the sinking of the *Titanic*.

9.31 Titanic: Survival and sex, analysis summary. Using your answers from Exercises 9.28 and 9.29, write a short paragraph that summarizes your analysis of the relationships between the *Sex* of a passenger and the passenger's *Survival*.

9.32 Titanic: Survival and age, analysis summary. Using your answers from Exercises 9.27 and 9.30, write a short paragraph that summarizes your analysis of the relationships between the *Age* of a passenger and the passenger's *Survival*.

9.33 Flight response of Pacific Brant, altitude: CHOOSE and FIT.[3] A 1994 study collected data on the effects of air traffic on the behavior of the Pacific Brant (a small migratory goose). Each fall, nearly the entire population of 130,000 of this species uses the Izembek Lagoon in Alaska as a staging area, where it feeds and stores fat for its southerly migration. Because offshore drilling near this estuary had increased the necessity of air traffic, an impact study was timely. The data represent the flight response to helicopter "overflights" to see what the relationship between the proximity of a flight, both lateral and altitudinal, would be to the propensity of the Brant to flee the area. For this experiment, air traffic was restricted to helicopters because a previous study had ascertained that helicopters created more radical flight response than other aircraft.

The data are in **FlightResponse**. Each case represents a flock of Brant that has been observed during one overflight in the study. Flocks were determined observationally as contiguous collections of Brants, flock sizes varying from 10 to 30,000 birds. For this study, the variables we investigate are: **FltResp**

Altitude	The experimentally determined altitude of the overflight by the helicopter. Units are in 100 m, with the variable range being 0.91 to 12.19, recorded at 9 distinct values: 0.91, 1.52, 3.05, 4.57, 6.10, 6.71, 7.62, 9.14, and 12.19.
Lateral	The perpendicular or lateral distance (in 100 m) between the aircraft and flock, as determined from studying area maps to the nearest 0.16 km.
AltCat	A categorical variable, derived from the altitude variable. The range [0, 3) is category "low," the range [3, 6] is "mid," and the range (6, infinity) is "high."
LatCat	A categorical variable, derived from the *Lateral* variable. The range [0, 10) is category 1; [10, 20) is 2; [20, 30) is 3; and [30, infinity) is 4.
Flight	This is a binary variable in which 0 represents an outcome where fewer than 10% of a flock flies away during the overflight and 1 represents an outcome where more than 10% of the flock flies away. This is the response variable of interest in this study.

a. Create a two-way table of *AltCat* by *Flight*. For each level of *AltCat*, calculate the odds and log(odds) for a flight response. Does there appear to be a relationship between altitude and flight, based solely on this table? Describe the nature and intensity of this response.

b. Calculate a logistic regression model using *Flight* as the response variable and *Altitude* as the explanatory variable. Does this model confirm your suspicion from part (a) about the existence and direction of a relationship between flight and altitude? Explain. Report model estimates and interpret the estimated slope coefficient.

9.34 Flight response of Pacific Brant, lateral distance: CHOOSE and FIT. Refer to the data in **FlightResponse**, discussed in Exercise 9.33. **FltResp**

a. Create a two-way table of *LatCat* by *Flight*. For each level of *LatCat*, calculate the odds and log(odds) for a flight response. Does there appear to be a relationship between lateral distance and flight, based solely on this table? Describe the nature and intensity of this response.

b. Calculate a logistic regression model using *Flight* as the response variable and *Lateral* as the explanatory variable. Does this model confirm your suspicion from part (a) about the existence and direction of a relationship between flight and lateral distance? Explain. Report model estimates and interpret the estimated slope coefficient.

9.35 Nikkei 225. If we know how the New York stock market performs today, can we use that information to predict whether the stock market in Japan will go up or down tomorrow? Or is the movement of the Japanese stock market today better predicted by how the Japanese market did yesterday? The file **Markets** contains data from two stock markets for 56 days. The variables recorded are *Date*; *Nik225ch* (the one-day change in the Nikkei 225, a stock index in Japan); *DJIAch* (the one-day change in the New York–based Dow Jones Industrial Average from the previous day); *Up* (1 or 0 depending on whether or not the Nikkei 225 went up on a date); and *lagNik* (the one-day change in the Nikkei 225 from the previous day). Thus if we want to predict whether the stock market in Japan will go up or down on a Tuesday, we might use the Monday result from Japan (*lagNik*) or the Monday result from New York (*DJIAch*)—remembering that when it is Monday evening in New York, it is Tuesday morning in Japan. **Market**

a. Fit a logistic model with *Up* as the response and *DJIAch* as the predictor. Is *DJIAch* a significant predictor of the direction the Nikkei 225 will go the next day? Explain the basis for your conclusion.

b. Fit a logistic model with *Up* as the response and *lagNik* as the predictor. Is *lagNik* a significant predictor of the direction the Nikkei 225 will go the next day? Explain the basis for your conclusion.

c. Compare the models in part (a) and part (b). Which variable, *DJIAch* or *lagNik*, is a more effective predictor of where the Nikkei 225 is going the next day? Explain the basis for your decision.

9.36 Red states and blue states in 2016: Compare models. Can we use state-level variables to predict whether a state votes for the Democratic versus the Republican presidential nominee? The file **Election16** contains data from 50 states plus the District of Columbia. The variables recorded are: **Elect16**

State	state
Abr	abbreviation for the state
Income	per capita income as of 2007
HS	percentage of adults with at least a high school education
BA	percentage of adults with at least a college education
Adv	percentage of adults with advanced degrees
Dem.Rep	%Democrat–%Republican in a state
	including those who lean toward either party according to a 2015 Gallup poll
TrumpWin	1 or 0 indicating whether the Republican candidate Donald Trump did or did not win a majority of votes in the state

Fit separate logistic regression models to predict *TrumpWin* using each of the predictors *Income*, *HS*, *BA*, and *Dem.Rep*. Which of these variables does the most effective job of predicting this response? Which is the least effective? Explain the criteria you use to make these decisions.*

9.37 Red states and blue states in 2016, income: Odds ratio. Refer to the data in **Election16** that are described in Exercise 9.36. Run a logistic regression model to predict *TrumpWin* for each state using the per capita *Income* of the state. **Elect16**

a. Use the estimated slope from the logistic regression to compute an estimated odds ratio and write a sentence that interprets this value in the context of this problem.

b. Find a 95% confidence interval for the odds ratio in (a).

9.38 Red states and blue states in 2016, Income/1000: Odds ratio. Refer to the data in **Election16** that are described in Exercise 9.36. In Exercise 9.37 you fit a logistic regression model to predict *TrumpWin* using *Income*. The units of the *Income* variable are dollars, with values ranging from \$39,665 (Mississippi) to \$74,551 (Maryland). The odds ratio and interval in Exercise 9.37 are awkward to interpret since they deal with the change in the odds when state income changes by \$1, a very trivial amount! To get an odds ratio that may be more meaningful, create a new variable (call it *IncomeTh*) using *Income*/1000 to express the state per capita incomes in \$1000s. **Elect16**

a. Run the logistic regression using *IncomeTh* as the predictor of *TrumpWin*. How does the fitted prediction equation change? How (if at all) does the predicted probability of Trump winning a state change?

b. Use the estimated slope from the logistic regression to compute an estimated odds ratio, and write a sentence that interprets this value in the context of this problem.

c. Find a 95% confidence interval for the odds ratio in (b).

9.39 Gunnels. The dataset **Gunnels**[4] comes from a study on the habitat preferences of a species of eel, called a gunnel. Biologist Jake Shorty sampled quadrats along a coastline and recorded a binary variable, *Gunnel*, where 1 represents the presence of the species found in the quadrat and a 0 its absence. Below is output from a logistic model fit with *Time* as the explanatory variable, where *Time* represents the time of day as "minutes from midnight," with a minimum of 383 minutes (about 6:23 a.m.) and a maximum of 983 minutes (about 1:23 p.m.). The output is given below: **Gunnel**

```
Coefficients:
             Estimate  Std. Error z value  Pr(>|z|)
(Intercept)  0.371980  0.644047   0.578   0.564
Time        -0.005899  0.001049  -5.624   1.86e-08 ***
---
Signif. codes: 0 '***' 0.001 '**' 0.01 '*' 0.05 '.' 0.1 ' ' 1
```

*Note: We will consider models that include multiple predictors for logistic regression in the next chapter.

Answer the questions that follow:

a. State the null hypothesis that the *P*-value of 1.86*e*-08 allows you to test.

b. What happens to the probability of finding a gunnel as you get later in the day? Does it get smaller or larger?

c. Find a 95% confidence interval for the slope parameter in the logistic model.

d. What quantity does $0.371980 + (-0.005899)600 = -3.16742$ estimate?

e. Find the estimated odds ratio for an additional minute of time after midnight.

f. Give a 95% confidence interval for the odds ratio found in part (e).

g. Find an estimate for the odds ratio for an additional hour or 60 minutes of time passing. Give a sentence interpreting the meaning of this number in simple language using the term "odds."

9.40 Levee failures. Research was done to study which factors are related to levee failure on the lower Mississippi River. One variable considered as a possible explanatory variable was the amount of constriction factor in the river at that location. The data are in the file **LeveeFailures**.[5] The output is given below: **LevFail**

```
Coefficients
Term                 Coef  SE Coef  Z-Value  P-Value
Constant            0.571    0.350     1.63    0.102
ConstrictionFactor -0.691    0.346    -2.00    0.046
```

a. State the null hypothesis that the *P*-value 0.046 allows you to test.

b. What happens to the probability of a levee failure as the constriction factor gets larger? Explain.

c. Find a 95% confidence interval for the slope parameter in the logistic model.

9.41 Leukemia treatments.[6] A study involved 51 untreated adult patients with acute myeloblastic leukemia who were given a course of treatment, after which they were assessed as to their response. The variables recorded in the dataset **Leukemia** are: **Leukem**

Age	at diagnosis in years
Smear	differential percentage of blasts
Infil	percentage of absolute marrow leukemia infiltrate
Index	percentage labeling index of the bone marrow leukemia cells
Blasts	absolute number of blasts, in thousands
Temp	highest temperature of the patient prior to treatment, in degrees Fahrenheit
Resp	1 = responded to treatment and 0 = failed to respond
Time	the survival time from diagnosis, in months
Status	0 = dead and 1 = alive

a. Fit a logistic model using *Resp* as the response variable and *Age* as the predictor variable. Interpret the results and state whether the relationship is statistically significant.

b. Form a two-way table that exhibits the nature of the relationship found in (a).

c. Redo parts (a) and (b), but using *Temp* as the single predictor variable.

9.42 First-Year GPA. In Example 4.2,we considered relationships between a number of variables in the **FirstYearGPA** dataset to help explain the variability of GPA for first-year students. One of those variables was an indicator, *FirstGen*, for whether the student was a first-generation college attendee (1 if so, 0 if not). In this exercise, you will compare several different ways of seeing if there is some association between *GPA* and *FirstGen*. **Y1GPA**

a. Use a two-sample *t*-test to see if there is a significant difference in the average GPA between students who are and are not first-generation college attendees. Report the *P*-value of the test and the direction of the relationship, if significant.

b. Use a simple linear regression to predict *GPA* using *FirstGen*. Report a *P*-value and indicate the nature of the relationship if a significant relationship is present.

c. Do the comparison once more, this time using a logistic regression with *GPA* as the predictor and *FirstGen* as the response. Compare the conclusion (and *P*-value) you would draw from this model to the results from parts (a) and (b).

9.43 THC versus prochlorperazine. An article in the *New England Journal of Medicine* described a study on the effectiveness of medications for combatting nausea in patients undergoing chemotherapy treatments for cancer. In the experiment, 157 patients were divided at random into two groups. One group of 78 patients were given a standard antinausea drug called prochlorperazine, while the other group of 79 patients received THC (the active ingredient in marijuana). Both medications were delivered orally, and no patients were told which of the two drugs they were taking. The response measured was whether or not the patient experienced relief from nausea when undergoing chemotherapy. The results are summarized in the 2×2 table shown below (and stored in **ChemoTHC**): **Chemo**

Drug	Effective	Not Effective	Patients
THC	36	43	79
Prochlorperazine	16	62	78

a. Find the proportion of patients in each of the sample groups for which the treatment was effective.

b. Fit a binary logistic regression model to predict *Effectiveness* (Yes or No) using type of *Drug* as a binary predictor. Give the logit form of the fitted model.

c. Use the model to predict the odds and the probability of effectiveness for each of the two drugs. Compare predicted proportions to the sample proportions in (a).

d. Find the odds ratio comparing the effectiveness of THC to prochlorperazine based on the binary logistic regression model. Write a sentence that interprets this value in the context of this problem and find a 95% confidence interval for the odds ratio.

e. The table shows that THC was effective in more cases, but is this difference statistically significant? Use information from the logistic regression output to do a formal test at a 1% level to see if there is evidence that THC is more effective than prochlorperazine in combatting nausea for chemotherapy patients.

9.44 Suicide. Researchers were concerned with the high rate of suicide in China. In particular, not only is there the obvious concern for loss of life, but there is a resource concern with respect to medical facilities. According to the source article ". . . in the USA, it is estimated that for every suicide death there are 13 emergency department (ED) visits and 45 hospitalisations due to non-fatal suicide attempts." The researchers collected data from three counties in the Shangdong province of China. The data are in the file **SuicideChina**.[7] Several different questions can be researched using this data. We will focus on how the age of the victim predicts the probability of death from the suicide attempt. **SuiChi**

a. Produce an empirical logit plot to check for linearity of the logit with respect to age. What conclusion do you reach about the appropriateness of logistic regression? Hint: Combine ages 12–14 and ages 90–100 since these sample sizes are very small and lead to probabilities of 0 or 1.

b. Fit the logistic model. State whether the relationship is statistically significant.

c. What happens to the probability of the suicide being successful as age increases? Explain.

9.45 Kershaw Pitches: Speed versus Result. Clayton Kershaw is a baseball player who won the 2013 Cy Young Award as the best pitcher in the National League while pitching for the Los Angeles Dodgers. The file **Kershaw** contains information on each of 3,204 pitches he threw during that season.[8] One of the variables (*StartSpeed*) is the speed of the ball (in mph) when it leaves the pitcher's hand. Will faster pitches tend to be more successful? The variable *Result* broadly characterizes the outcome of each pitch from the pitcher's perspective as either positive ("Pos" = a strike or an out) or negative ("Neg"= a ball or a hit). **Kershaw**

a. Fit a logistic regression model to explain the *Result* based on the *StartSpeed* of each pitch. Give both the logit and probability forms of the fitted model.

b. From the signs of the coefficients, does it look like faster pitches tend to give more or less positive results?

c. According to the fitted model, what proportion of 95 mph Kershaw pitches should have a positive result? How about his pitches at 75 mph?

d. Does it appear that *StartSpeed* is a useful predictor of the results of Kershaw's pitches? Justify your answer.

Open-ended Exercises

9.46 Backpack weights. Is your back aching from the heavy backpack you are lugging around? A survey of students at California Polytechnic State University (San Luis Obispo) collected data to investigate this question. The dataset **Backpack**[9] includes the following variables: **BckPck**

Back Problems (0 = No, 1 = Yes)
Backpack Weight
Body Weight
Ratio

Use the data to explore the possibility of backpacks being responsible for back problems. What are the relative merits of using the backpack weight alone as a predictor compared to using the ratio of backpack to body weight? Which would you recommend and why?

9.47 Risky youth behavior. What happens when someone who is underaged needs a ride home after a night of drinking? The concern of the media today focuses on those who drive home in this situation. However, what about those who ride with someone who has been drinking? We'll denote such an event as RDD, which stands for riding with a drunk driver. This choice can be as devastating as driving drunk, resulting in a large number of RDD traffic deaths. A recent study examines who accepts these rides and why.[10] **Youth**

We'll examine RDD behavior with only a few variables from a very large dataset to examine questions concerning youth who ride with drivers who have been drinking (RDD) so as to identify where safety messages might best be targeted. We'll use a dataset **YouthRisk** derived from the Youth Risk Behavior Surveillance System (YRBSS), which is an annual survey conducted by the Centers for Disease Control and Prevention (CDC) to monitor the prevalence of health-risk youth behaviors.[11] The following variables are included in **YouthRisk**:

ride.alc.driver	1 = respondent rode with a drinking driver within past 30 days, 0 = did not
female	1 = female, 0 = male
grade	A factor variable with levels "9," "10," "11," "12"
age4	14, 15, 16, 17, or 18 years of age
smoke	1 = Yes to "have you ever tried cigarette smoking, even one or two puffs?," 0 = No

Investigate the following hypotheses:

a. Some researchers proposed that the odds of riding with a driver who has been drinking are higher for young women compared to young men. Do these data provide evidence of this?

b. One study suggests that youth become more reckless about drinking and driving after they obtain their driver's license.[12] On the other hand, Poulin and co-workers[13] reported that a driver's license was found to be protective among Canadian youth. What does the evidence in this survey suggest about the effect of obtaining a driver's license in the United States?

c. Ryb and colleagues[14] have reported that smokers are more likely to RDD, citing that injury-prone behaviors were also more common among smokers than nonsmokers. Is smoking associated with an increased risk of RDD?

9.48 Good movies. Statisticians enjoy movies as much as the next person. One statistician movie fan decided to use statistics to study the movie ratings in his favorite movie guide, *Movie and Video Guide*, by Leonard Maltin. He was interested in discovering what features of Maltin's Guide might correlate to his view of the movie. Maltin rates movies on a one-star to four-star system, in increments of half-stars, with higher numbers indicating a better movie. Our statistician has developed, over time, the intuition that movies rated 3 or higher are worth considering, but lower ratings can be ignored. He used a random number generator to select a simple random sample of 100 movies rated by the Guide. For each movie, he measured and recorded these variables:

Title	the movie's title
Year	the year the movie was released (range is 1924–1995)
Time	the running time of the movie in minutes (range is 45–145)
Cast	the number of cast members listed in the guide (range is 3–13)
Rating	the movie's Maltin rating (range is 1–4, in increments of 0.5)
Description	the number of lines of text Maltin uses to describe the movie (range is 5–21)
Origin	the country where the movie was produced (0 = USA, 1 = Great Britain, 2 = France, 3 = Italy, 4 = Canada)

The data are in the file **Film**.

For the purposes of his study, the statistician also defined a variable called Good?, where 1 = a rating of 3 stars or better and 0 = any lower rating. He was curious about which variables might be good predictors of his personal definition of a good movie. Analyze the data to find out. Write a short report of your findings. (*Note:* Restrict your explanatory variables to *Year*, *Time*, *Cast*, and *Description*.) **Flim**

9.49 Lost letter. In a 1965 article in *Public Opinion Quarterly*, psychologist Stanley Milgram reported on a study where he intentionally lost letters to see if the nature of the address on the letter would affect return rates when people found the letters lying on a public sidewalk. For example, he found that letters addressed to a subversive organization such as Friends of the Nazi Party were far less likely to be returned than those addressed to an individual or to a medical research organization.

In 1999 Grinnell College students Laurelin Muir and Adam Gratch conducted an experiment similar to Milgram's for an introductory statistics class. They intentionally "lost 140 letters" in either the city of Des Moines, the town of Grinnell, or on the Grinnell College campus. Half of each sample were addressed to Friends of the Confederacy and the other half to Iowa Peaceworks. Do the return rates differ depending on address? Get the data in **LostLetter**. Use logistic regression and analyze the data to decide, location by location, if there is a difference. Summarize your findings. **Lost**

9.50 Political activity of students. Students Jennifer Wolfson and Meredith Goulet conducted a survey of Grinnell College students to ascertain patterns of political behavior. They took a simple random sample of 60 students who were U.S. citizens and conducted phone interviews. Using several "call backs," they obtained 59 responses. The dataset for the study is **Political** and the variables are:

Year	the class year of the student (1 = first year, 2 = second year, 3 = junior, 4 = senior)
Sex	the sex of the student (0 = male, 1 = female)
Vote	voting pattern of student (0 = not eligible, 1 = eligible only, 2 = has registered, but not voted, 3 = has voted)
Paper	"How often do you read newspapers or news magazines, not including comics?" (0 = never, 1 = less than once/week, 2 = once/week, 3 = 2 or 3 times/week, 4 = daily)
Edit	"Do you read the editorial page?" (0 = No, 1 = Yes)
TV	"How often do you watch TV for news?" (0 = never, 1 = less than once/week, 2 = once/week, 3 = 2 or 3 times/week, 4 = daily)
Ethics	"Should politics be ruled by ethical considerations or by practical power politics?" (scale of 1–5, with 1 being "purely ethical" and 5 being "purely practical")
Inform	"How informed do you consider yourself to be about national or international politics?" (scale of 1–5 with 1 = generally uninformed, 5 = very well informed)
Participate	a variable derived from *Vote*, with *Vote* = 0 transformed to missing value, *Vote* = 1 or 2 transformed to 0, and *Vote* = 3 transformed to 1

One question we can ask of these data is a classical concern in democracies: Do the better-informed citizens tend to be those who vote? Use *Participate* as the response and use simple logistic regression to investigate this question. Summarize your findings. **Poli**

Johanna Goodyear/Shutterstock

CHAPTER 10

Multiple Logistic Regression

In this chapter you will learn to:

- Extend the ideas of the previous chapter to logistic regression models with two or more predictors.
- Choose, fit, and interpret an appropriate model.
- Check conditions.
- Do formal inference both for individual coefficients and for comparing competing models.

In the last chapter, we found that GPA is a useful predictor of admission to medical school. What if you also know a student's MCAT score: Does that additional information lead to a better prediction? Or can you ace the MCAT and not worry about your GPA's effect on your chance of admission? Notice that each of these last two questions involves a pair of predictors and a binary response. Here's a second example with similar structure: Does the amount of pupil dilation when looking at nude photos (first predictor) help predict sexual orientation (coded as a binary response)? If so, is the relationship the same for males and females (second predictor)?

dismiss the number of applications as a predictor, but with many variables and few cases, we don't have what we need to tease apart the issues. *(Sometimes, the best you can do as a good data analyst is to recognize that "Without better data, I can't tell.")*

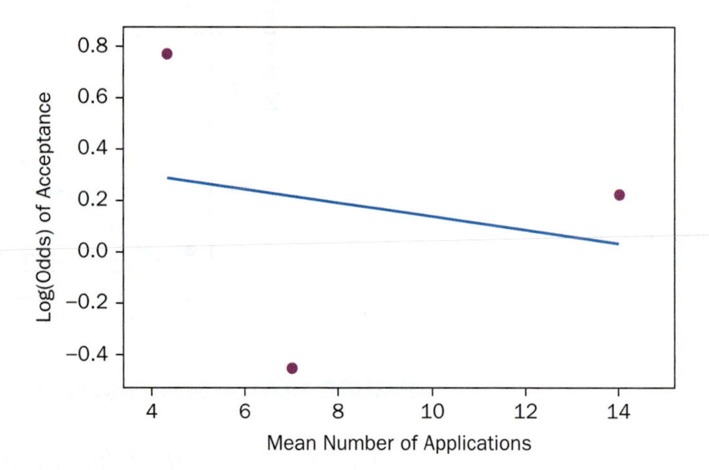

FIGURE 10.8 Log(odds) of acceptance versus number of applications

4. *Acceptance and Sex*. Table 10.4 summarizes the admission counts by sex. The acceptance rate for men was 12/27, or 44%, for women 18/28, or 64%. Here, also, it is reasonable to think we are seeing a pattern that involves interaction of some sort. If not, the data suggest discrimination against male applicants.

	Accepted?		
Female?	0	1	Total
0	15	12	27
1	10	18	28
Total	25	30	55

TABLE 10.4 Acceptance rates for male and female applicants

5. *Interaction of GPA and Sex* (Figure 10.9). If we decide that all four of the predictors above (*GPA*, *MCAT*, *Apps*, and *Sex*) belong in the model, then there are six possible two-way interactions, and four possible three-way interactions.* With enough cases, we might turn to automated methods to choose a set of reasonable models by deciding which terms to include. Here, and all too often, this automated approach is an unaffordable luxury. With fewer cases, our models must rely more on plots and context-based simplifying assumptions. The next few paragraphs look at a few of these interactions, starting with *GPA* and *Sex*.

   The graph shows lines for males and females that are clearly separated, with the solid line for females above the dashed line for males—a better chance of acceptance for women. The slopes of the two lines are not the same but are similar—close enough that, give or take chancelike variation, the true relationships might well correspond to parallel lines. The fact that the lines are so clearly separated raises a question: Are women given an advantage in the selection process, or is there some sex-related confounding factor that belongs in the model?

* Remember Francis Anscombe's caution that the ratio of cases to parameters should be at least 5 to 1.

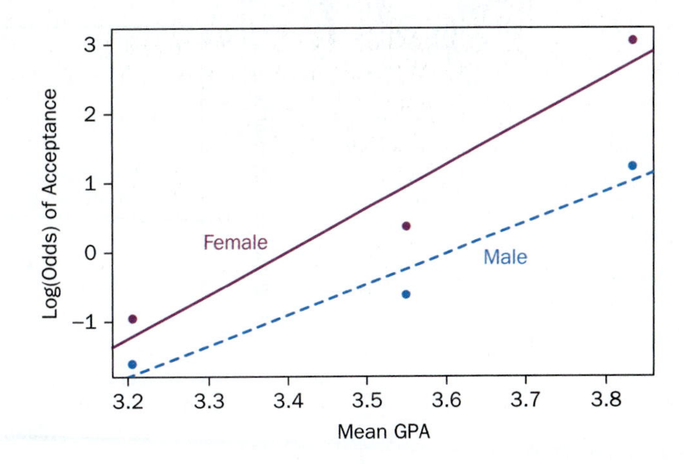

FIGURE 10.9 Log(odds) of acceptance versus *GPA* for female and male applicants

6. *Interaction of GPA and MCAT*. The relationships in Figure 10.10 all look strongly linear, and the graph suggests a difference in slope that depends on *MCAT* scores. Specifically, for those with a very high *MCAT* score (≥ 40), the line is steep: *GPA* makes a big difference. For those with mid-range scores (35–39) or low scores (< 35), the lines are nearly parallel and less steep: A higher *GPA* helps your chances, but not as much.

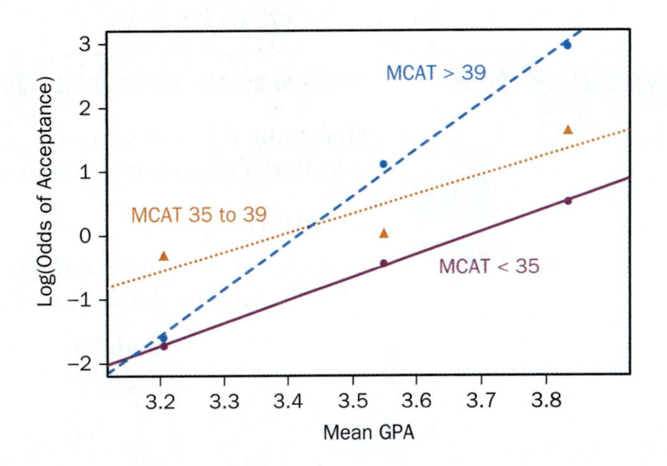

FIGURE 10.10 Log(odds) of acceptance versus *GPA* for three groups of *MCAT* scores

7. *Interaction of GPA and number of applications*. According to this graph (Figure 10.11), among students with a really high *GPA*, these who applied to more places were less likely to be accepted; among students with low *GPA*s, those who applied to more places were more likely to be accepted; and for students with *GPA*s in the middle, there was no clear relationship between the number of applications and chance of acceptance. However, caution is in order. Not one of the lines fits really well. In fact, for each of the three *GPA* groups, the relationship looks curved in the same way, higher at the extremes and lower in the middle. Finally, think about sample size and variability. We have $3(3) = 9$ categories, and only 55 cases total.

   There are plenty of reasons to be skeptical.

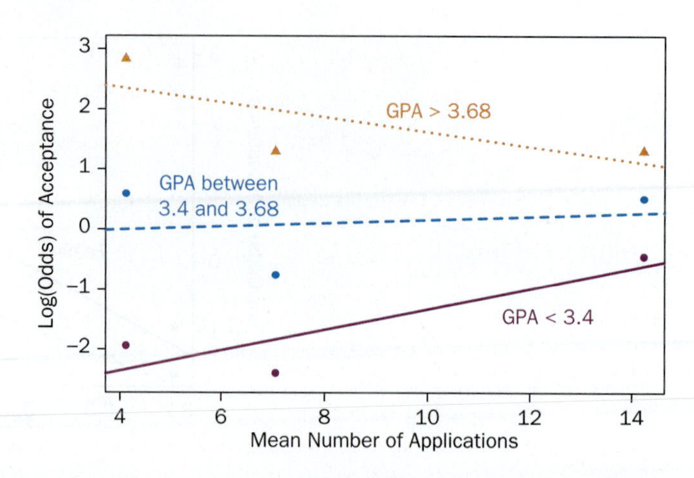

FIGURE 10.11 Log(odds) of acceptance versus number of applications for the three groups of *GPA*s

As we move on, remind yourself that for the **Eyes** data, with 106 cases and just two predictors (*DilationDiff* and *Sex*), we enjoyed the luxury of 53 cases per predictor. For the admissions data, with many variables and few cases, we don't have nearly enough data to find an unambiguous single comprehensive model. Instead, we can choose and fit models that focus on particular questions of interest. The example below offers just two instances. There are many more possibilities (as the exercises will demonstrate).

EXAMPLE 10.3

MedGPA

Medical school admissions: Models from context

Question 1. If you know an applicant's *MCAT* score, does their *GPA* provide additional information about their chances of acceptance?

CHOOSE

The question suggests comparing a pair of nested models, both containing *MCAT*: a null model without *GPA*, and an alternative with *GPA*.

$$\text{Model 0:} \quad \text{logit}(\pi) = \beta_0 + \beta_1 MCAT$$

$$\text{Model 1:} \quad \text{logit}(\pi) = \beta_0 + \beta_1 MCAT + \beta_2 GPA$$

FIT

Here are the fitted models:

$$\text{Model 0:} \quad \text{logit}(\hat{\pi}) = -8.712 + 0.246 MCAT$$

$$\text{Model 1:} \quad \text{logit}(\hat{\pi}) = -22.373 + 0.165 MCAT + 4.677 GPA$$

Notice that when *GPA* is added to the model, the coefficient for *MCAT* drops substantially, from roughly 0.25 down to less than 0.20. This drop reflects the correlation of 0.54 between *GPA* and *MCAT*. As *GPA* takes on some of the explanatory burden, differences in *MCAT* scores matter much less. Notice also that the coefficient for *GPA* is large because of the scale. For *MCAT* scores, a difference of 5 is large; for *GPA* a difference of 0.5 is large.

ASSESS

Is *GPA* a useful second predictor when you already know a student's *MCAT* score? A drop-in-deviance test (see Section 10.4) gives a *P*-value of 0.001, indicating that yes, *GPA* is definitely useful as a second predictor.

Question 2. For students with the same *GPA* and *MCAT* score, does applying to more schools increase the chance of admission?

CHOOSE

Before getting to the models, two cautions are in order. First, we don't have nearly enough data to make direct comparisons of students "with the same *GPA* and *MCAT*," so we resort to regression as a statistician's way to "equalize by adjusting."* Second, as worded, the question asks about a cause-and-effect relationship, even though our data are observational. The best we can do is to check whether the observed relationships are consistent with a causal explanation. With these cautions in mind, you can think about how to choose models that are suggested by the question. Here, as for Question 1 above, a pair of nested models seems useful—a base (null) model with terms for *GPA* and *MCAT*, together with an augmented model that also contains a term for the number of applications.

FIT

Here are the fitted models:

$$\text{Model 0:} \quad \text{logit}(\hat{\pi}) = -22.373 + 0.165MCAT + 4.677GPA$$

$$\text{Model 1:} \quad \text{logit}(\hat{\pi}) = -23.689 + 0.173MCAT + 4.861GPA + 0.044APPS$$

The coefficients for both *GPA* and *MCAT* are comparatively stable. Adding the number of applications to the model changes the coefficient for *GPA* by only 0.2, from 4.7 to 4.9, and the coefficient for *MCAT* by less than 0.01, from 0.165 to 0.173.

ASSESS

For these two models the drop-in-deviance test is unambiguous, with a *P*-value of 0.56. There is no evidence that *APPS* is a useful predictor, *provided GPA* and *MCAT* are already in your model.

What next? In the examples so far, we have shown ways to choose models, and shown fitted coefficients that came from maximizing the likelihood. We have also assessed model utility using a drop-in-deviance test that is essentially the same as the one in Chapter 9. In the next section, we look more closely at model assessment.

10.3 Checking Conditions

In this section, we show you two ways to think about whether to trust the formulas that give us the probabilities we use for hypothesis tests and confidence intervals. As you read this section, be a bit of a skeptic. Keep in mind that the formulas of the next section will give you values for tests and intervals even if there is no justification for using them. It is your job to decide whether and how far you can trust the results to mean what they appear to say. In this context, you have two main things to look at: how to assess linearity and how to assess the probability part of the model. Both are required for formal inference. But also keep in mind that your skepticism has logical limits: *Even if you can't justify formal inference, there's still a lot you can do with statistical models.*

*Just as with multiple regression, a model allows you to compare the fitted responses for two individuals who have the same values for all but one of the predictor variables. This model-based comparison is often called "adjusting for" those predictors.

Checking Linearity

To check whether a relationship is linear, we use empirical logit plots, as in Section 10.2. You have already seen several such plots that did show linear relationships (Figures 10.2, 10.4, and 10.6). What if the relationship is *not* linear? We consider two general possibilities: strong *nonlinear* relationships and "points all over the map." If your points are all over the map (Example 10.4 below), there's not much you can do, but the absence of a relationship is a strong indication that your predictor is not of much use in the model. On the other hand, if you have a strong but nonlinear relationship, you can often find a transformation that makes the relationship more nearly linear (Examples 10.6 and 10.7).

EXAMPLE 10.4

MedGPA

Medical school admissions: Points all over the map Figure 10.12(a) shows an empirical logit plot for the number of applications. As you can see, there is no obvious relationship. Granted, if you "connect the dots" as in Figure 10.12(b), you can create an up-and-down pattern, but with so few cases, we can't tell whether the pattern has meaningful substance or is just an illusion due to chancelike variation. Regardless, there is no simple transformation that will give a strong linear pattern. Bottom line: Don't include the number of applications in your model for this dataset.

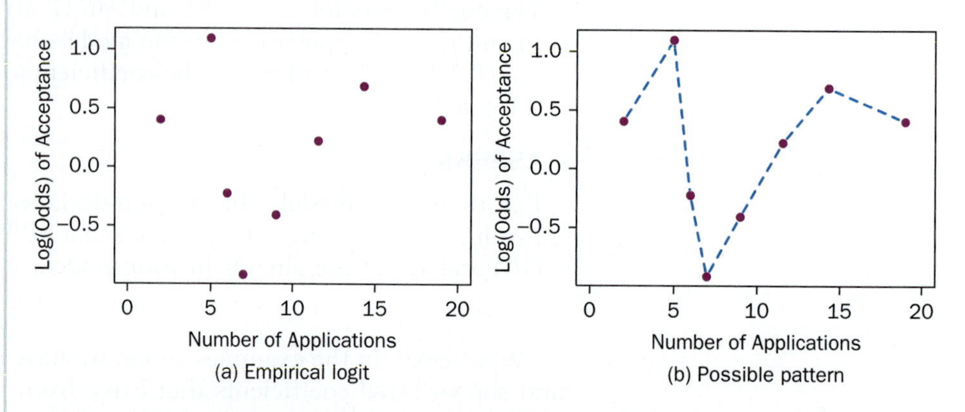

FIGURE 10.12 Medical school admissions data: Empirical logit plot for number of applications (left) and possible pattern (right)

EXAMPLE 10.5

CredRsk

Drinking and drawing Researchers[2] conducted a survey of 450 undergraduates in large introductory courses at either Mississippi State University or the University of Mississippi. We are interested in modeling whether a student overdraws a checking account (*Overdrawn*). Data for several possible predictors from the study are stored in **CreditRisk**, which includes the following variables.

Age	"What is your current age (in years)?"
Sex	"What is your sex?" 0 = male, 1 = female
DaysDrink	"During the past 30 days, on how many days did you drink alcohol?"
Overdrawn	"In the last year, I have overdrawn my checking account." 0 = no, 1 = yes

You will consider all of the predictors in Exercise 10.49, but in this example we focus on whether students who drink a lot (*DaysDrink*) are more likely to overdraw their checking accounts. Before fitting a logistic model, we explore the relationship between *Overdrawn* and *DaysDrink* with an empirical logit plot. Figure 10.13 shows some curvature, which suggests that a transformation is called for. But which one?

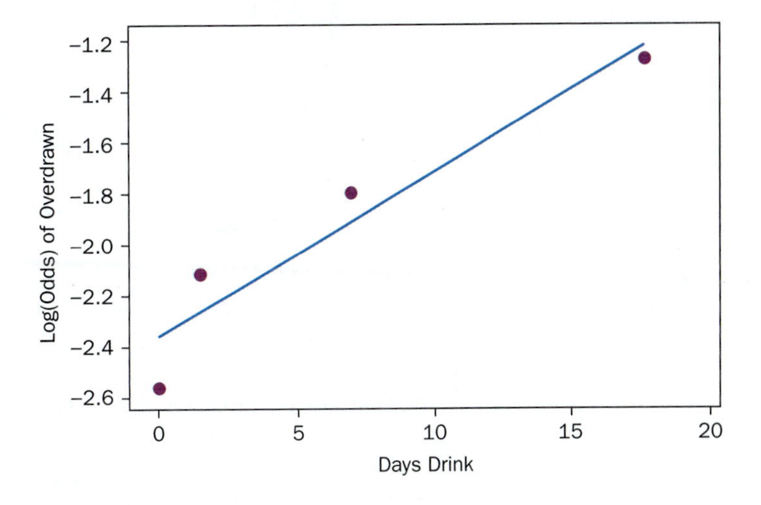

FIGURE 10.13 Empirical logit plot for predicting *Overdrawn* using *DaysDrink* in the original scale

The Power Family of Transformations

For many variables, if you need to transform to get a linear plot, you will be able to find a suitable choice by considering transformations from the "power family"—those of the form "x-to-the-*power*." These transformations take each value of x and change it to x^{power} for some fixed value of *power*. Squares (*power* = 2) transform x to x^2. Square roots (*power* = 1/2) transform x to $\sqrt{x}$. Reciprocals (*power* = −1) transform x to $1/x$. For technical reasons, the log transformation corresponds to *power* = 0. *If your empirical logit plot is concave down, you should "transform down," that is, try powers less than 1. If your plot is concave up, you should "transform up," that is, try powers greater than 1.*

If your plot is concave down, transforming to logs (*power* = 0) is often a good first choice to try. If you get lucky, your plot will be linear, or nearly so. If not, the shape of the curve will suggest other transformations.

EXAMPLE 10.6

CredRsk

Drinking and drawing: Linearity with a power transformation Figure 10.13 shows a concave-down pattern in the empirical logit plot for modeling *Overdrawn* with the original scale for *DaysDrink*. This suggests using a log transform, but a number of students reported zero days of drinking, which causes trouble when computing the log. For this reason we use *LogDrink* = *log*(*DaysDrink* + 1). Note that this transformation sends zero values for *DaysDrink* to zero values for *LogDrink*. Figure 10.14 shows that the empirical logit plot in the log scale looks better for linearity.

If by using a log transformation we had gone too far so that the curvature was now in the opposite direction with a plot that is concave up, then we could try a power between 0 (logs) and 1 (the original scale). The log transformation looks good, but for comparison purposes we can look at Figure 10.15, which shows empirical logits when using a square root transformation (power = 0.5). Notice that the square root transformation "pulls down" the curve in Figure 10.13, but not quite as much as does the log transformation shown in Figure 10.14. The square root transformation would also be acceptable for obtaining linearity in this situation.

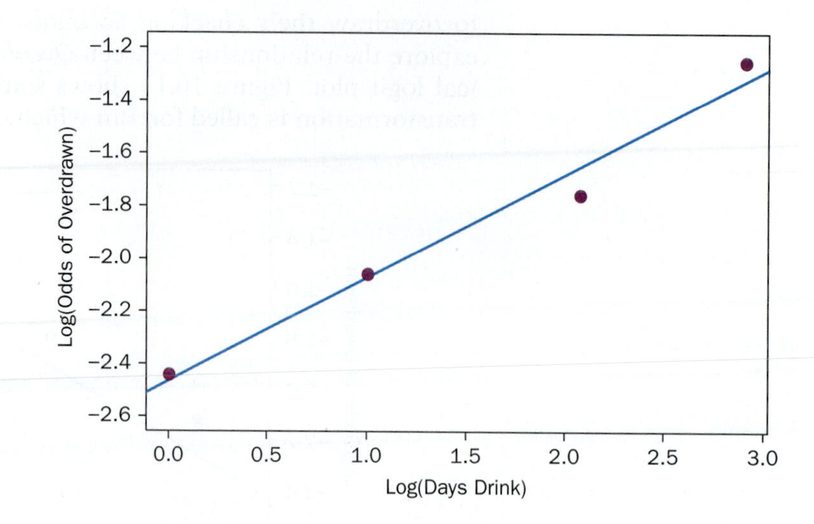

FIGURE 10.14 Empirical logit plot for *Overdrawn* with a log scale for days drinking

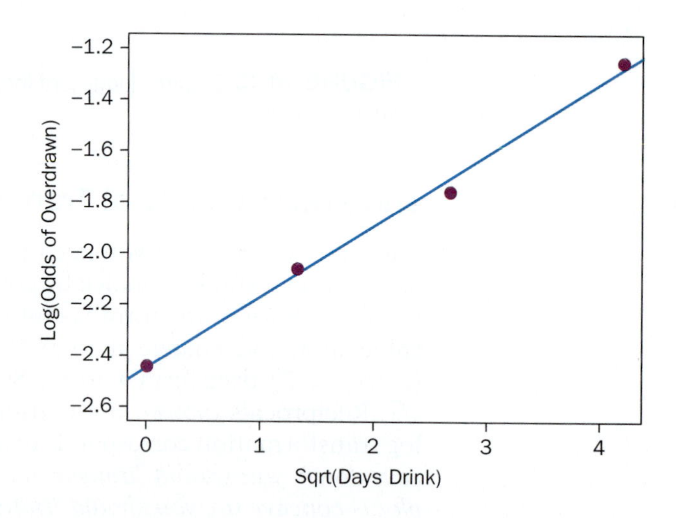

FIGURE 10.15 Empirical logit plot for *Overdrawn* with square root of days drinking

End-effects: Transformations from the power family typically work best when the original scale is not bounded by an endpoint. Occasionally, though, you need to correct for an "end-effect." This is, in fact, what the logistic transform does. Values of π must be between 0 and 1, and probabilities often tend to bunch up near one or both endpoints. The logistic transform stretches out the ends, changing 0 to $-\infty$ and 1 to ∞. The same bunching at an endpoint happens with *GPA* in the medical school data: *GPA* has a cap of 4.0, and grades of applicants tend to cluster at the high end.

EXAMPLE 10.7

MedGPA

Medical school admissions: Transforming to avoid an "end-effect" Figure 10.16 shows the empirical logit plot for *GPA*. If you ignore the right-most point, the relationship of the remaining four points is strongly linear (dashed line), but with all five points included the linearity is not nearly as strong (solid line).

The *GPA* ceiling at 4.0 suggests the possibility that the apparent nonlinearity may be due to an "end-effect." The fact that *GPA* is capped at 4.0

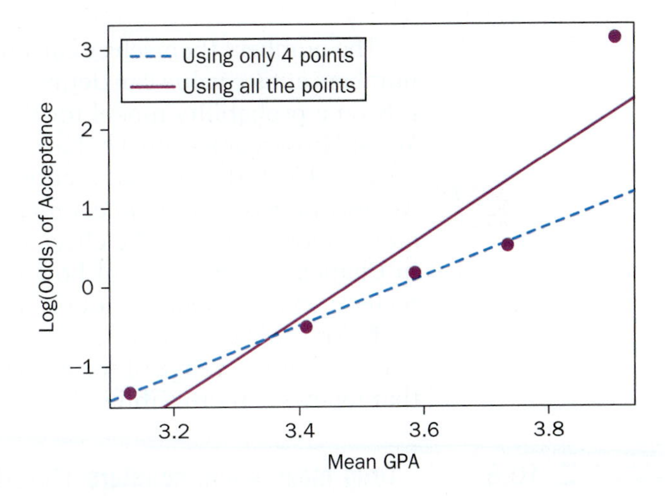

FIGURE 10.16 Empirical logit plot with *GPA* in the original scale

suggests rechecking linearity with *GPA* transformed to log(*GPA*/(4 – *GPA*)).* In the new scale, the empirical logit plot (Figure 10.17) shows a strong linear relationship for all five points.

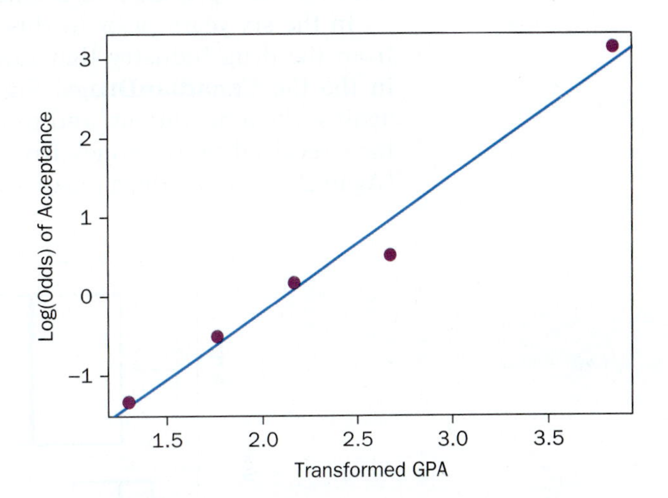

FIGURE 10.17 Empirical logit plot with *GPA* in the transformed scale

Is it worth transforming *GPA*? For this example, probably not, at least for most purposes. The plot in the original scale is reasonably close to linear, and it is much easier to think about changes in *GPA* than about changes in log(*GPA*/(4 – *GPA*)). However, if the upper end of the range of *GPA*s is what you care most about, it might be worth trying an analysis using the transformed scale. At the very least, the new plot does tend to confirm the presence of an end-effect.

Checking the Probability Model

Suppose you have chosen a scale that makes relationships linear. What next? Typically, we want our analysis to head toward standard errors, hypothesis tests, and confidence intervals. Your software stands ready to spit out the numbers at the click of a mouse. Why not go ahead?

* The transformation is the log of the ratio of distances from the endpoints, the log of (*GPA* – *min*)/(*max* – *GPA*), with *min* = 0 and *max* = 4. Note that this is also what the logit transform does for probabilities, *min* = 0 and *max* = 1.

The numbers themselves can't hurt you, but if you're not careful, taking the numbers at face value can definitely hurt you. The methods of formal inference rely on a probability model for the data. Specifically, just as in Chapter 9, the Yes or No outcomes must behave according to the Bernoulli (spinner) model of Section 9.3. Just as in that section, *before we take SEs or P-values at face value, we need to check randomness and independence, by thinking about the process that created our data.* The check of these conditions is essentially the same as in Chapter 9, regardless of how many predictors we have, because the main focus is on the response, not on the predictors.

Before going on to the next example, you may want to revisit Example 9.14, which covered a range of scenarios from clearly Yes to clearly No. The example that follows is more subtle.

EXAMPLE 10.8

CADrug

Drug makers and senators The Affordable Care Act, widely known as ObamaCare, was unpopular with Republican legislators. In early January of 2017 the newly seated 115th U.S. Senate crafted a budget bill as preparation for repealing the Affordable Care Act. The "Klobuchar amendment" to that bill was introduced with the purpose of lowering drug prices by allowing prescription drugs to be imported from Canada. The drug manufacturing industry was opposed to this amendment.

In the six years prior to this vote, senators had accepted contributions from the drug industry that ranged as high as $342,310. These data* are in the file **CanadianDrugs**. Figure 10.18 shows that senators who voted against the amendment, and thus were "With" the drug industry, tended to have received more money from the industry than did senators who voted "Against" what the drug makers wanted.

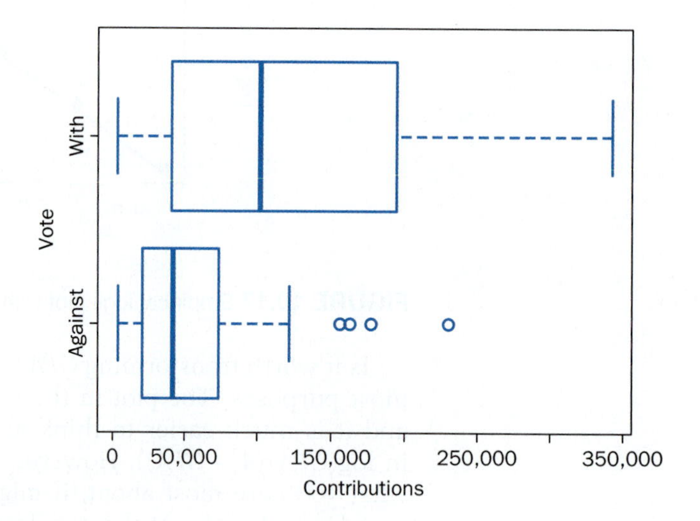

FIGURE 10.18 *Contributions* versus *Vote*

CHOOSE

In Figure 10.19 we see that there is a reasonably linear trend between empirical logits for *Vote* and size of contributions from drug makers.

* The data exclude two senators who did not vote on the amendment and four senators who were new to Congress and thus had received no money from the drug industry. The remaining 94 senators represent 49 states (every state except California), and each of these senators had received at least $3,000.

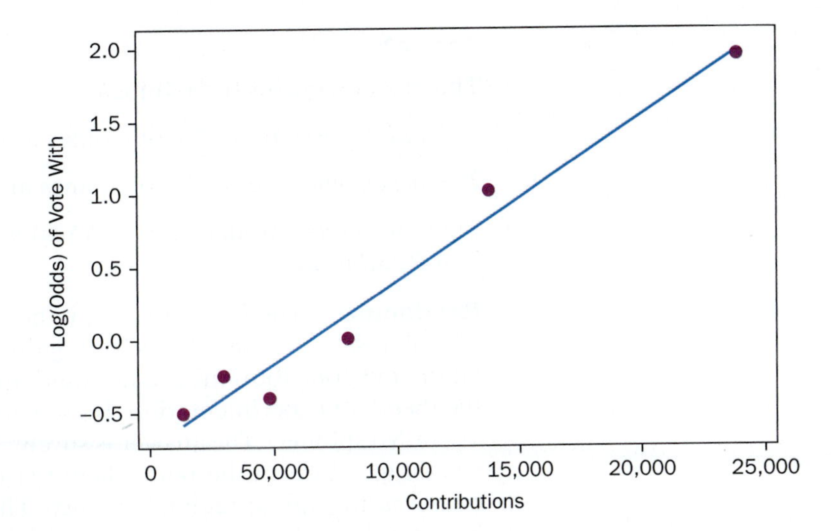

FIGURE 10.19 Empirical logits for *Vote* at different contribution levels

If we allow for a *Party* effect, the empirical logit plot (Figure 10.20) shows a difference between Democrats and Republicans. The lines in Figure 10.20 are not perfectly parallel, but there is no significant evidence of an interaction (the *P*-value for interaction is 0.55), so a "parallel curves" model is appropriate. The output for that model, below, shows that as *Contributions* increase, so does the likelihood of voting with the drug industry. There is also a strong *Party* effect. (Republicans were more likely to vote with the drug industry.) We discuss these sorts of inferences more thoroughly in the next section.

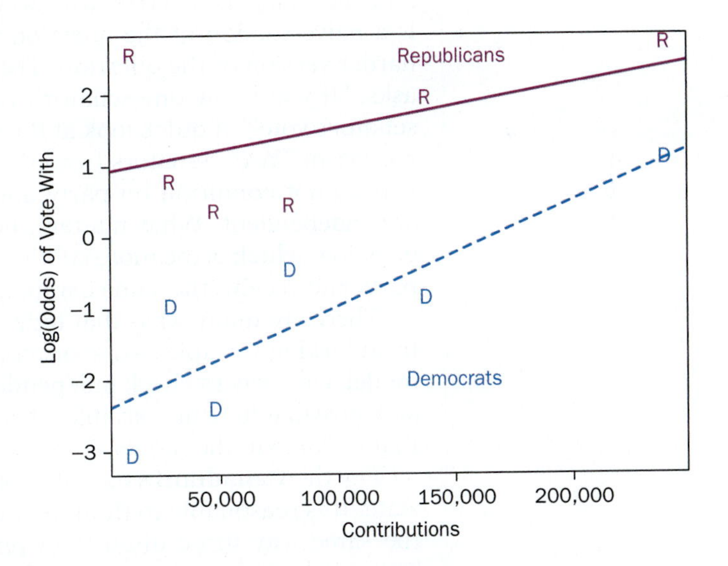

FIGURE 10.20 Empirical logits for *Vote* at different contributions levels

```
Coefficients:
                  Estimate   Std. Error  z value    Pr(>|z|)
(Intercept)    -2.31942366   0.57773602    -4.01  0.0000595 ***
Contributions   0.00001469   0.00000411     3.58    0.00035 ***
Party:Rep       2.50992529   0.56286044     4.46  0.0000082 ***

Null deviance: 129.246 on 93 degrees of freedom
Residual deviance: 88.922 on 91 degrees of freedom
```

ASSESS

The model requires three things:

1. *Randomness*: Yes or No outcomes are random.
2. *Independence*: Yes or No outcomes are independent.
3. *Linear model*: Model for $\pi = P(\text{Yes})$ is $\text{logit}(\pi) = \beta_0 + \beta_1 Contributions + \beta_2 Republican$.

Randomness: The U.S. Senate is in no way a random sample, so we cannot appeal to sampling as a basis for regarding outcomes as random. Party affiliation and contributions are not randomly assigned, so we cannot appeal to the theory of experimentation. Do senators decide which way to vote using a Yes or No spinner? The answer is surely "No" (cynics to the contrary notwithstanding). More to the point, how reasonable is it to think of the spinner model as an approximation to reality? This is a vastly more difficult question, because the right version to ask is *conditional: Given* a party affiliation and contribution amount, do senators' Yes or No (Against or With) votes behave like outcomes from a spinner? Data can't help us here because, for any given combination of party and contribution, the sample size is tiny, typically $n = 1$. Bottom line: Although there is no clear basis for saying that the spinner model does apply, there is also no clear basis for saying it does not apply.

Independence: If outcomes are not random, then independence is not an issue because independence has no meaning without probabilities. However, to the extent that randomness might be reasonable as an approximate model, it is important to think also about independence. Without independence, the formulas for the standard errors will be wrong, the *P*-values will be wrong, and the margins of error will be wrong. Here, as with randomness, there is a naive version of the question with a simple answer, and a subtler and harder version of the question. The simple version ignores conditioning and asks, "If you know one senator's vote, does that help you predict any other senator's vote?" A quick look at the votes state-by-state gives an equally quick answer of "Yes." Senators from the same state tend to vote the same way. If you do not condition on party and contribution amount, votes are clearly *not* independent. What matters, however, is the *conditional* version of the question, which is the more subtle version: If two senators belong to the same party and receive the same campaign amounts, are their votes independent?

There are many ways that independence can fail, but most failures result from lurking variables—one or more shared features that are not part of the model. One way to check independence is to use the applied context to guess at a possible lurking variable, then compare predictions with and without that variable in the model.

For the **CanadianDrug** data, one possible lurking variable is a senator's state. It is reasonable to think that senators from the same state tend to vote the same way more often than predicted by the independence condition. This prediction is one we can check: For each senator, we use our model to compute a fitted *P*(With) given party and contribution level. We then use these fitted probabilities to compute the chance that both senators from a state vote the same way, *assuming that votes are independent*. This lets us compute the expected number of states where both senators vote the same way, if the independence condition holds. We can then compare the expected number with the observed number. If votes are in fact independent, observed and expected should be close. Because in some states either a senator did not vote or was new to Congress (and thus excluded from our data), there are 45 states with pairs of voting senators in our data who can be compared. Table 10.5 shows the actual numbers.

	Together	Split	Total
Actual	31.0	14.0	45
If independent	26.2	18.8	45

TABLE 10.5 Observed and expected numbers of states with split votes in **CanadianDrugs**

Observed ("Actual") and expected ("If independent") are close enough to each other that the condition of independence would be considered plausible.*

The linear logit model: Randomness and independence are required for formal inference, but they are not enough. The theory also requires that, conditional on the values of the predictors, $P(\text{With})$ is constant. Less formally, there is a single spinner for all cases with the same combination of predictor values. Or to put it negatively, for the theory to work, there must be no lurking variables. For the **CanadianDrug** data, we considered the lurking variable *State*, but other variables are out there; for example, maybe it matters how long a senator has been in Congress, and thus receiving money from the drug industry. Moreover, if we wanted to use the categorical variable *State*, it has so many categories that building it into the model is unworkable. We would need 48 indicator variables, with only 94 cases.

Where does this leave us in relation to the probability model? Agnostic at best. We have no clear basis to justify the probability part of the model. At the same time, we have no compelling basis for declaring the model to be so clearly wrong as to make tests and intervals also clearly wrong. It is reasonable to go ahead with formal inference but prudent to regard the results as tentative guidelines only. This caveat aside, it is important to recognize that, even without formal inference, the logistic model still has great value. It fits well, and it provides a simple, accurate, and useful description of a meaningful pattern. In particular, the fitted model is consistent with the research hypothesis that contributions from the drug industry and votes on the Klobuchar amendment are related.

Checking conditions is the hard part of inference, as the last example illustrates. You really have to think. By comparison, getting the numbers for formal inference is a walk in the park. Keep in mind, however, as you go through the next section, that what matters is not the numbers themselves, but what they do tell you, and what they cannot tell you, even if you want them to.

10.4 Formal Inference: Tests and Intervals

Formal inference for logistic regression is in many ways quite similar to formal inference for ordinary regression. For both kinds of regression, we have tests and intervals for individual coefficients in the model (see Table 10.6). These are based on relating a standardized regression coefficient either to a t-distribution (ordinary regression) or to a normal distribution (logistic regression). For both kinds of regression, we also have tests to compare nested pairs of models. The test statistics and reference distributions are different (F for ordinary, χ^2 for logistic), but in most other respects, you can rely on ordinary regression methods as a guide to the new methods. Moreover, you already know about z-tests

* A chi-square test statistic has a value of 1.34.

	Ordinary Regression	Logistic Regression
Test or interval for β	$t = \frac{\hat{\beta}-\beta}{SE(\hat{\beta})}$	$z = \frac{\hat{\beta}-\beta}{SE(\hat{\beta})}$
	t-distribution	*z-distribution*
Test for nested models	$F = \frac{\Delta SSE/\Delta df}{MSE_{Full}}$	$G = \Delta(-2\log(L))$
	F-distribution	*χ^2-distribution*

TABLE 10.6 Comparing inference methods for ordinary and logistic regression

and intervals for individual coefficients from Chapter 9. These will be almost exactly the same as before. Only the interpretation of the coefficients and the computation of their standard errors are different, and the computer will take care of the computation for you. Finally, you already know about the drop-in-deviance test based on the difference in $-2\log(L)$ for a nested pair of models. With more than one predictor, you have more choice about the models to compare, but you have already seen examples of that in Example 10.3. The only other new thing is counting degrees of freedom, but that can be put in a single sentence.

Preview of Multipredictor Logistic Models

Putting all this together means that none of the methods of this section are truly new. What is new is their use for analyzing multipredictor logistic models. Accordingly, we first summarize the methods with formulas, then illustrate with examples, starting first with tests and intervals for single coefficients before going on to drop-in-deviance tests. In what follows, you will see a sequence of five examples chosen to illustrate a variety of key points. By way of preview, these examples are outlined below.

z-tests and Intervals

Example 10.9. Simple test and interval using z.
Data: Eyes, with *Gay* as response, *DilateDiff* and *Sex* as predictors.
Focus: Mechanics of the z-test, of the z-interval, and the interval for the odds ratio.

Example 10.10. More complicated tests using z.
Data: MedGPA, with separate quadratic models for women and men relating *GPA* to chance of admission.
Focus: The question the z-test answers may not be the one you want.

Nested Likelihood Ratio Tests

Example 10.11. Simple LRT using χ^2.
Data: MedGPA

a. If *GPA* is already in your model, do you also need *MCAT* score?
b. Model utility for the model with both *GPA* and *MCAT*.

Focus: Mechanics of the LRT.

Example 10.12. Simple LRT using χ^2.
Data: MedGPA
Focus: LRT and z-test may not give the same P-values.

Example 10.13. A sequence of LRTs using χ^2.
Data: MedGPA as in Example 10.10.
Focus: Using several LRTs to choose predictors.

Tests and Intervals for Individual Coefficients Using a Normal Distribution

If you remind yourself "z instead of t," the tests and intervals here are basically the same as for individual coefficients in ordinary multiple regression.

z-TEST AND INTERVALS FOR A SINGLE PREDICTOR

For a multiple binary logistic regression model with k predictors,

$$\log\left(\frac{\pi}{1-\pi}\right) = \beta_0 + \beta_1 X_1 + \beta_2 X_2 + \cdots + \beta_k X_k$$

A test for the contribution of the predictor X_i, given the other predictors in the model, H_0: $\beta_i = 0$ versus H_a: $\beta_i \neq 0$, uses the test statistic (Wald z-statistic)

$$z = \frac{\hat{\beta}_i}{SE_{\hat{\beta}_i}}$$

with a P-value from a standard normal distribution.

A confidence interval for an individual coefficient is found by

$$\hat{\beta}_i \pm z^* SE_{\hat{\beta}_i}$$

The estimated odds ratio for a one-unit change in the predictor X_i is

$$e^{\hat{\beta}_i} \text{ with confidence interval } (e^{\hat{\beta}_i - z^* SE_{\hat{\beta}_i}}, e^{\hat{\beta}_i + z^* SE_{\hat{\beta}_i}})$$

EXAMPLE 10.9

Eyes

The eyes have it: Simple tests and intervals using z The lines below show partial output from fitting the model (in Example 10.1) to predict being gay based on pupil dilation and sex for the data in **Eyes**.

Coefficients:

	Estimate	Std. Error	z value	Pr(>|z|)
(Intercept)	−1.266	0.382	−3.31	0.00092
DilateDiff	3.291	0.818	4.02	0.000058
SexMale	1.290	0.495	2.60	0.0092

Each z-value (fourth column) equals an estimated coefficient (second column) divided by its estimated standard error (third column). Each (two-tailed) P-value (fifth column) comes from referring z to the standard normal distribution.

For the **Eyes** data, these P-values tell us that both *DilateDiff* and *SexMale* belong in the model. More formally, we have evidence to reject the hypothesis that there is no association between *Gay* and either of the two predictor variables.

If we use Minitab instead of R, we get confidence intervals automatically. With R, we have to construct them using the formula

$$\hat{\beta} \pm z^* SE_{\hat{\beta}}$$

SexMale. For the sex indicator *SexMale*, this formula gives a 95% interval of

$$1.29 \pm 1.96(0.4953) \quad \text{or} \quad 0.32 \text{ to } 2.26$$

We can translate this confidence interval for β_2 to an interval for the odds ratio, by using the two values as exponents: The actual odds ratio should be between $e^{0.32} = 1.38$ and $e^{2.26} = 9.58$. More specifically, a difference of 1 in the predictor (*SexMale*) corresponds to multiplying the odds ratio by a number between 1.38 and 9.58. If a male (*SexMale* = 1) and female (*SexMale* = 0) have the same dilation difference, the interval, taken at face value, says that the odds of the male being gay is between 1.38 and 9.58 times the odds of the female being gay.

DilateDiff. For the dilation difference, the formula gives a 95% interval of

$$3.29 \pm 1.96(0.8180), \quad \text{or} \quad 1.69 \text{ to } 4.89$$

Again, we can translate the confidence interval for β_1 to an interval for the odds ratio, by using the two values as exponents: The population value of the odds ratio is estimated to be between $e^{1.69} = 5.42$ and $e^{4.89} = 132.95$. We see that an increase of 1 in the dilation difference (recall, from Figure 10.1 or Figure 10.5, that a dilation difference of 1 is quite large) corresponds to multiplying the odds ratio by a number between 5.42 and 132.95.

Almost surely, at this point you are ready to agree: Getting the numbers is the easy part; giving them meaning is the hard part. The only way to master this is to practice.

The next example illustrates one of the shortcomings of the z-test: It may not answer the question you want it to answer.

EXAMPLE 10.10

MedGPA

Medical school data: Multiple dependence and conditional *P*-values In Example 10.2(5), an empirical logit plot (Figure 10.9) suggests that quadratic curves may fit better than lines, with separate curves for men and women. (Note that we now switch from the categorical variable *Sex* to the indicator variable *FEM*, which takes the value 1 for females and 0 for males.) We can explore that suggestion using a model like the following:

$$logit(\pi) = \beta_0 + \beta_1 FEM + \beta_2 GPA + \beta_3 GPA^2 + \beta_4 FEM \cdot GPA + \beta_5 FEM \cdot GPA^2$$

In this model, the intercepts are β_0 for male applicants and $(\beta_0 + \beta_1)$ for female applicants. The linear terms are β_2 and $(\beta_2 + \beta_4)$, respectively; the quadratic terms are β_3 and $(\beta_3 + \beta_5)$. Put differently, for men we fit the model $\beta_0 + \beta_2 GPA + \beta_3 GPA^2$. For women we fit $(\beta_0 + \beta_1) + (\beta_2 + \beta_4)GPA + (\beta_3 + \beta_5)GPA^2$.

Here is the information the computer gives us about the coefficients for this model:

	Estimate	Std. Error	z value	Pr(>|z|)
(Intercept)	53.4790	55.2279	0.968	0.333
FEM	9.7948	162.0299	0.060	0.952
GPA	-37.0418	32.8472	-1.128	0.259
GPA2	6.1276	4.8571	1.262	0.207
FEM:GPA	-5.9204	93.4748	-0.063	0.949
FEM:GPA2	0.9993	13.4642	0.074	0.941

Taking the results at face value, it may seem that not even one of the predictors belongs in the model: All the *P*-values are 0.2 or larger! How can this be, when we already know that *GPA* belongs in the model? The important lesson here is that *the z- and P-values are conditional. The questions they address may not be the ones you want to ask.* To be concrete, the *P*-value for *FEM* is 0.952. Imagine the following exchange between a data analyst and a friend:

Data analyst: Does *FEM* belong in the model?
Friend: No way: The *P*-value of 0.952 is quite large.

In fact, the P-value answers a very different question:

Data analyst: If you already have all the other predictors in the model, do you also need *FEM*?

Friend: No, not even close. Whatever useful information *FEM* provides is already provided by the other variables.

Abstractly, the key point is that these z-tests are conditional, just as with ordinary regression: Each P-value assumes that all the other predictors are in the model. On a practical level, if you are trying to choose predictors, the P-values in a "big" model like the one in this example are probably not a very helpful thing to look at. If not, then what? We'll address this question in Example 10.13, in connection with nested likelihood ratio tests.

Nested Likelihood Ratio Tests (LRTs) Using χ^2

You may want to review the nested pairs of models in Example 10.3 of Section 10.2. In that example, the emphasis was on ways to choose nested pairs of models and how to interpret the results of the corresponding LRT. In the next example, the emphasis is on the mechanics of the LRT, as summarized below.

DROP-IN-DEVIANCE TEST (NESTED LIKELIHOOD RATIO TEST)

A test for the overall effectiveness of the model

$$H_0 : \beta_1 = \beta_2 = \cdots = \beta_k = 0 \text{ versus } H_a : \text{At least one } \beta_i \neq 0$$

uses the test statistic

$$G = -2\log(L_0) - (-2\log(L))$$

where L_0 is the likelihood of the sample under a constant model and L is the likelihood under the larger model. The P-value comes from the upper tail of a χ^2-distribution with k degrees of freedom.

The **nested likelihood ratio test** works the same way, but with more general hypotheses.

$$H_0 : \text{Reduced model} \quad \text{versus} \quad H_a : \text{Full model}$$

The nested G-statistic is the difference in $-2log(L)$ between the two models. If the reduced model has k_1 predictors and the full model has $k_1 + k_2$ predictors, then the P-value comes from the upper tail of a χ^2-distribution with k_2 degrees of freedom.

In the three examples that follow, Example 10.11 illustrates the mechanics of the LRT in a simple situation; then Examples 10.12 and 10.13 illustrate issues that are less straightforward.

EXAMPLE 10.11

MedGPA

Medical school data: A simple LRT using χ^2 In this example, we illustrate two basic uses of the nested LRT: (1) a test for a single predictor in the presence of one or more other predictors, and (2) a test for two or more predictors at once, as in assessing overall model utility.

1. In Example 10.3, we used a nested LRT to answer the question, "If *MCAT* is already in your model, can you get reliably better predictions if you also include *GPA*?" In that example, we merely summarized the results of the test using the P-value. Now we present some of the details of the method.

This time, just for variety, we reverse the roles of *MCAT* and *GPA*, so that the new question is, "If *GPA* is already in your model, can you get reliably better predictions if you also include *MCAT* in your model." For this question, our two nested models are

$$H_0: \ logit(\pi) = \beta_0 + \beta_1 GPA$$
$$H_a: \ logit(\pi) = \beta_0 + \beta_1 GPA + \beta_2 MCAT$$

Here, H_0 corresponds to the reduced model and H_a to the full model. To carry out the LRT, we need the values of $-2\log(L)$ for both models. We compute the difference in values and refer that difference to a chi-square distribution with 1 degree of freedom. The $df = 1$ comes from subtracting the numbers of parameters in the two models: 3(full) – 2(reduced). Here is some output for the models:

Reduced model:

```
            Estimate Std.  Error  z value   Pr(>|z|)
(Intercept)   -19.207  5.629  -3.412   0.000644 ***
GPA             5.454  1.579   3.454   0.000553 ***
---
    Null deviance: 75.791 on 54 degrees of freedom
Residual deviance: 56.839 on 53 degrees of freedom
```

Full model:

```
            Estimate Std.   Error  z value  Pr(>|z|)
(Intercept)   -22.3727  6.4538  -3.467   0.000527 ***
GPA             4.6765  1.6416   2.849   0.004389 **
MCAT            0.1645  0.1032   1.595   0.110786
---
    Null deviance: 75.791 on 54 degrees of freedom
Residual deviance: 54.014 on 52 degrees of freedom
```

The value of $-2\log(L)$ is called "residual deviance" in the R output: 56.839 for the reduced model and 54.014 for the full model. Subtracting gives a "drop-in-deviance" of 2.825. Referring this to a chi-square distribution with $df = 1$ gives a *P*-value of 0.093:

$-2\log(L)$ = residual deviance			
Reduced	**Full**	**Difference**	***P*-value**
56.839	54.014	2.825	0.093

The conclusion here is not clear-cut. The *P*-value is moderately small. If the goal is a formal test of H_0 versus H_a, we fail to reject H_0 (at the 0.05 level), but that "failure" could easily be a consequence of the small sample size. If the goal is to find a useful model, it may well make sense to include *MCAT* in the model. As an additional check, we can use a nested LRT to check overall utility for the model that includes both *GPA* and *MCAT*.

2. To assess the utility of the model $logit(\pi) = \beta_0 + \beta_1 GPA + \beta_2 MCAT$, we compare the residual deviance of 54.014 for that model with a residual deviance of 75.791 for the null model $logit(\pi) = \beta_0$. The drop-in-deviance is 75.791 – 54.014 = 21.777, with degrees of freedom = 2 because the full model has 3 coefficients and the null model has 1. To get a *P*-value, we refer the drop-in-deviance to a chi-square distribution on 2 df. This *P*-value is 0.000019, a strong confirmation of model utility.

P-values and Model Choosing

If you want a formal test of a null hypothesis that $\beta = 0$ and your data satisfy the required conditions, you can interpret *P*-values in the usual way, regarding 0.01 and 0.05 as benchmarks. On the other hand, for some purposes it may make sense to work with a model that includes variables for which the *P*-value is in the rough neighborhood of 0.10. Part of the logic is that with smaller sample sizes, *P*-values tend to be larger, even for predictors that truly matter, because the smaller samples give larger standard errors. Moreover, for some purposes, it may be useful to work with two or three models, rather than a single model that you think of as "best."

EXAMPLE 10.12

MedGPA

Medical school data: LRT and *z*-test may not give the same *P*-values If you look back at the computer output for the full model in the last example, you will find that the *P*-value for *MCAT* is 0.111. As always, this is a conditional *P*-value, conditional on all the other predictors being in the model. For this example, "all the other predictors" refers just to *GPA*, the only other predictor. Thus the *P*-value of 0.111 refers to the question, "If *GPA* is already in the model, do you get reliably better predictions if you also include *MCAT* in the model?" In other words, the *P*-value refers to the same question as the one we addressed using the nested LRT in Example 10.11. The LRT gave us a *P*-value of 0.093, not 0.111. What's up?

Both the *z*-test and the LRT are large-sample approximations in that even if all the required conditions are satisfied perfectly, the distributions for the *z*-statistics and for the drop-in-deviance are not exactly equal to the normal and chi-square distributions we use to get *P*-values. In short, the *P*-values are only approximations to the exact *P*-values. The approximations tend to be better for larger samples, and it can be proved mathematically that in the limit as the sample size goes to infinity, the approximations converge to the exact values. Less formally, the approximations get more accurate as your sample size increases. One consequence is that for smaller samples, your *P*-values from a *z*-test and the corresponding LRT may not agree. In particular, the difference here between 0.093 and 0.111 is a consequence of the comparatively small sample size.

In Example 10.10, we saw how fitting one big model with many predictors and then looking at *z*-tests may not help much in deciding which predictors to keep and which ones to leave out. A better strategy is based on looking at sequences of nested tests, as in the next example.

EXAMPLE 10.13

MedGPA

Medical school data: Using several nested LRTs to choose models One way to build a model is to start with the null model $logit(\pi) = \beta_0$ and add predictors one step at a time. As you will see, this method, though logical, is not easy to make automatic. For this example, using the data for medical school admissions, we start by using nested LRTs to choose one predictor from among *GPA*, *MCAT*, and *FEM*. At each step after that, we use a new set of nested tests to find the most useful predictor from among those available, and decide whether to add it to the model. If at some point we include a quantitative predictor such as *GPA*, at the next step we include *GPA2*, that is, the square of *GPA*, as a possible predictor; if at any point we have included a pair of terms, at the next step we include their interaction as a possible predictor.

First predictor? Our null model is $logit(\pi) = \beta_0$. To choose a first predictor, we carry out three nested tests, one for each of *GPA*, *MCAT*, and *FEM*:

	$-2\log(L)$			
Predictor	**Reduced**	**Full**	**Difference**	***P*-value**
GPA	75.791	56.839	18.952	1.34×10^{-5}
MCAT	75.791	64.697	11.094	8.66×10^{-4}
FEM	75.791	73.594	2.197	0.138

The biggest drop-in-deviance is for *GPA*, with a tiny *P*-value, so we add that predictor to our model to get a new reduced model, $logit(\pi) = \beta_0 + \beta_1 GPA$.

Second predictor? Now that *GPA* is in the model, we allow GPA^2 as a possible predictor. We have three new nested LRTs, one each for *MCAT*, *FEM*, and GPA^2:

	Additional	$-2\log(L)$			
	Predictor	**Reduced**	**Full**	**Difference**	***P*-value**
GPA +	*MCAT*	56.839	54.014	2.825	0.090
	FEM	56.839	53.945	2.894	0.089
	GPA^2	56.839	55.800	1.039	0.308

Clearly, we don't want to add GPA^2 at this point. Even though the *P*-values for *MCAT* and *FEM* are not tiny, they are small enough to make it worth seeing what happens with one or both of them in the model. Which one to include is somewhat arbitrary, but because the *MCAT* was designed to predict success in medical school, choosing to include *MCAT* is reasonable on the basis of context. This choice gives a new reduced model, $logit(\pi) = \beta_0 + \beta_1 GPA + \beta_2 MCAT$.

Third predictor? With *GPA* and *MCAT* in the model, we now expand our list of possible predictors to include both quadratic terms GPA^2 and $MCAT^2$ and the interaction $GPA \cdot MCAT$, along with *FEM*. The drop-in-deviance tests reveal a mild surprise:

Additional	$-2\log(L)$			
Predictor	**Reduced**	**Full**	**Difference**	***P*-value**
GPA^2	54.014	53.235	0.779	0.377
$MCAT^2$	54.014	54.005	0.009	0.925
$GPA \cdot MCAT$	54.014	53.186	0.828	0.363
FEM	54.014	50.786	3.228	0.072

The *P*-value for *FEM* is lower with *MCAT* in the model than it was with *MCAT* *not* in the model. None of the other possible predictors are worth including, but adding *FEM* to the model seems reasonable. As a check, we can also use a nested LRT to compare the resulting model $logit(\pi) = \beta_0 + \beta_1 GPA + \beta_2 MCAT + \beta_3 FEM$ with the null model $logit(\pi) = \beta_0$. The drop-in-deviance is $75.791 - 50.786 = 25.005$. The null model has 1 coefficient and the new full model has 4, so we refer 25.005 to a chi-square on 3 df to get a *P*-value, which turns out to be 0.000015, which implies the full model is a significant improvement over the null model.

Fourth predictor? With all three of *GPA*, *MCAT*, and *FEM* in the reduced model, we have five possibilities for an additional predictor:

Additional **Predictor**	**Reduced**	$-2\log(L)$ **Full**	**Difference**	***P*-value**
GPA^2	50.786	49.580	1.207	0.272
$MCAT^2$	50.786	50.781	0.005	0.939
$GPA \cdot MCAT$	50.786	50.226	0.560	0.454
$FEM \cdot GPA$	50.786	50.702	0.084	0.771
$FEM \cdot MCAT$	50.786	48.849	1.937	0.164

An argument could be made that $FEM \cdot MCAT$ is worth including in the model. For example, the data suggest that female applicants may have a greater chance of acceptance than do males with the same *GPA* and *MCAT* score. If you want to pursue this, you might choose to explore models with interaction terms of *FEM* with other predictors. On the other hand, the *P*-value for $FEM \cdot MCAT$ is rather large, and for some purposes it makes sense not to include the term in the model, and instead to stop with the three-predictor model.

As the new material in this chapter ends, it is important to remind yourself: Statistical tests and *P*-values may come at the end of the chapter, but they are not the be-all and end-all of statistical modeling. Even without a single *P*-value, models can be useful for description and for prediction. Even if the conditions required for inference are not satisfied, *P*-values can still be useful as an informal guide. Finally, for formal tests and intervals that rely on a probability interpretation of *P*-values and confidence levels, the required conditions of randomness and independence must be satisfied. If the probability part of the model is wrong, there is no basis for the probability part of the conclusion.

10.5 Case Study: Attractiveness and Fidelity

Can you judge sexual unfaithfulness by looking at a person's face?

College students were asked to look at a photograph of an opposite-sex adult face and to rate the person, on a scale from 1 (low) to 10 (high), for attractiveness. They were also asked to rate trustworthiness, faithfulness, and sexual dimorphism (i.e., how masculine a male face is and how feminine a female face is). Overall, 68 students (34 males and 34 females) rated 170 faces (88 men and 82 women).

Faithful

The data[3] are in the file **FaithfulFaces**, which includes the variables *Attract*, *Trust*, *Faithful*, *SexDimorph*, *FaceSex* (M for males and F for females), and *RaterSex* (M for male raters and F for female raters). Figure 10.21 is a scatterplot of *Faithful* versus *Attract* with points coded by sex of the face being rated and with regression lines added. There is a significant interaction between *FaceSex* and *Attract*, so an analysis that includes both men and women would be complicated by that fact. We will analyze data for just the 82 women (leaving the male faces data for an exercise).

Let's begin by looking at relationships among some of the possible predictor variables. Figure 10.22 shows a scatterplot matrix. We can see that assessment of faithfulness (*Faithful*) is negatively associated with being attractive (*Attract*) and being feminine (*SexDimorph*), and positively associated with trustworthiness (*Trust*). There is also a strong, positive association between *Attract* and *SexDimorph*.

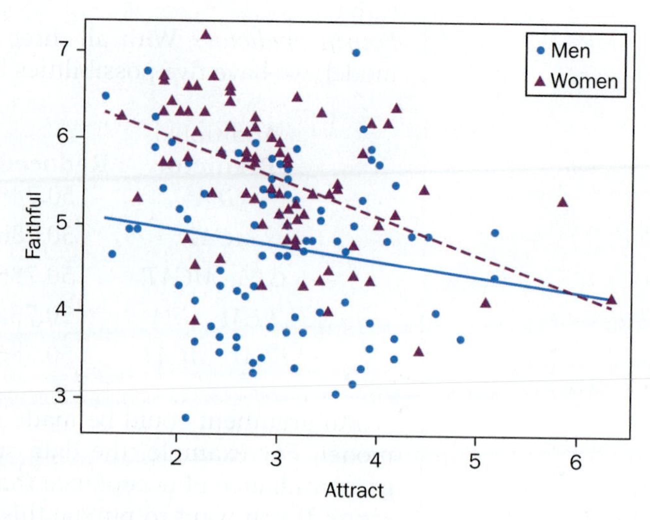

FIGURE 10.21 Faithfulness rating is related to attractiveness rating, but the relationship is different for women (a steeper line) than for men (a more shallow line)

If we want to predict *Faithful*, then we might consider using *Attract*, *SexDimorph*, and *Trust* together in an ordinary (not logistic) multiple regression model. The output shown below suggests that *SexDimorph* doesn't add much to the predictive power of *Attract* and *Trust*. Given how strongly *SexDimorph* is correlated with *Attract*, it is not surprising that only one of them is needed in a multiple regression model.

```
Coefficients:
             Estimate Std.  Error    t value  Pr(>|t|)
(Intercept)   4.5957        0.3998   11.50    < 2e-16 ***
Attract      -0.5058        0.1203   -4.21    6.9e-05 ***
SexDimorph   -0.0938        0.1058   -0.89    0.38
Trust         0.6330        0.0857    7.38    1.5e-10 ***
---
Residual standard error: 0.547 on 78 degrees of freedom
Multiple R-squared: 0.552, Adjusted R-squared: 0.535
F-statistic: 32.1 on 3 and 78 DF, p-value: 1.31e-13
```

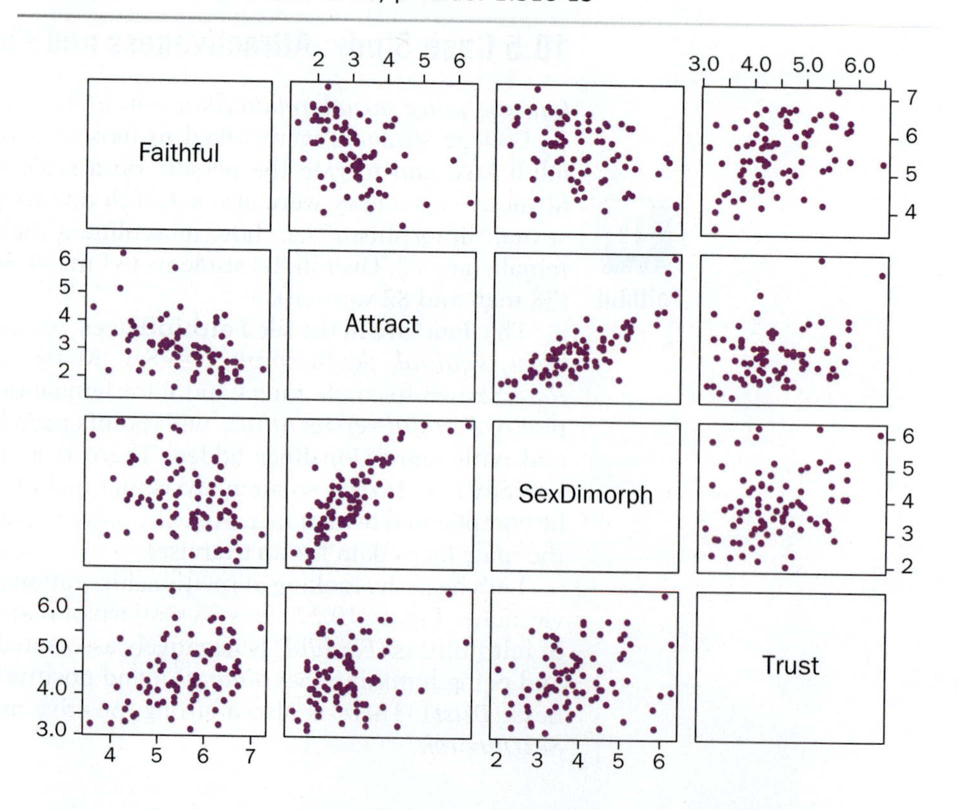

FIGURE 10.22 Scatterplot matrix of ratings given to photographs of faces of 82 women

When we pare things down we end up with two highly significant predictors of *Faithful*. The more attractive a woman is, the less likely she is judged to be faithful. The more trustworthy a woman is deemed to be, the more likely she is judged to be faithful.

```
Coefficients:
             Estimate Std. Error  t value  Pr(>|t|)
(Intercept)   4.4899     0.3810   11.78    < 2e-16 ***
Attract      -0.5917     0.0712   -8.31    2.1e-12 ***
Trust         0.6318     0.0856    7.38    1.4e-10 ***
—
Residual standard error: 0.546 on 79 degrees of freedom
Multiple R-squared: 0.548, Adjusted R-squared: 0.536
F-statistic: 47.8 on 2 and 79 DF, p-value: 2.45e-14
```

But are attractive women actually less faithful? And are women who are seen as trustworthy actually more faithful? The women who were rated had reported whether or not they had ever been unfaithful to a partner, so the researchers were able to check on the accuracy of the ratings. This led to the creation of the variable *Cheater*, which is 1 for those who had been unfaithful and 0 for others.

CHOOSE

Before fitting logistic regression models, let's explore the data using boxplots. The boxplots in Figure 10.23(a) show that the distribution of faithfulness ratings is pretty much the same whether or not a woman has actually cheated, as are the ratings for attractiveness. The biggest difference appears to be in trustworthiness, where cheaters tend to get lower ratings than noncheaters.

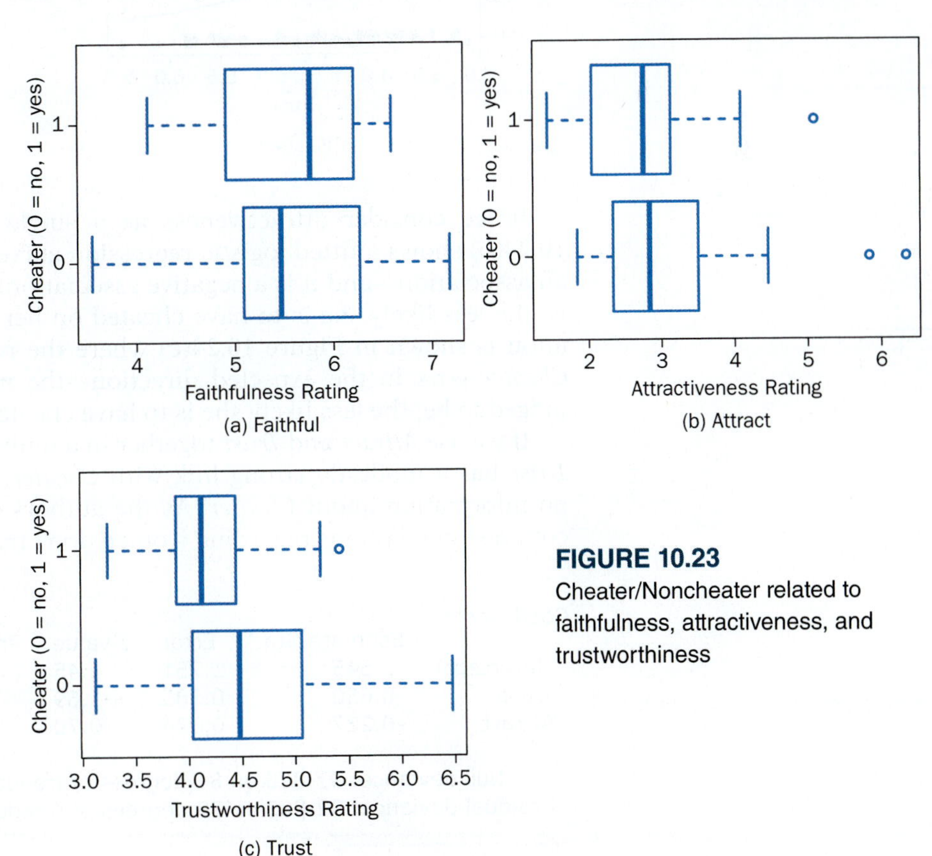

FIGURE 10.23 Cheater/Noncheater related to faithfulness, attractiveness, and trustworthiness

FIT

Now let's use logistic regression to see how well we can predict actual unfaithfulness (*Cheater* = 1). One might think that *Faithful* would be a useful predictor of *Cheater*, but the slope between the two variables is almost exactly zero, which is what the boxplots in Figure 10.23(a) suggested. Figure 10.24(a) shows a jittered plot of *Cheater* against *Faithful*, with the fitted logistic regression curve added. The logistic curve is essentially a flat line in this case.

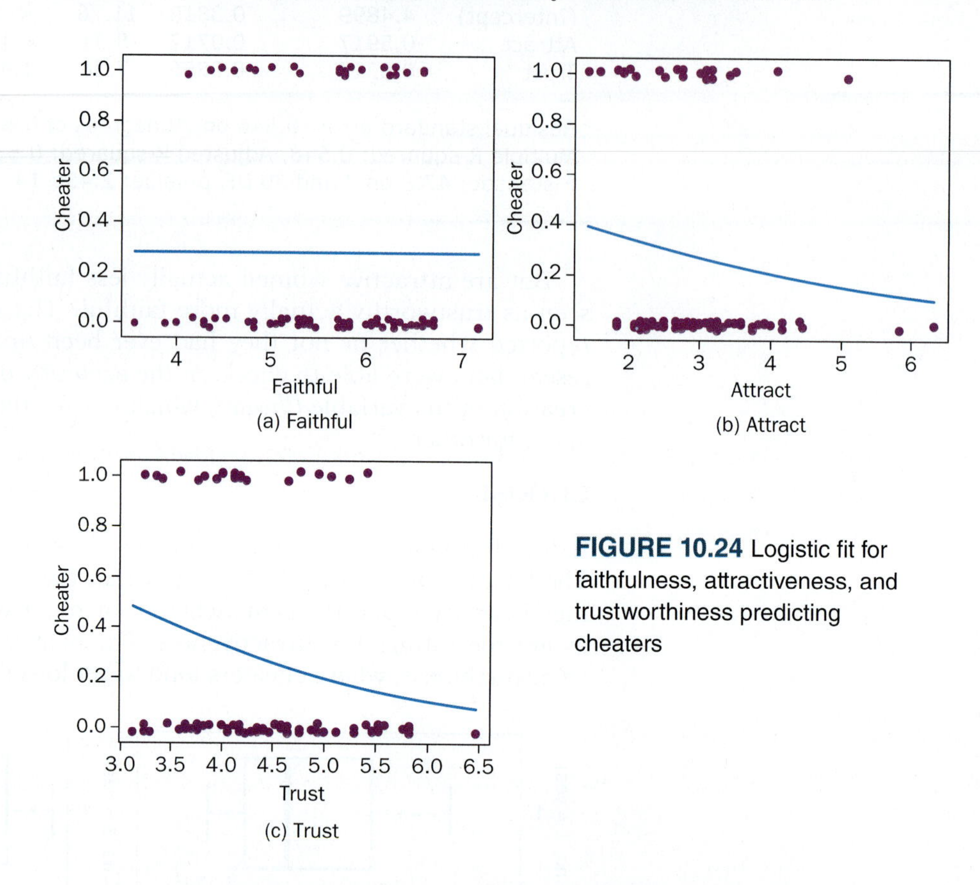

FIGURE 10.24 Logistic fit for faithfulness, attractiveness, and trustworthiness predicting cheaters

If we consider attractiveness as a single predictor of *Cheater*, Figure 10.24(b) shows a fitted logistic regression curve with only a very weak hint of an association—and it is a negative association! The more attractive a woman is, the less likely she is to have cheated on her partner. The strongest association is shown in Figure 10.24(c) where the relationship between *Trust* and *Cheater* goes in the expected direction—the more trustworthy a woman is judged to be, the less likely she is to have cheated on her partner.

If we use *Attract* and *Trust* together in a multiple logistic model, we see that *Trust* has a modestly strong link with *Cheater*, while *Attract* adds essentially no information about *Cheater*. As the authors of the research paper say, the common perceptions regarding women don't match the truth.

```
Coefficients:
            Estimate Std.  Error  z value  Pr(>|z|)
(Intercept)   2.545        1.751    1.45     0.146
Trust        -0.650        0.385   -1.69     0.091
Attract      -0.227        0.324   -0.70     0.483
---
     Null deviance: 97.320 on 81 degrees of freedom
Residual deviance: 92.867 on 79 degrees of freedom
```

ASSESS

We have been led to a model that has *Trust* as a single predictor of *Cheater*. Here is some output for fitting that model.

```
Coefficients:
              Estimate Std.  Error   z value   Pr(>|z|)
 (Intercept)   2.136         1.617    1.32     0.187
 Trust        -0.706         0.373   -1.89     0.059 .
 ---
     Null deviance: 97.320 on 81 degrees of freedom
 Residual deviance: 93.378 on 80 degrees of freedom
```

The z-test given in the output is not completely reliable, so we can conduct an LRT. The drop-in-deviance is $97.320 - 93.378 = 3.942$, which we compare to a chi-square distribution with one degree of freedom: $P(\chi^2_1 > 3.942) = 0.047$. Thus the P-value is 0.047, which is not much different for the z-test and indicates some modest, but not overly convincing evidence of the effectiveness of *Trust* as a predictor of *Cheater*.

As a next step, we explore whether adding a quadratic term improves the model. The output below shows that there is no statistically significant quadratic effect.

```
Coefficients:
              Estimate Std.  Error   z value   Pr(>|z|)
 (Intercept)  -9.228        10.196   -0.91     0.37
 Trust         4.564         4.708    0.97     0.33
 Trustsq      -0.597         0.537   -1.11     0.27
 ---
     Null deviance: 97.320 on 81 degrees of freedom
 Residual deviance: 91.963 on 79 degrees of freedom
```

In our final model (based on *Trust* alone), we see that the coefficient of *Trust* is -0.706. Thus we can say that as perceived trustworthiness goes up by 1 point, the odds of having cheated go down by a factor of $e^{-0.706} = 0.494$.

USE

If $Trust = 4$, then the predicted log(odds) of *Cheater* are $2.136 - 0.706(4) = -0.688$; the odds of having cheated are $e^{-0.688} = 0.503$; the probability of having cheated is $e^{-0.688}/(1 + e^{-0.688}) = 0.503/1.503 = 0.335$.

If $Trust = 5$, then the predicted log(odds) of *Cheater* are $2.136 - 0.706(5) = -1.39$; the odds of having cheated are $e^{-1.39} = 0.249$; the probability of having cheated is $e^{-1.39}/(1 + e^{-1.39}) = 0.199$. (We could estimate these probabilities by reading the fitted logistic curve in Figure 10.24(c).)

CHAPTER SUMMARY

Our look at logistic regression reveals that a **binary response**, like a continuous normally distributed response, can be modeled as a function of one or more explanatory variables. Unlike a continuous normally distributed response, we model the logit—log(odds)—of a binary response as a linear function of explanatory variables. This approach provides consistency with the modeling approach that has been emphasized throughout this book. Where appropriate, logistic regression allows for inferences to be drawn: A *P*-value assesses whether the association between the binary response and explanatory variables is significant, and a confidence interval estimates the magnitude of the population odds ratio. Logistic regression can accommodate multiple explanatory variables, including both categorical and continuous explanatory variables. When we have a categorical predictor with multiple categories, we can use several indicator variables to include information from that predictor in a logistic regression model. Logistic regression extends the set of circumstances where you can use a model and does so within a familiar modeling framework.

EXERCISES

Conceptual Exercises

10.1 Risky youth behavior. In Exercise 9.47, we learned that Ryb and colleagues have reported[4] that smokers are more likely to ride with drivers who have been drinking (RDD):

> Injury-prone behaviors were more common among smokers than nonsmokers: riding with drunk driver (38% versus 13%). In multiple logistic regression models adjusting for demographics, SES (socio-economic status), and substance abuse, smoking revealed significantly higher odds ratios (OR) for riding with drunk driver (OR = 2.2).[5]

a. Based on the percentages, compute the odds for riding with a drinking driver for both smokers and nonsmokers.

b. Using the odds from (a), find the odds ratio for riding with a drinking driver for smokers versus nonsmokers.

c. Explain why this odds ratio differs from the given odds ratio of 2.2.

10.2 Backpack weights. In Exercise 9.46, we studied backpacks and back problems. Overall, females are more likely to have back problems than are males: 44% versus 18%. In a multiple logistic regression model that adjusts for *BackpackWeight* and *BodyWeight*, the odds ratio of back problems for females versus males is 4.24.

a. Based on the percentages, compute the odds for having back problems for both females and males.

b. Using the odds from (a), find the odds ratio for having back problems for females versus males.

c. Explain why this odds ratio differs from the given odds ratio of 4.24.

10.3 Empirical logits. Figure 10.25 shows a plot of empirical logits for a dataset with two predictors: *X*, a continuous variable, and *Group*, a categorical variable with two levels, 1 and 2. The circles are for *Group* = 1 and the triangles are for *Group* = 2. What model is suggested by this plot?

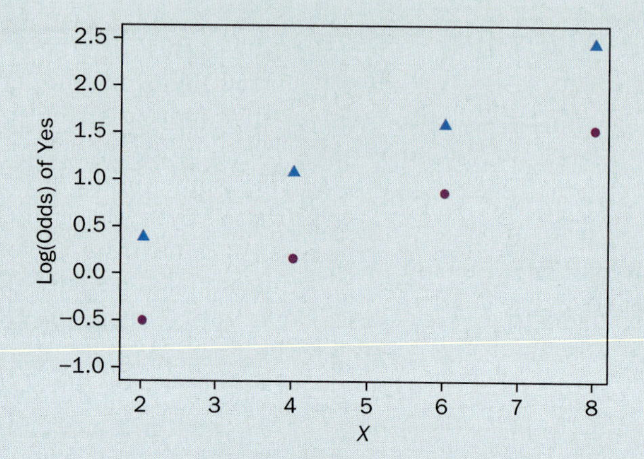

FIGURE 10.25 Empirical logit plot for Exercise 10.3

10.4 More empirical logits. Figure 10.26 shows plot of empirical logits for a dataset with two predictors: *X*, a continuous variable, and *Group*, a categorical variable with three levels, 1, 2, and 3. The circles are for *Group* = 1, the triangles are for *Group* = 2, and the squares are for *Group* = 3. What model is suggested by this plot?

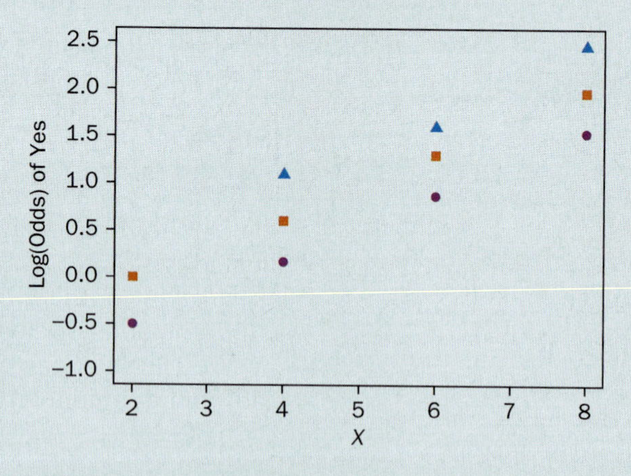

FIGURE 10.26 Empirical logit plot for Exercise 10.4

10.5 Empirical logits again. Figure 10.27 shows a plot of empirical logits for a dataset with two predictors: *X*, a continuous variable, and *Group*, a categorical variable with two levels, 1 and 2. The circles are for *Group* = 1 and the triangles are for *Group* = 2. What model is suggested by this plot?

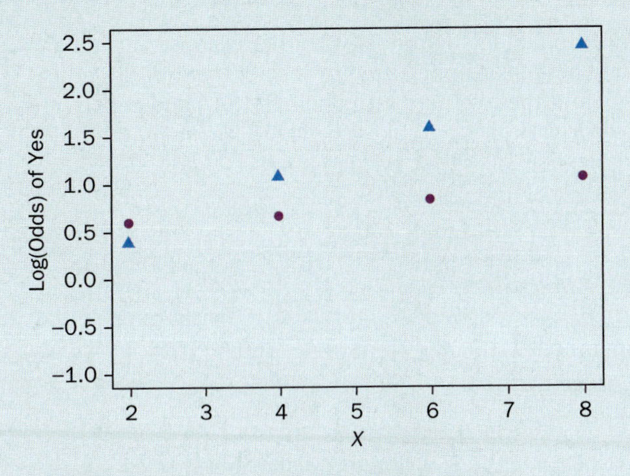

FIGURE 10.27 Empirical logit plot for Exercise 10.5

10.6 Yet more empirical logits. Figure 10.28 shows a plot of empirical logits for a dataset with two predictors: *X*, a continuous variable, and *Group*, a categorical variable with three levels, 1, 2, and 3. The circles are for *Group* = 1, the triangles are for *Group* = 2, and the squares are for *Group* = 3. What model is suggested by this plot?

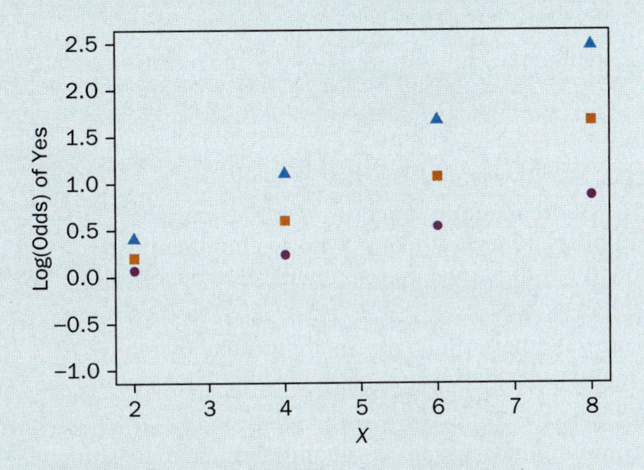

FIGURE 10.28 Empirical logit plot for Exercise 10.6

10.7 Transformations based on empirical logits. Figure 10.29 shows a plot of empirical logits. What transformation is suggested by this plot?

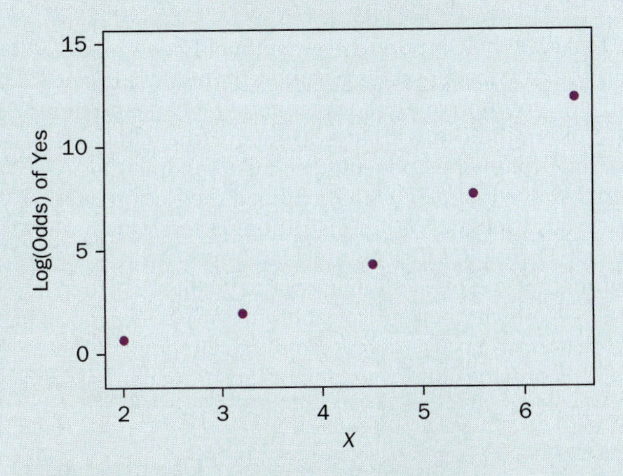

FIGURE 10.29 Empirical logit plot for Exercise 10.7

10.8 More transformations based on empirical logits. Figure 10.30 shows plot of empirical logits. What transformation is suggested by this plot?

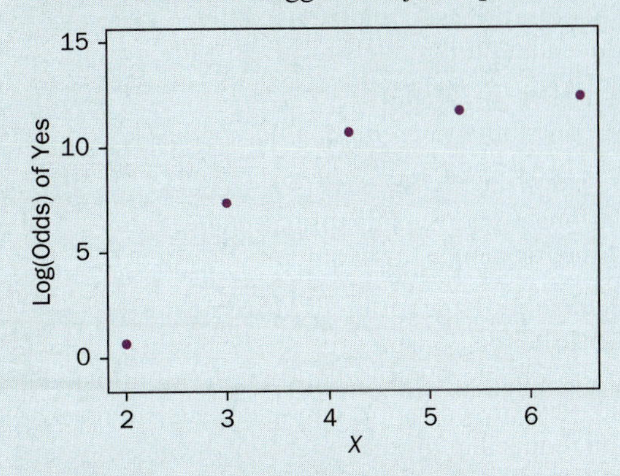

FIGURE 10.30 Empirical logit plot for Exercise 10.8

10.9 Model building. Figure 10.31 shows parallel boxplots. Suppose we want to model *Y* as depending on the levels of *A* and on the continuous variable *X*. What logistic model is suggested by this plot?

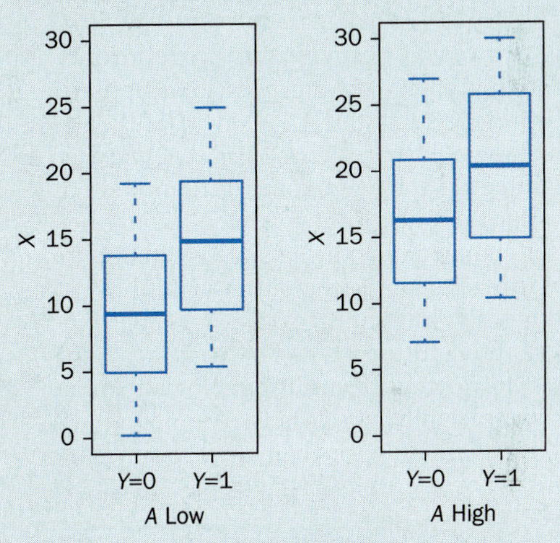

FIGURE 10.31 Parallel boxplots for Exercise 10.9

10.10 More model building. Figure 10.32 shows parallel boxplots. Suppose we want to model *Y* as depending on the levels of *A* and on the continuous variable *X*. What logistic model is suggested by this plot?

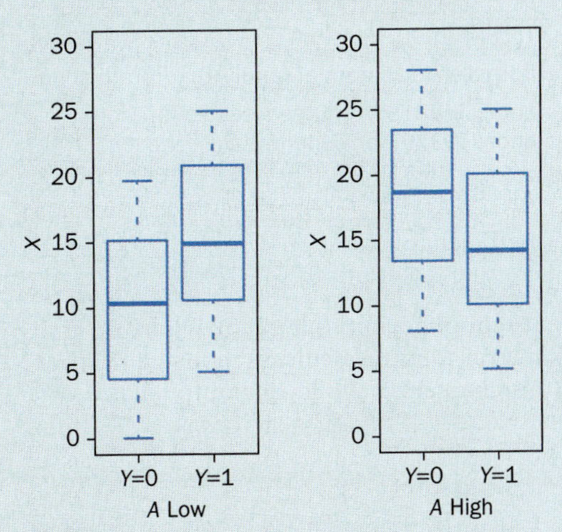

FIGURE 10.32 Parallel boxplots for Exercise 10.10

10.11 Medical school acceptance: Nested likelihood ratio tests. Consider the medical school data presented in this chapter. For each of the questions listed below, a nested likelihood ratio test could be used. In each case, state the reduced and full models for the likelihood ratio test. **MedGPA**

a. Does *GPA* contribute to predicting *Acceptance* beyond the effects of *MCAT* and *Sex*?

b. Is the relationship between logit(*Acceptance*) and *MCAT* the same for men as for women?

10.12 Medical school acceptance: More nested likelihood ratio tests. Consider the medical school data presented in this chapter. For each of the questions listed below, a nested likelihood ratio test could be used. In each case, state the reduced and full models for the likelihood ratio test. **MedGPA**

a. Does *MCAT* contribute to predicting *Acceptance* beyond the effects of *GPA* and *Sex*?

b. Is the relationship between logit(*Acceptance*) and *GPA* the same for men as for women when controlling for the effect of *MCAT*?

10.13 Putting prowess: Coefficient interpretation. In Chapter 9 we studied the putting prowess of a statistician and found a fitting logistic model that says that the log(odds) of making a putt are $3.26 - 0.57 Length$, where *Length* (given in feet) is the length of the attempted putt.

Sometime after the **Putts1** data were collected, the statistical golfer changed his putting style, switching from the traditional way of gripping the putter, with the left hand higher than the right hand, to a new grip with the left hand lower on the putter.

The output below comes from fitting a logistic model with the response variable being whether or not a putt was made and the two predictors being *Length* and *LeftHandLow* (Low or High).

```
Coefficients:
              Estimate Std.  Error z value  Pr(>|z|)
(Intercept)    3.2262 0.3513    9.18 <2e-16 ***
Length        -0.5604 0.0641   -8.75 <2e-16 ***
LeftHandLow    0.4616 0.2642    1.75 0.081 .
—
Signif. codes: 0 '***' 0.001 '**' 0.01 '*' 0.05 '.' 0.1 ' ' 1
(Dispersion parameter for binomial family taken to be 1)
    Null deviance: 90.2157 on 9 degrees of freedom
Residual deviance: 2.3792 on 7 degrees of freedom
```

Interpret the coefficient of 0.46 for *LeftHandLow*, in the context of this setting. **Putts1**

10.14 Medical school acceptance: Coefficient interpretation. Consider the medical school data presented in this chapter. The following output comes from fitting a logistic model with the response variable being whether or not a student was accepted into medical school and the two predictors being *MCAT* and *Sex_F* (an indicator variable that takes the value 1 for females and 0 for males).

Coefficients

Term	Coef SE	Coef	VIF
Constant	-10.02	3.50	
MCAT	0.2669	0.0941	1.05
Sex_F	1.066	0.633	1.05

Interpret the coefficient of 1.066 for *Sex_F*, in the context of this setting. **MedGPA**

10.15 Putting prowess: Estimated probability. Refer to the model for estimating the probability of making a putt shown in Exercise 10.13. **Putts1**

a. Compute the estimated probability that the statistical golfer will make a putt that is 5 feet using the "low left hand" grip.

b. Compute the estimated probability that the statistical golfer will make a putt that is 5 feet using his original grip.

c. Are your answers to parts (a) and (b) consistent with your answer to Exercise 10.13?

10.16 Medical school acceptance: Estimated probability. Refer to the model for estimating the probability of making a putt shown in Exercise 10.14. **MedGPA**

a. Compute the estimated probability that a female student with an MCAT score of 40 will be accepted into medical school.

b. Compute the estimated probability that a male student with an MCAT score of 40 will be accepted into medical school.

c. Are your answers to parts (a) and (b) consistent with your answer to Exercise 10.14?

Guided Exercises

10.17 CAFE. The Corporate Average Fuel Economy (CAFE) bill was proposed by Senators John McCain and John Kerry to improve the fuel economy of cars and light trucks sold in the United States. However, the bill was, in effect, indefinitely postponed when an amendment was passed, by a vote of 62–38, that charged the National Highway Traffic Safety Administration to develop a new standard. The file **CAFE** contains data on this vote, including the vote of each senator (*Vote*, which is 1 for Yes and 0 for No), whether or not each senator is a Democrat (*Dem*, which is 1 for Democrats and 0 for Republicans),* monetary contributions that each of the 100 senators

* Jim Jeffords, an Independent, caucused with the Democrats.

received over his or her lifetime from car manufacturers (*Contribution*), and *logContr*, which is the natural log of (1 + Contribution).[6] As we might expect, there is a strong positive association between contributions from car manufacturers and voting Yes on the CAFE amendment. But is the effect the same for Democrats as for Republicans? That is, we wonder whether the slope between logit(*Vote*) and *logContr* is the same for the two parties. **CAFE**

a. Fit the model in which logit(*Vote*) depends on *logContr* and on party, allowing for different coefficients of *logContr* for Democrats (*Dem*) and for Republicans. Use a Wald z-test to check for a difference. What is the P-value?

b. Repeat part (a) but this time use nested models and the drop-in-deviance test (the nested likelihood ratio test). What is the P-value?

c. How do the P-values from parts (a) and (b) compare? If they are the same, why is that? If they are different, why are they different?

10.18 Eyes.
Consider the **Eyes** data presented in this chapter, where we considered the model of logit(*Gay*) depending on *dilateDiff* and *Sex*. Now we want to ask whether the slope between logit(*Gay*) and *dilateDiff* is the same for the two sexes. **Eyes**

a. Fit the model in which logit(*Gay*) depends on *dilateDiff* and on *Sex*, allowing for different coefficients of *dilateDiff* for males and for females. Use a Wald z-test to check for a difference. What is the P-value?

b. Repeat part (a) but this time use nested models and the drop-in-deviance test (the nested likelihood ratio test). What is the P-value?

c. How do the P-values from parts (a) and (b) compare? If they are the same, why is that? If they are different, why are they different?

10.19 Blue jay morphology: Sex and bill characteristics. Biologists took several measurements of the heads of blue jays. Among the variables measured were *BillLength*, *BillWidth*, and *BillDepth* (where *BillDepth* is the distance between the upper surface of the upper bill and the lower surface of the lower bill, measured at the nostril). All measurements were in millimeters. The data are in the file **BlueJays**.[7] We want to study the relationship between sex (coded as M/F in the variable KnownSex and as 1/0 in the variable Sex) and measurements of the blue jays. **BluJay**

a. Make parallel boxplots of *BillLength* by *KnownSex*, *BillWidth* by *KnownSex*, and *BillDepth* by *KnownSex*. Which of these three predictors has the weakest relationship with *KnownSex*? Which has the strongest relationship?

b. Fit a multiple logistic regression model on Sex depending on *BillLength*, *BillWidth*, and *BillDepth*. Which predictor has the largest P-value?

c. Fit a simple logistic regression model of Sex depending on the predictor from part (b) that has the weakest relationship with Sex. What is the P-value for the coefficient of that predictor?

d. Comment on the results of parts (b) and (c). Why are the P-values so different between parts (b) and (c)?

10.20 Blue jay morphology: Sex and size. The dataset in the file **BlueJays** includes *Head* (head length, in mm) and *Mass* (in gm) for a sample of blue jays. We want to study the relationship between sex (coded as M/F in the variable *KnownSex* and as 1/0 in the variable *Sex*) and these measurements. **BluJay**

a. Make parallel boxplots of *Head* by *KnownSex* and *Mass* by *KnownSex*. Which of these two predictors has the strongest relationship with *KnownSex*?

b. Fit a simple logistic regression model on *Sex* depending on *Mass*. What is the P-value for testing that *Mass* has no predictive power?

c. Fit a multiple logistic regression model of *Sex* depending on *Mass* and *Head*. Which is the P-value for the coefficient of *Mass* in the multiple logistic model?

10.21 Medical school acceptance: GPA and sex. Consider the medical school data presented in this chapter. Is the relationship between logit(*Acceptance*) and *GPA* the same for men as for women? That is, is the slope between logit(*Acceptance*) and *GPA* the same for the two sexes? **MedGPA**

a. Fit the model in which logit(*Acceptance*) depends on *GPA* and on *Sex*, allowing for different coefficients of *GPA* for men and for women. Use a Wald z-test to check for a difference. What is the P-value?

b. Repeat part (a) but this time use nested models and the drop-in-deviance test (the nested likelihood ratio test). What is the P-value?

c. How do the P-values from parts (a) and (b) compare? If they are the same, why is that? If they are different, why are they different?

10.22 Medical school acceptance: MCAT and sex. Consider the medical school data presented in this chapter. Is the relationship between logit(*Acceptance*) and *MCAT* the same for men as for women? That is, is the slope between logit(*Acceptance*) and *MCAT* the same for the two sexes? **MedGPA**

a. Fit the model in which logit(*Acceptance*) depends on *MCAT* and on *Sex*, allowing for different coefficients of *MCAT* for men and for women. Use a Wald z-test to check for a difference. What is the P-value?

b. Repeat part (a) but this time use nested models and the drop-in-deviance test (the nested likelihood ratio test). What is the P-value?

c. How do the P-values from parts (a) and (b) compare? If they are the same, why is that? If they are different, why are they different?

10.23 Sinking of the *Titanic*. In Exercises 9.27–9.32, we considered data on the passengers who survived and those who died when the oceanliner *Titanic* sank on its maiden voyage in 1912. The dataset in **Titanic** includes the following variables: **Titanic**

Age	which gives the passenger's age in years
Sex	which gives the passenger's sex (male or female)
Survived	a binary variable, where 1 indicates the passenger survived and 0 indicates death
SexCode	which numerically codes male as 0 and female as 1

a. In Exercises 9.27–9.31, you fit separate logistic regression models for the binary response *Survived* using *Age* and then *SexCode*. Now fit a multiple logistic model using these two predictors. Write down both the logit and probability forms for the fitted model.

b. Comment on the effectiveness of each of the predictors in the two-predictor model.

c. According to the fitted model, estimate the probability and odds that an 18-year-old man would survive the *Titanic* sinking.

d. Repeat the calculations for an 18-year-old woman and find the odds ratio compared to a man of the same age.

e. Redo both (b) and (c) for a man and woman of age 50.

f. What happens to the odds ratio (female to male of the same age) when the age increases in the *Titanic* data? Will this always be the case?

10.24 Sinking of the *Titanic*: Add interaction term. Refer to the situation described in Exercise 10.23. Perhaps the linear relationship between log(odds) of survival and *Age* is much different for women than for men. Add an interaction term to the two-predictor model based on *Age* and *SexCode*. **Titanic**

a. Explain how the coefficients in the model with *Age*, *SexCode*, and *Age* · *SexCode* relate to separate linear models for males and females to predict log(odds) of survival based on *Age*.

b. Is this model a significant improvement over one that uses just the *SexCode* variable (and not *Age*)? Justify your answer by showing the details of a nested likelihood ratio test for the two terms involving *Age*.

10.25 Red states or blue states in 2008. In Exercises 9.36 and 9.37, you considered some state-level variables for the data in **Election08** to model the probability that the Democratic candidate, Barack Obama, won a state (response = *ObamaWin*). Among the potential predictors were the per capita *Income*, percentages of adults with high school (*HS*) or college (*BS*) degrees, and a measure (*Dem.Rep*) of the %Democrat−%Republican leaning in the state. **Elect08**

a. Fit a logistic model with *ObamaWin* as the response and *Dem.Rep*, *HS*, *BA*, and *Income* as the predictors. Which predictor has the strongest relationship with the response in this model?

b. Consider the model from part (a). Which predictors (if any) are not significantly related to *ObamaWin* in that model?

c. Identify the state that has a very large positive deviance (residual) and the state that has a very large negative deviance (residual). (*Note*: Deviance residuals are available in computer output. They can be treated similarly to residuals in the regression setting; we discuss them further in the next chapter.)

d. Consider applying a backward elimination process, starting with the four-predictor model in part (a). At each "step," find the least significant predictor, eliminate it, and refit with the smaller model—unless the worst predictor is significant, say, at a 10% level, in which case you stop and call that your final model. Describe what happens at each step in this situation.

10.26 Red states or blue states in 2016. The file **Election16** contains data from 50 states related to the 2016 U.S. presidential election. One variable is *TrumpWin*, which is 1 for the states won by Donald Trump, the Republican candidate, and 0 for states that Trump lost. Among the potential predictors are the per capita *Income*, percentages of adults with high school (*HS*), or college (*BS*) degrees, and a measure (*Dem.Rep*) of the %Democrat−%Republican leaning in the state. **Elect16**

a. Fit a logistic model with *TrumpWin* as the response and *Dem.Rep*, *HS*, *BA*, and *Income* as the predictors. Which predictor has the strongest relationship with the response in this model?

b. Consider the model from part (a). Which predictors (if any) are not significantly related to *TrumpWin* in that model?

c. Identify the state that has the largest positive deviance (residual) and the state that has the largest negative deviance (residual). (*Note*: Deviance residuals are available in computer output. They can be treated similarly to residuals in the regression setting; we discuss them further in the next chapter.)

d. Consider applying a backward elimination process, starting with the four-predictor model in part (a). At each "step," find the least significant predictor, eliminate it, and refit with the smaller model—unless the worst predictor is significant, say, at a 10% level, in which case you stop and call that your final model. Describe what happens at each step in this situation.

10.27 Nikkei 225. In Exercise 9.35, you considered models to predict whether the Japanese Nikkei 225 stock index would go up ($Up = 1$) or down ($Up = 0$) based on the previous day's change in either the New York–based Dow Jones Industrial Average (*DJIAch*) or previous Nikkei 225 change (*lagNik*). The data are in **Markets**. **Market**

a. Run a multiple logistic regression model to use both *DJIAch* and *lagNik* to predict *Up* for the Nikkei 225 the next day. Comment on the importance of each predictor in the model and assess the overall fit.

b. Suppose we had a remarkable occurrence and *both* the Nikkei 225 and Dow Jones Industrial average were unchanged one day. What would your fitted model say about the next day's Nikkei 225?

c. Based on the actual data, should we be worried about multicollinearity between the *DJIAch* and *lagNik* predictors? (*Hint*: Consider a correlation coefficient and a plot.)

10.28 Leukemia treatments. Refer to Exercise 9.41 that describes data in **Leukemia** that arose from a study of 51 patients treated for a form of leukemia. The first six variables in that dataset all measure pretreatment variables: *Age*, *Smear*, *Infil*, *Index*, *Blasts*, and *Temp*. Fit a multiple logistic regression model using all six variables to predict *Resp*, which is 1 if a patient responded to treatment and 0 otherwise. **Leukem**

a. Based on values from a summary of your model, which of the six pretreatment variables appear to add to the predictive power of the model, given that other variables are in the model?

b. Specifically, interpret the relationship (if any) between *Age* and *Resp* and also between *Temp* and *Resp* indicated in the multiple model.

c. If a predictor variable is nonsignificant in the fitted model here, might it still be possible that it should be included in a final model? Explain why or why not.

d. Despite your answer above, sometimes one gets lucky, and a final model is, simply, the model that includes all "significant" variables from the full additive model output. Use a nested likelihood ratio (drop-in-deviance) test to see if the model that excludes precisely the nonsignificant variables seen in (a) is a reasonable choice for a final model. Also, comment on the stability of the estimated coefficients between the full model from (a) and the reduced model without the "nonsignificant" terms.

e. Are the estimated coefficients for *Age* and *Temp* consistent with those found in the Exercise 9.41 using these data? Consider both statistical significance and the value of the estimated coefficients.

10.29 Intensive care unit: Quantitative predictors. The data in **ICU**[8] show information for a sample of 200 patients who were part of a larger study conducted in a hospital's Intensive Care Unit (ICU). Since an ICU often deals with serious, life-threatening cases, a key variable to study is patient survival, which is coded in the *Survive* variable as 1 if the patient lived to be discharged and 0 if the patient died. Among the possible predictors of this binary survival response are the following:

Age	= age (in years)
AgeGroup	= 1 if young (under 50), 2 if middle (50-69), 3 if old (70+)
Sex	= 1 for female, 0 for male
Infection	= 1 if infection is suspected, 0 if no infection
SysBP	= systolic blood pressure (in mm of Hg)
Pulse	= heart rate (beats per minute)
Emergency	= 1 if emergency admission, 0 if elective

Consider a multiple logistic regression model for *Survive* using the three quantitative predictors in the dataset *Age*, *SysBP*, and *Pulse*. **ICU**

a. After running the three-predictor model, does it appear as though any of the three quantitative variables are not very helpful in this model to predict survival rates in the ICU? If so, drop one or more of the predictors and refit the model before going on to the next part.

b. The first person in the dataset (ID #4) is an 87-year-old man who had a systolic blood pressure of 80 and a heart rate of 96 beats per minute when he checked into the ICU. What does your final model from part (a) say about this person's chances of surviving his visit to the ICU?

c. The patient with ID #4 survived to be discharged from the ICU ($Survive = 1$). Based on your answer to (b), would you say that this result (this patient surviving) was very surprising, mildly surprising, reasonably likely, or very likely?

10.30 Intensive care unit: Binary predictors. Refer to the **ICU** data on survival in a hospital ICU as described in Exercise 10.29. The dataset has three binary variables that could be predictors of ICU survival: *Sex*, *Infection*, and *Emergency*. **ICU**

a. First, consider each of the three binary variables individually as predictors in separate simple logistic regression models for *Survive*. Comment on the effectiveness of each of *Sex*, *Infection*, and *Emergency* as predictors of ICU survival on their own.

b. A nice feature of the multiple linear model for the log(odds) is that we can easily use several binary predictors in the same model. Do this to fit the three-predictor model for *Survive* using *Sex*, *Infection*, and *Emergency*. How does the effectiveness of each predictor in the multiple model compare to what you found when you considered each individually?

c. The first person in the dataset (ID #4) is an 87-year-old man who had an infection when he was admitted to the ICU on an emergency basis. What does the three-predictor model in part (b) say about this person's chances of surviving his visit to the ICU?

d. The patient with ID #4 survived to be discharged from the ICU (*Survive* = 1). Based on your answer to (c), would you say that this result (this patient surviving) was very surprising, mildly surprising, reasonably likely, or very likely?

e. Does the model based on the three binary predictors in this exercise do a better job of modeling survival rates than the model based on three quantitative predictors that you used at the start of Exercise 10.29? Give a justification for your answer.

10.31 Sinking of the *Titanic*: Categorical predictor. The **Titanic** data considered in Exercises 10.23 and 10.24 also contain a variable identifying the travel class (1st, 2nd, or 3rd) for each of the passengers. **Titanic**

a. Create a 2 × 3 table of *Survived* (Yes or No) by the three categories in the passenger class (*PClass*). Find the proportion surviving in each class. Make a conjecture in the context of this problem about why the proportions behave the way they do.

b. Use a chi-square test for the 2 × 3 table to see whether there is a significant relationship between *PClass* and *Survive*.

c. Create indicator variables (or use *PClass* as a factor) to run a logistic regression model to predict *Survived* based on the categories in *Pclass*. Interpret each of the estimated coefficients in the model.

d. Verify that the predicted probability of survival in each passenger class based on the logistic model matches the actual proportion of passengers in that class who survived.

e. Compare the test statistic for the overall test of fit for the logistic regression model to the chi-square statistic from your analysis of the two-way table. Are the results (and conclusion) similar?

10.32 Flight response of Pacific Brant. The data in Exercise 9.33 dealt with an experiment to study the effects of nearby helicopter flights on the flock behavior of Pacific Brant. The binary response variable is *Flight*, which is coded as 1 if more than 10% of the flock flies away and 0 if most of the flock stays as the helicopter flies by. The predictor variables are the *Altitude* of the helicopter (in 100 m) and the *Lateral* distance from the flock (also in 100 m). The data are stored in **FlightResponse**. **FltResp**

a. Fit a two-predictor logistic model with *Flight* modeled on *Altitude* and *Lateral*. Does this model indicate that both variables are related to *Flight*, controlling for the other variable? Give the fitted model and then give two interpretive statements: one about the relationship of *Flight* to *Altitude* and the other about the relationship of *Flight* to *Lateral*. Incorporate the estimated slope coefficients in each statement.

b. The model fit in (a), the "additive model," assumes that the relationship between, say, flight response and lateral distance is the same for each level of altitude. That is, it assumes that the slope of the model regarding the *Flight*-versus-*Lateral* relationship does not depend on altitude. Split your dataset into three subsets using the *AltCat* variable. For each subset, fit a logistic model of *Flight* on *Lateral*. Report your results. How, if at all, does the relationship between *Flight* and *Lateral* distance appear to depend on the altitude?

c. The lack of "independence" found in (b) suggests the existence of an interaction between *Altitude* and *Lateral* distance. Fit a model that includes this interaction term. Is this term significant?

d. Give an example of an overflight from the data for which using the model from (a) and the model from (c) would give clearly differing conclusions for the estimated probability of a flight response.

10.33 March Madness. Each year, 64 college teams are selected for the NCAA Division I Men's Basketball tournament, with 16 teams placed in each of four regions. Within each region, the teams are seeded from 1 to 16, with the (presumed) best team as the #1 seed and the (presumed) weakest team as the #16 seed; this practice of seeding teams began in 1979 for the NCAA tournament. Only one team from each region advances to the Final Four, with #1 seeds having advanced to the Final Four 63 times out of 156 possibilities during the years 1979–2017. Of the 156 #2 seeds, only 33 have made it into the Final Four; the lowest seed to ever make a Final Four was a #11 (this has happened three times). The file **FinalFourLong17** contains data on *Year* (1979–2017), *Seed* (1–16), and *Final*4 (1 or 0 depending on whether a team did or did not make it into the Final Four). The file **FinalFourShort17** contains the same information in more compact form, with *Year*, *Seed*, *In* (counting the number of teams that got into the Final Four at a given *Seed* level and *Year*), and *Out* (counting the number of teams that did not get into the Final Four at a given *Seed* level and *Year*). We are interested in the relationship between *Seed* and *Final*4—or equivalently, between *Seed* and *In*/*Out*. **Long17** **Short17**

a. Fit a logistic regression model that uses *Seed* to predict whether or not a team makes it into the Final Four. Is there a strong relationship between these variables?

b. Fit a logistic regression model that uses *Seed* and *Year* to predict whether or not a team makes it into the Final Four. In the presence of *Seed*, what is the effect of *Year*? Justify your answer.

10.34 March Madness and Tom Izzo. Consider the Final Four data from Exercise 10.33. A related dataset is **FinalFourIzzo17**, which includes an indicator variable that is 1 for teams coached by Tom Izzo at Michigan State University and 0 for all other teams. **Izzo17**

a. Fit a logistic regression model to use *Seed* and the *Izzo* indicator to predict whether or not a team makes it into the Final Four.

b. Is the *Izzo* effect consistent with chance variation, or is there evidence that Izzo-coached teams do better than expected?

10.35 Alito Confirmation vote. When someone is nominated to serve on the Supreme Court, the Senate holds a confirmation hearing and then takes a vote, with senators usually voting along party lines. However, senators are also influenced by the popularity of the nominee within the senator's state. The file **AlitoConfirmation** has data from for each of the 50 states on *StateOpinion*, which is a measure of the level of support within the state[9] for Samuel Alito when he was nominated. The variable *Party* is D for Democrats and R for Republicans. Vote is 1 if the senator voted in favor of conformation and 0 if not. **Alito**

a. Conduct a logistic regression of *Vote* on *StateOpinion*. What does the fitted model say about the relationship between the two variables?

b. Conduct a regression of *Vote* on *StateOpinion* and *Party*. Are *StateOpinion* and *Party* useful predictors of *Vote*? Which predictor has a stronger effect?

10.36 Thomas Confirmation vote. When someone is nominated to serve on the Supreme Court, the Senate holds a confirmation hearing and then takes a vote, with senators usually voting along party lines. However, senators are also influenced by the popularity of the nominee within the senator's state. The file **ThomasConfirmation** has data from each of the 50 states on *StateOpinion*, which is a measure of the level of support within the state for Clarence Thomas when he was nominated. The variable *Party* is D for Democrats and R for Republicans. Vote is 1 if the senator voted in favor of confirmation and 0 if not. **Thomas**

a. Conduct a logistic regression of *Vote* on *StateOpinion*. What does the fitted model say about the relationship between the two variables?

b. Conduct a regression of *Vote* on *StateOpinion* and *Party*. Are *StateOpinion* and *Party* useful predictors of *Vote*? Which predictor has a stronger effect?

c. Fit an appropriate model to determine whether there is an interaction between *StateOpinion* and *Party*. How strong is the evidence of an interaction?

10.37 Health care vote. On November 7, 2009, the U.S. House of Representatives voted, by the narrow margin of 220–215, to enact health insurance reform. Most Democrats voted Yes, while all Republicans voted No. The file **InsuranceVote** contains data for each of the 435 representatives and records several variables: *Party* (D or R); congressional district; *InsVote* (1 for Yes, 0 for No); *Rep* and *Dem* (indicator variables for Party affiliation); *Private*, *Public*, and *Uninsured* [the percentages of non–senior citizens who have private health insurance, public health insurance (e.g., through the Veterans Administration), or no health insurance]; and *Obama* (1 or 0 depending on whether the congressional district voted for Barack Obama or John McCain in November 2008). **InsVote**

a. Fit a logistic regression model that uses *Uninsured* to predict *InsVote*. As the percentage of uninsured residents increases, does the likelihood of a Yes vote increase? Is this relationship stronger than would be expected by chance alone (i.e., is the result statistically significant)?

b. Fit a logistic regression model that uses *Dem* and *Obama* to predict *InsVote*. Which of the two predictors (being a Democrat or representing a district that voted for Obama in 2008) has a stronger relationship with *InsVote*?

c. (Optional) Make a graph of the Cook's Distance values for the fitted model from part (b). Which data point is highly unusual?

10.38 AHCA2017 vote. On May 4, 2017, the U.S. House of Representatives voted, by the narrow margin of 217–213, to pass the American Health Care Act. Most Republicans voted Yes, while all Democrats voted No. The file **AHCAvote2017** contains data for each of the 430 representatives who voted and records several variables, including *Party* (D or R); *AHCAVote* (1 for Yes, 0 for No); *Rep* and *Dem* (indicator variables for Party affiliation); *uni*2015 (the percentage of citizens without health insurance in 2015), *uni*2013 (the percentage of citizens without health insurance in 2013), *uniChange* (*uni*2015 – *uni*2013), and *Trump* (1 or 0 depending on whether the congressional district voted for Donald Trump or Hillary Clinton in November 2016). **AHCA**

a. Fit a logistic regression model that uses *uniChange* to predict *AHCAVote*. As the change in percentage of uninsured citizens decreases, does the likelihood of a Yes vote decrease? Is this relationship stronger than would be expected by chance alone (i.e., is the result statistically significant)?

b. Fit a logistic regression model that uses *uniChange* and *Trump* to predict *AHCAVote*. Which of the two predictors (representing a district that saw a large change in uninsured rate or a district that voted for Trump in 2016) has a stronger relationship with *AHCAVote*?

c. (Optional) For the fitted model from part (b), which data points have the largest and smallest deviance residuals?

10.39 Gunnels: Time and water. Consider again the data describing the habitat preferences for the species of eel known as gunnels (see Exercise 9.39). Recall that *Gunnel* is binary, with 1 indicating a presence of the species in a sample quadrat and 0 an absence. In Exercises 9.39 and 11.21, we looked at the relationship to *Gunnel* for the variables *TIME* and *Water*, separately, one at a time. Now we consider a model with both predictors. **Gunnel**

a. Recall that we generally prefer to have predictors that are independent of one another. Investigate the strength and nature of the relationship between these two predictors by first using boxplots and comparative sample statistics and then by writing a brief summary of what you find.

b. Fit the two-predictor linear logit model, include the summary table of coefficients, and summarize the statistical significance of each predictor as well as the nature of the relationship between these two predictors and the presence or absence of gunnels.

c. Use the two-predictor model to estimate the odds in favor of finding a gunnel in standing water at 10 a.m., so *Time* = 600.

d. Now use the two-predictor model to estimate the odds in favor of finding a gunnel when there is no standing water in the quadrat at 10 a.m.

e. Use the previous two answers to find an estimate for the 10 a.m. odds ratio of finding a gunnel in a quadrat with standing water versus no standing water. Comment on the size of the odds ratio.

10.40 Gunnels: Quadratic model. In Exercise 10.39, we fit an additive two-predictor model relating *Gunnel* to *Time* and *Water*. We now want to consider whether this simple model is adequate by comparing it to a full quadratic model, taking into account that *Water* is binary. **Gunnel**

a. Fit the linear logit model that includes the linear terms along with an interaction term for *Time* and *Water*, a quadratic term for *Time*, and an interaction term for *Water* and the quadratic *Time* term. Summarize the results, commenting on the nature of the coefficients and their statistical significance.

b. Now use (a) to perform a nest likelihood ratio test to decide if the simpler additive model dealt with in Exercise 10.39 is adequate.

10.41 Gunnels: Many predictors. Construct a model that predicts the presence of gunnels using as predictors seven variables that relate to the geology and timing of the observation: *Time*, *Fromlow*, *Water*, *Slope*, *Rw*, *Pool*, and *Cobble*.
These predictor variables have the following interpretations: **Gunnel**

Time	minutes from midnight
Fromlow	time in minutes from low tide (before or after); as an indicator of how low in the intertidal the biologist was working (always worked along the water)
Water	is there any standing water at all in the quadrat? Binary 1 = YES, 0 = NO
Slope	slope of quadrat running perpendicular to the waterline, estimated to the nearest 10 degrees
Rw	estimated percentage of cover in quadrat of rockweed/algae/plants, to the nearest 10%
Pool	is there standing water deep enough to immerse a rock completely? Binary 1 = YES, 0 = NO (always NO when water = NO)
Cobble	does the dominant substratum involve rocky cobbles? Binary 1 = YES, 0 = NO

a. Give a summary table of the model that lists the statistical significance or lack thereof for the seven predictors. Using the 0.05 significance level, which variables are statistically significant?

b. Explain in plain language why we might not trust a model that simply eliminates from the model the nonsignificant predictors.

c. Use a nested likelihood ratio test to ascertain whether the subset of predictors deemed nonsignificant in (a) can be eliminated, as a group, from the model. Give a summary of the calculations in the test: a statement of the full and reduced models, the calculation of the test statistic, the distribution used to compute the *P*-value, the *P*-value, and a conclusion based on the *P*-value.

d. Would you recommend using a series of nested LRT tests to find a final model that is more than the reduced model and less than the full model? Explain why or why not?

e. Using the reduced model from (c), state the qualitative nature of the relationship between each of the predictors and the response variable, *Gunnel*; that is, state whether each has a positive or negative relationship, controlling for all other predictors.

10.42 Faithful faces of men. The case study of Section 10.5 features an analysis of data recorded after college students looked at photographs of 82 women. Here we analyze similar data on 88 men. The file **FaithfulFaces** contains several variables measured on a scale from 1 (low) to 10 (high), including *Trust* (a rating a trustworthiness), *Faithful* (a judgment of how likely the man was to be faithful), *Attract* (attractiveness), and *SexDimorph* (how masculine the male face was). We also know the value of the variable *Cheater* (1 for men who had been unfaithful, 0 for others). **Faithful**

a. Give a summary table of the model that lists the statistical significance or lack thereof for the four predictors. Using the 0.10 significance level, which variables are statistically significant?

b. Use a nested likelihood ratio test to ascertain whether the subset of predictors deemed nonsignificant in (a) can be eliminated, as a group, from the model. Give a summary of the calculations in the test: a statement of the full and reduced models, the calculation of the test statistic, the distribution used to compute the *P*-value, the *P*-value, and a conclusion based on the *P*-value.

c. Using the reduced model from (b), state the qualitative nature of the relationship between each of the predictors and the response variable, *Cheater*; that is, state whether each has a positive or negative relationship, controlling for all other predictors.

Open-Ended Exercises

10.43 Risky youth behavior. Use the Youth Risk Behavior Surveillance System (YRBSS) data (**YouthRisk**) to create a profile of those youth who are likely to ride with a driver who has been drinking. What additional covariates do you think would be helpful? **Youth**

10.44 Backpack weights. Consider the **Backpack** data from Exercise 9.46 on page 451. The dataset contains the following more extensive set of variables than we introduced in that exercise: **BckPck**

Back Problems	0 = no, 1 = yes
BackpackWeight	in pounds
BodyWeight	in pounds
Ratio	backpack weight to body weight
Major	code for academic major
Year	year in school
Sex	Female or Male
Status	*G* = graduate or *U* = undergraduate
Units	number of credits currently taking

a. Use this more extensive dataset to investigate which factors seem to figure into the presence of back pain for these students. Perform a thorough exploratory analysis. You may want to explore the relationships between potential predictors and consider interaction terms. In writing up your report, provide rationale for your results and comment on possible inferences that can be made.

b. One study that piqued the students' interest said that a person should not carry more weight in a backpack than 10% to 15% of body weight. Use these data to see which of the measured variables appears to relate to the likelihood that a student carries too much weight.

10.45 Math placement: Does it work? Students at a small liberal arts college took a placement exam prior to entry in order to provide them guidance when selecting their first math course. The dataset **MathPlacement** contains the following variables: **MthPlc**

ID	Identification number
Gender	0 = female or 1 = male
PSATM	PSAT math score
SATM	SAT math score
ACTM	ACT math score
Rank	High school rank (adjusted to size class)
Size	Number of students in high school graduating class
GPAadj	Adjusted HS GPA
PlcmtScore	Score on math placement test
Recommends	Placement recommendation
RecTaken	Student took the recommended course? 1 = yes
Grade	Grade in the course

Note: The *Recommends* variable comprises the following possible values:

- R0: Stat, 117 or 210
- R12: R0 course or 120
- R2: 120 or 122
- R4: 122
- R6: 126
- R8: 128

Use these data to decide how well the math placement process is working. Consider using only the placement score to place students into a course. If they take the recommended course, how do they do? Define "success" as a grade of "B" or above. Does it improve the model to add other variables?

10.46 Film. Consider the dataset **Film** giving information of a random sample of 100 films taken from the 19,000+ films in the Maltin film guide, which was fully introduced in Exercise 9.48. Build a multiple logistic regression model for the binary response variable *Good?* using possible explanatory variables *Year*, *Time*, *Cast*, or *Description*. Explain your results in a report. **Film**

10.47 Lost letter. Recall from Exercise 9.49 that in 1999 Grinnell College students Laurelin Muir and Adam Gratch conducted an experiment where they intentionally "lost 140 letters" in either the city of Des Moines, the town of Grinnell, or on the Grinnell College campus. Half of each sample were addressed to Friends of the Confederacy and the other half to Iowa Peaceworks. Would the return rates differ depending on address? Get the data in **LostLetter**. Use multiple logistic regression to ascertain the effect of both address and location on the return rate. Report on your findings. **Lost**

10.48 Basketball.[10] Since 1991, David Arseneault, men's basketball coach of Grinnell College, has developed a unique, fast-paced style of basketball that he calls "the system." In 1997 Arseneault published a book entitled *The Running Game: A Formula for Success* in which he outlines the keys to winning basketball games utilizing his fast-paced strategy. In it, Arseneault argues that: (1) making opponents take 150 trips down the court, (2) taking 94 shots (50% of those from behind the three-point arc), (3) rebounding 33% of offensive missed shots, and (4) forcing 32 turnovers would ultimately result in a win. It is these four statistical goals that comprise Arseneault's "keys to success." The dataset **Hoops** comes from the 147 games the team played within its athletics conference between the 1997–1998 season through the 2005–2006 season. Use these data to investigate the validity of Coach Arseneault's "keys to success" for winning games in the system. Then use the variables available to, if possible, improve on his keys. Write a report of your findings. The variables are: **Hoops**

Game	an ID number assigned to each game
Opp	name of the opponent school for the game
Home	dummy variable where 1 = home game and 0 = away game
OppAtt	number of field goal attempts by the opposing team
GrAtt	number of field goal attempts by Grinnell
Gr3Att	number of three-point field goal attempts by Grinnell
GrFT	number of free throw attempts by Grinnell
OppFT	number of free throw attempts by the opponent
GrRB	total number of Grinnell rebounds
GrOR	number of Grinnell offensive rebounds
OppDR	number of defensive rebounds the opposing team had
OppPoint	total number of points scored in the game by the opponent
GrPoint	total number of points scored in the game by Grinnell
GrAsst	number of assists Grinnell had in the game
OppTO	number of turnovers the opposing team gave up
GrTO	number of turnovers Grinnell gave up
GrBlocks	number of blocks Grinnell had in the game
GrSteal	number of steals Grinnell had in the game
40*Point*	dummy variable that is 1 if some Grinnell player scored 40 or more points
BigGame	dummy variable that is 1 if some Grinnell player scored 30 or more
WinLoss	dummy variable that is 1 if Grinnell wins the game
PtDiff	point differential for the game (Grinnell score minus opponent's score)

10.49 Credit cards and risk.[11] Researchers conducted a survey of 450 undergraduates in large introductory courses at either Mississippi State University or the University of Mississippi. There were close to 150 questions on the survey, but only four of these variables are included in the dataset **CreditRisk**. (You can consult the paper to learn how the variables beyond these four affect the analysis.) These variables are as follows: **CredRsk**

Age	age of the student, in years
Sex	0 = male, 1 = female
DaysDrink	"During the past 30 days, on how many days did you drink alcohol?"
Overdrawn	"In the last year, I have overdrawn my checking account." 0 = no, 1 = yes

The primary response variable of interest to the researchers was *Overdrawn*: What characteristics of a student predict his or her overdrawing their checking account? Explore this question and report your findings. (*Note*: There are several missing values, most notably for *Age*.)

© motive56/Shutterstock

CHAPTER 11

Additional Topics in Logistic Regression

In this chapter you will learn to:

- Apply maximum likelihood estimation to the fitting of logistic regression models.
- Assess the logistic regression model.
- Use computer simulation techniques to do inference for the logistic regression parameters.
- Analyze two-way tables using logistic regression.
- Recognize Simpson's Paradox and understand how the relationship between two variables can change when we control for a third variable.

TOPICS

In Chapters 9 and 10, we introduced you to what we consider to be the basics of logistic regression. Those chapters contain the topics we think you must understand to be equipped to follow someone else's analysis and to be able to adequately perform a logistic regression on your own dataset. This chapter takes a more in-depth look at the subject of logistic regression and introduces you to ideas that, while not strictly necessary for a beginning analysis, will substantially strengthen an analysis of data.

We prefer to think of the sections of this chapter as topics rather than subdivisions of a single theme. Each topic stands alone and they can be read in any order. Because of this, the exercises at the end of the chapter have been organized by topic.

TOPIC 11.1 Fitting the Logistic Regression Model

The main goal of this section is to show you how statisticians fit a logistic regression model. Before we get into details, there are two preliminaries: What's new and what's not.

What's New?

For a concrete way to see what's new, compare the data for two examples from Chapter 9, losing sleep (Example 9.1) and medical school admission (Example 9.4).

In Figure 11.1(a), there are few X-values (only 5), and many Y-values for each X. This makes it possible to compute the observed proportion ($\hat{p}$) for each value of X. In turn, that makes it possible to compute logit($\hat{p}$), and to plot these values versus X. All this creates a strong temptation to use least squares with

$$\text{Observed} = \text{logit}(\hat{p})$$
$$\text{Fitted} = \beta_0 + \beta_1 X$$

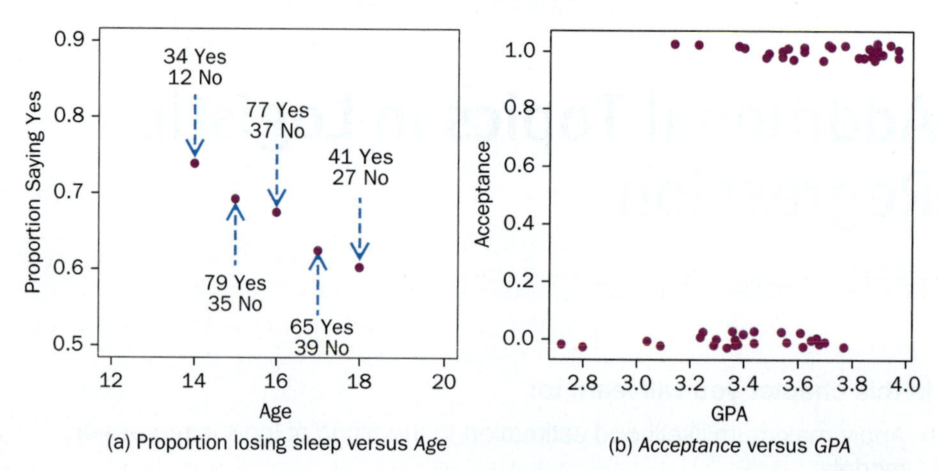

FIGURE 11.1 Visual summaries of logistic regression data

Resist! Least squares won't work! Why not? Look at Figure 11.1(b), which plots observed values versus X. For this dataset, there are lots of X-values, not just a few. For the raw data, each Y is a single observed value, either 0 or 1. For $Y = 0$ or $Y = 1$, corresponding logit values do not exist because either $Y = 0$ or $1 - Y = 0$ and log(0) is not defined. We have no way to compute logit(Y), and so no way to compare logit(Y) with logit(model) $= \beta_0 + \beta_1 X$. If we can't compute residuals, we can't use least squares. What can we do instead?

What's Not New?

Clearly, we need a new method for fitting the model. That's the main focus of this section. The new method may seem quite different on the surface. Fortunately, however, there's reason to think that by the end of the section you will start to get a sense that "the more things change, the more they stay the same." As with least squares, we'll end up with a numerical measure of how well a model fits. As with least squares, we'll choose parameter values that make the fit of the model to the data as good as possible. As with least squares, we'll be able to use the measure of fit to compare two nested models. All this is a preview of where we'll end up, but to get there we need to start with a new method called maximum likelihood estimation, which chooses parameter values that maximize a quantity called likelihood.

LIKELIHOOD

Definition: The **likelihood** is the probability of the observed data, regarded as a function of the parameters.

Question: If the likelihood is just the probability, why do you give it a new name?
Answer: The difference between "probability" and "likelihood" depends on what is fixed and what is variable. For probabilities, the parameters are constants and the data values are variable. For likelihood, it's the other way around: The data values are fixed, and the parameters are variable.

Question: What good is that?
Answer: Your question needs two answers, how and why. One answer tells *how* we'll use the likelihood to fit the model: Instead of choosing β-values to minimize the sum of squared errors (SSE), we'll choose β-values to maximize the likelihood. For ordinary regression, the smaller the SSE, the better the fit. For the logistic regression, the larger the likelihood, the better the fit.

Question: And the why?
Answer: Concrete examples may help you build intuition, but meanwhile, here's the abstract answer, called the "maximum likelihood principle."

MAXIMUM LIKELIHOOD PRINCIPLE

For any estimates of the parameters we can measure how probable it is to obtain the actual data. The more likely the data, the better the estimates. The **maximum likelihood estimates** are chosen to make this likelihood as large as possible.

Using maximum likelihood is different from least squares, but in some ways is quite similar:

Ordinary Regression Using Least Squares	Logistic Regression Using Maximum Likelihood
SSE = sum of squared errors = sum of (obs – fit)^2	L = probability of observed data
SSE depends on the parameter values β_0 and β_1.	L depends on the parameter values β_0 and β_1.
Choose parameter values that minimize SSE.	Choose parameter values that maximize L.

Why Is Maximum Likelihood Reasonable? A Simple Example

To illustrate the maximum likelihood principle, we start with a very simple example where a probability of success has only three possible values.

EXAMPLE 11.1

Tale of three spinners Imagine a guessing game with three choices: 1/4, 1/2, and 3/4. Each choice corresponds to the probability $\pi = P(\text{Yes})$ for a spinner, as in Figure 11.2. In what follows, we give you three tiny datasets and ask you to use the data to guess the value of π. We hope your intuition will lead you to use the maximum likelihood principle.

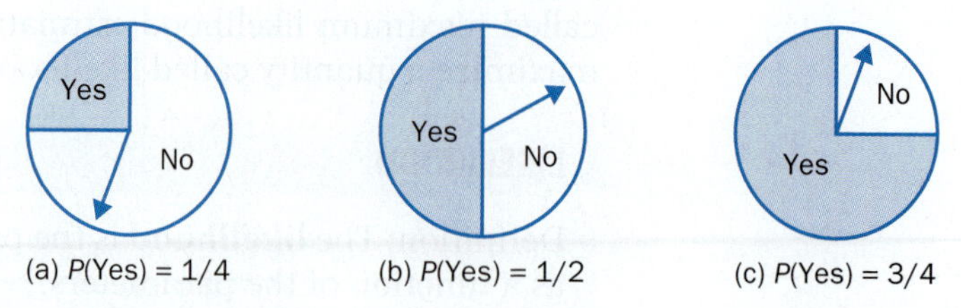

FIGURE 11.2 Three spinners

a. *Single spin*. A spinner is chosen, spun once, and the outcome is Yes. If you have to guess which spinner was used to get the Yes, what is your choice: 1/4, 1/2, or 3/4? (Answer before you go on.)

Notice that each value of π assigns a probability to the actual data. Taken together, these probabilities define the likelihood:

π = parameter value	1/4	1/2	3/4
$P(Yes)$ = likelihood = $L(\pi)$	1/4	1/2	3/4

We can show the likelihood function in a graph. See Figure 11.3.

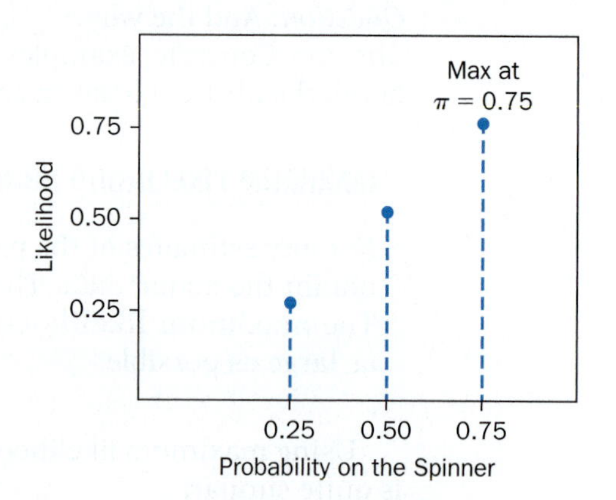

FIGURE 11.3 Discrete likelihood for single spin example

If, like most people, you think $\pi = 3/4$ is the best choice, you have chosen the parameter value that assigns the largest possible probability to what actually happened.

For this simplest version of the example, there was only one observed value. In the next version, there are two.

b. *Two spins*. Now suppose that a spinner is chosen, spun twice, and the outcome is (No, No). As before, you are told the outcome, but not which spinner it came from. What is your best guess?

Here, as before, each value of π assigns a probability to the actual data. Because the spins are independent, the probabilities multiply:

π = parameter value	1/4	1/2	3/4
$P(\text{No, No})$ = likelihood = $L(\pi)$	$\left(\frac{3}{4}\right)\left(\frac{3}{4}\right)$	$\left(\frac{1}{2}\right)\left(\frac{1}{2}\right)$	$\left(\frac{1}{4}\right)\left(\frac{1}{4}\right)$
	9/16	4/16	1/16

Taken together, the values for P(data) define the likelihood function shown in Figure 11.4.

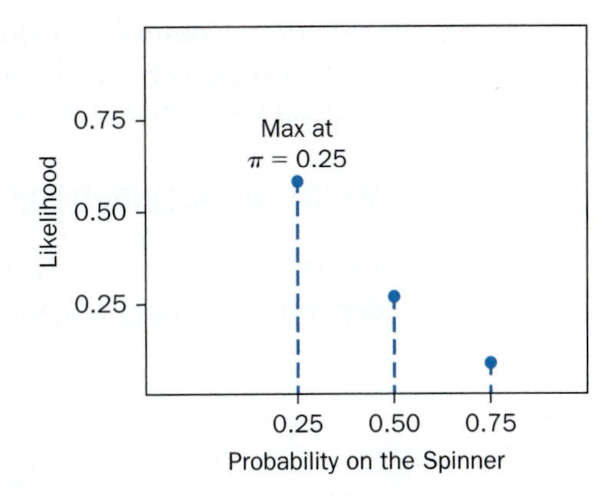

FIGURE 11.4 Discrete likelihood for two spins example

If your intuitive guess for the value of π was 1/4, then once again you were choosing the maximum likelihood estimate.

c. *Four spins*. Finally, suppose that a spinner is chosen, spun four times, and the result is (No, Yes, Yes, No). Which spinner is it this time?
   If your guess is $\pi = 1/2$, then once again you have chosen the maximum likelihood estimate (see Figure 11.5):

π = parameter value	1/4	1/2	3/4
P(Yes, No, No, Yes)	$\left(\frac{3}{4}\right)\left(\frac{1}{4}\right)\left(\frac{1}{4}\right)\left(\frac{3}{4}\right)$	$\left(\frac{1}{2}\right)\left(\frac{1}{2}\right)\left(\frac{1}{2}\right)\left(\frac{1}{2}\right)$	$\left(\frac{1}{4}\right)\left(\frac{3}{4}\right)\left(\frac{3}{4}\right)\left(\frac{1}{4}\right)$
= likelihood = $L(\pi)$	9/256	16/256	9/256

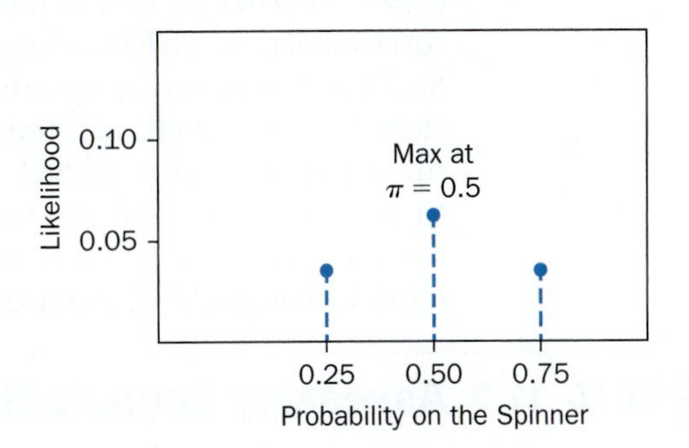

FIGURE 11.5 Discrete likelihood for four spins example

If you think back through the example, you can identify two main steps needed for maximum likelihood estimation.

1. *Likelihood*. For each value of the parameter, compute the value of each of the Yes and No outcomes. Then multiply these individual probabilities to assign an overall probability to the observed data. There will be one *P*(data) value for each parameter value. Taken together, these give the likelihood function.

2. *Maximize*. Find the parameter value that gives the largest possible *P*(data).

Maximum Likelihood for Logistic Regression

For logistic regression, we have two parameters, β_0 and β_1, and each parameter has infinitely many possible values. To fit the model by the method of maximum likelihood, the computer finds the values of $\hat{\beta}_0$ and $\hat{\beta}_1$ that make the likelihood (as described below) as large as possible. The mathematical details of maximizing the likelihood belong in a different course, so we will rely on statistical software to compute the estimates.

MAXIMUM LIKELIHOOD FOR LOGISTIC REGRESSION

Likelihood. For any $\hat{\beta}_0$ and $\hat{\beta}_1$ estimates, assign probability for each case depending on its predictor value and Yes/No response:

$$\text{if Yes} \qquad \pi = \frac{e^{\hat{\beta}_0+\hat{\beta}_1 x}}{1+e^{\hat{\beta}_0+\hat{\beta}_1 x}}$$

$$\text{if No} \qquad 1-\pi = \frac{1}{1+e^{\hat{\beta}_0+\hat{\beta}_1 x}}$$

Multiply the individual probabilities to get the likelihood, which will be a function of $\hat{\beta}_0$ and $\hat{\beta}_1$.

Maximize. Choose the values of $\hat{\beta}_0$ and $\hat{\beta}_1$ that make the likelihood as large as possible.

A Useful Parallel and a Look Ahead

Although the method of fitting is called maximum likelihood, and we have described it here in terms of maximizing the likelihood, in practice the computer software does something different, but equivalent. Instead of maximizing the likelihood, it minimizes $G = -2\log(\text{likelihood})$.* Working with logs offers many advantages, but one advantage in particular will be important for what follows. For maximum likelihood problems, the quantity $G = -2\log(\text{likelihood})$ behaves a lot like the sum of squared residuals (sum of squared errors, *SSE*) for least squares problems. We fit a model by minimizing *SSE* for regression or by minimizing $G = -2\log(\text{likelihood})$ for logistic regression. For both methods, the smaller the value, the better the fit, so we can use the minimum value of *SSE* or the minimum value of $G = -2\log(\text{likelihood})$ to measure how well the model fits. Better yet, just as we can compare two nested regression models using ΔSSE, we can compare two nested logistic models using ΔG. There are many other parallels as well.†

TOPIC 11.2 Assessing Logistic Regression Models

The previous two chapters have introduced you to the basics of logistic regression. You've seen important information to help you understand someone's use of logistic regression in a report or journal or to perform a basic logistic analysis of your own. This chapter takes a deeper look at logistic regression. Though not necessary for a basic analysis, knowing more about how logistic regression works can allow you to understand more and get more out of your logistic regression analyses.

As we have done with OLS—ordinary least squares—models, we assess our fitted logistic regression model with an eye toward the following general concerns:

1. **Utility:** Is it worthwhile to even bother with the explanatory variable(s)? That is, do the explanatory variables appear to be related to the probability of interest?

*It is a mathematical property of logarithms that the values of β_0 and β_1 that maximize the likelihood are always the same as the values of β_0 and β_1 that minimize $G = -2\log(\text{likelihood})$.

†In fact, least squares is a special case of maximum likelihood. It can be proved that for regression and ANOVA the two quantities *SSE* and *G* are one and the same.

2. **Fit:** Are there indications that the model does not fit well? Do the observed proportions differ substantially from the modeled proportions? Are there unusual observations or a need for transforming explanatory variables?
3. **Conditions:** Are there any concerns about the plausibility of the conditions for a logistic regression model?
4. **Predictions:** How well does the model perform? Is it good at prediction?

Assessing the Utility of a Logistic Regression Model

We have already investigated some ways in which to assess the utility of a logistic regression model in Chapters 9 and 10. Under the utility Question 1 above, we consider: Does the model under consideration indicate that there is an association between the explanatory variable(s) and the probability of interest? Is this model better at describing the probability of success than a model with no predictors, that is, a model that predicts a constant response value regardless of explanatory variable values?

There are a couple of ways we can assess the utility of a logistic regression model, including performing a z-test for a model with a single explanatory variable or using a nested Likelihood Ratio Test (LRT) for a model with several explanatory variables. Like an OLS t-statistic, a z-statistic for an explanatory variable coefficient (also known as a Wald statistic) in logistic regression is one tool we can use to find such an association and thereby assess the utility of the model. Another way to look for an association is to compare models by comparing likelihoods. Specifically, the difference in $-2\log(L)$ for two nested models is the test statistic for the nested LRT. (Recall that L represents the likelihood.) The value of this test statistic can be compared to a χ^2 distribution to assess the utility of a model with additional explanatory variables.

Just as we have done earlier, we use this full-versus-reduced approach as a way to assess the overall utility of a model (see Example 10.3, page 464). Here, we consider the simple case where the reduced model has no explanatory variables and the full model has one or more explanatory variables. More formally, we are comparing these two models:

$$H_0: \beta_1 = \beta_2 = \cdots = \beta_k = 0 \qquad \text{(reduced model)}$$

versus

$$H_a: logit(\pi) = \beta_0 + \beta_1 X_1 + \beta_2 X_2 + \cdots + \beta_k X_k, \text{ at least one } \beta_i \neq 0 \qquad \text{(full model)}$$

The reduced model here is the *constant model* that predicts the same probability of success regardless of predictor values. The full model here is the model we are considering: *the linear logit model*. With a model that is useful, we would expect rejection of H_0.

Assessing the Fit of a Logistic Regression Model

We now change topics. Suppose we are satisfied that we have a useful model; that is, we are satisfied that the linear logit model captures the nature of the association between the probability of interest and the explanatory variable(s) better than a model of constant probability. Realizing that no model is perfect, we might ask whether there is evidence of a problem with the fit of the model. Might the logit model be too simple to capture all the nuance in the association? Might there be unexplained variation in probability of success *beyond* the linear logit fit? This question is another facet of assessing a model and we again attack it with the full-versus-reduced model approach, leading to a test we call a **goodness-of-fit test**. In this test—in contrast to the previous test—the linear logit model becomes the null or reduced model, while the full model is a **saturated model**. "Saturated" suggests a model as full as can be, which makes

goodness-of-fit test

saturated model

sense here given that every possible combination of the predictors has its own, separately fitted probability estimate; a less saturated model would fit fewer parameters and would smooth out the probabilities across these combinations. Here are the full and reduced models in this context:

$$H_0 : logit(\pi) = \beta_0 + \beta_1 X_1 + \beta_2 X_2 + \cdots + \beta_k X_k \text{ (reduced model)}$$
$$H_a : logit(\pi) = p_i, \qquad i = 1, 2, \ldots, n \qquad \text{(full model)}$$

In the full model, $p_1, p_2, \ldots, p_n$ represent separate probabilities for each value of the explanatory variable or each combination of the collection of explanatory variables; the notation assumes n distinct values of the explanatory variable or n distinct combinations of the collection of explanatory variables. The full (saturated) model specifies a probability for each level (or combination of levels) of the explanatory variable(s), and in doing so, it provides the best fit possible. Implementing this approach will only be feasible if there are enough replicates* at each level (or combination of levels) of the explanatory variable(s) in the observed data. We'll see that this will work well for the putting example but not the MCAT data.

EXAMPLE 11.2 Putting

Putts2

Length of putt (in feet)	3	4	5	6	7
Number of successes	84	88	61	61	44
Number of failures	17	31	47	64	90
Total number of putts	101	119	108	125	134

TABLE 11.1 Putting success versus distance of the putt

binomial random variables

The data for this application (stored in the file **Putts2** and shown in Table 11.1) can be modeled as a set of **binomial random variables**. A binomial random variable counts the number of successes for a fixed number of independent trials. For example, we have $n_3 = 101$ putts for the first distance of 3 feet. Our model states that these putts can be considered the result of 101 independent trials, each of which has a specified probability (call it p_3—we will index from 3 to 7, rather than 1 to 5) of success. For these data, the value of the binomial variable is 84. For the second distance of 4 feet, there was another set of $n_4 = 119$ independent trials, with some fixed success probability (call it p_4). For these data, the value the binomial variable takes on is 88. Similar calculations can be completed for putt lengths of 5, 6, and 7 feet. For each of the 5 putt lengths, we have a fixed number of independent trials with some fixed "success probability" that changes with each putt length.

This is the underlying model used when fitting the saturated model. It is important to keep in mind the assumptions of the binomial model:

- Each trial results in a success or failure. Here each putt results in making the putt or not.
- The probability of success is the same for each trial at a given level (or combination of levels) of the predictor(s). This implies in the putting example that the probability of making the putt is the same for every putt from a given distance.

*Replicates at each level mean multiple observations with the same value of the explanatory variable.

- The trials are independent of one another at a given level of the predictor. When putting, this implies that what happens on one putt does not affect the probability of success or failure of other putts.

If any of these conditions are suspect, the goodness-of-fit test is suspect. (For example, might you have reason to suspect that the set of all 3-foot putts are independent of one another? This is a question you can consult your golfing friends about.)* If the assumptions do seem reasonable, we can fit the saturated model using maximum likelihood estimates, which turn out to be nothing more than the empirical proportions for each putt length. These proportions are the "saturated probability estimates" of Table 11.2.

Length of putt (in feet)	3	4	5	6	7
Saturated probability estimates $\hat{p}$	0.832	0.739	0.565	0.488	0.328
Logits of saturated estimates	1.600	1.040	0.260	−0.050	−0.720
Probability estimates: linear logit model	0.826	0.730	0.605	0.465	0.330
Logit estimates: linear logit model	1.558	0.992	0.426	−0.140	−0.706

TABLE 11.2 Goodness-of-fit test: Saturated model is "full"; linear logit model is "reduced"

The reduced model is the linear logit model that expresses the logit of π_d as a linear function of distance, and from the computer output (as in Example 9.17 on page 441), we see that the fitted model has the form

$$logit \frac{\widehat{\pi}}{1-\widehat{\pi}} = 3.26 - 0.566(distance)$$

or, equivalently,

$$\widehat{\pi} = \frac{e^{3.26-0.566(distance)}}{1+e^{3.26-0.566(distance)}}$$

These give the "Probability estimates: linear logit model" of Table 11.2; their logits are also given there and it is these logits that fall exactly on a line, as the model specifies.

The goodness-of-fit test compares the probability estimates from these two models: The closer they are to one another, the better the goodness of fit. As before, the comparison is made using a likelihood ratio test. Recall from Chapter 10 that in the nested likelihood ratio test the test statistic is

$$G = -2\ln(L_{Reduced}) - (-2\ln(L_{Full}))$$

When the full model is the saturated model, this becomes

$$G = -2\ln(L_{Reduced}) - (-2\ln(L_{Saturated}))$$

residual deviance
deviance

This last expression is called the **residual deviance** by R and is called the **deviance** by Minitab. Small residual deviance values indicate a reduced model that fits well; large values indicate poorer fit. This is intuitive if we

*In this case, only the first putt on a given green was recorded, which lends credence to the independence assumption.

consider the deviance as representing variation unexplained by the model (although this is not in exactly the same form as we encountered in OLS). We assess statistical significance using a χ^2 distribution with degrees of freedom equal to the difference in the number of parameters estimated in the two models, in this case, $5 - 2 = 3$. (Note that this df is given by Minitab and R output.) In this case, the residual deviance is 1.0692 on 3 df, which gives a *P*-value of 0.78. This result suggests that there is no compelling evidence that the fit of the model is problematic. So the null model is not rejected and we conclude that there are no detectable problems with the fit of the linear logistic regression model.

We can rewrite the likelihood ratio test statistic as

$$G = [-2\ln(L_{Reduced}) - (-2\ln(L_{Saturated}))] - [(-2\ln(L_{Full})) - (-2\ln(L_{Saturated}))]$$

or more simply as

$$G = Deviance_{Reduced} - Deviance_{Full}$$

This derivation explains again why the nested likelihood ratio test can be called the "drop-in-deviance" test.

Before leaving the topic of residual deviance and the goodness-of-fit test, we consider one more way of understanding the logic of this test. The following R output fits two logit models. The first model—"model with linear distance"—fits the probability of success to a linear logistic model of distance. The second model—"saturated model"—fits separate estimates to the logit for each putt distance, treating distance as a factor with five unordered levels ("3", "4", "5", "6", "7"). The residual deviance for the saturated model is 0 (i.e., −2.3315e-14) because there is no better fit of a model to these data: The fitted values of probability are precisely the empirical probabilities. Hence, the drop-in-deviance statistic is simply 1.0692 – 0, the 1.0692 in the linear logit model already being a comparison to the saturated model.

Thus the residual deviance appearing on R output is a likelihood ratio statistic comparing that model to a saturated model.

```
### Model with linear distance
glm(formula = cbind(n.made, n.missed) ~ distance, family = binomial)
             Estimate  Std. Error  z value    Pr(>|z|)
(Intercept)  3.25684   0.36893     8.828    <2e-16 ***
distance    -0.56614   0.06747    -8.391    <2e-16 ***

    Null deviance: 81.3865 on 4 degrees of freedom
Residual deviance: 1.0692 on 3 degrees of freedom
AIC: 30.175

### Saturated Model
# Model with Proportions for each Distance
glm(formula = cbind(n.made, n.missed) ~ factor(distance), family = binomial)
                   Estimate  Std. Error  z value   Pr(>|z|)
(Intercept)         1.5976    0.2659     6.007    1.89e-09 ***
factor(distance)4  -0.5543    0.3382    -1.639    0.101
factor(distance)5  -1.3369    0.3292    -4.061    4.90e-05 ***
factor(distance)6  -1.6456    0.3205    -5.134    2.84e-07 ***
factor(distance)7  -2.3132    0.3234    -7.154    8.46e-13 ***

    Null deviance: 81.387 on 4 degrees of freedom
Residual deviance: -2.3315e-14 on 0 degrees of freedom
AIC: 35.106
```

EXAMPLE 11.3

MedGPA

MCAT Table 11.3 illustrates the type of data found in the **MedGPA** dataset. It is apparent that the MCAT data are not in the form that the putting data are. That is, the MCAT data consist of many different possible predictor (GPA) levels and there are not replicates at each level. While we could think of the observations of the number of successful applicants at each GPA level as binomial variables, the number of trials is small and often just 1. (When counting successes for just a single trial, the variable is referred to as a binary observation or a Bernoulli trial.) In the table, we have 3 cases when GPA is 3.54, 2 when GPA is 3.56, only 1 when GPA is 3.58, and so on. The residual deviance test is only appropriate when we have more than 5 cases for all distinct values of the explanatory variable (or all combinations of the set of explanatory variables). The MCAT example is more typical than the putting example, that is, it is more common to have few, if any, replicate values or combinations of values of the explanatory variables. Therefore, goodness-of-fit tests based on the residual deviance can rarely be used in logistic regression.

GPA	Admitted	Denied	$\hat{p}$
3.54	2	1	0.67
3.56	1	1	0.50
3.58	1	0	1.00
3.61	0	1	0.00
3.62	2	1	0.67
3.65	0	1	0.00

TABLE 11.3 Some of the MCAT data

There is a test proposed by Hosmer and Lemeshow[1] that groups data in order to compute a goodness-of-fit measure that can be useful in more general circumstances. When this test is conducted, the cases are divided into several (up to 10) groups based on similar logistic predicted probabilities and the sample proportions in those groups are compared to the logistic fits—again, with a chi-square procedure. This test performs better with large samples so that there are at least five observations in each group. In this case, the df will be the number of groups minus 2. This option may be preferable when there are a large number of observations with little replication in the data. There are several other tests for goodness-of-fit whose results are provided routinely by software packages that we do not discuss here.

Reasons for Lack of Fit

While there was no problem detected with fit in the putting data, when a lack-of-fit is discovered, it is essential to investigate it. Here are three possible reasons for lack of fit, which we give *in the order in which we suggest you consider them*:

1. **Omitted explanatory variables or transformation needed**
2. **Unusual observations**
3. **Overdispersion**

We consider each of these possibilities below.

Omitted Explanatory Variables or Transformation Needed

This is the classic challenge in modeling, and it should be considered before any of the other reasons for lack-of-fit. Are there additional predictors you could include in your model? Do you need a quadratic or interaction term? Much of this book has provided ways in which to address just these kinds of concerns. Note that applications where there are many replicates also allow us to compute empirical logits and look graphically at the assumption of linearity. Without replicates, near-replicates, or large samples, graphical checks are not as easily interpreted with binary data. A common approach in logistic modeling is to compare models using a likelihood ratio test to determine whether quadratic or interaction terms are useful (see Chapter 10).

Unusual Observations

In the following case, we are using the goodness-of-fit statistic as a guide to finding extreme or unusual observations; we are not using it to make formal inferences about fit. Nonetheless, it can be very helpful to construct such diagnostic plots.

In addition to looking at summary statistics for goodness-of-fit, it is often useful to examine each individual observation's contribution to those statistics. Even if the summary statistic suggests there are no problems with fit, we recommend that individual contributions be explored graphically. We examine individual contributions in the following way: Each observation is removed one at a time from the dataset and the summary goodness-of-fit statistic, G, is recalculated. The change in G ($\Delta\chi^2$) provides an idea of how each particular observation affects the G. A large goodness-of-fit G suggests problems with the fit of the model. An observation contributing a proportionately large amount to the goodness-of-fit G test statistic requires investigation.

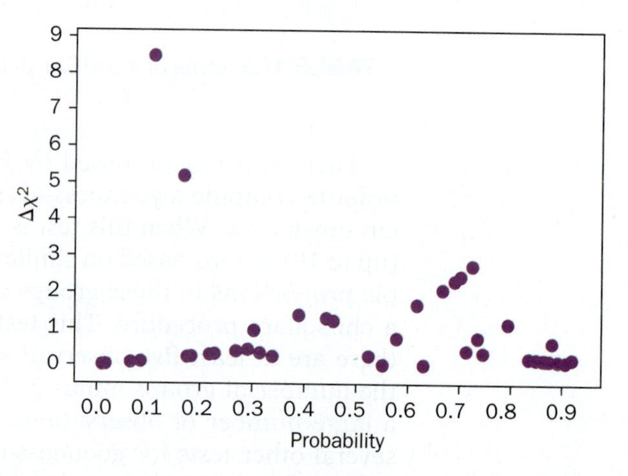

FIGURE 11.6 Contributions to G using $\Delta\chi^2$ criterion

Figure 11.6 shows the contribution of each point to the goodness-of-fit statistic for the medical school acceptances based on GPA. It is quite clear that there are two observations strongly affecting goodness-of-fit. These two students have relatively low GPAs (3.14, 3.23), yet both were accepted to medical school. This does not fit the pattern for the rest of the data.

Overdispersion

It is easier to understand overdisperson in the context of binomial data, for example, with $n > 5$ for most combinations of explanatory variables, although it is possible for overdipersion to occur in other circumstances as well. As established above, we may model such data using a binomial variable with n trials and with probability of success π for that combination. Intuitively, we expect the number of successes to be $n\pi$, on average, and from the theory of binomial variables, the variance is known to be $n\pi(1 - \pi)$.

When R fits a logistic model, part of the standard output is the sentence `(Dispersion parameter for binomial family taken to be 1)`, which implies that the variance is assumed to be $n\pi(1-\pi)$. In practice, it could happen that *extra-binomial dispersion* is present, meaning that the variance is larger than expected for levels of the predictor; that is, the variance is greater than $n\pi(1-\pi)$. This could happen for a number of reasons, and there are a number of ways to take this into account with your model. One approach that is often helpful is to fit a model identical to the usual linear-logistic model, but with one extra parameter, ϕ, that inflates the variance for each combination of the predictors by a factor of ϕ, so that the model variance is $\phi n\pi(1-\pi)$ rather than $n\pi(1-\pi)$, for some ϕ greater than 1. When we inflate the variance in this way, we are no longer working with a true likelihood but rather we use a *quasi-likelihood*, the details of which are left to our computer software.

In the case of binomial data, the deviance goodness-of-fit test, discussed earlier in this section, provides a check on dispersion. If the residual deviance is much greater than the degrees of freedom, then overdispersion may be present. For example, the putting data produce a residual deviance of 1.0692 on 3 degrees of freedom, so there is no evidence of overdispersion here. But consider the following similar example.

EXAMPLE 11.4

Putts3

Another putting dataset Table 11.4 gives another dataset (**Putts3**) that leads to a similar fit, but with an overdispersion problem.

Length of putt (in feet)	3	4	5	6	7
Number of successes	79	94	60	65	40
Number of failures	22	25	48	60	94
Total number of putts	101	119	108	125	134
Sample proportion ($\hat{p}$)	0.782	0.790	0.556	0.520	0.299
Logistic model $\hat{\pi}$	0.826	0.730	0.605	0.465	0.330

TABLE 11.4 Another putting dataset to illustrate overdispersion

Fitting the logistic regression of successes on the length of putt in R gives the same fit as for the original data but with a much larger residual deviance, due to the fact that the sample variances for the different lengths of putts do not match very well with what a binomial distribution predicts.

```
Coefficients:
              Estimate  Std. Error  z value   Pr(>|z|)
(Intercept)    3.25684     0.36893    8.828   <2e-16 ***
Length        -0.56614     0.06747   -8.391   <2e-16 ***

(Dispersion parameter for binomial family taken to be 1)

    Null deviance: 87.1429 on 4 degrees of freedom
Residual deviance:  6.8257 on 3 degrees of freedom
```

Overdispersion might be evident, for example, when our golfer has good or bad days. In those cases, data from the same type of day may be more similar or correlated so that the counts of successes are more spread out than we would expect if every putt was made with the same probability of success for each length of putt. If we had a binary variable measuring good and bad days, we could include that in the model as a second predictor. We don't have such a variable, but by using an overdispersion parameter, we take into account unmeasured variables such as good and bad days.

Comparing the two sets of R output illustrates the problems of fitting a regular linear logistic model when an overdispersed model is more appropriate. The slope and intercept estimates are identical, which will always occur. The differences lie in the standard errors. Not taking overdispersion into account leads to standard error estimates that are too small and thereby inflates the chance of statistical significance in the z-tests for coefficients. Notice here that with the usual model, the coefficient for *Length* appears highly statistically significant (P-value ≈ 0), whereas with the overdispersion model fit the standard errors are, appropriately, larger and, in this case, the P-value increases to 0.011, a less statistically significant result. In some cases, the change in P-value could be even more pronounced.

In general, if we do not take the larger variance into account, the standard errors of the fitted coefficients in the model will tend to be too small, so the z-tests for terms in the model are significant too often.

Note: To fit the overdispersion model in R change the `family=binomial` part of the glm command to `family=quasibinomial`. Here is output from using `family=quasibinomial` in R for the data in the previous table:

```
              Estimate  Std. Error  t value     Pr(>|t|)
(Intercept)    3.2568     0.5552     5.866    0.00988 **
Length        -0.5661     0.1015    -5.576    0.01139 *

(Dispersion parameter for quasibinomial family taken to be 2.264714)

    Null deviance: 87.1429 on 4 degrees of freedom
Residual deviance:  6.8257 on 3 degrees of freedom
```

Assessing the Conditions of a Logistic Regression Model

Recall that the conditions for a logistic regression model include:

1. The observations are independent of one another.
2. There is a linear relationship between the logits and the explanatory variables.
3. No important explanatory variables have been omitted.
4. There are no unusual observations that differ greatly from the predicted pattern.

Some ways to check these conditions follow:

1. *The observations are independent of one another.* As in the case of linear regression, this condition is best checked by scrutinizing the way in which the data were gathered. If the data come from a survey, was there a cluster sampling scheme? If so, the observations cannot be considered independent. Are relatives included? If so, the observations cannot be considered independent. Are observations that were gathered closer in time similar to each other? If so, the observations cannot be considered independent. Other considerations may apply as well.
2. *There is a linear relationship between the logits and the explanatory variables.* If there are repeated observations at each level of the explanatory variable, construct empirical logits and plot against the explanatory variable. Perform a goodness-of-fit test, which again requires multiple observations at each level of X. Residual analyses may be helpful as described below.

3. *No important explanatory variables have been omitted.* Add other explanatory variables and assess the improvement in fit.

4. *There are no unusual observations that differ greatly from the predicted pattern.* Residual analyses may be helpful as described below.

Are there any concerns about the model conditions? We began this section assessing the fit of a logistic regression model for binomial data with $n > 5$ using the residual deviance, which is equivalent to a likelihood ratio test comparing the model to a saturated model. If we have concerns about the fit, we check out potential reasons for lack-of-fit. We assess the condition that the model includes the correct explanatory variables in forms that are not in need of a transformation. If we have binomial data with $n > 5$, we can construct empirical logits and examine this question graphically. If there are no repeat observations (the more usual case), we can compare increasingly complex models, models with quadratic terms or interactions, for example, using nested LRTs.

If we are satisfied with the predictors in the model, we can use the $\Delta\chi^2$ plot, as in Figure 11.6, to identify unusual observations. We can also use residuals. There are a number of different kinds of residuals used in logistic regression, including Pearson residuals (which you may have encountered in χ^2 analysis) and deviance residuals (whose sum of squares gives the residual deviance).

Beyond simply identifying unusual observations, many other diagnostic methods have been proposed for assessing the conditions of a logistic model. Recall that when we studied ordinary least squares regression, we examined plots of residuals, leverages, and Cook's distance* values to determine which points in a dataset deviate from what the model predicts and which points have high influence on the fitted model. Analogous ideas apply to logistic regression.

EXAMPLE 11.5

Eyes

Eye dilation: Residuals Consider the **Eyes** data from Example 10.1 on page 457, where we examine dilation difference (pupil dilation when looking at same-sex nudes versus looking at opposite-sex nudes) and sex as predictors of whether or not a person is gay. The logistic model that uses *DilateDiff* and *SexMale* provides useful predictions, but there is an unusual point present.

A plot of Pearson residuals (Figure 11.7(a)) shows that one person has a residual close to 5. Since Pearson residuals have, roughly, mean 0 and standard deviation 1, a value as far from zero as 5 is quite extreme. A plot of Cook's distance values (Figure 11.7(b)) shows that this same person is

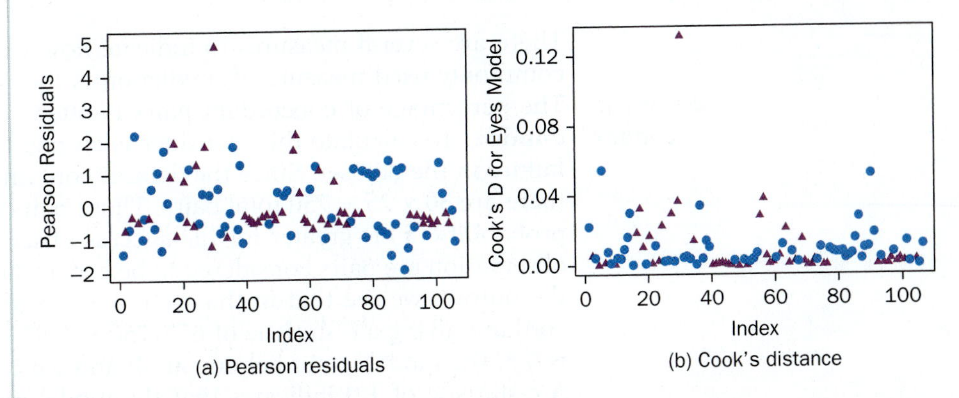

FIGURE 11.7 Finding unusual points in the eyes model, male points are circles, female points are triangles

*See Section 4.4.

extreme on this measure. Both plots are color-coded, so we can see that the unusual person is a female (triangle). Given that her dilation difference is negative (*DilateDiff* $= -0.58$) the model predicts only a 4% chance that she would be gay, but she is. Thus she has a very large residual.

Assessing Prediction

In the previous example, our objective was to describe a relationship, not to predict a new outcome. If our purpose in fitting a logistic regression model is prediction, we would like to know more than whether it seems to fit well or not. It would be helpful to know how well the model performs at prediction.

Many of the ways to assess prediction are intuitively appealing but not all are particularly effective at judging the predictive capabilities of a model. A natural and easily understood statistic is the percentage of points that are correctly classified. The model estimates $\hat{\pi}$ for each observation and we can say that the model predicts success if $\hat{\pi} > 0.5$. To find the percentage of correctly classified points, we count up the number of successes predicted out of the actual number of successes and the number of failures predicted out of the actual failures in the data. Some packages allow the user to specify the cutoff predicted probability to be different from 0.5. Consequently, this percentage can differ markedly even when using the same data. Users should be aware of this sensitivity to small changes in the cutoff. As an example, the output below shows the model of medical school acceptance on MCAT and GPA10. We used this model to construct Table 11.5, a contingency table of whether a student is denied or accepted to medical school (row headings) versus whether the model predicts being denied or accepted based on a 0.5 cutoff of estimated acceptance probability from the fitted model. We see that of the 55 cases, 41 ($= 18 + 23$) are correct predictions from the model, a success rate of 74.5%.

		Predicted	
		$\widehat{Deny}$	$\widehat{Accept}$
Actual	Deny	18	7
	Accept	7	23

TABLE 11.5 Rows are the actual admissions outcome: Deny or Accept; columns are predicted by the model: $\widehat{Deny}$ or $\widehat{Accept}$; based on a 0.5 cutoff

There are several measures to indicate how well a model is performing. One commonly used measure of prediction is the percentage of *concordant pairs*. The percentage of concordant pairs is often referred to as the **c-statistic** or **c-index**. To calculate this, consider each possible pairing of a success and a failure in the dataset. Since the dataset contains 30 successes and 25 failures, there are $30 \times 25 = 750$ total pairs. Those pairs for which the model-predicted probability ($\hat{\pi}$) is greater for the success observation than it is for the failure observation are pairs considered to be concordant with the fitted model. From the output, we see that in this example 625 of the 750 total pairs were concordant, giving a c-statistic of $625/750 = 0.833 = 83.3\%$. When the c-statistic is 0.5, the model is not helping at all and we can do just as well by guessing. A c-statistic of 1.0 indicates that the model perfectly discriminates between successes and failures. Our 83.3% indicates a good degree of accuracy in the model predictions.

c-statistic
c-index

If our goal is prediction, we will want to see good performance, as indicated by the c-statistic or other similar measures. Different packages report one or more of the many logistic regression versions of R^2. Unfortunately, these

pseudo-R^2 **pseudo-R^2s** are controversial and each suffers inadequacies. In general, this group of "measures of association" is not recommended or discussed further here. The output below, from the Minitab package, includes the c-statistic, but we have expunged much of the output that the package provided:

```
Response Information

 Variable    Value   Count
 Acceptance  1          30  (Event)
             0          25
             Total      55

Logistic Regression Table
                                                          Odds     95% CI
 Predictor       Coef    SE Coef      Z      P   Ratio   Lower   Upper
 Constant   -22.3727     6.45360  -3.47  0.001
 MCAT       0.164501    0.103154   1.59  0.111    1.18    0.96    1.44
 GPA10      0.467646    0.164155   2.85  0.004    1.60    1.16    2.20

Log-Likelihood = -27.007
Test that all slopes are zero: G = 21.777, DF = 2, P-Value = 0.000

Measures of Association:
(Between the Response Variable and Predicted Probabilities)

 Pairs        Number  Percent  Summary Measures
 Concordant      625     83.3  Somers' D                0.67
 Discordant      124     16.5  Goodman-Kruskal Gamma    0.67
 Ties              1      0.1  Kendall's Tau-a          0.34
 Total           750    100.0
```

Issues

We discuss here three issues that are important to logistic regression analysis, the first of which is unique to the logistic setting, the other two of which are analogous to OLS (ordinary least squares) regression:

1. *Perfect separation.* Strangely enough, when you include a predictor in the model that can perfectly discriminate between successes and failures, you will run into computational difficulties. You can easily anticipate this situation by doing a thorough exploratory data analysis in which you construct tables of each predictor by the response. Even so, it is still possible to have a combination of covariates completely separate successes from failures. A clue that this may be happening is the presence of extraordinarily large SEs for the coefficients. In general, if you spot very large SEs, you may be dealing with some kind of computational issue. For example, the algorithm that searches for a maximum likelihood may not have converged for any number of reasons. We should not use models with these large SEs. A number of sophisticated methods have been proposed for modeling in this situation, but none are included here. A simpler approach, that of a randomization (or permutation) test, is described in the next section.

2. *Multicollinearity.* Like OLS, we need to remain vigilant about the possibility of multicollinearity in logistic regression. As before, highly correlated predictors will probably not be a big problem for prediction, but we must be very cautious in interpreting coefficients when explanatory variables are

highly correlated. Again for categorical predictors, we should construct and examine simple two-way tables.

3. *Overfitting*. Overfitting—that is, closely fitting the idiosyncrasies of a particular dataset at the cost of producing generalizable results—can also be a problem with logistic regression. As with OLS, there are several ways in which to avoid overfitting, including cross-validation and bootstrapping.

TOPIC 11.3 Randomization Tests for Logistic Regression

In logistic regression modeling, randomization tests can be used for testing how well a model fits and whether an explanatory variable is associated with the response. Why would we bother with randomization tests? The tests and confidence intervals we have constructed using logistic regression results rely on approximations to distributions that require relatively large samples. For example, when we compare models by examining the difference in the deviances, we assume that when the null hypothesis is true, the difference would look as if it is coming from a chi-square distribution. When the samples are small, this approximation may not be correct and it may be better to perform a randomization test. Here we will examine the simple case of a single binary explanatory variable with a binary response. With a little ingenuity, you can take these methods and extend them to more complex situations.

EXAMPLE 11.6

Arch

Archery Heather Tollerud, a Saint Olaf College student, undertook a study of the archery scores of students at the college who were enrolled in an archery course. Students taking the course record a score for each day they attend class, from the first until the last day. Hopefully, the instruction they receive helps them improve their game. The data in **ArcheryData** contain the first and last day scores, number of days attended, and sex of the student. Let's begin by determining whether males and females exhibit similar improvements over the course of the semester.

Both variables we are considering, *Improve*, the response, and *Sex*, a possible explanatory variable, are binary variables. Hence the information we have can be summarized in a 2×2 table. See Table 11.6.*

	Improved	Did Not Improve
Males	9	1
Females	6	2

TABLE 11.6 Archery *Improve* by *Sex*

```
Coefficients:
            Estimate  Std. Error  z value  Pr(>|z|)
(Intercept)   1.0986      0.8165    1.346     0.178
Sexm          1.0986      1.3333    0.824     0.410

(Dispersion parameter for binomial family taken to be 1)

    Null deviance: 16.220 on 17 degrees of freedom
Residual deviance: 15.499 on 16 degrees of freedom
  (2 observations deleted due to missingness)
AIC: 19.499
```

* The *Improve* variable is binary, with value 1 when the last day's score exceeds the first day's score and value 0 otherwise.

Using R to fit a logistic regression to this data, we estimate that males improve $e^{1.0986} = 3$ times what females improve on average, so the observed odds ratio is 3. However, we have a very small number of students so it might be better to use a randomization test here, as opposed to typical logistic regression. Our approach will be to randomly assign the 15 improvements and 3 that are not improvements to the 8 women and 10 men. After each random assignment, we will use the "new data," fit a logistic regression, and take note of the odds ratio and residual deviance. These ORs (odds ratios) are a measure of the extent of the differences between males and females produced by chance. Repeating many times—1000 or 5000 or 10,000 times—we'll look at the distribution of ORs and determine the proportion of random assignments for which the odds ratio equals or exceeds the observed odds ratio of 3. With 10,000 samples, we found 1964 greater than the OR of 3 that we observed. This is consistent with the logistic regression results. Nearly 20% of the samples generated by chance yielded odds ratios larger than what we observed. It is difficult to argue that our observed result is unusually large and deserving of "statistically significant" status.

One of the reasons logistic regression is so useful is that it takes us beyond analyzing tables of counts and allows us to incorporate continuous variables. For example, we could determine how attendance affected the students' improvement.

```
Coefficients:
             Estimate  Std. Error  z value  Pr(>|z|)
(Intercept)    5.5792     11.0819    0.503     0.615
attendance    -0.1904      0.5278   -0.361     0.718

(Dispersion parameter for binomial family taken to be 1)

    Null deviance: 16.220 on 17 degrees of freedom
Residual deviance: 16.071 on 16 degrees of freedom
  (2 observations deleted due to missingness)
AIC: 20.071
```

The coefficient of -0.1904 is surprising: Does it really pay to skip class? A closer look at the output reveals that this negative association is not statistically significant. We might have considered a randomization (or permutation) test for this problem as well. And there's no reason not to proceed in a manner similar to the previous analysis. Randomly assign an attendance to each person and record attendance and improvement status for each person. Once again, repeat this many times and determine the proportion of the samples with an OR smaller than what we observed, $e^{-0.1904} = 0.83$.

Despite the fact that the point estimate suggests that each additional class period reduces your improvement by a factor of 17%, the *P*-values from the logistic regression (*P*-value = 0.718; two-sided) and the randomization test ($p = 0.5150$) do not provide convincing evidence that the observed reduction in improvement is significant, keeping in mind that this is an observational study. Nonetheless, you might be able to come up with reasons for why those who do not attend as much still show improvement. Possibly, they are excellent archers at the start of the course and know they will not benefit from more instruction, which in turn suggests that it is the poorer archers attending class and struggling with improvement.

TOPIC 11.4 Analyzing Two-Way Tables with Logistic Regression

In Chapter 9, we learned how to analyze a problem with a binary response variable and a binary predictor, in other words, how to analyze a 2×2 table using logistic regression. In this section, we extend the logistic method to cover the general two-way table and then relate all of this to topics you probably studied in your introductory statistics course for analyzing two-way tables: the z-test when the table is 2×2 and the more general chi-square test for two-way tables.

In the last section of Chapter 9, we saw how logistic regression can be used with a binary response and binary predictor to analyze the information in a 2×2 table. Now that we have a model for logistic regression with multiple predictors, we can extend those ideas to work with larger $2 \times k$ tables. While we still assume a binary categorical response, the categorical predictor may now have any number of categories. The key to this analysis is to create 0–1 indicator variables for each of the categories in the predictor. If the predictor has k categories, we build a multiple logistic regression model using any $k - 1$ of the indicator variables. The next example illustrates why we need to omit one of the indicators and how to interpret the coefficients to learn about the probability of the response in each of the k predictor categories.

EXAMPLE 11.7

Tip

Joking for a tip Can telling a joke affect whether or not a waiter in a coffee bar receives a tip from a customer? A study investigated this question at a coffee bar at a famous resort on the west coast of France.[2] The waiter randomly assigned coffee-ordering customers to one of three groups: When receiving the bill one group also received a card telling a joke, another group received a card containing an advertisement for a local restaurant, and a third group received no card at all. Results are stored in **TipJoke** and are summarized in the following 2×3 table:

	Joke Card	Advertisement Card	No Card	Total
Left a Tip	30	14	16	60
Did Not Leave a Tip	42	60	49	151
Total	72	74	65	211

The explanatory variable here is the type of card (if any) given to the customer. This variable is categorical but not binary, because it has three categories. The response variable is whether or not the customer left a tip. We will regard leaving a tip as a success, coded with a 1.

We can calculate the observed success (tipping) rates to be $30/72 = 0.417$ for the "joke" group, $14/74 = 0.189$ for the "advertisement" group, and $16/65 = 0.246$ for the "no card" group. Figure 11.8 shows a segmented bar graph of the proportion of tips within each card category. It appears that perhaps the joke card does produce a higher success rate, but we need to investigate whether the differences are statistically significant.

To use the categorical information of the card type in a logistic model, we construct new indicator variables for each of the card groups:

$$Joke = \begin{cases} 1 & \text{if Card = Joke} \\ 0 & \text{otherwise} \end{cases} \quad Ad = \begin{cases} 1 & \text{if Card = Ad} \\ 0 & \text{otherwise} \end{cases} \quad None = \begin{cases} 1 & \text{if Card = None} \\ 0 & \text{otherwise} \end{cases}$$

Note that if we know the value of any two of these indicator variables, we can find the value of the third, since

$$Joke + Ad + None = 1$$

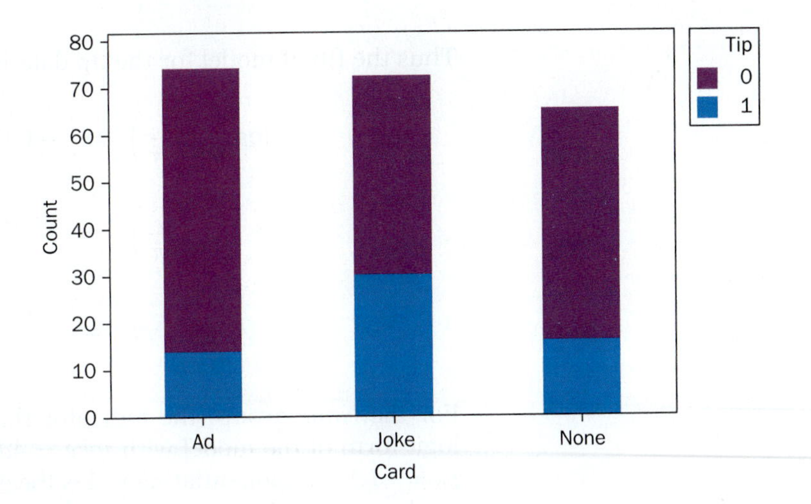

FIGURE 11.8 Proportion of tips within each card group

Thus we should not put all three indicator variables in the same regression model since any one is exactly collinear with the other two. Besides the information for three indicators being redundant, mathematically having all three in a single model will cause computational problems. Specifically, it is not possible to find a unique solution, so most software packages will either produce an error message or automatically drop one of the predictors if we try to include them all in a model.

In general, when we want to use indicator variables to code the information for a predictor with k categories in a regression model, we should include all but one of the indicators, that is, use any $k - 1$ indicator predictors. It doesn't really matter which indicator is omitted, although you may have a preference depending on the context. As we shall see, we can recover information about the omitted category from the constant term in the model; the other coefficients show how each of the other categories differs from the one that was left out. For this reason, we often call the group that was omitted the *reference* category. In the tipping example, we can choose to omit the *None* indicator if we are interested in assessing how the other two treatments relate to doing nothing. While it may seem odd at first to leave out one category, note that we have been doing that quite naturally with binary variables; for example, we did this when coding sex with a single indicator (1 = Female, 0 = Male) for the two groups.

We are now in a position to fit a logistic regression model to predict tip success based on the type of card. We omit the indicator for *None* and fit a multiple logistic model to predict *Tip* (1 = Yes, 0 = No) using the *Joke* and *Ad* indicators:

$$\log(\text{odds})\ \text{tips} = \beta_0 + \beta_1 Joke + \beta_2 Ad$$

Computer output includes the following fit and assessment of the model.

Logistic Regression Table

					Odds	95% CI	
Predictor	Coef	SE Coef	Z	P	Ratio	Lower	Upper
Constant	-1.11923	0.287938	-3.89	0.000			
Joke	0.782759	0.374234	2.09	0.036	2.19	1.05	4.56
Ad	-0.336056	0.413526	-0.81	0.416	0.71	0.32	1.61

Thus the fitted model for the tip data is

$$\log\left(\frac{\hat{\pi}}{1-\hat{\pi}}\right) = -1.12 + 0.78Joke - 0.34Ad$$

$$\left(\frac{\hat{\pi}}{1-\hat{\pi}}\right) = e^{-1.12} \cdot e^{0.78Joke} \cdot e^{-0.34Ad}$$

$$\hat{\pi} = \frac{e^{-1.12+0.78Joke-0.34Ad}}{1+e^{-1.12+0.78Joke-0.34Ad}}$$

For the *None* group (the indicator that was omitted from the model), the logit form of the model with $Joke = Ad = 0$ gives the estimated log(odds) of $\hat{\beta}_0 = -1.12$. Exponentiating yields the estimated odds of getting a tip with no card as $e^{-1.12} = 0.326$, and from the probability form of the fitted model, the estimated probability is $\hat{\pi} = 0.326/1.326 = 0.246$. Notice that this matches the sample proportion of 16/65.

For the *Joke* group, the fitted equation indicates that the log(odds) for the *Joke* group can be found by increasing the log(odds) for *None* by $\hat{\beta}_1 = 0.78$. Exponentiating to get to the odds scale implies that we multiply the odds for *None* ($e^{-1.12} = 0.326$) by $e^{0.78}$ ($= 2.181$) to get the estimated odds for *Joke*, that is, $0.326(2.181) = 0.711$. Nonetheless, it is the odds ratio that is often of interest with logistic regression and this is easily determined by exponentiating the coefficient. Here, the odds of getting a tip with a joke is $e^{0.78} = 2.181$ times the odds of getting a tip with no card. In addition, from the estimated odds of 0.711 we can get the probability of a tip with a joke to be $\hat{\pi} = 0.711/1.711 = 0.416$. Again, this matches (up to round-off) the sample proportion, which is 30/72.

The fitted equation also provides an estimate of the odds ratio for *Ad* ($e^{-0.34} = 0.712$). To estimate the probability of a tip with an *Ad*, first compute $\log\left(\frac{\hat{\pi}}{1-\hat{\pi}}\right) = -1.12 - 0.34 = -1.46$. Exponentiating, we get $\left(\frac{\hat{\pi}}{1-\hat{\pi}}\right) = e^{-1.46} = 0.232$. Solving, we get $\hat{\pi} = 0.232/1.232 = 0.188$, the sample proportion (within round-off) of tips given when an *Ad* card was used.

From the computer output, we see that the indicator variable for the joke card has a small *P*-value (0.036) and a positive coefficient (0.783), so we can conclude that the joke card does have a higher success rate than the reference (no card) group. But the *P*-value for advertisement card is not small (0.416), so these data provide insufficient evidence that an advertisement card leads to a higher rate of tipping than no card at all. From the estimated odds ratio for the *Joke* indicator, we can conclude that the odds of receiving a tip are roughly doubled ($e^{0.78} = 2.181$) when a joke is presented with the bill. Because this study was a randomized experiment, we can conclude that the joke card caused an increase in the tipping rate.

Review: Two-sample z-test and Chi-square Test for a Two-way Table

An alternative (and possibly more familiar) method for assessing whether the proportions differ between two groups is to use a two-sample test based on the normal distribution. This procedure is often covered in an introductory statistics course, so we only review it briefly here and compare the results to what we would see in binary logistic regresion.

z-TEST FOR DIFFERENCE IN TWO PROPORTIONS

Assume we have two independent samples of sizes n_0 and n_1, respectively. Let X_0 and X_1 be the counts of successes in each group so that $\hat{\pi}_0 = X_0/n_0$ and $\hat{\pi}_1 = X_1/n_1$. To test $H_0 : \pi_0 = \pi_1$ versus $H_a : \pi_0 \neq \pi_1$,

$$z = \frac{\hat{\pi}_1 - \hat{\pi}_0}{\sqrt{\hat{\pi}(1-\hat{\pi})\left(\frac{1}{n_1} + \frac{1}{n_0}\right)}}$$

where $\hat{\pi} = \frac{X_1 + X_0}{n_1 + n_0}$ is the proportion obtained by pooling the two samples. Assuming the sample sizes are not small, the distribution of the test statistic z is approximately standard normal under H_0, so we use the normal distribution to compute a *P*-value.

For the TMS (1) versus placebo (0) data in Example 9.18 on page 441, we have these sample proportions:

$$\hat{\pi}_1 = \frac{39}{100} = 0.39 \qquad \hat{\pi}_0 = \frac{22}{100} = 0.22 \qquad \hat{\pi} = \frac{39 + 22}{100 + 100} = 0.305$$

and the test statistic is

$$z = \frac{0.39 - 0.22}{\sqrt{0.305(1 - 0.305)\left(\frac{1}{100} + \frac{1}{100}\right)}} = 2.61$$

The *P*-value = $2P(Z > 2.61) = 0.009$, where $Z \sim N(0, 1)$, which again indicates that the proportion of patients who got relief from migraines using TMS was significantly different from the corresponding proportion for those using the placebo. Note that the test statistic and *P*-value for this two-sample z-test for a difference in proportions are similar to the test statistic (2.58) and *P*-value (0.010) from the computer output (page 441) for testing whether the slope of the logistic regression model differs from zero.

Review: Chi-square Test for a 2 × 2 Table

In an introductory statistics class, you might also have studied a chi-square procedure for testing relationships between categorical variables using a two-way table. How does this test relate to the assessment of the binary logistic regression with a single binary predictor?

CHI-SQUARE TEST FOR A 2 × 2 TABLE

Start with data of *observed* counts (O_{ij}) in a 2 × 2 table. Compute an *expected* count for each cell using $E_{ij} = \frac{RowTotal \cdot ColumnTotal}{n}$:

Group	Success		Failure	
0	O_{00}	(E_{00})	O_{01}	(E_{01})
1	O_{10}	(E_{10})	O_{01}	(E_{11})

The test statistic compares the *observed* and *expected* counts within each of the cells of the table and sums the results over the four cells:

$$X^2 = \sum \frac{(Observed - Expected)^2}{Expected}$$

If the sample sizes are not small (expected counts at least 5), the test statistic under H_0 follows a chi-square distribution with 1 degree of freedom.

Here are the observed and expected counts for the 2 × 2 table based on the TMS and placebo data:

Group	Success	Failure	Total
Placebo	22 (30.5)	78 (69.5)	100
TMS	39 (30.5)	61 (69.5)	100
Total	61	139	200

The chi-square test statistic is

$$ts = \frac{(22-30.5)^2}{30.5} + \frac{(78-69.5)^2}{69.5} + \frac{(39-30.5)^2}{30.5} + \frac{(61-69.5)^2}{69.5} = 6.817$$

This gives a P-value $= P(\chi^2_1 > 6.817) = 0.009$, which matches the two-sample z-test exactly (as it always must in the 2×2 case). While this chi-square statistic is not exactly the same as the chi-square test statistic based on $-2\log(L)$ in the logistic regression (6.885), the P-values are very similar and the conclusions are the same.

On to the 2 × *k* Case

While the two-sample z-test does not easily generalize to the $2 \times k$ case, the chi-square test does. We now review the basics of the chi-square test for these larger tables.

CHI-SQUARE TEST FOR A 2 × *k* TABLE

Start with data of observed counts (O_{ij}) in a $2 \times k$ table. Compute an *expected* count for each cell using $E_{ij} = \frac{RowTotal \cdot ColumnTotal}{n}$:

Response	Group #1	Group #2	...	Group # k	Total
0 = No	O_{01} (E_{01})	O_{02} (E_{02})	...	O_{0k} (E_{0k})	# No
1 = Yes	O_{11} (E_{11})	O_{12} (E_{12})	...	O_{1k} (E_{1k})	# Yes

The test statistic compares the *observed* and *expected* counts within each of the cells of the table and sums the results over the $2k$ cells:

$$X^2 = \sum \frac{(Observed - Expected)^2}{Expected}$$

If the sample sizes are not small (expected counts at least 5), the test statistic under H_0 follows a chi-square distribution with $k - 1$ degrees of freedom.

Note: This procedure works the same way for a $k \times 2$ table where the binary response determines the columns.

The expected counts are just the frequencies we would see if the counts in each cell exactly matched the overall proportion of successes and failures for the entire sample. This is essentially the "constant" model that we use as a base for comparison when computing the G-statistic to test the overall fit of a logistic regression model. Indeed, while not identical, the results of the chi-square test for a $2 \times k$ table are very similar to the overall G-test for a model with $k - 1$ indicator variables as predictors.

EXAMPLE 11.8

Joking for a tip The output below is for a chi-square test on the table of results from the tipping experiment, along with the G-test of $H_0 : \beta_1 = \beta_2 = 0$ for the model that uses *Joke* and *Ad* indicators in predicting *Tips*. Note that setting both of those coefficients equal to zero in the model is equivalent to saying that the tip proportion should be the same for each of the three groups (in this case, estimated as $60/211 = 0.284$)—which is the same assumption that generates the expected counts in the chi-square test for the table.

Here is output for a chi-square test of the 2×3 table:

Rows: Tip Columns: Card

	Ad	Joke	None	All
0	60	42	49	151
	52.96	51.53	46.52	151.00
1	14	30	16	60
	21.04	20.47	18.48	60.00
All	74	72	65	211
	74.00	72.00	65.00	211.00

Cell Contents: Count
Expected count

Pearson Chi-Square = 9.953, DF = 2, P-Value = 0.007

Here is output for logistic regression to predict *Tips* based on *Joke* and *Ad*:

Predictor	Coef	SE Coef	Z	P	Odds Ratio	95% CI Lower	95% CI Upper
Constant	-1.11923	0.287938	-3.89	0.000			
Joke	0.782759	0.374234	2.09	0.036	2.19	1.05	4.56
Ad	-0.336056	0.413526	-0.81	0.416	0.71	0.32	1.61

Log-Likelihood = -121.070
Test that all slopes are zero: G = 9.805, DF = 2, P-Value = 0.007

We see that the chi-square statistics for the two procedures are quite similar (9.953 and 9.805) with the same degrees of freedom (2) and essentially the same P-value (0.007). So the tests are consistent in showing that the tip rate differs depending on the waiter's action (*Joke*, *Ad*, or *None*). To understand the nature of this difference in the chi-square test, we see that the observed number of tips in the joke group (30) is quite a bit higher than expected (20.47), while the observed counts for both the advertising and no card groups are lower than expected. This is consistent with the coefficients we observed in the logistic model, which showed that the odds of a tip were significantly better for the joke group, but not much different between the advertising and no card groups. So if you happen to find yourself waiting on tables at a coffee bar in western France, you probably want to learn some good jokes!

white victims and having a white victim greatly increases the chance of conviction, at first it appears that the conviction rate is higher among white defendants because of the defendant's race. But again, it is the race of the victim, the third variable, that is more important.

Figure 11.9 shows these relationships graphically. The diamonds represent minority defendants and the triangles represent white defendants. White victims are represented by the points at the right of the graph; minority victims are represented by the points at the left of the graph. The open diamonds are above the open triangles, but the solid triangle, showing the overall conviction rate of white defendants, is higher than the solid diamond, showing the overall conviction rate of minority defendants.

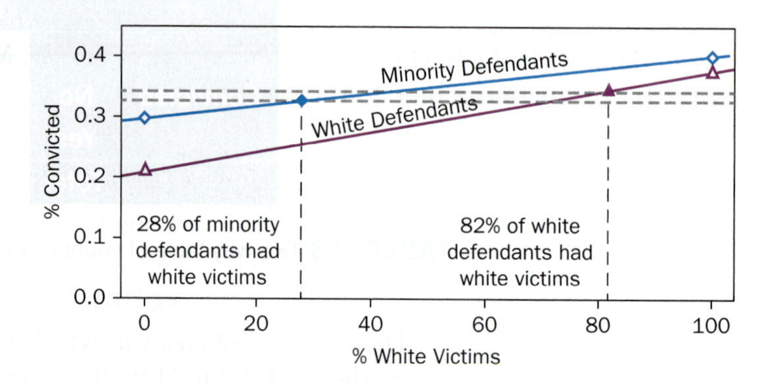

FIGURE 11.9 A graphical display of Simpson's Paradox

To explore the inter-relationships among the three variables, we fit a multiple logistic model. The response variable *Convicted* records convicted/not convicted and the two predictor variables are *IndWhiteDefendant* = 1 for white defendants and 0 for minority defendants and *IndWhiteVictim* = 1 for white victims and 0 for minority victims. The fitted model is that the log(odds) of conviction equal $-0.950 + 0.624$*IndWhiteVictim* $-$ 0.253*IndWhiteDefendant*, as shown in the following output:

```
Coefficients:

                   Estimate  Std. Error  z value  Pr(>|z|)
(Intercept)          -0.915       0.255    -3.59   0.00033
IndWhiteVictim        0.624       0.357     1.75   0.08013
IndWhiteDefendant    -0.253       0.350    -0.72   0.46990
```

The logistic fit shows that the race of the victim is a much stronger predictor of conviction than is the race of the defendant. Moreover, the coefficient on *IndWhiteDefendant* is negative, meaning that being a white defendant decreases the odds of conviction, but the coefficient on *IndWhiteVictim* is positive, meaning that having a white victim increases the odds of conviction. This is evidence that crimes against whites are considered to be much more serious than crimes against minorities.

Simpson's Paradox can arise in many situations. You will probably never be involved with an assault case, as in Example 11.9, but you might fly somewhere, so consider the next example.

EXAMPLE 11.10

On-time airlines Which airline, Delta or American, is more likely to have a flight arrive on time? This depends on the airport. There are more delays on flights into New York's LaGuardia airport than there are on flights into Chicago's O'Hare airport. Since Delta has many flights into LaGuardia but few flights into O'Hare, we might expect a lot of flight delays for Delta. American, on the other hand, has more flights into O'Hare and thus we might expect American to have a good on-time arrival rate.

Let's compare on-time arrival proportions using the same approach we used in the previous example. The data stored in **Airlines** are all flights by Delta or by American to either O'Hare or to LaGuardia in March of 2016. There are 10,333 observations in this dataset.

Table 11.10 shows the cross-classification of *Airline* with *OnTime*, for both airports combined. This allows us to find the overall on-time arrival rate for each airline. Note that the on-time rate was higher for American ($5880/7330 = 0.802$) than for Delta ($2334/3003 = 0.781$).

		Airline		
		Delta	American	Total
OnTime	**No**	659	1450	2109
	Yes	2344	5880	8224
	Total	3003	7330	10,333

TABLE 11.10 Two-way table of counts for both airports combined

Now let's consider the third variable *airport*. Most of the American flights were to O'Hare airport, where on-time arrivals are quite common. The breakdown by airline is shown in Table 11.11.

		Airline		
		Delta	American	Total
OnTime	**No**	92	825	917
	Yes	527	4229	4756
	Total	619	5054	5673

TABLE 11.11 Two-way table of counts for planes flying to O'Hare

Most of the Delta flights were to LaGuardia airport, where on-time arrivals are less common. The breakdown by airline is shown in Table 11.12.

		Airline		
		Delta	American	Total
OnTime	**No**	567	625	1192
	Yes	1817	1651	3468
	Total	2384	2276	4660

TABLE 11.12 Two-way table of counts for planes flying to LaGuardia

Now compare the combined data in Table 11.10 with the airport-by-airport data in Tables 11.11 and 11.12. We see another example of Simpson's Paradox here: Delta has a higher on-time arrival rate for each of the two airports (O'Hare: $527/619 = 0.851 > 4229/5054 = 0.837$; Laguardia: $1817/2384 = 0.762 > 1651/2276 = 0.725$), but overall Delta has a lower on-time rate ($0.781 < 0.802$). As noted above, the airport matters more than the airline. Within each airport the Delta advantage over American is small. The difference between O'Hare (Table 11.11) and LaGuardia (Table 11.12) is more substantial.

Here is the output from fitting a multiple logistic regression model in which on-time arrival depends on airport (*IndOHare* = 1 for O'Hare and 0 for LaGuardia) and on airline (*IndDelta* = 1 for Delta and 0 for American). The fitted model is that the log(odds) of an on-time arrival equals $0.98 + 0.647 IndOHare + 0.174 IndDelta$, as shown in the following output:

```
Coefficients:
             Estimate  Std. Error  z value     Pr(>|z|)
(Intercept)   0.9809     0.0443     22.16   <2e-16 ***
IndOHare      0.6471     0.0544     11.89   <2e-16 ***
IndDelta      0.1737     0.0588      2.95   0.0031 ***
```

Flying to O'Hare (rather than to LaGuardia) increases the odds of being on-time. Flying on Delta (rather than American) also increases the odds of being on-time, but not by as much.

EXERCISES

Topic 11.1 Exercises: Fitting the Logistic Regression Model

11.1 Looking at likelihoods. You are trying to estimate a true proportion (π) using the results of several independent Bernoulli trials, each with the same probability of success (π). Open the Excel file labeled **likelihood-play.xls**. You will find a graph of a likelihood function for a series of Bernoulli trials. You can specify different values for the numbers of successes and failures and you will see a corresponding change in the likelihood function. What happens as the number of trials increases? How does the graph change as the proportion of successes changes? How does the corresponding graph of the log of the likelihood compare?

11.2 Compute a likelihood. In Example 9.18 (page 441) on zapping migraines, we showed how to fit the logistic model, by hand, using sample proportions from the 2×2 table. One can also use these proportions to directly compute the *G*-statistic given in the Minitab output. Show this calculation and verify that your answer agrees with the Minitab output.

11.3 Spinners. George has three spinners like those in Example 11.1, giving outcomes of Yes or No. The first has $\pi = P(\text{Yes}) = 1/4$, the second spinner has $\pi = P(\text{Yes}) = 1/3$, and the third has $\pi = P(\text{Yes}) = 1/2$.

a. George chooses a spinner, it is spun once, and the outcome is Yes. What is the likelihood, $L(\pi)$, for each of the three spinners? Which spinner has the maximum likelihood?

b. George chooses a spinner, it is spun twice, and the outcome is (Yes, No). What is the likelihood, $L(\pi)$, for each of the three spinners? Which spinner has the maximum likelihood?

c. George chooses a spinner, it is spun three times, and the outcome is (Yes, No, No). What is the likelihood, $L(\pi)$, for each of the three spinners? Which spinner has the maximum likelihood?

11.4 Spinners again. Ann has three spinners that are similar to the spinners described in Exercise 11.3. The first has $\pi = P(\text{Yes}) = 1/4$, the second spinner has $\pi = P(\text{Yes}) = 1/2$, and the third has $\pi = P(\text{Yes}) = 2/3$.

a. Ann chooses a spinner, it is spun once, and the outcome is Yes. What is the likelihood, $L(\pi)$, for each of the three spinners? Which spinner has the maximum likelihood?

b. Ann chooses a spinner, it is spun twice, and the outcome is (Yes, No). What is the likelihood, $L(\pi)$, for each of the three spinners? Which spinner has the maximum likelihood?

c. Ann chooses a spinner, it is spun three times, and the outcome is (Yes, No, No). What is the likelihood, $L(\pi)$, for each of the three spinners? Which spinner has the maximum likelihood?

11.5 Likelihoods. For a logistic regression model with $\hat{\beta}_0 = 0.5$ and $\hat{\beta}_1 = 0.7$, the likelihood of Yes, for a given value of X, is $\pi = e^{0.5+0.7X}/(1+e^{0.5+0.7X})$ and the likelihood of No is $1-\pi$.

a. Suppose $X = 1$. What is the likelihood of Yes?

b. Suppose $X = 2.2$. What is the likelihood of Yes?

c. Suppose that two observations are made and that $X = 1$ for the first and $X = 2.2$ for the second. What is the likelihood of the pair of outcomes (Yes, Yes)?

11.6 Likelihoods. For a logistic regression model with $\hat{\beta}_0 = 0.5$ and $\hat{\beta}_1 = -1$, the likelihood of Yes, for a given value of X, is $\pi = e^{0.5-X}/(1+e^{0.5-X})$ and the likelihood of No is $1-\pi$.

a. Suppose $X = 1$. What is the likelihood of Yes?

b. Suppose $X = 2$. What is the likelihood of Yes?

c. Suppose that two observations are made and that $X = 1$ for the first and $X = 2$ for the second. What is the likelihood of the pair of outcomes (Yes, No)?

Topic 11.2 Exercises: Assessing Logistic Regression Models

11.7 Two kinds of deviance. In the archery example (Example 11.6 on page 514), we gave logistic output from R that models archery improvement on attendance. Some computer output for fitting this model to the data in **ArcheryData** is reproduced below: **Arch**

```
              Estimate  Std. Error  z value  Pr(>|z|)
(Intercept)     5.5792     11.0819    0.503     0.615
attendance     -0.1904      0.5278   -0.361     0.718

    Null deviance: 16.220 on 17 degrees of freedom
Residual deviance: 16.071 on 16 degrees of freedom
```

a. Write down the full and reduced models corresponding to the null deviance, using notation related to the context of these data.

b. Write down the full and reduced models corresponding to the residual deviance, using notation related to the context of these data.

11.8 Related significance tests. Refer to Exercise 11.7 that includes some computer output for fitting a logistic model to predict archery improvement based on class attendance. **Arch**

a. Describe a test based on the deviances that tests the same hypothesis as the one tested by the z-statistic of -0.361.

b. Do the two tests in part (a) necessarily lead to the same P-values? Explain.

c. Explain why, in this example, you might not want to trust either of the tests referred to in (b).

d. What does the text suggest for the remedy to the situation described in (c)?

11.9 The "other" putting data. Consider Example 11.4 and the data in Table 11.4 on page 509. Some output based on these data for modeling the chance of making a golf putt based on its length is reproduced below: **Putts3**

```
             Estimate  Std. Error  z value     Pr(>|z|)
(Intercept)   3.25684     0.36893    8.828  <2e-16 ***
Lengths      -0.56614     0.06747   -8.391  <2e-16 ***

(Dispersion parameter for binomial family taken to be 1)

    Null deviance: 87.1429 on 4 degrees of freedom
Residual deviance: 6.8257 on 3 degrees of freedom
```

a. Which rows of Table 11.4 are compared by the residual deviance?

b. The null deviance compares the row of logistic model $\hat{\pi}$'s to something. What is that thing?

c. Compute a P-value using the null and residual deviances and explain what other piece of the output tests the same hypotheses.

d. Do the two P-values discussed in (c) lead to the same conclusion? Are they guaranteed to always do so?

e. What piece of the R output does the text suggest should lead to our considering an overdispersion, quasi-likelihood fit?

11.10 Lost letters. Consider the data from Exercise 9.49 on page 452, which describes a study of intentionally losing letters to ascertain how different addresses affect return rates. The data are in **LostLetter**. **Lost**

a. Fit a logistic model that models *Returned* on *Location* and *Address*. Summarize your findings.

b. Use the residual deviance to test for goodness of fit for the logistic model in part (b).

c. Write down explicitly the empirical proportions for the saturated model being considered by the goodness-of-fit test and the corresponding linear logit estimated probabilities that these saturated logits are being compared to.

d. Does the output suggest the need to fit an overdispersion model here? Explain.

11.11 Bird nests revisited. Birds construct nests to protect their eggs and young from predators and from harsh environmental conditions. The variety of nest types varies widely, including some that are open and some that are closed. Amy R. Moore, a student at Grinnell College, did a study where she collected several variables in an attempt to understand what predicts whether a nest will be open or closed. The response variable *Closed* is 1 for closed nests and 0 for open nests. One model to consider is the simple additive model to predict closed and open nests using linear terms for the duration of nest use (*TotCare*) and the size of the bird species (*Length*). The logit form of the fitted model is

$$logit(\hat{\pi}) = -4.351 - 0.3036Length + 0.3061TotCare$$

Assess the predictive power of this model using the "percentage correctly classified" approach discussed in Topic 11.2 (the data are in **BirdNest**). Try it two different ways: **Bird**

a. Use 0.5 as the cutoff point. If the estimated probability is above 0.5, predict that the bird builds a closed nest; otherwise, predict it's an open nest.

b. In the original sample, 57 of the 83 species (68.7%) build open nests. Use this as a cutoff, so that the predicted probability of being closed needs to exceed 0.687 before we classify the predicted nest type to be closed.

c. Discuss the effectiveness of this model to predict a closed nest and discuss the relative merits of the two choices of cutoff.

11.12 Field goals in the NFL.[3] Data in **FGByDistance** summarize all field goals attempted by place kickers in the National Football League during regular season games for the 2000 through 2008 seasons. We derived this dataset from a larger dataset, where each case represents one attempted kick and includes information on the game, date, and kicker, team of the kicker, and whether the kick was made, missed, or blocked. **FGByDistance** summarizes the 8520 attempted kicks by distance. Distances of kicks ranged from 18 yards to 76 yards. The minimum possible kick distance is 18 yards—because the shortest possible kick is from the opponent's 1-yard line, when the kicker would stand at the 8-yard line, and then another 10 yards would be added to the distance of the kick because of the 10-yard depth of the end zone. **FGDist**

The variables for **FGByDistance** are:

Dist	The distance of the attempted kick from the goal posts
N	The number of attempted kicks, of the 8520, from that distance
Makes	The number of those attempted kicks that were successful
PropMakes	The proportion of successes
Blocked	The number of those attempted kicks that were blocked
PropBlocked	The proportion of attempted kicks that were blocked

The first 10 cases of the dataset are printed here:

	Dist	N	Makes	PropMakes	Blocked	PropBlocked
1	18	15	15	1.0000000	0	0.000000000
2	19	107	105	0.9813084	1	0.009345794
3	20	204	201	0.9852941	1	0.004901961
4	21	197	194	0.9847716	0	0.000000000
5	22	244	239	0.9795082	2	0.008196721
6	23	298	291	0.9765101	2	0.006711409
7	24	241	233	0.9668050	1	0.004149378
8	25	211	202	0.9573460	5	0.023696682
9	26	237	226	0.9535865	1	0.004219409
10	27	238	226	0.9495798	1	0.004201681

a. Create a model to describe the relationship between the probability of a make as the response variable and the distance of the attempted kick as the explanatory variable. Report your findings, including a summary of your model, a description of your model-building process, and relevant computer output and graphics.

b. Repeat the previous exercise, except using the probability of a kick being blocked as the response variable.

11.13 Blue jay morphology. The data in the file **BlueJays** include *Head* (head length, in mm) and *Mass* (in gm) for a sample of blue jays. We want to study the relationship between sex (coded as M/F in the variable *KnownSex* and as 1/0 in the variable *Sex*) and these measurements. **BluJay**

a. Fit a simple logistic regression model of *Sex* depending on Head. Then use the fitted model to make predictions of male/female for each of the 123 blue jays in the sample; predict "male" if the estimated probability is above 0.50. What proportion of the blue jays are correctly classified?

b. Now fit the multiple logistic regression model of *Sex* depending on *Head* and *BillDepth*. Then use the fitted model to make predictions of male/female for each of the 123 blue jays in the sample; predict "male" if the estimated probability is above 0.50. What proportion of the blue jays are correctly classified?

Topic 11.3 Exercises: Randomization Tests

11.14 Small sample of NFL field goals. We extracted 30 cases at random from the 8520 cases in the NFL field goal data described in Exercise 11.12. The data are in **SampleFG**. Suppose we had only this much data and wanted to model the probabilty of a make as a function of distance. **SampleFG**

a. Create such a model using the maximum likelihood approach to fit a linear logistic model with distance as the one predictor. What is the estimated slope for the linear logistic model and what is its level of statistical significance?

b. Repeat part (a), except using a randomization test to ascertain the statistical significance of the slope parameter.

c. Compare the approaches in (a) and (b), especially with respect to the significance of the relationship of probability of make to distance. At the common significance levels such as 5% or 1%, would conclusions be the same using either approach? Do you have any reason to prefer one approach to the other?

The data are printed here:

ID	Yards	Results2	ID	Yards	Results2
5255	35	1	5372	36	1
1802	51	1	226	51	0
7941	42	1	4332	51	1
5836	44	1	2196	36	1
7168	28	1	2387	33	1
3763	41	1	5319	49	0
5256	22	1	5133	22	1
7445	48	0	5919	45	0
2992	43	0	4486	44	1
6523	19	1	5402	52	0
8285	33	1	4134	48	1
7612	27	1	1198	20	1
6659	48	0	1212	24	1
1505	40	1	8484	26	1
2190	26	0	6127	43	0

11.15 Gunnels: Amphiso. In Exercise 9.39, we learned of a study about the presence of a type of eel, called the gunnel, in a quadrat. In this exercise, we investigate whether this presence might be related to a variable that describes the density of a type of gunnel food in the quadrat.

The relevant response variable from the dataset **Gunnels** is the binary variable *Gunnel*, where 1 indicates the presence of gunnels in the quadrat and 0 represents the absence of any gunnels. The explanatory (ordinal) variable we consider here is *Amphiso*, a measure of the density of a type of crustacean that the gunnel eats. Here is the code for the *Amphiso* variable: $0 = 0$, $1 =$ "≤ 10," $2 = 10 - 30$, $3 = 31 - 50$, $4 =$ "≥ 50." So the values of this ordinal variable increase as the density of the crustaceans increases, with the 0 category indicating a negligible density. **Gunnel**

a. Create a two-way table that uses *Amphiso* for the row variable and *Gunnel* for the column variable. Then compute row percentages or proportions and use them to describe the nature of the relationship.

b. Now perform a randomization test to ascertain the *P*-value of the result found in (a). The logic of this test is this. The null hypothesis to be tested says that the *Amphiso* values for each case have no relationship to the response variable, *Gunnel*. Thus the *Amphiso* values can be viewed as meaningless labels and the value of any test statistic we use to measure the strength of relationship between *Amphiso* and *Gunnel* should be nothing more than one of the many possible values that would be obtained by randomly rearranging these *Amphiso* values and recomputing the test statistic.

This logic suggests the following scheme for our randomization test:

i. First, we must select some test statistic that measures the strength of a relationship between *Amphiso* and *Gunnel*. For this exercise, we will use a test statistic from a chi-square test for a two-way table, a test statistic you probably learned about in a first statistics course. When there is no relationship between the explanatory and response variables, the value of the chi-square statistic will be 0, with larger values resulting as the relationship grows stronger.

ii. Randomly permute the vector of 1592 values of the *Amphiso* vector, that is, rearrange these values.

iii. For the "new dataset" created by the original *Gunnel* vector alongside the rearranged *Amphiso* vector, compute the test statistic, and record it into a vector called "TestStat."

iv. Repeat steps (ii) and (iii) many times, for example, 100,000.

The *P*-value can now be approximated as the proportion of the 100,000 samples in which the test statistic obtained is greater than or equal to the test statistic computed from the original two-way table, from (a).

Carry out this process and report the estimate of the *P*-value found.

11.16 Gunnels: Substratum. In Exercise 9.39, we learned of a study about the presence of a type of eel, called a gunnel, in a quadrat. In this exercise, we investigate whether this presence might be related to a variable that describes the substratum of the quadrat. The relevant response variable from the dataset **Gunnels** is the binary variable *Gunnel*, where 1 indicates the presence of gunnels in the quadrat and 0 represents the absence of any gunnels.
The explanatory variable will be *Subst*, which gives the substratum information for a quadrat with these codes: **Gunnel**

1	solid rock
2	rocky cobbles
3	mixed, pebbles/sand
4	fine sand
5	mud
6	mixed mud/shell detritus
7	cobbles on solid rock
8	cobbles on mixed, pebbles/sand
9	cobbles on fine sand
10	cobbles on mud
11	cobbles on mixed mud/shell detritus
12	cobbles on shell detritus
13	shell detritus

a. Create a two-way table that uses *Subst* for the row variable and *Gunnel* for the column variable. Then compute row percentages or proportions and use them to describe the nature of the relationship.

b. Now perform a randomization test to ascertain the *P*-value of the result found in (a). The logic of this test is this. The null hypothesis to be tested says that the *Subst* values for each case have no relationship to the response variable, *Gunnel*. Thus the *Subst* values can be viewed as meaningless labels and the value of any test statistic we use to measure the strength of relationship between *Subst* and *Gunnel* should be nothing more than one of the many possible values that would be obtained by randomly rearranging these *Subst* values and recomputing the test statistic.

This logic suggests the following scheme for our randomization test: First, we must select some test statistic that measures the strength of a relationship between *Subst* and *Gunnel*. For this exercise, we will use a test statistic from a chi-square test for a two-way table, a test statistic you probably learned about in a first statistics course. When there is no relationship between the explanatory and response variables, the value of the chi-square statistic will be 0, with larger values resulting as the relationship gets stronger.

i. Randomly permute the vector of 1592 values of the *Subst* vector, that is, rearrange these values.

ii. For the "new dataset" created by the original Gunnel vector alongside the rearranged *Subst* vector, compute the test statistic, and record it into a vector called "TestStat."

iii. Repeat steps (i) and (ii) many times; for example, 100,000.

The *P*-value can now be approximated as the proportion of the 100,000 randomizations in which the test statistic obtained is greater than or equal to the test statistic computed from the original two-way table, from (a).

Carry out this process and report the estimate of the *P*-value found.

For parts (c) and (d): Suppose we had expected that the gunnels would prefer substrata involving cobbles. Notice, also, that the two-way table shows support for this expectation in the data: Categories 2, 7, 8, and 9 are the substrata categories with the highest incidence of gunnels and these all involve cobbles.

With this expectation of preference for cobbles in mind, we can redo the randomization test by using a test statistic more sensitive to the preference for cobbled substrata. This test statistic is defined very simply as follows: Take the two-way table for Subst by Gunnel and form two subtables:

Subtable 1 comprises rows 2 and 7–12;
Subtable 2 comprises the other rows.

The test statistic is then the overall incidence proportion of gunnels in Subtable 1 minus the overall incidence proportion of gunnels in Subtable 2.

c. Show that the value of the test statistic for the original data is 0.0773.

d. Now perform the randomization test using the new test statistic. Report the *P*-value from this test and state a conclusion about statistical significance.

11.17 Lost letter. Recall that in Exercise 9.49, we learned of a study about whether the address on a letter would affect the probability of return when people found the letter lying on a public sidewalk. The dataset **LostLetter** contains the data from the study by Muir and Gratch, who "lost" letters in three distinct locations: a city, a town, and a college campus. For this exercise, we investigate just the 40 letters lost in the town of Grinnell, Iowa. We use *Returned* as the response variable and *Peaceworks* for the explanatory variable. *Returned* is an indicator for whether the letter is returned to a U.S. postbox when found, and *Peaceworks* is an indicator for the address on the letter. (Recall the two addresses used were for fictitious organizations called Iowa Peaceworks and Friends of the Confederacy.) You will also need the variable *GrinnellTown*, an indicator for whether a case is for a letter lost in the town of Grinnell, to subset the dataset. **Lost**

a. Using the subset of the data for the 40 cases in the town of Grinnell, create a two-way table of *Peaceworks* by *Returned*. Use simple percentages to summarize the nature of the relationship you find there.

b. Fit a linear logistic model to the data and comment on the nature of the estimated relationship, including a statement that uses odds ratio, and comment on statistical significance.

c. Because of the small sample size here, a randomization test is a sensible alternative procedure to the significance test from the logistic model. Perform such a randomization test using the odds ratio as the test statistic. Summarize the process of computing the test, report the *P*-value, and compare it to what you found in (b).

Topic 11.4 Exercises: Analyzing Two-Way Tables with Logistic Regression

11.18 Gunnels: Binary predictor. Consider again the **Gunnels** dataset. We again try to predict the presence of a gunnel, this time using a binary variable derived from *Amphiso*. *Amphiso* gives a measure of the density of small, shrimplike crustaceans that can be classified as either amphipod or isopod. These crustaceans are a source of food for the gunnels. **Gunnel**

a. Create a binary variable, called *Crust*, that is valued 1 for a density value 1 or greater and 0 otherwise. Create a 2×2 table with *Crust* as the row identifier and *Gunnel* as the column identifier. Using simple row proportions or percentages, describe the nature of the relationship you find.

b. Use a chi-square test to confirm or deny the statistical significance of the relationship found in (a).

c. Use a logistic model to confirm the results found in (b).

d. Give the odds ratio for finding gunnels when there is a nonnegligible density of the crustaceans, and use the odds ratio in an interpretive sentence. Also, include a 95% confidence interval.

11.19 Gunnels: Ordinal predictor. In Exercise 11.18, we found that gunnels prefer quadrats where there is some food. We used as the predictor there a binary variable, *Crust*, derived from an ordinal variable, *Amphiso*, that gave more complete information about the density of the food source (the crustaceans) in the quadrat. The values for *Amphiso* are $0 = 0$, $1 =$ "≤ 10," $2 = 10 - 30$, $3 = 31 - 50$, $4 =$ "≥ 50." So the values of this ordinal variable increase as the density of the crustaceans increases, with the 0 category indicating a negligible density. **Gunnel**

a. Create a two-way table that uses *Amphiso* as the row identifier and *Gunnel* as the column identifier. Compute row proportions or percentages and use these to describe the nature of the relationship between the probability of finding a gunnel and the density of these crustaceans.

b. Use a chi-square test to confirm or deny that there is a statistically significant relationship between the two variables. *Note*: The expected counts for some of the cells are small, that is, much less than the 5.0 that often leads to erroneous results. First, perform the chi-square test on the original 5×2 table. Your software will likely give you an error message. Combine the categories 2, 3, or 4 into a new category 2 and then perform the chi-square test again. (You might still get an error message because one expected value is 4.97, slightly below 5. But we view this discrepancy as not worth worrying about.)

c. Use a logistic regression model with indicator variables to predict gunnel presence with the density of the crustaceans. Use the 0 density category as the omitted category. Include the summary table from the logistic regression and interpret what the results tell you about the relationship.

11.20 Gunnels: Nested likelihood ratio test. In Exercise 11.19 , we saw that the differences between the 2, 3, and 4 categories appear small. We can formally test this observation using a nested LRT test. **Gunnel**

a. Define a variable called *Aisum* as $Aisum = AI2 + AI3 + AI4$. Now fit a linear logistic model that predicts *Gunnel* using *AI1* and *Aisum*. Produce the summary table of output for this model.

b. Now fit the model that predicts *Gunnel* using *AI1*, *Aisum*, *AI2*, and *AI3*. Produce the summary table.

c. Explain why the model fit in (a) can be considered a reduced model to that fit in (b).

d. Now use the nested likelihood-ratio test for this instance of a full and reduced model. State a conclusion and interpret the results in terms of a preferred model for predicting the presence of gunnels from the density of these crustaceans.

11.21 Gunnels: Water. The *Water* variable is a binary variable that is 1 when there is standing water in the quadrat and 0 if there is not. One might expect that the presence of standing water would increase the chances of finding this water-loving creature. Does it? **Gunnel**

a. Form a 2×2 table of counts and state your conclusion to the main question using a simple comparison of percentages.

b. Use a two-proportion z-test to ascertain the statistical significance of your conclusion from (a).

c. Use a chi-square test to ascertain the statistical significance of your conclusion from (a).

d. Now use a linear-logistic model fit to ascertain the statistical significance of the conclusion from (a).

e. Relate the slope estimate from (d) to the two-way table in (a).

11.22 Pines: Binary predictor. The dataset **Pines** contains data on an experiment conducted by the Department of Biology at Kenyon College at a site near the campus at Gambier, Ohio. In April 1990, student and faculty volunteers planted 1000 white pine (*Pinus strobes*) seedlings at the Brown Family Environmental Center. These seedlings were planted in two grids, distinguished by 10- and 15-foot spacings between the seedlings. Several other variables were measured and recorded for each seedling over time. In the dataset are all cases for which complete data are available.

One question of interest was whether the spacing seemed to affect the propensity of deer to browse the trees. Browsing involves destroying parts of the tree, such as the bark, and can damage the trees.

The variable *Spacing* refers to the experimental condition and takes on the values 10 and 15. The variable *Deer95* is an indicator for deer browsing for the year 1995, where a tree gets a value of 1 if the tree was browsed and 0 if it was not. *Deer97* is the similar variable for the year 1997. **Pines**

a. Form a two-way table using *Spacing* as the predictor variable and *Deer95* as the response variable and describe the nature of the relationship, using simple percentage comparisons.

b. Find the odds-ratio from the two-way table in (a) and interpret it in simple prose.

c. Use a two-sample z-test to decide if the difference in browsing is significantly different for the two values of tree spacing.

d. Confirm the results of (c) using a simple linear logistic model. Summarize the results.

e. Repeat (a–d) for *Deer97*, and then comment on the main difference within the data between 1995 and 1997.

11.23 Pines: Ordinal predictor. Consider again the **Pines** dataset. This time we investigate the relationship between the amount of thorny cover in the forest and the propensity of deer to browse a tree. One thought is that the more thorny cover on the forest floor, the less likely the deer will be to browse trees there. The amount of thorny cover was measured and recorded in 1995 and becomes the variable *Cover95* in the **Pines** dataset. This ordinal, categorical variable has four values, which are defined as 0 = no thorny cover, 1 = some thorny cover but less than 1/3 of the land covered by thorny brush, 2 = between 1/3 and 2/3 covered with thorny brush, and 3 = greater than 2/3 of the land covered with thorny brush. **Pines**

a. Use a two-way table to investigate the question for 1995, that is, use *Deer95* as the response variable and *Cover95* as the predictor variable. Describe what you see in this table, using simple percentage comparisons for your description.

b. Use a chi-square test to decide if the relationship described in (a) is statistically significant.

c. Now use a linear logistic model to predict *Deer95* with the categorical predictor *Cover95*. Use the absence of cover, that is, *Cover95* = 0, as the omitted category. Summarize the results and compare them to what you found in (b).

d. Use the model from (c) to estimate the odds ratios for each of the categories of *Cover95* compared to the absence of thorny cover.

e. Compute the percentage of correct predictions of browsings that the model makes, defining a predicted browse to be any case in which the model predicts greater than a 0.244216 probability of browsing. The 0.244216 value is simply the baseline rate of browsing for 1995, that is, 0.244216 = 190/778.

f. Repeat (a–c) for the *Deer97* response variable and summarize how 1995 and 1997 deer browsing differs.

11.24 Pines: Nested likelihood ratio test. We now investigate the relationship between the pair of predictors *Cover95* and *Spacing* on deer browsing. **Pines**

a. Using *Deer95* as the response variable, fit a linear logistic model using *Cover95* and *Spacing* as the predictors. Summarize the results: Which predictors are statistically significant and what is the nature of the relationship of each to *Deer95* (i.e., to deer browsing of the trees)?

b. Use the model to estimate the odds ratio for a deer browsing a tree for the 15-foot versus the 10-foot spacing and include a 95% confidence interval for this odds ratio.

c. Use a nested likelihood-ratio test to decide if the categorical variable *Cover95* is important to the predictive power of the model, beyond the *Spacing* predictor. Report the relevant computer output and interpret the results.

d. Fit a logistic model using *Deer97* as the response variable and *Spacing* and *Cover95* as the predictors and comment on the results.

11.25 Lost letters. Consider the data from Exercise 9.49 on page 452, which describes a study of intentionally losing letters to ascertain how different addresses affect return rates. The data are in **LostLetter**. **Lost**

a. The study dropped letters in three locations (Des Moines, Grinnell town, and Grinnell College campus). Do a chi-square analysis to decide if there is a relationship between *Location* and *Returned*. Report a *P*-value and state a conclusion. If there is a significant relationship, explore the tables of observed and expected counts and describe the nature of the relationship in simple terms.

b. Fit a logistic model of *Returned* on *Location*. Describe the statistical significance of the coefficients and interpret the results in the context of the dataset.

c. Test whether the model in (b) shows that *Location* is helpful for explaining *Returned*. State a conclusion, making clear what two models you are comparing.

d. Redo the chi-square analysis of part (a) separately for both addresses: Iowa Peaceworks and Friends of the Confederacy. Summarize your findings.

e. Now fit a logistic model that models *Returned* on *Location* and *Address*. Summarize your findings.

f. What advantages or disadvantages does the logistic analysis of part (e) have over the chi-square analysis of part (d)?

g. Write down explicitly the empirical proportions for the saturated model being considered by the goodness-of-fit test and, for comparison, the corresponding linear logit estimated probabilities that these saturated logits are being compared to.

11.26 Tipping study. Consider Example 11.7, the tipping study on page 516, where we consider three indicator variables for the type of card. Some output for that model is shown below. Answer the following questions using just this output, and *not* using any computer software: **Tip**

Predictor	Coef	SE Coef	Z	P	Odds Ratio	95% CI Lower	95% CI Upper
Constant	-1.11923	0.287938	-3.89	0.000			
Joke	0.782759	0.374234	2.09	0.036	2.19	1.05	4.56
Ad	-0.336056	0.413526	-0.81	0.416	0.71	0.32	1.61

a. Write down the linear logit equation that includes *Joke* and *None* as the two explanatory variables.

b. Write down the linear logit equation that includes *Ad* and *None* as the two explanatory variables.

11.27 Titanic. Recall the **Titanic** data that describe fatalities from the sinking of the famed British luxury liner in 1912. In Exercise 9.28, we investigated the relationship between *Survived*, the binary variable that is valued 1 for a person's surviving and 0 otherwise, and *SexCode*, an indicator for femaleness. **Titanic**

a. Use software to fit a logistic model of *Survived* predicted by *SexCode* to decide whether there is a statistically significant relationship between sex and survival.

b. Confirm the results of the significance of the slope coefficient by using a chi-square test on the two-way table and comparing *P*-values.

11.28 THC versus prochlorperazine. Exercise 9.43 compared two drugs for combating nausea in chemotherapy recipients. The two drugs were THC and prochlorperazine, the latter being the standard drug and THC being a new drug. **Chemo**

a. Use a linear logistic model to ascertain, at the 1% significance level, if THC is the more effective drug.

b. Address the question in part (a) using a two-sample *z*-test to compare the proportions of effective cases between the THC and prochlorperazine groups.

c. Try one more test, this time using a chi-square test based on the observed and expected counts in the 2×2 table.

d. Are the results of the three tests in parts (a), (b), and (c) consistent with each other? Are any of them identical in the strength of evidence? Comment.

11.29 Intensive care unit. Refer to the **ICU** data on survival in a hospital ICU as described in Exercise 10.29. In addition to the quantitative *Age* variable, the data have a categorical variable that classifies patients into three broad groups based on age. We'll call patients "Young" if they are under 50 years old ($AgeGroup = 1$), "Middle" if they are in their 50s or 60s ($AgeGroup = 2$), and "Old" if they are 70 or older ($AgeGroup = 3$). **ICU**

a. Produce a two-way table showing the number of patients who lived and died in each of the three age groups. Also compute the sample proportion who survived in each age group. What do those proportions tell us about a possible relationship (if any) between *AgeGroup* and *Survive*?

b. Suppose that we ignore the fact that *AgeGroup* is really a categorical variable and just treat it as a quantitative predictor of *Survive*. Fit that logistic regression model. Compute the estimated proportion surviving in each age group from this model.

c. Use the logit form of the model in (b) to find the log(odds) for survival in each of the age groups. Since the value coding the "Middle" group ($AgeGroup = 2$) is exactly between the coding for the other two groups ($AgeGroup = 1$ and $AgeGroup = 3$), what does that imply about estimated log(odds) for the "Middle" group? Given what you observed in the two-way table in part (a), is that reasonable for the ICU data and these age groups?

d. Create indicator variables for the age groups and use any two of them in a multiple logistic regression model to predict *Survive*. Explain what the coefficients of this model indicate about how the odds of ICU survival are related to the age groups.

e. Check that the survival proportions estimated for each age group using the model in (d) match the sample proportions you found part (a).

f. Perform a chi-square test on the two-way table from part (a). Compare the results to the chi-square test for the effectiveness of the logistic model you fit using the indicator variables in part (d). Are the results consistent? What do they tell you about the relationship (or lack of a relationship) between the age groups and ICU survival?

g. Try a logistic model using the actual ages in the **ICU** dataset, rather than the three age groups. Does this provide a substantial improvement? Give a justification for your answer.

Topic 11.5 Exercises: Simpson's Paradox

11.30 Death penalty in Florida. The table that follows shows data for defendants convicted of murder in Florida. There are three variables present: the race of the defendant, the race of the victim, and whether or not the defendant was sentenced to death. **FLDP**

	White victim			Black victim	
	Defendant			Defendant	
	White	Black		White	Black
Death penalty	19	11	Death penalty	0	6
No DP	132	52	No DP	9	97
Total	151	63	Total	9	103

a. Construct a two-way table that shows how defendant's race is related to the death penalty. That is, aggregate over victim's race.

b. Show that Simpson's Paradox is present here: Confirm that the death penalty percentage is higher for white defendants than for black defendants in the aggregate table (from part (a)) but is lower when the victim is white and when the victim is black.

c. Explain how the relationship between defendant's race and victim's race causes the paradox to be present. Use everyday language, as if you were explaining the situation to someone who has not studied statistics.

11.31 Smoking status. Smoking status and age were recorded for 1314 women in Whickham, England. A follow-up study 20 years later made note of mortality. The following table shows the data. **Whick2**

	Baseline age 18-64			Baseline age 65+	
	Smoker	Nonsmoker		Smoker	Nonsmoker
Alive	437	474	Alive	6	28
Dead	95	65	Dead	44	165
Total	532	539	Total	50	193

a. Construct a two-way table that shows how smoking status is related to mortality. That is, aggregate over age.

b. Show that Simpson's Paradox is present here: Confirm that the percentage alive is higher for smokers than for nonsmokers in the aggregate table (from part (a)) but is lower when the woman is young and when the woman is old.

c. Explain how the relationship between smoking status and age causes the paradox to be present. Use everyday language, as if you were explaining the situation to someone who has not studied statistics.

11.32 Death penalty in Florida: Logistic model. The raw data from Exercise 11.30 are available in the file **FloridaDP**. Use the data to fit a logistic regression model that uses race of the victim and race of the defendant to predict whether or not the death penalty is imposed. **FLDP**

a. Is the coefficient of *White Victim* positive or negative? What does that mean?

b. Is the coefficient of *DefendantWhite* positive or negative? What does that mean?

c. Which predictor, victim's race or defendant's race, has a stronger relationship with the response variable?

11.33 Smoking status: Logistic model. The raw data from Exercise 11.31 are available in the file **Whickham2**. Use the data to fit a logistic regression model that uses *AgeGroup* and *Smoker* to predict *Alive*. **Whick2**

a. Is the coefficient of *AgeGroup* positive or negative? What does that mean?

b. Is the coefficient of *Smoker* positive or negative? What does that mean?

c. Which predictor, age group or smoking status, has a stronger relationship with the response variable?

11.34 Graduation rates. Are athletes more likely or less likely to graduate from college than nonathletes? The table that follows shows 6-year graduation data for white and black football and basketball players and for male nonathletes at NCAA Division I schools. **AthGrad**

	White			Black	
	Athlete	Non-athlete		Athlete	Non-athlete
Graduate	1,088	112,906	Graduate	970	7895
Nongraduate	659	75,270	Nongraduate	1,106	14,661
Total	1747	188,176	Total	2,076	22,556

a. Calculate the graduation rate for white athletes and for white nonathletes. Then do the same for black athletes and black nonathletes. For each race, which rate is higher?

b. Here is the table that combines data across race. Calculate the graduation rate for athletes and for nonathletes. Which is higher?

	Athlete	Non-athlete
Graduate	2058	120,801
Non-graduate	1765	89,930
Total	3823	210,732

c. Explain how the relationship between race and being an athlete causes the Simpson's Paradox to be present.

11.35 Cleveland Cavaliers basketball. During the 2016–17 season, Kyrie Irving and Channing Frye both played basketball for the Cleveland Cavaliers. Who was the better shooter? Basketball players can attempt "2-point" shots or "3-point" shots. Following are the data for each player for 2-point shots and for 3-point shots. **Cavs**

	2-point shots			3-point shots	
	Irving	Frye		Irving	Frye
Hit	494	101	Hit	177	137
Miss	485	84	Miss	264	198
Total	979	185	Total	441	335

a. Calculate the Hit percentage for Irving on 2-point shots and for Fry on 2-point shots. Then do the same for 3-point shots. For each type of shot, who had the higher success rate?

b. Here is the table that combines data across both kinds of shots. Calculate the Hit percentage rate for Irving and for Frye. Which is higher?

	Irving	Frye
Hit	671	238
Miss	749	282
Total	1420	520

c. Explain how the relationship between type of shot and player causes the Simpson's Paradox to be present.

11.36 Graduation rates: Logistic model. The raw data from Exercise 11.34 are available in the file **AthleteGrad**. **AthGrad**

a. Use the data to fit a logistic regression model that uses *Student* to predict whether or not a student graduates within 6 years. Is the coefficient of *StudentNonAthlete* (i.e., the indicator that the student was not an athlete) positive or negative? What does that mean?

b. Now use the data to fit a logistic regression model that uses *Race* and *Student* to predict whether or not a student graduates within 6 years. Is the coefficient of *RaceWhite* (i.e., the indicator that the student was white) positive or negative? What does that mean?

c. Is the coefficient of *StudentNonAthlete* (i.e., the indicator that the student was not an athlete) positive or negative in the model from part (b)? What does that mean?

d. Which predictor, *Race* or *Student*, has a stronger relationship with the response variable?

11.37 Cleveland Cavaliers basketball: Logistic model. The raw data from Exercise 11.35 are available in the file **CavsShooting**. Cavs

a. Use the data to fit a logistic regression model that uses *Player* to predict whether or not a shot is hit. Is the coefficient of *PlayerIrving* (i.e., the indicator that the player was Irving) positive or negative? What does that mean?

b. Now use the data to fit a logistic regression model that uses *ShotType* and *Player* to predict whether or not a shot is hit. Is the coefficient of *ShotTypeTwo* (i.e., the indicator that a shot was a 2-point shot) positive or negative? What does that mean?

c. Is the coefficient of *PlayerIrving* (i.e., the indicator that the player was Irving) positive or negative in the model from part (b)? What does that mean?

d. Which predictor, *ShotType* or *Player*, has a stronger relationship with the response variable?

© Gary M. Hart, Gary Hart Photography, www.EloquentImages.com

UNIT D

Unit D: Time Series Analysis

CHAPTER 12: TIME SERIES ANALYSIS

Fit models to time series based on time and past history of the series. Measure relationships within a series using autocorrelation and use differencing to obtain stationarity. Model seasonal patterns in time series. Build ARIMA models that use past history of a times series and residuals.

Rapax/Shutterstock

CHAPTER 12

Time Series Analysis

In this chapter you will learn to:

- Fit regression models based on time to explain patterns found in time series data.
- Use autocorrelations to measure dependence within a time series.
- Use differencing to help obtain a stationary series.
- Fit models to account for seasonal trends.
- Build ARIMA models based on the past history of a time series and residuals.

In Section 1.2 on page 27 we considered several conditions for regression models. One of these is the condition of *independence*, stating that the residual for any one data point should be independent of the residuals for other data points. When this was first introduced, we advised considering how the data were collected (for example, with a random sample) as a way to assess whether the condition is met. In this chapter we consider a specific type of data, a **time series**, where the independence condition is often violated.

time series

TIME SERIES

A **time series** is a quantitative variable that is measured at regular intervals in time.

For example, we might be interested in the

- Number of new hits to cnn.com each hour
- Daily closing price of the S&P 500 stock index
- Average monthly temperature over several years in Minneapolis, Minnesota
- Quarterly heating oil use for a residence in Lisbon, Iowa
- Yearly sales at Amazon.com

In each of these data situations we may have *historical data* from past time periods (hourly, daily, monthly, etc.) and may be interested in building models to forecast values for future time periods. In such settings we *want* to have some dependence between adjacent data values that we can exploit to help improve those predictions. We can use the time variable as a predictor (Section 12.1) or the past history of the series itself (Sections 12.2 and 12.3).

12.1 Functions of Time

EXAMPLE 12.1

SeaIce

Arctic ice Climatologists have used satellite images to estimate the amount of ice present in the Arctic Ocean each year since 1979. The measurements[1] (in millions of square kilometers) are taken each September when the ice coverage is at a minimum. Yearly data from 1979 to 2015 are stored in **SeaIce**. One of the variables (*Extent*) measures the area inside the outer boundaries created by the ice (which may include some interior pockets of open water). The *Area* variable estimates the area of just the ice itself. Some of these values are shown in Table 12.1. For this example we will focus on models to forecast the *Extent* in future years.

Year	1979	1980	1981	1982	...	2011	2012	2013	2014	2015
Extent	7.22	7.86	7.25	7.45	...	4.63	3.63	5.35	5.29	4.68
Area	4.54	4.83	4.38	4.38	...	3.18	2.37	3.75	3.70	3.37

TABLE 12.1 Yearly *Extent* and *Area* for September sea ice in the Arctic Ocean

Time Series Plot

time series plot

We generally display the structure of a historical time series with a **time series plot**, which is just a modified scatterplot with time on the horizontal axis and the time series values on the vertical axis. Since each time point has a single value, we use a line to connect adjacent points (and often we omit the points entirely and show only the lines). Figure 12.1 shows a time series plot for the *Extent* variable in the **SeaIce** dataset. The plot shows a generally decreasing trend with the amount of Arctic sea ice tending to go down over this time period.

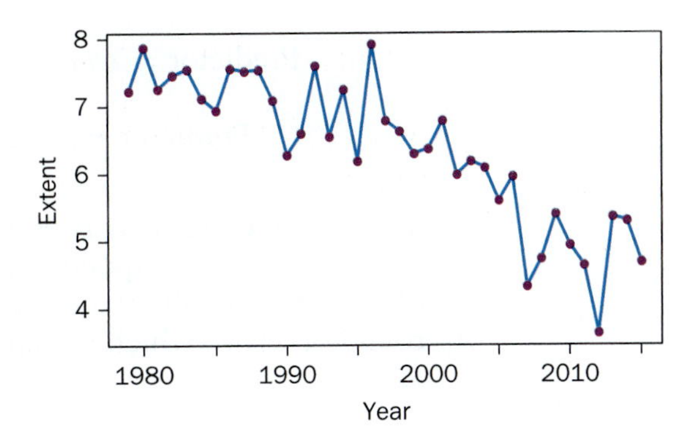

FIGURE 12.1 Time series plot of *Extent* for Arctic ice from 1979–2015

EXAMPLE 12.2 Linear fit for Arctic ice

SeaIce

CHOOSE/FIT

One way to model this trend is to use a simple linear model (as in Chapter 1) with a time variable as the predictor.

$$Extent = \beta_0 + \beta_1 Year + \epsilon$$

Some output for fitting a regression line based on *Year* to explain sea ice *Extent* is shown below along with a visual display in Figure 12.2.

```
Coefficients:
              Estimate  Std. Error  t value  Pr(>|t|)
(Intercept) 180.728966   17.396966    10.39  3.10e-12
Year         -0.087321    0.008711   -10.02  7.97e-12
---

Residual standard error: 0.5658 on 35 degrees of freedom
Multiple R-squared: 0.7417, Adjusted R-squared: 0.7343
F-statistic: 100.5 on 1 and 35 DF, p-value: 7.968e-12
```

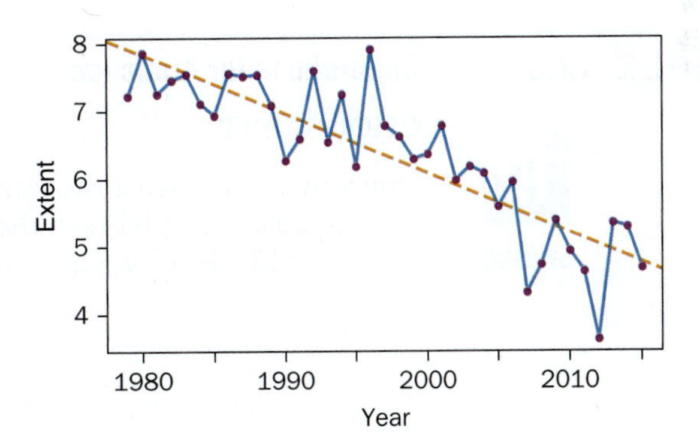

FIGURE 12.2 Arctic sea ice with linear fit

The *Year* variable is a convenient numeric predictor in the sea ice example, but some time scales (e.g., monthly data over several years) may not have an obvious numeric analog. For this reason, we commonly use a generic time scale of $t = 1, 2, 3, ..., n$ to code the time variable. For a single predictor model, this use of t only changes the intercept of the fitted model. For example, the slope, regression diagnostics, and visual display of the fit (other than labeling the horizontal scale) are unchanged when using the generic time scale, rather than *Year* in the Arctic sea ice example.

$$\text{Predictor} = Year: \quad \widehat{Extent} = 180.73 - 0.0873Year$$

$$\text{Predictor} = t: \quad \widehat{Extent} = 8.0080 - 0.0873t$$

For example, when $Year = 2000$ we get $180.729 - 0.08732(2000) = 6.089$. But $Year = 2000$ corresponds to $t = 22$ (since the first year is 1979) and when $t = 22$ we get $8.0080 - 0.08732(22) = 6.087$. The slight difference is just due to round-off error, which is more pronounced when the larger $Year = 2000$ is used.

ASSESS

We can also plot the residuals from the linear model for Arctic ice extent (using t as the predictor) as a time series as shown in Figure 12.3. We see that the residuals tend to be negative in the early years, mostly positive in the middle, generally turning negative for later years. Such a pattern in a residual plot indicates that the residuals are not randomly distributed and suggests that we might want to account for some curvature in the series over time.

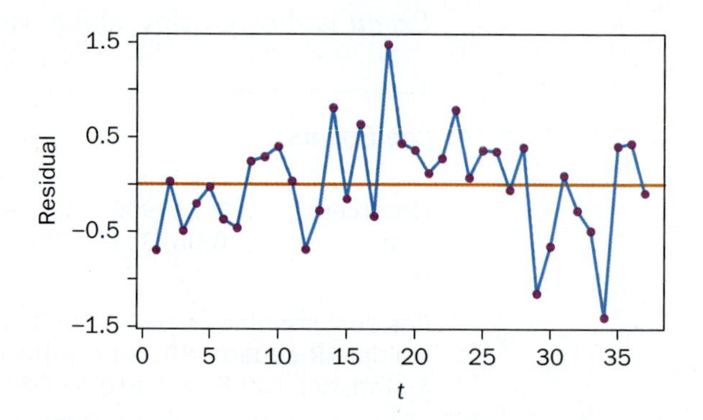

FIGURE 12.3 Residuals for linear fit of sea ice extent

EXAMPLE 12.3 **Quadratic fit for Arctic ice**

SeaIce

CHOOSE/FIT

One way to deal with the curvature in the time series of Arctic sea ice is to fit a quadratic model (see the discussion on polynomial regression starting on page 113). Here we use t and t^2 as predictors for *Extent* with the **SeaIce** data.

```
Coefficients:
             Estimate  Std. Error  t value  Pr(>|t|)
(Intercept)  7.470476  0.273792    27.285   <2e-16
t           -0.004622  0.033225    -0.139   0.8902
tsq         -0.002176  0.000848    -2.566   0.0149
---
Residual standard error: 0.5254 on 34 degrees of freedom
Multiple R-squared: 0.7836, Adjusted R-squared: 0.7708
F-statistic: 61.55 on 2 and 34 DF, p-value: 5.016e-12
```

ASSESS

We see that the quadratic term is fairly significant (P-value $= 0.0149$), the value of R^2 increases to 78.4% (from 74.2% in the linear model), the standard error of the regression is smaller (down from 0.57 to 0.53), and the curved fit in Figure 12.4 does a better job of capturing the general trend of the data. (Exercise 12.20 asks you to plot the residuals for this model to see if the curved pattern in the residuals has been removed.) The negative coefficients indicate that the Arctic sea ice is diminishing over time at an accelerating rate—a finding that has unfortunate implications for those concerned with possible effects of climate change.

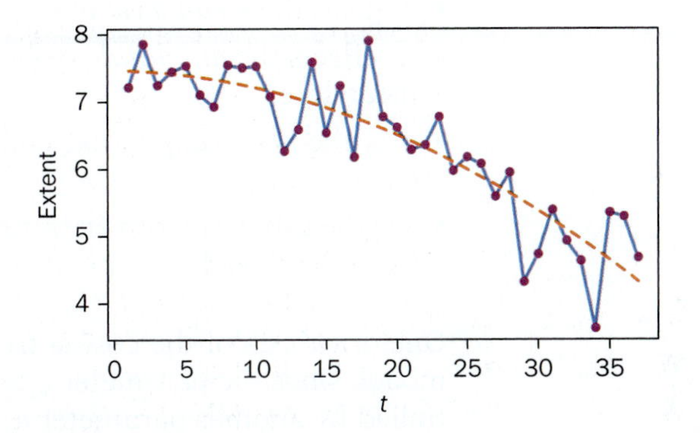

FIGURE 12.4 Quadratic fit for yearly Arctic ice extent

EXAMPLE 12.4

Bridge12

Peace Bridge traffic Figure 12.5 shows the monthly traffic (in thousands of vehicles) over the Peace Bridge[2] connecting the United States and Canada near Niagara Falls. The data for this time period (2012–2015) are stored in **PeaceBridge2012**. The dominant feature in the time series plot is a periodic pattern with high traffic volume in the summer months (especially July and August) and fewer vehicle crossings in the winter months. This is known as a *seasonal* time series, which exhibits a regular, repeating pattern over a fixed, predictable time period (S). Monthly data usually have a seasonal period of $S = 12$, quarterly data have $S = 4$, and daily data might have $S = 7$ or $S = 5$ depending on whether weekends are included. We'll consider two different models for dealing with such seasonal patterns.

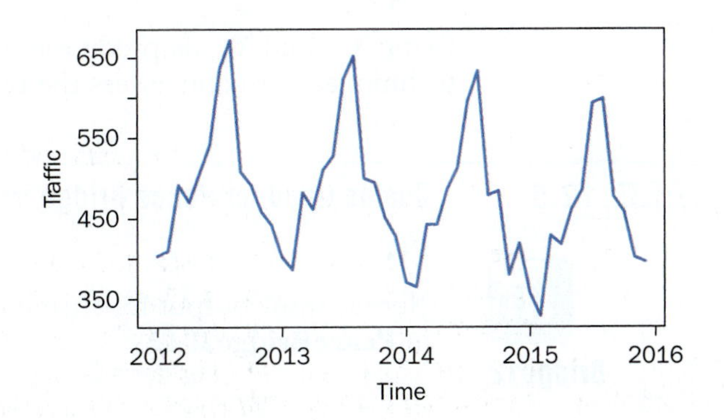

FIGURE 12.5 Monthly traffic (in thousands) over the Peace Bridge (2012–2015)

Cosine Trend Model

In mathematics we often use trigonometric functions, particularly sin and cos, to produce a periodic pattern. To model the structure seen in a seasonal time series we need to add some parameters to allow the model to fit the pattern of the data. One such model* for a series with time denoted by t and seasonal period S is

$$Y = \beta_0 + \alpha \cos\left(\frac{2\pi t}{S} + \phi\right) + \epsilon$$

where

- β_0 gives the *mean level* of the series
- α reflects the *amplitude* or how far the peaks and valleys extend beyond the mean level
- ϕ is the *phase angle* that controls the time at which the peaks and valleys occur
- ϵ is the usual random error term accounting for deviations in the actual data from the model

One drawback of the cosine trend model in this form is that it is not a linear model, since the parameter ϕ is buried within the cosine function that is multiplied by another parameter α. This means we cannot use our usual multiple regression techniques to estimate the parameters and fit the model when the model has that form. Fortunately, with a bit of algebraic manipulation we can obtain an equivalent form of this model.†

$$Y = \beta_0 + \beta_1 \cos\left(\frac{2\pi t}{S}\right) + \beta_2 \sin\left(\frac{2\pi t}{S}\right) + \epsilon$$

Since we know the value of the time variable t and period S, we create two new predictors

$$X_{\cos} = \cos\left(\frac{2\pi t}{S}\right) \qquad \text{and} \qquad X_{\sin} = \sin\left(\frac{2\pi t}{S}\right)$$

to obtain the equivalent multiple regression model

$$y = \beta_0 + \beta_1 X_{\cos} + \beta_2 X_{\sin} + \epsilon$$

Using $X_{\cos}$ and $X_{\sin}$ as predictors we can use ordinary least squares regression techniques to fit and assess the two-predictor model.

EXAMPLE 12.5 **Cosine trend for Peace Bridge traffic**

Bridge12

FIT

Here is some output for fitting the cosine trend model to the *Traffic* variable in **PeaceBridge2012**.

*You can find an online app to explore how the parameters of this cosine model affect the shape of the fitted curve at *http://shiny.stlawu.edu:3838/sample-apps/CosineTrend/*.

†If we need to recover the original parameters, $\alpha = \sqrt{\beta_1^2 + \beta_2^2}$ and $\phi = \tan^{-1}\left(\frac{-\beta_2}{\beta_1}\right)$.

```
Coefficients:
            Estimate  Std. Error  t value   Pr(>|t|)
(Intercept)  478.325       6.061   78.916    < 2e-16
Xcos         -78.217       8.572   -9.125   8.45e-12
Xsin         -61.688       8.572   -7.197   5.21e-09
---

Residual standard error: 41.99 on 45 degrees of freedom
Multiple R-squared: 0.7501, Adjusted R-squared: 0.739
F-statistic: 67.53 on 2 and 45 DF, p-value: 2.824e-14
```

This gives a fitted model of

$$\widehat{Traffic} = 478.3 - 78.2\cos\left(\frac{2\pi t}{12}\right) - 61.7\sin\left(\frac{2\pi t}{12}\right)$$

ASSESS

Note that t-tests show that the coefficients of both the $X_{\cos}$ and $X_{\sin}$ terms are clearly nonzero; however, in general, we don't worry much about those tests of individual terms since we need both to get the equivalent form of the cosine trend model. To judge effectiveness of the cosine model we rely more on the R^2 value (75.0%) and F-test for the overall model ($F = 67.53$, P-value ≈ 0), which is quite significant. We also look at a time series plot of the traffic series along with the fitted cosine trend as in Figure 12.6. This shows a pretty good fit to the seasonal pattern in bridge traffic, although it tends to consistently fall short of the highest peaks in the summer months.

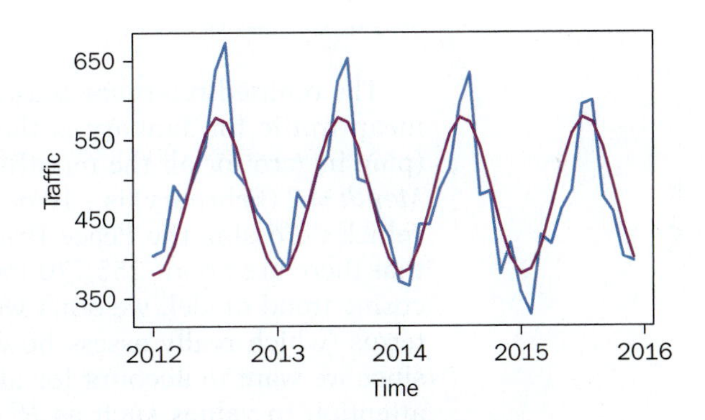

FIGURE 12.6 Cosine fit (in purple) for Peace Bridge traffic

Seasonal Means Model

An alternative to the cosine trend model for seasonal data is to allow for a different mean value within each season. One way to accomplish this within a multiple regression framework is to treat the seasonal variable (e.g., month or quarter) as a categorical predictor. As we consider in Topic 4.5, the basic idea is to create indicator variables for each seasonal period and then include all but one of these variables in the regression model. For example, with monthly data we might exclude the indicator for January and use

$$Y = \beta_0 + \beta_1 Feb + \beta_2 Mar + \beta_3 Apr + \beta_4 May + \beta_5 June + \cdots + \beta_{10} Nov + \beta_{11} Dec + \epsilon$$

Recall that β_0 represents the mean level for the omitted season (January) and the other coefficients reflect how the mean for that season differs from the mean of the reference season.

EXAMPLE 12.6 Seasonal means for Peace Bridge traffic

Bridge12

FIT

Here is some output for fitting the seasonal means model to the *Traffic* variable in **PeaceBridge2012**. The months are coded as $1, 2, \ldots, 12$ in the *Month* variable so the as.factor(Month) notation in the output arises from having the software (R) treat *Month* as a categorical predictor.

```
Coefficients:
                    Estimate  Std. Error  t value   Pr(>|t|)
(Intercept)           383.08       12.95   29.576    < 2e-16
as.factor(Month)2     -10.68       18.32   -0.583   0.563676
as.factor(Month)3      78.22       18.32    4.271   0.000136
as.factor(Month)4      64.90       18.32    3.543   0.001116
as.factor(Month)5     108.02       18.32    5.897   9.53e-07
as.factor(Month)6     133.47       18.32    7.287   1.39e-08
as.factor(Month)7     228.97       18.32   12.500   1.17e-14
as.factor(Month)8     255.77       18.32   13.963   4.21e-16
as.factor(Month)9     107.77       18.32    5.884   9.94e-07
as.factor(Month)10     99.05       18.32    5.407   4.31e-06
as.factor(Month)11     39.55       18.32    2.159   0.037579
as.factor(Month)12     37.92       18.32    2.070   0.045642
---
Residual standard error: 25.9 on 36 degrees of freedom
Multiple R-squared: 0.9239, Adjusted R-squared: 0.9007
F-statistic: 39.74 on 11 and 36 DF, p-value: < 2.2e-16
```

The omitted reference season is $Month = 1$ (January), so the estimated mean traffic for January is the constant term, 383.08 (thousand) vehicles (plug in zero for all the months in the model). The estimated coefficient of $Month = 2$ (February) is -10.68, so, on average, there are about 10,680 fewer vehicles crossing the Peace Bridge in February than in January; but notice that there are about 255,770 more vehicle crossings in August. As with the cosine trend model, we don't worry too much about the tests for individual terms (which really assess how each month's mean differs from January), since we want to account for all months in the model. However, we do pay attention to values such as R^2 (92.4%) and the overall *F*-test ($F = 39.74$, *P*-value ≈ 0), which show that the seasonal means model is quite effective at explaining the bridge traffic.

ASSESS

Figure 12.7 shows a time series plot of the seasonal means fit over this time period. It does a much better job than the cosine trend model at explaining the sharp summer peaks and has a much better R^2 (92.4% to 75.0%), adjusted R^2 (90.1% to 73.9%), and standard deviation of the error term (25.9 to 42.0). This is not surprising since the seasonal means model is using 12 parameters (including the constant term) for just 48 data points compared to only 3 parameters for the cosine trend model. Also note that the estimate for each month in the seasonal means model uses only the four values in the dataset at that month and completely ignores the fact that adjacent months might have similar or related values.

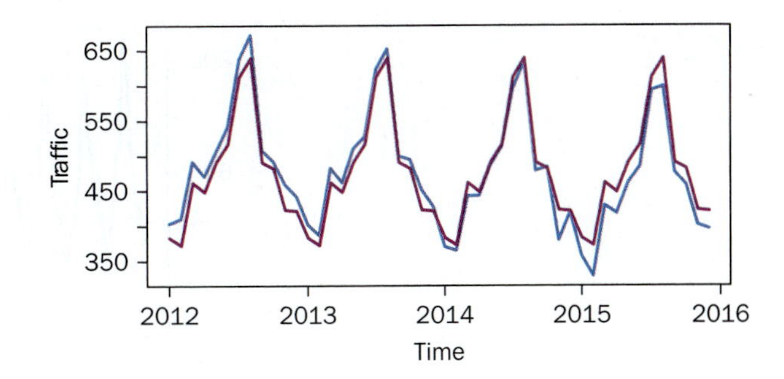

FIGURE 12.7 Seasonal means fit (in purple) for Peace Bridge traffic

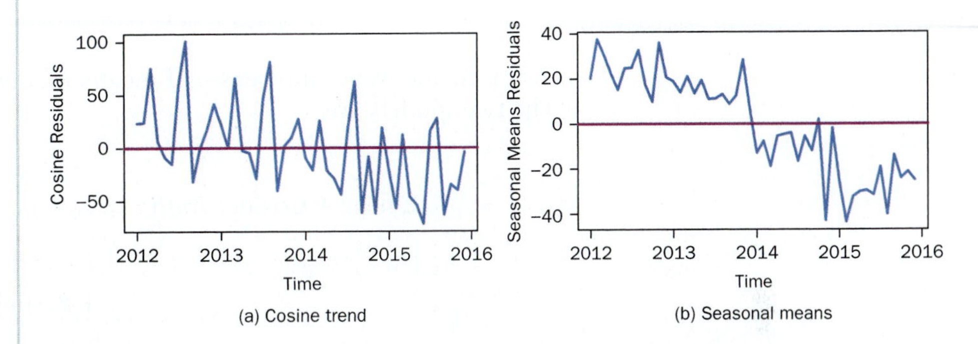

FIGURE 12.8 Residuals for models to predict Peace Bridge traffic

Figure 12.8 shows time series plots of the residuals for both the cosine trend (Figure 12.8(a)) and seasonal means (Figure 12.8(b)) models to predict Peace Bridge traffic. While the trend is more obvious in the seasonal means residuals, both plots show a tendency for residuals to decrease from generally positive in the first two years to more negative in the second two years. Thus the residuals may not be independent and there may still be some structure that we can extract from the time series to improve the model.

One of the benefits to the "function of t" regression models we consider in this section is that they may be combined in the same model to account for different features of a time series. For example, we can include a t term to handle an increasing or decreasing trend while using seasonal means or cosine terms to reflect seasonal patterns.

EXAMPLE 12.7

Bridge03

Peace Bridge traffic (2003–2015) Figure 12.9 shows a longer series of monthly traffic over the Peace Bridge from 2003 to 2015. The data for these 156 months are stored in **PeaceBridge2003**. The regular seasonal pattern with high peaks in the summer and low values in the winter is still very apparent and the generally decreasing trend as time goes along is more visible.

CHOOSE

We consider two models to account for both the decreasing trend and seasonal pattern by adding a linear term for the time variable ($t = 1, 2, \ldots, 156$)

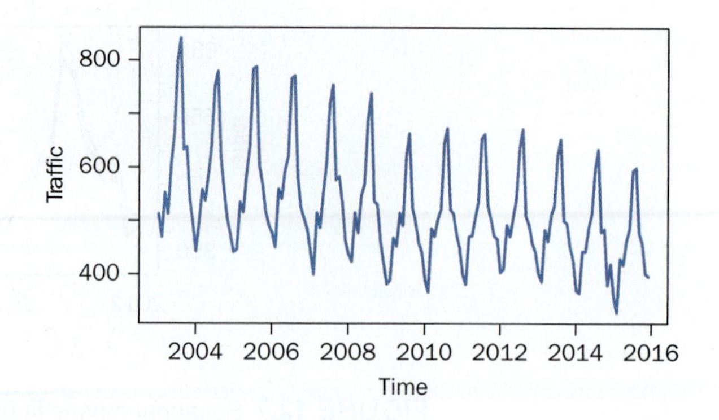

FIGURE 12.9 Monthly Peace Bridge traffic from 2003–2015

to the cosine trend and seasonal means models of Examples 12.5 and 12.6. The two models are

$$\text{Linear + Cosine: } \mathit{Traffic} = \beta_0 + \beta_1 t + \beta_2 \cos\left(\frac{2\pi t}{S}\right) + \beta_3 \sin\left(\frac{2\pi t}{S}\right) + \epsilon$$

$$\text{Linear + Seasonal Means: } \mathit{Traffic} = \beta_0 + \beta_1 t + \beta_2 \mathit{Month2} + \beta_3 \mathit{Month3} + \cdots + \beta_{12} \mathit{Month12} + \epsilon$$

FIT

Fitting the two models gives the following prediction equations:

$$\widehat{\mathit{Traffic}} = 601.50 - 0.981t - 86.33\cos\left(\frac{2\pi t}{12}\right) - 77.77\sin\left(\frac{2\pi t}{12}\right)$$

$$\widehat{\mathit{Traffic}} = 490.86 - 0.979t - 10.2\mathit{Month2} + 82.2\mathit{Month3} + 64.5\mathit{Month4} + 119.6\mathit{Month5} + 150.7\mathit{Month6} + 268.6\mathit{Month7} + 296.7\mathit{Month8} + 133.0\mathit{Month9} + 115.1\mathit{Month10} + 63.4\mathit{Month11} + 42.2\mathit{Month12}$$

ASSESS

Figure 12.10 shows the fitted series for both the cosine (Figure 12.10(a)) and seasonal means (Figure 12.10(b)) models with the extra linear t term to account for the decreasing trend. We see that the model with a cosine trend still has trouble reaching the high peaks in the summer months. Measures of overall fit (R^2, adj. R^2, and s_ϵ) also look considerably better for the model based on seasonal means with a linear trend. Even though that model has more parameters to estimate (13 compared to 4), it is probably the better choice in this situation.

Model	R^2	Adjusted R^2	s_ϵ
Linear + Cosine	81.2%	80.8%	45.0
Linear + Seasonal Means	95.1%	94.7%	23.7

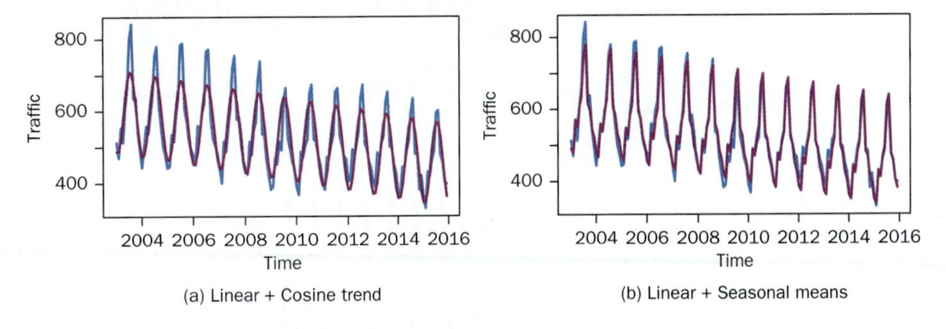

FIGURE 12.10 Adding a linear trend to predict Peace Bridge traffic

Finally, we check a time series plot of the residuals (Figure 12.11) for the linear trend + seasonal means model with the **PeaceBridge2003** data. This shows no obvious increasing, decreasing, or seasonal patterns, so we've probably done an adequate job of summarizing the main structure of the series.

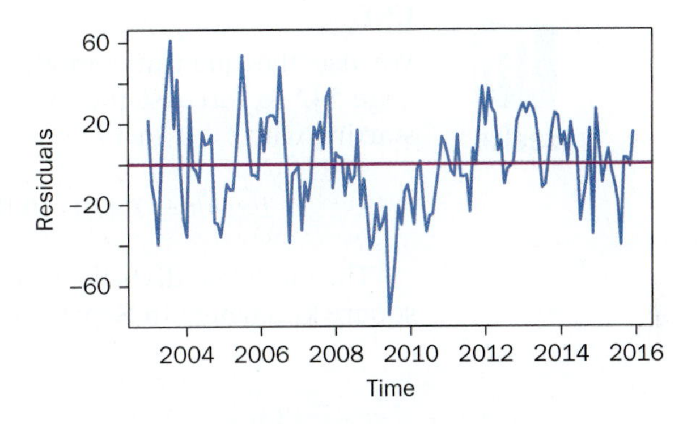

FIGURE 12.11 Residuals for seasonal means + linear trend model for Peace Bridge traffic

The Peace Bridge traffic example demonstrates a common approach to fitting a time series model by breaking the model down into "components" that reflect various types of structure. We see that the overall traffic series is mostly the sum of a linear component, a seasonal component, and fairly random noise. This *decomposition* is shown graphically in Figure 12.12.

Time Series Forecasts

While one purpose of creating a time series model is to help understand the structure of the data (e.g., the decreasing trend in sea ice or the seasonal pattern in bridge traffic), another common goal is to forecast future values of the series. For the regression models of this section, this is easily accomplished by substituting the future time values into the fitted prediction equation.

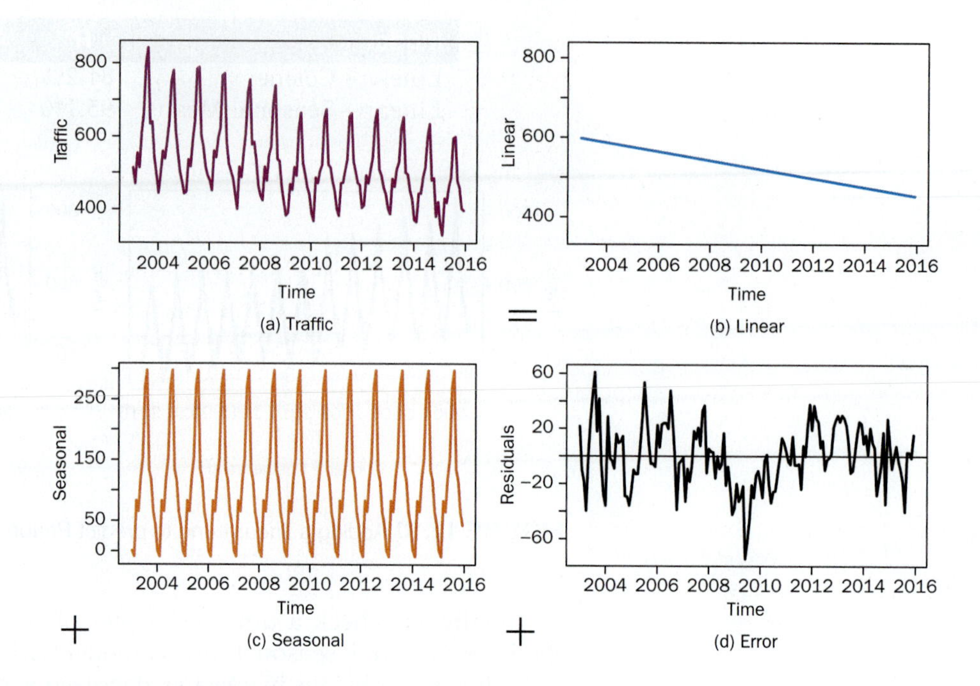

FIGURE 12.12 Decomposition of Peace Bridge traffic series

EXAMPLE 12.8

Arctic sea ice in 2017

SeaIce

USE

We use the quadratic model for the Arctic sea ice in Example 12.3 on page 542 to forecast the extent of the ice in September 2017. For the data starting with $t = 1$ in 1979, we see that $t = 39$ in 2017, so the forecast is

$$\widehat{Extent} = 7.470 - 0.004622(39) - 0.002176\left(39^2\right) = 3.98$$

The model predicts the extent of the Arctic sea ice to be about 3.98 million square kilometers in September 2017.

But how accurate is that estimate? In Section 2.4 on page 70 we introduce the ideas of a confidence interval for mean response and prediction interval for individual responses in a regression setting. Those ideas apply to the regression models in this section, although in a time series setting we are almost always interested in a prediction interval, since we are looking to capture a specific value at a future point in time. We often refer to these prediction intervals as **forecast intervals**.

forecast intervals

EXAMPLE 12.9

Forecast interval for Arctic sea ice in 2017

SeaIce

USE

Because we fit the quadratic model for sea ice extent with multiple regression, we rely on technology to find a 95% prediction interval for year 2017 when $t = 39$.

```
       fit        lwr        upr
1  3.980082   2.746408   5.213756
```

This confirms our forecast of 3.98 million sq. km. for the sea ice extent on September 2017 and lets us be 95% sure that the actual measurement that year will be somewhere between 2.75 and 5.21 sq. km. By the time you are reading this book, the September 2017 value will probably be available on the website, so you can check if the interval works!

Figure 12.13 shows the last few values of the sea ice time series along with forecasts for the next five years with 95% forecast bounds. Generally, the intervals get wider as we go further into the future (although it's hard to see in this figure that the bounds around the 2020 forecast are slightly wider than in 2016), and *we should use extra caution when forecasting values that are well beyond the time frame where the model was built.*

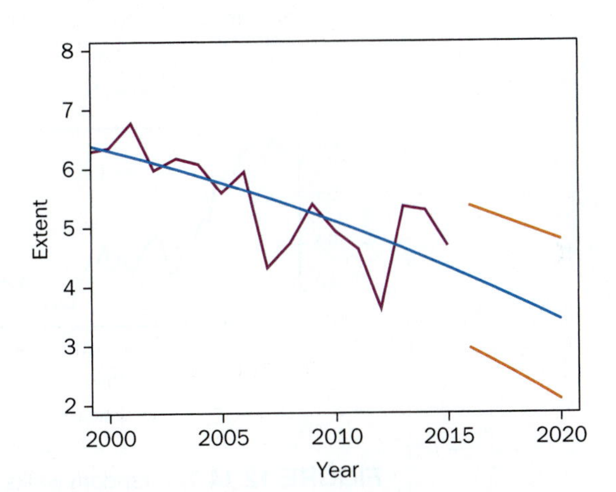

FIGURE 12.13 Historical sea ice extent with forecast and bounds for the next five years

12.2 Measuring Dependence on Past Values: Autocorrelation

We generally expect that values in a time series may be related to values at other times in the series and can use this fact to build forecast models. For example, a data value that is unusually high or low at time $t-1$ might be followed by a similarly high or low value at the adjacent time t. This suggests using the value at the previous time, Y_{t-1}, to predict the current value of Y_t. Here is one such model.

RANDOM WALK

In a **random walk** model, the value at time t is a random change from the value at $t-1$.

$$Y_t = Y_{t-1} + \epsilon_t$$

where $\epsilon_t \sim N(0, \sigma_\epsilon)$ is the usual independent error term.

In words, we think of a random walk as *new value equals old value plus error*.

EXAMPLE 12.10

Apple

Hypothetical stock prices Suppose that a stock starts with a price of \$100, and each day the price changes by a random amount from the previous day's price. What might such a random walk look like if those daily changes come from a normal distribution with mean 0 and standard deviation = \$1.50?

Figure 12.14 shows two time series simulated from this model over a period of three months, along with a third series that shows prices from an actual stock. Can you guess which is the real stock?

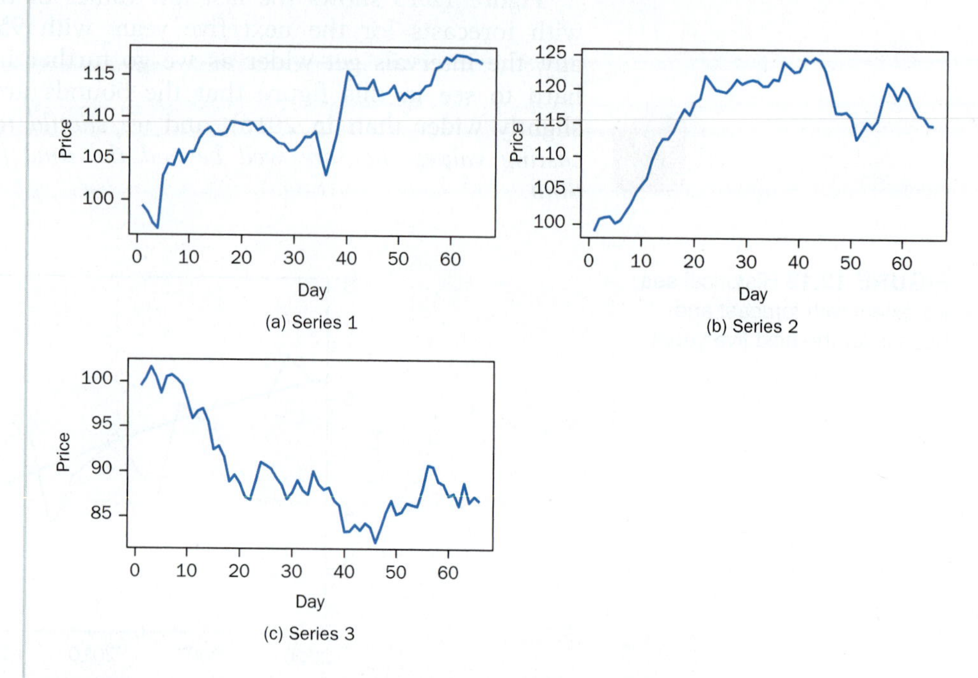

FIGURE 12.14 Two random walks and one real stock price series

The real stock prices are daily closing prices for Apple Inc. between July 21, 2016, and October 21, 2016. The file **AppleStock** contains these prices along with the daily price change and trading volume. That series is shown in Figure 12.14(a). The other two series were computer generated from a random walk model starting at \$100 and assuming a N(0,1.5) error term.

So how can we tell when a series we see might be a random walk? Note that the random walks model says that

$$Y_t - Y_{t-1} = \epsilon_t$$

differenced series

Thus the *changes* in the series should be random. We refer to this as a **differenced series** or the *first differences* of the original series.

DIFFERENCES

The *first difference* of a time series is

$$\Delta Y_t = Y_t - Y_{t-1}$$

This is another time series that has one fewer value than the original series.

Figure 12.15 shows the differenced series for each of the series shown in Figure 12.14. We expect the computer-generated series (b and c) to be random fluctuations of a N(0,1.5) variable. The differences for the actual Apple stock prices (a) also look relatively random.

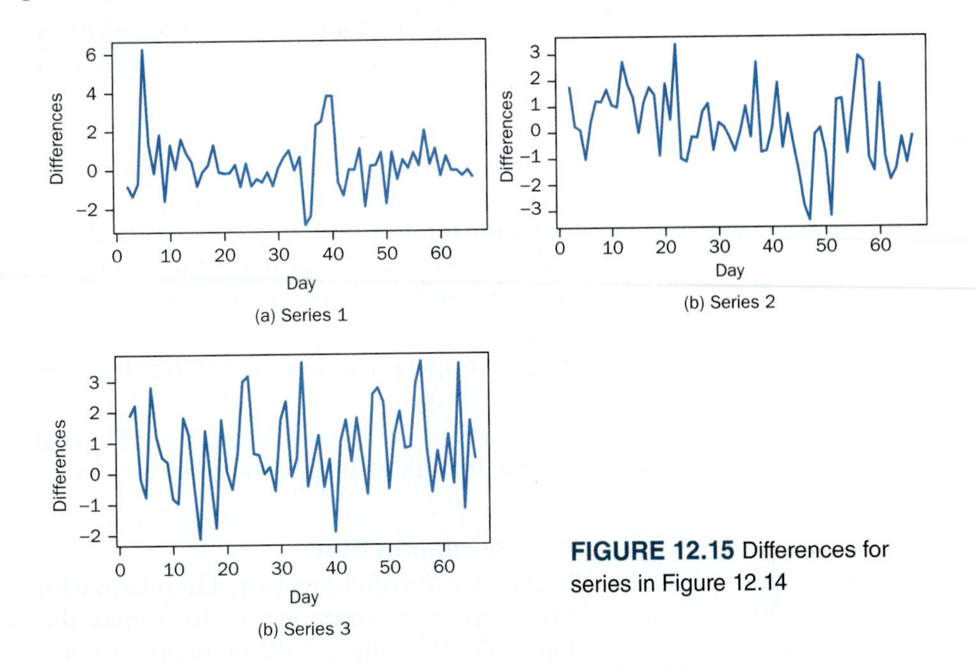

FIGURE 12.15 Differences for series in Figure 12.14

Autocorrelation

How can we tell that the series displayed in Figure 12.15 (at least parts b and c) are really "random" and there is not some underlying dependence in the series values? One way is to produce a scatterplot of values in the series versus the previous value. This is known as a **lagplot**. Figure 12.16 shows lagplots for Apple stock price series from Figure 12.14(a) and its differences series from Figure 12.15(a). We see a clear positive association in the lagplot for the original Apple price series, but not much relationship between adjacent values in the lagplot of the differences.

lagplot

FIGURE 12.16 Lagplots for Apple stock prices and their differences

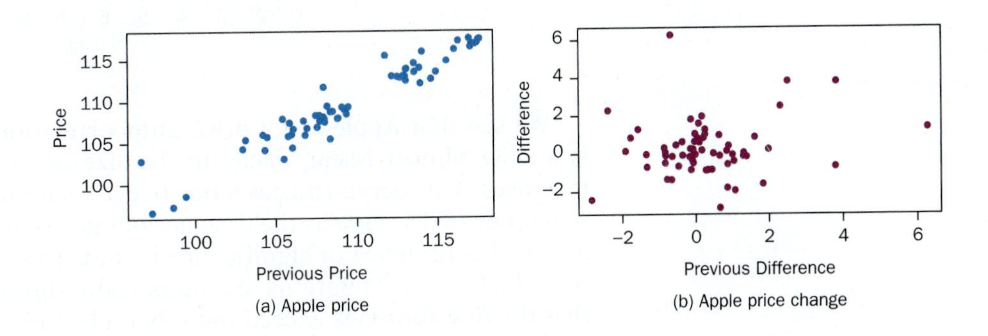

We could quantify a measure of the association between adjacent values in a time series by computing the correlation of the pairs of values in a lagplot, but in some cases we might also be interested in the relationship between values two, three, or even more time periods apart. For example, with monthly data, such as the Peace Bridge traffic of Example 12.7 on page 547, we might be interested in the relationship between values 12 months apart. In time series we measure such relationships between values of the same series with a quantity known as the **autocorrelation**.

autocorrelation

SAMPLE AUTOCORRELATION

The *sample autocorrelation* at lag k measures the strength of association between series values that are k time periods apart.

$$r_k = \frac{\sum(y_t - \bar{y})(y_{t+k} - \bar{y})}{\sum(y_t - \bar{y})^2}$$

This is similar to simply finding the ordinary correlation between $\{y_1, y_2, \ldots, y_{t-k}\}$ and the shifted $\{y_{1+k}, y_{2+k}, \ldots, y_t\}$, except that we use the mean, $\bar{y}$, and the total variability $\sum(y_t - \bar{y})^2$ for the entire sample. In practice, we generally rely on technology to compute autocorrelations for a sample series. For example, with the Apple stock prices we find

Original Apple price series $r_1 = 0.915$; Differenced Apple price series $r_1 = 0.161$.

For the Peace Bridge traffic counts in **PeaceBridge2003**, we find the autocorrelation for values 12 months apart is $r_{12} = 0.85$.

Autocorrelation Plot

Because we are often interested in relationships at several different lags within a time series, a common way to display the autocorrelations is with a plot that shows the values of the autocorrelations at many different lags. We sometimes call this an **ACF plot** (for autocorrelation function) or a **correlogram**. Figure 12.17 shows the ACF plot for the first 12 lags of the Apple stock price series.

ACF plot
correlogram

FIGURE 12.17 ACF of Apple stock prices

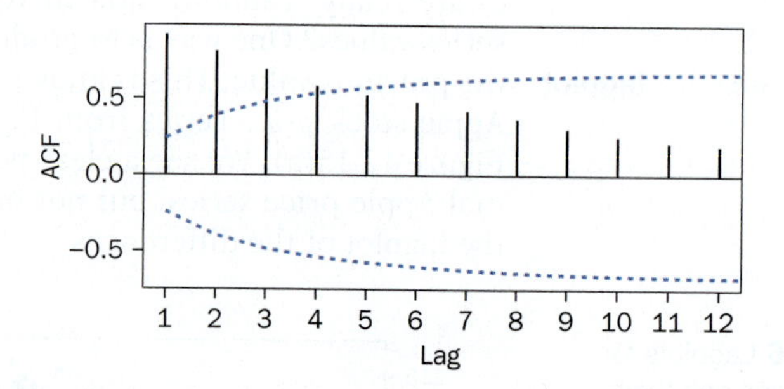

We see that Apple stock price autocorrelations are all positive, and there is a slow, almost linear, decay in the size of the autocorrelations as the lag increases. This demonstrates a positive association between stock price values that tends to decrease as the time periods get farther apart. The plot also shows a pair of confidence or significance bounds (dotted blue lines) that help identify which autocorrelations are statistically significant (at a 5% level). We see that the first four lags exceed these bounds, indicating convincing evidence of a positive autocorrelation at those lags. The rest of the lags don't exceed the significance bounds, but the regular pattern suggests that there may still be some relationships that the sample size is not large enough to confirm statistically.

Figure 12.18 shows the ACF plot of the *differences* of the Apple stock prices. The plot is simply showing the autocorrelations for the daily changes in Apple stock prices. At lag 1 we are comparing today's change (ΔY_t) to yesterday's change (ΔY_{t-1}). Here we see that none of the autocorrelations exceed the significance bounds, so we don't find convincing evidence for association among any of the lags of the differenced series. This is the pattern we would expect

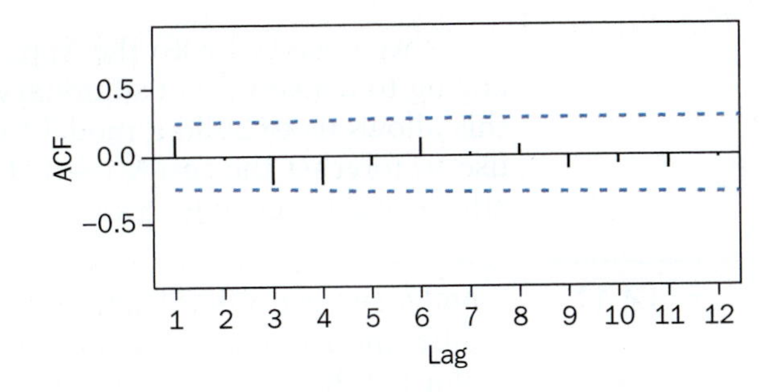

FIGURE 12.18 ACF of the differences of Apple stock prices

to see if the values of the time series are independent of each other. Thus a possible model for the Apple stock prices that shows no strong relationships between residuals would be

$$\Delta Y_t = Y_t - Y_{t-1} = \epsilon_t$$

This is simply a random walk model, $Y_t = Y_{t-1} + \epsilon_t$, which states that each day's new stock price is just a random change from the previous day.

Stationarity

When building a model to forecast values of a time series based on past values, we would like the nature of the relationship to stay stable over time. For example, if the variability of the series or autocorrelation between adjacent values consistently increased over time, we shouldn't use a model built from early values to try to forecast relationships later in the series. This leads to the idea of a **stationary series**, for which the *mean*, *variance*, and *autocorrelation structure* are all constant over time. While there are a number of ways to check stationarity for a theoretical time series model or a sample series, we consider two relatively informal methods for detecting nonstationarity (departures from stationarity) here.

stationary series

- *Look for long generally increasing or decreasing trends in the time series plot.* This is usually a clear indication of a mean that is not constant over time. For example, the extent of Arctic sea ice in Figure 12.1 and the Peace Bridge traffic in Figure 12.9 both show a consistent decreasing trend, while the Apple stock prices in Figure 12.14 are generally increasing. On the other hand, the differences of the random walk series in Figure 12.15 are more typical of a series with a constant mean.
- *Look for a slow linear decay in the ACF plot.* Sometimes called a "picket fence," we have seen this pattern in the ACF plots of the Apple stock prices (Figure 12.17) and it also appears for the ACF of the Arctic sea ice shown in Figure 12.19, so we have evidence of nonstationarity. The slow linear decay is not present in the ACF for the differences of the Apple stock prices, which are more stationary (Figure 12.18).

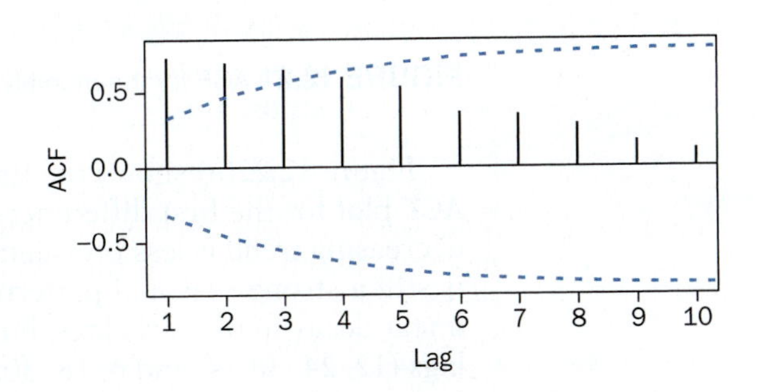

FIGURE 12.19 ACF for the Arctic sea ice series

As we observed with the Apple Stock price series, we can often use differencing to convert a nonstationary series to one that is stationary. Essentially this allows us to build a model for the *changes* in the series that we can then use to forecast the series itself. The next two examples further illustrate the effects of differencing on stationarity.

EXAMPLE 12.11

SeaIce

Arctic sea ice: Differences Figure 12.20(a) shows a time series plot of the differences of the Arctic sea ice extent series. Figure 12.20(b) shows the ACF plot for those differences. We see that taking a first difference has done a good job of producing a more stationary series, with no consistent growth or decay over time and an ACF plot with no slow linear decay. We do see evidence of significant autocorrelation at the first lag of the differenced series, which indicates that the change in Arctic sea ice extent over one year might be helpful in forecasting the change over the next year. We look at a model to exploit this fact in the next section.

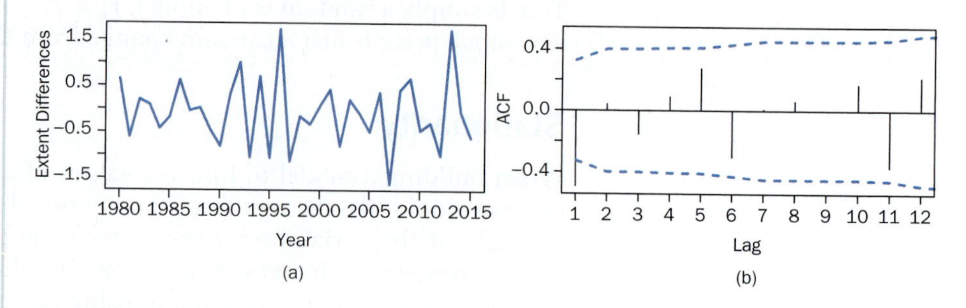

FIGURE 12.20 Time series plot and ACF for the differences of the Arctic sea ice series

EXAMPLE 12.12

Bridge03

Differences for Peace Bridge traffic The time series plot of the Peace Bridge traffic from **PeaceBridge2003** in Figure 12.9 on page 548 shows a strong seasonal pattern and a general declining trend that indicates a lack of stationarity. Figure 12.21 is the ACF plot for that series. The dominant feature in the ACF plot is the seasonal pattern with large positive autocorrelations near lags 12, 24, 36, . . . and negative autocorrelations near lags 6, 18, 30, . . . when time periods are at opposite points of the calendar year.

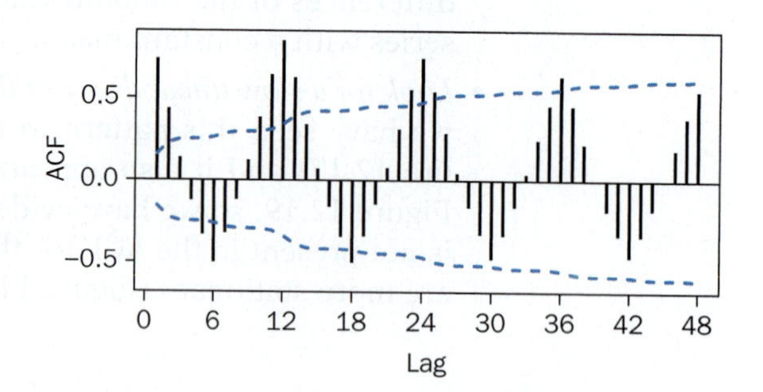

FIGURE 12.21 ACF for the monthly Peace Bridge traffic (2003–2015)

Figure 12.22(a) shows the time series plot and Figure 12.22(b) shows the ACF plot for the first differences in the bridge traffic series. We see that the decreasing trend is less prevalent than what we see in Figure 12.21, but there is still a strong seasonal pattern. Note that the ACF does not have the slow linear decay in the early lags, but this phenomenon is present at the *seasonal* lags (12, 24, 36, 48 and 6, 18, 30, 42).

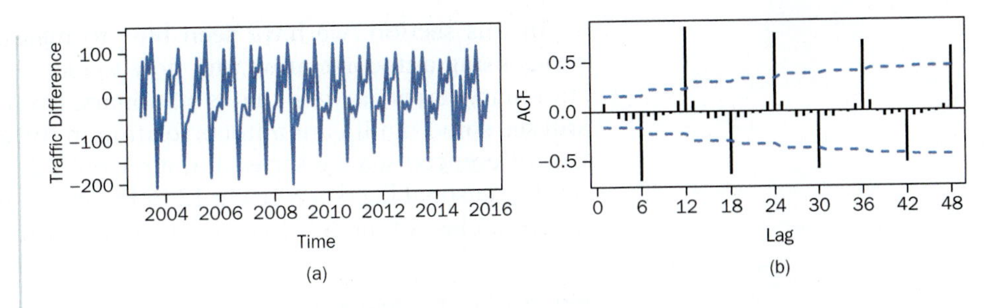

FIGURE 12.22 Time series plot and ACF for the differences of the Peace Bridge traffic

This suggests trying a *seasonal difference* to look at the change in bridge traffic from the same month the previous year.

$$\Delta_{12}y_t = y_t - y_{t-12}$$

Figure 12.23(a) shows the time series and Figure 12.23(b) shows the ACF plots for the seasonal differences of the Peace Bridge traffic data. Notice that the nonstationarity at the seasonal time periods has been diminished, but now we see the linear decay picket fence at the early lags!

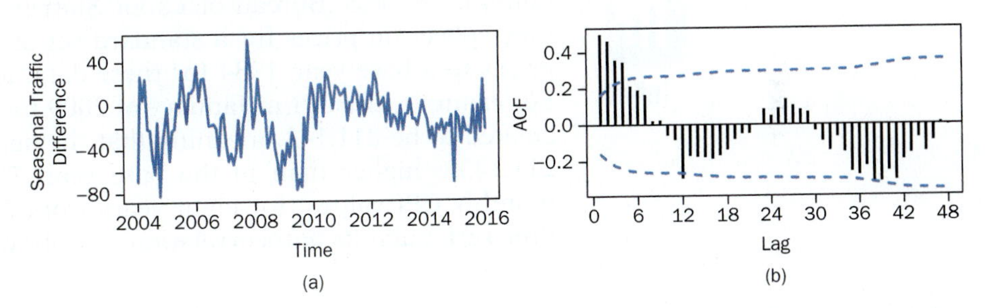

FIGURE 12.23 Time series plot and ACF for seasonal differences of the Peace Bridge traffic

The graphs suggest we need differencing at both the regular (lag 1) and seasonal (lag 12) levels, so why not try both? Note that we can do it in either order, take a lag 1 difference of the seasonal differences or vice versa, and the results are the same (see Exercise 12.21).

$$\Delta(\Delta_{12}y_t) = \Delta_{12}(\Delta y_t)$$

Figure 12.24(a) shows the time series and Figure 12.24(b) shows the ACF plots for the bridge traffic data after taking both differences. The constant mean and variability conditions look fine in the time series plot, and there is no slow, linear decay in either the early lags or seasonal lags of the ACF plot, so it looks like taking both differences produces a series that is relatively stationary.

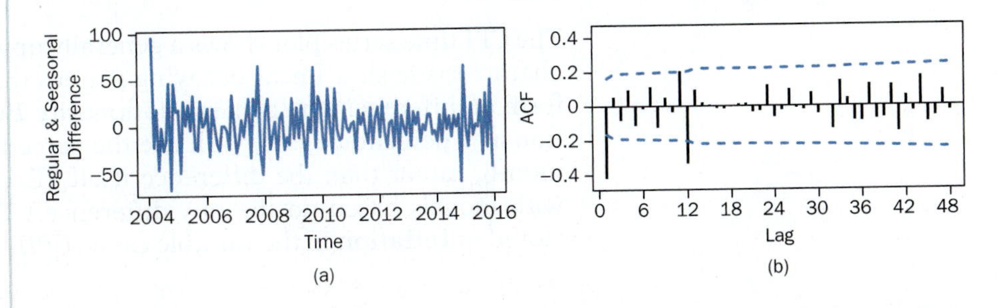

FIGURE 12.24 Time series and ACF after one regular and one seasonal difference for bridge traffic

In this section, we have seen how to measure association within a time series using autocorrelation and how to use differencing to transform a nonstationary series into one that is stationary. Even after this differencing we may still see some significant autocorrelation in the early lags (e.g., at lag 1 in both the differenced sea ice series and after both differences for the bridge traffic) or at a seasonal lag (such as lag 12 in Figure 12.24). In the next section we look at a new class of models to take advantage of those relationships.

12.3 ARIMA Models

ARIMA

In this section we consider a general class of time series models, known as **ARIMA*** models, that exploit relationships at different lags within a time series. When fitting these models, we generally follow a first step (if needed) of using differences or a transformation to get a stationary series, before fitting a model.

EXAMPLE 12.13

Inflat

Consumer Price Index: Checking stationarity The Consumer Price Index (CPI) is a common tool for tracking how fast prices are rising (or more rarely, falling). The U.S. Bureau of Labor Statistics constructs the CPI from data they collect on prices for a standard set of consumer goods. They scale all prices to a base year, 1984 for these data, to allow comparisons over time. For example, the CPI in January of 2009 for U.S. urban consumers was calculated to be 211.143, meaning that the aggregate prices at this time were 211.143% higher than in the base year. The data file **Inflation** contains monthly CPI values[3] for eight years from 2009 through 2016. Plots of the time series and its autocorrelations are shown in Figure 12.25.

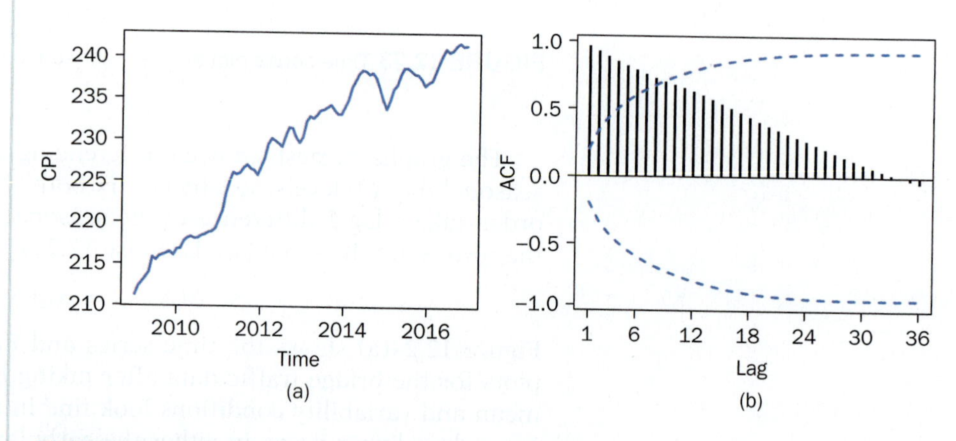

FIGURE 12.25 Time series and ACF for the CPI series

CHOOSE

The CPI time series plot shows a generally upward trend and the ACF has the characteristic slow linear decaying pattern that both indicate we need some form of differencing to achieve stationarity. Because we often consider inflation as a percentage rate, we'll use the percentage difference from month to month, rather than the difference itself. (Exercise 12.22 on page 584 deals with models based on the raw difference.) The percentage differences are stored in **Inflation** in the variable called *CPIPctDiff*. Figure 12.26 shows plots

*ARIMA stands for *autoregressive, integrated, moving average*. Each of these terms will be described in this section.

of the CPI percent difference time series and their autocorrelations. We see a more stable mean level and significant ACF spikes at only a few lags (1, 8, 12, and 24), so stationarity looks much better for the percent differences of the monthly CPI data.

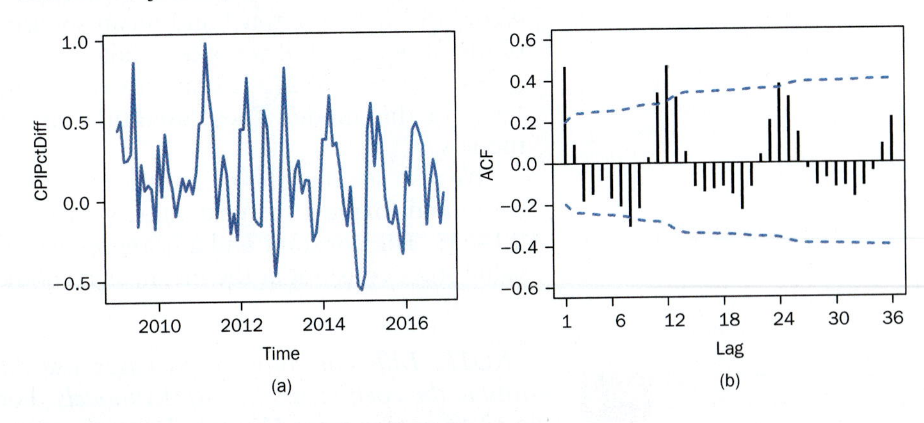

FIGURE 12.26 Time series and ACF for CPI percent differences

Autoregressive Models

The autocorrelation ($r_1 = 0.462$) is significant for the first lag of the *CPIPctDiff* series, indicating that one month's CPI change is positively related to the next. We can exploit this relationship in the model by using the previous month's value as a predictor.

$$Y_t = \delta + \phi_1 Y_{t-1} + \epsilon_t$$

first order autoregressive

This is known as a **first order autoregressive** model, AR(1), and is slightly different from an ordinary least squares regression model* because the predictor and response values both come from the same time series. Fortunately we can use software to fit the model, as the next example does for the *CPIPctDiff* series.

EXAMPLE 12.14

Inflat

Consumer Price Index: AR(1)

FIT

Here is some computer output for fitting the AR(1) model to the percent differences in the CPI series.

Type	Coef	SE Coef	T	P
AR 1	0.4743	0.0908	5.22	0.000
Constant	0.07694	0.02794	2.75	0.007
Mean	0.14635	0.05315		

```
Number of observations: 96
Residuals:    SS = 7.04583
              MS = 0.07496    DF = 94
```

The fitted model is $\widehat{y}_t = 0.07694 + 0.4743 y_{t-1}$. The last value in the series is $y_{96} = 0.033$ for December 2016, so the forecast for the percent change in the Consumer Price Index for January 2017 is $\widehat{y}_{97} = 0.07694 + 0.4743(0.033) = 0.093$.

*This is one reason we use δ and ϕ_1 in the autoregressive model rather than the usual β_0 and β_1 of ordinary least squares regression.

We can interpret the information labeled "SE Coeff," "T," and "P" in the computer output similar to ordinary least squares regression to show that both the constant term (P-value = 0.007) and first-order autoregressive term (P-value = 0.000) are important to keep in this model. The sum of squared errors (SS in the output) and mean square error (MS) reflect the accuracy of the fit. As usual, if we take a square root of the mean square error, we get $s_\epsilon = \sqrt{0.07496} = 0.2738$ as an estimate of the standard deviation of the error term for this model. These quantities are useful for comparing competing models.

We also see that the mean is listed as 0.14635. Note that if we are at the mean and forecast the next value we get $\widehat{y}_t = 0.07694 + 0.4743(0.14635) = 0.14635$. The constant and autoregressive coefficient are chosen so that the value stays constant at the mean, as it should for a stationary series.

NOTE: Different software packages use different optimization methods to estimate the coefficients in ARIMA models. For example, the output in Example 12.14 comes from Minitab. Below is output using R to fit the same AR(1) model to the *CPIPctDiff* series. Note that the $\hat{\phi}_1$ estimate (under `ar1`) and its standard error are slightly different, and R reports the mean for the model, rather than an estimate of the constant term.

```
Series: CPIPctDiff
ARIMA(1,0,0) with non-zero mean
          Coefficients:
                 ar1     mean
              0.4694   0.1463
        s.e.  0.0897   0.0517
sigma^2 estimated as 0.07511: log likelihood=-11.07
AIC=28.14   AICc=28.4   BIC=35.84
```

Of course, if the previous Y_{t-1} value helps predict Y_t, we might expect other values in the history of the series to be useful in the model. This leads to the more general p^{th}-order autoregressive model that uses information from the past p values in the series.

AUTOREGRESSIVE MODEL

A pth-*order autoregressive* model depends on a linear combination of the past p values in the series.

$$Y_t = \delta + \phi_1 Y_{t-1} + \phi_2 Y_{t-2} + \cdots + \phi_p Y_{t-p} + \epsilon_t$$

EXAMPLE 12.15

Inflat

Consumer Price Index: AR(2) Would a second order autoregressive model be more appropriate for describing the percent change in the CPI series?

$$Y_t = \delta + \phi_1 Y_{t-1} + \phi_2 Y_{t-2} + \epsilon_t$$

Here is some output for fitting the AR(2) model to the *CPIPctDiff* variable in **Inflation**.

Type	Coef	SE Coef	T	P
AR 1	0.5493	0.1023	5.37	0.000
AR 2	-0.1633	0.1028	-1.59	0.115
Constant	0.08948	0.02774	3.23	0.002
Mean	0.14572	0.04517		

Number of observations: 96

Residuals: SS = 6.86763 (backforecasts excluded)
MS = 0.07385 DF = 93

We see that the test for the lag 2 coefficient (P-value = 0.115) indicates that it is not very important in this model, so we probably don't need to use the information from two months ago to predict the current value. The sum of squared errors is better (6.876 compared to 7.046), as it will always be when we add additional terms to the model, and the mean square error is also a bit lower (0.07385 compared to 0.07496). Note that, as usual, the degrees of freedom drops from 94 to 93 when we add the additional term, so the change in the MSE is not as impressive as the change in SSE.

Should we keep the second-order term in the model? We have somewhat conflicting information and could make a case for either "yes" or "no." The P-value for that term is not significant, even at a generous 10% level, but the MSE (and hence s_ϵ) is smaller for the AR(2) model. As we have seen in past examples, all arrows don't always point in the same direction and statisticians may disagree about which is the "best" model. In this case, we would probably prefer the AR(1) model since it is simpler, all terms are significant, and the MSE is very close to what is achieved by the more complicated AR(2) model.

Moving Average Models

Another way to use historical information from a series to predict future values is to look at the past residuals. For example, if economic conditions cause a larger than expected increase in the CPI (a positive residual), we might expect the next month to also be above what is expected. Similarly, an unusually small value (negative residual) might be followed by other below-normal values. A **moving average*** time series model uses the past residuals as predictors for future values of the series.

moving average

MOVING AVERAGE MODEL

A qth-*order moving average* model depends on a linear combination of the past q values in the random error series, $\epsilon_{t-1}, \epsilon_{t-2}, \ldots, \epsilon_{t-q}$.

$$Y_t = \delta + \epsilon_t - \theta_1\epsilon_{t-1} - \theta_2\epsilon_{t-2} - \cdots - \theta_q\epsilon_{t-q}$$

Note: By convention, we write the model by subtracting the moving average terms, although the coefficients $\theta_1, \theta_2, \ldots, \theta_q$ can be positive or negative.

*The term *moving average* is also commonly used to denote forecasts based on an average of past values of the series. For example, $\widehat{Y}_t = (Y_{t-1} + Y_{t-2} + Y_{t-3})/3$. This is actually a special case of an autoregressive model. Take care to not confuse this with the way we use the term *moving average* in this section.

EXAMPLE 12.16

Consumer Price Index: MA(1) The ACF plot for CPI percent differences (Figure 12.26(b) on page 559) shows a significant relationship at lag 1 in the series, so let's try an MA(1) model.

$$Y_t = \delta + \epsilon_t - \theta_1 \epsilon_{t-1}$$

Here is some computer output for fitting the MA(1) model to the *CPIPctDiff* variable in **Inflation**.

Type	Coef	SE Coef	T	P
MA 1	-0.4219	0.0936	-4.51	0.000
Constant	0.14536	0.04023	3.61	0.000
Mean	0.14536	0.04023		

Number of observations: 96
Residuals: SS = 7.23212
MS = 0.07694 DF = 94

We see that both the moving average and constant terms have significant P-values. However, the sum of squared errors (7.232) and mean square error (0.07694) are both slightly worse than the AR(1) model for the same data. The forecast equation is

$$\widehat{y}_t = 0.14536 + 0.4219\hat{\epsilon}_{t-1}$$

Why isn't ϵ_t in the forecast equation? We won't know the residual at time t until we have the value of the series, y_t. So, as is the case with all of our regression models, we use the expected value of zero when predicting that error term. The residual at time $t-1$ can be estimated by comparing the known value of y_{t-1} to the value that would have been predicted from the previous data, $\hat{\epsilon}_{t-1} = y_{t-1} - \hat{y}_{t-1}$. If we want to forecast further into the future, the value of any future residual is set to zero. Note also how the sign for the moving average coefficient switches from what appears in the computer output. This is consistent with the positive lag 1 autocorrelation that we see in Figure 12.26.

USE

Forecasts by hand are a bit trickier with moving average models since we don't see the past residuals as readily as we do the past series values for an autoregressive model. The model is fit by forecasting all of the historical values (and minimizing SSE) so those residuals can generally be obtained from the computer. For example, this MA(1) model forecasts a value of 0.0259 for the CPI percent difference in December 2016 and the actual value is 0.033, so the residual at that last time point is 0.0071. This means the forecast for January 2017 is

$$\begin{aligned}
\text{Model:} \quad y_t &= \delta + \epsilon_t - \theta_1 \epsilon_{t-1} \\
\text{Forecast:} \quad \hat{y}_t &= \hat{\delta} - \hat{\theta}_1 \hat{\epsilon}_{t-1} \\
\text{For January 2017:} \quad \widehat{y}_{97} &= 0.14536 + 0.4219(0.0071) = 0.1483
\end{aligned}$$

ARIMA Models

We have now considered three important tools for fitting a time series model based on the past history.

- *Differences* to help with stationarity
- *Autoregressive* terms to reflect dependence on past series values
- *Moving average* terms to reflect dependence on past residuals

Autoregressive Integrated Moving Average

Why not include all three in the same model? This leads to the general **Autoregressive Integrated Moving Average** model, or ARIMA for short.* We need to choose three parameters for an ARIMA, the number of autoregressive terms (p), the number of differences (d), and the number of moving average terms (q). Thus we often use the notation ARIMA(p, d, q) to specify a particular model of this form.

ARIMA(p, d, q) MODEL

An ARIMA(p, d, q) model uses p autoregressive and q moving average terms applied to the dth difference of a series.

$$\Delta^d Y_t = \delta + \phi_1 \Delta^d Y_{t-1} + \cdots + \phi_p \Delta^d Y_{t-p} + \epsilon_t - \theta_1 \epsilon_{t-1} - \cdots - \theta_q \epsilon_{t-q}$$

In practice, any of p, d, or q might be zero and we try to keep the model as simple as possible. We generally choose the number of differences† d first to remove any obvious departures from stationarity and then experiment with choices for p and q. The autocorrelation patterns in the ACF can help in choosing p and q, although a full description of ACF patterns associated with these models is beyond the scope of this course.

EXAMPLE 12.17

SeaIce

Arctic sea ice: ARIMA models We return to the yearly extent of Arctic sea ice introduced in Example 12.1 on page 540 for the data in **SeaIce**.

CHOOSE

The original series has a generally decreasing trend and Figure 12.19 on page 555 shows a slow linear decay in the ACF for the original series, so we need some differencing to achieve better stationarity. Example 12.11 shows that a single regular difference is relatively stationary, so setting $d = 1$ looks like a reasonable choice for an ARIMA model.

The ACF of the first differences in Figure 12.19 shows a significant negative spike at lag 1, but nothing significant after that. This suggests using $p = 1$, $q = 1$, or both $p = q = 1$ as tentative models. Here is the computer output for fitting all three of these ARIMA models.

*You may wonder why the term *integrated* is used here. In mathematics, we often think of integrate as the process of reversing a difference.

†Take care to distinguish between the dth difference (Δ^d) and a difference (Δ_d) at lag d. For example, $\Delta^2 Y_t = \Delta(\Delta Y_t) = (Y_t - Y_{t-1}) - (Y_{t-1} - Y_{t-2})$, while $\Delta_2 Y_t = Y_t - Y_{t-2}$.

FIT

ARIMA(1, 1, 0):

Type	Coef	SE Coef	T	P
AR 1	-0.5204	0.1476	-3.52	0.001
Constant	-0.1098	0.1081	-1.02	0.317

Differencing: 1 regular difference
Number of observations: Original series 37, after differencing 36

Residuals: SS = 14.3153
MS = 0.4210 DF = 34

ARIMA(0, 1, 1):

Type	Coef	SE Coef	T	P
MA 1	0.8152	0.0987	8.26	0.000
Constant	-0.08391	0.01908	-4.40	0.000

Differencing: 1 regular difference
Number of observations: Original series 37, after differencing 36

Residuals: SS = 11.2173
MS = 0.3299 DF = 34

ARIMA(1, 1, 1):

Type	Coef	SE Coef	T	P
AR 1	-0.0722	0.2132	-0.34	0.737
MA 1	0.7782	0.1339	5.81	0.000
Constant	-0.08902	0.02229	-3.99	0.000

Differencing: 1 regular difference
Number of observations: Original series 37, after differencing 36

Residuals: SS = 11.2108
MS = 0.3397 DF = 33

ASSESS

We see that the *P*-value for the constant term in the ARIMA(1, 1, 0) model is not significant, so we might drop that term, but the SSE for that model (14.32) is already the worst of the three and would only get larger without the constant term. Note that the AR term is quite significant in the ARIMA(1, 1, 0) model, but not at all so in the ARIMA(1, 1, 1). This is similar to the situation in ordinary least squares regression where predictors are related to each other. The first-order autoregressive and moving average terms both reflect relationships at lag 1. In this case, if the MA term is not present, the AR term shows up as being needed, but the AR term does not seem to be needed if the MA term is also in the model. Even though the SSE (11.21) is smallest for the ARIMA(1, 1, 1) model (as it must be among these three), we eliminate it from consideration since the AR term is probably not needed. This leaves the ARIMA(0, 1, 1) model as our choice to describe and forecast yearly Arctic sea ice extent. Notice that the SSE only increases slightly to 11.22 for the ARIMA(0, 1, 1) model.

Recall that one goal of fitting an ARIMA model is to get residuals that no longer exhibit dependence at any lags. We can check this by examining the ACF of the *residuals* after fitting the model, as shown in Figure 12.27. We see no problems with a lack of independence in the residuals for the ARIMA(0, 1, 1) model.

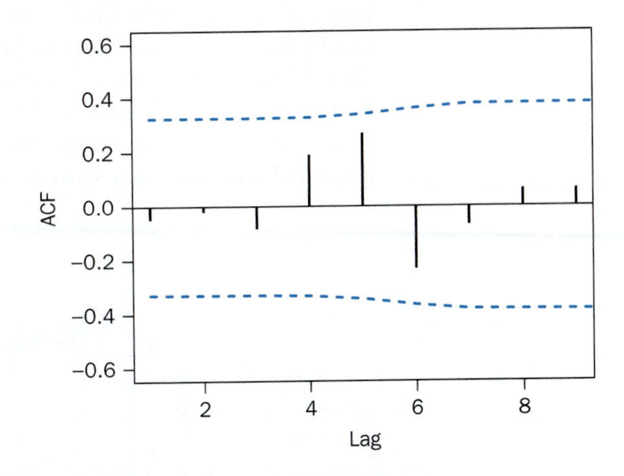

FIGURE 12.27 ACF for the residuals of the ARIMA(0,1,1) model for Arctic sea ice extent

We get some indication of the effectiveness of the model from the small *P*-value of the moving average coefficient, but we can also compare this ARIMA(0, 1, 1) fit to other models such as the linear and quadratic trends of Examples 12.2 and 12.3 from earlier in this chapter. We see that the models using ordinary least squares regression on t and t^2 work a bit better for the Arctic sea ice extent series. Both least squares models have smaller s_ϵ than the ARIMA model. The quadratic model has the smallest MSE.

	# parameters	SSE	MSE	s_ϵ
Linear	2	11.20	0.320	0.566
Quadratic	3	9.39	0.276	0.525
ARIMA(0,1,1)	2	11.22	0.330	0.574

USE

Although the quadratic model looks better for this series, we'll look at forecasts using the ARIMA(0, 1, 1) model to get some more experience with this kind of model. The fitted equation is

$$\Delta\hat{y}_t = -0.0839 - 0.8152\hat{\epsilon}_{t-1}$$

If we expand the difference $\Delta\hat{y}_t = \hat{y}_t - \hat{y}_{t-1}$ this becomes

$$\hat{y}_t = \hat{y}_{t-1} - 0.0839 - 0.8152\hat{\epsilon}_{t-1}$$

When working with this sort of forecast equation the basic rules are

- If you need a past value of the series use it, but if you need a future value forecast it first.
- If you need a past residual use the historical fit, but if you need a future residual use zero.

$$\text{Forecast for 2016: } \hat{y}_{2016} = y_{2015} - 0.0839 - 0.8152\hat{\epsilon}_{2015}$$
$$= 4.68 - 0.0839 - 0.8152(0.0179) = 4.582$$
$$\text{Forecast for 2017: } \hat{y}_{2017} = 4.582 - 0.0839 = 4.498$$

After the first year, each subsequent forecast is just a reduction of 0.0839 from the previous year's forecast. This is very similar to the linear model that has a slope of −0.0873. In practice, we let software do the calculations for predicting future values and bounds around those predictions. Figure 12.28 shows the last few years of the Arctic sea ice extent series together with the historical fits, forecasts (in blue), and forecast bounds (in red) for the next four years using the ARIMA(0, 1, 1) model. Compare this to Figure 12.13 on page 551 that shows a similar graph for the quadratic fit.

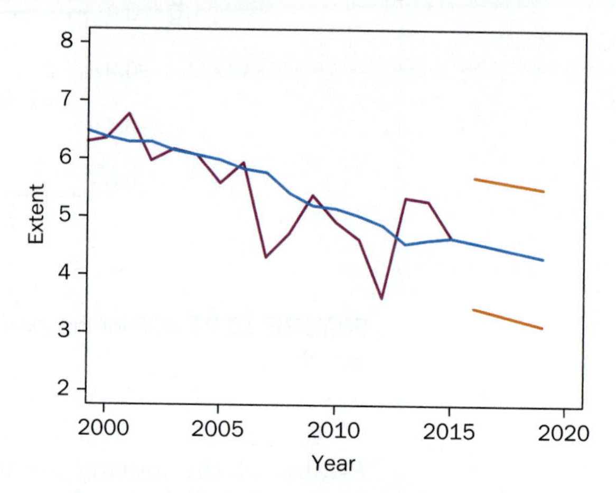

FIGURE 12.28 Recent historical, forecast values, and bounds for the ARIMA(0,1,1) model for Arctic sea ice extent

Seasonal ARIMA

EXAMPLE 12.18

Bridge03

Peace Bridge traffic: ARIMA models In Example 12.12 on page 556 we see that taking one regular and one seasonal difference for the Peace Bridge traffic data in **PeaceBridge2003** does a good job of producing a reasonably stationary series. We reproduce the ACF for the differenced series in Figure 12.29. We see big spikes at lags 1 and 12 and one that is barely significant at lag 11.

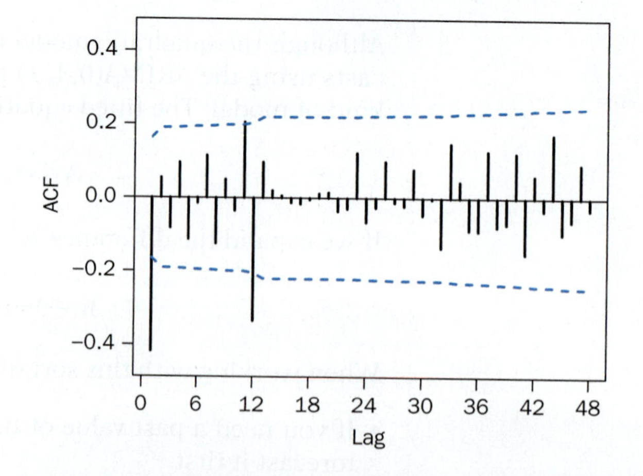

FIGURE 12.29 ACF for one regular and one seasonal difference of the Peace Bridge traffic

To handle the lag 1 autocorrelation, we can try ARIMA models for the differenced series with some combination of $p = 1$ or $q = 1$. The best option turns out to be a single moving average term with no constant.

```
Type          Coef   SE Coef      T       P
MA      1   0.5847    0.0670   8.72   0.000
Differencing: 1 regular, 1 seasonal of order 12
Number of observations: Original series 156, after differencing 143
Residuals:   SS = 68339.2
             MS = 481.3 DF = 142
```

The model is $\Delta\Delta_{12}Y_t = \epsilon_t - \theta_1\epsilon_{t-1}$ which gives a fitted forecast function (after expanding the differences) of

$$\widehat{y}_t = y_{t-1} + y_{t-12} - y_{t-13} - 0.5847\hat{\epsilon}_{t-1}$$

Figure 12.30 shows the ACF of the residuals for this moving average model on the differences. We see that the big spike at lag 1 is gone, but there still is significant autocorrelation in the residuals at lag 12.

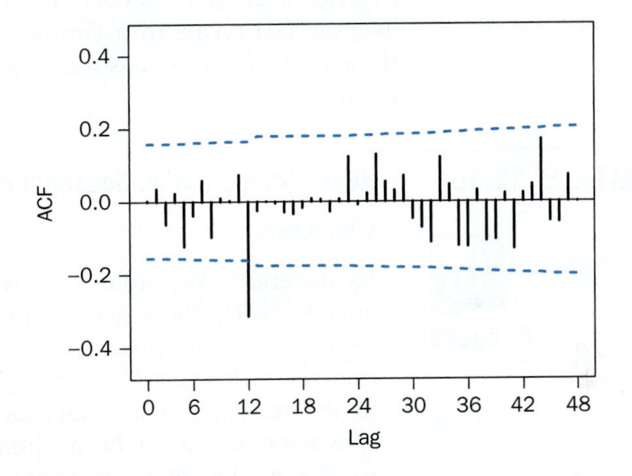

FIGURE 12.30 ACF for the residuals of an MA(1) model fit to the differenced bridge traffic

It makes sense that monthly data might show a strong relationship between the same months one year apart. How should we deal with an autocorrelation at a seasonal point like lag = 12? We could try an ARIMA model with $p = 12$ or $q = 12$, but that would require estimating 12 parameters when most of the lower lag terms are probably not important at all! A better option is to use autoregressive or moving average terms at just the seasonal lags. For example, if the seasonal period is S, we use a model like the one below, which has two seasonal autoregressive terms and one seasonal moving average term.*

$$Y_t = \delta + \Phi_1 Y_{t-S} + \Phi_2 Y_{t-2S} + \epsilon_t - \Theta_1\epsilon_{t-S}$$

*By convention we use capital Φ_i and Θ_j to denote the coefficients of the seasonal terms, reserving the lowercase ϕ_i and θ_j for coefficients of regular (nonseasonal) terms.

In practice we might need any combination of regular differences, autoregressive, and moving average terms along with seasonal differences, autoregressive, and moving average terms. This requires a more general ARIMA model to accommodate both regular and seasonal components.

SEASONAL ARIMA MODEL

An ARIMA$(p,d,q) \times (P,D,Q)_S$ model combines p regular autoregressive terms, d regular differences, and q regular moving average terms with P seasonal autoregressive terms, D seasonal differences, and Q seasonal moving average terms at a seasonal period of S.

Writing down a full ARIMA$(p,d,q)\times(P,D,Q)_S$ model can get quite involved, since the terms interact multiplicatively.* For example, we have already seen that choosing $d = 1$ and $D = 1$ for the Peace Bridge traffic series introduces the term Y_{t-13} into the model from the interaction of the regular and seasonal differences. At this point, we will rely on technology to sort out the details, fit the model, and produce forecasts from a seasonal ARIMA.

We have already seen how we choose the regular (d) and seasonal (D) differences to help with stationarity and how to deal with significant low lag autocorrelations by experimenting with regular autoregressive (p) and moving average(q) terms. We use seasonal autoregressive (P) and moving average (Q) terms in a similar manner to deal with significant autocorrelations exclusively at seasonal lags ($S, 2S, \ldots$). The next example illustrates this process.

EXAMPLE 12.19 **Peace Bridge traffic: Seasonal ARIMA models**

Bridge03

CHOOSE

By the end of Example 12.18 we determined that we need one regular difference ($d = 1$), one seasonal difference ($D = 1$), one regular moving average term ($q = 1$), and no constant term. One flaw in this ARIMA$(0,1,1)\times(0,1,0)$ model† is the significant autocorrelation shown in Figure 12.30 at lag 12 in its residuals. This suggests adding either a single seasonal autoregressive term ($P = 1$) or a single seasonal moving average term ($Q = 1$) to the model. Here is some computer output for fitting both of these models.

FIT

ARIMA$(0,1,1)\times(1,1,0)$ with no constant:

```
Type        Coef   SE Coef      T      P
SAR   12  -0.4903   0.0740  -6.63  0.000
MA     1   0.6311   0.0638   9.89  0.000
Differencing: 1 regular, 1 seasonal of order 12
Number of observations: Original series 156, after differencing 143
Residuals:   SS = 53805.0
             MS = 381.6  DF = 141
```

*A complete description of the model generally uses "backward shift" notation with polynomials to code each of the six components.

†For simplicity, we often omit the S subscript from the notation for a seasonal ARIMA model when the context is clear.

ARIMA(0, 1, 1) × (0, 1, 1) with no constant:

```
Type          Coef   SE Coef     T      P
MA     1    0.5536    0.0691  8.02  0.000
SMA   12    0.6129    0.0655  9.35  0.000
Differencing: 1 regular, 1 seasonal of order 12
Number of observations: Original series 156, after differencing 143
Residuals:    SS = 49199.3
              MS = 348.9   DF = 141
```

ASSESS

Both coefficients in both models are clearly significant, but the MSE for the model with a seasonal autoregressive term ($MSE = 381.6$) is quite a bit larger than with a seasonal moving average term ($MSE = 348.9$), so we will go with the ARIMA(0, 1, 1) × (0, 1, 1) model.

Does the new seasonal moving average term help address the issue with residual autocorrelation at lag 12? Figure 12.31, the ACF for the residuals of this model, shows that it does, and now we see no evidence of significant autocorrelation at any lag of the residuals.

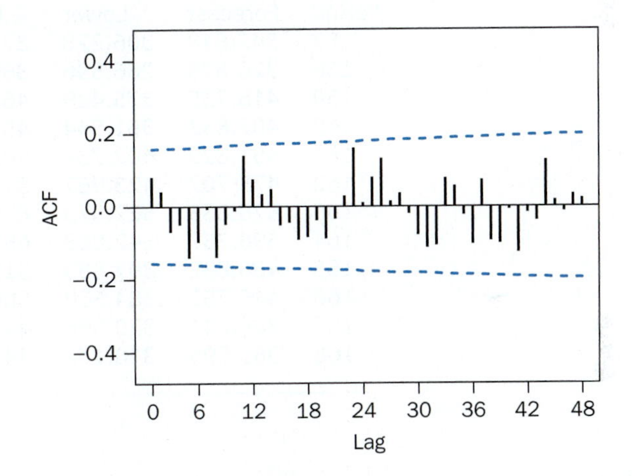

FIGURE 12.31 Residual ACF for the ARIMA(0,1,1) × (0,1,1) model fit to Peace Bridge traffic

Recall in Example 12.17 on page 563 that the ARIMA model for extent of sea ice was not as effective as the linear or quadratic trend models. How does the seasonal ARIMA model for Peace Bridge traffic compare to the linear + cosine trend and linear + seasonal means models that we fit in Example 12.7 on page 547?

	# parameters	SSE	MSE	s_ϵ
Cosine + Linear	4	307,907	2,025.7	45.01
Seasonal means + Linear	13	80,633	563.9	23.75
ARIMA(0, 1, 1) × (0, 1, 1)	2	49,199	348.9	18.68

We need to temper our excitement about the big drop in SSE for the seasonal ARIMA model a bit by realizing that the ARIMA model is working with

13 fewer residuals. Since we took one regular and one seasonal difference at the outset of the ARIMA process, that model is fit using only 156 − 13 = 143 differences. However, the MSE and s_ϵ take this into account, since the degrees of freedom reflect the sample size after differencing as well as the number of parameters in the model. Thus we see pretty clear evidence that the seasonal ARIMA model with just two parameters fits the historical Peace Bridge traffic data much better than either of the ordinary least squares regression models.

USE

Based on the historical data and the ARIMA(0, 1, 1) × (0, 1, 1) model, what do we predict for monthly traffic (in thousands of vehicles) over the Peace Bridge in 2016? Computer output with the forecasts and 95% prediction intervals are shown below and a graphical display of these values (forecasts in purple, bounds in blue) with the last few years of historical data (in black) are shown in Figure 12.32.*

Forecasts from period 156

		95% Limits	
Period	Forecast	Lower	Upper
157	342.848	306.228	379.468
158	326.699	286.596	366.801
159	418.735	375.429	462.040
160	407.832	361.544	454.120
161	451.829	402.739	500.918
162	475.707	423.967	527.446
163	576.833	522.573	631.093
164	598.757	542.088	655.426
165	456.832	397.853	515.811
166	445.752	384.550	506.955
167	384.244	320.896	447.592
168	383.595	318.172	449.017

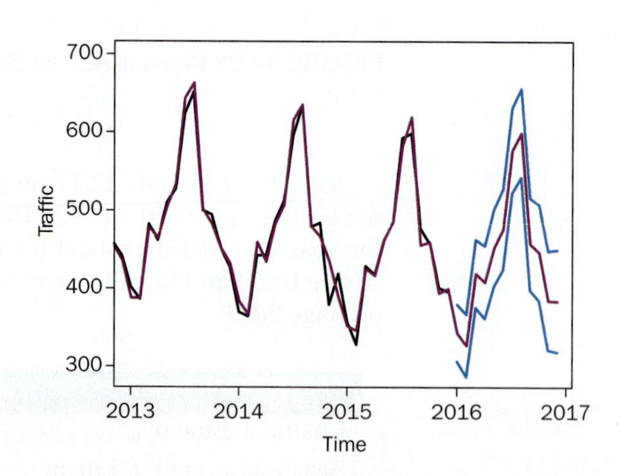

FIGURE 12.32 Forecasts (purple) for ARIMA(0,1,1) × (0,1,1) model and Peace Bridge traffic

*Note: By the time you read this, the monthly data for 2016 should be available at *http://www.peacebridge.com/index.php/historical-traffic-statistics/yearly-volumes* so you can check how the forecasts worked!

12.4 Case Study: Residual Oil

The U.S. Energy Information Administration[4] tracks the production and distribution of various types of petroleum products. One category, *residual oil*, contains heavier oils (often called No. 5. and No. 6) that remain after lighter oils (such as No. 4 home heating oil) are distilled away in the refining process.* It is used in steam-powered ships, power plants, and other industrial applications. The data in **ResidualOil** track the amount of residual oil delivered from U.S refineries (in millions of gallons/day) each quarter from 1983 through 2016. Figure 12.33 shows a time series plot of the $n = 136$ values in this series. The dominant pattern is a general decreasing trend. Our goal is to develop models for forecasting the residual oil deliveries for each quarter of 2017 and 2018.

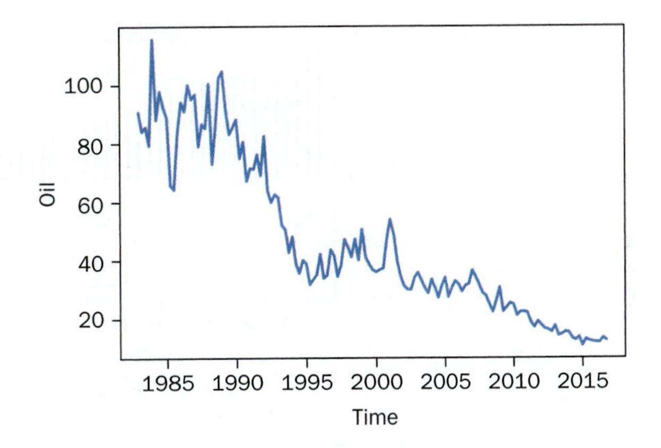

FIGURE 12.33 Quarterly U.S. residual oil deliveries (millions of gallons, 1983–2016)

EXAMPLE 12.20

ResOil

Residual oil: Initial choices We first consider two alternatives for dealing with the decreasing trend in the data.

- If we are thinking of a "function of t"-type model, we might start with a linear trend.
- If we are thinking of an ARIMA model, we should look at the autocorrelation function and the first differences of the series to try to achieve stationarity.

Figure 12.34(a) shows the linear fit on the original series and Figure 12.34(b) shows a time series plot of the residuals. We see some curvature in the data with mostly positive residuals in the earlier years, negative residuals in the middle, and residuals growing to positive again in the most recent years. This suggests using some sort of transformation (such as a logarithm of the *Oil* values) or a quadratic term in the model.

Figure 12.35 shows the ACF of the original *Oil* series and a plot of the first differences. As expected we see the slow linear decay in the ACF that is typical of a nonstationary series and a more stationary appearance to the first differences. However, we also see clear evidence of changing variability, with more volatility in the differences in early years and less variability in more recent years. This also suggests trying a transformation, such as *log*(*Oil*), which we consider in the next example.

*Note that the use of the term *residual* is similar to how we use it in a statistical model as the part that is "left over" after the model has been fit to extract the important structure of the data.

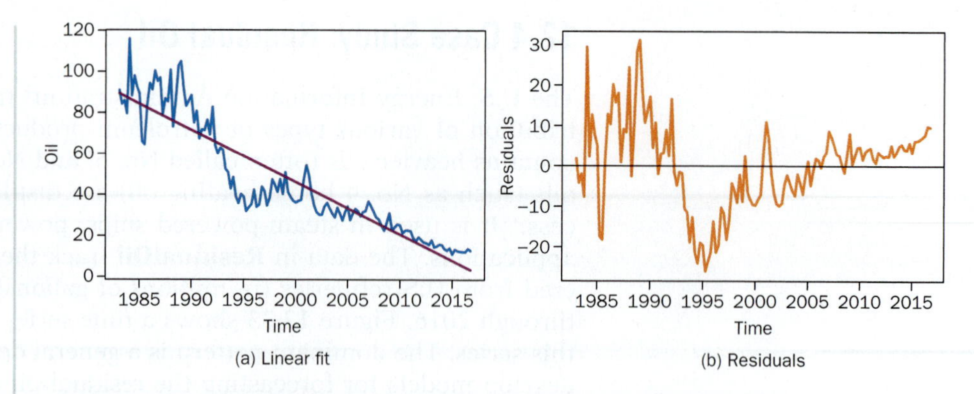

(a) Linear fit (b) Residuals

FIGURE 12.34 Linear fit and residuals for the original *Oil* series

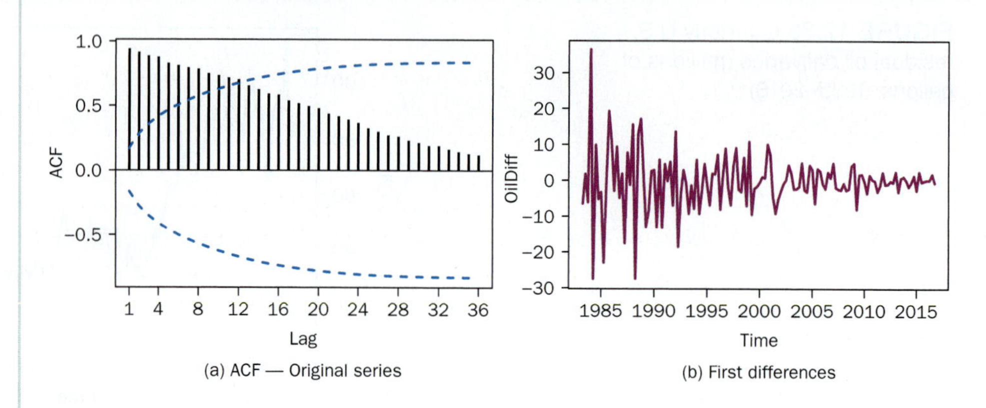

(a) ACF — Original series (b) First differences

FIGURE 12.35 ACF and first differences for the original *Oil* series

EXAMPLE 12.21

ResOil

Residual oil: Using a log transformation Both the curvature in the original time series plot and the decreasing variability in the plot of the first differences indicate considering a transformation of the original *Oil* values. Figure 12.36 shows a time series plot of the (natural) logarithm of the *Oil* variable (stored as *LogOil* in **ResidualOil**) along with a linear fit. We see a less curved pattern.

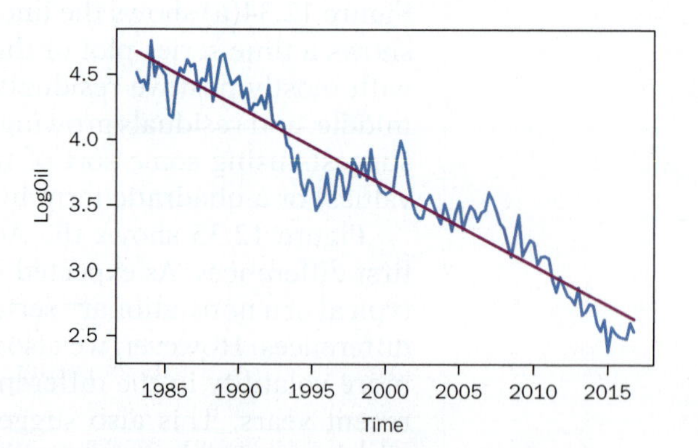

FIGURE 12.36 Logarithm of residual oil deliveries with a linear fit

The following ouptput is for fitting the linear model for *LogOil* using time alone. This model explains about 91.5% of the variability in *LogOil*.

```
Coefficients:
              Estimate   Std. Error  t value   Pr(>|t|)
(Intercept)  4.6825974   0.0313643   149.30    <2e-16
t           -0.0150975   0.0003973   -38.01    <2e-16
---
Residual standard error: 0.1819 on 134 degrees of freedom
Multiple R-squared: 0.9151, Adjusted R-squared: 0.9145
F-statistic: 1444 on 1 and 134 DF, p-value: < 2.2e-16
```

EXAMPLE 12.22

ResOil

Residual oil: Seasonal means? Should we include a seasonal component in the model? One might expect deliveries of some petroleum products (such as home heating oil) to be higher in the fall and winter quarters, but lower in the spring and summer (e.g., see the monthly heating oil data in Exercises 12.42 and 12.43). Because the seasonal period ($S = 4$) is fairly short, we may have trouble detecting a pattern in the time series plot of Figure 12.36, especially in the presence of the strong decreasing trend. An alternative is to look at boxplots of the residuals within each quarter after fitting the linear trend (as shown in Figure 12.37) to see if there are differences between the quarters.

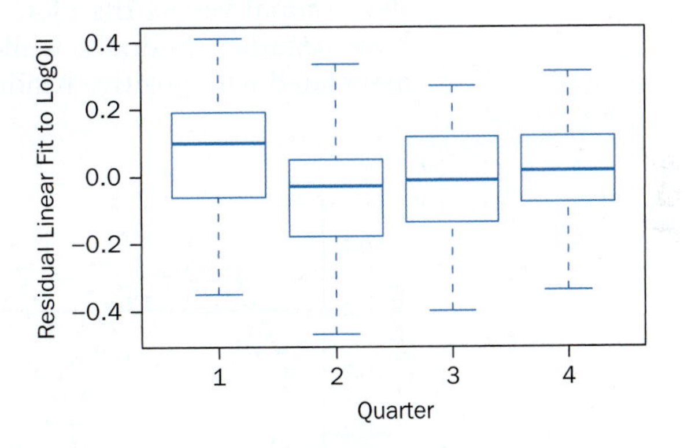

FIGURE 12.37 Residuals by *Quarter* for linear fit to *LogOil*

The boxplots of residuals from the linear trend are all relatively symmetric, but the median is quite a bit higher in the first quarter (indicating values that tend to be above the line) and somewhat lower in quarters two and three. Although we could fit a cosine trend to the quarterly data, that model would not be as reasonable in this case, because the high and low periods are adjacent to each other rather than at opposite points in the year. However, a seasonal means model (with linear trend) would account for differences between the quarters and only use five parameters (as compared to four for the cosine + linear trend).

Here is some output for fitting the seasonal means + linear trend model.

```
Coefficients:
                 Estimate   Std. Error   t value   Pr(>|t|)
(Intercept)     4.7524859   0.0400767    118.585   < 2e-16
t              -0.0150846   0.0003884    -38.842   < 2e-16
as.factor(Qtr)2 -0.1222160  0.0431082     -2.835   0.00531
as.factor(Qtr)3 -0.1006655  0.0431135     -2.335   0.02107
as.factor(Qtr)4 -0.0602074  0.0431222     -1.396   0.16501
---

Residual standard error: 0.1777 on 131 degrees of freedom
Multiple R-squared: 0.9207, Adjusted R-squared: 0.9183
F-statistic: 380.4 on 4 and 131 DF, p-value: < 2.2e-16
```

We see that the coefficients for Qtr2 and Qtr3 have significant *P*-values, indicating a difference from the reference quarter (Qtr1), while Qtr4 is not as significant. This is consistent with what we observed in the comparative boxplots. We also see slight improvement in both the adjusted R^2 (from 91.45% up to 91.83%) and the estimated standard deviation of the error term (from 0.1819 down to 0.1777).

Figure 12.38 shows the usual residual plots (residuals versus fits and normal quantiles) for checking the conditions of a regression model. The normality of the residuals looks pretty good, but we see some problems with the residual versus fits plot. Most of the smallest fitted values (below 3.0) have negative residuals, while predictions between 3.0 and 3.5 tend to be associated with positive residuals.

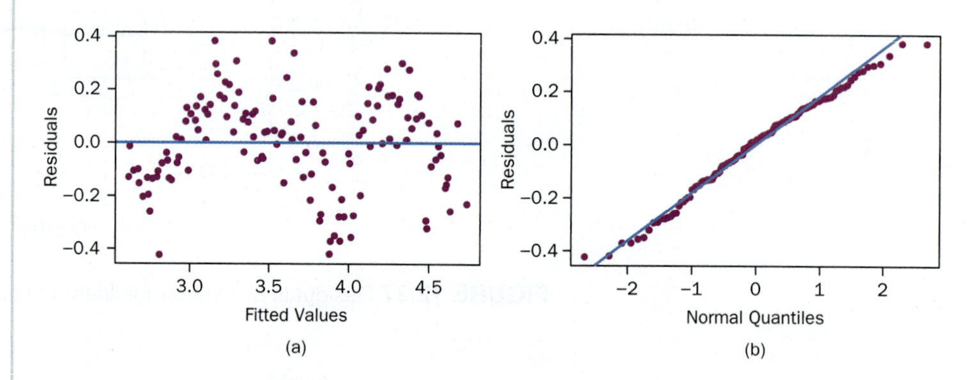

FIGURE 12.38 Residual plots for seasonal means + linear trend model of *logOil*

Finally, since this is time series data, we can look at the residuals as a time series and see if they appear to be independent of each other or still exhibit some patterns. Figure 12.39 shows this plot. We see that the residuals stay positive or negative for long stretches of time (e.g., all the quarters from 1993 to 1997) and show signs of generally increasing and decreasing patterns. In fact, the lag 1 autocorrelation of the residuals is 0.801, which shows very strong evidence of dependence in the residuals from the seasonal means + linear trend model for *LogOil*.

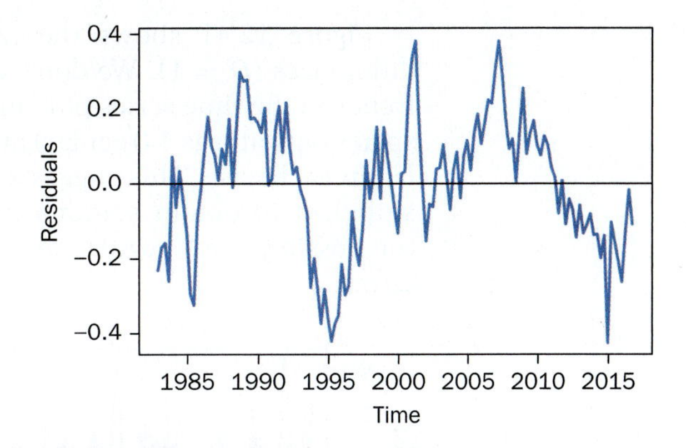

FIGURE 12.39 Time series of residuals for seasonal means + linear model

EXAMPLE 12.23

ResOil

Residual oil: Seasonal ARIMA We now consider ARIMA models for the *LogOil* series. In choosing an ARIMA$(p, d, q) \times (P, D, Q)$ model, we first consider the number of regular differences (d) and seasonal differences (D) we might need to make the series reasonably stationary. After that, we need to select numbers of regular autoregressive (p) or regular moving average (q) terms to account for any remaining low lag autocorrelations, along with seasonal autoregressive (P) or seasonal moving average terms (Q) to reflect possible autocorrelations at seasonal lags.

Figure 12.36 shows a clear decreasing trend for the logarithms of the residual oil series and the ACF for *LogOil* (not shown here) looks a lot like the ACF in Figure 12.35 for the original *Oil* series—the classic slow linear decay with many positive autocorrelations at early lags. These facts suggest the need for differencing: either a regular difference ($d = 1$) or a seasonal difference at lag four ($D = 1$), or possibly both.

Figure 12.40 shows the time series of first differences for the *LogOil* series and their autocorrelations. The time series, Figure 12.40(a), looks relatively stationary (at least no long increasing or decreasing trends) and the variability looks more constant across the time frame (especially when compared to the first differences of the original *Oil* series in Figure 12.35). However, there are lots of significant autocorrelations (see Figure 12.40(b), especially the positive spikes at seasonal lags 4, 8, 12, 16, and 32 (with positive but not quite significant values at lags 20, 24, 28, and 36). This suggests the possible need for a seasonal difference at lag = 4 to account for the quarterly data (perhaps with or without the regular difference).

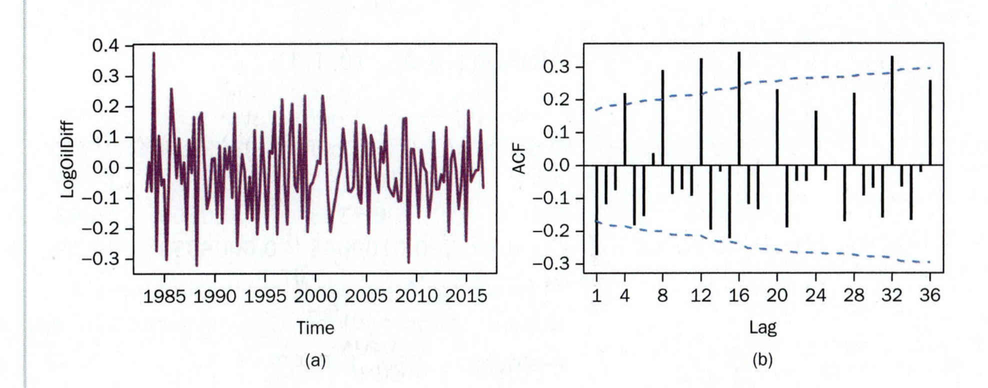

FIGURE 12.40 Plot and ACF for first differences ($d = 1$) of *LogOil*

Figure 12.41 shows the time series and ACF for the first seasonal differences ($D = 1$). We don't see any serious problems with nonstationary trends in the time series plot and the autocorrelations show large significant spikes only at lags 1 (regular) and 4 (seasonal) with smaller spikes at the next lag in each case. This suggests that a single seasonal difference is probably sufficient to obtain stationarity, and we might consider models with one (or possibly two) regular and seasonal moving average or autoregressive terms.

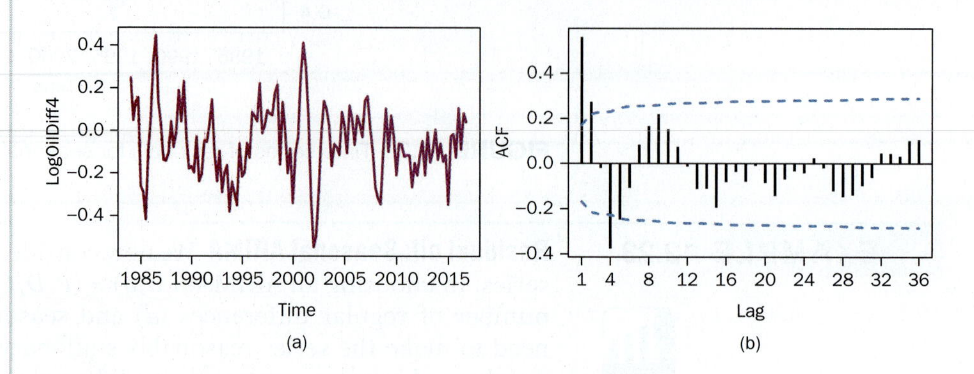

FIGURE 12.41 Plot and ACF for first seasonal differences ($D = 1$) of *LogOil*

Here is some output for seasonal ARIMA models with $d = 0, D = 1, p = 1$ or $q = 1$, and $P = 1$ or $Q = 1$. The constant terms are significant so they are retained in each of the models.

ARIMA$(1, 0, 0) \times (1, 1, 0)$:

Type	Coef	SE Coef	T	P
AR 1	0.6868	0.0677	10.14	0.000
SAR 4	-0.5492	0.0766	-7.17	0.000
Constant	-0.02819	0.01070	-2.63	0.009

Differencing: 0 regular, 1 seasonal of order 4
Number of observations: Original series 136, after differencing 132
Residuals: SS = 1.95020
MS = 0.01512 DF = 129

ARIMA$(1, 0, 0) \times (0, 1, 1)$:

Type	Coef	SE Coef	T	P
AR 1	0.8165	0.0527	15.48	0.000
SMA 4	0.9313	0.0560	16.63	0.000
Constant	-0.0109643	0.0008853	-12.39	0.000

Differencing: 0 regular, 1 seasonal of order 4
Number of observations: Original series 136, after differencing 132
Residuals: SS = 1.48475
MS = 0.01151 DF = 129

ARIMA$(0, 0, 1) \times (1, 1, 0)$:

Type	Coef	SE Coef	T	P
SAR 4	-0.4318	0.0820	-5.27	0.000
MA 1	-0.5841	0.0733	-7.97	0.000
Constant	-0.08648	0.01844	-4.69	0.000

Differencing: 0 regular, 1 seasonal of order 4
Number of observations: Original series 136, after differencing 132
Residuals: SS = 2.30699
MS = 0.01788 DF = 129

ARIMA$(0, 0, 1) \times (0, 1, 1)$:

Type	Coef	SE Coef	T	P
MA 1	-0.6810	0.0649	-10.49	0.000
SMA 4	0.5487	0.0752	7.30	0.000
Constant	-0.061289	0.008789	-6.97	0.000

Differencing: 0 regular, 1 seasonal of order 4
Number of observations: Original series 136, after differencing 132
Residuals: SS = 2.22389
MS = 0.01724 DF = 129

We see that all of the *P*-values are very small for testing each of the terms in all four models, so we don't see any terms that should be dropped. Also, each of the models has the same differencing and same number of parameters, so we can use either the sum of squared errors or mean squared error for comparisons. This shows a clear "favorite" with the second model, ARIMA$(1, 0, 0) \times (0, 1, 1)$, having the smallest value for both measures. But is this model adequate or should we consider adding additional terms? We could try models with more terms, but we could also look at the residuals to see if there is any remaining structure to be modeled.

Figure 12.42 shows the ACF of the residuals for each of the four ARIMA models. Remember that we'd like to see little or no evidence of autocorrelation in the residuals after fitting a model.

We see that the two models with a regular MA term (Figure 12.42(c) and (d)) both show a large positive autocorrelation in the residuals at lag 2, along with less significant spikes at lags 3, 7, and 15. The large early autocorrelation at lag 2 is particularly worrisome. Fortunately the models with a regular AR term do a good job of addressing the autocorrelation at lag 2. The ACF plots in Figure 12.42(a) and (b) show a first spike at lag 7, with fairly strong spikes at lags 8 and 15 for the first model and only one other barely significant spike at lag 15 for the second model. Among these four models, the most independent residual structure appears to be found with the ARIMA$(1, 0, 0) \times (0, 1, 1)$ model. Fortunately, this model also had the smallest SSE and MSE so would be a good choice at this point.

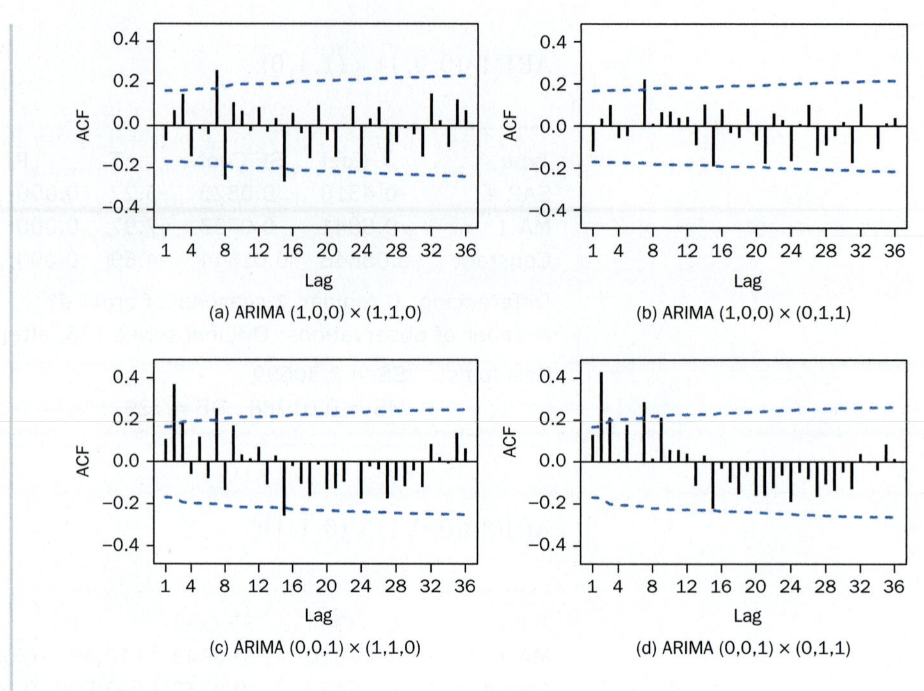

FIGURE 12.42 Residual ACFs for four ARIMA models for *logOil*

Should we try additional terms to address the remaining autocorrelations? Probably not. An autocorrelation at lag 8 might suggest adding a second seasonal term, but the one at lag 7 is more problematic. There might be some economic reason why values seven quarters apart are uniquely related, but it is more likely just a random occurrence. Note that we are actually doing 36 different tests when assessing the residual autocorrelations at lags 1 through 36. Even if there are no relationships, when we are doing lots of tests at a 5% significance level, about 5% of them will show up as "significant" just by random chance. That is why we pay particular attention to residual autocorrelations at early lags or seasonal lags and are less concerned about spikes at more random locations. In practice, we rarely find a "perfect" model that optimizes all the various criteria for building a model, so we often need to be happy with models that are "pretty good," even if they have a few flaws.

EXAMPLE 12.24

ResOil

Residual oil: Forecasts for 2017–2018 Recall that our goal in this case study is to forecast the residual oil values for each quarter of 2017 and 2018. The best "function of t" model we found for the residual oil data is the seasonal means + linear model for *LogOil* in Example 12.22. That model requires estimating five parameters and produces an estimated standard deviation for the error term of $\widehat{\sigma}_\epsilon = 0.1777$. However, the residuals for that model are still highly correlated. The ARIMA$(1, 0, 0) \times (0, 1, 1)$ model from the previous example uses only three parameters (1 regular AR, 1 seasonal MA, and the constant), has an estimated standard deviation of $\widehat{\sigma}_\epsilon = \sqrt{MSE} = \sqrt{0.01151} = 0.1073$, and shows little evidence of autocorrelation in the residuals. Looks like the ARIMA model would be the best choice at this point.

We use technology to compute forecasts and 95% prediction bounds for the next eight quarters (2017 and 2018). The results are shown in Table 12.2 and displayed along with the last few years of the time series for *LogOil* in Figure 12.43. We see that the forecasts maintain the expected pattern of highest values in Qtr1, dropping to the lowest values in Qtr2, and gradually climbing back in Qtr3 and Qtr4. Also, notice that the intervals tend to be wider for each quarter in 2018 than the corresponding quarter in 2017. This generally occurs as we forecast further into the future.

Time	Forecast	Lower	Upper
2017 Q1	2.575	2.363	2.787
2017 Q2	2.479	2.206	2.752
2017 Q3	2.491	2.183	2.798
2017 Q4	2.526	2.198	2.855
2018 Q1	2.558	2.211	2.905
2018 Q2	2.454	2.095	2.812
2018 Q3	2.459	2.093	2.825
2018 Q4	2.489	2.119	2.860

TABLE 12.2 Forecasts and bounds for *LogOil* in 2017–2018

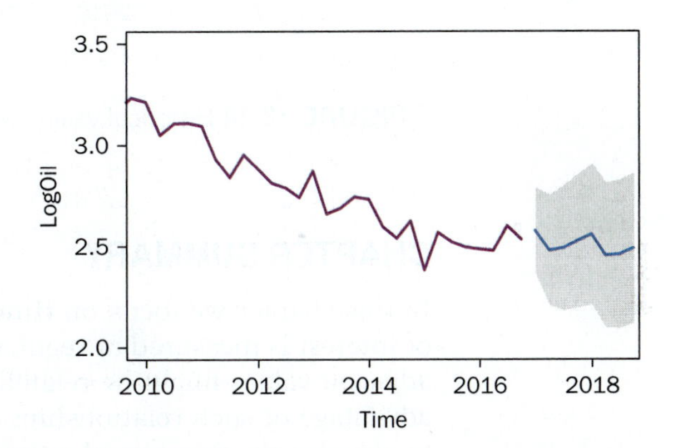

FIGURE 12.43 Historical values with forecasts and bounds for *LogOil* in 2017–2018

Finally, note that our task is to forecast the quarterly residual oil values for 2017 and 2018 and, so far, we have only produced forecasts and bounds for the logarithm of the *Oil* values. Fortunately, all that remains is to exponentiate those forecasts and bounds to obtain results on the original *Oil* scale (millions of gallons/day). These are shown in Table 12.3 and Figure 12.44.*

*By the time you are reading this, you should be able to go to the EIA website and check the accuracy of some of these forecasts!

Time	Forecast	Lower	Upper
2017 Q1	13.14	10.63	16.23
2017 Q2	11.93	9.08	15.67
2017 Q3	12.07	8.87	16.42
2017 Q4	12.51	9.00	17.37
2018 Q1	12.91	9.13	18.26
2018 Q2	11.63	8.13	16.64
2018 Q3	11.69	8.11	16.86
2018 Q4	12.05	8.32	17.47

TABLE 12.3 Forecasts and bounds for *Oil* in 2017–2018

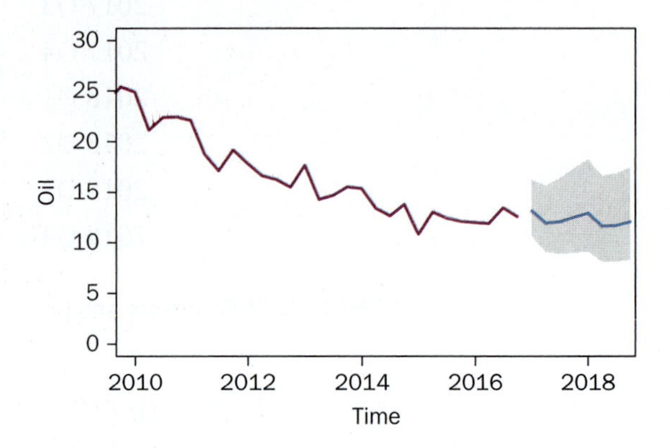

FIGURE 12.44 Historical values with forecasts and bounds for *Oil* in 2017–2018

CHAPTER SUMMARY

In this chapter we focus on **time series models** where the response variable of interest is measured at regular time intervals. In this setting we expect that adjacent values might be related (not independent) and, in fact, we may take advantage of such relationships to improve a model. We first start with tools to *visualize the structure* of a time series that include

- Looking at a **time series plot**, with time on the horizontal axis and the response value on the vertical axis, to show increasing/decreasing trends, seasonal patterns, and variability
- Looking at **autocorrelations** at various lags (for example, with an **ACF plot**) to reveal dependencies within the series

Depending on what the plots show, we may choose models to handle *increasing/decreasing trends*, with techniques such as

- A **linear model in** t: $Y_t = \beta_0 + \beta_1 t + \epsilon_t$
- A **quadratic model in** t: $Y_t = \beta_0 + \beta_1 t + \beta_2 t^2 + \epsilon_t$
- A **log transformation** for exponential growth or decay: $log(Y_t)$
- A model for the **differences**: $\Delta Y_t = Y_t - Y_{t-1}$ (or second differences)

To handle models with *seasonal trends* at a seasonal period S, we might use

- A **seasonal means model** that specifies a different mean at each seasonal period: $Y_t = \mu_i + \epsilon_t$
- A **cosine trend model**: $Y = \beta_0 + \alpha \cos\left(\frac{2\pi t}{S} + \phi\right) + \epsilon = \beta_0 + \beta_1 \cos\left(\frac{2\pi t}{S}\right) + \beta_2 \sin\left(\frac{2\pi t}{S}\right) + \epsilon$
- **Seasonal terms in an ARIMA model** (see below)

We can also choose to combine some of these models, for example, by adding a linear term to a seasonal means or cosine trend model to account for both a seasonal pattern and increasing/decreasing trend.

Finally, we consider the general class of **ARIMA models** that can reflect a greater variety of interdependence structures within a time series. The general steps to fitting a time series model include

1. Take sufficient differences (either regular or seasonal) or use a transformation (such as $log(Y_t)$) to make the series **stationary**. For stationarity, we look for no regular increasing/decreasing trends in the series and an ACF plot that does not show a slow linear decay at either regular (lags $1, 2, 3, \ldots$) or seasonal (lags $S, 2S, 3S, \ldots$) periods. This is where we specify d (and D in the case of seasonal differences).
2. Choose one or more regular autoregressive (p) and/or moving average (q) terms to account for dependence (autocorrelation) on past series values or past residuals at low lags.
3. If there is dependence at seasonal periods, choose one or more seasonal autoregressive (P) and/or seasonal moving average (Q) terms.

At this stage we choose p, q, P, and Q somewhat by trial and error, although a full course in time series analysis will offer more guidance on how to match various autocorrelation patterns with different ARIMA models.

When deciding on a final model, just as with many of the modeling situations in this book, we have several (possibly competing) goals to help guide our choices.

- Try to use only significant terms in the model (e.g., by checking tests of individual coefficients).
- Try to minimize the size of the residuals (e.g., by checking SSE or s_ϵ).
- Try to keep the model simple. (Avoid adding lots of extra terms that may work against each other and not really contribute much to the model.)
- Try to have no significant autocorrelations in the residuals.

The last item is particularly relevant to time series, where autocorrelation in the residuals may signal valuable structure in the data that has not yet been fully exploited by the model. However, we need to recognize that it is often infeasible or even impossible to satisfy all these criteria with a single model. We need to use them to help guide the process of selecting and assessing models, but not reject a model that might be useful simply because of a minor flaw (such as a random "significant" autocorrelation at lag 9 of the residuals).

EXERCISES

Conceptual Exercises

12.1 Yearly and seasonal? Suppose that we have a time series that has values at yearly intervals.

a. Explain why we would generally *not* see a seasonal model applied to data that has just one value for each year.

b. Give an example of a yearly time series where you might expect to find a seasonal pattern. Hint: Think about politics.

12.2 *Year* or *t*? In Example 12.2 on page 541 we consider using a linear model to predict *Extent* for the Arctic sea ice data. The linear model could use either the *Year* (1979 to 2015) as the predictor or a generic time scale t (from 1 to 37). The two prediction equations had the same slope ($\hat{\beta}_1 = -0.08732$), but different intercepts ($\hat{\beta}_0 = 180.73$ for *Year* and $\hat{\beta}_0 = 8.008$ for t). How do the residuals compare between those two models? **SeaIce**

12.3 Specifying cosine trend models. We describe some seasonal time series patterns in Example 12.4 on page 543 and fit a cosine trend model for monthly traffic over the Peace Bridge in Example 12.5. Write down the cosine trend models for each of the following settings.

a. Quarterly sales data for a large company.

b. Daily closing prices for a stock, with a linear term for time t. You may ignore holidays and assume that market data are available for all weekdays (but not weekends).

c. Daily riders on a metropolitan transit system, allowing for a quadratic trend over time t.

d. After surgery, you get blood pressure checked every four months (three times a year).

12.4 Specifying seasonal means models. In Example 12.6 on page 546 we fit a seasonal means model for the monthly traffic over the Peace Bridge. Write down the seasonal means model for each of the following settings.

a. Quarterly sales data for a large company.

b. Daily closing prices for a stock, with a linear term for time t. You may ignore holidays and assume that market data are available for all weekdays (but not weekends).

c. Daily riders on a metropolitan transit system, allowing for a quadratic trend over time t.

d. After surgery, you get your blood pressure checked every four months (three times a year).

12.5 Guessing regular differencing from a plot. Figure 12.45 shows times series plots for five different series. Based on just the plots for each series, indicate whether or not at least one *regular* difference would help achieve stationarity. Note: The next exercise asks a similar question about seasonal differences.

12.6 Guessing seasonal differencing from a plot. As in Exercise 12.5 we are interested in guessing whether differencing might be needed to help with stationarity for the time series that are plotted in Figure 12.45. In this exercise, comment on whether you think a *seasonal* difference might be required.

12.7 Guessing regular differencing from an ACF. Figure 12.46 shows autocorrelation plots for five different series. Based on just the plots, which series do you think would need at least one *regular* difference to help achieve stationarity? Note: The next exercise asks a similar question about seasonal differences.

12.8 Guessing seasonal differencing from an ACF. As in Exercise 12.7 we are interested in guessing whether differencing might be needed to help with stationarity for

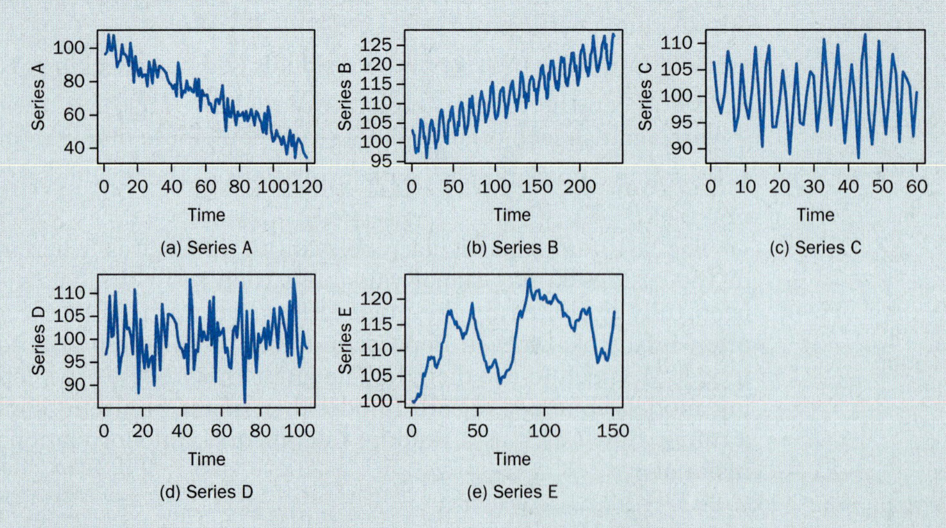

FIGURE 12.45 Five time series plots

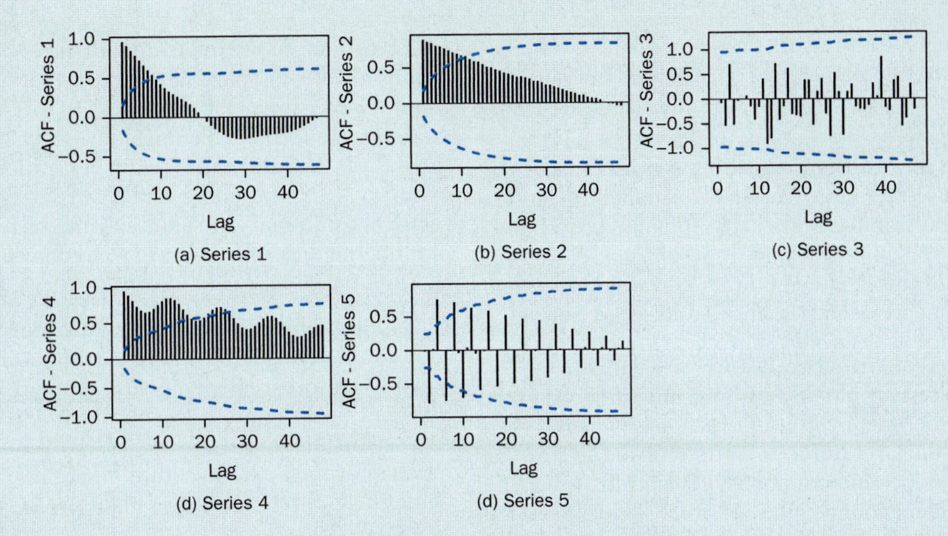

FIGURE 12.46 Five time series ACF plots

the time series that are plotted in Figure 12.46. In this exercise, comment on whether you think a *seasonal* difference might be required.

12.9 Matching time series plot to its ACF. The five time series (Series A through Series E) shown in the time series plots of Figure 12.45 for Exercises 12.5 and 12.6 are the same as the five series (Series 1 through Series 5) shown in the ACF plots of Figure 12.46 for Exercises 12.7 and 12.8, but the order has been scrambled. Match each series with its ACF and explain your reasoning.

12.10 Not seasonal. Which variable in the list below would you expect to *not* have a seasonal component? Each variable is measured on a monthly basis.

a. Ice cream sales

b. Reservations at a beach hotel in Virginia

c. Cholesterol levels for an aging author

d. New memberships in Orange Theory Fitness

12.11 Seasonal. Which variable in the list below would you expect to contain a seasonal component?

a. The cost of a dental cleaning

b. Household spending on gifts

c. An Internet provider's stock price

d. The price of milk

12.12 Investment considerations. When you invest in a stock, you are willing to accept a certain amount of risk, but you also have certain expectations. Let's assume that you want to purchase a stock and hold onto it for at least a decade. Mark each statement as *True* or *False*.

a. Over a short period of time, the stock price will go up and down.

b. Over a long period of time, you are expecting the stock to go down.

12.13 Training considerations. Athletes use a variety of training methods to improve their performance over time. Mark each statement as *True* or *False*.

a. Performance, as measured by a fitness test, after 2 weeks of a training program is independent of performance after 1 week of the program.

b. A seasonal component in performance is likely, even if the athlete continues the training program all year.

c. A decreasing trend in weight is likely if calorie intake is not increased as the intensity of the program increases.

Guided Exercises

12.14 Arctic sea ice: linear fit for *Area*. In Example 12.2 on page 541 we consider a linear model to predict the yearly *Extent* of Arctic sea ice. Another variable in the **SeaIce** dataset measures the *Area* of the ice. This is a smaller amount that deducts interior pockets of open water within the outer boundaries of the ice that determine the *Extent*. SeaIce

a. Obtain a plot of the *Area* time series and compare it to the *Extent* series from Example 12.2.

b. Fit a linear model based on t (1 to 37), to predict *Area*.

c. The linear fit for the sea ice *Extent* variable shows some concern with curvature. Use a plot (or plots) to see if there is a similar concern with the *Area* variable.

12.15 Arctic sea ice: quadratic fit for *Area*. In Example 12.3 on page 542 we see that a quadratic model is useful for predicting the *Extent* of Arctic sea ice. Does this also hold for the *Area* variable in the **SeaIce** dataset? SeaIce

a. Fit a quadratic model for *Area* and add the fitted model to a time series plot of *Area*.

b. Is the quadratic term needed in the model?

c. Plot the residuals from the quadratic model for *Area* as a time series. Does it look like the concern with curvature has been addressed?

12.16 Arctic sea ice: autocorrelation for linear residuals. Exercise 12.14 looks at a linear model to predict the *Area* of Arctic sea ice each September, based on either the *Year* or *t*. Find the lag 1 autocorrelation for the residuals of this model and comment on what this tells you about independence of the residuals. **SeaIce**

12.17 Arctic sea ice: autocorrelation for quadratic residuals. Exercise 12.15 looks at a quadratic model based on *t* to predict the *Area* of Arctic sea ice each September. Find the lag 1 autocorrelation for the residuals of this model and comment on what this tells you about independence of the residuals. **SeaIce**

12.18 Arctic sea ice: linear forecasts. Consider the linear model for predicting the *Area* of Arctic sea ice in Exercise 12.14. **SeaIce**

a. Find forecasts and 95% prediction bounds for values in September of 2017, 2019, and 2021. Describe how those forecasts and bounds change as we go further out in time.

b. If we assume the model remains accurate into the future, at what time would it predict the Arctic sea ice to be completely gone in September?

12.19 Arctic sea ice: quadratic forecasts. Consider the quadratic model for predicting the *Area* of Arctic sea ice in Exercise 12.15. **SeaIce**

a. Find forecasts and 95% prediction bounds for values in September of 2017, 2019, and 2021. Describe how those forecasts and bounds change as we go further out in time.

b. If we assume the model remains accurate into the future, at what time would it predict the Arctic sea ice to be completely gone in September?

12.20 Arctic sea ice: residuals for quadratic (*Extent*). In Example 12.3 on page 542, we consider a quadratic model due to curvature shown in Figure 12.3 for the residuals versus fits plot from the linear model for the Arctic sea ice *Extent* variable in **SeaIce**. Plot the residuals from the quadratic model in *t* as a time series and comment on whether or not the curvature issue has been addressed. **SeaIce**

12.21 Combining differences. In Example 12.12 we note that, when using both a regular and seasonal difference for the Peace Bridge traffic data, the order of the differences does not matter. This means we can take a regular difference (ΔY_t) and then compute seasonal (lag 12) differences for those values, or do the seasonal ($\Delta_{12} Y_t$) difference first and then do a regular (lag 1) difference. Show, algebraically, that these methods give the same result.

12.22 Consumer Price Index: differences. In Example 12.13 on page 558 we see that the data on the monthly Consumer Price Index (CPI) are clearly not stationary, but the percentage differences (*CPIPctDiff* in the dataset **Inflation**) are relatively stationary. Would the ordinary first differences of the CPI series itself also be relatively stationary? Produce a time series plot and ACF plot for the differences (ΔCPI) and comment stationarity. **Inflat**

12.23 Consumer Price Index: autoregressive models. In Examples 12.14 and 12.15 starting on page 559 we consider autoregressive models for the percentage differences of the monthly CPI data (*CPIPctDiff* in the dataset **Inflation**). In this exercise we consider similar models for the actual CPI differences (ΔCPI, not the percentage differences). **Inflat**

a. Fit a first-order autoregressive model to the differences of the *CPI* series (i.e., an ARIMA$(1, 1, 0)$ model). Check if the constant term is important to keep in the model. If not, drop it and refit. Write down the fitted ARIMA$(1, 1, 0)$ model.

b. Is the first-order autoregressive term needed in this model? Justify your answer.

c. Comment on anything interesting you see in the ACF of the residuals for this model.

d. Now add a second autoregressive term to run an ARIMA$(2, 1, 0)$ for the *CPI* series. Is the second-order autoregressive term helpful in the model?

12.24 Consumer Price Index: moving average models. In Example 12.16 on page 562 we consider an MA(1) model for the percentage differences of the monthly CPI data (*CPIPctDiff* in the dataset **Inflation**). In this exercise we consider moving average models for the actual CPI differences (ΔCPI, not the percentage differences). **Inflat**

a. Fit a first-order moving averge model to the differences of the *CPI* series (i.e., an ARIMA$(0, 1, 1)$ model). Check if the constant term is important to keep in the model. If not, drop it and refit. Write down the fitted ARIMA$(0, 1, 1)$ model.

b. Is the first-order moving average term needed in this model? Justify your answer.

c. Comment on anything interesting you see in the ACF of the residuals for this model.

d. Now add a second moving term to run an ARIMA$(0, 1, 2)$ for the *CPI* series. Is the second-order moving average term helpful in the model?

12.25 Hawaii CO2: time series plot. The data in **CO2Hawaii** shows the monthly carbon dioxide level (CO2 in ppm) measured at the Mauna Loa Observatory in Hawaii[5] from 1988 through 2017. Create time series plot of the data and comment on any interesting features. **CO2HI**

12.26 South Pole CO2: time series plot. The data in **CO2SouthPole** shows the monthly carbon dioxide level (CO2 in ppm) measured at the South Pole[6] from 1988 through 2016. Create time series plot of the data and comment on any interesting features. **CO2SP**

12.27 Hawaii CO2: linear or quadratic? Consider the times series of *CO2* measurements from Hawaii introduced in Exercise 12.25, but ignore possible monthly seasonal trends for the moment. **CO2HI**

a. Fit a linear model using t for the *CO2* data in *CO2Hawaii*. Write down the fitted prediction equation.

b. Does the linear model explain a substantial portion of the variability in the Hawaii *CO2* series?

c. Would it be better to replace the linear model with a quadratic one? Give a justification your answer.

d. Plot the residuals from both the linear and quadratic models as time series. Comment on what these plots tell you.

12.28 South Pole CO2: linear or quadratic? Consider the times series of CO2 measurements from the South Pole introduced in Exercise 12.26, but ignore possible monthly seasonal trends for the moment. **CO2SP**

a. Fit a linear based on t for the *CO2* data in *CO2SouthPole*. Write down the fitted prediction equation.

b. Does the linear model explain a substantial portion of the variability in the South Pole *CO2* series?

c. Would it be better to replace the linear model with a quadratic one? Give a justification for your answer.

d. Plot the residuals from both the linear and quadratic models as time series. Comment on what these plots tell you.

12.29 Hawaii CO2: seasonal component. In Exercise 12.27 you are asked about linear and quadratic trends for the Hawaii *CO2* series. Now it's time to work on possible seasonal trends in the series. Assume for this exercise that a quadratic model in t will account for the overall increasing trend. Here we decide between a cosine trend model and seasonal means for handling the seasonal pattern. **CO2HI**

a. Fit a cosine trend model, along with terms for a quadratic trend in t, for the *CO2* data in **CO2Hawaii**. Record the adjusted R^2 value.

b. Fit a seasonal means model, along with terms for a quadratic trend in t, for the *CO2* data in **CO2Hawaii**. Record the adjusted R^2 value.

c. What CO2 level does each model predict for July 2018 ($t = 370$)?

d. Do you prefer the model in (a) or (b)? Give some justification for your answer.

e. Do we really need the terms for a quadratic trend in this model? Take whichever model you chose in (d) and delete the terms for the quadratic trend to leave only the seasonal terms. Find the percentage of variability in the Hawaii *CO2* series explained by just the seasonal means and produce a time series plot for the fitted (seasonal only) model on the same graph as the original series. Comment on what you see.

12.30 South Pole CO2: seasonal component. In Exercise 12.28 you are asked about linear and quadratic trends for the South Pole *CO2* series. Now it's time to work on possible seasonal trends in the series. Assume for this exercise that a quadratic model in t will account for the overall increasing trend. Here we decide between a cosine trend model and seasonal means for handling the seasonal pattern. **CO2SP**

a. Fit a cosine trend model, along with terms for a quadratic trend in t, for the *CO2* data in **CO2SouthPole**. Record the adjusted R^2 value.

b. Fit a seasonal means model, along with terms for a quadratic trend in t, for the *CO2* data in **CO2SouthPole**. Record the adjusted R^2 value.

c. What CO2 level does each model predict for December 2017 ($t = 360$)?

d. Do you prefer the model in (a) or (b)? Give some justification for your answer.

e. Do we really need the terms for a quadratic trend in this model? Take whichever model you chose in (d) and delete the terms for the quadratic trend to leave only the seasonal terms. Find the percentage of variability in the South Pole CO2 series explained by just the seasonal means and produce a time series plot for the fitted (seasonal only) model on the same graph as the original series. Comment on what you see.

12.31 Google stock prices: function of *t* models. In Example 12.10 on page 552 we see a series based in a few months of daily closing prices for Apple Stock. The dataset **TechStocks** has longer series (two years from December 1, 2015, to December 1, 2017) of daily closing prices for several tech stocks:[7] Apple (APPL), Alphabet—the parent company of Google (GOOG), and Microsoft (MSFT). Use only the Google prices below. **Tech**

a. Create a time series plot for the Google price data. You may use a generic t axis or the *Date* variable included in the dataset. Comment on any important features of the time series.

b. Consider a linear model in t for describing the Google price series. What proportion of variability in the series does this model explain?

c. Use either a plot with series and linear model or a time series plot of the residuals to show why there might be a problem with the linearity condition of this model.

d. Two possible ways to address the linearity issue are to add a quadratic term to the model or to use a log transformation on the original data. Fit both of those models and report the fitted equations.

e. Does either change in (d) help with the linearity issue?

f. Use the model you prefer from (d) to find a forecast (with bounds) for the Google stock price on December 15, 2017. Hint: Counting only stock trading days gets to $t = 514$.

12.32 Microsoft stock prices: function of *t* models. Answer the parts (a) through (f) of Exercise 12.31 for the Microsoft (MSFT) stock prices in **TechStocks**. Tech

12.33 Google stock prices: ARIMA models. Exercise 12.31 considers function of t models for the daily Google stock prices in **TechStocks**. This exercise deals with ARIMA models for the same series. Tech

a. Plot the Google stock price (*GOOG*) and its ACF. What do you think about stationarity of the original series?

b. Plot the first differences for the Google stock prices and find their ACF. What do you think about stationarity of the differences?

c. Find the mean of the differences from (b). What does this value tell you about Google stock prices over this two-year period?

d. Using the ACF from (b), which lags (if any) show a significant autocorrelation for the differences?

e. What does the ACF for the differences suggest about possible AR or MA terms for an ARIMA model with one difference for the Google stock prices? Hint: Focus on autocorrelation at the low lags (1, 2, or 3) and ignore random spikes at unusual lags. Note that about 1 in 20 autocorrelations will be flagged as "significant" at a 5% level, even if the data were generated at random.

f. Fit the ARIMA model suggested by (e) using one difference and including a constant term. Report the forecast equation. Note: Some software might require at least one AR or MA term to fit the model. If so, include one.

g. How does the constant term in your model from (f) compare to the mean difference you found in (c)?

h. Use the ARIMA model from (f) to find a forecast (and bounds) for the closing Google stock price on December 15, 2017 (which is 10 trading days after the last date in the series).

12.34 Microsoft stock prices: ARIMA models. Answer parts (a) through (h) of Exercise 12.33 for the Microsoft (*MSFT*) stock prices in **TechStocks**. Tech

12.35 Key West water temperature: plots. Obtaining time series data from a variety of sources is getting much easier. A student[8] obtained hourly water temperatures in the Gulf of Mexico near Key West from October 3, 2016, to October 3, 2017, from NOAA.* These hourly temperatures, in degrees Fahrenheit, are in **KeyWestWater**. KeyW

a. Construct a time series plot of *WaterTemp* and comment on what you see.

b. We might expect a seasonal pattern repeating each day, but that is difficult to see in a plot that shows hourly data over 365 days. Create a time series plot for any 4-day (96-hour) period and comment on whether you see a seasonal pattern with $S = 24$.

c. Create an ACF plot for the full *WaterTemp* series and comment on what you see.

12.36 Key West water temperatures: differencing. The ACF plot in Exercise 12.35 clearly shows that a regular difference may be helpful for the temperatures in the Gulf of Mexico near Key West. Use software to create a series of the differences of the *Watertemp* variable in **KeyWestWater**. KeyW

a. If everything works as you hope, the mean of the differences should be zero. What is the mean of the differences that you created?

b. The largest difference is 4.3. Would you tag this observation as an outlier? Explain.

c. Create a time series plot of the differences. Does this time series plot look similar to the time series plot of *WaterTemp*? Explain.

d. Do you think the regular difference was helpful in obtaining stationarity? Explain.

e. Create an ACF plot for the differences and comment on what you see.

12.37 Key West water temperatures: linear, quadratic, and cosine trend models. We considered a variety of basic function of t models in Section 12.1. Let's see if any of those models are useful for the water temperatures in the Gulf of Mexico off of Key West. KeyW

a. Fit a simple linear model using t, the hourly time index, as the predictor variable and then create a time series plot of the residuals for this model. Is t a significant predictor of temperature? How much of the variation in water temperatures is explained by using this model? Comment on the residual plot.

b. Fit a quadratic model using t, the hourly time index, as the predictor variable and then create a time series plot of the residuals for this quadratic model. Is the quadratic term t^2 a significant predictor of temperature? How much of the variation in water temperatures is explained by using this model? Comment on the residual plot.

c. Fit a cosine trend model by adding the terms $X_{\cos}$ and $X_{\sin}$ with $S = 24$ to your quadratic model. Are these two predictors helpful here? How much of the variation in water temperatures is explained by using this model?

*The National Oceanographic and Atmospheric Administration website https://www.nodc.noaa.gov provides lots of data that can be used for course projects.

12.38 Key West water temperatures: smoothing with running averages. When introducing the moving average model (page 561) we noted that it should not be confused with simply averaging recent past values. To distinguish these, we'll use the term *running average* to refer to using the average of some fixed number of past values to help smooth a series. But how many past values should we use? That is what you will explore in this exercise. Create plots of the Key West water temperature series along with running averages using orders 24 (one day), 48 (two days), 168 (1 week), 336 (2 weeks), and 720 (30 days or 1 month). Comment on what happens as you increase the number of past days in the running average. Looking only at the graphs, what order do you think summarizes thr general pattern of the series best? **KeyW**

12.39 Key West water temperatures: comparing ARIMA output. Compare and contrast the output from two statistical software packages for an ARIMA(1, 1, 0) × ARIMA(0, 1, 1) model with $S = 24$ and no constant term fit to the *WaterTemp* series. **KeyW**

From Minitab:

```
Final Estimates of Parameters
Type          Coef   SE Coef         T       P
AR     1   -0.0641   0.0123      - 5.19   0.000
SMA   24    0.9796   0.000105   9303.2   0.000

Differencing: 1 regular difference
Number of observations: Original series 6572, after
differencing 6547
Residuals:   SS = 314.513
             MS = 0.04805   DF = 6545
```

From R:

```
ARIMA(1,1,0)(0,1,1)[24]
 Coefficients:
          ar1      sma1
      -0.0695   -0.9660
s.e.   0.0126    0.0055

sigma^2 estimated as 0.04822: log likelihood=605.55
AIC=-1205.11 AICc=-1205.1 BIC=-1184.75
```

12.40 Key West water temperatures: comparing ARIMA models. Exercise 12.39 shows output from R for fitting an ARIMA(1, 1, 0) × $(0, 1, 1)_{24}$ to *WaterTemp* series in **KeyWestWater**. Output for three other models for this series follows that modify some of the seasonal parts of that model. Based only on the output shown for these four models, which would you prefer? **KeyW**

ARIMA(1, 1, 0)—no seasonal component:

```
ARIMA(1,1,0)
 Coefficients:
          ar1
      -0.0375
s.e.   0.0123

sigma^2 estimated as 0.04897: log likelihood=587.38
AIC=-1170.75  AICc=-1170.75  BIC=-1157.17
```

ARIMA(1, 1, 0) × $(0, 0, 1)_{24}$—seasonal AR (no seasonal difference)

```
ARIMA(1,1,0)(1,0,0)[24]
 Coefficients:
          ar1    sar1
      -0.0687  0.1286
s.e.   0.0127  0.0126

sigma^2 estimated as 0.04822: log likelihood=638.89
AIC=-1271.77  AICc=-1271.77  BIC=-1251.4
```

ARIMA(1, 1, 0) × $(0, 1, 0)_{24}$—no seasonal moving average

```
ARIMA(1,1,0)(0,1,0)[24]
 Coefficients:
          ar1
      -0.1731
s.e.   0.0122

sigma^2 estimated as 0.08461: log likelihood=-1203.67
AIC=2411.34 AICc=2411.34 BIC=2424.91
```

Open-ended Exercises

12.41 Arctic sea ice: ARIMA for Area. Example 12.17 on page 563 looks at ARIMA models for the *Extent* of Arctic sea ice using the September data found each year from 1979 to 2015. The **SeaIce** dataset also has information on the *Area* of the ice in the same years. Do a similar analysis to look at ARIMA models for the *Area*. Once you have settled on a reasonable ARIMA model, use it to forecast the Arctic sea ice *Area* in 2017, 2019, and 2021 (as you may have done in Exercises 12.18 and 12.19). **SeaIce**

12.42 Heating oil: functions of *t*. The case study in Section 12.4 looked at models for quarterly data on residual oil production. We also have data[9] in **HeatingOil** on the monthly residential consumption of fuel oil (in 1,000s of barrels) from January 1983 through December 2016. For this exercise consider models that are functions of *t* (e.g., linear, quadratic, cosine trend, seasonal means) as in Section 12.1. We'll save the ARIMA models for the next exercise. Once you have settled on a model, use it to forecast the *FuelOil* consumption for each month in 2018. **HeatOil**

12.43 Heating oil: ARIMA. Exercise 12.42 looks at function of *t* model for the monthly *FuelOil* series *HeatingOil*. In this exercise you should consider ARIMA models for forecasting this series. First decide on possible differences (regular or seasonal) to help with stationarity, then investigate adding some (regular or seasonal) autoregressive or moving average terms. Try to keep the model relatively simple. When you have checked residuals and settled on a model, use it to forecast the *FuelOil* consumption for each month in 2018. **HeatOil**

Answers to Selected Exercises

CHAPTER 0

0.1 Response : Explanatory

(a) Sleep (Q): Major (C). **(b)** Final (Q): Exam1 (Q). **(c)** Final time (Q): Gender (C, B). **(d)** Handedness (C, B): Major (C), Gender (C, B), Final (Q).

0.3 **(a)** Units = MLB games; Response = Game time (Q); Explanatory = Runs (Q), Margin (Q), Pitchers (Q), Attendance (Q), League (C). **(b)** Units = putts; Response = Made/Missed (C); Explanatory = Distance (Q). **(c)** Units = Brees games; Variables = Passing Yards (Q), Pass Attempts (Q), Completions (Q), Touchdown Passes (Q).

0.5 **(a)** Nutrition experts. **(b)** Experiment. **(c)** Serving Size (Q). **(d)** Bowl (C), Spoon (C).

0.7 **(a)** *WineQuality* (Q). **(b)** *WinterRain* (Q), *AverageTemp* (Q), *HarvestRain* (Q). **(c)** More. **(d)** Less. **(e)** More. **(f)** Observational.

0.9 **(a)** Members of an entering class. **(b)** Parameters. **(c)** Students who want to take math. **(d)** Statistics.

0.11 **(a)** Yes. **(b)** Yes. **(c)** Spoon size. **(d)** Similar.

0.13 **(a)** 9.825. **(b)** 0.175. **(c)** 7.96.

0.15 **(a)** 54 mph. **(b)** 61.6 mph. **(c)** 7.6 mph.

0.17 **(a)** Not a short answer. **(b)** $y = \mu + \epsilon$; $\mu > 50$. **(c)** $\hat{y} = 62.52$.

0.19 **(a)** Approximately normally distributed. **(b)** *P*-value ≈ 0, reject H_0. **(c)** Subjects' guesses significantly higher than 50%.

0.21 **(a)** $y = \mu_i + \epsilon$. **(b)** Women: $\hat{y} = 66.459$; Men: $\hat{y} = 57.802$. **(b)** Residuals normal, *P*-value ≈ 0, reject H_0. **(c)** Mean percent correct are significantly different for men and women students.

0.23 Not convincing evidence that financial incentives help increase mean weight loss after seven months.

0.25 No evidence that female reading mean is higher than male reading mean.

0.27 **(a)** The mean running pace is clearly slower in later years. **(b)** The average distance run per day is different in later years.

0.29 The mean difference for *Both* between men and women students is −5.95%, which is statistically different from 0% (*P*-value ≈ 0).

0.31 **(a)** Effect size = 0.23, small. **(b)** Effect size = 1.07, large.

CHAPTER 1

1.1 **(c)**.

1.3 $\hat{\beta}_1 = 0.4674$.

1.5 $\hat{\beta}_0 = 1.3655$.

1.7 Not a short answer.

1.9 $\hat{\sigma}_\epsilon = 1.39959$.

1.11 $df = 114$.

1.13 5.

1.15 **(a)** $\widehat{Width} = 37.72 - 0.01756\ Year$. **(b)** Not a short answer. **(c)** 3.197 mm.

1.17 **(a)** $\widehat{MaxGripStrength} = 36.16 + 4.705\ Attractive$. **(b)** As *Attractive* increases by 1, *MaxGripStrength* increases by 4.7 kg, on average. **(c)** 50.3.

1.19 **(a)** Some positive association. **(b)** $\widehat{Calories} = 87.43 + 2.48\ Sugar$. **(c)** For every additional gram of sugar in a serving of cereal, the expected calories increase by 2.48 calories.

1.21 **(a)** $\widehat{Calories} = 112.23$. **(b)** Residual = 20.09. **(c)** Not a short answer.

1.23 Not a short answer.

1.25 **(a)** There is a weak negative relationship. **(b)** $\widehat{Adj2007} = 388.204 - 54.427\ Distance$. **(c)** 92.13; hard to interpret because of nonconstant variance. **(d)** nonconstant variance.

1.27 **(a)** Decreasing, but curved trend. **(b)** Clear curved pattern. **(c)** Very linear for *logVoltage* vs *Time*. **(d)** $\widehat{logVoltage} = 2.19 - 2.059\ Time$. **(e)** Curved pattern in the residual plot.

1.29 **(a)** Some curvature and increasing variability for large *Mass*. **(b)** Strong positive association. **(c)** $\widehat{LogWetFrass} = -0.739 + 1.054\ LogMass$. **(d)** Not a short answer. **(e)** Not a short answer.

1.31 **(a)** Not a linear pattern. **(b)** Better positive association, but still not linear. **(c)** No, for both.

1.33 **(a)** Relatively linear after the first four points. **(b)** $\widehat{Price} = -1647.17 + 0.841\ Year$. **(c)** Linear fit looks good. **(d)** Conditions are reasonably met. **(e)** 1958: residual = 4.5, standardized = 2.95.

1.35 **(a)** Not a short answer. **(b)** $\widehat{Hgt96} = 241.3 + 2.250\ Hgt90$. **(c)** Not a short answer.

1.37 **(a)** Yes. **(b)** $\widehat{Hgt97} = 40.6 + 1.10\ Hgt96$. **(c)** Yes.

1.39 **(a)** Strong positive association. **(b)** $\widehat{Cassim} = 0.00379 + 0.0639\ Intake$. **(c)** Problems with normality and equal variance.

1.41 **(a)** Positive, but weak. **(b)** *Gdiam03* is stronger. **(c)** $\widehat{Wall03} = -1.0521 + 0.36821\ Gdiam03$. **(d)** $\widehat{Wall03} = 20.7$, residual = −0.57. **(e)** $\hat{\sigma}_\epsilon = 1.50$.

1.43 **(a)** Both models show strong linear patterns, but the 3-D is tighter. **(b)** The 2-D model has 1.18 standard error for regression; not as good as the 0.65 for 3-D.

1.45 **(a)** Reasonably symmetric. **(b)** *Runs* has the strongest linear relationship with *Time*. **(c)** $\widehat{Time} = 148.0 + 4.18\ Runs$. **(d)** Normal plot shows departure from normality.

1.47 **(a)** $\widehat{SRA} = -1,732,400 + 868\ Year$ 2003: $-5,642.7$, 2011: $-12,201$.
(b) $\widehat{SRA} = -2,257,997 + 1131\ Year$.

1.49 Linear trend is consistent across *Instar*.

1.51 Mean of residuals is always zero when $\hat{\beta}_0 = \bar{y} - \hat{\beta}_1\bar{x}$.

CHAPTER 2

2.1 False.

2.3 True.

2.5 False.

2.7 True.

2.9 **(a)** No, could have high r^2 but a curved relationship. **(b)** No, could have low r^2 even for data simulated from a linear model.

2.11 **(a)** $t = 4.56$, *P*-value < 0.001, reject H_0. **(b)** (8.62, 22.38).

2.13 **(a)** Very weak linear relationship. **(b)** *P*-value is 0.87. Fail to reject H_0.

2.15 **(a)** $t = 3.51$, *P*-value $= 0.001$, reject H_0. **(b)** (1.044, 3.918).

2.17 **(a)** Yes. $t = 3.18$, *P*-value $= 0.005$. **(b)** No pattern. **(c)** 1.344, 0.422. **(d)** (0.61, 2.08).

2.19 **(a)** -54.427. **(b)** $(-70.46, -38.39)$. **(c)** Lack of constant variance.

2.21 $r^2 = 0.733$.

2.23 **(a)** $r^2 = 98.5\%$. **(b)** $t = 35.68$, *P*-value ≈ 0, reject H_0. **(c)** $F = 1273.1$.

2.25 **(a)** $t = 13.46$, *P*-value ≈ 0, Reject H_0. **(b)** $r^2 = 61.4\%$. **(c)** $F = 181.25$, *P*-value ≈ 0, Reject H_0. **(d)** $\sqrt{F} = \sqrt{181.25} = 13.46 = t$.

2.27 **(a)** $\widehat{Yards} = 86.1 + 5.69\ Attempts$. **(b)** mean yards/attempt = 7.7, slope = 5.69. **(c)** 33.9%.

2.29 **(a)** $t = 18.38$, *P*-value ≈ 0, Reject H_0. **(b)** $\hat{\beta}_1 = 0.368$, $SE_{\hat{\beta}_1} = 0.0200$. **(c)** $\hat{\sigma}_\epsilon = 1.50$. **(d)** No, $r^2 = 36.3\%$. **(e)** 6.19 to 6.43 mm. **(f)** $r = 0.602$, $t = 18.4$, *P*-value ≈ 0, Reject H_0.

2.31 **(a)** $t = 4.72$, *P*-value ≈ 0, Reject H_0. **(b)** $r^2 = 2.7\%$. **(c)** $F = 22.28$. **(d)** $r^2 = 2.7\%$. **(e)** No (small r^2).

2.33 **(a)** (1.079, 1.113). **(b)** No. **(c)** No.

2.35 **(a)** $\widehat{Eggs} = 29.56 + 79.24\ BodyMass$. **(b)** Not a short answer. **(c)** 79.86 to 79.24. **(d)** 19.5% to 26.6%.

2.37 $\widehat{Gesell} = 105.63 - 0.779\ Age$, $r^2 = 0.11$, not significant.

2.39 $\widehat{Gesell} = 102.1 - 0.554\ Age$, $r^2 = 0.064$, not significant.

2.41 **(a)** (1.767, 2.735). **(b)** $(-0.594, 5.097)$. **(c)** Not a short answer.

2.43 **(a)** 10.714 grams. **(b)** 10.15 to 11.28 grams. **(c)** 7.88 to 13.54 grams. **(d)** Not a short answer.

2.45 **(a)** (0.285, 2.221). **(b)** $(-5.780, 8.287)$. **(c)** Same, $\hat{y} = 1.253$. **(d)** PI is wider than CI for mean. **(e)** $x^* = \bar{x} = 99.07$. **(f)** (3.680, 18.017), but $\bar{x} = 0$ is much smaller than the markets in the sample.

2.47 **(a)** \$360,990. **(b)** (\$207,017, \$514,964). **(c)** The variance in the residual plot is not constant. **(d)** (\$211,866, \$499,049). **(e)** Logging has led to constant variance in the residual plot.

2.49 **(a)** $\widehat{Spring} = 244$. **(b)** (223.7, 264.3). **(c)** (183, 305). **(d)** (c).

2.51 **(a)** Weak, positive linear relationship. **(b)** $\widehat{Survey2} = 40.417 + 0.395\ Survey1$. **(c)** lack of normality. **(d)** The relationship is not $Survey1 = Survey2$.

2.53 **(a)** (\$263,854, \$310,121). **(b)** Wider.

2.55 **(a)** *WetFrass*, $r = 0.990$. **(b)** $\widehat{Nfrass} = 0.000297 + 0.00967\ WetFrass$. **(c)** $t = 113.36$, *P*-value ≈ 0, reject H_0. **(d)** Not a short answer.

2.57 Not a short answer.

2.59 Not a short answer.

2.61 **(a)** $\widehat{Gate} = 22,576.6 + 111.826\ Enroll$. **(b)** $r^2 = 49.1\%$. **(c)** $\widehat{Gate} = 184,165$. **(d)** $Residual = -138,594$.

CHAPTER 3

3.1 **(a)** $\widehat{Final} = 100$. **(b)** $\widehat{Final} = 82.31$. Residual $= -2.13$.

3.3 No. A larger coefficient does not (necessarily) indicate a stronger relationship.

3.5 As the score on the project goes up by 1, after accounting for the midterm grade, the average final score goes up by 1.2 points.

3.7 **(a)** True. **(b)** False.

3.9 **(a)** Yes, positive. **(b)** Negative. **(c)** Negative.

3.11 **(a)** Negative. **(b)** $Price = \beta_0 + \beta_1 Year + \beta_2 Mileage + \epsilon$. **(c)** Negative.

3.13 **(a)** $Arsenic = \beta_0 + \beta_1 Year + \beta_2 Miles + \beta_3 Year \cdot Miles + \epsilon$. **(b)** $Lead = \beta_0 + \beta_1 Year + \beta_2 Iclean + \beta_3 Year \cdot Iclean + \epsilon$. **(c)** $Titanium = \beta_0 + \beta_1 Miles + \beta_2 Miles^2 + \epsilon$. **(d)** Not a short answer.

3.15 **(a)** 194. **(b)** 194. **(c)** 195. **(d)** 191.

3.17 **(a)** $t = 1.08$, *P*-value $= 0.282$, Fail to reject H_0. **(b)** $(-0.0182, 0.08656)$. **(c)** 96.13.

3.19 **(a)** $\widehat{Animus} = 124.3 + 0.564 Black - 0.578 Hispanic - 2.054 BachPlus + 1.519 Age65Plus$. **(b)** Problems with normality and constant variance.

3.21 **(a)** $R^2 = 87.1\%$. **(b)** $\hat{\sigma}_\epsilon = 13.4$. **(c)** $F = 23.64$, P-value $= 0.001$, reject H_0. **(d)** Fall: $t = -4.93$, P-value $= 0.002$, reject. H_0 Ayear: $t = 4.57$, P-value $= 0.003$, reject H_0.

3.23 **(a)** Linear relationship not evident. **(b)** $t = -0.17$, P-value $= 0.866$, linear relationship not significant. **(c)** M: $\widehat{pH} = 6.89 - 0.00045Age$, F: $\widehat{pH} = 6.903 - 0.00045Age$.

3.25 **(a)** $t = -16.6$, P-value $= 0.003$, Clinton did worse in states with a paper trail. **(b)** $t = -6.15$, P-value $= 0.13$. Controlling for the percentage of African Americans in each state, the effect of having a paper trail is negative but is not statistically significantly different from zero. **(c)** *PaperTrail* only model has P-value $= 0.001$. Using both variables, *PaperTrail* has a P-value $= 0.053$.

3.27 **(a)** Scatterplot shows a clear positive trend with two clusters of points. **(b)** $t = 14.88$, P-value ≈ 0, reject H_0. **(c)** Model: $NetSupport = \beta_0 + \beta_1 Months + \beta_2 Late + \epsilon$
Fit: $\widehat{NetSupport} = -70.04 + 0.26875Months + 39.689Late$.
(d) $t = 4.166$, P-value $= 0.000951$, reject H_0.

3.29 **(a)** Negative relationship with 2000 points being higher on the percent report scale than 1998. **(b)** $t = -3.44$, P-value $= 0.001$, linear relationship is significant. **(c)** 1998: $\widehat{PctReport} = 77.08 - 0.717Period$; 2000: $\widehat{PctReport} = 94.91 - 0.717Period$; methods working. **(d)** 1998: $\widehat{PctReport} = 76.43 - 0.668Period$; 2000: $\widehat{PctReport} = 95.98 - 0.765Period$; P-value for interaction term $= 0.699$, fail to reject H_0.

3.31 **(a)** Relationship is similar. Women's intercept is higher. **(b)** Gender is significant, indicating different intercepts. **(c)** Intercept is no longer significant at the 0.05 level.

3.33 **(a)** The plot shows several clusters of points, but no linear relationship. **(b)** $t = 1.366$, P-value $= 0.1921$, fail to reject H_0. **(c)** $t = -1.96$, P-value $= 0.07$, reject H_0. **(d)** Coefficient of *Unemployment* switches sign between (b) and (c).

3.35 **(a)** Substantial curvature to plot. **(b)** $\widehat{Height} = 100.2 + 13.383Age - 0.2643Age^2$. **(c)** 241.5 cm.

3.37 **(a)** There is curvature. **(b)** $\widehat{TotalPrice} = -522.7 + 2386.0Carat + 4498.2Carat^2$; $R^2 = 0.9257$, adjusted $R^2 = 0.9253$. **(c)** Non-normal residuals; nonconstant variance. **(d)** $\widehat{TotalPrice} = -723 + 2942Carat + 4078Carat^2 + 88Carat^3$; $R^2 = 0.9257$, adjusted $R^2 = 0.9251$. **(e)** Non-normal residuals; nonconstant variance.

3.39 **(a)** Neither the equal variance nor normality conditions are met for this model. **(b)** The complete second-order model is still a good choice for *ln(TotalPrice)*. **(c)** Both equal variance and normality are reasonable.

3.41 **(a)** $\widehat{ProteinProp} = 0.48 - 0.2532\ Calcium - 0.0278\ Calcium^2$. **(b)** Not a short answer. **(c)** Not a short answer. **(d)** $t = -5.12$, P-value ≈ 0, reject H_0. **(e)** $R^2 = 89.4\%$.

3.43 **(a)** $R^2 = 0.3495$, $SSE = 899$. **(b)** $R^2 = 0.4170$, $SSE = 806$. **(c)** $R^2 = 0.3239$, $SSE = 935$. **(d)** (b).

3.45 **(a)** $\widehat{NetSupport} = -66.628 + 0.2104Months + 13.115Late + 0.1740Months \cdot Late$. **(b)** $t = 1.36$, P-value $= 0.1959$, interaction term is not needed. **(c)** $F = 10.14$, P-value $= 0.0022$, reject H_0.

3.47 **(a)** $F = 62.9$, P-value $= 10^{-8}$, reject H_0. **(b)** $t = 11.15$, $F = 124.3$, both P-values $= 3 \times 10^{-9}$, reject H_0.

3.49 **(a)** All three predictors are statistically significant; coefficients for *logdistance*, *logsquarefeet*, and *no_full_baths* are, respectively: −0.04883, 0.59328, and 0.05667. **(b)** Model conditions appear to be met. **(c)** The R^2 for the more complicated model has made a tiny increase: from 78.34% to 80.07%. **(d)** The P-value from the nested F-test is about 0.09, suggesting the added complexity may not be worth it.

3.51 **(a)** $\widehat{Height} = 102.48 + 12.566Age - 0.2763Age^2$. **(b)** $t = 6.07$, P-value ≈ 0, the effect is statistically significant. **(c)** Model from (c).

3.53 **(a)** $\widehat{GPA} = 2.97$. **(b)** PI: (2.213, 3.727). **(c)** $\widehat{GPA} = 2.985$, PI: (2.230, 3.741).

3.55 Different intercepts, similar slopes.

3.57 Not a short answer.

CHAPTER 4

4.1 **(a)** Strong, positive trend. **(b)** Strong, negative trend. **(c)** Weak, negative trend.

4.3 **(a)** Forward selection gives *Runs*, *ERA*, *Saves*, and *WHIP*; $R^2 = 88.63\%$. **(b)** Backwards elimination gives *BattingAverage*, *Runs*, *Saves*, and *WHIP*; $R^2 = 88.36\%$. **(c)** Best subsets gives *Runs*, *Doubles*, *Saves*, and *WHIP*; $R^2 = 88.85\%$. **(d)** (a) 11.16, (b) 11.88, (c) 10.54. **(e)** Best subsets (c) has the lowest C_p and the highest R^2.

4.5 **(a)** Strongest: *MeanDailyGn*, $r = -0.432$. Weakest: *E2*, $r = -0.125$. **(b)** P-value $= 0.023$, reject H_0. **(c)** *E2*, *MaxDailyGn*, *Oocytes*, $R^2 = 27.1\%$. **(d)** Not a short answer.

4.7 **(a)** *Runs, Pitchers,* and *Attendance* ($R^2 = 64.5\%$). **(b)** *Runs*, *Pitchers*, and *Attendance* ($R^2 = 53.8\%$). **(c)** *Runs* alone ($Cp = 1.3$). **(d)** Model in (c).

4.9 **(a)** $\widehat{GPA} = 1.1475 + 0.46605HSGPA + 0.015328HU + 0.19917White$. P-values are 0.000 for *HSGPA*, 0.000 for *HU*, and 0.010 for *White*. Standard error of regression is 0.377 and $R^2 = 28.4\%$. **(b)** Not a short answer. **(c)** Mean error $= -0.0595$; std. dev. of errors 0.4066 is close to 0.377. **(d)** $r = 0.596$. **(e)** shrinkage $= -0.071$.

4.11 **(a)** Plot shows a negative association. **(b)** 53.9%. **(c)** *LogGDP* decreases by about 1.4 for an increase of one in *Religiosity*. **(d)** Kuwait studres = 3.99. **(e)** 72.4%. **(f)** After accounting for region, *LogGDP* decreases by about 1 for an increase of one in *Religiosity*. **(g)** F = 4.95, P-value $= 0.001438$. Yes, region indicators help the model. **(h)** Kuwait studres = 3.37.

4.13 $\bar{x}_{White} = 117.87$, $\bar{x}_{Black} = 110.56$, $\bar{x}_{Hispanic} = 118.52$, $\bar{x}_{Other} = 117.14$.

4.15 Single predictor: $\widehat{SystolicBP} = 136.23 + 8.437 Overwt$
Indicators: $\widehat{SystolicBP} = 136.32 + 8.050 Overweight + 16.87 Obese$. The results for the two models are very similar.

4.17 P-value ≈ 0.0014.

4.19 **(a)** Original: $R^2 = 0.5307$ or $F = 6.219$ or $\hat{\sigma}_\epsilon = 16.45$. **(b)** Answers vary. **(c)** Answers vary. **(d)** Not a short answer. **(e)** P-value $= 0.0168$. **(f)** P-value $= 0.0156$.

4.21 Z-interval from SE: (1.54, 2.10)
Percentile interval: (1.48, 2.03)
Reverse percentile: (1.61, 2.16).

4.23 **(a)** Bootstrap correlations look normal. **(b)** Z-interval from SE: (−0.111, 0.599)
Percentile interval: (−0.094, 0.610)
Reverse percentile: (−0.122, 0.582). **(c)** All intervals include zero, so not much evidence of association.

CHAPTER 5

5.1 True.

5.3 False.

5.5 **(c)**.

5.7 **(d)**.

5.9 Protects against bias; permits conclusions about cause; justifies using a probability model.

5.11 **(a)** Ethnicity cannot be assigned, Inference about cause not justified. **(b)** Samples must be random; yes.

5.13 Major: Categorical, ANOVA; Sex: Categorical, Two-sample t-test; Class year: Categorical, ANOVA; Political Inclination: Categorical, ANOVA; Sleep time: Quantitative; Study time: Quantitative; BMI: Quantitative; Money spent: Quantitative.

5.15 **(a)** Explanatory: font; response: final exam score. **(b)** Randomized experiment; treatment randomly assigned to subjects. **(c)** Subjects randomly assigned to treatments.

5.17 **(a)** Students. **(b)** Fonts. **(c)** Assign 10 people to each font.

5.19 3, 36, 39.

5.21 **(a)** If the four groups all have the same mean score, there is a 0.003 probability of collecting sample data in which the sample mean scores are as or more different than those displayed by this sample. **(b)** We have only shown there is some difference among the four types of fonts. **(c)** We have only concluded that at least one difference exists. **(d)** We do have evidence that at least one difference exists.
(e) Treatments were randomly assigned to subjects. **(f)** We can only generalize to the population of students like those in this instructor's class.

5.23 **(a)** $H_0 : \alpha_1 = \alpha_2 = \alpha_3 = 0$. **(b)** The actual values of response variable. **(c)** The residuals.

5.25 **(a)** Answers will vary. **(b)** Answers will vary.

5.27 **(a)** Minus group has fewer contacts on average. **(b)** Conditions met.

5.29 Concern about equal variances condition.

5.31 **(a)**

Variable	Group	N	Mean	StDev
ABeta-42	mAD	17	761.3	426.7
	MCI	21	341.0	406.4
	NCI	19	336.3	435.6

(b) Skewed distributions; normality not met.

5.33 **(a)** $DF = 4$; SS $= 10{,}998$; $F = 13.609$. **(b)** 5. **(c)** reject H_0.

5.35 **(a)** $df_{\text{type}} = 2$, $df_E = 9$, $df_{\text{Total}} = 11$, $SSE = 33.09$, $MS_{\text{type}} = 18.755$, $F = 5.101$. **(b)** Measures the amount of variability between the means of the three county types. **(c)** 0.033. **(d)** $H_0 : \alpha_1 = \alpha_2 = \alpha_3 = 0$, reject the null hypothesis.

5.37 $F = 5.66$, P-value $= 0.006$, reject H_0.

5.39 **(a)** $H_0 : \alpha_B = \alpha_H = \alpha_W = 0$, H_a : at least one $\alpha_k \neq 0$. **(b)** $F = 13.60$, P-value ≈ 0, reject H_0.

5.41 **(a)**

Group	Mean	St. Dev.
A	1.0	0.1
B	10.0	1.0
C	100.0	10.0
D	1000.0	100.0

(b) 1000, yes, slope of 1.
(c) 0, logarithm.

(d)

Group	Mean	St. Dev.
A	−0.0015	0.0436
B	0.9985	0.0436
C	1.9985	0.0436
D	2.9985	0.0436

(e) 1, yes.

5.43 **(a)** Yes. **(b)** Approximately 0.5. **(c)** $p = 0.5$, square root.

5.45 Mixed and plus are indistinguishable from each other, but both have significantly more contacts, on average, than minus.

5.47 Whites have a significantly higher mean score than both of the other groups, but there is no significant difference between blacks and Hispanics.

5.49 **(a)** Sample means: 20.144, 32.940, 11.922; sample SDs: 3.861, 3.063, 2.422. **(b)** Weights from largest to smallest: RT, CH, SS. **(c)** SDs now similar, still skewed distributions and outliers.

5.51 **(a)** Sample means: 26.4, 14.7, and 14.2; Sample SDs: 8.2, 11.5, and 11.9. **(b)** Abeta levels tend to be higher in the mAD group than in the other two groups. **(c)** Normality condition for ANOVA now met. **(d)** $F = 7.36$, P-value $= 0.001$, reject H_0.

5.53 **(a)** There is a difference. **(b)** P-value is 0.006, reject H_0. **(c)** P-value is 0.006, reject H_0. **(d)** Yes.

5.55 **(a)** There is not a difference. **(b)** P-value is 0.771, do not reject H_0. **(c)** P-value is 0.771, do not reject H_0. **(d)** Yes.

5.57 **(a)** $H_0 : \alpha_a = \alpha_f = \alpha_v = 0$, H_a : at least one $\alpha_k \neq 0$. **(b)** Condition of equal variance is not met.

5.59 **(a)** Similar centers; whites and blacks: skewed to the left with more variability; Hispanics and others: symmetric with less variability.
(b)

Race	N	Mean	StDev
black	332	110.56	23.40
hispanic	164	118.52	18.17
other	48	117.15	17.60
white	906	117.87	22.52

(c) The difference between means is small compared to amount of overlap between the observations of all groups.

5.61 **(a)** Black mothers have babies with a significantly different mean birth weight than white or Hispanic mothers. **(b)** Answers may vary.

5.63 $F = 20.99$, P-value ≈ 0.

5.65 LSD = 0.213, don't conclude Houston and Boston are different.

5.67 **(a)** ANOVA not appropriate; residuals not normally distributed. **(b)** ANOVA appropriate; $F = 12.99$, P-value ≈ 0, reject H_0.

5.69 Normality and equal variance conditions not met for original data; all conditions are met using a natural log transformation; $F = 0.81$, P-value $= 0.713$, do not reject H_0.

5.71 Conditions are met; $F = 5.96$, P-value $= 0.001$, reject H_0; using Fisher's LSD, time 0 had a significantly higher mean than all other time periods; time 5 is also significantly higher than time 25.

5.73 **(a)** Conditions are met; Reject H_0; using Fisher's LSD, all three sizes of cars have significantly different mean noise levels. **(b)** Previous exercise could result in cause-and-effect conclusion; the results of this exercise could be generalized to larger population, but not a cause-and-effect conclusion.

CHAPTER 6

6.1 **(a)** Calcium concentration in plasma. **(b)** Sex, additive. **(c)** Sex: observational, 2 levels; Additive: experimental, 2 levels. **(d)** No.

6.3 **(a)** Adaption score. **(b)** Subject, shock. **(c)** Subject: observational, 18 levels; Shock: experimental, 3 levels. **(d)** Yes.

6.5 **(a)** Running speed. **(b)** Dogs, diet. **(c)** Dogs: observational, 5 levels; Diet: experimental, 3 levels. **(d)** Yes.

6.7 **(a)** Not significant. **(b)** Significant.

6.9 **(a)** 3. **(b)** 4. **(c)** 12.

6.11 **(a)** Large. **(b)** Less.

6.13 True.

6.15 True.

6.17 **(a)** Version B is the block design. **(b)** Having groups of similar units.

6.19 Treatment or measurement changes the subject and treatment or measurement takes too long. Examples will vary.

6.21 Answers will vary.

6.23 **(a)** Protect against bias, permit conclusions about cause, justify using a probability model. **(b)** Reduce residual variation.

6.25 **Reusing:** A cloud cannot be reused. **Subdividing:** To create blocks by subdividing, you would have to be able to seed half of each cloud and measure rainfall from each half of the cloud. Neither is practical. **Grouping:** No practical way to seed some clouds in a group.

6.27 **(a)** Complete Block: Diets, 2; Blocks, 4; Residual, 8. Completely Randomized: Diets, 2; Residual, 12. **(b)** The block design uses fewer dogs but takes longer, has fewer *df* for residual, and splits off between-dog differences from residual error.

6.29 **(a)** For each rat, randomly assign one treatment to each time slot. **(b)** The treatment may change the subject.

6.31 **(a)** $F_{Shrew} = 19.34$, P-value ≈ 0.000 $F_{Phase} = 3.79$, P-value $= 0.059$. **(b)** No, plot shows reasonable linear trend. **(c)** Yes, outlier, possible linearity.

6.33 **(a)** Not a short answer. **(b)** 1.32, 0.88, 0.65; no transformation.

6.35 Food cooked in iron pots has the most iron. Food in clay pots has a slightly higher amount of iron than food in aluminum pots. Two low outliers for the meat and vegetable dishes cooked in the clay pots. Much smaller spread for poultry. Change of scale is not likely to help.

6.37 3.0 HCL is significantly different from water.

6.39 **(a)**

124.93	0.97	−223.54	97.64
−31.28	−0.24	55.97	−24.45
−93.64	−0.72	167.56	−73.19

(b) Scatterplot. **(c)** Slope approx. 0.75; $p = 0.25$; Fourth root.

6.41 Conditions are met; both *Moon* and *Month* are significant in predicting the number of admissions.

CHAPTER 7

7.1 **(a)** 3. **(b)** 4. **(c)** 12. **(d)** 40.

7.3 c

7.5 **(a)** large. **(b)** less.

7.7 Yes

7.9 True

7.11 No

7.13 **(a)** (ii). **(b)** (iii). **(c)** (ii). **(d)** (v). **(e)** (v).

7.15 $MSB < MSE < MSA < MSAB$

7.17 Factor A: #3 < #2 < #1
Factor B: #2 < #3 < #1
Interaction AB: #3 < #2 < #1

7.19 Met: Comparison, replication, crossing, randomization; not met: blocking

7.21 There is an interaction

7.23 Yes

7.25 No

7.27 **(a)** A two-way factorial design would have several servers of each sex. **(b)** Male/Female: 1; can't distinguish between Male/Female and server effects; interaction: $df = 0$. **(c)** The design *should* have used more than one server of each sex. **(d)** For each of the six servers, randomly assign "face" or "no face" to a sequence of tables.

7.29 **(a)** As a two-way factorial with one observational factor and one experimental factor. **(b)** Yes. **(c)** Because each subject is tested under both conditions.

7.31 Interaction present

7.33 $df_F = 1, df_G = 1, df_{Inter} = 1, df_{Residuals} = 96$, $df_{Total} = 99$, $SSF = 12{,}915$, $SSG = 2{,}500$, $SSR = 9{,}600$, $MSF = 12{,}915$, $MS_{Inter} = 400$, $F_F = 129.15$, $F_G = 25.0$, $F_{Inter} = 4.0$

7.35 **(a)** Slope ≈ 0.5, $p = 0.5$, square root. **(b)** Slope ≈ 0.8, $p = 0.2$, fifth root.

7.37 **(a)** Oil, Ultrasound not different, interaction might be. **(b)** Yes. **(c)** Conditions met; oil significant, ultrasound and interaction not significant.

7.39 **(a)** female, yes. **(b)** increasing variability. **(c)** 7.51 is more than twice 2.12. **(d)** square root.

7.41 **(a)** female, no. **(b)** log transformation. **(c)** log_{10} versus natural log.

7.43 Conditions are met; both the two-way additive and two-way nonadditive models account for less than 4% of the variation in *Lifespan*

7.45 When there is more oil in the sample, less ultrasound is better and freshwater is better. When there is less oil in the sample, more ultrasound is better and salt water is better. Also, if the sample has freshwater, less ultrasound is better and when the water is salty, more ultrasound is better.

7.47 Not a short answer

CHAPTER 8

8.1 False.

8.3 Test statistic = 1.12, *P*-value = 0.341, fail to reject H_0.

8.5 Test statistic = 1.22, *P*-value = 0.297, fail to reject H_0.

8.7 Test statistic = 7.17, *P*-value = 0.007, reject H_0.

8.9 The multiplier of the SE.

8.11 **(a)** JW takes significantly longer than all of the others, and TS takes a significantly shorter time than all others. The rest are not significantly different from each other. **(b)** Answers may vary.

8.13 **(a)** All three categories are significantly different from each other. **(b)** All three categories are significantly different from each other. **(c)** Results are the same.

8.15 **(a)** $H_0 : \frac{1}{2}(\mu_D + \mu_E) = \frac{1}{2}(\mu_D + \mu_E)$ $H_a : \frac{1}{2}(\mu_D + \mu_E) \neq \frac{1}{2}(\mu_D + \mu_E)$. **(b)** $\frac{1}{2}\mu_D + \frac{1}{2}\mu_E - \frac{1}{2}\mu_F - \frac{1}{2}\mu_G$.

8.17 **(a)** $H_0 : \frac{1}{2}(\mu_{1p} + \mu_{8p}) = \mu_N$ $H_0 : \frac{1}{2}(\mu_{1p} + \mu_{8p}) \neq \mu_N$. **(b)** $\frac{1}{2}(\mu_{1p}) + \frac{1}{2}(\mu_{8p}) - \mu_N$ Estimated value is 0.52. **(c)** 3.625. **(d)** $t = 0.143$, $df = 120$, *P*-value = 0.8865, fail to reject H_0.

8.19 **(a)** $H_0 : \frac{1}{2}(\mu_s + \mu_m) = \mu_l$ $H_0 : \frac{1}{2}(\mu_s + \mu_m) \neq \mu_l$. **(b)** $\frac{1}{2}(\mu_s) + \frac{1}{2}(\mu_m) - \mu_l$ Estimated value is 56.46. **(c)** 3.808. **(d)** $t = 14.83$, $df = 33$, *P*-value ≈ 0; reject the H_0.

8.21 False.

8.23 **(a)** Normality condition violated. **(b)** Test statistic = 613.5, *P*-value = 0.0089; the median number of beds are significantly different.

8.25 **(a)** Either test would be okay. **(b)** $t = 4.50$, *P*-value = 0.000; the mean number of steps is significantly different between weekdays and weekend days. **(c)** Test statistic = 2125, *P*-value = 0.0085; the median number of steps is significantly different between weekdays and weekend days. **(d)** The conclusions are the same.

8.27 **(a)** Equal variance condition violated. **(b)** Test statistic = 34.8, *P*-value = 0.001; the group medians of the egg lengths are significantly different for the different types of bird nest.

8.29 **(a)** Kruskal-Wallis is a more robust procedure because of lack of equal variances. **(b)** Test statistic = 9.06, *P*-value = 0.17; we cannot conclude that the median number of steps per day is significantly different for different days of the week. **(c)** Given the variability in the medians from day-to-day, the number of steps per day on weekend days could be due to random chance. However, when all weekdays are combined and both weekend days are combined, the difference is bigger than chance would suggest.

8.31 0.086.

8.33 **(a)** 0.03. **(b)** 0.06. **(c)** Randomization test accounts for a large difference in *either* direction. **(d)** (a). **(e)** A *t*-test can be either one-sided or two-sided. The two-sided *t*-test and *F*-test are equivalent.

8.35 *P*-values should be similar.

8.37 2 within-subjects: *Style*(typed/hand) × *Sound*(music/quiet).

8.39 **(a)** Not a short answer. **(b)** $H_0 : \mu_a = \mu_v$. **(c)** Some evidence of longer response with visual.

8.41 Significant, *P*-value ≈ 0.

8.43 Not a short answer.

8.45 **(a)** Fingers, significant, *P*-value = 0.018. **(b)** Objects, significant, *P*-value ≈ 0. **(c)** Interaction, significant, *P*-value ≈ 0.

8.47 Not a short answer.

8.49 **(a)** $Y = \mu + \alpha_k + \epsilon$. **(b)** $Y = \beta_0 + \beta_1 lemon + \beta_2 paper + \epsilon$. **(c)** β_0 : mean shelf life for control; β_1 : change in mean shelf life for lemon treated; β_2 : change in mean shelf life for paper towel treated.

8.51 **(a)** 3. **(b)** $\hat{\beta}_0 = 117.15$, $\hat{\beta}_1 = -6.58$, $\hat{\beta}_2 = 1.37$, $\hat{\beta}_3 = 0.73$. **(c)** The ANOVA tables are identical.

8.53 **(a)** Conditions are met; constant term and Oil are significant. **(b)** Not a short answer. **(c)** Not a short answer.

8.55 **(a)** Size, type, and interaction are significant. **(b)** Not a short answer. **(c)** Medium and small cars are significantly different from large cars in terms of noise. Filter type is not significant for large cars or small cars but is significant for medium-sized cars. **(d)** Not a short answer.

8.57 **(c)**.

8.59 **(a)** The conditions have all been met; both treatment and size of the thorax are significant in explaining the longevity. **(b)** ANCOVA.

8.61 **(a)** Conditions are met; $F = 22.98$, P-value ≈ 0; reject H_0. **(b)** Conditions are not met; there is an interaction between sex and height. **(c)** ANOVA.

8.63 Not a short answer.

CHAPTER 9

9.1 Regression line could go below zero or above one, which is not possible for a proportion.

9.3 **(a)** 1. **(b)** 9. **(c)** 1/9.

9.5 **(a)** 0.67. **(b)** 0.91. **(c)** 0.20.

9.7 3.86.

9.9 **(a)** Curve does not rise so steeply. **(b)** Horizontal shift to left. **(c)** Curve falls.

9.11 **(a)** Might be true. **(b)** Must be true. **(c)** Must be true. **(d)** Might be true. **(e)** Cannot be true.

9.13 **(a)** Might be true. **(b)** Cannot be true. **(c)** Must be true. **(d)** Might be true. **(e)** Might be true.

9.15 −4.605; 0; 0.693; 1.

9.17 **(a)** $(1, 0)$. **(b)** $(1, -1)$. **(c)** $(0, 0)$. **(d)** $(-1, -1)$. **(e)** $(-1, 0)$. **(f)** $(-1, 1)$. **(g)** $(0, -1)$. **(h)** $(0, 1)$. **(i)** $(1, 1)$.

9.19 **(a)** Logit and probability forms:

$$logit(\hat{\pi}) = -8.712 + 0.246MCAT$$

$$\hat{\pi} = \frac{e^{-8.712 + 0.246MCAT}}{1 + e^{-8.712 + 0.246MCAT}}$$

(b) 1.279. **(c)** Approximately 76%. **(d)** About 35.4.

9.21 **(a)** 2.676. **(b)** 0.728. **(c)** 1.67. **(d)** 0.817, 0.089 more than for 6 cm.

9.23 6.79 cm.

9.25 slope $= -0.23$.

9.27 **(a)** Age boxplots are similar.
(b) $logit(\hat{\pi}) = -0.0814 - 0.008795Age$, P-value $= 0.0928$, do not reject H_0.

9.29 **(a)** $OR = 9.986$. **(b)** $(7.67, 13.01)$. **(c)** $OR = 9.986$, $log(9.986) = 2.301$. **(d)** $\hat{p}_f = 0.667$. **(e)** Independence and random are problematic.

9.31 Not a short answer.

9.33 **(a)** Odds: high (3.47); mid (1.57); low (1.24)
Log(odds): high (1.24); mid (0.452); low (0.211)
(b) The odds of flight increase by a factor of $e^{0.11503} = 1.12$, or 12% (p = 0.01) for each additional 100 m of altitude.

9.35 **(a)** The coefficient for *DJIAch* is 0.013 ($p = 0.001$). **(b)** The coefficient for *lagNik* is −0.004 ($p = 0.09$). **(c)** *DJIAch*.

9.37 **(a)** OR = 0.9998 for each $1 of income. **(b)** (0.9997, 0.9999).

9.39 **(a)** $H_0 : \beta_1 = 0$. **(b)** The probability decreases. **(c)** $(-0.007955, -0.003843)$. **(d)** log-odds of presence at 600 (10 a.m.). **(e)** 0.9941. **(f)** (0.9921, 0.9962). **(g)** 0.7019.

9.41 **(a)** OR $= 0.95$, $p = 0.017$. **(b)** Answers may vary depending upon the cutoff chosen. **(c)** OR $= 0.96$, $p = 0.069$; answers for 2-way table vary depending on cutoff chosen.

9.43 **(a)** $\hat{p}_{THC} = 0.456$, $\hat{p}_{Pro} = 0.205$.
(b) $log\left(\frac{\hat{\pi}}{1-\hat{\pi}}\right) = -1.35455 + 1.17686THC$.
(c) for the THC group: probability = 0.456
for the Prochlorperazine group, probability = 0.21.
(d) 3.24, (1.60, 6.57). **(e)** P-value≈ 0.0005 (one-tail), reject H_0.

9.45 **(a)** $\widehat{logit} = -1.4098 + 0.021235StartSpeed$. **(b)** more positive. **(c)** 0.647, 0.546. **(d)** yes, P-value≈ 0.000.

9.47 **(a)** OR $= 2.71$, $p < 0.0001$. **(b)** OR $= 1.20$, $p < 0.0001$. **(c)** OR $= 3.22$, $p < 0.0001$.

9.49 Des Moines: P-value $= 0.015$, reject H_0
Grinnell town: P-value $= 0.039$, reject H_0
Grinnell campus: P-value $= 0.147$, do not reject H_0.

CHAPTER 10

10.1 **(a)** smokers = 0.613, nonsmokers = 0.149. **(b)** 4.11. **(c)** 4.11 does not take into account other differences between smokers and nonsmokers that may be associated with RDD.

10.3 $\log\left(\frac{\pi}{1-\pi}\right) = \beta_0 + \beta_1 X + \beta_2 Group1$, *Group*1 is an indicator variable for *Group*.

10.5 $\log\left(\frac{\pi}{1-\pi}\right) = \beta_0 + \beta_1 X + \beta_2 Group1 + \beta_3 X \cdot Group1$, *Group*1 is an indicator variable for *Group*.

10.7 Squares.

10.9 $\log\left(\frac{\pi}{1-\pi}\right) = \beta_0 + \beta_1 Ahigh + \beta_2 X$; *Ahigh* is an indicator variable for *A*.

10.11 **(a)** Restricted model: $\text{logit}(\pi) = \beta_0 + \beta_1 MCAT + \beta_2 Sex$; Full model: $\text{logit}(\pi) = \beta_0 + \beta_1 MCAT + \beta_2 Sex + \beta_3 GPA$.
(b) Restricted model: $\text{logit}(\pi) = \beta_0 + \beta_1 MCAT + \beta_2 Sex$; Full model: $\text{logit}(\pi) = \beta_0 + \beta_1 MCAT + \beta_2 Sex + \beta_3 Sex \cdot MCAT$.

10.13 Having "left hand low" increases the log(odds) by 0.46.

10.15 **(a)** 0.7080. **(b)** 0.6045. **(c)** Yes.

10.17 **(a)** 0.472. **(b)** 0.4567. **(c)** Similar.

10.19 **(a)** *BillWidth*; *BillLength*. **(b)** *BillWidth*. **(c)** 0.046. **(d)** Short answer not appropriate.

10.21 **(a)** 0.645. **(b)** 0.642. **(c)** Similar.

10.23 **(a)** $logit(\hat{\pi}) = -1.160 - 0.00635Age + 2.466SexCode$, $\hat{\pi} = \frac{e^{-1.160 - 0.00635Age + 2.466SexCode}}{1+e^{-1.160 - 0.00635Age + 2.466SexCode}}$. **(b)** *SexCode* is effective (P-value ≈ 0), *Age* is not effective (P-value $= 0.305$). **(c)** $\widehat{odds} = 0.2796$, $\hat{\pi} = 0.219$. **(d)** $\widehat{odds} = 3.293$, $\hat{\pi} = 0.767$,

OR = 11.78. **(e)** $\widehat{odds}_m = 0.2282$, $\hat{\pi}_m = 0.186$, $\widehat{odds}_f = 2.687$, $\hat{\pi}_f = 0.729$, OR = 11.77. **(f)** Gender OR is same at all ages.

10.25 **(a)** *Dem.Rep* *P*-value = 0.020. **(b)** *HS* *P*-value = 0.697, *BA* *P*-value = 0.186, *Income* *P*-value = 0.182. **(c)** Indiana (1.938), Missouri (−1.746). **(d)** $logit(\hat{\pi}) = -21.94 + 0.5116 Dem.Rep + 0.6986 BA$.

10.27 **(a)** $logit(\pi) = -0.0273 + 0.0135 DJIAch - 0.00382 lagNik$, *DJIAch* *P*-value = 0.001, *lagNik* *P*-value = 0.094, Overall fit is good, *P*-value ≈ 0. **(b)** $\hat{\pi} = 0.493$. **(c)** $r = -0.068$, multicollinearity is not an issue.

10.29 **(a)** *Pulse*. **(b)** $\widehat{odds} = 0.85$ and $\hat{\pi} = 0.46$. **(c)** Reasonably likely.

10.31 **(a)** $\hat{p}_1 = 0.599$, $\hat{p}_2 = 0.427$, $\hat{p}_3 = 0.194$. **(b)** $\chi^2 = 172.5$, *P*-value ≈ 0, reject H_0. **(c)** Not a short answer. **(d)** Predicted proportions match those in (a). **(e)** $\chi^2 = 173.1$, *P*-value ≈ 0, reject H_0, results are similar to part (b).

10.33 **(a)** $logit(\pi) = -0.2227 - 0.4848 Seed$, *P*-value ≈ 0, reject H_0. **(b)** $logit(\pi) = -0.2227 - 0.4848 Seed + 0 Year$, *P*-value = 1 for *Year*, do not reject H_0.

10.35 **(a)** Strong, positive. **(b)** Both useful, *Party* stronger.

10.37 **(a)** $\hat{\beta}_1 = 0.9875 > 0$, so likelihood increases, but *P*-value = 0.486, so not significant. **(b)** Both are very strong predictors. **(c)** LA-2 has large Cook's D.

10.39 **(a)** Not a short answer. **(b)** Not a short answer. **(c)** 0.0667584. **(d)** 0.0199147. **(e)** 3.352.

10.41 **(a)** *Time*, *Fromlow*, *Rw*, and *Cobble*. **(b)** Not a short answer. **(c)** *P*-value is 0.1207; reduced model. **(d)** The reduced model was adequate. **(e)** Negative for *Time* and *Fromlow*; positive for *Rw* and *Cobble*.

10.43 Not a short answer.

10.45 Not a short answer.

10.47 Not a short answer.

10.49 Not a short answer.

CHAPTER 11

11.1 Max is at $\hat{\pi}$, less variability as n increases.

11.3 **(a)** 1/4, 1/3, 1/2. **(b)** 3/16, 2/9, 1/4. **(c)** 9/64, 4/27, 1/8.

11.5 **(a)** 0.769. **(b)** 0.885. **(c)** 0.680.

11.7 **(a)** Full: $logit(\pi) = \beta_0 + \beta_1 Attendance$; Reduced: $logit(\pi) = \beta_0$. **(b)** Full: $logit(\pi) = p_i$, where i ranges over all attendance values; Reduced: $logit(\pi) = \beta_0 + \beta_1 Attendance$.

11.9 **(a)** Logistic model $\hat{\pi}$ to Sample proportion $\hat{p}$. **(b)** Overall proportion of success, 0.576. **(c)** $\chi^2 = 80.3172$, *P*-value ≈ 0. Similar to z-test for slope, $z = -8.391$, *P*-value ≈ 0. **(d)** Both tests have very small *P*-values, evidence that putt success is related to length. Methods won't always give the same conclusion. **(e)** Check if residual deviance (6.83) is much larger than its degrees of freedom (3).

11.11 **(a)** 72% are predicted correctly. **(b)** 78% are predicted correctly. **(c)** The 0.687 cutoff has a slightly lower error rate.

11.13 **(a)** 78.9%. **(b)** 83.7%.

11.15 **(a)** Incidence rate for gunnels is lowest when there are no crustaceans in the quadrat. For higher densities the first density (1) shows a clearly higher incidence rate of about 8.5% than the higher densities, which show incidence rates between 5 and 5.8%. **(b)** *P*-value ≈ 0.00041, reject H_0.

11.17 **(a)** Peaceworks address elicits a higher estimated return rate: 95% versus 65%. **(b)** Not a short answer. **(c)** Not a short answer.

11.19 **(a)** The table is

	Gunnel	
Amphiso	0	1
0	1165	21
1	151	14
2	160	9
3	49	3
4	19	1

(b) There is a clear statistically significant relationship between these variables. **(c)** Model confirms answer to part (b).

11.21 **(a)** In standing water there is an increased probability of the gunnel being present: 4.4% chance versus 1.5% chance. **(b)** Significant with *P*-value of 0.00069. **(c)** Significant with *P*-value of 0.00069. **(d)** Significant with *P*-value of 0.0012. **(e)** Not a short answer.

11.23 **(a)**

	Cover95			
Deer95	0	1	2	3
0	0.716	0.675	0.801	0.859
1	0.284	0.325	0.199	0.141

(b) Results are statistically significant; *P*-value = 0.00002. **(c)** Results agree. **(d)** 1.21; 0.626; 0.415. **(e)** 56.1%. **(f)** Not a short answer.

11.25 **(a)** *P*-value = 0.000, reject H_0. **(b)** *P*-values vary depending on reference location, but all are small. **(c)** *G*-statistic = 17.771, *P*-value = 0.000. **(d)** Confederacy: *P*-value = 0.016, Peaceworks: *P*-value = 0.005. **(e)** Each coefficient and overall fit are significant. **(f)** Answers vary.

11.27 **(a)** $z = -17.1$, *P*-value ≈ 0, reject H_0. **(b)** $\chi^2 = 329.8$, *P*-value ≈ 0, reject H_0.

11.29 **(a)** Young 0.915, Middle 0.779, Old 0.719. **(b)** Young 0.893, Middle 0.814, Old 0.698. **(c)** 2.117, 1.477, 0.837. **(d)** Answers vary. **(e)** The proportions match. **(f)** Two-way: $\chi^2 = 7.75$, *P*-value = 0.0208 Logistic: $G = 8.57$, *P*-value = 0.0138, the results are consistent. **(g)** Not a substantial improvement.

11.31 **(a)**

	Smoker	Nonsmoker
Alive	443	502
Dead	139	230
Total	582	732

(b) For both younger and older women, separately, the percentage alive is higher for nonsmokers. But with both groups combined, the percentage alive is higher for smokers.

(c) Younger women are more likely to have been smokers and to still be alive.

11.33 **(a)** Negative. Controlling for smoking, probability of being alive is lower in the older age group. **(b)** Negative. Controlling for age group, probability of being alive is lower for smokers. **(c)** *AgeGroup*.

11.35 **(a)** 0.505, 0.546, 0.401, 0.409; Frye had higher Hit rate. **(b)** 0.473, 0.458; Irving had higher Hit rate. **(c)** Not a short answer.

11.37 **(a)** Positive. **(b)** Positive. **(c)** Negative. **(d)** *ShotType*.

CHAPTER 12

12.1 **(a)** Pattern would have to repeat at regular yearly intervals. **(b)** One example: yearly spending on political ads.

12.3 **(a)** $S = 4$. **(b)** linear $+ S = 5$. **(c)** quadratic $+ S = 7$. **(d)** $S = 3$.

12.5 **(a)** Yes. **(b)** Yes. **(c)** No. **(d)** No. **(e)** Yes.

12.7 **(a)** Yes. **(b)** Yes. **(c)** No. **(d)** Yes. **(e)** No.

12.9 **(a)** Series A = Series 2. **(b)** Series B = Series 4. **(c)** Series C = Series 5. **(d)** Series D = Series 3. **(e)** Series E = Series 1.

12.11 **(b)**.

12.13 **(a)** False. **(b)** False. **(c)** True.

12.15 **(a)** $\widehat{Area} = 4.45426 + 0.057315t - 0.0027141t^2$
(b) Yes, *P*-value = 0.00114
(c) No consistent curvature in residuals.

12.17 $r_1 = 0.174$ is not significant.

12.19 **(a)** 2017: $\widehat{Area} = 2.56$, bounds $= (1.45, 3.67)$
2019: $\widehat{Area} = 2.24$, bounds $= (1.06, 3.42)$
2021: $\widehat{Area} = 1.90$, bounds $= (0.64, 3.16)$. **(b)** $t \approx 53$ (2031).

12.21 Not a short answer.

12.23 **(a)** $\widehat{CPI}_t = 0.1695 + 1.4759CPI_{t-1} - 0.4759CPI_{t-2}$. **(b)** *P*-value ≈ 0, ϕ_1 is needed. **(c)** Residual ACF spikes at lags 3, 11, 12, 24. **(d)** *P*-value = 0.081, not significant at 5%.

12.25 Increasing trend and monthly seasonal pattern.

12.27 **(a)** $\widehat{Y}_t = 347.4 + 0.1579t$. **(b)** Yes, $R^2 = 97.45\%$. **(c)** Yes. **(d)** Not a short answer.

12.29 **(a)** Adjusted $R^2 = 99.76\%$. **(b)** Adjusted $R^2 = 99.88\%$. **(c)** Cosine trend: $\widehat{Y}_{370} = 406.7$.
Seasonal means: $\widehat{Y}_{370} = 410.2$. **(d)** Seasonal means.
(e) $R^2 = 1.25\%$.

12.31 **(a)** Increasing trend. **(b)** $R^2 = 84.12\%$. **(c)** Curvature in the series. **(d)** Quadratic: $\widehat{y}_t = 723.5 - 0.09277t + 0.001418t^2$, log transformation: $\widehat{log(y_t)} = 6.515 + 0.0007449t$.
(e) Quadratic: yes; log: no. **(f)** $\widehat{y}_{514} = \$1050.36$, PI = (\$992.97 to \$1107.75).

12.33 **(a)** Increasing trend in plot, slow decay in ACF, not stationary. **(b)** Differences look stationary. **(c)** Mean difference = \$0.48. **(d)** Lag 8. **(e)** No AR or MA terms, ARIMA(0, 1, 0). **(f)** $\hat{y}_t = y_{t-1} + 0.483$. **(g)** Essentially the same. **(h)** $\hat{y}_{514} = \$1015.00$, PI = (\$957.74 to \$1072.27).

12.35 Not a short answer.

12.37 **(a)** Yes, $R^2 = 37.89\%$. **(b)** Yes, $R^2 = 58.94\%$. **(c)** No, $R^2 = 58.96\%$.

12.39 Not a short answer.

12.41 Not a short answer.

12.43 Not a short answer.

Notes and Data Sources

CHAPTER 0

1. K. Volpp, L. John, A.B. Troxel, L. Norton, J. Fassbender, and G. Lowenstein (2008), "Financial Incentive-based Approaches for Weight Loss: A Randomized Trial", *JAMA*, *300*(22): 2631–2637.

2. In a study reported in the *American Journal of Preventive Medicine*

3. Dansinger et al., 2005

4. *Source*: **www.baseball-reference.com**.

CHAPTER 1

1. *Source:* Cars.com, February 2017.

2. *Sources:* Physicians–American Medical Association, Chicago, IL, Physician Characteristics and Distribution in the U.S., accessed May 17, 2006. Community Hospitals–Health Forum LLC, an American Hospital Association (AHA) Company, Chicago, IL, Hospital Statistics, e-mail accessed May 4, 2006. Related information from **http://www.healthforum.com**.

3. Source: Lawrence R. Heaney, (1984), Mammalian Species Richness on Islands on the Sunda Shelf, Southeast Asia, *Oecologia* vol. 61, no. 1, pages 11–17.

4. *Source:* **http://trackandfield.about.com/od/longjump/qt/olymlongjumpmen.htm**

5. We thank Priscilla Erickson and Professor Robert Mauck from the Department of Biology at Kenyon College for allowing us to use these data.

6. The data are courtesy of Melanie Shoup and Gordon Gallup, Jr.

7. We thank Professors Harry Itagaki, Drew Kerkhoff, Chris Gillen, Judy Holdener, and their students for sharing this data from research supported by NSF InSTaRs grant No. 0827208.

8. The data were obtained from Wikipedia; the URL is **http://en.wikipedia.org/wiki/History_of_United_States_postage_rates**.

9. The data were obtained from **http://Registrar.Kenyon.edu** on June 1, 2012.

10. Thanks to the Kenyon College Department of Biology for sharing these data.

11. Thanks to Suzanne Rohrback for providing these data from her honors experiments at Kenyon College.

12. Thanks to the Kenyon College Department of Biology for sharing these data.

13. *Source:* Cal Poly Student project.

14. We thank Professor Itagaki and his research students for sharing these data.

CHAPTER 2

1. Data scraped from the MLB GameDay website (**http://gd2.mlb.com/components/game/mlb/**) using pitchRx.

2. These data are from the paper "High-Artic Butterflies Become Smaller with Rising Temperatures" by Bowden, J. et al. (2015) *Biology Letters* 11: 20150574.

3. Seth Stephens-Davidowitz. "The Cost of Racial Animus on a Black Candidate: Evidence Using Google Search Data." *Journal of Public Economics* 118, October 2014.

4. Data downloaded from **http://www.pro-football-reference.com/players/T/TomlLa00/gamelog/2006/**

5. We thank Professor Itagaki and his students for sharing this data from experiments on *Manduca sexta*.

6. *Source*: Cal Poly students using a horse sale website.

7. *Source*: CDC National Vital Statistics Reports at **http://www.cdc.gov/nchs/data/nvsr/nvsr57/nvsr57_14.pdf**

CHAPTER 3

1. Data downloaded from **www.nfl.com**.

2. *Source:* JSE Data Archive, **http://www.amstat.org/publications/jse/jse_data_archive.htm**, submitted by Juha Puranen.

3. Refer to O'Connor, Spotila (1992) "Consider a Spherical Lizard: Animals, Models, and Approximations", Am Zool 32:179–193.

4. This example is based on Kleisner K, Chvatalova V, Flegr J (2014); "Perceived Intelligence Is Associated with Measured Intelligence in Men but Not Women." *PLoS ONE 9(3)*: e81237. doi:10.1371/journal.pone.0081237. We are using "crystallized IQ" as actual IQ, as Kleisner et al. write "Crystallized intelligence is the ability to use skills, knowledge, and experience. This sort of reasoning improves with age and reflects the lifetime achievement of an individual." Also, for ease of interpretation we have back-transformed their data from the z-scores they give to IQ scores with a mean of 125 and an SD of 15.

5. The funnel dropping experiment was originally described in Gunter, B. "Through a Funnel Slowly with Ball Bearing and Insight to Teach Experimental Design," *The American Statistician*, Vol. 47, Nov. 1993.

6. Data obtained from *Zillow.com* in April 2017.

7. *Source:* Personal communication from clothing retailer David Cameron.

8. *Source:* There are advanced methods for handling collinearity, see Belsley, Kuh, and Welsch (1980) *Regression Diagnostics: Identifying Influential Data and Sources of Collinearity*.

9. *Source:* Franklin County Municipal Court.

10. B. Lantry, R. O'Gorman, and L. Machut (2008), "Maternal Characteristics versus Egg Size and Energy Density," *Journal of Great Lakes Research 34*(4): 661–674.

11. Data are from Phyllis Lee, Stirling University, and are related to Lee, P., et al. (2013), "Enduring Consequences of Early Experiences: 40-year Effects on Survival and Success Among African Elephants (*Loxodonta Africana*)," *Biology Letters*, 9: 20130011.

12. Data are from Phyllis Lee, Stirling University, and are related to Lee, P., et al. (2013), "Enduring Consequences of Early Experiences: 40-year Effects on Survival and Success Among African Elephants (*Loxodonta Africana*)," *Biology Letters*, 9: 20130011.

CHAPTER 4

1. *Source*: Student survey in an introductory statistics course.

2. We thank Dr. Priya Maseelall and her research team for sharing these data.

CHAPTER 5

1. C. P. Wilsie, Iowa State College Agricultural Station (1944) via Snedecor and Cochran, *Statistical Methods*.

2. "Survival and Behavioral Responses of the Potato Leafhopper, Empoasca Fabae (Harris), on Synthetic Media," MS thesis by Douglas Dahlman (1963), Iowa State University. The data can be found in *Analyzing Experimental Data by Regression* by David M. Allen and Foster B. Cady, Belmont, CA: Lifetime Learning (Wadsworth).

3. R. L. Stevenson (1886), *Dr. Jeckyl and Mr. Hyde*.

4. https://www.census.gov/programs-surveys/ahs/data/2007/ahs-2007-public-use-file--puf-.html

5. Cobb, George W. (1984), "An Algorithmic Approach to Elementary ANOVA," *The American Statistician, 38*(2). Hoaglin, David C., Frederick Mosteller, and John W. Tukey (1991), *Fundamentals of Exploratory Analysis of Variance*, Wiley Interscience.

6. Diaconis, Persi (1988), *Group Representations in Probability and Statistics*, Institute of Mathematical Statistics, IMS Notes, vol. 11.

7. Personal communication.

8. Box, G. E. P., and D. R. Cox (1964), "An Analysis of Transformations," *JRSS B* (26), pp. 211–252.

9. Ionnidis, John P. A. (2005), "Why Most Published Research Findings Are False," *PLoS Medicine, 2*(8).

10. Jeffrey Arnett (1992), "The Soundtrack of Recklessness: Musical Preferences and Reckless Behavior Among Adolescents," *Journal of Adolescent Research, 7*(3): 313–331.

11. Beis, D., Holzwarth, K., Flinders, M., Bader, M., Wöhr, M., and Alenina, N. (2015), "Brain Serotonin Deficiency Leads to Social Communication Deficits in Mice." *Biology Letters 11*:20150057. **http://dx.doi.org/10.1098/rsbl.2015.0057**

12. Data and documentation downloaded from **http://www.amstat.org/publications/jse/jse_data_archive.htm**

13. Violetta N. Pivtoraiko, Eric E. Abrahamson, Sue E. Leurgans, Steven T. DeKosky, Elliott J. Mufson, and Milos D. Ikonomovic (2015), "Cortical Pyroglutamate Amyloid-β Levels and Cognitive Decline in Alzheimer's Disease," *Neurobiology of Aging* (36) 12–19. Data are read from Figure 1, panel d.

14. Many thanks to the late Professor Bob Black at Cornell College for sharing these data with us.

15. Data from: Vosteen I., Gershenzon, J., and Kunert G. (2016), "Hoverfly Preference for High Honeydew Amounts Creates Enemy-free Space for Aphids Colonizing Novel Host Plants," *Dryad Digital Respository*. **http://dx.doi.org/10.5061/dryad.37972**. Original article: Vosteen I., Gershenzon, J., and Kunert G. (2016), "Hoverfly Preference for High Honeydew Amounts Creates Enemy-free Space for Aphids Colonizing Novel Host Plants," *Journal of Animal Ecology 85*(5): 1286–1297. **http://dx.doi.org/10.1111/1365-2656.12564**

16. Data from the Annie E. Casey Foundation, KIDS COUNT Data Center, **http://www.datacenter.kidscount.org**

17. Data explanation and link can be found at **http://lib.stat.cmu.edu/DASL/Datafiles/tastedat.html**.

18. Data from: **http://www.stat.ufl.edu/~winner/data/meniscus.txt**. Original source article: P. Borden, J. Nyland, D.N.M. Caborn, and D. Pienkowski (2003), "Biomechanical Comparison of the FasT-Fix Meniscal Repair Suture System with Vertical Mattress Sutures and Meniscus Arrows," *American Journal of Sports Medicine, 31*:3, pp. 374–378.

19. Data provided by Rosemary Roberts and discussed in "Persistence of Fenthion Residues in Olive Oil," Chaido Lentza-Rizos, Elizabeth J. Avramides, and Rosemary A. Roberts, (January 1994), *Pest Management Science, 40*(1): 63–69.

20. Data from experiment run by Las Vegas high school student Chris Mathews for a science fair project in spring 2016. Shared with Ann Cannon via email.

21. Data were used as a case study for the 2003 Annual Meeting of the Statistical Society of Canada. See **Chttp://www.ssc.ca/en/education/archived-case-studies/case-studies-for-the-2003-annual-meeting-blood-pressure**.

22. Data provided by Allan Rossman.

23. Data explanation and link can be found at **http://www.stat.ucla.edu/projects/datasets/seaslug-explanation.html**.

24. Data explanation and link can be found at **http://lib.stat.cmu.edu/DASL/Stories/airpollutionfilters.html**.

CHAPTER 6

1. Thanks to Dr. Jeff Chiarenzilli of the St. Lawrence University Geology Department.

2. Camner, P. and Philipson, K. (1973), "Urban Factor and Tracheobronchial Clearance," Archives of Environmental Health, 27(2).

3. Katherine Ann Maruk (1975), "The Effects of Nutrient Levels on the Competitive Interaction between Two Species of *Digitaira*," unpublished master's thesis, Department of Biological Sciences, Mount Holyoke College.

4. D. A. Daly and E. B. Cooper (1967), "Rate of Stuttering Adaptation under Two Electro-shock Conditions," *Behavior Research and Therapy, 5*(1): 49–54.

5. S. Zentall and J. Shaw (1980), "Effects of Classroom Noise on Performance and Activity of Second-grade Hyperactive and Control Children," *Journal of Educational Psychology, 72*(6): 630–840.

6. (July 20, 2002),"Antioxidants for Greyhounds? Not a Good Bet" *Science News, 162*(2): 46.

7. (July 20, 2002),"Gender Differences in Weight Loss," *Science News, 162*(2): 46.

8. D. A. Daly and E. B. Cooper (1967), "Rate of Stuttering Adaptation under Two Electro-shock Conditions," *Behavior Research and Therapy, 5*(1): 49–54.

9. The experiment is described in J. Simpson and W. L. Woodley (1971), "Seeding Cumulus in Florida: New 1970 Results," *Science, 172*: 117–126.

10. F. Loven (1981), "A Study of the Interlist Equivalency of the CID W-22 Word List Presented in Quiet and in Noise," unpublished master's thesis, University of Iowa.

11. Thomas E. Bradstreet and Deborah L. Panebianco (2017), "An Oral Contraceptive Drug Interaction Study," *Journal of Statistics Education, 12*:1, DOI: 10.1080/10691898.2004.11910719.

12. A. A. Adish, et al. (1999), "Effect of Food Cooked in Iron Pots on Iron Status and Growth of Young Children: A Randomized Trial," *The Lancet 353*: 712–716.

13. The original discussion of the study appears in S. Blackman and D. Catalina (1973), "The Moon and the Emergency Room," *Perceptual and Motor Skills 37*: 624–626. The data can also be found in Richard J. Larsen and Morris L. Marx (1986), *Introduction to Mathematical Statistics and Its Applications*, Prentice-Hall: Englewood Cliffs, NJ.

CHAPTER 7

1. S. M. Kosslyn (1980) *Image and Mind*, Cambridge, MA: Harvard University Press.

2. Chester Bliss (1970), *Statistics in Biology*, McGraw-Hill.

3. Iowa Agricultural Experiment Station, Animal Husbandry Swine Nutrition Experiment No. 577 (1952) via George W. Snedecor and William G. Cochran (1967), *Statistical Methods*, 6th edition, The Iowa State University Press, Ames, IA, p. 345.

4. Fred L. Ramsey and Daniel W. Schafer (2002), *The Statistical Sleuth*, 2nd ed., Pacific Grove, CA, Duxbury, pp. 405–407.

5. W. Alvarez and F. Asaro (1990), "What Caused the Mass Extinction? An Extraterrestrial Extinction," *Scientific American 263*(4): 76–84. E. Courtillot (1990), "What Caused the Mass Extinction? A Volcanic Eruption." *Scientific American 263*(4): 85–93.

6. The original article is T. Yamada, S. Takahashi-Abbe, and K. Abbe (1985), "Effects of Oxygen Concentration on Pyruvate Formate-Lyase In Situ and Sugar Metabolism of *Streptocucoccus mutans* and *Streptococcus samguis*," *Infection and Immunity, 47*(1): 129–134. Data may be found in J. Devore and R. Peck (1986), *Statistics: The Exploration and Analysis of Data*, St. Paul, MN: West.

7. Steven Swanson and Graham Caldwell (2001), "An Integrated Biomechanical Analysis of High Speed Incline and Level Treadmill Running," *Medicine and Science in Sports and Exercise, 32*(6): 1146–1155.

8. B. Rind and P. Bordia (1996). "Effect on Restaurant Tipping of Male and Female Servers Drawing a Happy Face on the Backs of Customers' Checks," *Journal of Social Psychology, 26*: 215–225.

9. S. S. Zentall and J. H. Shaw (1980), "Effects of Classroom Noise on Performance and Activity of Second-grade Hyperactive and Control Children," *Journal of Educational Psychology, 72*: 830–840.

10. B. Rind and P. Bordia (1996), "Effect on Restaurant Tipping of Male and Female Servers Drawing a Happy Face on the Backs of Customers' Checks," *Journal of Social Psychology, 26*: 215–225.

11. Thanks to Hamisi Babusa, visiting scholar at St. Lawrence University for the data.

CHAPTER 8

1. See Ewan Cameron and Linus Pauling (Sept. 1978), "Supplemental Ascorbate in the Supportive Treatment of Cancer: Reevaluation of Prolongation of Survival Times in Terminal Human Cancer," *Proceedings of the National Academy of Sciences of the United States of America, 75*(9), 4538–4542.

2. Many thanks to the late Professor Bob Black at Cornell College for sharing these data with us.

3. **http://lib.stat.cmu.edu/DASL/Stories/CalciumandBloodPressure.html**, the online data source, Data and Story Library.

4. Mary Ann DiMatteo (1972), "An Experimental Study of Attitudes Toward Deception," unpublished manuscript, Department of Psychology and Social Relations, Harvard University, Cambridge, MA.

5. The justification for the randomization test is different. For an explanation, see M. Ernst (2004), "Permutation Methods: A Basis for Exact Inference," *Statistical Science, 19*: 676–685.

6. "What Is the Use of Experiments Conducted by Statistics Students?" (online). The actual data can be found at the Australasian Data and Story Library (OzDASL) maintained by G. K. Smyth, **http://www.statsci.org/data**. The article can be found at **http://www.amstat.org/publications/jse/v2n1/mackisack.html**.

7. These data are not real, though they are simulated to approximate an actual study. The data come from John Grego, director of the Stat Lab at the University of South Carolina.

8. Data supplied by Robin Lock, St. Lawrence University.

9. Data come from the Data and Story Library (DASL), **http://lib.stat.cmu.edu/DASL/Stories/nursinghome.html**

10. Data were accessed from **http://www.statsci.org/data/oz/cloudtas.html**. This is the web home of the Australasian Data and Story Library (OzDASL). The data were discussed in A.J. Miller, D.E. Shaw, L.G. Veitch, and E.J. Smith (1979) "Analyzing the Results of a Cloud-seeding Experiment in Tasmania," *Communications in Statistics: Theory and Methods*, A8(10): 1017–1047.

11. Data supplied by Jeff Witmer.

12. Oswald H. Latter (January 1902), "The Egg of Cuculus Canorus: An Enquiry into the Dimensions of the Cuckoo's Egg and the Relation of the Variations to the Size of the Eggs of the Foster-Parent, with Notes on Coloration," *Biometrika 1*(2): 164–176.

13. M. Ernst (2004), "Permutation Methods: A Basis for Exact Inference," *Statistical Science, 19*: 676–685.

14. Data can be found in Siddhartha R. Dalal, Edward B. Fowlke, and Bruce Hoadley (1989), "Risk Analysis of the Space Shuttle: Pre-Challenger Prediction of Failure," *Journal of the American Statistical Association, 84*(408): 945–957.

15. Sarah Marshall-Pescini and Andrew Whiten (2008), "Social Learning of Nut-Cracking Behavior in East African Sanctuary-Living Chimpanzees (*Pan Troglodytes schweinfurthii*)," *Journal of Comparative Psychology, 122*(2):186–194.

16. Data collected by Oberlin College student Ksenia Vlasov.

17. Data collected by Oberlin College students Arjuna Pettit, Jr. and Jeremy Potterfield.

18. Data from a student at Oberlin College.

19. Kyriacos Kareklas, Daniel Nettle, and Tom V. Smulders (2013,) "Water-Induced Finger Wrinkles Improve Handling of Wet Objects," *Biology Letters,* **http://dx.doi.org/10.1098/rsbl.2012.0999**.

20. K. G. Volpp, L. K. John, A. B. Troxel, et al. (December 2008), "Financial Incentive-based Approaches for Weight Loss," *Journal of the American Medical Association 200*(22): 2631–2637.

21. Data supplied by Allan Rossman.

ONLINE SECTIONS 8.9–8.11

1. George W. Snedecor and William G. Cochran (1980), *Statistical Methods*, 7th edition, The Iowa State University Press, Ames, IA.

2. George W. Snedecor and William G. Cochran (1980), *Statistical Methods*, 7th edition, The Iowa State University.

3. G. Box, W. Hunter and S. Hunter (2005). *Statistics for Experimenters*, 2nd ed.

CHAPTER 9

1. Based on results in R. B. Lipton et al. (2010), "Single-pulse Transcranial Magnetic Stimulation for Acute Treatment of Migraine with Aura: A Randomised, Double-blind, Parallel-group, Sham-controlled Trial," *The Lancet Neurology* 9(4):373-380.

2. 2013 data scraped from the MLB GameDay website, **http://gd2.mlb.com/components/game/mlb/** using pitchRx.

3. Data from Roxy Peck, Larry D. Haugh, and Arnold Goodman, eds. (1998), *Statistical Case Studies: A Collaboration Between Academe and Industry*, Washington DC: SIAM and ASA.

4. Thanks to Jake Shorty, Bowdoin biology student, for this data.

5. Data came from **http://www.stat.ufl.edu/~winner/data/lmr_levee.txt**. Original source article is A. Flor, N. Pinter, and W. F. Remo. (2010). "Evaluating Levee Failure Susceptibility on the Mississippi River Using Logistic Regression Analysis," *Engineering Geology,* Vol. 116, pp. 139–148.

6. Data from Brian S. Everitt. (1994). *Statistical Analysis Using S-Plus*, 1st ed. New York: Chapman & Hall.

7. Data: J. Sun, X. Guo, J. Zhang, M. Wang, C. Jia, and A. Xu. (2015). Data from: "Incidence and Fatality of Serious Suicide Attempts in a Predominantly Rural Population in Shandong, China: A Public Health Surveillance Study." Dryad Digital Repository. **http://dx.doi.org/10.5061/dryad.r0v35**. Original article: J. Sun, X. Guo, J. Zhang, M. Wang, C. Jia, and A. Xu. (2015). "Incidence and Fatality of Serious Suicide Attempts in a Predominantly Rural Population in Shandong, China: A Public Health Surveillance Study." *BMJ Open 5*(2): e006762. **http://dx.doi.org/10.1136/bmjopen-2014-006762**

8. Data scraped from the MLB GameDay website, **http://gd2.mlb.com/components/game/mlb/** using pitchRx.

9. See J. Mintz, J. Mintz, K. Moore, and K. Schuh (2002),"Oh, My Aching Back! A Statistical Analysis of Backpack Weights," *Stats: The Magazine for Students of Statistics, 32*: 1719.

10. The article "Which Young People Accept a Lift From a Drunk or Drugged Driver?" in *Accident Analysis and Prevention* (July 2009, pp. 703–709) provides more details.

11. A more recent version of the full dataset is available at **https://www.cdc.gov/brfss/annual_data/annual_data.htm**. The full dataset would be a rich source of information for those interested in updating the results of the study.

12. Denis M. McCarthy and Sandra A. Brown (2004), "Changes in Alcohol Involvement, Cognitions and Drinking and Driving Behavior for Youth after They Obtain a Driver's License," *Journal of Studies on Alcohol 65*: 289–296.

13. See Christiane Poulin, Brock Boudreau, and Mark Asbridge (January 2007), "Adolescent Passengers of Drunk Drivers: A Multi-level Exploration into the Inequities of Risk and Safety," *Addiction 102*(1): 51–61.

14. Gabriel E. Ryb et al. (2007), "Smoking Is a Marker of Risky Behaviors Independent of Substance Abuse in Injured Drivers," *Traffic Injury Prevention, 8*(3): 248–52.

CHAPTER 10

1. These data come from the article, "The Eyes Have It: Sex and Sexual Orientation Differences in Pupil Dilation Patterns," by G. Rieger and R. C. Savin-Williams, published in 2012 in *PLoS ONE*. The full study included 325 students. Here we are analyzing a subset of the data that excludes White students.

2. S. L. Worthy, J. N. Jonkman, and L. Blinn-Pike (2010), "Sensation-Seeking, Risk-Taking, and Problematic Financial Behaviors of College Students," *Journal of Family and Economic Issues, 31*: 161–170.

3. This case study is based on the *Biology Letters* article, "Women Can Judge Sexual Unfaithfulness from Unfamiliar Men's Faces," by G. Rhodes et al., published in November 2012. All of the 68 raters were heterosexual Caucasians, as were the 170 persons who were rated. (We have deleted 3 subjects with missing values and 16 subjects who were over age 35.)

4. Ryb et al. (2007), "Smoking Is a Marker of Risky Behaviors Independent of Substance Abuse in Injured Drivers," *Traffic Injury Prevention, 8*(3): 248-52.

5. From the online summary of the article's results at **http://www.ncbi.nlm.nih.gov/pubmed/17710714**.

6. Data were collected by Scott Preston, a professor of statistics at SUNY, Oswego.

7. Data came from Keith Tarvin, Department of Biology, Oberlin College.

8. The Data and Story Library (DASL), **http://lib.stat.cmu.edu/DASL/Datafiles/ICU.html**.

9. These numbers are taken from J. P. Kastellec, J. R. Lax, and J. Phillips (2010), "Public Opinion and Senate Confirmation of Supreme Court Nominees," *Journal of Politics, 72*(3): 767–84. In this paper the authors used opinion polls and an advanced statistical method known as multilevel regression and poststratification to determine the StateOpinion levels.

10. These data were collected by Grinnell College students Eric Ohrn and Ben Johannsen.

11. S. L. Worthy, J. N. Jonkman, and L. Blinn-Pike (2010), "Sensation-Seeking, Risk-Taking, and Problematic Financial Behaviors of College Students," *Journal of Family and Economic Issues, 31*: 161–170.

CHAPTER 11

1. David W. Hosmer and Stanley Lemeshow (2000), *Applied Logistic Regression*, John Wiley and Sons, Hoboken, NJ.

2. Nicholas Gueguen (2002), "The Effects of a Joke on Tipping When It Is Delivered at the Same Time as the Bill," *Journal of Applied Social Psychology, 32*: 1955–1963.

3. We thank Sean Forman and Doug Drinen of Sports Reference LLC for providing us with the NFL field goal dataset.

CHAPTER 12

1. Sea ice data were updated from **ftp://sidads.colorado.edu/DATASETS/NOAA/G02135/north/monthly/data/N_09_extent_v3.0.csv** after originally being found in G. Witt (2013), "Using Data from Climate Science to Teach Introductory Statistics," *Journal of Statistics Education, 21*(1) at **http://www.amstat.org/publications/jse/v21n1/witt.pdf**.

2. **http://www.peacebridge.com/index.php/historical-traffic-statistics/yearly-volumes**

3. Data downloaded from the Bureau of Labor statistics at **https://www.bls.gov/data/**.

4. Data downloaded from the EIA website at **https://www.eia.gov/petroleum/data.php#consumption**.

5. Data downloaded for MOL (Mauna Loa) from the ESRL/GMD data page at **https://www.esrl.noaa.gov/gmd/ccgg/trends/data.html**.

6. Data downloaded for SPO (South Pole) from the ESRL/GMD data page at **https://www.esrl.noaa.gov/gmd/ccgg/trends/data.html**.

7. Data downloaded using the Quandl R package (12/2/2017).

8. Thanks to Kyle Johnston '18 from Kenyon College for providing these data that were used in his Senior Exercise, a capstone project for his mathematics major. A few missing values were interpolated to provide a complete series.

9. U.S. Energy Information Administration website, **https://www.eia.gov/totalenergy/data/monthly/index.php**

General Index

Bolded page numbers indicate material in Sections 8.9–8.11; available online.

ex